RIVER ECOSYSTEM ECOLOGY: A GLOBAL PERSPECTIVE

A DERIVATIVE OF ENCYCLOPEDIA OF INLAND WATERS

RIVER ECOSYSTEM ECOLOGY: A GLOBAL PERSPECTIVE

A DERIVATIVE OF ENCYCLOPEDIA OF INLAND WATERS

EDITOR

PROFESSOR GENE E. LIKENS
Cary Institute of Ecosystem Studies
Millbrook, NY, USA

AMSTERDAM • BOSTON • HEIDELBERG • LONDON • NEW YORK • OXFORD
PARIS • SAN DIEGO • SAN FRANCISCO • SINGAPORE • SYDNEY • TOKYO
Academic Press is an imprint of Elsevier

ACADEMIC
PRESS

Academic Press is an imprint of Elsevier
525 B Street, Suite 1900, San Diego, CA 92101-4495, USA
30 Corporate Drive, Suite 400, Burlington, MA 01803, USA
32 Jamestown Road, London, NW1 7BY, UK
Radarweg 29, PO Box 211, 1000 AE Amsterdam, The Netherlands

Material in this work originally appeared in *Encyclopedia of Inland Waters* by Gene E. Likens (Elsevier Inc. 2009)

British Library Cataloguing in Publication Data
A catalogue record for this book is available from the British Library

Library of Congress Cataloging-in-Publication Data
River ecosystem ecology : a global perspective : a derivative of Encyclopedia of inland waters / editor, Gene E. Likens.
p. cm.
Includes bibliographical references and index.
ISBN 978-0-12-810213-8 (alk. paper)
1. Stream ecology. I. Likens, Gene E., 1935- II. Encyclopedia of inland waters.
QH541.5.S7R585 2010
577.6′4–dc22 2010010996

ISBN: 978-0-12-810213-8

For information on all Academic Press publications
visit our website at elsevierdirect.com

10 11 12 10 9 8 7 6 5 4 3 2 1

EDITOR

Professor Gene E. Likens is an ecologist best known for his discovery, with colleagues, of acid rain in North America, for co-founding the internationally renowned Hubbard Brook Ecosystem Study, and for founding the Institute of Ecosystem Studies, a leading international ecological research and education center. Professor Likens is an educator and advisor at state, national, and international levels. He has been an advisor to two governors in New York State and one in New Hampshire, as well as one U.S. President. He holds faculty positions at Yale, Cornell, Rutgers Universities, State University of New York at Albany, and the University of Connecticut, and has been awarded nine Honorary doctoral Degrees. In addition to being elected a member of the prestigious National Academy of Sciences and the American Philosophical Society, Dr. Likens has been elected to membership in the American Academy of Arts and Sciences, the Royal Swedish Academy of Sciences, Royal Danish Academy of Sciences and Letters, Austrian Academy of Sciences, and an Honorary Member of the British Ecological Society. In June 2002, Professor Likens was awarded the 2001 National Medal of Science, presented at The White House by President G. W. Bush; and in 2003 he was awarded the Blue Planet Prize (with F. H. Bormann) from the Asahi Glass Foundation. Among other awards, in 1993 Professor Likens, with F. H. Bormann, was awarded the Tyler Prize, The World Prize for Environmental Achievement, and in 1994, he was the sole recipient of the Australia Prize for Science and Technology. In 2004, Professor Likens was honored to be in Melbourne, Australia with a Miegunyah Fellowship. He was awarded the first G. E. Hutchinson Medal for excellence in research from The American Society of Limnology and Oceanography in 1982, and the Naumann-Thienemann Medal from the Societas Internationalis Limnologiae, and the Ecological Society of America's Eminent Ecologist Award in 1995. Professor Likens recently stepped down as President of the International Association of Theoretical and Applied Limnology, and is also a past president of the American Institute of Biological Sciences, the Ecological Society of America, and the American Society of Limnology and Oceanography. He is the author, co-author or editor of 20 books and more than 500 scientific papers.

Professor Likens is currently in Australia on a Commonwealth Environment Research Facilities (CERF) Fellowship at the Australian National University.

CONTRIBUTORS

R Abell
WWF-United States, Washington, DC, USA

J H Aldstadt III
University of Wisconsin-Milwaukee, Milwaukee, WI, USA

J D Allan
School of Natural Resources and Environment, University of Michigan, Ann Arbor, MI, USA

J L Ammerman
SEAL Analytical, Inc., Mequon Technology Center, Mequon, WI, USA

M E Benbow
University of Dayton, Dayton, OH, USA

A C Benke
University of Alabama, Tuscaloosa, AL, USA

S Blanch
WWF-Australia, Darwin, NT, Australia

H A Bootsma
University of Wisconsin-Milwaukee, Milwaukee, WI, USA

J P Bravard
University Lumière-Lyon 2, France

J Bruce Wallace
University of Georgia, Athens, GA, USA

P A Bukaveckas
Virginia Commonwealth University, Richmond, VA, USA

D T Chaloner
University of Notre Dame, Notre Dame, IN, USA

J Chen
Peking University, Beijing, P. R. China

C E Cushing
Streamside Programs, CO, USA

H Décamps
Centre National de la Recherche Scientifique and Université Paul Sabatier, Toulouse, France

J A Downing
Iowa State University, Ames, IA, USA

S L Eggert
USDA Forest Service, Northern Research Station, Grand Rapids, MN, USA

M C Feller
University of British Columbia, Vancouver, BC, Canada

B Finlayson
University of Melbourne, VIC, Australia

S G Fisher
Arizona State University, Tempe, AZ, USA

D Gann
Florida International University, Miami, FL, USA

J W Grubaugh
University of Memphis, Memphis, TN, USA

S K Hamilton
Michigan State University, Hickory Corners, MI, USA

A E Hershey
University of North Carolina at Greensboro, Greensboro, NC, USA

B R Hodges
University of Texas at Austin, Austin, TX, USA

J A Hubbart
University of Missouri, Columbia, MO, USA

J R Jones
University of Missouri, Columbia, MO, USA

H-P Kozerski
Institute of Freshwater Ecology and Inland Fisheries, Berlin, Germany

G A Lamberti
University of Notre Dame, Notre Dame, IN, USA

B G Laub
University of Maryland, College Park, MD, USA

M E McClain
Florida International University, Miami, FL, USA

W H McDowell
University of New Hampshire, Durham, NH, USA

M D McIntosh
University of Dayton, Dayton, OH, USA

T McMahon
University of Melbourne, VIC, Australia

M Meybeck
Université Pierre et Marie Curie, Paris, France

J L Meyer
University of Georgia, Athens, GA, USA

R J Naiman
University of Washington, Seattle, WA, USA

T N Narasimhan
University of California, Berkeley, CA, USA

M A Palmer
University of Maryland Center for Environmental Sciences, Solomons, MD, USA

M Peel
University of Melbourne, VIC, Australia

F D Peter
Swiss Federal Institute of Aquatic Science and Technology (Eawag), Duebendorf, Switzerland

F Petit
University of Liège, Liège, Belgium

M E Power
University of California, Berkeley, CA, USA

C Revenga
The Nature Conservancy, Arlington, VA, USA

B L Rhoads
University of Illinois at Urbana-Champaign, Urbana, IL, USA

C T Robinson
Swiss Federal Institute of Aquatic Science and Technology (Eawag), Duebendorf, Switzerland

G G Sass
Illinois Natural History Survey, Havana, IL, USA

R Siber
Swiss Federal Institute of Aquatic Science and Technology (Eawag), Duebendorf, Switzerland

R A Sponseller
Arizona State University, Tempe, AZ, USA

R J Stevenson
Michigan State University, East Lansing, MI, USA

K M Stewart
State University of New York, Buffalo, NY, USA

A N Sukhodolov
Institute of Freshwater Ecology and Inland Fisheries, Berlin, Germany

M Thieme
WWF-United States, Washington, DC, USA

J H Thorp
University of Kansas, Lawrence, KS, USA

K Tockner
Institute of Biology, Freie Universität Berlin, Germany

D Tonolla
Swiss Federal Institute of Aquatic Science and Technology (Eawag), Duebendorf, Switzerland

U Uehlinger
Swiss Federal Institute of Aquatic Science and Technology (Eawag), Duebendorf, Switzerland

F Wang
University of Manitoba, Winnipeg, MB, Canada

M R Whiles
Southern Illinois University, Carbondale, IL, USA

E Wohl
Department of Geosciences, Colorado State University, Ft. Collins, CO, USA

CONTENTS

HUMAN IMPACTS ON STREAMS AND RIVERS

RIVERS OF THE WORLD

INTRODUCTION TO RIVER ECOSYSTEM ECOLOGY: A GLOBAL PERSPECTIVE

Rivers, streams, brooks, runs, forks, kills, creeks, are among the many names for lotic (running or fluvial) ecosystems within the landscapes of the Earth. These systems facilitate the gravitational transport of water, dissolved substances, and large and small particulate materials downstream through a diversity of types of drainage networks from relatively simple channels to highly complicated "braided" channels, both above and below ground (e.g. Allan and Castillo, 2007). The tight connection in terms of structure and function between the river and its drainage basin (catchment=European usage or watershed=American usage) has been the subject of detailed study for many decades (e.g. Hynes, 1975; Likens, 1984; Allan and Castillo, 2007). The drainage area bordering the stream is called the riparian zone and is of critical importance to the function, as well as the protection and management of a river (e.g. Naiman et al., 2005).

Nevertheless, rivers and streams are far more than channels transporting water, chemicals and sediments downstream. They function as ecosystems (e.g. Fisher and Likens, 1972, 1973) with all of the varied and complicated activities and interactions that occur among their abiotic and biotic components, which are characteristic of all ecosystems (e.g. Allan and Castillo, 2007). Thus, they are not functioning just as "Teflon pipes" in the landscape that many have assumed in the past.

Rivers and streams comprise about 0.006% of the total fresh water on the Earth (Likens, 2009b), but like lakes, reservoirs and wetlands are valued by humans far out of proportion to their small size, as these systems supply diverse drinking, irrigation, waste removal, food, recreation, tourism, transportation and aesthetic services. Rivers with the largest volume of fresh water in the world, like the Amazon, Congo, Yangtze and Orinoco, are located in the tropics or semi-tropics. In fact, some 25% of the freshwater flow to the oceans of the world comes from two rivers, the Congo and the Amazon Rivers, both at approximately the same latitude (Likens, 2009b).

This volume consists of 5 sections: 1. Introduction to River Ecosystems; 2. Physical and Chemical Processes Influencing Rivers; 3. Ecology of Flowing Waters; 4. Human Impacts on Streams and Rivers; and, 5. Rivers of the World.

The articles in this volume are reproduced from the Encyclopedia of Inland Waters (Likens, 2009a). I would like to acknowledge and thank the authors of the articles in this volume for their excellent and up-to-date coverage of these important riverine topics.

Gene E. Likens
Cary Institute of Ecosystem Studies
Millbrook, NY
December 2009

References

Allan JD and Castillo MM (2007) *Stream Ecology: Structure and Function of Running Waters,* 2nd edn., p. 436. Dordrecht, The Netherlands: Springer.

Fisher SG and Likens GE (1972) Stream ecosystem: organic energy budget. *BioScience* 22(1): 33–35.

Fisher SG and Likens GE (1973) Energy flow in Bear Brook, New Hampshire: an integrative approach to stream ecosystem metabolism. *Ecol. Monogr.* 43(4): 421–439.

Hynes HBN (1975) The stream and its valley. *Verh. Internat. Verein. Theoret. Ange Limnol.* 19: 1–15.
Likens GE (1984) Beyond the shoreline: a watershed-ecosystem approach. *Verh. Internat. Verein. Limnol.* 22: 1–22.
Likens GE (2009a) (Editor-in-Chief) *Encyclopedia of Inland Waters* 3 Volumes). Elsevier, Academic Press.
Likens GE (2009b) Inland waters. In: Likens GE (ed.) *Encyclopedia of Inland Waters,* volume 1, pp. 1–5. Oxford: Elsevier.
Naiman RJ, Décamps H, and McClain ME (eds.) Riparia: Ecology, Conservation, and Management of Streamside Communities. Elsevier Academic Press, Inc.
Wetzel RG (2001) *Limnology. Lake and River Ecosystems,* 3rd edn. Academic Press.

PHYSICAL AND CHEMICAL PROCESSES INFLUENCING RIVERS

Contents

Physical Properties of Water

K M Stewart, State University of New York, Buffalo, NY, USA

Introduction

Water is an indispensable and remarkable substance that makes all forms of life possible. Speculation about possible past or present life on other planets within our solar system, or on any extraterrestrial body somewhere within the universe, is conditioned on the evidence for or against the existence of past or present water or ice. Humans can and did survive and evolve without petroleum products (gas and oil) but cannot survive and evolve without water. Water is the most important natural resource.

By far the greatest volume (~76%) of water on Earth is in the oceans. A smaller fraction (~21%) is found within sediments and sedimentary rocks. A still smaller fraction (~1% of the overall volume) is freshwater, and of that 1%, about 73% is in the form of ice (mostly contained within the Greenland and Antarctic ice caps), and only about 23% of that 1% is liquid freshwater. If we consider further that about one-fifth of the world's liquid freshwater is contained within the five St. Lawrence Great Lakes in North America, and another approximately one-fifth is contained within the deepest freshwater lake on Earth, Lake Baikal, in Russia, we are left with an unevenly distributed resource. It is obvious that if the expanding human populations around the world do not conserve and manage this precious resource very carefully, they put themselves at great peril.

Liquid water can be formed through some hydrogen bonding and electrostatic attraction of two slightly positively charged atoms of the gaseous hydrogen (H) and one slightly negatively charged atom of the gaseous oxygen (O) to form one molecule of water (H_2O). **Figure 1** provides two views of that polar molecule. **Figure 1(a)** and **1(b)** show the somewhat lopsided or asymmetrical arrangement of two smaller hydrogen atoms, separated by an angle of ~105°, and a larger oxygen atom. **Figure 1(a)** is a simple 'ball and spoke' representation whereas **Figure 1(b)** shows the shared electron orbits, positive (+) and negative (−) poles, and the number (eight each) of protons and neutrons in the nucleus of the oxygen atom.

The relative elemental simplicity of water is somewhat deceptive because of the great influence that some of the unusual properties of water have on the physics, chemistry, and biology of the world generally, and on the distribution of life specifically. The following discussion will describe briefly some of these unusual properties and provide examples of how these properties may help us understand the world of inland waters.

Density

Density may be simply defined as the amount of weight or mass contained in a specific volume. If the volumes of all substances could be standardized to one size, e.g., one cubic centimeter (cm^3), then a measure of the weight or mass in that fixed volume gives the density. **Table 1** lists a few comparative densities (rounded to two decimals) of two liquids (water and mercury) and some selected solids.

Density differences in inland waters may be caused by variations in the concentrations of dissolved salts,

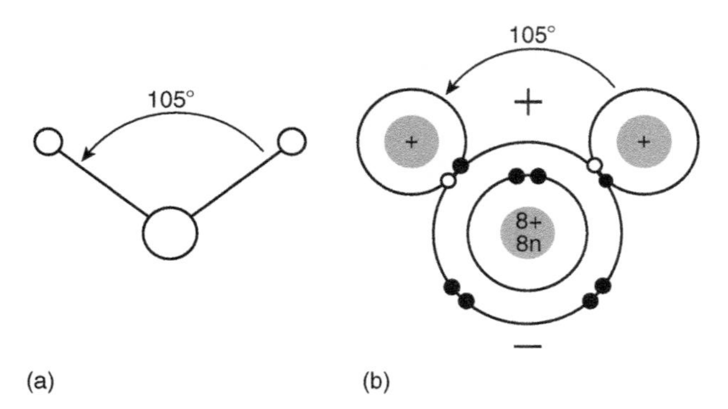

Figure 1 Two schematic representations (a) and (b) of a water molecule. (Modified from various sources.)

Table 1 Some comparative densities of water and other substances or elements

Substance	*Densities (g cm^{-3})*
Wood	
Seasoned balsa	0.11–0.14
Seasoned maple	0.62–0.75
Seasoned ebony	1.11–1.33
Water	1.00
Calcium	1.55
Aluminum	2.70
Iron	7.87
Lead	11.34
Mercury	13.55
Uranium	18.95
Platinum	21.45

Information from multiple sources.

Table 2 Comparative densities of average ocean water (salinity ∼35%), freshwater ice, and pure distilled water at different temperatures

Temperatures (°C)	*Densities (g cm^{-3})*
20.0	1.02760, ocean water (salinity 35%)
0.0	0.9168, freshwater ice
0.0	0.99987, pure water (from here on)
2.0	0.99997
3.98–4.00	1.00000
6.0	0.99997
8.0	0.99988
10.0	0.99973
12.0	0.99952
14.0	0.99927
16.0	0.99897
18.0	0.99862
20.0	0.99823
22.0	0.99780
24.0	0.99733
26.0	0.99681
28.0	0.99626
30.0	0.99568
32.0	0.99505

Values from Hutchinson (1957), Pinet (1992), and Weast and Astle (1979).

by changes in the water temperature, and in pressure. For the vast majority of inland lakes, only vertical differences in salt concentrations and temperatures are of significant influence to mixing processes. Fixed or uniform additions of salts to the water tend to cause *linear increases* in the density of water. In contrast, fixed or uniform changes in the temperature (both below and above 4 °C) of water cause *nonlinear changes* in the density of water (see **Table 2**). The density of pure water is maximum at a temperature of 4 °C (3.98 °C to be precise). It is at this temperature that the interatomic and intermolecular motions and intermolecular distances of water molecules are least. One consequence of this reduction is that more molecules of H_2O can fit into a fixed space at 4 °C than at any other temperature. This compaction allows the most mass per unit volume and thus the greatest density. It is especially noteworthy that the temperature at which water has the maximum density is above its freezing point.

Because the differences in densities, within a few degrees above and below 4 °C, are very slight, it takes relatively little wind energy to induce substantial vertical mixing when water temperatures are within those ranges. An example period, for those lakes that become covered with ice in the winter, would be shortly before an ice cover develops and shortly after the ice cover departs. However, it takes much more energy to cause extensive mixing when the density differences are high, such as is common between the usually warm upper waters and colder lower waters of Temperate Zone lakes during summer. The greater the top-to-bottom differences in temperature, the greater the top-to-bottom differences in density and, consequently, greater are the energies required for wind-induced mixing.

There is an old, but still valid, cliché in the northern hemisphere that '... it is cold up north and warm down south.' Water temperatures in more northerly Temperate Zone lakes tend to average cooler than those of more southerly tropical lakes. Interestingly, although the upper-water summer temperatures in tropical lakes are somewhat higher than those of Temperate Zone lakes, the lower-water temperatures in tropical lakes are substantially higher than those ordinarily found in the lower waters of Temperate Zone lakes. It might there fore seem that there would be an easy top-to-bottom mix of the water in tropical lakes. Indeed some shallow tropical lakes, with only slight top-to-bottom temperature differences, may have this. However, because of the nonlinear increases in water density with temperature, tropical lakes can be surprisingly stable and resistant to much vertical mixing.

Table 2 provides a listing of some comparative densities. Let us consider two hypothetical lakes with just a 2 °C spread between their lower and upper waters. For example, if a Temperate Zone lake in the spring, not long after the ice departed, had lower and upper waters of 4.0 and 6.0 °C, respectively, the density difference would be $1.00000-0.99997=0.00003\,g\,cm^{-3}$. In contrast, a warmer tropical lake whose lower and upper water temperatures may be 26.0 and 28.0 °Cs would have density differences that are much greater ($0.99681-0.99626=0.00055\,g\,cm^{-3}$). Thus, the top-to-bottom ratio or density difference of these two lakes with a temperature difference of just 2 °C would be 55/3 or ~18 times as great in the tropical lake as in the Temperate Zone lake. The example above is only hypothetical but it shows the nonlinear influence of density changes with temperature, a property of water that influences, to varying degrees, the stratification and mixing of lakes around the world.

Heat Capacity/Specific Heat

Heat is a form of energy and, as such, we can measure changes in the temperature of a given volume of a substance and determine its heat capacity. Water is the common standard used and its heat capacity (arbitrarily defined as the heat needed to increase the temperature of 1 g of water by 1 °C) is comparatively large. When the mass is also considered then the number of calories needed to raise 1 g of a substance by 1 °C is termed its *specific heat*. For water, the value is $1\,cal\,g^{-1}$. That quantity may not seem like much but, compared to other materials, the heat capacity or specific heat of water ($1.00\,cal\,g^{-1}$) and ammonia ($1.23\,cal\,g^{-1}$) are much greater than that of most other substances (**Table 3**). Consequently, these two liquids are commonly used to exchange heat in refrigerators and air conditioners.

Along with its ever changing and mesmerizing aesthetic qualities, inland waters are of immense importance in the storage and release of heat. In terms of freshwater lakes, the influence of their heat capacity can be seen most easily around very large lakes located in Temperate Zone latitudes and more inner continental areas. It is in these areas that even larger swings in seasonal air temperature would ordinarily occur in the absence of those lakes. Parts of the immediate surrounding areas of Lake Baikal in Russia (this is actually the world's deepest freshwater lake as well as one with the greatest volume of water) and the five St. Lawrence Great Lakes of North America are prime examples of the 'thermal buffering' these large lakes provide to their surroundings because of their large heat capacity.

Table 3 The specific heat ($cal\,g^{-1}$) of selected substances compared to that of ice, pure water, and ammonia

Aluminum	0.215
Copper	0.092
Gold	0.030
Lead	0.030
Silver	0.056
Zinc	0.092
Ethyl alcohol	0.60
Ice (at 0 °C)	0.51
Water	1.000
Ammonia	1.23

Information from multiple sources.

For humans, this may mean some 'beneficial economic consequences' as portions of a lake's heat capacity are slowly released or 'shed' to down-wind regions as the fall and winter seasons progress. The immense thermal capacity of Lake Baikal is such that the lake and its immediate environments are roughly 10 °C warmer in December and January, and about 7 °C cooler in June and July, than in the cities of Irkutsk (about 50 km to the west of the southern half of Lake Baikal) and Ulan-Ude (about 70 km to the east of the lake). Several coastal and near-coastal regions of the St. Lawrence Great Lakes also provide impressive beneficial evidence of the influence of the Great Lake's heat capacity. There may be reduced costs associated with home and business heating in some coastal regions. An extended or milder autumnal period permits greater production in near-shore plantations of fruit trees and vineyards. Economic benefits may also accrue in some coastal regions of higher terrain during winter, when enhanced snows permit additional winter skiing, snowmobiling, and other winter sports.

However, some influences of a lake's heat capacity have 'detrimental economic consequences'. There are costs involved with snow removal, increased vehicular accidents (because of slippery roads), the corrosion of cars (attributable to road salts), and the potential long-term ecological changes associated with lake and stream salinization. There are also greater heating costs in spring as cooler water bodies extend their cooling influence inland. In late fall and winter, before an ice-cover develops, heavy snows may result when water vapor, being formed by evaporative processes off a relatively warm lake, is buoyed into much colder Arctic air (northerly Temperate Zone) crossing the lake. The rising water vapors may freeze, coalesce to ice crystals, and be carried down wind to shore areas where they fall out as snow. Perhaps the most dramatic of all the detrimental consequences is seen following the sometimes paralyzing effect of occasional, but intense, 'lake-effect' snow storms of mesoscale proportions. The lake-effect snow storms tend to have their greatest impact at the downwind end of the St. Lawrence Great Lakes after very cold Arctic air ($\geq$13 °C colder than the temperature of the lake) has moved across a long axis of the lakes and deposited its snows. These deposits or drops of snow may be in a broader synoptic pattern, but sometimes they are in very narrow bands of thick snow that may bring auto traffic, schools, and businesses to a stop. In the St. Lawrence Great Lakes region of North America, three of the better known areas where unusually heavy deposits of lake-effect snows may occur are (1) portions of the Upper Peninsula of Michigan on the southeastern shore of Lake Superior, (2) the southeasterly and easterly shores of Lake Ontario, especially the Tug Hill Plateau area of New York State, and (3) the easterly end of Lake Erie, around Buffalo, NY. Indeed, the St. Lawrence Great Lakes have been considered 'weather factories' capable of causing twists of climate found in few other parts of the world.

Heat of Fusion/Melting

This is just the amount of heat exchanged during a phase shift from either liquid water to solid ice, or from solid ice to liquid water. One gram of water at 0.0 °C can be converted to ice at 0.0 °C if 80 cal (79.72 cal g^{-1} to be precise) are released in the process. The same quantity, i.e., 80 cal, is required to melt that 1 g of ice back to 1 g of water. No further caloric additions or subtractions are needed to effect the phase shift.

Because of the heat needed to melt ice, researchers might intuitively expect to see a brief but substantial drop in the mean or weighted lake-water temperature when the ice cover of a lake melts in the spring season. For example, assume there is a hypothetical northerly latitude and a 20-m deep lake in late winter (March). Consider that the lake is covered with 50 cm of ice at 0.0 °C. Consider further that the weighted mean temperature of the 1950 cm (essentially 1950 g) water column below the ice is 3.0 °C. The heat content of that water column would be 5850 cal (1950 g × 3 cal g^{-1} = 5850 cal). Assuming that there are no further gains or losses of heat to the lake, the amount of heat required to melt the ice would be 3680 cal (80 cal g^{-1} × 50 cm of ice × 0.92 g cm^{-1}, allowing for density of pure ice rounded to two decimals = 3680 cal). If some of the caloric content of the water column could be used to melt all the ice, the total caloric content would drop to 2170 cal (5850 cal - 3680 cal = 2170 cal). If those 2170 cal were now equally distributed within a 1-cm^2 square and 20-m (2000 cm, essentially 2000 g) deep water column, the mean water temperature would need to drop from 3 to 1.08 °C (2170 cal/2000 cal = 1.08 °C). A drop of about 2 °C during the melting of ice would be large!

As it turns out, the hypothetical example in the above paragraph is not realistic. Some background follows. Many years ago as a graduate student, I took daily measurements of ice thickness and top-to-bottomwater temperatures for two winters and right through the spring ice break up in a Midwestern U.S. lake. From conversations with others, I was told to expect, and did anticipate, a substantial drop in mean water temperature as the ice melted... especially in the last few days of ice cover when the ice thinned rapidly. However, I did not measure any big drops in lake temperature and, in retrospect, should not have anticipated them. The reasons researchers do not see large decreases in lake temperatures with ice loss reflect some interacting physics. For example, there may be somewhat differing weather patterns each spring. The ice generally melts over an extended period of time, from several days to several weeks, not suddenly. Half or more of the total ice thickness may be lost from the top of the ice by melting from warming air temperatures above the ice, not necessarily from waters that are just above freezing below the ice. Because of its *albedo* (*percent of incoming solar radiation that is reflected back into space*) dark or open water generally reflects only a small fraction of the incoming solar radiation, whereas white snow cover on a frozen lake can reflect a large fraction of incident radiation. Indeed, snow cover extending into the spring period can delay the date the ice disappears. However, with increasing amounts of solar radiation, rising air temperatures, melting snows, and darkening ice, the water below the ice may be gaining some heat from solar inputs at the same time it is losing some heat in melting an overlying ice cover. Moral of the story: Do not expect a big drop in mean water temperature as an ice cover melts on a lake.

Heat of Vaporization/Condensation

As was the case for 'Heat of Fusion/Melting,' the heat of vaporization/condensation also represents the amount of heat exchanged during a phase shift. For vaporization, it is the quantity of heat (540 cal g^{-1}) needed to convert 1 g of water to 1 g of water vapor. The same amount of heat is exchanged or released in the phase shift during the condensation of 1 g water vapor to 1 g of water.

Aquatic scientists may be naturally impressed with the large amount of heat exchanged (80 cal g^{-1}) in the phase shift from water to ice, or from ice to water, but the amount of heat exchanged ($540\,cal\,g^{-1}$) in the phase shift from water to water vapor, or water vapor to water is 6.75 times larger ($540/80 = 6.75$). Although the importance of this large amount of heat exchange via vaporization or condensation may be underappreciated by humans, it is huge. On a small but critical scale for life, water evaporating off perspiring warm-blooded animals, including humans, helps maintain body temperatures within narrow survivable limits. On a global scale, the seemingly endless phase shifts between liquid water and water vapor in the atmosphere are key determinants in the redistribution of water and heat within the hydrological cycle around the world.

Isotopes

An isotope is one of two or more forms of the same chemical element. Different isotopes of an element have the same number of protons in the nucleus, giving them the same atomic number, but a different number of neutrons giving each elemental isotope a different atomic weight. Isotopes of the same element have different physical properties (melting points, boiling points) and the nuclei of some isotopes are unstable and radioactive. For water (H_2O), the elements hydrogen (atomic number 1) and oxygen (atomic number 16) each have three isotopes: ^1H, ^2H, and ^3H for hydrogen; ^{16}O, ^{17}O, and ^{18}O for oxygen. In nature, the ^1H and ^{16}O (usually just given as O) isotopes are by far the most common. In water, the water molecule may be given as ^1H_2O or hydrogen oxide, ^2H_2O or deuterium oxide, and ^3H_2O or tritium oxide, the radioactive one. Both of the latter two are sometimes called heavy water because of their increased mass. However, the phrase 'heavy water' gained notoriety primarily because of the association of ^2H_2O or deuterium oxide, also called the deuterated form of water, in the development of nuclear weapons. Many elements have isotopes, but the isotopes of hydrogen and oxygen are of particular interest because fractionation occurs in vapor–liquid–solid phase changes. Heavier molecular 'species' tend to be enriched in the condensation phase and lighter molecular 'species' in the vapor phase. Some isotopes can be used to great advantage as tracers in understanding water movements and exchanges within atmospheric, oceanic, lake, stream, and ground water systems.

Sublimation

Water is said to be sublimated, sublimed, or undergo sublimation when it passes directly from a solid (ice) stage to a gas (vapor stage) without becoming a liquid in between. The latent heat of sublimation, i.e., the heat required to make the form of water change from ice to a water vapor, is $679\,cal\,g^{-1}$. This quantity is larger than the heat required to melt ice ($80\,cal\,g^{-1}$) and vaporize water ($540\,cal\,g^{-1}$) combined ($80 + 540 = 620\,cal\,g^{-1}$). Because there may be multiple heat sources and sinks (e.g., the air above the ice and the water below the ice) associated with changing ice thickness on frozen Temperate Zone lakes, it is a challenge to assess the quantitative role that sublimation may play in those changes.

Some practical effects of sublimation may be visualized by observing a reduction in the volume of some dry ice (solid CO_2) or camphor. In another example, after several weeks of continuing subfreezing temperatures and deep frost, and assuming that no deicing salts were used, sublimation is most likely responsible for the slow disappearance of an ice sheet over the surface of a frozen sidewalk. Sublimation is also the main process by which wet clothes, which were hung out to dry in subfreezing temperatures, may dry. In the latter case, the water on the clothing quickly freezes to ice, but then slowly vaporizes through sublimation, and the clothes dry. In more recent years, freeze-dried vegetables, fruits, and other products (including instant coffee) provide other examples where the practical application of sublimation is utilized to both market and preserve food.

Surface Tension and Cohesiveness

Surface tension may be regarded as the resistance offered by liquid water to forces attempting to deform or break through the surface film of water. It is an interesting property and, for water, the surface tension measured in Newton's per meter ($N\,m^{-1}$), is high and shows a slight *increase* as the temperature falls from 100 ($0.0589\,N\,m^{-1}$) to 0 °C ($0.0765\,N\,m^{-1}$). The molecules of water are strongly attracted to each other through their *cohesiveness* (attraction of like substances). The properties of surface tension and cohesiveness work together in water in shaping the small rounded water droplets seen on a table top or a car windshield. The same properties help to form the slightly flattened to spherically-shaped raindrops as they fall through the air.

The primary force for restoring larger wind-generated surface and internal waves of lakes is gravity, but the primary force for restoring the much smaller capillary waves or ripples on a lake's surface seems to be surface tension of the water itself.

The surface tension of water is sometimes used to advantage in parlor games in which someone claims that he/she can float a more dense (than water) steel needle on less dense water. When the needle is lowered slowly and carefully with its long axis paralleling the surface of the water, it may be possible to 'float the needle' because the high surface tension of the water may prevent the needle from sinking. Do not try this by lowering one of the sharp ends of the needle first because a point application of the needle will exceed the surface tension of the water film, and the needle will sink rapidly.

When responding to a 'fire call' in fire trucks, water is the most common and practical substance used by firemen. Water is cool, it suppresses heat, it puts out fires and sometimes there is much water to spare. However, the *high surface tension of water can reduce its effectiveness in suppressing some fires.* Surfactants are compounds that reduce the surface tension of water. In their response to a 'fire call' firemen often quickly attach hoses to street fire hydrants and spray water from that source on a burning structure. Although the addition of tiny quantities of surfactants to water may help put out fires, it is not practical (or safe) to add surfactants to an entire distribution system of a city. However, the addition of tiny quantities of surfactants to the volume (roughly 1.89 m^3 or 500 gallons in the United States) of water being carried in the fire truck would make that truck water 'wetter.' Some combustibles could be penetrated more easily by this wetter water of reduced surface tension and selected fires could be put out more rapidly.

There is a specialized community of organisms, sometimes called neuston, associated with the surface film. For many observers of nature, it is always fascinating to see small insects such as pond skaters or water-striders (*Gerris* sp., within the insect Order Hemiptera), and whirligig beetles (*Gyrinus* sp. and *Dineutes* sp., within the insect Order Coleoptera), running around on the surface of ponds, sheltered lakes, and some streams. Because of padded ends to the long middle and hind feet of water striders, and the much shortened but paddle-like feet of the whirligig beetles, the high surface tension of the water is such that the insects may dimple, but not break through, the surface film.

One of the easiest ways of getting popcorn into your mouth is by touching your tongue to some popcorn in a container. Here again it is the surface tension of the water on your tongue that lets you 'hold on' to the light popcorn easily.

Viscosity

This property may be thought of as the internal friction or resistance exerted on one substance (gas, liquid, or solid) as that substance tries to flow or move through the same or another liquid. One way of visualizing the influence that liquids or semiliquids of progressively greater viscosities might exert would be to take three glass marbles (same diameter and density) and drop one in each of three similar-sized glasses, one glass containing water, one light oil, and one honey, all at the same temperature. The marble would descend quite rapidly in water, more slowly in the light oil, and very much more slowly in the glass of honey. In this example, honey would obviously exert the most friction or resistance to movement through it and have the greatest viscosity. Viscosity is usually measured in poises ($N\,s\,m^{-2}$) or centipoises (= 0.01 P). Water at 20 °C has a viscosity of 0.01002 P or 1.002 cP.

The rate of passive descent through a liquid reflects the density of the liquid itself as well as the surface area and density of the substance moving through it. Viscosity changes with water temperature in that viscosities decrease as water temperatures rise and increase as water temperatures fall. Many fish are powerful enough, slippery from mucous on their skin, and shaped so they can 'slip through' water relatively easily. In contrast tiny zooplankton, with multiple projections on their body, are ordinarily challenged as they attempt to move in any direction and particularly so when moving in cool waters.

Colligative Properties

These are the four special properties of water that are significantly altered or modified when solutes are added to and dissolve in water. The alterations or modifications of a colligative property (regarded as a binding property) may be predictable in dilute solutions when the number of solute particles is known. It is the number of solute particles, not their chemical nature, that determines the extent to which a property is modified.

The four colligative properties of water are *vapor pressure* (when water is in equilibrium with its own vapor), *osmotic pressure* (the pressure controlling the diffusion of a solvent across a semipermeable membrane), *boiling point* (the temperature at which water undergoes a phase shift to a gas), and *freezing point* (the temperature at which water undergoes a phase shift to a solid). Even at standardized pressures and temperatures, the extent to which a property is modified depends on the number of solute particles added. Generally, if we add a fixed number of solute particles of a sugar or salt to a liter of pure water, there would be some consequences. The vapor pressure would be lowered but the osmotic pressure would rise. The boiling point (also termed *boiling-point elevation*) would be elevated a bit above the usual boiling point of 100.0 °C. In the latter case, a watery mixture with solutes (e.g., a well-mixed soup being heated for a meal) would have to get hotter than the boiling point of pure water before it would boil. The freezing point would be lower than 0.0 °C. A practical application of this (also termed *freezing-point depression*) is easily seen, in parts of the northerly Temperate Zone in winter, following the application of deicing salts to melt the ice and snow on roads and sidewalks. Although not a colligative property as such, a simple increase in physical pressure also lowers the melting point of ice (~0.007 °C/atm) and helps form snowballs (when the snow is not too cold) and form a lubricating layer of water under the blade of an ice skate.

Further Reading

Eichenlaub VL (1979) *Weather and climate of the Great Lakes Region*, pp. 1–27. Notre Dame, IN: University of Notre Dame Press.

Hutchinson GE (1957) Geography, Physics, and Chemistry. *A Treatise on Limnology*, vol. I. Chap. 3. New York, NY: John Wiley.

Klaff J (2002) *Limnology, Inland Water Ecosystems*, pp. 35–40. Upper Saddle River, NJ: Prentice-Hall.

Kozhov MM (1963) *Lake Baikal and Its Life*, pp. 15–19. The Hague, Netherlands: Dr. W. Junk Publishers.

van der Leeden F, Troise FL, and Todd DK (1990) *The Water Encyclopedia,* 2nd edn., pp. 774–777, Chelsea, MI: Lewis Publishers.

Lemmon EW, McLinden MO, and Friend DG (2003) Thermophysical properties of fluid systems. In: Linstrom PJ and Mallard WG (eds.) *NIST Chemical Webbook, NIST Standard Reference Database Number 69*. Gaithersburg, MD: National Institute of Standards and Technology. http://webbok.nist.gov.

Morgan JJ and Stumm W (1998) Properties. In: Kroschwitz JI and Howe-Grant M (eds.) *Encyclopedia of Chemical Technology,* 4th edn., vol. 25, pp. 382–405. New York, NY: John Wiley.

Mortimer CH (2004) Lake Michigan in motion. Responses of an Inland Sea to Weather, Earth-Spin, and Human Activities, p. 143. Madison, WI: University of Wisconsin Press.

Pinet PR (1992) *Oceanography, An Introduction to the Planet Oceanus*, pp. 120–128. St. Paul, MN: West Publishers.

Scott JT and Ragotzkie RA (1961) The heat budget of an ice-covered inland lake. *Tech. Rep. 6. ONR Contract 1202 (07)*. Madison, WI: Deptartment Meteorology, University of Wisconsin.

Voet D, Voet JG, and Pratt CW (1999) *Fundamentals of Biochemistry,* pp. 23–27. New York, NY: John Wiley.

Wallace RA, Sanders GP, and Ferl RJ (1991) *Biology. The Science of Life*. ch. 2. New York, NY: Harper Collins.

Weast RC and Astle MJ (1979) *CRC Handbook of Chemistry and Physics,* 60th edn. Boca Raton, FL: CRC Press.

Chemical Properties of Water

J H Aldstadt III and H A Bootsma, University of Wisconsin-Milwaukee, Milwaukee, WI, USA
J L Ammerman, SEAL Analytical, Inc., Mequon Technology Center, Mequon, WI, USA

Water is H_2O, hydrogen two parts, oxygen one, but there is also a third thing, that makes it water and nobody knows what it is.

—D.H. Lawrence (1929)

Introduction

Water is the most abundant molecule on Earth. In spite of being so common, water is quite unusual – from its high melting and boiling points to its tremendous solvating power, high surface tension, and the largest dielectric constant of any liquid. In this article, we present an overview of the chemical properties of water. The phrase 'chemical property' is context dependent, which we define in general as a description of the way that a substance changes its identity in the formation of other substances. A universally accepted set of chemical properties does not exist in the same way that there is, more or less, a standard set of physical properties for a given substance. Whereas a given substance has intrinsic physical properties (such as melting point), by our definition chemical properties are clearly tied to change. In addition to reactivity, a substance's 'chemical properties' also typically include its electronegativity, ionization potential, preferred oxidation state(s), coordination behavior, and the types of bonding (e.g., ionic, covalent) in which it participates. Because these properties are extensively studied in general chemistry courses, we will not further discuss them here. Rather, we move beyond the basic general chemistry concepts and focus upon water in a limnologic context – particularly, its bulk fluid structure and aspects of its chemical reactivity in the hydrosphere.

In the following pages, we begin by briefly reviewing the molecular structure of water and then discuss models for its structure in 'bulk' solution. We then turn our attention to the hydration of ions and an overview of important reactions that involve water, including acid–base, complexation, precipitation, and electron transfer. We conclude with a look at trends in the chemical composition of freshwater that are fundamental to the field of limnology.

The Structure of Water

Knowledge of the structure of water is the basis for understanding its unique chemical and physical properties. Like the other nonmetallic hydrides of the Group 16 elements, water is a triatomic molecule that forms a nonlinear structure. In terms of group theory, water has two planes of symmetry and a twofold rotation axis and is therefore assigned point-group C_{2v}. The H–O–H angle is 104.5°, formed as a result of the distortion of the O–H bond axes by the two pairs of nonbonding electrons on the oxygen atom. Although water is often described as having four sp^3-hybridized molecular orbitals in a slightly distorted tetrahedral geometry, models based solely upon that configuration fail to accurately predict the properties of liquid water, particularly the extent and influence of hydrogen bonding on the structure of the bulk fluid state. However, a tetrahedral geometry is in fact present in the solid state, giving rise to the sixfold axis of symmetry that is characteristic of ice, and in large part as the basis of the networks that form in the bulk liquid, though in a rapidly fluctuating dynamic state.

Models for the bulk fluid structure of water are a function of the noncovalent van der Waals forces that exist between water molecules. There are five major types of van der Waals forces that occur between neutral molecules and ions in solution: (1) London (or dispersion) forces, in which transient dipoles form by variations in electron density between neutral molecules; (2) Debye forces, in which the dipole of a molecule induces the formation of a dipole in an adjacent neutral molecule; (3) Keesom forces, which form between neighboring dipoles; (4) Coulombic forces, the electrostatic attraction (and repulsion) of ions; and (5) hydrogen bonds, which involve the electrophilic attraction of a proton to electronegative atoms such as oxygen and nitrogen. All of these forces are present in aqueous solution to varying degrees – hydrogen bonding being the most dominant. The high negative charge density of the oxygen atom relative to the high positive charge density of the hydrogen atom creates a large (1.84 D) electric dipole moment for the water molecule (**Figure 1**). Because of the large dipole moment, the partial positive charge on the H atom is attracted to electron density, while the partial negative charge on the O atom causes the attraction of electrophilic H atoms. In this way, hydrogen bonds are formed, representing the strongest of the van der Waals forces that exist between neutral molecules. While each hydrogen bond is ~20 times weaker than a typical covalent bond, each water molecule can

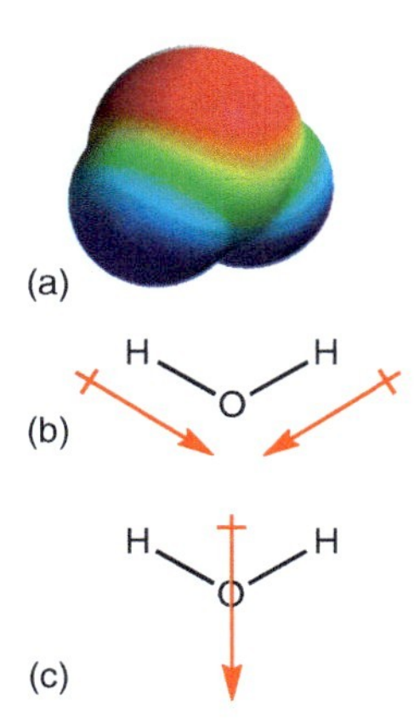

Figure 1 (a) The distribution of electron density in molecular water (red = high, blue = low). Representation of the electric dipolar nature of molecular water, as contributing dipoles along each O–H axis (b) and as a net dipole (c).

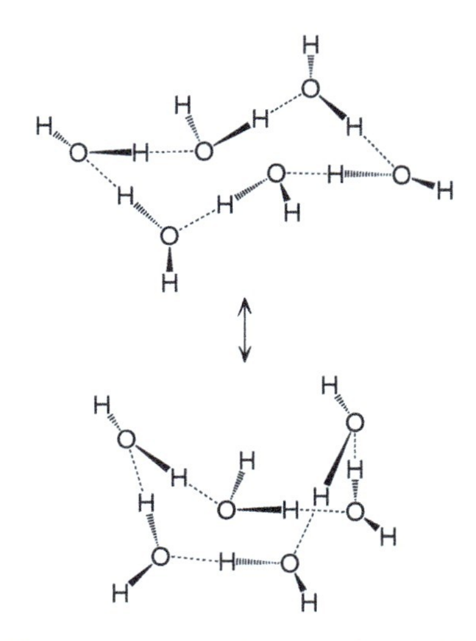

Figure 2 The arrangement of hexagonal water into a 'chair' conformation (top) and less stable 'boat' conformation (bottom).

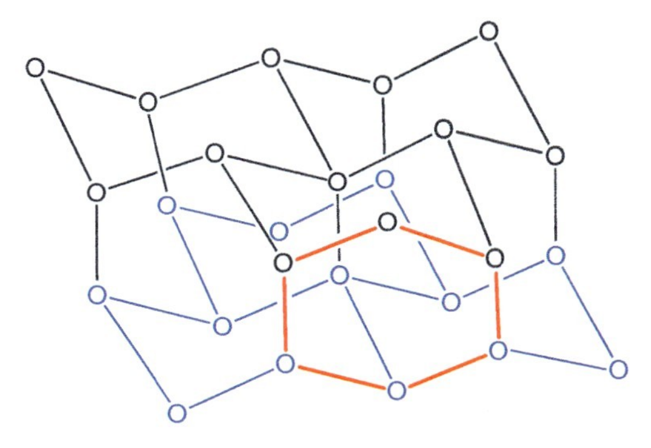

Figure 3 The structure of the most common form of ice (hexagonal ice), an arrangement based upon the HOH 'chair' hexamer. Each oxygen atom is at the approximate center of a tetrahedron formed by four other oxygen atoms. The sixfold axis of symmetry is shown in red for a layer of water 'chairs' (black) overlaying another layer (blue). (Hydrogen atoms are not shown for clarity.)

participate in multiple hydrogen bonds – one to each H atom and one (or more) to each nonbonded pair of electrons on the O atom.

The key to understanding the structure of bulk water – and its abnormal properties – is understanding the way that noncovalent hydrogen bonds affect its intermolecular interactions. Although one might expect that the random translational motion of molecules in a liquid results in an amorphous structure, the extensive network of hydrogen-bonded molecules in the liquid state of water gives rise to a surprisingly very high degree of order. Water has considerable short-range order that continues to a distance of at least ~10–15 Å from the 2.75 Å diameter water molecule. Hydrogen bonds are certainly not peculiar to water, but in water they form such elaborate, extensive, and strong networks that they create a 'bulk' structure with significant order, order that is in fact maintained up to its boiling point.

A great deal of research has been devoted to improving our understanding of water's structure in condensed phases – broadly divided into studies of short-range and long-range order, the latter defined as beyond ~15 Å. These research endeavors have been both theoretical and empirical, with theoreticians employing advanced computational tools for molecular modeling, and experimentalists armed with a wide variety of spectroscopic techniques. Models for the structure of water in the solid phase (i.e., in the various ices that can form) generate little controversy because theoretical models can be directly verified by crystallographic and neutron-scattering techniques. Because of the much more limited atomic motion in the solid state, crystallographic methods have provided an accurate picture of the various ices that form as a function of temperature and pressure. The most common type of ice under ambient conditions is hexameric ice, in which six water molecules are hydrogen bonded to form a hexagonal ring, as shown in **Figure 2**. The most stable state for this structure is a so-called 'chair' conformation (analogous to cyclohexane), in which H–O···H bonds alternate around the ring (where '–' is a covalent bond and '···' is a hydrogen bond). Also shown in **Figure 2** is the 'boat' conformation, an energetically less stable conformation than the 'chair' structure. Each O atom has a nearly tetrahedral arrangement of H atoms surrounding it, in which two H atoms are covalently bonded and two noncovalently as hydrogen bonds. The sixfold axis of symmetry found in ice (**Figure 3**) is the result of the building blocks of cyclic hexamers.

Unlike models for ice, much controversy continues to surround models for the structure of liquid water. This may be somewhat surprising given that water is a simple molecule, yet general agreement on a realistic model remains elusive despite the application of powerful computational and experimental approaches. Predicting the precise arrangements of hydrogen-bonded neighboring water molecules is challenging because the structures are in a state of rapid flux (at subpicosecond timescales). Some insight into the structure of bulk water can be gleaned by examining the structural changes that occur upon the melting of ice. When ice melts, the increase in temperature causes a slight disruption of the hydrogen-bonded network, thereby initially causing the ice crystalline lattice to collapse. Whereas the structure of ice is >80% ordered, only an ~10% decrease in order occurs upon transition to the liquid phase. In this way, much if not most of the short-range order is maintained, which in fact continues to persist in part all of the way to the boiling point at 100 °C, where the order is essentially lost completely. The partial collapse of the ordered environment during melting results in slightly more compact hexameric chairs. Consequently, water has the very unusual property of maximal density at a temperature that is higher than its melting point. Above 4 °C, further disruption of the intricate networks of cyclic hexamers by more intensive thermal agitation causes the structures to become more open with a consequent decrease in water's density.

Water forms clusters in the liquid state. The presence of 'ice-like' structures in water, based on not only hexameric but also pentameric and octameric building blocks, along with 'free' swimming water molecules in more amorphous regions, is the generally accepted model (**Figure 4**). However, there have been intriguing studies that suggest that there are regions that are far more complex than the structures analogous to ice. Curiously, one of the earliest is found in Plato's dialogue *Timaeus*, where the ancient Greek's classification of matter – Earth, Fire, Air, and Water – is described in mathematical (geometric) terms. In the Platonic conception of 'substance,' matter is intrinsically composed of triangles. Earth is cubic (i.e., two equilateral triangles each comprising six faces), Fire is tetrahedral (four triangles), and Air is octahedral (eight triangles). In Plato's view, water is the most complex structure, taking the form of an icosahedron. A regular icosahedron has 20 faces, with five equilateral triangles meeting at each of the 12 vertices. Thus, along with the dodecahedron, these regular convex polyhedra comprise the famous 'Platonic Solids.' This ancient conception of water may seem quaint, yet it is strikingly similar in concept to several recent theoretical models of the structure of water in the bulk liquid phase. Clusters based on dodecahedra and icosahedra have been proposed by molecular modeling and supported by experiment to exist in water – though the evidence remains somewhat controversial. Early work by Searcy and Fenn

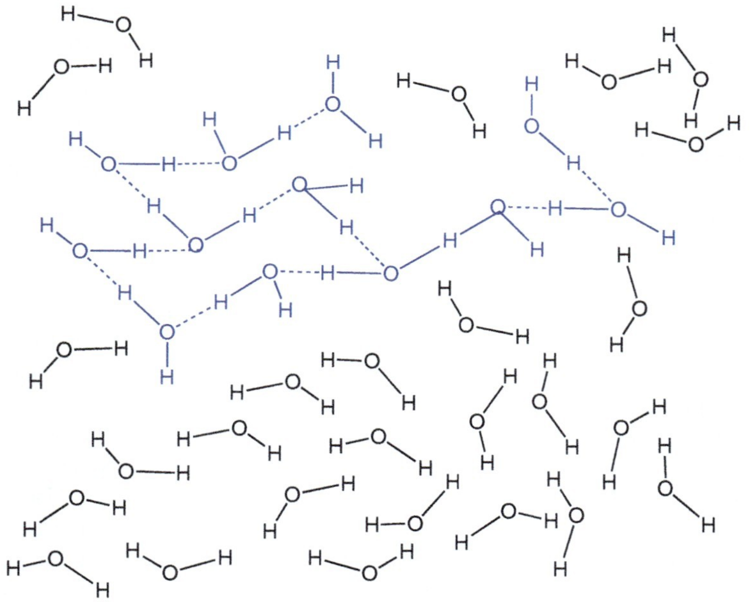

Figure 4 Proposed models for the structure of bulk water. (Top) The "flickering cluster" model, with ice-like ordered regions (high-lighted in blue) surrounded by amorphous regions where little short-range order is present. Molecular modeling and some experimental evidence suggests that quite complex structures, such as dodecahedra (bottom left) and icosahedra (bottom right), may also exist.

on protonated water clusters by molecular beam mass spectrometry found that a large peak in the spectrum, which corresponded to 21 water molecules (a so-called 'magic number') was present, that is, for a cluster of unusual stability. Speculation arose that the structure of this 'magic' cluster was a dodecahedral complex of 20 water molecules, each vertex occupied by an oxygen atom and a hydronium ion trapped within (e.g., as in clathrates). Recent work by Dougherty and Howard has indeed found evidence for dodecahedral clusters, and Chaplin has proposed a theoretical model for the formation of icosahedral clusters, a model that has been supported by recent neutron scattering experiments.

Solvation by Water

Ions in aqueous solution interact with one another and with other nonelectrolytes, and their presence in water's dipolar electronic field creates relatively strong noncovalent bonds such that the hydrated ion is the form that undergoes further interactions and chemical reactions, and has consequent implications for the rates of these processes. Only in the gas phase do 'bare' (unsolvated) ions exist; in the liquid phase, all ions are hydrated to some degree.

To appreciate the solvating power of water, the solubility parameter (δ) provides a useful measure, defined as the ratio of the energy required to completely break all intermolecular forces that maintain the liquid state. We represent δ quantitatively as

$$\delta = \sqrt{\left(\frac{\Delta E_V}{V}\right)}$$

where ΔE_V is the total energy required to vaporize a solute. One can think of δ as the 'cohesive energy density' of a substance. Of course δ correlates strongly with polarity, with water not surprisingly having the highest value of δ when compared to other common solvents (**Figure 5**).

Before studying an example of the structure of a hydrated metal ion, we recognize that each water molecule is already 'solvated' to a very high degree of structural complexity. And because of the autoionization reaction of water, which we can represent as a net reaction:

$$H_2O \rightleftharpoons H^+ + OH^-, \qquad [1]$$

protons and hydroxide ions are formed that also become hydrated. Realistic structures of the reaction [1] products continue to be the subject of debate, but much evidence suggests that a more realistic way to describe the autoionization of water is

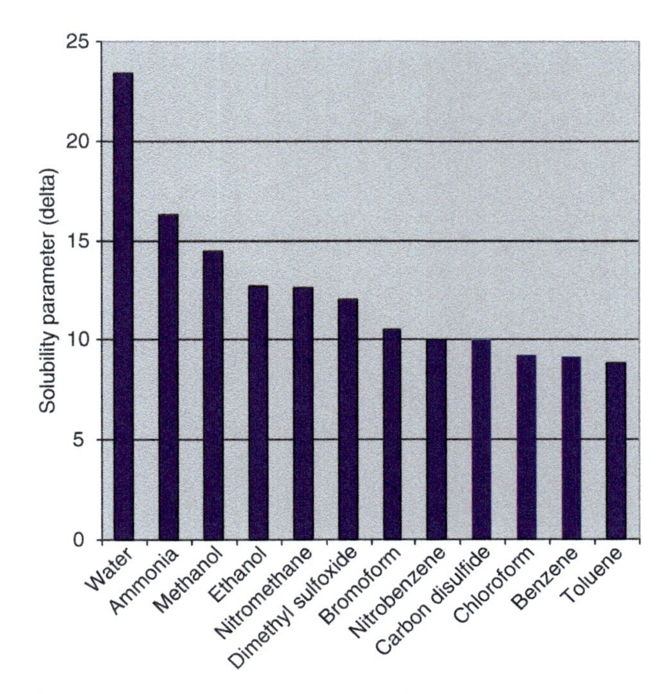

Figure 5 A comparison of Hildebrand's solubility parameter (δ) for various liquids (25 °C).

$$6H_2O \rightleftharpoons (H_2O)_2H^+ + (H_2O)_3OH^- \qquad [2]$$

Proposed structures for these ions are shown in **Figure 6**. For convenience, the simplistic products of reaction [1] are commonly used in the literature. However, more complex structures, such as those depicted in **Figure 6**, are themselves not yet fully accepted as realistic.

For ions in aqueous solution, the structures formed by hydration reactions are driven by geometric and electronic factors. The number of water molecules that coordinate as ligands to an ion typically varies from four to nine, and is a function of factors that include ion size, the number of vacant orbitals present, and the degree of ligand–ligand repulsion. Given the great interest in pollution by toxic metals, our understanding of cation hydration is more extensive than for anions, yet hydration of the latter should not be surprising given the dipolar nature of water as a ligand.

In **Figure 7**, the 'concentric shell' model for the hydration of an ion is illustrated for aluminum ion, which exists under ambient conditions in the +3 oxidation state. Three regions form the shells – an inner layer, known as the primary (1°) shell, an intermediate layer known as the secondary (2°) shell, and a third region comprised of the bulk fluid. The structure of the 1° shell is highly ordered, as shown in **Figure 7** for the tricapped trigonal prismatic

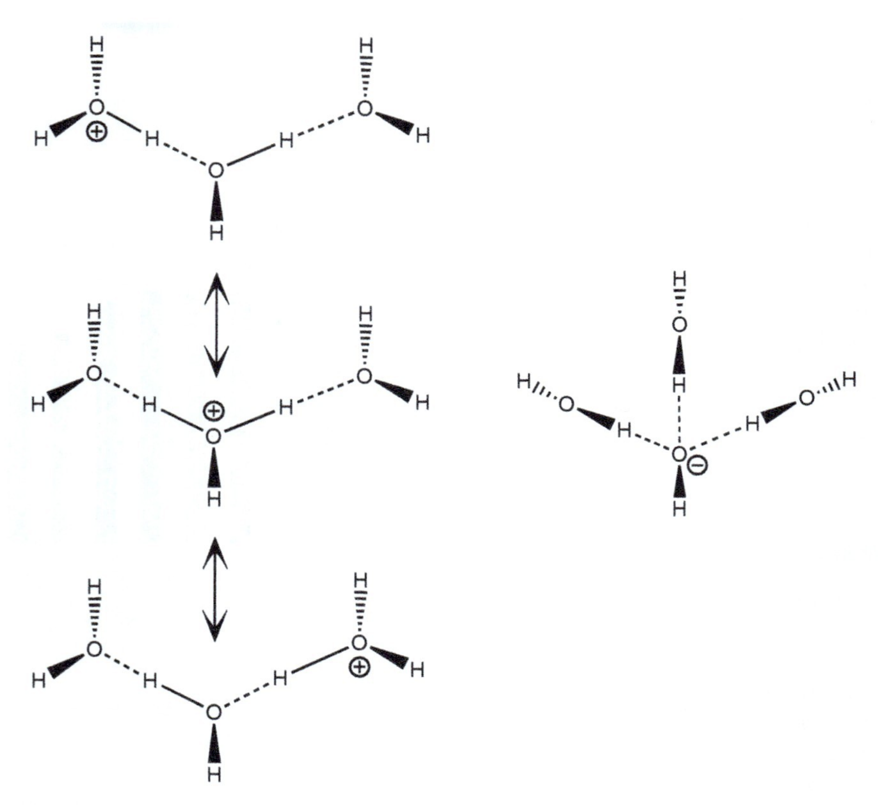

Figure 6 'Proton hopping' among three water molecules which together constitute a more accurate representation of a hydrated proton ($H_5O_2^+$). The center structure is the most energetically stable of the three shown. A more realistic structure for solvated hydroxide ion ($H_7O_4^-$) is also shown (right). Hydrogen bonds are denoted by dashes (---).

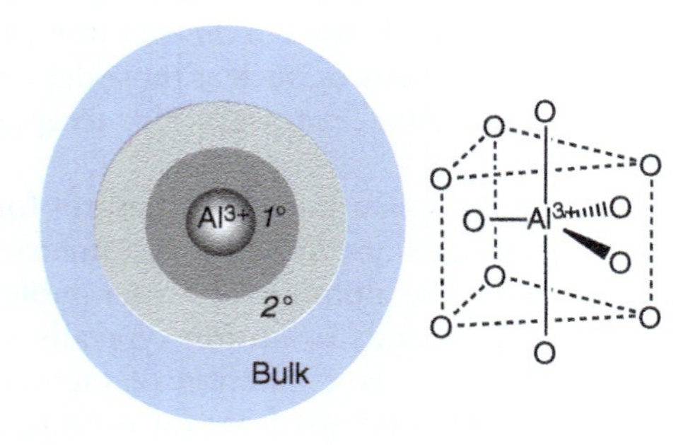

Figure 7 The 'concentric shell' model (left) for the hydration spheres surrounding a cation, showing the primary, secondary, and bulk solution shells. The primary hydration shell of aluminum ion (right), a tricapped trigonal prismatic geometry in which only the O atom positions for the 11 coordinating water molecules are shown.

arrangement of 11 water molecules closely surrounding the trivalent cation. In the 2° shell, the influence of the Al(III) ion's high charge density would create a more loosely held though structurally defined layer. The bulk fluid extends beyond the 2° shell where the range of the ion's force field has no apparent effect on the fluid structure. It is important to note that the concentric shell model is simplistic, focusing on the strongest inner layers that are present. That is, the model ignores long-range ordering effects, which, because of their weakness, are inherently difficult to study. For example, molecular modeling (theoretical) studies have suggested that for heavy metal ions in aqueous solution, the surrounding water would be affected by the electronic field of the ion to a distance corresponding to several dozen or more layers of water molecules. Only beyond these layers would the bulk water reflect the 'undisrupted' structural state of a pure solution of water.

The Reactivity of Water

While we may tend to think of water as relatively inert, it is actually a very reactive molecule, with the oxygen atom behaving as a strong electrophile and the protons involved in autoionization reactions. However, water's reactivity is attenuated by its extensive hydrogen bonding. The eightfold ratio between water's single relatively heavy (O) atom and two light (H) atoms, and the charge inequity that exists between them, gives rise to a rapid exchange of protons between adjacent water molecules (proton hopping). In a pure solution of water, proton hopping among water molecules is constantly occurring at a high rate – even at pH 7 where it is slowest, it occurs on the order of 1000× per second (**Figure 5**). In studies of hydrogen bonding and the solvation of ions by water, the exchange of protons is even faster than the millisecond timescale observed for a bulk solution of pure water. Nevertheless, water is treated as a stable molecule because the net structure (H–O–H) is maintained in spite of its intrinsic dynamic state.

Fundamentally, chemical reactions occur as means for a species (atom, molecule, or ion) to increase its thermodynamic stability. We can generally classify chemical reactions into two broad categories: (1) those that involve changes in oxidation state, and (2) those that involve changes in coordination environment. While the former redox processes stand alone, the latter type of reaction can be divided into acid–base, complexation, and precipitation reactions. We can illustrate these three subcategories of coordination reactions by the example of a series of hydrolytic reactions involving the Al(III) ion:

$$Al^{3+}_{(aq)} + H_2O \rightleftharpoons Al(OH)^{2+}_{(aq)} + H^+_{(aq)} \quad pK_1 = 8.2$$

$$Al(OH)^{2+}_{(aq)} + H_2O \rightleftharpoons Al(OH)^{+}_{2(aq)} + H^+_{(aq)} \quad pK_2 = 19.0$$

$$Al(OH)^{+}_{2(aq)} + H_2O \rightleftharpoons Al(OH)_{3(s)} + H^+_{(aq)} \quad pK_3 = 27.0$$

$$Al(OH)_{3(s)} + H_2O \rightleftharpoons Al(OH)^{-}_{4(aq)} + H^+_{(aq)} \quad pK_4 = 31.4$$

(The subscript 'aq' denotes 'in aqueous solution,' a reminder that all of these species are hydrated, the structures of which are not shown.) While none of the reactions above cause changes in oxidation states, all are acid–base reactions because of the generation of a (hydrated) proton. All can be classified as complexation reactions as well because of hydroxide ion acting as a ligand in its coordination to the metal cation, with the formation of complex cations and anions (with the exception of the third reaction). For the third reaction, because of the formation of a solid product, we classify it as a precipitation reaction. Chemical reactions in the environment that involve water as a reactant or product – i.e., each type of reaction illustrated above as well as redox reactions – represent an enormous volume of scholarly work; the interested reader is therefore referred to the 'Further Reading' listed at the end of this article and elsewhere in this Encyclopedia.

Trends and Patterns in Limnology

The chemistry that is mediated by water in natural aquatic systems varies in space and time. Often this variability is expressed in the form of trends and patterns, and by understanding their causes it is possible to gain insight into the mechanisms that control water chemistry. Ultimately, variation in the chemistry of lakes and rivers can be attributed to three controlling factors: (1) physical processes and properties, including lake morphometry, weather, and climate; (2) geologic setting; and (3) biological factors, including the abundance and composition of biota within the water body and its watershed. Each of these factors may in turn be influenced by human activities. A discussion of how these factors influence water chemistry is best facilitated by examining some observed patterns for three important classes: dissolved gases, major ions, and nutrients.

Dissolved Gases

The dissolved gases of primary interest in most aquatic ecosystems are oxygen and carbon dioxide. Both of these molecules are nonpolar, therefore, as they partition at the air–water interface their hydration by water is minimal and consequently their solubility is very low. The only van der Waals forces that act upon them are very weak Debye forces, in which water's strong dipolar field induces a transient dipole in the nonpolar molecule's electronic field. These gases are of primary importance because they both influence and reflect biological processes. As a result, they serve as tracers of electron flow (i.e., energy flow) in an ecosystem. Reactions that convert energy into an organic form will reduce CO_2. In the case of photosynthesis, energy is derived from light and water is the electron donor, with the resultant production of O_2. CO_2 can also be reduced by chemoautotrophic bacteria, using other alternate electron donors, such as ammonium (NH_4^+), methane (CH_4), and hydrogen sulfide (H_2S). In each case, anabolic processes result in a loss of dissolved CO_2. Conversely, the decomposition of organic material results in the production of CO_2 and the loss of O_2, if that gas is available. In general, the balance between carbon reduction and oxidation in lakes and rivers is controlled by light-driven photosynthesis. This, and the physical exchange of gases between water and the atmosphere, results in deep waters having higher CO_2 concentrations and lower dissolved O_2 concentrations than surface waters. In lakes that are chemically or thermally stratified, the combination of decomposition and reduced vertical mixing can result in anoxia in the hypolimnion. In lakes that are well mixed, anoxia will occur in the sediment. Under these conditions, bacteria will use other electron donors in the metabolism of organic carbon. The electron donor used depends on the relative availability and the Gibbs' free energy of reaction resulting from the use of that donor. As a result, a vertical redox gradient is created, in which the various electron acceptors serially decrease with depth.

For lakes of a given size and within a geographic/climatic region, dissolved gas concentrations can vary according to the loading of nutrients and organic carbon. Lakes with high nutrient loads will exhibit large diurnal fluctuations in surface dissolved O_2 and CO_2 concentrations, because of high photosynthetic rates during the day and high respiration rates at night. Lakes with high organic carbon loads may be persistently supersaturated with CO_2 and undersaturated with O_2.

Temperature is a key property that determines the solubility of gases in water (**Figure 8**). This has ramifications both for the distribution of dissolved gases within lakes, and for the relationship between climate and dissolved gases, especially O_2. Within large temperate lakes in which plankton metabolism is generally slow, there is usually sufficient dissolved O_2 at all depths to support aerobic organisms. Smaller lakes that stratify may develop an anoxic hypolimnion, with the probability of anoxia increasing with the duration of stratification and lake productivity. In tropical lakes and rivers, warm temperatures result in lower dissolved O_2 saturation concentrations, and higher decomposition rates, making these systems more prone to anoxia than their temperate counterparts.

Major Ions

Major ions are those that contribute significantly to the salinity of water. Major cations generally include Ca^{2+}, Mg^{2+}, Na^{+}, and K^{+}, while major anions may include HCO_3^-, CO_3^{2-}, Cl^-, SO_4^{2-}, and sometimes NO_3^- All of these species are of course solvated by water, and the concentric shells that are formed may extend relatively far into the 'bulk' water. The absolute and relative abundance of the hydrated major ions in rivers and lakes are controlled by three factors: basin geology, rainfall, and evaporation–crystallization processes. Hence geographic variations in major ion composition can be related to one or more of these factors. For example, the relatively low Ca^{2+} concentrations in lakes and rivers of Precambrian Shield regions of North America and northern Europe are because of the dominance of igneous granite in their watersheds, while the high sodium and chloride concentrations of lakes in many dry regions is because of evaporative concentration of these salts. Because the above three factors differentially influence various major ions, the salinity and relative abundance and distribution of ions can be used to infer which of these processes is most significant for a given water body (**Figure 9**). Some exceptions to the pattern shown in **Figure 9** occur, especially in Africa, where a combination of intense weathering, low Ca^{2+} concentrations in rock, and evaporative concentration can result in moderately high salinities that are dominated by Na^+ and HCO_3^-.

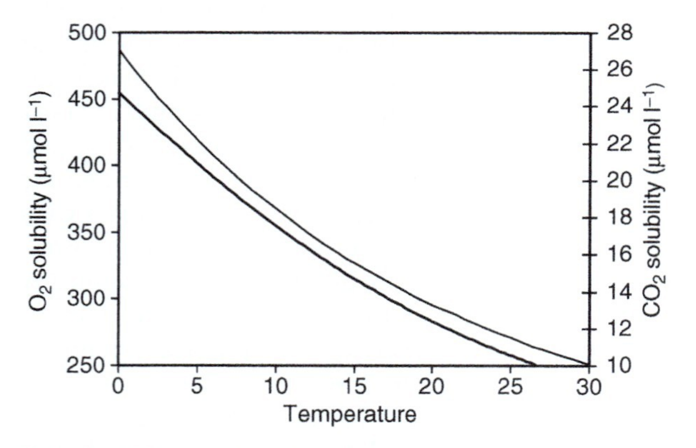

Figure 8 The solubility of oxygen (——) and carbon dioxide (——) in freshwater at a pressure of one atmosphere and an atmospheric CO_2 partial pressure of 380 μatm.

Nutrients

Most algae require a minimum of 14 essential nutrients to grow. The nutrient that limits algal growth in a water body depends on the availability of these nutrients relative to algal demand. In most water bodies, phosphorus or nitrogen is the limiting nutrient, but trace elements such as iron and molybdenum may also be limiting in some systems.

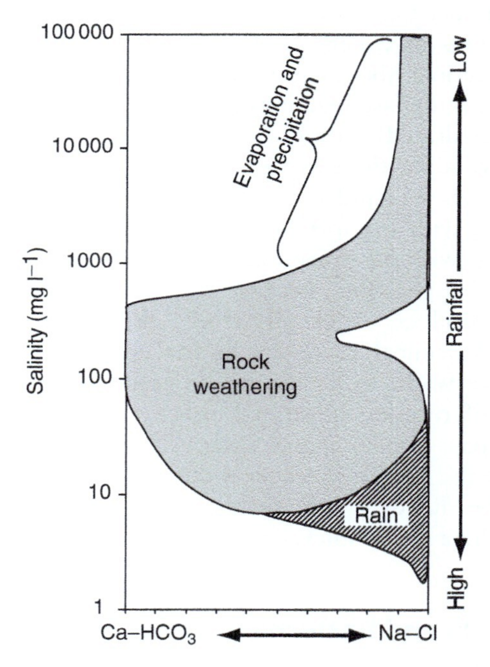

Figure 9 Influence of geology and climate on salinity and major ion composition of inland waters.

The effects of accelerated nutrient loading to lakes and rivers resulting from human activities, referred to as eutrophication, are well documented in the scientific literature, and are not addressed here. Phosphorus input to lakes and rivers is controlled primarily by rock composition and weathering intensity, but the availability of phosphorus to algae is influenced by the availability of other elements, and by biologically mediated processes. In iron-rich waters, inorganic phosphorus is bound as insoluble ferric phosphate or adsorbed onto ferric oxides and oxyhydroxides, and such systems tend to be unproductive and phosphorus limited. In calcareous regions, including the Laurentian Great Lakes, calcium minerals may serve as a source of phosphorus through weathering, but this phosphorus is often biologically unavailable because of adsorption to minerals such as calcium carbonate and precipitation with calcium to form apatite. The equilibrium between dissolved and particulate phosphorus is influenced by redox potential, with phosphorus dissolution being accelerated under anoxic conditions. Such conditions also promote denitrification, through which biologically available nitrate is ultimately reduced to nitrogen gas, which cannot be assimilated by most algae. Over an annual cycle, water column anoxia is more prevalent in tropical lakes than in temperate lakes, increasing phosphorus availability while promoting nitrogen loss. As a result, nitrogen limitation of algae tends to be more common in the tropics. These patterns can be modified by lake depth. In deep lakes, phosphorus is sequestered more efficiently into sediment, and as a result these lakes tend to have a lower concentration of phosphorus in the water column relative to shallow lakes with similar external phosphorus loads.

Conclusion

Primarily because of an extensive network of hydrogen bonding, water is structurally complex and has very unusual properties. Ions and molecules are solvated by water, and the resulting structures affect their reactivity and hence their toxicity, transport, and fate. Understanding how nutrients and pollutants are transformed by their interaction with water is essential to understanding the dynamics of the Earth's hydrosphere. Furthermore, these chemical transformations affect how these compounds are transported to other environmental compartments (e.g., the lithosphere and biosphere).

Water is said to be the most studied molecule. Yet while theory and experiment have greatly improved our knowledge of water's structure, important questions remain only partially answered. In particular, questions concerning the structure of water in the liquid state, specifically how hydrogen bonding determines long-range ordering effects, continue to intrigue researchers. Given the astonishing properties of such a simple molecule, one might conclude that hydrogen bonding is indeed that 'third thing' to which D.H. Lawrence was alluding.

See also: Physical Properties of Water.

Further Reading

Baird C and Cann M (2005) *Environmental Chemistry,* 3rd edn. New York, NY: W.H. Freeman.

Barrett J (2003) *Inorganic Chemistry in Aqueous Solution.* Cambridge, UK: Royal Society of Chemistry.

Chaplin MF (2000) A proposal for the structuring of water. *Biophysical Chemistry* 83: 211–221.

Cotton FA, Wilkinson G, Murillo CA, and Bochmann M (1999) *Advanced Inorganic Chemistry,* 6th edn. New York, NY: Wiley-Interscience.

Dougherty RC and Howard LN (1998) Equilibrium structural model of liquid water: evidence from heat capacity, spectra, density, and other properties. *Journal of Chemical Physics* 109: 7379–7393.

Kusalik PG and Svishchev IM (1994) The spatial structure in liquid water. *Science* 265: 1219–1221.

Manahan SE (2005) *Environmental Chemistry,* 8th edn. Boca Raton, FL: CRC Press.

Marcus Y (1985) *Ion Solvation.* Chichester, UK: Wiley-Interscience.

Martell AE and Motekaitis RJ (1989) Coordination chemistry and speciation of Al(III) in aqueous solution. In: Lewis TE (ed.) *Environmental Chemistry and Toxicology of Aluminum,* pp. 3–19. Chelsea, MI: Lewis Publishers.

Searcy JQ and Fenn JB (1974) Clustering of water on hydrated protons in a supersonic free jet expansion. *Journal of Chemical Physics* 61: 5282–5288.

Stumm W and Morgan JJ (1996) *Aquatic Chemistry: Chemical Equilibria and Rates in Natural Waters,* 3rd edn. New York, NY: Wiley-Interscience.

VanLoon GW and Duffy SJ (2000) *Environmental Chemistry: A Global Perspective.* New York, NY: Oxford University Press.

Wallqvist A and Mountain RD (1999) Molecular models of water: Derivation and description. *Reviews in Computational Chemistry* 13: 183–247.

Wetzel RG (2001) *Limnology: Lake and River Ecosystems,* 3rd edn. San Diego, CA: Academic Press.

Zwier TS (2004) The structure of protonated water clusters. *Science* 304: 1119–1120.

Relevant Websites

http://www.lsbu.ac.uk/water/ – "Water Structure and Science" by Professor Martin Chaplin, London South Bank University, London, England, UK.

http://witcombe.sbc.edu/water/chemistry.html – "The Chemistry of Water" by Professor Jill Granger, Sweet Briar College, Sweet Briar, Virginia, USA.

http://webbook.nist.gov/chemistry/ – The National Institute of Standards and Technology's "Chemistry WebBook," Gaithersburg, Maryland, USA.

Hydrological Cycle and Water Budgets

T N Narasimhan, University of California, Berkeley, CA, USA

Introduction

The Earth's geology, its atmosphere, and the phenomenon of life have been profoundly influenced by water through geological time. One manifestation of this influence is the hydrological cycle, a continuous exchange of water among the lithosphere, the atmosphere, and the biosphere. The present-day hydrological cycle is characterized by a vigorous circulation of an almost insignificant fraction, about 0.01%, of the total water existing on the Earth. Almost all beings on land require freshwater for sustenance. Yet, remarkably, they have evolved and proliferated depending on the repeated reuse of such a small fraction of available water. The partitioning of water among the components of the hydrological cycle at a given location constitutes water balance. Water-balance evaluations are of philosophical interest in comprehending the geological and biological evolution of the Earth and of practical value in environmental and natural-resource management on various scales. The purpose here is to outline the essential elements of the hydrological cycle and water budgets relevant to inland waters and aquatic ecosystems.

Hydrological Cycle

The hydrological cycle is schematically shown in **Figure 1**. Atmospheric water vapor condenses and precipitates as rain or snow. A small portion of this is intercepted by vegetation canopies, with the rest reaching the ground. A portion of this water flows over land as surface water toward the ocean or inland depressions, to be intercepted along the way by ponds, lakes, and wetlands. Another portion infiltrates to recharge the soil zone between the land surface and the water table, and the groundwater reservoir below the latter. Pulled by gravity, groundwater can move down to great depths. However, because of the presence of low permeability earth layers, the downward movement is resisted, and water is deflected up toward the land surface to be discharged in streams, lakes, ponds, and wetlands. Water escaping the influence of resistive layers and moving to greater depths encounters geothermal heat. Geothermal heating too has the effect of countering downward movement and impelling groundwater toward the land surface. At the land surface, surface water and discharging groundwater are subject to evaporation by solar radiation and to transpiration by plants as they consume water for photosynthesis. Collectively referred to as evapotranspiration, this transfer of water back to the atmosphere completes the hydrological cycle.

The components of the hydrological cycle, namely, atmosphere, surface water, and groundwater (including soil water), are intimately interlinked over a variety of spatial scales (meters to thousands of kilometers) and temporal scales (days to millions of years). Information on the volume of water stored in each component of the hydrological cycle, and the relevant spatial and temporal scales are summarized in **Table 1**.

Water is a slightly compressible liquid, with high specific heat capacity and latent heats of melting and evaporation. It exists in solid, liquid, and gaseous phases within the range of temperatures over which life, as we know it, can sustain. Its bipolar nature enables it to form cage-like structures that can trap nonelectrolyte molecules as well as charged ions. For these reasons, water is an active chemical agent, efficient transporter of mechanical energy and heat, and a carrier of dissolved and suspended substances. These attributes render water to be an extraordinary geological and biological agent that has endowed the Earth with features no other celestial object is known to possess.

The hydrological cycle is driven mostly by solar energy and to a minor extent by geothermal heat. The Earth's erosional and geochemical cycles exist due to water's ability to do mechanical work associated with erosion, chemically interact with rocks and minerals, and transport dissolved and suspended materials. Collectively, the hydrological, erosional, and geochemical cycles constitute the vital cycles that sustain life. The interrelationships among these vital cycles can be conveniently understood by examining the lithospheric components of the hydrological cycle.

Hydrological Cycle: Lithospheric Components

Surface water On the Earth's surface, water breaks down rocks physically and chemically through weathering, aided by solar energy and by actions of microbes, plants, and animals. The products of weathering are transported as sediments (bedloads and suspended loads) and dissolved chemicals. In addition, water also transports leaf litter and other

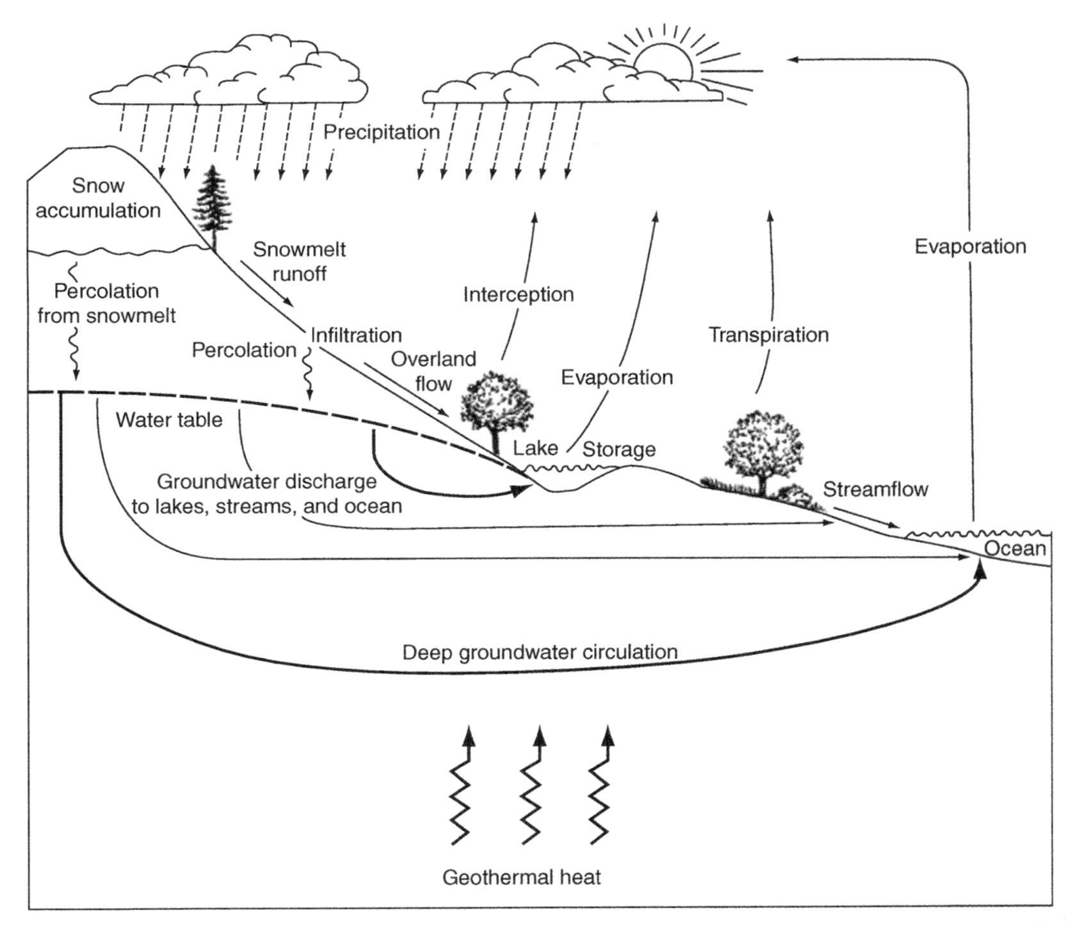

Figure 1 Schematic description of the hydrological cycle: adapted from T. Dunne and L. B. Leopold, 1978, *Water in Environmental Planning*, p. 5.

Table 1 Hydrological cycle: spatial and temporal scales

	Storage, % of Total[a,b]	*Spatial scale*[c]	*Residence Time*[d]
Atmosphere	0.001	Km to thousands of km	Days
Surface water[e]	0.01	Meters to hundreds of km	Weeks to years
Soil water	0.05	Meters to tens of meters	Weeks to years
Groundwater	2.1	Tens of meters to hundreds of km	Days to millions of years
Oceans and seas	95.7	Km to thousands of km	Thousands of years
Ice caps and glaciers	2.1	Km to thousands of km	Tens of thousands of years

[a]Total volume of water on Earth, 1.43×10^9 km^3.
[b]From Unesco, 1971, *Scientific framework for World water balance*, p. 17.
[c]Distance over which cycle is completed.
[d]From Unesco, 1971, *Scientific framework for World water balance*, p. 17.
[e]Includes lakes and reservoirs, river and stream channels, swamps.

decaying vegetation and animal matter. The sediments and organic matter together contribute to the cycling of life-sustaining nutrients. The habitats of flora and fauna along the course of a river depend, in very complex ways, on the texture of sediments as well as on their chemical makeup. A glimpse into the intricate influence of physical nature of sediments and the aquatic chemical environment on an organism's life cycle is provided by salmon, an anadromous fish. In the wild, salmon is hatched in gravelly stream beds that provide protection from predators and abundant supplies of oxygen to the eggs. Once hatched, the young fingerlings must have narrowly constrained aquatic chemical and thermal environment to survive as they migrate from a freshwater environment to a marine environment where they will spend their adult life.

Soil water The soil zone lies between land surface and the water table, where water and air coexist. Soil water, which is held in the pores by capillary forces, is not amenable for easy extraction by humans. However, plants have the ability to overcome capillary forces and extract water for their sustenance. Microbial populations constitute an integral part of the soil biological environment. With abundant availability of oxygen and carbon dioxide in the air, the soil is an active chemical reactor, with microbially mediated aqueous reactions.

In the soil zone, water movement is dominantly vertical, and a seasonally fluctuating horizontal plane separates vertically upward evaporative movement from downward directed gravity flow. Water moving down by gravity reaches the water table to recharge the groundwater reservoir. The journey of water from the time it enters the groundwater reservoir to the time it emerges back at the land surface may be referred to as regional groundwater motion. Regional groundwater motion constitutes a convenient framework for an integrated understanding of the formation of sedimentary rocks and minerals, and the areal distribution of soils and aquatic ecosystems on land.

Groundwater Infiltrating water enters the groundwater reservoir at high elevations, and driven by gravity, moves vertically down in areas of groundwater recharge. Depending on topographic relief and the distribution of permeable and impermeable layers, the vertically downward movement is resisted sooner or later, and the movement becomes subhorizontal. With further movement, flow is deflected up toward the land surface in areas of groundwater discharge. Groundwater discharge typically occurs in perennial stream channels, wetlands, low-lying areas, and springs. In these discharge areas, surface water and groundwater directly interact with each other, with important geological and biological consequences. For example, the spectacular tufa towers of Mono Lake in California represent precipitates of calcium carbonate resulting from a mixing of subaqueous thermal springs with the lake water. Hyporheic zones, which play an important role in stream ecology, are groundwater discharge areas where stream flow is augmented by groundwater discharge.

Regional groundwater flow provides a framework to interpret patterns of chemical processes in the subsurface. Water in recharge areas is rich in oxygen and carbon dioxide and has a significant ability to chemically break down minerals through corrosive oxidation reactions. However, available oxygen is consumed as water chemically interacts with the minerals along the flow path, and the oxidation potential of groundwater progressively decreases along the flow path. In swamps and wetlands of discharge areas, water exists under strong reducing (anaerobic) conditions. Between these two extremes, ambient conditions of acidity (pH) and redox state (Eh) govern the chemical makeup of water as well as the types of minerals and microbial populations that are compatible with ambient water chemistry. In general, the cation content of groundwater reflects the chemical make up of the rocks encountered along the flow path, and the anion content is indicative of the progress of chemical reactions.

The concept of hydrochemical facies denotes the diagnostic chemical aspect of aqueous solutions reflecting the progress of chemical processes within the framework of regional groundwater motion. Given the concept of regional groundwater motion, and that of hydrogeochemical facies, one can readily see how the spatial distribution of various types of soils, and the distribution of different types of ecosystems over a watershed, must represent the profound influence of the lithospheric segment on the hydrological cycle. Regional groundwater flow pattern in the Atlantic Coastal Plain as deciphered from hydrochemical information is shown in **Figure 2**.

Nutrient Cycling and Energy Balance

A discussion of the hydrological cycle is incomplete without examining its connections to nutrient cycling and solar energy balance.

A glimpse into connections to nutrient cycling can be gained by examining the role of water in the cycling of carbon, sulfur, and phosphorus. Almost all biological carbon originates in atmospheric carbon dioxide through photosynthesis by plants and phytoplankton. Water, essential for the photosynthesis process, is transferred as water by plants from the soil via leaves to the atmosphere, completing the hydrological cycle. In the lithosphere, water plays a dominant role in the decomposition and mineralization of organic carbon on diverse time scales, ultimately producing carbon dioxide or methane to be returned to the atmosphere. Sulfur is a multivalent, redox-controlled chemical species which plays an important role in metabolic reactions of plants.

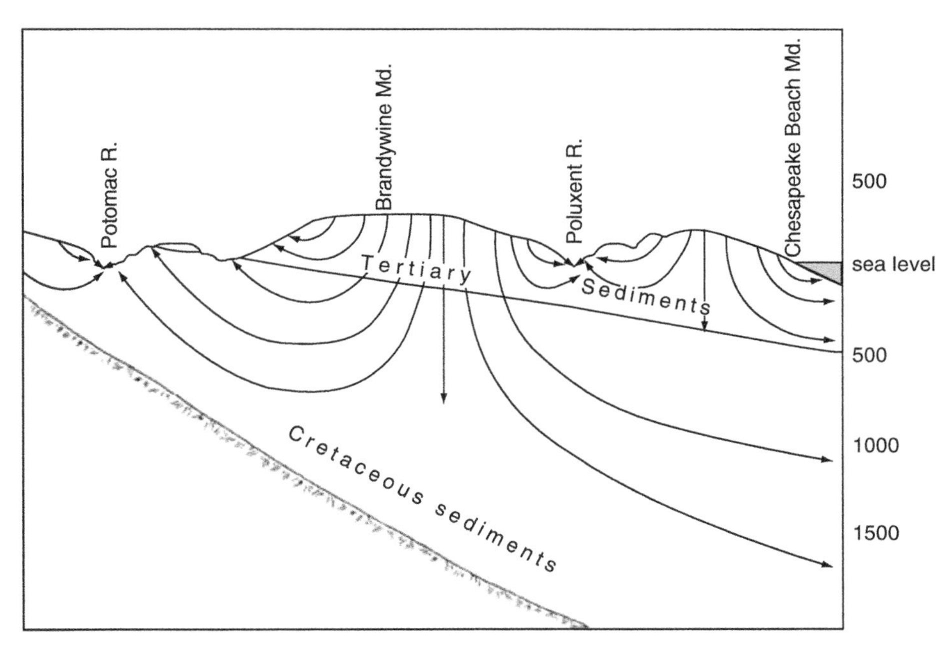

Figure 2 Groundwater flow patterns inferred from hydrochemical facies in the Atlantic coastal plain (W. Back, 1960, *Origin of Hydrochemical Facies of Groundwater in the Atlantic Coastal Plains*).

Under reducing conditions, sulfur is insoluble in water. Sulfate, its most oxidized form, is water soluble, and it is in this form that sulfur usually enters plant roots. Sulfide minerals constitute the principal source of sulfur in the lithosphere, and they are oxidized in the presence of bacteria to sulfate and become available for uptake by plants. In plants, sulfur is fixed in a reduced form. Thus, sulfur of dead organic matter is mobilized by oxidizing waters to sulfate to sustain the sulfur cycle. Phosphorus, which plays several important roles in the biological processes of plants and animals, is water soluble only under very narrow ranges of redox and pH. It does not readily form gaseous compounds. Therefore, phosphorus cycling is almost entirely restricted to the lithosphere. Phosphorus cycling effectively maintains biological habitats despite the severe aqueous constraints that limit its mobility.

The hydrological cycle is driven largely by solar energy. Just like water, solar energy is also subject to cyclic behavior. On the land surface, the energy received as incoming solar radiation (insolation) is balanced partly by outgoing longwave radiation, partly as sensible heat by convecting air columns and partly as latent heat transferred by water from the land to the atmosphere. Of the total solar radiation received from the sun on land, the amount of energy returned by water to the atmosphere amounts to about 46%, a major fraction. Any significant perturbation of this contribution will have influence global climate.

Summary

The concept of hydrological cycle is elegantly simple. But, its importance in the functioning of the geological and biological Earth is profound, transcending water itself. It plays an overarching role in the cycling of solar energy, sediments, and chemical elements vital for the sustenance of life. Although it is clear that contemporary ecosystems reflect an evolutionary adaptation to the delicate linkages that exist among the various components of the hydrological cycle, it is also apparent that evolving life must have influenced the evolution of the hydrological cycle over geological time. Life, it appears, is simultaneously a product of the hydrological cycle and its cause.

Water Budgets

Framework

Despite advances in science and technology, climate remains outside human control and manipulation. Humans, just as plants and animals, have to pattern their existence submitting to the variability of climate. However, surface water, soil water, and groundwater lie within reach of human control, to be managed for human benefit. In this context, the concept of water budgets becomes relevant.

Given a volume element of the Earth with well-defined boundaries, water budget consists in quantifying the relationships among inflow, outflow, and change in storage within the element. This simple concept is as valid over the Earth as a whole treated as a volume element, over a river basin, or over a small rural community. In a world of stressed water resources, water budget is assuming an ever increasing importance as a framework for wise and equitable water management.

Water is always in a state of motion, and its budget is governed by the simple notion that inflow must equal change in storage plus outflow. Symbolically, this may be stated as

$$P = E + R_{\mathrm{Su}} + R_{\mathrm{Gw}} + \Delta\mathrm{Su} + \Delta\mathrm{So} + \Delta\mathrm{Gw} + D_H \qquad [1]$$

where P is precipitation, E is evapotranspiration, R is runoff, Su is surface water, Gw is groundwater, So is soil water, D_H is diversion by humans, and Δ denotes change in storage. If the time interval of interest is smaller than a season, the terms involving change in storage cannot be neglected, the system being under transient conditions. If, however, the time interval of interest is a year or several years, seasonal increases and decreases in storage will effectively cancel out, and the water budget equation reduces to a steady-state balancing of inflow and outflow

$$P = E + R_{\mathrm{Su}} + R_{\mathrm{Gw}} + D_H \qquad [2]$$

Implicit here is the assumption that precipitation constitutes the only inflow into the volume element, which is reasonable if one considers a watershed enclosed by a water divide, without any water import. Clearly, if the volume element of interest is defined by open boundaries, terms representing water import and export have to be added to the equations.

Assessment of Components

The simplicity of the above equations belies the difficulties inherent in assessing the different components involved. Perhaps the most widely measured quantities in water budgets are precipitation and surface runoff. Rainfall data from aerially distributed rain-gauging stations are integrated over space to arrive at the total volume of water falling over an area during a period of interest. Runoff estimated at a given location on a stream with flow meters or river-stage data supplemented by rating curves, represents outflow from the watershed upstream of that location.

Evapotranspiration Experience has shown that evapotranspiration constitutes a significant percentage of precipitation over the land surface. Yet, quantification of evapotranspiration is a difficult task. The gravimetric lysimeter provides a way of experimentally estimating evapotranspiration from a soil mass of the order of a few cubic meters in size. Although of much value as tools of research, lysimeters are helpful in estimating evapotranspiration only over small areas. For watersheds and river basins, it is customary to use a combination of empirical and theoretical methods. In one such approach, the concept of potential evaporation plays a central role. Potential evaporation is understood to be the height of column of water that would be evaporated from a pan at a given location, assuming unlimited supply of water, as from a deep lake. If precipitation at the location exceeds potential evaporation, the soil is assumed to hold a maximum amount of water in excess of gravity drainage. If precipitation is less than potential evaporation, then the actual evapotranspiration will be limited to what precipitation can supply. In this case, empirical curves are used to estimate soil moisture storage based on precipitation deficit and the maximum water-holding capacity of the soil. With the availability of instruments of increased sophistication and super computers, energy methods are increasingly sought after to estimate evapotranspiration from watershed scale to continental scale. In these methods, the goal is to carry out a solar energy budget and isolate the amount of energy that is transferred by water from the land surface to the atmosphere as latent heat. This estimate is then converted to evapotranspiration. To support this model, data are generated from detailed micrometeorological measurements such as short-wave and long-wave radiation, temperature, humidity, cloud cover, and wind velocity. Another method for estimating evapotranspiration is to carry out an atmospheric water balance in a vertical column overlying the area of interest. In this method, evapotranspiration is set equal to the sum of precipitation and change in water vapor content of the column, less the net flux of water laterally entering the column.

Soil-water storage In the field, water content of soils can be profiled as a function of depth with the help of neutron logs or by Time Domain Reflectometry. In principle, one can empirically estimate change in soil-water storage by carrying out repeat measurements with these instruments. However, these methods are of limited value when estimates are to be made over large areas.

The concept of field water capacity, used widely by soil scientists and agronomists, denotes the quantity of water remaining in a unit volume of an initially wet soil from which water has been allowed to drain by gravity over a day or two, or the rate of drainage has become negligible. The water that remains is held by the soil entirely by capillary forces. Field capacity depends on soil structure, texture, and organic content and is commonly measured to help in scheduling irrigation. Empirical curves presented by Thornthwaite and Mather (1957) provide correlations among field water capacity, water retained in soil, and the deficit of precipitation with reference to potential evaporation. These curves can be used to estimate change in soil-water storage.

Groundwater storage Changes in groundwater storage occur due to two distinct physical processes. At the base of the soil zone, as the water table fluctuates, change in storage occurs through processes of saturation or desaturation of the pores. In this case, change in groundwater storage per unit plan area is equal to the product of the magnitude of the water-level fluctuation and the specific yield of the formation, a parameter that is approximately equal to porosity. In the case of formations far below the water table, water is taken into storage through slight changes in the porosity, depending on the compressibility of the formations. In this case, change in groundwater storage can be estimated from the product of water-level fluctuation and the storage coefficient of the formations.

Groundwater runoff The movement of water in the subsurface is quantified with Darcy's Law, according to which the volume of water flowing through a given cross sectional area per unit time is equal to the product of the hydraulic conductivity of the formation, the gradient of hydraulic head, and the cross-sectional area. In the field, hydraulic gradients can be obtained from water table maps. These, in conjunction with the known hydraulic conductivity of the geological formations can be used to estimate groundwater runoff.

Two Examples

Global water balance Between 1965 and 1974, the International Hydrological Decade Program of UNESCO did much to focus attention on the imperative to judiciously manage the world's freshwater resources. An important contribution to the efforts of IHD by the Russian National Committee was the publication, *World Water Balance and Water Resources of the Earth* (Unesco, 1978), which provided detailed estimates of water balance for the different continents, and for the Earth as a whole. The general finding was that for the world as a whole, total annual precipitation is of the order of 113 cm, or about 5.76×10^5 km^3. Globally, this precipitation is balanced by an equal magnitude of evapotranspiration. However, an imbalance exists between precipitation and evapotranspiration, if land and the oceans are considered separately. Over land, average annual precipitation is about 80 cm, or 1.19×10^5 km^3. Of this, evapotranspiration constitutes 48.5 cm (60.6%) and runoff constitutes 31.5 cm (39.4%), indicating a deficit of precipitation in comparison to evapotranspiration. Over the oceans, the average annual precipitation is about 127 cm, or 4.57×10^5 km^3, while evaporation is about 140 cm. The excess of evaporation over precipitation over the oceans is equal to the runoff from the land to the oceans.

California With a land area of 409 500 km^2, and spanning 10° of latitude and longitude, California exhibits remarkable diversity of physiography, climate, flora, and fauna. The Department of Water Resources of the State of California periodically prepares water balance summaries to aid state-wide water planning. The DWR's latest water balance estimates are instructive in that they provide comparison

Table 2 Statewide water balance, California – m^3 (maf[a])

	Water year (Percent of normal precipitation)		
	1998 (171%)	*2000 (97%)*	*2001 (72%)*
Precipitation	4.07×10^{11}(329.6)	2.32×10^{11} (187.7)	1.72×10^{11}(139.2)
Imports: Oregon/Nevada/Mexico	9.00×10^9 (7.3)	8.63×10^9 (7.0)	7.77×10^9 (6.3)
Total inflow	*4.16×10^{11} (336.9)*	*2.40×10^{11} (194.7)*	*1.79×10^{11} (145.5)*
Evapotranspiration[b]	2.58×10^{11} (208.8)	1.62×10^{11} (131)	1.53 (124.2)
Exports: Oregon/Nevada/Mexico	1.85×10^9 (1.5)	1.11×10^9 (0.9)	8.63×10^8 (0.7)
Runoff	1.49×10^{11} (120.8)	8.46×10^{10}(68.6)	4.30×10^{10} (34.9)
Total outflow	*4.08×10^{11} (331.1)*	*2.48×10^{11} (200.8)*	*1.97×10^{11}(159.8)*
Change in surface water storage	8.88×10^9 (7.2)	-1.60×10^9 (−1.3)	-5.67×10^9 (−4.6)
Change in groundwater storage	-1.72×10^9 (−1.4)	-5.55×10^9 (−4.5)	-1.20×10^{10} (−9.7)
Total change in storage	*7.15×10^9 (5.8)*	*-7.15×10^9 (−5.8)*	*-1.76×10^{10} (−14.3)*

[a]Million acre feet.

[b]Includes native plants and cultivated crops.

of water budget for an average year with those of a surplus year and a deficit year (California Department of Water Resources, 2005). Salient features are summarized in **Table 2**. It is interesting to note from the table that (1) groundwater is being over pumped even during surplus years, (2) California experiences a deficit of about 3% even during an average year, and (3) evapotranspiration varies from 62% during a surplus year to as much as 85% during a drought year.

Epilogue

Modern science has shown that the observed behavior of the hydrological cycle can be understood and explained in terms of the laws of mechanics and thermodynamics. However, the ability of modern science to describe the hydrological cycle in precise detail and to predict the future behavior of components of the hydrological cycle with confidence is severely limited. The limitation arises from the many spatial and temporal scales in which the components interact, the complexity of processes, difficulties of access to observation, and sparsity of data, not to mention the role of living beings that defy quantification. Yet, we have to draw upon our best science so as to use the world's limited supplies of freshwater wisely and equitably. This goal will be best achieved if we recognize the limitations of science, moderate our social and economic aspirations, and use science to help us adapt to the constraints imposed by the hydrological cycle.

Throughout history, humans have been fascinated with water. Although modern science has been successful in elucidating the details of the functioning of the hydrological cycle, its essential features were astutely recognized and viewed with awe centuries (perhaps even millenniums) before Christ in China, India, Greece, and Egypt. It is therefore fitting to conclude this discussion of the hydrological cycle with a psalm from the Hindu scripture:

"The waters which are from heaven, and which flow after being dug, and even those that spring by themselves, the bright pure waters which lead to the sea, may those divine waters protect me here" (Rig-veda, VII 49.2).

See also: Chemical Properties of Water; Hydrology: Streams; Physical Properties of Water.

Further Reading

Encyclopedia Britannica (1977) Hydrological Cycle 9: 102–116.

Back W (1960) Origin of hydrochemical facies of groundwater in the Atlantic Coastal Plains, Report, *21st Session, Int. Geol. Congress*, Copenhagen, Pt. 1, pp. 87–95.

Freeze RA and Cherry JA (1979) *Groundwater.* Englewood Cliffs, New Jersey: Prentice Hall, 604 pp.

Narasimhan TN (2005) Pedology: A hydrogeological perspective. *Vadose Zone J.* 4: 891–898.

Thornthwaite CW and Mather JR (1957) Instructions and tables for potential evapotranspiration and water balance. *Publication in Climatology,* Vol. 10, No. 3. Centerton, New Jersey: Thornthwaite and Associates.

U.S. Geological Survey (2007) The Water Cycle, Complete Summary, http://ga.water.usgs.gov/edu/watercyclesummary.html.

Hydrology: Streams

E Wohl, Department of Geosciences, Colorado State University, Ft. Collins, CO, USA

Introduction

Every point on the Earth's landmass lies within a drainage network formed of stream channels tributary to one another that eventually drain to an inland reservoir or to an ocean. The spatial arrangement of channels into a drainage network, the water and sediment moving from hillslopes and down streams, and the geometry of streams, all reflect climatic and geologic factors within the drainage basin.

Spatial Organization of Streams in Drainage Networks

A drainage network includes all the stream channels that drain toward a reference point. The network is bounded by a topographically defined drainage divide; precipitation falling on the far side of the divide flows down slope into an adjacent drainage network. A drainage network begins with first-order streams to which no other stream is tributary. In the most commonly used method of stream orders, a second-order stream begins at the junction of two first-order streams, a third-order stream begins at the junction of two second-order streams, and so on (**Figure 1**).

Patterns of drainage networks. The spatial distribution of streams within the network can be descriptively classified using terms including dendritic, rectangular, radial, and others. Dendritic drainages are the most widespread, taking their name from a resemblance to the outline of a tree (**Figure 2**). A dendritic drainage is commonly interpreted to reflect a relatively homogeneous substrate of moderate down slope gradients. A rectangular drainage, in contrast, has many right-angle tributary junctions that reflect a strong underlying control, such as joints in the bedrock, which influences the location of stream channels. A radial drainage network more likely reflects the underlying topographic control, such that individual streams radiate outward and down from a central high point such as a volcanic cinder cone. This descriptive classification for drainage networks is useful because it is readily apparent in aerial photographs, topographic maps, or digital elevation models of a landscape, and because the categories of the classification imply something about the geologic controls on the spatial arrangement of stream channels across a landscape.

Drainage density. Drainage networks can be quantitatively described using parameters such as drainage density, which is the ratio of total length of streams within a network to the surface area of the network (stream km/km^2 of drainage area). Drainage density reflects climatic controls, substrate on which the drainage network is formed, and age of the drainage network. The highest values of drainage density tend to occur in semiarid regions and in the seasonal tropics. In each of these regions, high-intensity rainfalls create sufficient erosive force to overcome the surface resistance of hillslopes and form stream channels. High values of drainage density can also be associated with very steep topography, with erodible substrates, and with patterns of land use such as deforestation that reduce hillslope resistance to surface erosion. Drainage networks initially form relatively rapidly on newly exposed landforms such as glacial or volcanic deposits. The rate of increase in drainage density then levels off with time as the network becomes fully integrated and the spacing of stream channels reflects the minimum surface area needed to produce sufficient runoff to support a channel.

Formation of stream channels. A stream channel can form as the result of either surface or subsurface processes, or some combination of the two. Heterogeneities in the surface and subsurface properties of hillslopes create zones of preferential flow during downslope movement of water. As water preferentially concentrates on the surface, the force exerted against the surface by the flowing water increases proportionally to the depth of the water. A self-enhancing feedback occurs such that an initial surface irregularity slightly concentrates surface flow on the hillslope, and the slightly deeper flow in this irregularity exerts more erosive force against the surface, thus deepening the irregularity, which then concentrates yet more flow as it widens and deepens. Eventually, the irregularity creates a spatially continuous downslope flow of water in the form of a rill. If one of a series of parallel rills enlarges faster than the neighboring rills, the master rill creates a secondary side slope between its channel and that of adjacent rills. This secondary slope facilitates shifting of the smaller rill so that it becomes tributary to the larger rill, and a drainage network begins to form.

An analogous process occurs in the subsurface, where differences in porosity and permeability create localized zones of greater flow that dissolve or physically erode material to create subsurface cavities. These cavities can form surface channels if the overlying material collapses into the cavity. The resulting sapping and

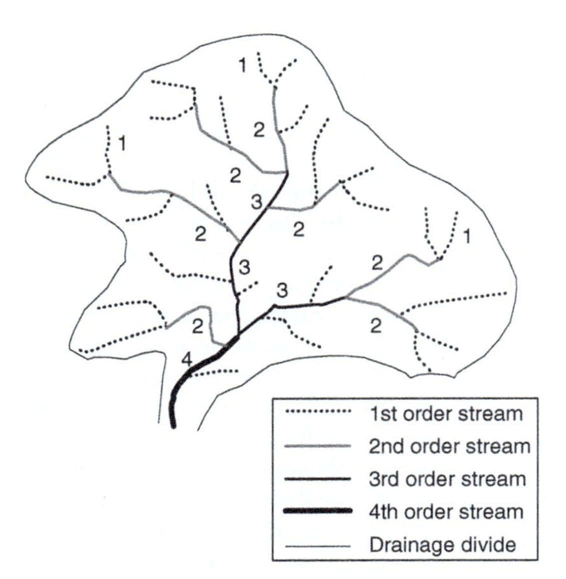

Figure 1 Schematic drawing of a drainage network, showing the ordering of streams, and delineation of the drainage divide.

Figure 2 Dendritic drainage network formed on a gently sloping surface with homogeneous underlying sediment, northwestern Australia. The trees in the photo are approximately 8–10 m tall.

piping networks have distinctive channels in which surface flow begins abruptly at an amphitheater-shaped depression in the ground surface.

Because the area of hillslope contributing flow to a stream channel increases downslope, thresholds for erosion and channel formation can be crossed at downstream portions of a slope first, and the stream channels then erode headward as the network of channels enlarges. If one set of channels erodes headward more rapidly than an adjacent network, the former channels can erode through the drainage divide and capture a portion of the adjacent network. This situation is occurring presently at the Casiquiare Canal, a naturally occurring channel along which a portion of the headwater drainage of the Amazon River of South America is capturing some of the headwater drainage of the adjacent Orinoco River.

Figure 3 Bedlands topography in Death Valley National Monument, California. Channels begin very close to the ridge crests, as can be seen most clearly along the dark brown ridge crest running across the lower third of the photograph.

The point along a hillslope at which stream channels begin to form depends on factors such as characteristics of precipitation, infiltration capacity of the surface, and erosional resistance of the surface (**Figure 3**). Regions with intense rainfall, low infiltration capacities, and highly erodible surfaces can have drainage networks that start very close to the crest of hillslopes, whereas other areas with less intense precipitation, higher infiltration, or greater surface resistance may have channel networks that begin much farther downslope.

Sources of Flow in Streams

The ultimate source of water flowing in any stream is snowmelt, rain-on-snow, or rainfall. Snowmelt generally produces regular seasonal patterns of stream flow during the onset of warmer temperatures when snow packs have accumulated during the winter melt. Snowmelt tends to be an important source of stream flow at higher latitudes and higher altitudes. Snow and rainfall can also enter streams after having been stored as ice in glaciers or icefields for periods of up to thousands of years. As with other forms of snowmelt, glacier melt is most pronounced during warmer seasons of the year, but can persist throughout the warm season (unlike snowmelt, which tends to be most pronounced during the early part of the warm season).

Rain-on-snow occurs when warmer temperatures cause rain to fall directly onto snow packs that have not yet completely melted. The warmer temperatures increase the melt rate of the snow which, when combined with rainfall, can create high rates of runoff and associated flooding. Rain-on-snow floods are particularly prevalent in low-to-moderate elevation catchments in coastal mountain ranges at middle latitudes.

The intensity, duration, and spatial extent of rainfall vary greatly among different types of climatic circulation patterns that generate rainfall. Convective storms create very intense rains that cover small areas ($1–10^2$ km^2) for periods of up to a few hours. Frontal storms that last for days can extend across 10^4 km^2. The most extensive rains are associated with cyclonic storms such as hurricanes that last for days to weeks and monsoonal circulation patterns that last for months at a time. Both cyclonic and monsoonal storms can cover large areas of $10^5–10^7$ km^2. Convective storms can generate enormously large stream flows within small drainage basins, but the effects of a small storm can be mitigated in large drainage basins where substantial portions of the basin remain unaffected by the storm. The more extensive frontal, cyclonic, and monsoonal storms can produce floods across much larger drainage basins.

The distribution of different types of precipitation reflects global-scale atmospheric circulation patterns, as well as regional topographic influences on the movement of air masses that bring moisture over a drainage basin. The regions with the greatest annual precipitation mostly lie within 30° north and south of the equator, where air masses moving across the warm surface of the tropical oceans pick up tremendous amounts of water vapor that is then transported inland to fall as precipitation. Smaller areas of very high precipitation can occur at higher latitudes where a mountain range forces moisture-bearing air masses to rise higher into the atmosphere, causing the water vapor within the air masses to condense and fall as precipitation, or where proximity to an ocean surface with relatively warm temperatures facilitates evaporation and inland transport of moisture from the ocean. Convective storms, which involve localized strong updrafts, are most common at latitudes 10° N–10° S. Frontal storms occur when the boundary between two air masses with different densities passes over a region and brings widespread precipitation. Monsoonal storms are associated with seasonal reversals of winds that draw moisture from adjacent oceans over land masses. Cyclonic storms, which have a strong rotational component, occur in two broad bands at approximately 10°–50° north and south of the equator.

Stream flow can also be dramatically affected by the failure of a natural or human-built reservoir. Lakes created when a landslide or debris flow blocks a stream commonly burst within a few days as the blockage is overtopped or weakened by seepage and piping. Water ponded upstream from glacial moraines or underneath glacial ice can also empty catastrophically when the moraine is overtopped or weakened within, or when the confined water builds sufficient pressure to lift the overlying glacial ice. Human-built fill and concrete dams can also fail by being overtopped or undermined. In each of these cases, sudden release of the ponded water initiates a catastrophic flood that continues until the reservoir is drained below the level of the remaining portions of the dam, or until the glacial ice once again shuts off the drainage path.

Patterns of stream flow reflect global and regional atmospheric circulation patterns and topography, as well as drainage area. Rivers in the equatorial and tropical latitudes commonly have the largest mean flow per unit drainage area because of the greater amounts of precipitation at these latitudes. Peak flow per unit drainage area tends to be greatest in relatively small rivers because the entire drainage area can be contributing runoff during intense precipitation events. Seasonal and interannual variability of flow tend to be largest in arid and semiarid regions, and in the seasonal tropics.

Movement of Water into Stream Channels

Precipitation falling across a landscape moves downward along various paths from hillslopes into stream channels. Precipitation can remain at the ground surface and move downslope relatively quickly as runoff or Hortonian overland flow. Precipitation can also infiltrate the ground surface and move downslope more slowly. Throughflow occurs when subsurface flow moves within the upper, unsaturated layers of sediment. Although the matrix as a whole remains unsaturated, concentrated zones of flow in pipes or macropores, or temporarily saturated zones, are particularly effective in moving water downslope into streams relatively rapidly. If the infiltrating water reaches the deeper, saturated layers of the subsurface, the water moves downslope with groundwater. Hillslopes tend to be heterogeneous environments as a result of small-scale variations in surface topography and the porosity and permeability of subsurface materials. Throughflow moving downslope can concentrate in topographic irregularities and zones of limited porosity and permeability along the hillslope and return to the surface to move downslope as saturation overland flow. Overland flow and shallow, concentrated subsurface flow in pipes or macropores usually move downslope most rapidly, and these sources of runoff are together sometimes referred to by the descriptive term quickflow. Other forms of

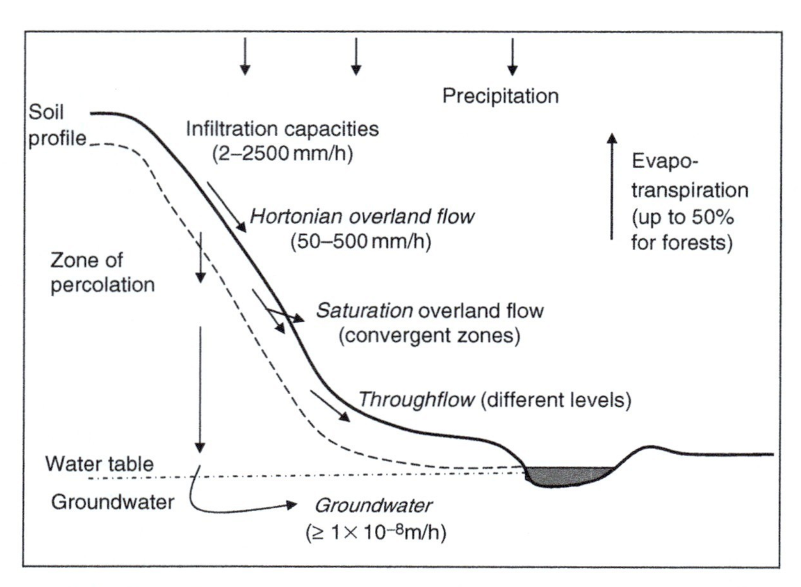

Figure 4 Schematic side view of hillslope illustrating four basic downslope pathways of water (italicized) and range of rates of movement.

throughflow, as well as groundwater flow, move at slower rates (**Figure 4**).

The distribution of water among these four basic flow paths commonly varies across time and space. Rainfall that initially produces throughflow can subsequently create overland flow, for example, if infiltration capacity declines following prolonged rainfall or an increase in rainfall intensity. Convex portions of a hillslope can produce dominantly throughflow, whereas concave portions of the slope have saturation overland flow during the same rainstorm.

Spatial and temporal variability in the downslope movement of water reflects the characteristics of precipitation inputs and hillslope pathways. Precipitation intensity and duration exert particularly important influences on downslope movement of water. Higher intensities of precipitation are more likely to overwhelm infiltration capacity and produce overland flow, but prolonged precipitation at any intensity has the potential to exceed infiltration capacity.

Hillslope characteristics including vegetation cover, downslope gradient, and the porosity and permeability of materials at the surface and in the subsurface also strongly influence the downslope movement of water. Vegetation cover intercepts some precipitation, allowing snow or rain to evaporate or sublimate directly from the plant without reaching the ground, or reducing the force of impact when raindrops bounce from the plant onto the ground. Vegetation also sheds dead leaves and branches that can build up over time in a surface layer of duff with high infiltration capacity. Linear cavities left in the subsurface when plant roots die and decay can create macropores that facilitate rapid downslope movement of water in the subsurface. Steeply sloping surfaces can create large subsurface pressure differences that facilitate more rapid subsurface flow. Hillslope materials with high porosity (percent of void space) and permeability (interconnectedness of void spaces) also facilitate rapid infiltration and downslope movement of subsurface water. Porosity and permeability can result from spaces between individual grains in unconsolidated materials. Sand and gravel tend to have lower porosity but higher permeability than finer silt and clay-sized particles, with the result that sand and gravel commonly have higher infiltration and downslope transmission of water. Larger cavities in the form of pipes or macropores in sediments, or fractures in bedrock, also facilitate downslope flow. Pipes and macropores can result from biological processes including animal burrows or decayed plant roots. They can also form by erosion when subsurface flow concentrated above a less-permeable unit builds sufficient force to remove particles and create a continuous subsurface cavity (**Figure 5**).

In general, hillslopes with limited vegetation cover or surfaces disturbed by humans are likely to have more overland flow, whereas subsurface flow paths become more important with greater vegetation cover and deeper, more permeable soils. However, even a catchment with continuous, dense forest cover can have rapid downslope transmission of precipitation during conditions of high rainfall intensity or where thin soils and preferential subsurface flow paths such as pipes and macropores are present.

Figure 5 Channel segments affected by subsurface piping along Cienega Creek in central Arizona. As subsurface pipes enlarge, the overlying sediment collapses into the cavity (upper photo) until eventually the collapse becomes longitudinally continuous, leaving a deeply cut channel with nearly vertical banks (lower photo).

Movement of Sediment into Stream Channels

Sediment transported downstream can come from adjacent hillslopes, floodplains, and valley bottoms, and from erosion of the bed and banks within the stream. Hillslope sediment enters streams via gradual processes of slope erosion that occur through slope wash, soil creep, rill erosion, and other movements of individual sediment particles. Large volumes of sediment can also be introduced to streams during mass movements such as landslides, debris flows, and rock falls. Mass movements become progressively more important sources of sediment to streams where adjacent slopes are steeper and where episodic triggers such as intense rainfalls, seismic shaking, or wildfires periodically destabilize the hillslopes. Mass movements are particularly important in bringing sediment directly into headwater streams in mountainous terrains where narrow valley bottoms and spatially limited floodplains leave little storage space for sediment between the hillslopes and stream channels. Mountainous terrains around the world produce an estimated 96% of the sediment that eventually reaches the ocean basins, but occupy only 70% of the land area within river basins (**Figure 6**).

Floodplains adjacent to streams provide a very important source of sediment to streams, although the dynamics of sediment movement between streams and floodplains are spatially and temporally complex. Overbank flows that inundate floodplains can deposit large volumes of sediment as particles carried in suspension settle from waters that move more slowly across floodplains. This sediment can remain in storage on the floodplain for periods ranging from hours to tens of thousands of years. The floodplain changes from a sink to a source of sediments when processes such as lateral stream migration cause the channel to move across the floodplain and reintroduce sediment from the floodplain into the stream. The rate and manner of floodplain deposition and

Figure 6 A massive rockfall coming from the right enters a stream channel in the Nepalese Himalaya, causing the channel to become braided downstream.

erosion vary with stream type. Meandering channels tend to erode the outer portion of each meander bend, for example, creating more predictable directions and rates of floodplain erosion, whereas braided channels can shift abruptly back and forth across the valley bottom in a much less predictable fashion. Because many nutrients and contaminants travel adsorbed to silt and clay particles, the storage and remobilization of floodplain sediments can exert a strong influence on stream chemistry and ecological communities.

Erosion of the stream bed and banks provides a third primary source of sediment in stream channels. This form of erosion can be very temporary; most floods erode the channel boundaries while discharge is increasing, but then redeposit sediment during the falling limb of the flood when discharge is decreasing once more. Bed and bank erosion can also be more sustained when a stream is progressively incising downward in response to an increase in discharge, a decrease in sediment supply from other sources, or a drop in the base level (the lowest point to which the stream flows; the ocean is the ultimate base level). Most streams are continually adjusting to changes in water and sediment supply and base level. As a result, erosion of stream bed and banks is also continual in most streams, although this erosion may be balanced by deposition elsewhere along the stream, as when a migrating meander bend has erosion of the outer bend and simultaneous deposition of a point bar on the inner bend.

Characteristics of Flow in Streams

Hydrology of streams. One of the simplest ways to characterize flow in a stream channel is to quantify discharge through time. Discharge, usually expressed in cubic meters or cubic feet per second, is volume of flow per unit time. Discharge is calculated by measuring the velocity, or rate of flow (meters per second) within a cross-sectional area (square meters) calculated from mean width and depth of the flow. Continuous records of discharge come from stream-gaging stations where calibrated rating curves are used to convert measurements of stage, or flow depth, into discharge. These continuous records can then be used to construct a hydrograph, which is a plot of discharge versus time. A flood hydrograph represents a discrete event, whereas an annual hydrograph represents variations in discharge over the course of a year (**Figure 7**).

Hydrographs can be used to differentiate base flow, which is the relatively constant input of water to the stream from groundwater sources, from runoff that results from snowmelt and rainfall entering the stream via throughflow and overland flow. The shape of the hydrograph can be characterized by the relative steepness of the rising and falling limbs of higher flow, as well as the magnitude, duration, and frequency of occurrence of higher flow.

Hydrograph shape is influenced by the precipitation mechanism, the paths of downslope movement of water, and location within the drainage network. Higher intensity precipitation and greater overland flow produce more peaked hydrographs. Convective rainfall that results in Hortonian overland flow will produce a flash flood, for example, whereas snowmelt or prolonged gentle rain created by a low-pressure trough that results in throughflow will produce a lower magnitude, and a more sustained flood peak. Other factors being equal, smaller basins tend to have more peaked hydrographs because the close connections between hillslopes and streams, the narrow valley bottoms with limited floodplains, and the relatively short stream networks all facilitate rapid movement of water through the channel network. Larger basins that have broad, longitudinally continuous floodplains and longer travel times from headwaters to downstream measurement points produce floods that are less peaked and more sustained. Longer travel times occur both because water must travel through a longer network of stream channels, and because more of the water travels slowly across floodplains rather than being concentrated within stream channels. The attenuation of flood discharges across floodplains is critical for depositing sediment and nutrients on floodplains, reducing flood hazards by limiting the magnitude of flood peaks, limiting channel erosion during floods by expending some of the flow energy, and nourishing floodplain wetlands and other ecosystems.

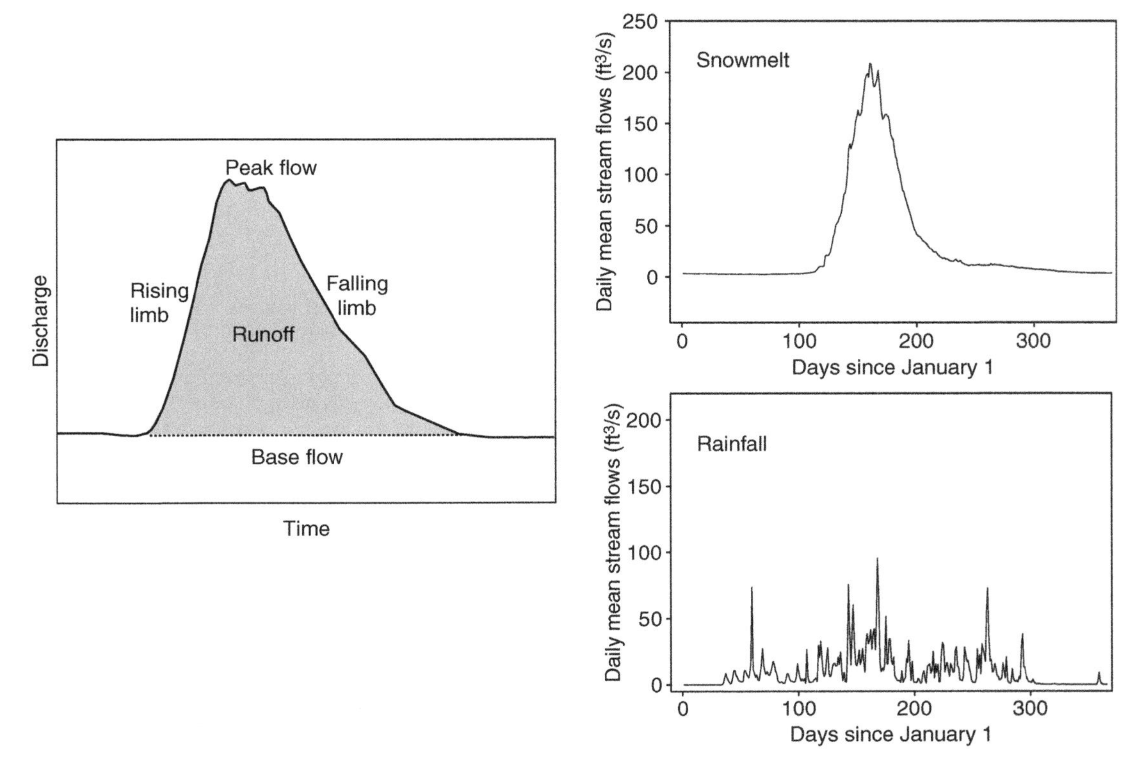

Figure 7 An idealized flood hydrograph (left), showing base flow, storm runoff (gray shading), and rising and falling limbs. Sample annual hydrographs (right) for a snowmelt-runoff stream (top) and a rainfall-runoff stream (bottom).

Flow duration curves, which plot magnitude of discharge versus the percent of time that discharge occurs, provide another means of characterizing the distribution of water in a stream through time. These curves graphically represent the variability of streamflow by the shape of the curve. Curves with low slope and high minimum values indicate a more ephemeral character and a quicker response to precipitation events. Flow duration curves are most frequently used for determining potential water supply for power generation, irrigation, or municipal use.

Flood-frequency curves indicate the average length of time, or recurrence interval, between floods of a similar magnitude. These curves are commonly used for predicting or mitigating flood hazards and for restoring streams in which the distribution of flow has been altered by dams, diversions, or other forms of flow regulation. Estimation of the recurrence interval of very large, infrequent floods, such as those that occur on average every hundred years, is particularly difficult because flood-frequency estimation is based on extrapolation from gage records that are commonly less than a century in duration. Supplementing gage records using information from historical, botanical, or geological sources can substantially improve the accuracy of estimated recurrence intervals for very large floods, or for streams with no gaging records.

Hydraulics of streams. Water flowing within a stream channel is converting potential energy to kinetic energy and heat. The amount of potential energy available for this conversion depends on the vertical drop as the water moves downstream, and on the mass of water moving downstream. Kinetic energy can be expended in overcoming external and internal resistance, and in transporting sediment. External resistance comes from roughness along the bed and banks of the stream. Individual grains that protrude into the flow create external roughness, as do bedforms such as ripples and dunes, coarse woody debris in the stream, irregularities in the channel banks, and downstream variations in channel shape such as meander bends or alternating pools and riffles. Internal resistance occurs when individual fluid elements do not follow all parallel flow paths and move at the same velocity (laminar flow), but instead move at different rates with components of vertical and lateral movement as well as downstream movement (turbulent flow). Flow in all natural stream channels is turbulent to some extent because the water moving along the stream bed and banks encounters more external resistance and moves more slowly than water toward the top center of the stream (**Figure 8**).

Sediment transport in streams. Sediment can be transported in solution within streams. This dissolved

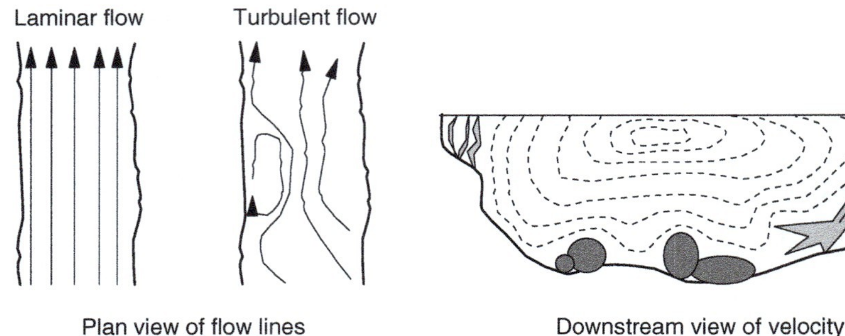

Figure 8 Simplified illustration of hydraulics in a natural channel. The plan view drawings illustrate laminar flow, in which all streamlines are parallel and water moves at equal rates, and turbulent flow, in which streamlines move at different rates, and flow has components of movement laterally across the channel and vertically within the channel, as well as downstream. The downstream view illustrates a natural channel with a slightly irregular cross-sectional form and sources of external resistance along the channel boundaries, including wood (right), cobbles and boulders (center), and submerged vegetation (left). The resulting isovels, or contours representing equal velocity distribution, are shown as dashed lines. The slowest velocities are along the sides and bottom of the channel.

Figure 9 A flood from the Paria River (mouth at upper left) joins the Colorado River at Lees Ferry, Arizona. The Colorado River, at right, flows relatively clear, whereas the Paria carries high concentrations of suspended sediment.

or solute load constitutes a greater proportion of sediment transport during periods of base flow, when water that has moved slowly through the subsurface and had longer periods of time to react with the surrounding matrix, constitutes a greater proportion of stream discharge. Dissolved load is also relatively large for streams draining rocks such as limestone, which is susceptible to chemical weathering, and for streams in tropical regions that tend to have higher rates of chemical weathering for all types of rocks.

Sediment that is not dissolved in stream water can move in suspension within the water column or in contact with the stream bed. Washload is the finest portion of the suspended material, predominantly silts and clays that do not form a substantial portion of the sediment on the streambed. Washload requires so little energy to be transported that it tends to remain in suspension for hours or days even in areas of still water. Suspended load refers to the slightly coarser sands and pebbles that alternate between periods of moving in suspension and periods of moving along the bed and can settle from suspension relatively rapidly when velocity decreases. Bedload moves in nearly continuous contact with the streambed as larger particles roll, slide, and bounce downstream. Because the larger particles that constitute suspended and bedload require greater amounts of energy to move, much of this transport occurs during floods (**Figure 9**).

Glossary

Bedforms – Regularly repetitive longitudinal alternations in streambed elevation, such as pools and riffles, steps and pools, or dunes.

Bedload – Sediment moving in nearly continuous contact with the streambed.

Dissolved load – Sediment transported in solution by stream flow.

Drainage density – A measure of the total length of stream channels per unit area of the drainage basin.

Drainage divide – A topographic high point or line that delineates the boundaries of a drainage network.

Drainage network – An integrated group of stream channels that drain toward a common point.

External resistance – Hydraulic resistance created by the channel boundaries.

Flood-frequency curve – A plot of flood magnitude versus recurrence interval.

Flow duration curve – A plot of discharge magnitude versus the percent of time that discharge occurs.

Glacier melt – Runoff created when glacial ice melts.

Groundwater flow – subsurface flow that occurs below the water table, or zone of saturation.

Hydraulics – The mechanical properties of liquids; for rivers, these properties are described by variables such as velocity.

Hydrograph – A plot of discharge versus time.

Internal resistance – Hydraulic resistance created by differences in the rate and direction of movement of individual fluid elements within a channel.

Laminar flow – Individual fluid elements follow parallel flow paths and move at the same velocity.

Overland flow (Hortonian, saturation) – Water moving across the ground surface; Hortonian overland flow has no infiltration, whereas saturation overland flow results from water that briefly infiltrates to shallow depths and then returns to the surface as the subsurface becomes saturated.

Piping – The processes whereby preferential flow in the unsaturated zone creates longitudinal cavities in the subsurface.

Rainfall – Liquid precipitation that results from different types of atmospheric circulation patterns.

Rain-on-snow – Rain falling directly on a snowpack, which increases the rate of snowmelt.

Reservoir failure – Collapse of a dam built by natural processes such as landslides, or by humans; the collapse results in rapid drainage of the water ponded behind the dam.

Rill – Channels that have no tributaries.

Sapping – The processes whereby preferential flow in the saturated zone creates longitudinal cavities in the subsurface.

Sediment transport – The movement of sediment in channels, includes dissolved, wash, suspended, and bedload.

Snowmelt – Runoff created when snowfall or snowpack melts.

Suspended load – Particulate material moving in suspension in stream flow and of a size that can settle relatively rapidly when velocity decreases.

Throughflow – Subsurface flow that occurs above the water table, or in the unsaturated zone.

Turbulent flow – Individual fluid elements move at different rates and exhibit lateral and vertical components of movement as well as moving downstream.

Washload – The smallest sizes of particulate material moving in suspension in stream flow; usually clay- and silt-sized particles that do not form a substantial portion of the sediment on the streambed.

See also: Climate and Rivers; Coarse Woody Debris in Lakes and Streams; Currents in Rivers; Ecology and Role of Headwater Streams; Flood Plains; Geomorphology of Streams and Rivers; Restoration Ecology of Rivers; Riparian Zones; Streams and Rivers as Ecosystems; Wetlands of Large Rivers: Flood plains.

Further Reading

Milliman JD and Syvitski JPM (1992) Geomorphic/tectonic control of sediment discharge to the ocean: The importance of small mountainous rivers. *The Journal of Geology* 100: 525–544.

Poff NL, Allan JD, Bain MB, *et al.* (1997) The natural flow regime. *BioScience* 47: 769–784.

House PK, Webb RH, Baker VR, and Levish DR (2002) *Ancient Floods, Modern Hazards: Principles and Applications of Paleoflood Hydrology.* American Geophysical Union Press.

Hirschboeck KK (1988) Flood hydroclimatology. In: Baker VR, Kochel RC, and Patton PC (eds.) *Flood Geomorphology,* pp. 525–544. New York: John Wiley and Sons.

Knighton D (1998) *Fluvial Forms and Processes: A New Perspective.* Oxford University Press.

Leopold LB, Wolman MG, and Miller JP (1964) *Fluvial Processes in Geomorphology.* Freeman and Company.

Wohl EE (2000) *Mountain Rivers.* American Geophysical Union Press.

Relevant Websites

http://www.usgs.gov/hazards/floods/ – U.S. Geological Survey floods.

http://www.fema.gov/hazard/flood/index.shtm – Federal Emergency Management Agency floods.

http://www.noaa.gov/floods.html – National Oceanic and Atmospheric Administration floods.

http://www.dartmouth.edu/~floods/ – Dartmouth Flood Observatory.

http://waterdata.usgs.gov/nwis/sw – U.S. Geological Survey surface-water records.

http://nrrss.nbii.gov/ – National River Restoration Science Synthesis.

http://www.willametteexplorer.info/ – Willamette Basin (Oregon) explorer.

http://water.usgs.gov/osw/streamstats/index.html – U.S. Geological Survey StreamStats.

Hydrology: Rivers

P A Bukaveckas, Virginia Commonwealth University, Richmond, VA, USA

What is a River?

There is no strict definition to distinguish rivers from streams and therefore the designation 'river' encompasses flowing waters of widely varying size. Flowing waters may be ranked in size by various metrics that include discharge (glossary), catchment area, and length of channel. For example, the discharge of the Amazon River is six orders of magnitude greater than that of a small river. This range of variation is comparable with the range in volume observed among lakes worldwide. Rivers are sometimes defined as 'non-wadeable' flowing waters since this delineation has practical implications for the way sampling activities are carried out. Along the continuum from headwater streams to large rivers, there are gradients in channel slope, width, and depth. Idealized gradients in geomorphology provide a basis for understanding differences in the structure and functioning of streams vs. rivers. For example, the greater width of river channels reduces the importance of riparian inputs while greater depth lessens the influence of benthic processes. Rivers in their natural settings exhibit complex geomorphologies that give rise to a rich variation in channel form and function and provide diverse habitats for aquatic biota.

Hydrology and Geomorphology

Water Sources and Discharge

Water sources to rivers are principally surficial inputs via tributary streams (**Table 1**). Owing to their small surface area, direct atmospheric inputs are usually minor though groundwater is important in some settings. For comparisons among river basins, discharge is converted to an areal water yield by dividing the volume of discharge by the area of the drainage basin. Water yields vary widely depending on the amount of precipitation relative to evapotranspiration (glossary). South American rivers such as the Amazon and Orinoco are notable for their high water yields, exceeding 1000 mm year^{-1} (**Table 2**). Arid and semiarid regions are characterized by low precipitation relative to evapotranspiration and water yields less than 100 mm year^{-1}. Arid regions occupy about one-third of the world's land area, including portions of several major river basins such as the Murray-Darling (Australia), Colorado (North America), Nile (Africa), and Ganges (Asia) Rivers.

Variation in river discharge arises from short-term, seasonal, and long-term variability in precipitation and evapotranspiration within the drainage basin. Over short time scales (days–weeks), discharge is affected by rain events associated with frontal passage. Though infrequent in occurrence, event-related discharge may account for a large proportion of the annual total. The frequency and magnitude of storm events is therefore an important factor influencing interannual variation in discharge. Event-driven and seasonal variations are superimposed upon long-term (decadal-scale) climatic cycles (e.g., El Niño Southern Oscillation), which may bring about extended periods of above- or below-average discharge. The combined effects of climatic variations occurring over multiple time scales results in a wide range of discharge conditions, which may exceed three orders of magnitude for a given site. Variation in discharge is typically larger that the variation in the concentration of dissolved and particulate substances such that the export of materials from the basin (flux rate) is principally determined by discharge.

Seasonal variation in rainfall and evapotranspiration give rise to predictable annual patterns in river discharge that are characteristic of climatic regions (**Figure 1**). In temperate-humid climates, rainfall may be distributed relatively uniformly throughout the year but seasonal changes in evapotranspiration give rise to variation in discharge. Warmer months are associated with high evapotranspiration, resulting in less runoff from the catchment and lower river discharge relative to colder months. Snowmelt may also contribute to a spring discharge pulse in climates that allow for winter accumulation of snow (including tropical rivers with mountainous catchments). The north-flowing rivers of Canada and Russia are representative of this hydrologic regime in exhibiting high year-round discharge but with a pronounced winter-spring peak. In tropical-humid climates, evapotranspiration is less variable throughout the year but rainfall is often strongly seasonal, particularly in regions affected by monsoons. Wet seasons are associated with elevated river stage and discharge and may be accompanied by extended periods of floodplain inundation. Most South American and African rivers are representative of the tropical unimodal hydrologic regime, which is characterized by an extended period of elevated discharge and floodplain inundation during the rainy season. Arid and

Table 1 Distinguishing characteristics of rivers, estuaries, and lakes

	Rivers	*Estuaries*	*Lakes*
Water movement	Unidirectional, horizontal	Bidirectional, horizontal	Vertical
Water forces	Gravitational	Tidal	Wind-induced
Water-level fluctuations	Large (seasonal)	Variable (daily, storm events)	Small (seasonal)
Water residence time	Days–weeks	Weeks–months	Months–years
Water sources	Runoff	Runoff, marine, precipitation	Runoff, groundwater, precipitation
Stratification	Rare	Common (salinity)	Common (thermal)
Transparency	Low (nonalgal particulates)	Variable (particulates, dissolved color)	High (algae, dissolved color)

Table 2 Water and sediment delivery from large river basins of the world

River	*Drainage area (10^6 km^2)*	*Discharge (km^3 year^{-1})*	*Water yield (mm year^{-1})*	*Sediment load (10^6 t year^{-1})*	*Sediment yield (t km^{-2} year^{-1})*
Amazon	6.15	6300	1024	1200	195
Colorado	0.64	20	31	0.01	0.02
Columbia	0.67	251	375	10	15
Congo (Zaire)	3.72	1250	336	43	12
Danube	0.81	206	254	67	83
Ganges–Brahmaputra	1.48	971	656	1060	716
Huang He (Yellow)	0.75	49	65	1050	1400
Indus	0.97	238	245	59	61
Mackenzie	1.81	306	169	42	23
Mekong	0.79	470	595	160	202
Mississippi	3.27	580	177	210	64
Niger	1.21	192	159	40	33
Nile	3.03	30	10	0	0
Orinoco	0.99	1100	1111	150	152
St. Lawrence	1.03	447	434	4	4

Source: Milliman JD and Meade RH (1983) Worldwide delivery of river sediment to the oceans. *Journal of Geology* 91: 1–21.

semiarid regions occur in both temperate and tropical climates and occupy about one third of the world's land area. They are characterized by low precipitation relative to evapotranspiration and include portions of several major river basins, including the Murray-Darling (Australia), Missouri (North America), Nile (Africa), and Ganges (Asia) Rivers. River basins in arid regions exhibit sustained periods of low discharge interspersed with short periods of elevated discharge. For example, the Murray-Darling River is fed by infrequent summer monsoons which, coupled with high rates of evapotranspiration, result in an annual discharge equivalent to only 3% of annual rainfall.

Large river basins may span climatic and topographic regions and exhibit complex hydrologic regimes. For example, the Rhone is a snowmelt-dominated river in its upper, mountainous sections but is influenced by a Mediterranean climate in its lower course. The river exhibits a complicated flow regime with low discharge periods shifting from winter in the upper course to autumn in the lower course and floods occurring in all seasons. Despite the problems inherent in categorizing this continuum of variation, hydrologic regimes are useful for facilitating comparisons among river basins (e.g., in response to land-use and climate change effects).

Flooding

Rivers experience large and rapid fluctuations in surface water elevation (i.e., 'stage') in response to runoff. The rate and magnitude of rise in river stage is dependent in part on the morphometry of the channel (**Figure 2**). Low banks enable the river to escape the active channel and inundate lateral areas (floodplain). During flooding, the widening of the river lessens the stage response to runoff and reduces water velocity because the force of the water is distributed over a wider area. Flood-prone rivers are common in both temperate and tropical climates and exhibit considerable variation in the extent, timing, and frequency of flooding events. In some settings (e.g., Amazon River) the annual flood pulse is a defining feature of the riverscape, important not only to the life cycles of

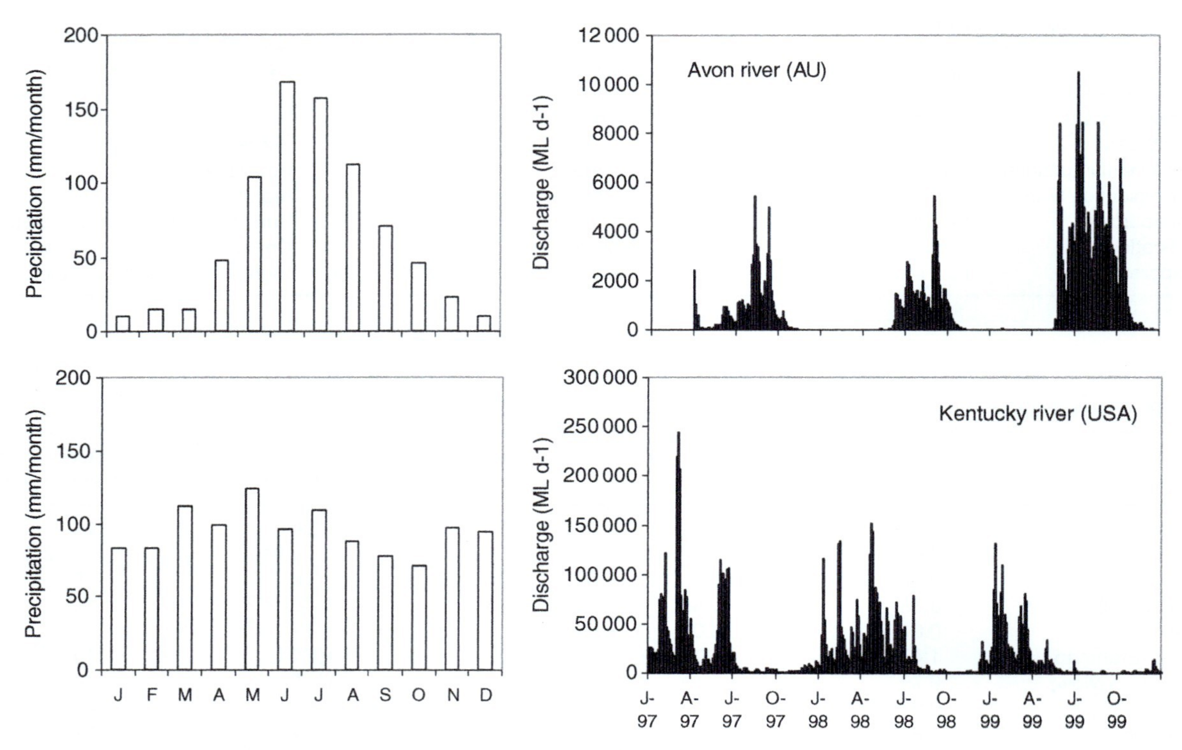

Figure 1 The hydrologic regimes of tropical and temperate rivers reflect differences in seasonal patterns of precipitation and evapotranspiration. The Avon River (Western Australia) experiences high evapotransipration throughout the year and variation in discharge is largely driven by seasonal patterns in rainfall. The Kentucky River (North America) receives similar rainfall throughout the year but variation in evapotranspiration results in similar seasonal patterns in river discharge (offset in northern vs. southern hemispheres).

riverine biota but also in shaping floodplain communities. Floodplains are rare in naturally constricted rivers; or may be disconnected if lateral water regulation structures (i.e., levees) are present. In constricted and levied channels, the effects of runoff on river velocity and stage are accentuated because the ratio of water volume to bottom area increases with rising stage. Thus, the influence of frictional resistance in dissipating energy is lessened with rising stage. Hydrodynamics of river channels are often depicted using simulation models that describe water movements in one, two, or three dimensions (longitudinal, lateral, vertical). These models typically rely on input data describing channel geomorphometry (cross-sectional depictions of river bed and bank elevation) and calibrated using measured surface water elevation and discharge. The models predict surface water elevation under various discharge scenarios and are used to forecast the timing, severity and location of flood events.

Water Movement

Energy is required to move water and in the case of rivers, this energy is derived from gravitational forces acting along an elevation gradient. Rivers are similar to estuaries in that both are flow-dominated (advective) systems; in estuaries, however, the movement of water is bidirectional and driven by tidal forces (**Table 1**). Water movement in lakes is driven by comparatively weak forces associated with wind-induced vertical mixing. The slope of the channel and the frictional resistance imposed by its boundaries determine the velocity with which water is carried down the elevation gradient. The roughness of the channel reflects the composition of bed and bank materials and the presence of natural and artificial structures (e.g., woody debris, wing dams). Turbulence arises as force is dissipated by frictional resistance. This mixing energy maintains particulate matter in suspension and is sufficient to overcome differential heating of surface and bottom layers. Consequently, thermal stratification is rarely observed in rivers except in cases where impoundments are present.

The length of time that water resides within a given segment of the river determines in part the potential for physical, chemical, and biological processes to act upon the dissolved and particulate constituents in through-flowing water. Because of the unidirectional

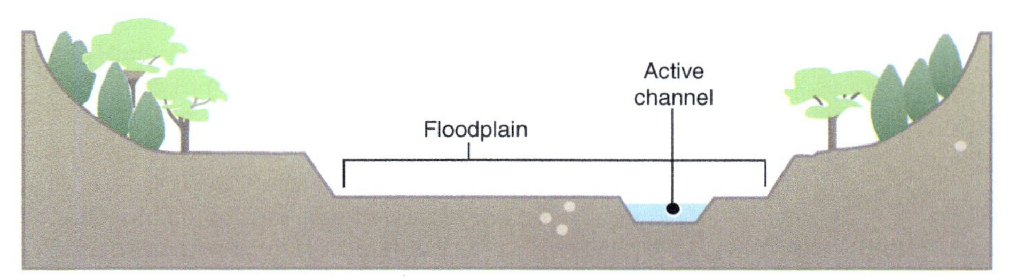

Floodplain river

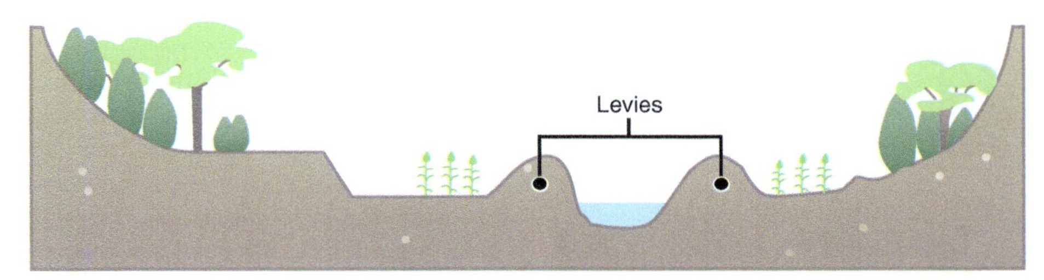

Floodplain river with levies

Constricted river

Figure 2 Cross-sectional morphologies of floodplain and constricted rivers. In floodplain rivers, rising river stage results in lateral inundation unless precluded by the presence of levees. Widening of the river during flood events increases frictional forces and reduces water velocity. In constricted rivers, lateral inundation is constrained by steep adjoining slopes resulting in rapidly increasing water velocity with rising river stage. (Illustration by Christopher O'Brion, VCU Design Services).

flow of water, transit time is a useful metric to characterize inter-river differences in the time required for water and materials to travel through a reach of specified length. Tracers such as dyes (rhodamine) or conservative solutes (chloride, bromide) are used to measure transit time by tracking the movement of labeled parcels of water. Tracer additions provide a reach-scale estimate that integrates longitudinal, lateral, and depth variations in water velocity. Application of this technique to larger rivers is problematic owing to the quantities of tracer required and the difficulty of achieving a laterally uniform addition. Transit time estimates may be obtained from hydrologic models using measured discharge and cross-sectional area to infer average (cross-sectional) velocity at multiple points along the channel. The coupling of transit time and nutrient uptake, termed nutrient spiraling (glossary), is a concept that has been widely used as a framework for understanding the interaction between hydrologic and biological processes in regulating nutrient retention. Transit time estimates are also used to design sampling programs in which a parcel of water is sampled repeatedly as it travels down the channel (termed LaGrangian sampling).

Geomorphology

At any point along a river course, channel morphology reflects the interplay between the force of water and the stability of bed and bank materials. Channel form is a quasi-equilibrium condition maintained by the dominant discharge and determined in part by the supply of sediment from upstream. Where rivers are not constrained by natural landforms or

Figure 3 Selective loss of fine materials may over time create channel reaches that are characterized by a predominance of large substrates such as the gravel bars illustrated here. Their presence in the river channel is important to the maintenance of biodiversity as some species colonize hard substrates or exploit interstitial spaces as a means of adapting to flowing environments. Gravel bars and other subsurface exchange zones are also important to ecosystem function such as nutrient retention. Photo of the Rio Apurimac in Peru by A. Aufdenkampe (see related paper by Aufdenkampe *et al.* (2007) *Organic Geochemistry* 38: 337).

water regulation structures, channels migrate laterally (meander) through erosion and redeposition of bank materials. Active channels are characterized by the ephemeral nature of their features (movement of bars and banks) and by their morphological complexity, which may include the presence of pools, riffles, side channels, and meanders (**Figure 3**). Constrained channels occur where natural landforms or water regulation structures limit lateral mobility. High discharge results in the erosion of bed materials leading to incised (entrenched) channels of low structural complexity and relatively uniform flow conditions. Channel forms and substrate conditions influence the structure and functioning of riverine food webs. For example, where flow conditions favor the deposition of fine materials, the accumulation of particulate organic matter enhances benthic microbial activity. Various schemes have been devised to categorize channel forms, though these efforts are often confounded by the continuous rather than discrete variation in channel features (e.g., width–depth ratio; size distribution of bed materials). Emerging technologies for sensing underwater environments hold much promise for linking biological and geophysical properties particularly in large rivers.

Water Regulation

Human activities have substantially affected the natural hydrologic cycles of rivers throughout the world. Land-use changes have indirect effects on river hydrology by altering the timing and quantity of runoff from the catchment. For example, urbanization creates impermeable surfaces that increase the volume and speed of storm runoff. Direct impacts include the abstraction (withdrawal) of river water for domestic supply and irrigation as well as the alteration of river channels by water regulation structures. Rivers have been altered through the construction of dams, levees and other channel modifications to accommodate local needs for flood protection, hydropower generation and navigation. Channelization (straightening) of river courses facilitates navigation but reduces channel and flow complexity thereby diminishing habitat diversity. Channelized rivers are subject to elevated flow velocities that cause erosion and necessitate bank stabilization. Levees preclude lateral exchange and thereby diminish the role of floodplains in material and energy cycles. In flood-prone rivers, biota are adapted to annual flood pulses that provide access to food and spawning areas within the floodplain. Among the most widespread of human impacts on rivers is the construction of dams, which currently number in excess of 45 000 worldwide. Together, their cumulative storage capacity is equivalent to 15% of global annual river runoff. Over half of the world's major rivers are affected by dams, most of which were constructed in the twentieth century. Dams induce pelagic conditions by increasing water storage and dissipating mixing energy. Pelagic conditions favor sediment deposition and biotic assemblages that differ from those occurring in flowing environments. The severity of water regulation effects varies according to the number and size of regulation structures along the river course. The cumulative effect of dams within a river basin can be gauged from their number and storage capacity expressed relative to river discharge (**Table 3**). Low dams (height < 10 m) are designed to maintain a minimum depth for navigation during low discharge and

Table 3 Water regulation effects vary according to the number and storage capacity of mainstem dams as illustrated by rivers of the central United States

River	*Mean annual discharge ($m^3\ s^{-1}$)*	*Storage capacity of mainstem impoundments (km^3)*	*Retention effect of impoundments (days)*
Kentucky	234	0.26	12.7
Green	314	1.11	41.0
Tennessee	1880	15.00	92.4
Cumberland	936	1.18	14.7
Wabash	800	0.19	2.8
Ohio	7811	8.92	13.2

The combined storage capacity of mainstem dams on the Tennessee River is equivalent to 26% of the river's annual discharge or approximately 92 days at average discharge. In contrast, the Wabash is a relatively free-flowing river with only a single mainstem impoundment that stores a volume less than 1% of its annual discharge.

thereby regulate stage but do not eliminate flowing conditions. High dams are designed for flood control and water storage. They inundate large areas and effectively create lake-like conditions, in some cases, resulting in thermal stratification of the water column.

Water Quality

Rivers integrate drainage waters from distant points in the landscape that may differ in topography, soils, vegetation, and land use. These differences give rise to widely varying water chemistry within river basins particularly where anthropogenic influences differ among sub-basins. Along the river course, water chemistry changes in response to inputs from these diverse sources and also reflects variable water residence times in channel, hyporheic, and lateral storage zones.

Particulate Matter

High concentrations of suspended particulate matter are a characteristic feature of rivers particularly during periods of elevated discharge. The upward component of water turbulence acts to maintain particulate matter in suspension, resulting in downstream transport. Particulate matter may originate within the channel through erosion of bed and bank materials, resuspension of sedimented materials, and biological production. Most particulate matter, however, is derived from sources outside the river channel that are transported via tributary streams. The rivers of Asia are particularly noted for their high sediment load. It is estimated that the Ganges, Brahmaputra, and Yellow Rivers contribute 20% of the total sediment load transported to the oceans (**Table 2**). High sediment production is attributed to natural factors affecting surface erosion (soil composition, steep slopes, and intensive rainfall) as well as anthropogenic effects associated with deforestation and urbanization. Riverine suspended matter is predominantly a fine-grained (<0.2 mm) mixture of mineral and organic particulates (e.g., clay and silt). Though recalcitrant, mineral particulates may undergo changes in their chemical composition through the selective sorption and desorption of dissolved substances. For example, proteins and other dissolved organic compounds adhere to the surfaces of mineral particulates, thereby altering both the bioavailability of these compounds and the chemical properties of particulate matter. Phosphate has a high sorption potential and is principally transported with the particulate fraction. The sorption capacity of particulate matter is determined by the number of available binding sites on the surfaces of the particles and their cumulative surface area (a function of particle density, shape, and size).

Particulate matter is the principal factor regulating water transparency in rivers, although light absorption by dissolved organic compounds may be important during periods of low discharge. When present in high concentrations, mineral particulates may have deleterious effects on filter-feeding organisms by interfering with feeding mechanisms or simply diluting the intake of the more nutritious organic fraction. This fraction includes phytoplankton and bacteria although these typically account for a small proportion of particulate organic matter. The bulk of the particulate organic matter is nonliving detrital material of terrestrial and aquatic origin. This material is of variable age and in varying stages of diagenesis, having been acted upon by both terrestrial and aquatic decomposers.

Dissolved Substances

River water contains dissolved inorganic and organic materials derived from mineral weathering and decomposition processes. Their concentration is largely determined by the types of soils and vegetation within the basin and the extent of interaction between runoff and soil. Low concentrations of dissolved substances occur where river basins are characterized by steep slopes and thin soils, particularly where soils are comprised of insoluble materials (e.g., sand, igneous rock). In these basins, river water is dilute (ion-poor) and similar in chemical composition to that of rain water. Gradual slopes and deeper soils allow for longer flowpaths and greater interaction between water and soil. In these settings, there is greater opportunity for biogeochemical processes to influence the chemistry of runoff especially where soils are dominated by easily-weathered materials (e.g., sedimentary rocks such as limestone). Temporal variation in dissolved ion concentrations is typically associated with rain and snowmelt events. High discharge is often characterized by lower concentrations of dissolved substances owing to rapid delivery of water via overland flow, shallow soil flowpaths, and short transit times in tributary streams. At the onset of rising discharge, rain or snow-melt waters may displace older groundwater, resulting in an initial increase in ion concentrations. Thus, the relationship between discharge and concentration is often nonlinear and ion-specific.

Geologic differences among river basins will influence both the total amount of ions present and their relative proportions. Despite these differences, major ions are generally similar and include bicarbonate, sulfate, chloride, and the base cations (Ca, Mg, Na, K). Climatic factors also influence ionic strength and composition particularly in arid regions where evapoconcentration effects are large. A tea-colored appearance is an apparent feature of some ('blackwater') rivers owing to elevated concentrations of dissolved organic compounds. Their presence is associated with characteristic types of vegetation that leach humic and tannic acids and in some cases (e.g., in coastal areas) by the predominance of sandy soils, which have limited capacity to retain these compounds.

Nutrients

Rivers are not simply conduits for transporting watershed-derived materials but rather, riverine processes may exert considerable influence on water chemistry, particularly for those elements whose abundance is low relative to biological demand. Nitrogen is transported in rivers in dissolved inorganic form (NO_3, NH_4) and in dissolved and particulate organic forms. The latter include living cells and detrital matter as well as a diverse array of dissolved organic compounds that are released through exudation, excretion, and decomposition. Nitrate is a highly mobile ion owing to its low sorption potential. Therefore, it is readily transported through soils and is typically the dominant form of N in rivers, where agriculture and urbanization are prevalent. Elevated NH_4 concentrations may occur below wastewater discharge points. Dissolved organic N assumes greater importance in rivers with minimal human influence. Unlike N, phosphorus is principally transported in the particulate fraction. Concentrations of dissolved P (including PO_4 and other reactive forms) are low owing to biotic uptake and high sorption affinity for mineral particulates (e.g., clay). Sorption processes are reversible such that particle-bound P may desorb and enter the bioavailable pool. Anthropogenic impacts are associated with increases in the total amount of P and the proportion that is in the dissolved fraction. Within rivers, inorganic forms of nitrogen, phosphorus, and silica may be transformed to particulate organic forms (e.g., in algal and bacterial cells). Dissolved silica is converted to its biogenic form by diatoms, a common component of benthic and pelagic algal communities in rivers (See **Algae of River Ecosystems**). Biogenic silica is relatively recalcitrant to remineralization (compared with N and P) such that autotrophic uptake results in progressive depletion of dissolved silica along the river course. Denitrification results in the loss of nitrogen to the atmosphere (as N_2) and is an important process determining N delivery from catchments.

Dissolved Gases

Dissolved gases, particularly oxygen and carbon dioxide, are of interest because their concentrations in river water are influenced by biological processes of photosynthesis and respiration. The solubility of dissolved gases is temperature dependent and therefore it is useful to express concentrations as a percent saturation; that is, relative to the expected concentration for a solution in atmospheric equilibrium. Departures from equilibrium concentrations occur when the rate at which gases are exchanged with the atmosphere is slow relative to rates at which gases are produced or consumed through biological activity. Atmospheric exchange is governed by the concentration gradient across the air–water interface, boundary layer thickness (a function of wind speed), the ratio of river surface area to volume, and factors related to agitation and turbulence of water (e.g., presence of waterfalls). Gas exchange occurs more rapidly in shallow and turbulent rivers relative to deeper, slow-moving rivers. In many rivers, dissolved oxygen is undersaturated while CO_2 is supersaturated (**Figure 4**). These departures from equilibrium reflect the heterotrophic nature of rivers in which community respiration exceeds autotrophic production. Respiration is supported in part by inputs of dissolved and particulate organic matter of terrestrial origin. Decomposition of terrestrial organic matter within the river results in a net production of CO_2 (i.e., in excess of photosynthetic C demand) and a net release of CO_2 from water to air. Diel variations in dissolved oxygen can be used to estimate production

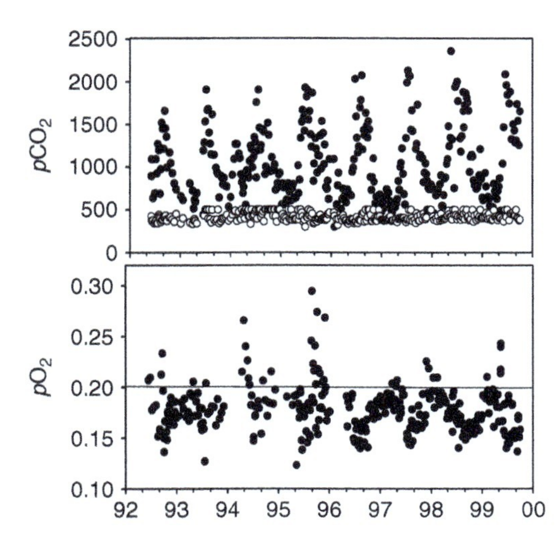

Figure 4 Partial pressure of dissolved carbon dioxide and oxygen in the Hudson River (NY, USA; atmospheric levels of CO_2 indicated by open circles). Persistent supersaturation of pCO_2 and undersaturation of pO_2 are indicative of net heterotrophic conditions whereby respiration exceeds net primary production (adapted from Cole and Caraco 2001, Marine and Freshwater Research).

and respiration provided that re-aeration rates can be reasonably estimated. Although undersaturation of dissolved O_2 is common, severe depletion (i.e., hypoxia – glossary) is rare in riverine environments because turbulent mixing promotes reaeration. Organic matter inputs from poorly-treated sewage effluent were once a wide-spread problem that resulted in chronic and severe oxygen depletion in rivers. Modern wastewater treatment plants are designed to minimize the biological and chemical oxygen demand of effluent.

Pollutants

Rivers integrate runoff over large areas of the landscape and therefore their pollutant loads reflect the cumulative effect of basin-wide releases. Macropollutants include a relatively short list of agents present in concentrations on the order of parts per million ($mg\ l^{-1}$) while micropollutants includes a much larger inventory of chemicals that occur at very low environmental concentrations (ppb or ppt; $\mu g\ l^{-1}$ or $ng\ l^{-1}$). The most common macropollutants are compounds of N and P, which originate in runoff from agricultural areas and from contamination by wastewater (including treated effluent and urban storm water overflow). Nitrogen and phosphorus often limit primary production in lakes and estuaries though their role in regulating the trophic state of rivers is less clear. Many rivers experience nutrient enrichment but biotic responses to elevated nutrient levels (i.e., eutrophication) may be muted by factors that constrain primary production (principally light and residence time). Other macropollutants include sulfate, chloride, and base cations; these are associated with atmospheric deposition, mining, wastewater, and de-icing. Their effects on river biota are less well studied compared with pollutants associated with eutrophication. Micropollutants are a diverse group of chemicals that have deleterious effects at low concentrations. They vary in their reactivity, mode of toxicity, and persistence in the environment and include inorganic pollutants such as metals as well as synthetic organic compounds (e.g., pharmaceuticals, detergents, pesticides). In rivers, the high throughput of water favors the rapid removal of pollutants in the dissolved form. Many pollutants, however, bind to particulates or enter the food chain, where they may persist over long periods of time in sediments and long-lived species such as fish. Regulatory policies aimed at mitigating pollution must take into account proximal effects on river biota as well as distant effects on receiving waters such as estuaries. In some cases (e.g., nutrients), the latter may exhibit greater sensitivity than rivers owing to their longer water residence time.

Biology of Rivers

Rivers owing to their diverse size, channel forms, and biogeographic settings differ greatly in their species assemblages. Constituent species include river specialists that rarely occur outside of flowing waters and habitat generalists that occur in both lentic and lotic waters. In coastal areas, marine species are seasonally important members of river food webs. Salmon and other anadromous fishes (glossary) serve as vectors for distributing marine-derived resources through drainage networks. Species inhabiting rivers face challenges imposed by the unidirectional flowing nature of their environment. Strategies include current avoidance in sheltered areas (along channel margins, behind debris dams or in interstitial spaces), and specialized adaptations such as attachment to hard substrates. Riverine species also share the benefits provided by water flow which supplies particulate matter to filter-feeding organisms, replenishes nutrients and oxygen at the cell boundary layer, and, during floods, allows periodic access to floodplain habitats.

Primary Producers

Attached algae (i.e., periphyton), phytoplankton, and macrophytes contribute to autotrophic production in rivers; their relative importance varies in accordance with river hydrogeomorphology. In shallow, fast-flowing rivers, benthic algae predominate particularly where rocks and woody debris provide stable substrates for colonization. Benthic algal abundance is determined by the availability of suitable substrates, light conditions (the extent of riparian shading), and flow regime (the frequency and severity of scour events). Nutrients and grazers may be important in some settings particularly where nutrient loading is associated with riparian disturbance and loss of canopy shading. In deep, slow-moving rivers, phytoplankton are often the dominant primary producers. Their abundance is principally determined by light availability. The average light intensity experienced by phytoplankton circulating within the river channel is determined by water transparency and the depth of the channel. Nutrients and grazing may be important particularly in regulated rivers and during low discharge conditions. Low flow velocities favor the accumulation of phytoplankton biomass owing to reduced washout (advective loss) and increased water transparency (due to sedimentation of nonalgal particulates). Phytoplankton communities are composed of taxa similar to those found in lentic environments but may also include detached benthic algae. Dominance by diatoms is often reported and may reflect their ability to tolerate the low light conditions in rivers (having a high light utilization efficiency) and the benefits of active mixing (to offset high sinking velocities). Channel morphometry is an important factor determining the species composition and areal coverage of submergent and emergent aquatic vegetation (**Figure 5**). Constricted and channelized rivers have steep shoreline areas, which provide little suitable habitat, whereas floodplain and low-gradient rivers allow for greater colonization in shallow-water areas. Substrate stability is likely a key factor determining the extent and persistence of macrophyte beds since perenniating structures (e.g., tubers, rhizomes) are vulnerable to displacement during periods of elevated discharge.

Figure 5 Aquatic macrophytes are common in rivers though usually they are restricted to channel margins and backwater areas, where flow conditions are reduced. Photo of Beaver River in the Adirondack Mountains of New York State (USA) by P. Bukaveckas.

Invertebrates

The diversity and productivity of invertebrates has received considerable attention in studies of riverine foods webs. For example, nearly one-third of Hynes' classic *Ecology of Running Waters* is devoted to benthic macroinvertebrates. Invertebrates are important to trophic energetics because they link primary sources of energy (autochthonous production and allochthonous inputs) to higher trophic levels such as fish. Pelagic invertebrates (zooplankton) are commonly found in regulated and deep rivers and at times in large numbers. Their abundance is determined by in situ production within the main channel and contributions from areas of reduced water velocity (**Figure 6**). Inputs from upstream reservoirs may also be important in some systems. River zooplankton assemblages are dominated by rotifers and small-bodied forms of cladocerans (e.g., *Bosmina*) and copepods, whereas larger zooplankters (*Daphnia*) are generally associated with lentic environments.

Benthic invertebrates are important components of river food webs and are widely used in habitat assessments owing to their sensitivity to water quality conditions. These include crustaceans such as amphipods and crayfish, mollusks (snails and bivalves), and a great variety of insects (dragonflies, damselflies, stoneflies, mayflies, midges, blackflies, caddisflies). The aquatic insects in particular draw attention to the productive nature of riverine environments through periodic emergence of adults in large numbers. Invertebrates may be grouped according to their feeding habits as predators, filtering and gathering collectors, deposit feeders, scrapers, and shredders. The River Continuum Concept predicts shifts in food resources and feeding habits along a gradient of stream order. In low- and middle-order streams, shredders and grazers rely on leaf litter inputs and benthic algal production whereas in rivers, collectors and filter-feeders utilize suspended particulate matter. Productivity is determined by water temperature, food quantity and quality and the presence of suitable habitat (e.g., hard substrates and snags). A variety of invertebrates, particularly oligochaetes, amphipods, chironomids and micro-crustaceans, occur in large numbers in the

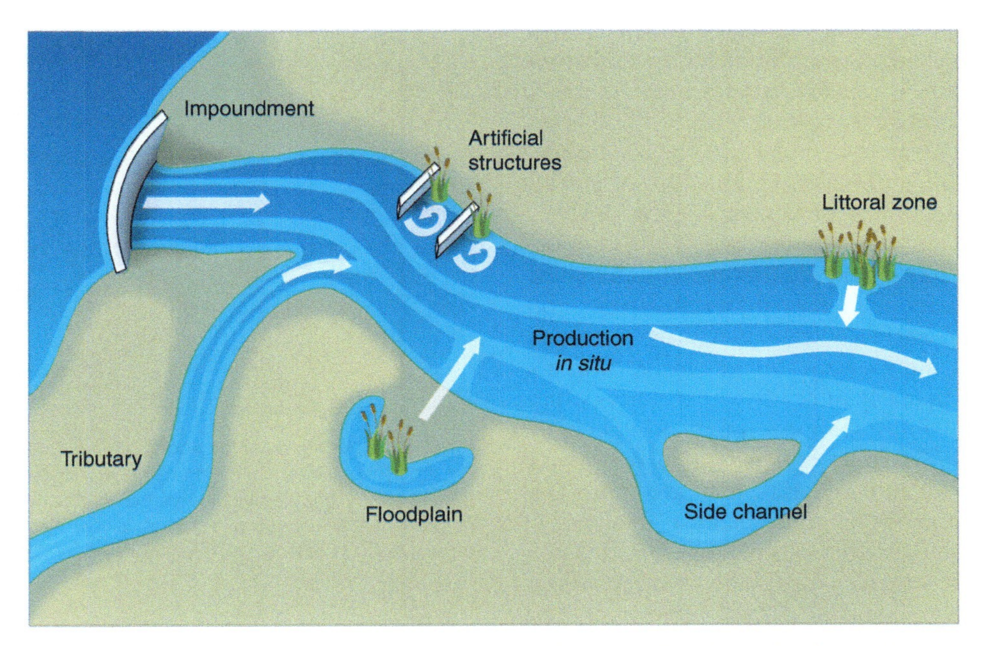

Figure 6 Sources of plankton to rivers include *in situ* production as well as inputs from tributaries, impoundments and near-shore areas of reduced water velocity (Illustration by John Havel and Christopher O'Brion).

subsurface zone (hyporheos; glossary) where they find refuge from predation and currents.

Fishes

Fish are typically the top predators in river food webs and, like macroinvertebrates, are often used as 'bioindicators' for habitat assessment. Many studies have focused on species that are important to commercial or recreational fisheries. However, quantitative estimates of abundance are difficult to obtain particularly in large and deep rivers. The lack of production and biomass estimates with which to compare against similar data for lower trophic levels greatly limits our understanding of food-web energetics. For example, the utility of using phosphorus or chlorophyll as a predictor of fish biomass, which is well-known for lakes, remains largely untested in rivers. In contrast, factors influencing the diversity and species composition of river fish communities are generally well studied. In both temperate and tropical rivers, the numbers of species increase with the size of the drainage basin. The dendritic form of river networks may foster high diversity (relative to contiguous water bodies of comparable area) by providing diverse habitat conditions and through isolation of populations in distant portions of the drainage basin. In many regions, rivers are ancient features of the landscape, thus providing opportunities for speciation among reproductively isolated populations. Anthropogenic influences generally act to make fish assemblages more similar within and among basins and lead to loss of biodiversity. In many rivers, the presence of water regulation structures has had a negative impact on species that prefer flowing conditions and in some cases, has restricted their ability to access former spawning areas. The introduction of nonnative fish species has also substantially altered fish communities in many rivers. As for other river biota, discharge is the key environmental factor structuring communities. In floodplain rivers, fish seek refuge from current velocities and utilize food resources in inundated areas. In levied and naturally constricted rivers, high discharge may cause high mortality, particularly of larval stages, due to elevated current velocities in the channel.

River Food Webs

Research on river food webs has focused on trophic energetics with the goal of understanding the sources of organic matter supporting secondary production. Several conceptual models have been advanced that relate the abundance of invertebrates and fishes to sources of organic matter from the catchment, the floodplain and the river itself. The most influential of these is the River Continuum Concept (RCC) published by Robin Vannote and his colleagues in 1980 and cited in over 1800 subsequent publications. The utility of the RCC model lies in its holistic view of drainage networks whereby changes in the physical template of the channel (morphometry and substrate composition) with increasing stream order is linked to corresponding changes in food resources and biotic communities. The model emphasizes the importance of terrestrial (allochthonous) inputs in supporting secondary production. Consumers in river environments are thought to benefit from allochthonous inputs to a greater extent than their lentic counterparts due to loading factors that reflect the large ratio of land to surface water area in river basins. Autochthonous inputs were thought to be of minor importance particularly in headwater reaches (where shading by the forest canopy limits primary production) and in large rivers (where turbidity and depth limit algal and aquatic plant growth). This viewpoint is supported by geochemical analyses of riverine particulate matter which show that it is predominantly of terrestrial origin. However, the utilization of allochthonous and autochthonous organic matter is determined not only by their relative availability but also by their suitability relative to consumer needs (e.g., edibility, digestibility, nutritional sufficiency). Allochthonous inputs are comprised of detrital materials low in nutritive value whereas organic matter of autochthonous origin is enriched in mineral nutrients (N, P) and important biochemicals (fatty acids, proteins, etc.). An alternative view of river food-web energetics (Riverine Productivity Model; RPM) is that higher trophic levels obtain a disproportionate fraction of energy (or key dietary factors) from autochthonous sources by selective feeding and preferential assimilation of the more nutritious algal component. Stable and radio isotopes of carbon are used to quantify inputs from various sources (e.g., aquatic vs. terrestrial) provided that the sources differ in their isotopic signatures. Stable isotope data have shown that various consumer groups in rivers rely on algal production despite the quantitative dominance of organic matter that is terrestrial in origin. While the RCC and RPM focus on transport and production within the main channel, the Flood-Pulse Concept (FPC) considers the contribution of floodplain areas in supporting riverine communities. The importance of floodplain resources depends on the duration, aerial extent and timing of floodplain inundation. In tropical regions, flooded areas may far exceed the size of the main channel, thereby allowing riverine consumers to utilize terrestrial resources over extensive areas (**Figure** 7). The growth of aquatic plants and algae in flooded areas may also augment terrestrial resources if the duration of flooding is sufficiently long and light-temperature conditions are favorable. Temperate rivers also experience periodic floods although these are typically of shorter

Figure 7 Inundation of the floodplain near a tributary of the Amazon River (Rio Unini). In many rivers, flood events follow a regular annual cycle to which riverine organisms and riparian communities are adapted. Flooding allows access by river organisms to terrestrial food resources in inundated areas. Photo by A. Aufdenkampe.

duration and occur during periods when water temperature is low (e.g., in association with winter rains or spring snowmelt). The three models differ by their emphasis on longitudinal transport of terrestrial organic matter (RCC), autochthonous production within the channel (RPM) and floodplain resources (FPC). They share the common view that an appreciation of river hydrogeomorphology is central to understanding variations in the quantity and quality of food resources and, in turn, the energetic efficiency of river foods webs.

Global Biogeochemical Cycling

Rivers account for only a small proportion of land area worldwide but play an important role in regional and global biogeochemical cycles. Rivers are the principal means by which terrestrial-derived materials are transported to the ocean. Over 90% of the earth's landmass is drained by rivers; the 100 largest rivers drain 65% of global land area. Rivers are the most powerful erosive force on the planet, substantially modifying landscape features and transporting 20 gigatons of sediment to the coastal margin annually. The input of dissolved and particulate organic carbon from rivers is sufficient to account for the estimated replacement times of oceanic dissolved organic carbon (ca. 4000–6000 year). Much progress has been made in recent years to assess material export from rivers, but few studies have examined within-river processes and their significance in regional and global biogeochemical cycles. Work by Jeff Richey and his colleagues has shown that waters of the Amazon release 13 times more carbon through out-gassing (evasion) of respired CO_2 than is exported to the ocean. The respired carbon originates from terrestrial sources and suggests that the overall carbon budget of the rainforest is more closely balanced than would be inferred from terrestrial biomass accumulation and fluvial export losses alone. The cumulative effects of human activities within river basins have given rise to global-scale alterations in water and material fluxes. Anthropogenic inputs have enhanced the delivery of nitrogen and phosphorus by rivers to coastal environments and led to widespread problems with eutrophication. The combined storage capacity of the world's dams has increased water storage and sediment retention thereby partially offsetting erosion losses associated with watershed disturbance.

Glossary

Discharge – The volume of water moving past a given point in the river per unit time (typically, $l\ s^{-1}$)

Evapotranspiration – The movement of water from the Earth's land surface to the atmosphere via evaporation and plant transpiration.

Nutrient spiraling – The uptake and release of dissolved nutrients during downstream transport.

Anadromous – Fishes that live predominantly in marine waters but are seasonal residents of freshwater streams and rivers during spawning and larval development.

Hypoxia – A reduced concentration of dissolved oxygen in a waterbody.

Hyporheos – The zone beneath and lateral to the river bed where river- and ground-water mix.

See also: Algae of River Ecosystems; Currents in Rivers; Restoration Ecology of Rivers; Streams and Rivers as Ecosystems.

Further Reading

Finlay JC (2001) Stable-carbon-isotope ratios of river biota: Implications for energy flow in lotic foodwebs. *Ecology* 82: 1052–1064.

Hynes HBN (1970) *The Ecology of Running Waters.* Toronto: University of Toronto Press.

Junk WJ, Bayley PB, and Sparks RE (1989) The flood-pulse concept in river-floodplain systems. In: Dodge DP (ed.) *Proceedings of the International Large Rivers Symposium. Can. Spec. Publ. Fish Acquat. Sci.* 106: 110–127.

Kalff J (2002) Rivers and the export of materials from drainage basins and the atmosphere. In: *Limnology*, pp. 94–121. Upper Saddle River, NJ: Prentice-Hall.

Meybeck M (1982) Carbon, nitrogen and phosphorus transport by world rivers. *American Journal of Science* 282: 401–450.

Milliman JD and Meade RH (1983) Worldwide delivery of river sediment to the oceans. *Journal of Geology* 91: 1–21.

Nilsson C, Reidy CA, Dynesius M, and Revenga C (2005) Fragmentation and flow regulation of the world's large river systems. *Science* 308: 405–408.

Richey JE, Hedges JI, Devol AH, Quay PD, Victoria R, Martinelli L, and Forsberg BR (1990) Biogeochemistry of carbon in the Amazon River. *Limnology and Oceanography* 35: 352–371.

Richey JE, Melack JM, Aufdenkampe A, Ballester VM, and Hess LL (2002) Outgassing from Amazonian rivers and wetlands as a large tropical source of atmospheric CO_2. *Nature* 416: 617–620.

Syvitski JPM, Vorosmarty CJ, Kettner AJ, and Green P (2005) Impact of humans on the flux of terrestrial sediment to the global coastal ocean. *Science* 308: 376–380.

Thorp JH and Delong MD (1994) The riverine productivity model: an heuristic view of carbon sources and organic processing in large river ecosystems. *Oikos* 70: 305–308.

Vannote RL, Minshall GW, Cummins KW, Sedell JR, and Cushing CE (1980) The river continuum concept. *Canadian Journal of Fisheries and Aquatic Sciences* 37: 130–137.

Vorosmarty CJ, Fekete BM, Meybeck M, and Lammers RB (2000) Global system of rivers: Its role in organizing continental land mass and defining land-to-ocean linkages. *Global Biogeochemical Cycles* 14: 599–621.

Vorosmarty CJ, Sharma KP, Fekete BM, Copeland AH, Holden J, Marble J, and Lough JA (1997) The storage and aging of continental runoff in large reservoir systems of the world. *Ambio* 26: 210–219.

Wetzel RG (2001) Rivers and lakes – Their distribution, origins and forms. In: *Limnology: Lake and River Ecosystems*, pp. 15–42. San Diego, CA: Academic Press.

Geomorphology of Streams and Rivers

J P Bravard, University Lumière-Lyon 2, France
F Petit, University of Liège, Liège, Belgium

Watersheds and River Networks

Rivers drain areas delimited by 'watershed divides,' which usually fit with topographic divides, except in karstic regions, where underground water fluxes are more complex. The watershed is an open system with zones of sediment and water production, transfer, and storage. It displays different levels of connectivity between slopes and streams, depending on a set of changing variables. Streams originate from overland flow and from saturated zones according to the watershed geology, topography, and vegetation, as well as to the precipitation type.

Horton (1945) and Strahler (1952) developed the 'stream order' concept. This ranking of rivers, from upstream individual reaches (low orders) to downstream large rivers (high orders), then became a widely used classification system (refer to 'see also' section). It has been demonstrated that river orders are strongly correlated to the watersheds characteristics in terms of climate, geology, and land cover. Furthermore, the density of present river networks (in km of river length km^{-2}) reflects a long history of changing climate conditions. Also, the mobility of networks through geological times may be of prime importance to explain the pattern of fish fauna: the changing sea level, the changing climate on continents, as well as river piracy (when a river is captured to the benefit of another watershed) may explain changes in the way rivers connect, and must be taken into account.

River Profile

The longitudinal stream profile reflects a balance between the transport capacity on one hand, the volume, and size of the bed material on the other. In general, for perennial rivers, the upper part of the profile tends to be concave and the discharge, increasing from upstream to downstream, makes the bedload transport possible on more and more gentle slopes. On the other hand, allogenic rivers, whose discharge does not increase in the downstream direction, do not display a marked concavity. Moreover, in semi-arid areas, the reduction of the discharge downstream (by evaporation, infiltration, and transmission losses) produces convex profiles.

Another basic principle is that the river slope is correlated to bedload size. Bedload size decreases downstream, by sorting and selecting deposits, by attrition (i.e., grinding of coarse particles), and by dissolution in the case of limestone. Thus, the profiles display an important concavity when the size of transported material decreases quickly.

Long profiles can sometimes present local convexities (steepening of channel gradient, which appears like a knickpoint) because of (i) tectonic activity, (ii) a more resistant bedrock outcropping, (iii) the injection of a coarser or larger load by a tributary or from the valley slopes, and (iv) consequences of past events like a lowering of the base level.

Particulate Material Transport in River Channels

Hydraulics of River Flow and Bankfull Discharge

The Reynolds number (Re) allows distinguishing laminar from turbulent flows. It is defined as

$$Re = ud/v \qquad [1]$$

where u is the velocity, d the depth, and v the kinematic viscosity (function of the temperature). The Reynolds number is dimensionless. For Reynolds numbers smaller than 500, viscous forces dominate and the flow is laminar. Laminar flows can be represented by a series of parallel layers without any mixing between them. For Reynolds numbers larger than 2000, turbulent forces are dominant: the flow is fully turbulent and there are lateral and vertical exchanges between the liquid veins. The flow is to some extent chaotic with fluctuations in instantaneous velocities; mixing of particles occurs with turbulent energy exchanges. Eddies and other forms of secondary flows are superimposed to the principal flow component. This is of great significance for the variations of instantaneous velocities and the mixing of sediments. The vertical flow components make it indeed possible to maintain particles in suspension. Most natural river flows have Reynolds numbers well in excess of 2000.

The Froude number (Fr) is defined as

$$Fr = u/(gd)^{0.5} \qquad [2]$$

with u, d as before, g being the acceleration of gravity. This number (also dimensionless) is used to differentiate between subcritical flows (where $Fr < 1$) and supercritical flows ($Fr > 1$). For supercritical flows, gravity waves cannot migrate upstream; surface waves are unstable and may break. This results in a considerable energy loss. In nature, most of the river flows are turbulent and subcritical.

In natural rivers, the banks and bed cause energy losses by friction; these losses are all proportional to their roughness. Roughness depends mainly on the nature of bed material, on the vegetation which can clutter the channel, on bedforms succession, and on the presence of sinuosity. Various formulae take into account roughness; the most usual one is the Manning's formula:

$$V=\left(R_h^{2/3} S_e^{1/2}\right)/n \quad [3]$$

where R_h is the hydraulic radius (i.e., the ratio of the wetted cross section on the wetted perimeter), S_e the energy slope, and n the Manning's roughness coefficient. The later can be determined by tables, where n_0 depends on the material constituting the bed bottom (it can be given from the median diameter of the grains, using the Strickler formula). For rivers whose cross sections display poorly differentiated longitudinal variations (i.e., in the absence of significant bedforms variations), the longitudinal water slope may be substituted to energy slope. Moreover, for broad and shallow channels, it is admitted that the hydraulic radius is very close to the average depth. Typical roughness values have been proposed for various river types: roughness generally varies between 0.020 and 0.100, but can exceed 0.150 and even reach a value of 0.42 for a small rivulet with aquatic vegetation at low water stage.

Bankfull Discharge

The discharge which is supposed to control the 3-dimensional morphology of the channel is called 'bankfull discharge' for temperate rivers. Furthermore, it has been proved that this discharge is the 'dominant' discharge, i.e., the discharge providing the maximum efficiency in bedload transport and leading to adjustments of channel morphology. Early studies by Leopold *et al.* (1964) suggested a recurrence interval of bankfull discharge varying between 1 and 2 years with 1.5 years on average, but a considerable scatter of results is observed with values ranging from 1.01 to 32 years. In fact, the recurrence of bankfull discharge depends on the basin size, on the bedload, and on hydrological regimes. The recurrence interval increases with basin size: it reaches only 0.5 years in the case of small rivers with pebble beds on impermeable substrata, but exceeds 1.5 years for large rivers. Base-flow dominated gravel-bed streams, and silty or sandy rivers experience less frequent bankfull discharges, along with a recurrence interval higher than 2 or even 3 years.

Suspended Load and Bedload Transport

Rivers transport dissolved load (which do not play any role in channel morphology), suspended load, and bedload. Concentration of suspended load during floods depends on the geology, on the vegetation cover of the watershed, and on the intensity of precipitations, at least for streams. Sedimentation occurs on the floodplain beyond bankfull discharge (Q_b), and in the channel during the recession of the flood (refer to 'see also' section). Sediment deposition affects aquatic habitats through the siltation of substrates.

The coarse particles lying on river beds are set into motion during floods when a critical shear stress is passed, according to the equation:

$$\tau = g\,\rho_w R_h S_e \quad [4]$$

In this equation, τ is the mean shear stress (expressed in $\mathrm{N m^{-2}}$), ρ_w is the density of water ($1000\,\mathrm{kg\,m^{-3}}$), g is the acceleration due to gravity ($9.81\,\mathrm{m\,s^{-2}}$), R_h the hydraulic radius, and S the energy line gradient or energy slope. Equation [4] represents the sum of the shear stresses created by the resistance of the particles (τ') and by the irregularities in the channels and the bank, i.e., the bedforms (τ''). τ'' is frequently referred to as the form drag. However, only the grain shear stress (τ') should be taken into consideration for the transport of sediments. Different methods exist to ascertain the distinction between τ' and τ''. One of the most commonly used and most efficient ones is based on the relationship between roughness due solely to resistance of particles (n_0), obtained using Strickler formula,

$$n_0 = 0.048\, D_{50}^{1/6} \quad [5a]$$

where D_{50} is the median diameter of the bed material (in m).

And the total roughness (n_t) in the Manning's formula (see eqn. [3], above)

$$\tau' = \tau(K)^{3/2} \quad [5b]$$

with $K = n_0/n_t$.

The Shields entrainment function is generally used to determine the threshold of bed sediment motion:

$$\theta_c = \tau_c/(\rho_s - \rho_w)gD = fctRe^* \quad [6]$$

where θ_c is the critical dimensionless shear stress, τ_c the critical boundary shear stress, ρ_s the density of sediment grains, ρ_w the density of water, g the acceleration due to gravity, and D the grain diameter (usually the D_{50}). θ_c is a function of the Reynolds' grain number $Re_* = u_* D/\nu$, where u_* is the shear velocity, and ν is the kinematic viscosity.

To resolve equation. [6], a value has to be assigned to θ_c. A value of 0.060 was initially proposed by Shields (1936) when $Re_* > 200$ (as it is generally the case in natural rivers and even in flumes). However, this value has been challenged in the case of naturally-sorted grains and different critical values of 0.045 and 0.030 have been proposed. Additionally, the constancy of θ_c has been questioned: an equation of the type $\theta_c = a\ (D_i/D_{50})^b$ has been proposed where the Shields factor (θ_c) varies as a function of the relationship between the size of the surface particles involved (D_i) and the diameter of the underlying material constituting the bed (D_{50}). In this type of equation, the coefficient a represents Shield's standard dimensionless coefficient (θ_c) in homogeneous sediment conditions when $D_i/D_{50} = 1$. Moreover, a negative sign of the exponent b indicates that θ_c values decrease as D_i increases. This equation includes both the hiding effects (when $D_i < D_{50}$) and the protrusion effects (when $D_i > D_{50}$). More recent studies confirm the validity of this approach, even if the values of coefficient and exponents sometimes differ significantly.

Other criteria are used for characterizing bedload mobilization. The critical erosion velocity introduced by Hjulström (1935) is the oldest and the most widely used in early studies. Critical values of velocity exist for deposition and for suspension. It appears that the lowest threshold mean velocity occurs for well-sorted 0.2–0.5 mm sands. Higher critical velocities are needed to move larger and heavier particles (gravel and pebbles) and also to erode the smaller particles, such as cohesive clays which are protected by submergence withing the laminar sublayer and because the cohesion of particles restrains the erosion. The velocity at which deposition occurs is lower than the critical velocity for gravels because of the inertia of particles. Usually, the cessation of transport corresponds with the fall velocity. Thus, the transport of suspended silt and clays is maintained over a wide range of flow velocities, between the threshold and fall velocities. Even if the critical mean velocity for given sediment size varies with flow depth and with sediment sorting and consolidation (Sundborg, 1956), this criterion can yet be used when for example the shear stresses are difficult to evaluate in the field.

More recently, the notion of specific stream power which represents the amount of work a river may perform was introduced:

$$\varpi = (\rho g\ Q_S)/w \qquad [7]$$

with ϖ as the stream power (W m^{-2}), Q the discharge, w the width of the water surface, and S the longitudinal slope. Specific power was initially used to evaluate bedload transport but, later on, other fields of applications were found, mostly to understand river activity, in particular regarding channel patterns and meander dynamics, or the possible reaction and adaptation of rivers to human interventions. More recently, the distance traveled by the bedload after its mobilization has been linked to excessive specific stream power in relation to critical specific stream power, the specific stream power which allows the displacement of particles. However, the major remaining problem is to determine a relation linking the critical specific stream power to the size of mobilized material. Various types of equations have been proposed from simple ones to quite complex with limited practical application and distinctions have been made for gravel-bed rivers, and for torrents with very coarse bedload. Compared to shear stress, specific stream power has the advantage of being easy to determine (discharge, slope, and width are easy to evaluate even after a flood event). However, specific stream power does not take into account the role of bedforms (hiding and protrusion effects, pool, and riffles sequences) and is therefore no more than a basic indicator of river dynamics.

For a long time, the discharge required to initiate bedload movement was commonly assumed as close to bankfull discharge. But recent studies conducted in temperate gravel-bed rivers, demonstrated that bedload movement is initiated for discharges lower than bankfull (0.4–0.7 Q_b) and occurs about fifteen days a year on average, but values may be highly variable between rivers as for the bankfull discharge recurrence interval. But in spite of this frequency, bedload downstream progression is slow – less than 4 km per century for Ardenne rivers (Belgium) – in comparison with values ranging between 10 and 20 km per century for high energy mountain rivers in Alpine and Mediterranean environments.

Bedload sediment yield (i.e., bedload discharge) is not important with respect to the material evacuated in suspension. For example, in temperate humid environments, bedload discharge (in terms of specific annual yield) varies from 1 to 5 t km^{-2} year^{-1} for gravel-bed rivers to more than 50 t/km^{-2} year^{-1} in Mediterranean mountains or in semi-arid areas. It is generally considered that bedload is about 10% of the total sediment load; it may be much higher in mountain streams and watersheds with hard rock.

Channel Morphology and Channel Adaptations

Channel Morphology

Steep long profiles of mountain channels are broken by waterfalls, rocky rapids, pebble, and log steps which dissipate energy (refer to 'see also' section). Log jams – which are unstable features – store bedload and give rise to various habitats for aquatic fauna (refuges, spawning grounds).

Downstream of mountain sectors, river patterns are classified on the basis of two descriptive variables: the sinuosity (Si, see **Table 1**) and the number of channels at the scale of the functional sector or reach with homogeneous geomorphology (a nested taxonomy at the scale of the sector may deal with the unit landforms (for instance: bars, riffles...), and the particles granulometry and setting. Each functional sector may be characterized by a specific fluvial pattern, by dominant processes of various intensities, and by associated specific landforms (**Figure 1**).

Table 1 Key variables used to describe channel morphology (after Petts and Amoros, 1995)

Variable	*Definition*
River channel cross-sectional form	
Size	
Channel capacity (Cc)	Cross-sectional area at bankfull stage equates to mean depth width (m^2)
Channel width (w)	Width of channel between the river banks (m)
Channel mean depth (d)	
Wetted perimeter (Wp)	Total length of channel bed and banks (m)
Shape	
Width depth ratio (w/d)	
Channel asymmetry (A^*)	$A^* = (A_r - A_l)/\mathrm{Cc}$ where A_r and A_l are the areas to the right and left of the channel center line and $\mathrm{Cc} = A_r + A_l$
Efficiency	
Hydraulic radius (R) Cc/Wp	A measure of channel efficiency in conveying water
River channel planform	
Sinuosity	
Channel sinuosity (Si)	Channel length/straight-line valley length
Meander form	
Meander wavelength (L)	Distance between two consecutive meander bends inflexion points (m)
Meander height (H), m	
Radius of curvature (r), m	
Channel multiplicity	
Braid intensity	Two times total bar length divided by reach length
Long profile	
Bed slope (Sb), m m^{-1}	
Profile gradient (S)	Gradient from 10% to 85% of length, upstream from river mouth (m m^{-1})
Bed roughness	
Bedform wavelength (l)	Use for pool-riffle spacing of riffle-dune forms (m)
Bedform amplitude (h), m	
Bed roughness (D84)	D 84 is a representative percentile of the bed particle size distribution curve

The three main geomorphic patterns are the following, considering that straight channels do not exist, except under structural control:

1. *The meander pattern.* Meanders are unique, narrow and deep channels, with a Si index exceeding 1.2 or 1.5. Meanders may be entrenched or ingrown into hard rock plateaus, or may migrate freely across flood plains. The morphological sets of alluvial meanders are lateral bars, riffles, and pools. A bar may display a chute across its inner part, as well as ridges and swales shaped by the lateral and downstream migration of the river during floods. Rivers transporting a considerable suspended load may build alluvial ridges and levees, which may be breached during floods, inducing crevasse splays over bottom flats. The velocity of lateral erosion displays a wide range of values. **Figure 2** presents the main geomorphic features of channels initiating sinuosities.

The cut-off of meanders depends on two main processes: the neck and the chute cut-offs.

Some morphometric relationships link geometric variables to the river discharge. For instance, the equation:

$$\mathrm{l} = 54.3 Q_{\mathrm{b}}^{0.5} \quad [8]$$

links the meander wavelength (l) and the bankfull discharge (Q_{b}). The multivariate equation:

$$\mathrm{l} = 618.Q_{\mathrm{b}}^{0.43}.M^{-0.7} \quad [9]$$

links l to Q_{b} and to the silt-clay index M (percentage of silt and clay in the river bank sediments): a high value of M protects the banks from lateral erosion and decreases sinuosity.

Alluvial meanders occur when transport capacity exceeds the river sediment load. Sinuosity has been interpreted as a means of dissipating specific stream power, which is usually comprised between 10 and 100 W m^{-2}.

2. *The braided pattern.* Braiding is a multichannel pattern, with low sinuosity (value of Si between 1 and 1.1), wide and flat channels, along with mobile gravel and/or sand bars. Sand bars are visible at low flow, but submerged during floods. The intensity of sediment transport is reflected in the degree of colonization of bars by vegetation. This conditions the permanence of terrestrial habitats, while aquatic

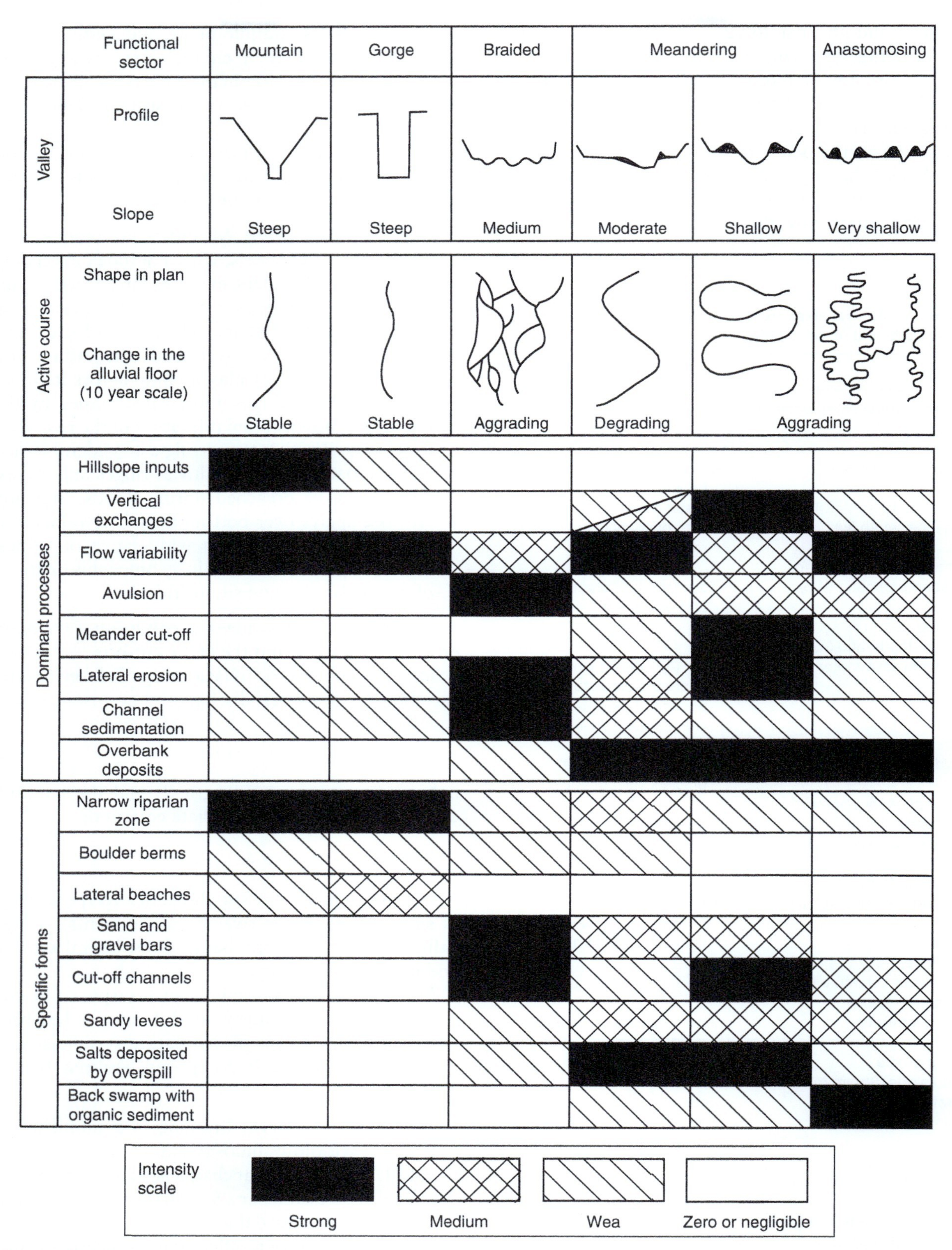

Figure 1 Types of functional sectors characterized by valley morphology, the style of the active course, the dominant processes, and the specific forms that make up the geomorphological patches within the hydrosystem (after Petts and Amoros (1996)).

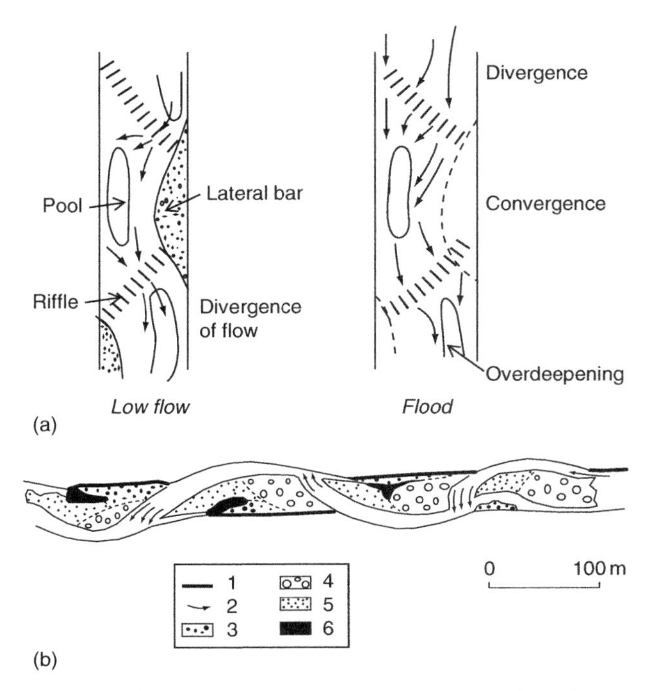

Figure 2 (a) Morphological development resulting from the initiations of alternated bars (after Ferguson (1987)). (b) Development of lateral bars and meander formation in an artificially straightened channel (after Lewin (1976)). 1. Initial banks, 2. Riffles in the channel at low flow, 3. Ancient gravelly deposits, 4. Recent gravelly deposits, 5. Gravel deposits in sheltered areas, 6. Sandy deposits.

ones are ephemeral due to the intense reworking of unit landforms and particles. The variables explaining braiding are: high values of bedload fluxes and of bank erodibility, high variability of discharges, and steep slopes. The braided pattern develops on steep alluvial valley floors (slope being usually more than 0.0007). The energy displayed in braided rivers commonly ranges between less than 100 W m^{-2} and more than 500 W m^{-2}.

3. *The anastomosed pattern.* Anastomosed rivers are multichannel rivers displaying stable high sinuosity channels on the active belt margins. Different types of anastomosed patterns have been described all over the world: aggrading patterns in flat post-glacial valleys (Western Canada), stable patterns in the humid tropics (Zaire), juxtaposed braided and anastomosed belts in arid regions (Australia), and also in cold regions (Lena).

However, recent studies stress the increasing diversity of geomorphic behavior of world rivers as most models have been defined in temperate regions. The originality of some tropical rivers due to the occurrence of extreme flood events is put forward. For instance, river beds of Southeast Asia display high stable landforms on the river margins, shaped by extreme events, and, in the inner part of the river tract, landforms adjustable for low and medium floods.

Discrimination between Geomorphic Patterns: Succession of Patterns along the River Continuum

Since Leopold and Wolman's successful work (1957), fluvial geomorphologists have proposed predictive equations for discriminating between fluvial patterns. These equations have progressively incorporated the size of bed sediment and the specific stream power (usually calculated for the bankfull discharge or for the 1.5 year recurrence interval discharge; see above), which integrates slope, bankfull discharge, and channel width.

From upland streams to the mouth of natural rivers, the slope and the size of both bed material and bedload decrease, while discharge and the silt-clay index increase. Downstream the production zone – with mountain torrents constrained by valley walls – rivers usually display braided patterns along the transit zone. The change in sediment size and in flow discharge explains the change towards meander pattern in the downstream reach. Braided patterns are characterized by high energy gradient, strong width/depth ratio as well as a high instability in space. The lower part of **Figure 3** describes the relative stability of variables involved in the functioning of fluvial patterns, including the size of particles, the stream power, as well as the texture of particles of the bed and of the floodplain.

However, this continuum may change over time, considering Holocene climate change and the impacts of human activities in the watershed. These may trigger an increase in flood discharge and bedload.

Channel Adjustments and Channel Metamorphosis

Even if river valleys and rivers may be considered as systems in quasi-equilibrium at an historical timescale (10–100 years), they undergo continual changes on longer terms. The variables are dependent or independent according to the time span considered. For instance, present channel morphology controls the flow hydraulics at a section (**Table 2**), but it is itself dependant of flow and sediment mean discharge at the historical timescale. These complex adaptations in time and space reflect the dynamic equilibrium.

The concepts of channel system and channel response (Schumm, 1977) are indeed efficient tools for understanding changes in river behavior (**Table 3**). At a timescale of decades or more, the relative

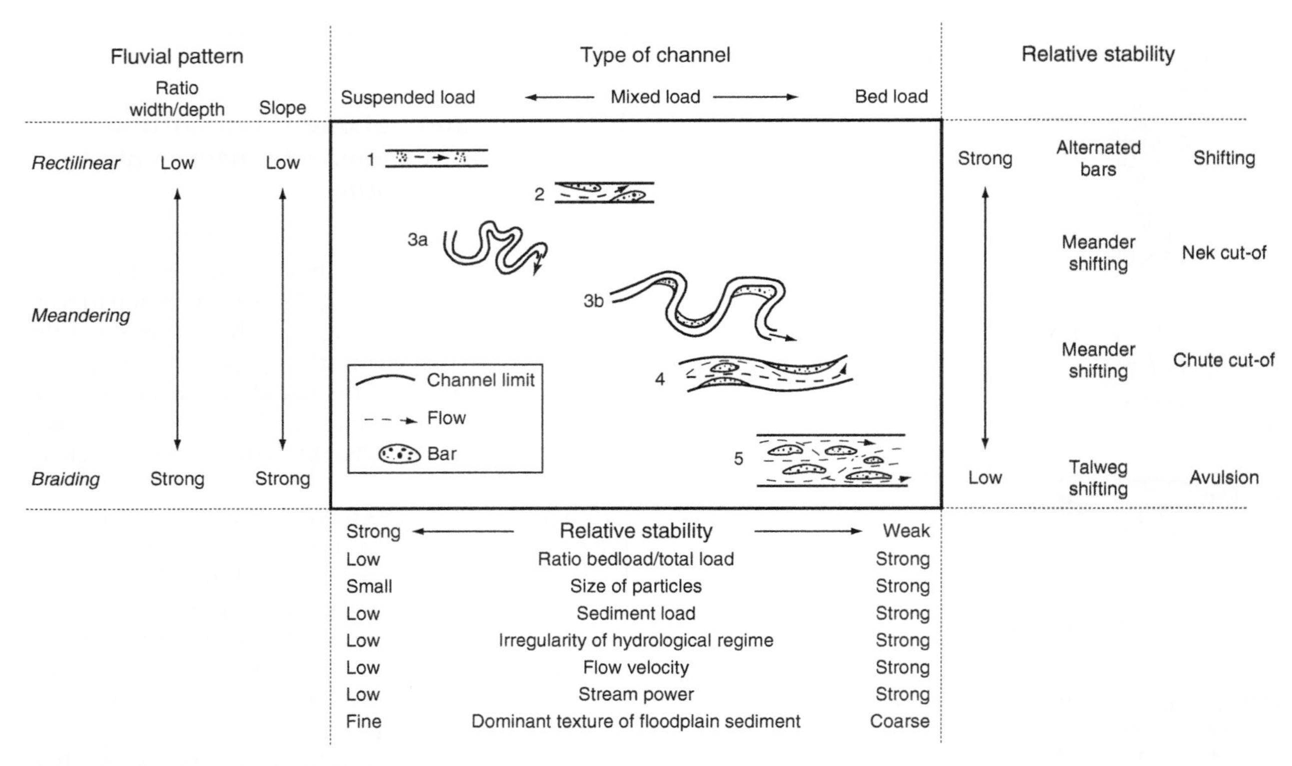

Figure 3 Classification of fluvial patterns (modified from Schumm (1977); Kellerhals and Church (1989).

Table 2 Status of river variables during timespans of decreasing duration (after Schumm and Lichty (1965))

Ranked variables	*Status of variables during designated time*		
	Geologic 10^4–10^6	Historical 10^2–10^3	Present 10^0–10^1
Geology (lithology, structure)	Independent	Independent	Independent
Palaeoclimate	Independent	Independent	Independent
Palaeohydrology	Independent	Independent	Independent
Relief or volume of system above base level	Dependent	Independent	Independent
Valley dimensions (width, depth, slope)	Dependent	Independent	Independent
Climate (precipitation, temperature, seasonality)		Independent	Independent
Vegetation (type and density)		Independent	Independent
Hydrology (mean discharge of water and sediment)		Independent	Independent
Channel morphology		Dependent	Independent
Hydraulic of flow (at-a-section)			Dependent

variations (+/−) of the two external variables, bankfull (Q_b) or dominant (Q_d) discharges, and Q_{sf} (bedload discharge) explain the adjustments of the geometric (internal) channel parameters at the reach scale. Changes from a meandering to a braided pattern do occur in response to the variations of the external variables along river reaches (at the scale of the functional sector).

However, the adjustment of hydraulic parameters varies in space and time. The reactivity to changes in external variables is higher for unit landforms belonging to the channel than for fluvial patterns, while long profiles (slopes at the river scale) are the less sensitive (**Figure 4**).

At a multidecadal or multicentennial scale, discrete reaches may experience a complete change of their geomorphic pattern (i.e., from braiding to meandering, for instance, the reverse change being possible), which is called 'fluvial metamorphosis.' This is due to long lasting changes in the external variables at the watershed scale. For instance, the Rhône River (France) and its main mountain tributaries displayed between 400 BC and ca. 1400 AD meander patterns from the upland 5th order rivers down to the Mediterranean Sea. This pattern changed towards braiding at the very beginning of the Little Ice Age (early 15th century), through the downstream progradation of 'sedimentation zones' from the uplands prone to increased sediment delivery, due to fragile cleared mountain slopes and to stronger (summer) storms. The main channel of the Rhone ('Grand Rhône') flowing across the Camargue delta to the sea was braiding during the 17th century.

Table 3 Influence of river discharge and bed load changes on the geometric variables describing river channels and on the adjustment of river behavior (pattern and long profile)

Balance between flow discharge and bedload discharge	*Adjustment of geometric variables*	*River responses*
$Q_l^- > Q_{sf}^-$	w^-, d^-, l^-, S^-, Si^+	Am: aggradation, meandering
$Q_l^- < Q_{sf}^-$	w^-, d^+, l^-, S^-, Si^+	Em: erosion, meandering
$Q_l^+ < Q_{sf}^+$	w^+, d^-, l^+, S^+, Si^-	Ab: aggradation, braiding
$Q_l^+ > Q_{sf}^+$	w^+, d^+, l^+, S^+, Si^-	Eb: erosion, braiding

Q_l is water discharge, Q_{sf} is bed load discharge, L is the meander wave length, and S is the profile gradient. Sign '+' means an increase while sign '–' means a decrease in the value of the variables (after Starkel (1983)).

Bedrock Channels

Bedrock channels are developed when the capacity of transportation exceeds the particle size on the long term, i.e., when sediment budgets are negative at the reach scale. Due to steep slope and to constriction by the valley walls, unit stream power is an order of magnitude larger than in alluvial channels (from 3×10^3 to $1 \times 10^4\,W\,m^{-2}$). Bed morphology is characterized by a high spatial variability due to turbulence, complexity of circulating cells of water on the margin of the main flow, and to critical and supercritical flows: rocky knickzones, incisional bedforms such as grooves and flutes, and reaches with boulder bars are the main features (refer to 'see also' section). Furthermore, slack water deposits are

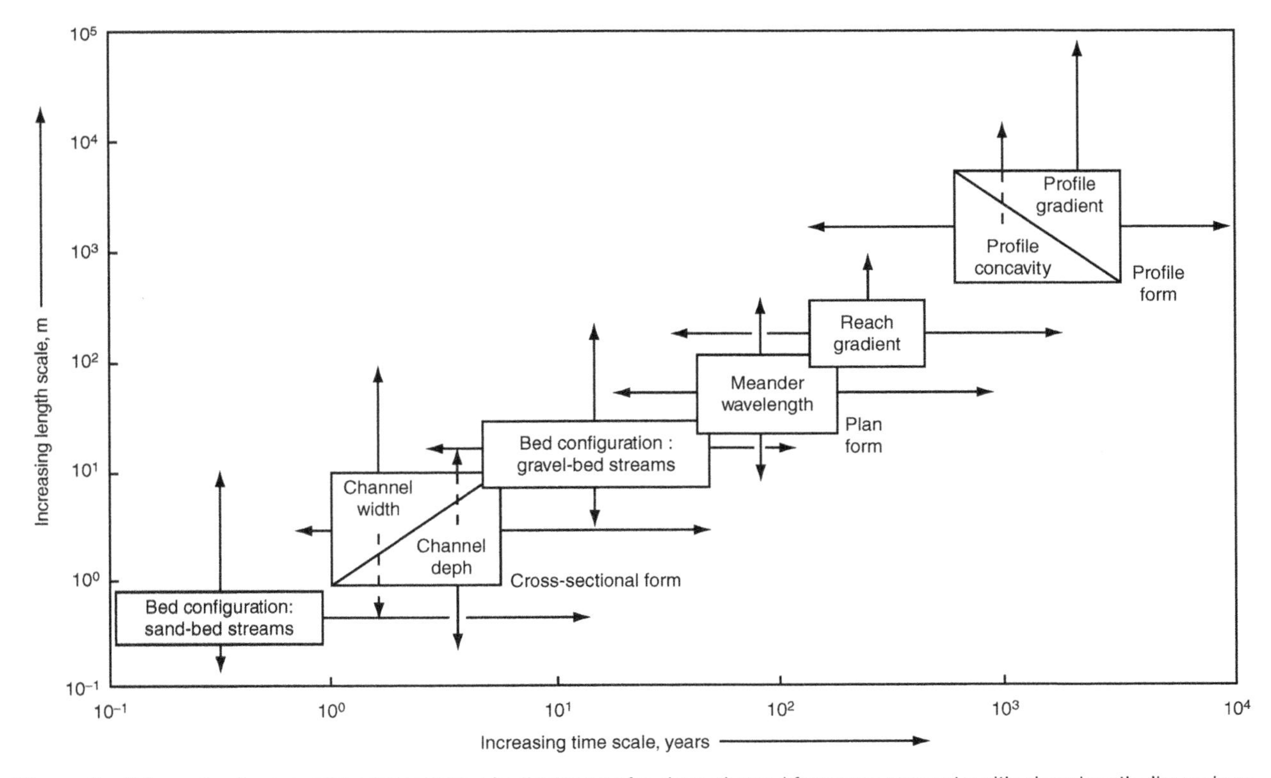

Figure 4 Schematic diagram of the timescales of adjustment of various channel forms components with given length dimensions in a hypothetical basin of intermediate size (after Knighton (1998)).

present at confluences in canyons, in caves and shelters, and in wider areas; they are markers of palaeofloods.

See also: Ecology and Role of Headwater Streams; Flood Plains.

Further Reading

Ferguson RI (1987) Hydraulic and sedimentary controls of channel pattern. In: Richard K (ed.) *River Channels, Environment and Process*, pp. 130–155. London: Blackwell.

Hjulström F (1935) Studies of the morphological activity of rivers as illustrated by the River Fyris. *Bulletin of the Geological Institute of Upsala,* pp. 221–527.

Horton RE (1945) Erosional development of streams and their drainage basins. Hydrophysical approach to quantitative geomorphology. *Geological Society America Bulletin* 56(3): 275–370.

Kellerhals R and Church M (1989) The morphology of large rivers: Characterization and management. *Canadian Special Publication on Fisheries and Aquatic Sciences* 106: 31–48.

Knighton D (1998) *Fluvial Forms and Processes: A New Perspective.* 383 pp. Oxford, UK: Oxford University Press.

Leopold LB and Wolman MG (1957) River channel patterns – Braided, meandering and straight. *US Geological Survey Professional Paper* 282B: 39–85.

Leopold LB, Wolman G, and Miller JP (1964) *Fluvial Processes in Geomorphology.* 522 pp. San Fransisco, CA: Freeman and Cie.

Lewin J (1976) Initiation of bed forms and meanders in a coarse-grained sediment. *Geological Society of America Bulletin* 87: 281–285.

Petts G and Amoros C (1996) *Fluvial Hydrosystems.* London: Chapman and Hall.

Rust BR and Nanson GC (1986) Contemporary and palaeochannel patterns and the Late Quaternary stratigraphy of Cooper Creek, Southwest Queensland, Australia. *Earth Surface Processes and Landforms* 11: 581–590.

Schumm A and Lichty RW (1965) Time, space and causality in geomorphology. *American Journal of Science* 263: 110–119.

Shields A (1936) Anwendung der Aehnlichkeitsmechanik und der turbulenzforschung auf die gesschiebebewegung. *Mitteilung der Preussischen versuchsansalt fuer Wasserbau und Schiffbau* Heft 26.

Starkel L (1983) The reflection of hydrologic changes in the fluvial environments of the temperate zone during the last 15,000 years. In: Gregory KJ (ed.) *Background to Palaeohydrology*, pp. 213–235. Chichester, UK: Wiley.

Strahler AN (1957) Quantitative analysis of watershed geomorphology. *Transactions American Geophysical Union* 38(6): 913–920.

Currents in Rivers

A N Sukhodolov and H-P Kozerski, Institute of Freshwater Ecology and Inland Fisheries, Berlin, Germany
B L Rhoads, University of Illinois at Urbana-Champaign, Urbana, IL, USA

Introduction

A characteristic feature of fluvial systems is the distinctive directed motion of water masses, or current, caused by gravitational forces. Currents in fluvial systems also differ from other geophysical flows (atmospheric, oceanic, and limnetic) primarily by a presence of irregular flow boundaries, or river channels, developed in bedrock or alluvium. In alluvial channels, cohesive or noncohesive materials are subjected to erosion, transport, and deposition, shaping channels by currents and causing alterations in the structure of the currents. For example, nonuniformities in flow structure produce local erosion at banks that with time evolve into the large-scale channel deformations – meander bends or loops. Flow curvature in bends produces centrifugal forces and counteracting pressure-gradient forces, thereby generating helical motion and secondary currents, which enhance channel deformation. Feedbacks among currents, transport of alluvium, and channel deformation result in self-regulating adjustments that dynamically sustain the form of natural fluvial systems.

Apart from their significance for processes of erosion and deposition, currents in rivers represent an abiotic component of fluvial ecosystems. Flow rates and patterns of currents determine transport and mixing of oxygen, nutrients, and pollutants. Moreover, distinctive flow patterns create specific habitats for various forms of aquatic life. On the other hand, biota can substantially influence currents, shaping the structure of habitat to favor conditions for dominant species. A good example is the interaction between biota and flow in vegetated river reaches.

Although the qualitative and quantitative assessment of river currents has attracted the attention of scientists for centuries, detailed understanding of these currents has proven elusive due in part to the lack of a general theory of turbulent flows. Therefore, available methods of characterizing river flow quantitatively are based either on case-specific computational models or purely empirical techniques. This article provides an abbreviated overview of the essential physical processes associated with river currents. The simple case of currents in wide and straight channels is considered first because it provides a theoretical framework and represents the basic (primary) class of currents. Then the effects of nonuniformity in morphology or composition of the riverbed that result in the development of secondary currents are considered along with the effects of channel curvature responsible for the secondary currents of centrifugal origin. Further, complications in the pattern of flow currents are considered using the example of flows through river confluences – essential components of river networks. The influence of human actions on river currents in the form of the complex structure of flow around groynes – transverse dikes that deflect the flow from erodible banks and promote navigability of river reaches – are discussed, followed by analysis of navigation-induced currents generated by commercial vessels. Conceptual and theoretical principles are illustrated with the examples of original field studies completed by the authors on rivers in Germany and the United States.

Controlling Factors and Classifications of Currents

Currents in rivers originate at a defined source (channel head or the junction of two streams) and evolve under the mutual influence of gravitational (G) and frictional (F) forces. At the most basic level, currents in rivers can be classified according to whether or not the bulk rate of flow remains constant over time (*steady flow*) or it changes over time (*unsteady flow*), and whether or not the bulk rate of flow remains constant over space (*uniform flow*) or it changes over space (*varied flow*). In the case of steady, uniform flow, gravitational and frictional forces are equal ($G = F$). However, in unsteady flow, the forces are unbalanced ($G \neq F$) over time, whereas in varied flow they are unbalanced over space. If $G > F$ the flow accelerates, whereas if $G < F$ the flow decelerates.

Motion of water in fluvial systems is a continuous physical process of energy transformation. Potential energy of liquid ρgh (where ρ is density of water, g is gravity acceleration, and h is flow depth) is transformed into kinetic energy ρU^2 (U is bulk flow velocity). The ratio between these two forces $Fr = U/\sqrt{gh}$, the Froude number, provides the basis for further classification. The river current can be in a *subcritical* ($Fr < 1$), critical ($Fr = 1$), or supercritical ($Fr > 1$) state. Subcritical flows are typical for lowland rivers and are characterized by smooth, undisturbed water surfaces. In the critical regime, the surface of the stream develops standing waves, and in supercritical conditions the surface of the water may become distorted into breaking waves. The critical and supercritical regimes are characteristic of mountain torrents and flow around or over engineering structures (dykes, dams), but can also develop in lowland rivers during large floods. Transitions between subcritical and supercritical regimes produce hydraulic drops, or abrupt decreases in flow depth, whereas transitions between supercritical and subcritical regimes result in hydraulic jumps, or abrupt increases in flow depth.

Rivers originate in uplands and flow downhill into lakes, seas, or oceans. Therefore, rivers flow within channels with longitudinal gradients, or slopes (S). The shear stress associated with the gravitational force per unit area oriented along the inclined plane of the channel bed is ρghS. A simple expression for the mean velocity of the flow current can be derived from assumptions of uniform flow as: $c_f = ghS/U^2$, where c_f is a friction factor. This equation can be rearranged as $U = C\sqrt{hS}$, which is known as the Chezy formula and C is the Chezy coefficient. A related empirical formula, $U = h^{2/3}S^{1/2}/n$, is known as the Manning–Strickler formula and n is the Manning coefficient ($C = h^{1/6}/n$). It can be seen that the empirical channel resistance coefficients are related to the friction factor as $c_f = g/C^2$ and $c_f = n^2 g/h^{1/3}$. Values of friction coefficients have been determined empirically and are summarized in standard manuals for open-channel flow computations. Values of the Chezy coefficient vary in rivers from 30 to 70, and the Manning coefficient ranges from 0.020 (lowland rivers) to 0.2 (flow over floodplains with terrestrial vegetation).

Although the theory of uniform flow is capable of describing bulk characteristics of currents in rivers, flow in rivers exhibits significant spatial variability because of zero local flow velocity near riverbeds and banks. This variability is a distinctive feature, providing diversity of habitat for aquatic life and is therefore a key factor determining patchiness in the community structure of aquatic organisms and plants communities. The following sections illustrate spatial patterns of currents in rivers and the possibilities of quantifying the processes producing these patterns.

Currents in Fluvial Channels

Because the permeability of riverbeds and banks is relatively low, velocity at these boundaries can be assumed to be zero. Therefore, velocities in a river cross-section reduce to zero values at the bottom and sides, and are maximal at the surface in the center of the channel (**Figure 1**). For steady uniform flow not close to the river banks, the gravitational shear-stress component ρghS is balanced by boundary friction

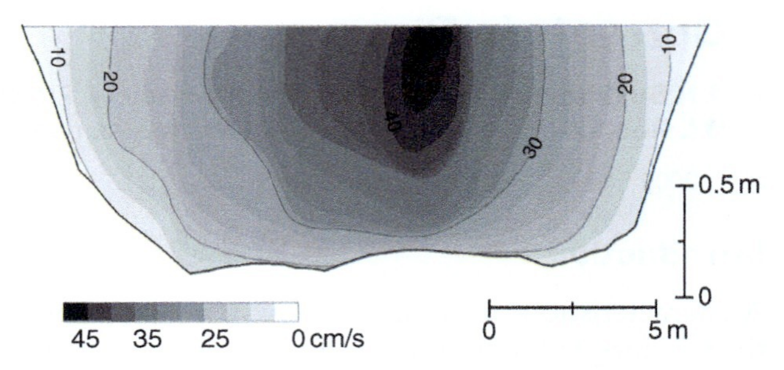

Figure 1 Distribution of time–mean streamwise velocity in a river cross-section (the Spree River, Germany).

causing shear stress within the water column that can be expressed as $\tau(z) = -\rho\overline{u'w'}$, where τ is shear stress, and u', w' are turbulent fluctuations of velocity in the streamwise and vertical directions, and z is the distance from the riverbed. It can be shown that in uniform flow shear stresses are linearly distributed over the flow depth $\tau = \tau_0(1 - z/h)$ with a maximum bed shear stress $\tau_0 = \rho ghS$, at the riverbed. A characteristic velocity scale, shear velocity, can be expressed respectively as $U_* = \sqrt{ghS} = \sqrt{\tau_0/\rho}$.

Relating turbulent velocity fluctuations $-\overline{u'w'}$ to time–mean velocity $U(z)$ at certain distance z from the bed provides a simple model of turbulence and allows shear stresses to be expressed as $\tau = \rho\nu_t\, dU/dz$, where ν_t is turbulent viscosity. This relationship can be integrated to obtain the vertical velocity distribution, but first requires an estimate of ν_t. The assumption of a parabolic distribution of turbulent viscosity over depth, $\nu_t = \kappa U_* z\sqrt{1 - z/h}$, when substituted into the expression for shear stress, yields a logarithmic distribution for mean velocity over river depth

$$U(z) = \frac{U_*}{\kappa}\ln\frac{z}{k_s} + B_0 \qquad [1]$$

where $\kappa = 0.4$ is an universal constant, k_s is a characteristic height of roughness elements, and $B_0 = 8.5$ is a constant of integration. Alternatively, eqn [1] can be expressed as

$$\frac{U(z)}{U_*} = \frac{1}{\kappa}\ln\left(\frac{z}{z_0}\right) \qquad [2]$$

where $z_0 = \exp(\ln k_s + \kappa B_0)$ is hydrodynamic roughness parameter. Integrating [2] over river depth provides a logarithmic function expressing the influence of riverbed resistance on the depth-averaged velocity (U_a):

$$\frac{C}{\sqrt{g}} = \frac{U_a}{U_*} = \frac{1}{\kappa}\left(\ln\frac{h}{z_0} - 1\right) \qquad [3]$$

To illustrate the performance of logarithmic law [2], experimental data from some rivers are presented in **Figure 2** in nondimensional coordinates $z/h = \exp[U\kappa/U_* - \ln(h/z_0)]$.

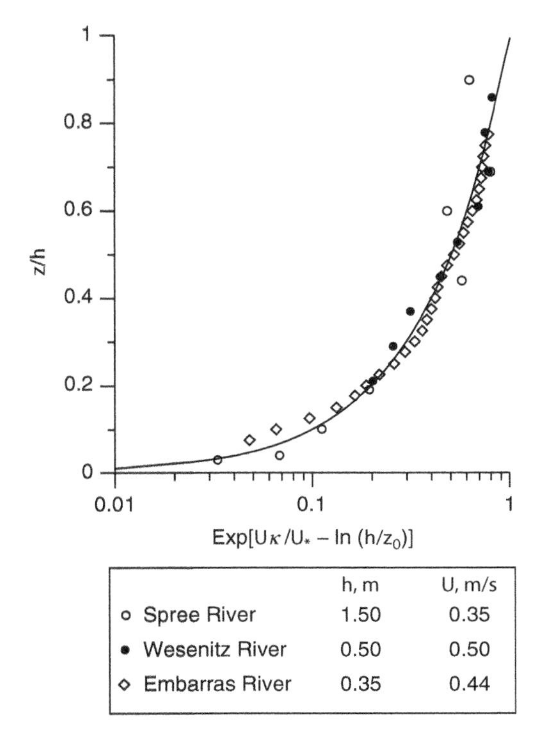

Figure 2 Comparison of measured time–mean streamwise velocities over the river depth (symbols) with predicted logarithmic law (line).

Secondary Currents

So far only the streamwise velocity, or the velocity component parallel to centerline of the channel, has been considered for straight river reaches with uniform cross-sectional geometry and riverbed material. However, natural stream channels usually meander and exhibit complex morphology and distributions of riverbed material. These variations produce components to currents that have significant magnitudes perpendicular to the streamwise component. These components, referred to as secondary currents, result in substantial three dimensionality of the overall pattern of currents in streams.

Depending on their genesis, secondary currents are classified into two categories: secondary currents of first and second kinds. Secondary currents of the first kind are produced by large-scale nonuniformities of channel pattern – for example, river bends. Centrifugal forces that develop in a curved channel produce superelevation of the water surface along the outer bank channel, which generates a counteracting pressure-gradient force. Local imbalances between these forces over the flow depth produce outward motion at the surface, downward motion along the outer bank, and upward, inward motion along the bed (**Figure 3**). The resulting pattern of helical motion redistributes momentum shifting the zone of maximum streamwise velocity towards the concave bank near the bend apex (**Figure 3**). In curvilinear systems of coordinates r (radial), θ (tangential), and z (vertical) dynamical equation of flow are presented in the following form:

$$u_r\frac{\partial u_r}{\partial r}+\frac{u_\theta}{r}\frac{\partial u_r}{\partial \theta}+w\frac{\partial u_r}{\partial z}-\frac{u_\theta^2}{r}=-gS_r+\frac{\partial}{\partial z}\left[\nu_t\frac{\partial u_r}{\partial z}\right] \quad [4]$$

$$u_r\frac{\partial u_\theta}{\partial r}+\frac{u_\theta}{r}\frac{\partial u_\theta}{\partial \theta}+w\frac{\partial u_\theta}{\partial z}+\frac{u_r u_\theta}{r}=gS_\theta+\frac{\partial}{\partial z}\left(\nu_t\frac{\partial u_\theta}{\partial z}\right) \quad [5]$$

where u_r, u_θ, and w are radial, tangential, and vertical mean velocities, and S_r, S_θ are radial and tangential slopes. Systems [4] and [5] can be solved analytically for the radial component, which represents the secondary current, if the distributions of tangential velocities and of turbulent viscosity are presented in an analytical form. For natural streams with large radii of channel curvature the distribution of mean velocities usually differs little from the logarithmic law and the parabolic distribution of turbulent viscosity applies $\nu_t=\kappa U_* z\sqrt{1-z/h}$, then

$$u_r=\frac{1}{\kappa^2}U\frac{h}{r}\left[F_1(\eta)-\frac{\sqrt{g}}{\kappa C}F_2(\eta)\right],$$

$$F_1(\eta)=\int_0^1\frac{2\ln\eta}{\eta-1}\mathrm{d}\eta,\text{ and }F_2(\eta)=\int_0^1\frac{\ln^2\eta}{\eta-1}\mathrm{d}\eta \quad [6]$$

where $\eta=z/h$. Comparison of values predicted with eqn [6] and measured in a typical lowland river meander bend show good agreement (**Figure 4**). The magnitude of secondary currents of the first kind can be up to 20–30% of the magnitude of streamwise velocity component. These currents are crucial for shaping riverbed relief in channels with loose alluvium.

Genesis of secondary currents of the second kind is attributed to the nonuniformity in distributions of roughness or morphology of the riverbed in straight river reaches. These secondary currents have much smaller magnitudes, about 5–10% of the primary current, and thus are similar to the magnitudes of turbulent fluctuations. Despite their small magnitudes, these currents are responsible for lateral redistributions of fine sediments on the channel bed, forming longitudinal ridges and thus shaping habitats of benthic invertebrates. Some researchers explicitly associate these currents with turbulent structures and use turbulence anisotropy terms as the driving force in models describing the formation of secondary currents in straight river reaches. However, quantitative methods describing such currents are still unavailable because of a lack of knowledge about river turbulence, and particularly about coherent structures.

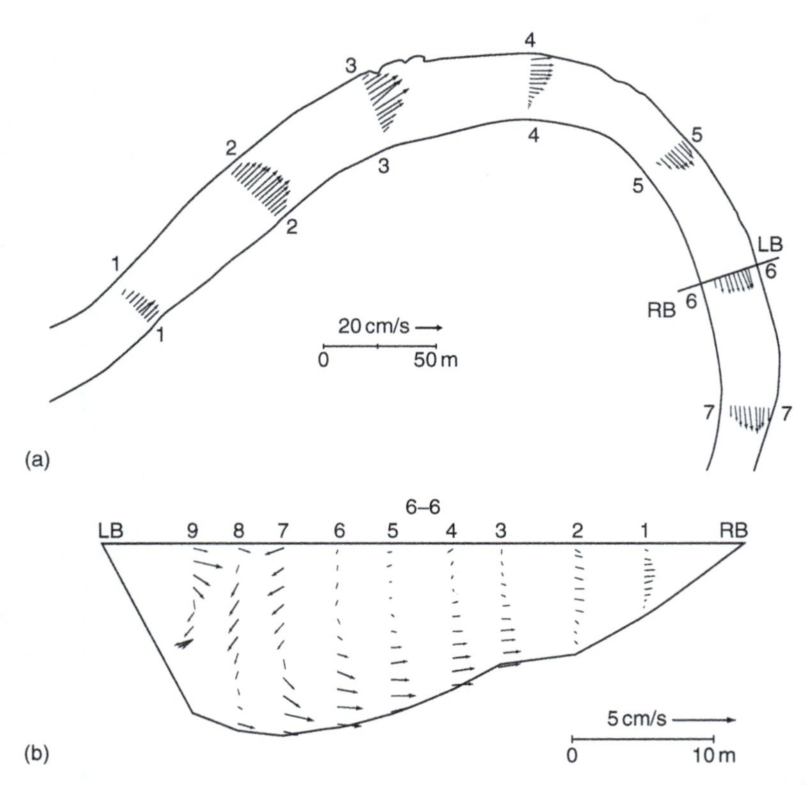

Figure 3 Distribution of time–mean velocity vectors near the free surface of the flow (a), and (b) secondary currents in the river cross-section, depth is enlarged five times (the Spree River, Germany).

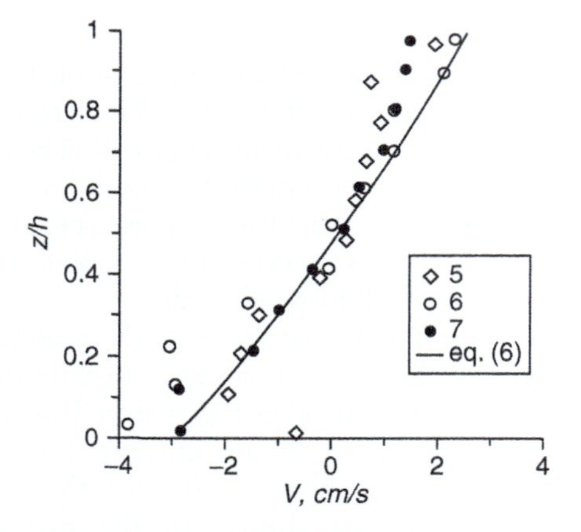

Figure 4 Comparison of measured (symbols mark verticals, **Figure 3(b)**), and predicted (line) time–mean radial velocities (the Spree River, Germany).

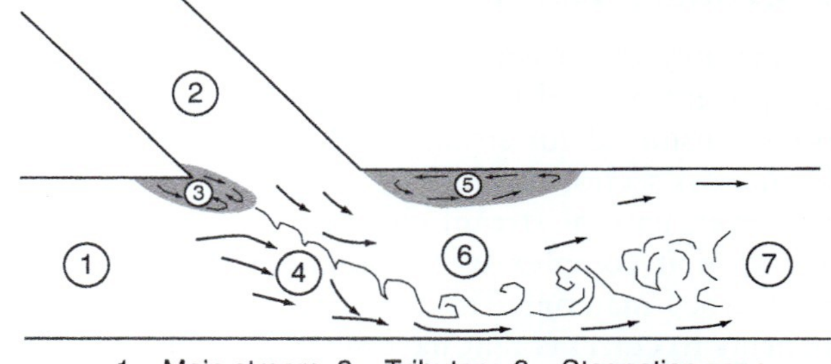

Figure 5 Schematic representation of a confluence.

Structure of Currents at River Confluences

River confluences, or the locations where two rivers join one another, are integral and ubiquitous features of river networks. Currents within confluences are marked by highly complex three-dimensional patterns of flow that include a zone of stagnant, recirculating flow at the upstream corner of the junction of the two rivers, a region of strong flow convergence within the center of the confluence, a shear layer between the merging flows, and in some cases a zone of separated flow near one or both of the banks (**Figure 5**). It is generally acknowledged that flow structure within confluences is influenced by the junction planform, junction angle, momentum flux ratio, and degree of concordance of the channel beds at the entrance to the confluence. A characteristic pattern of currents within a confluence is the presence of two discrete zones of maximum velocity associated with flow from the two

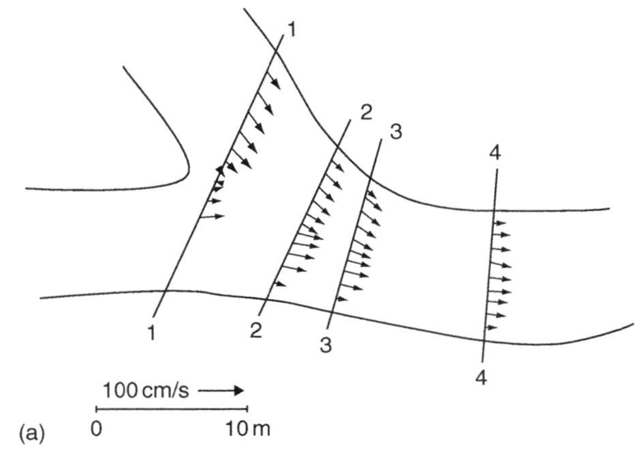

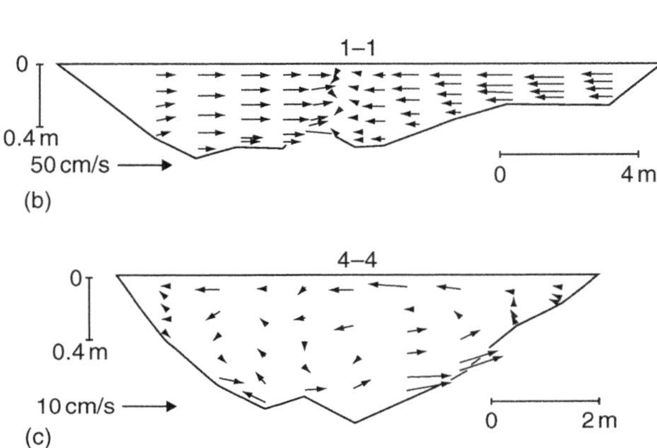

Figure 6 Distribution of time–mean velocity vectors near the free surface of the flow (a), and (b, c) secondary currents in the river cross-sections (confluence Kaskaskia-Copper Slough, USA).

upstream rivers (**Figure 6(a)** and **6(b)**). Between these zones is either a stagnation zone (a region of recirculating, separated flow characterized by negative downstream velocities) or farther downstream, a shear layer (a zone of intense turbulence along the interface between the converging flows).

The most prominent characteristic of the cross-stream velocity fields is the opposing orientation of transverse and vertical velocity vectors on each side the confluence (**Figure 6(b)**). The magnitudes of the cross-stream vectors reflect the momentum ratio of the two confluent streams with the largest vectors located on the side of the confluence corresponding to the dominant tributary. As the flow moves downstream, the two distinct zones of maximum downstream velocity gradually converge. Gradual convective acceleration of flow occurs within the low-velocity region between the two maxima until a uniform downstream velocity field with a single zone of maximum velocity develops downstream of the confluence. A pattern of helical flow, similar to secondary currents of the first kind developing in river bends, can also be present at confluences where curvature of flow from a lateral tributary into the downstream channel generates an effect similar to that which occurs in bends (**Figure 6(c)**).

Although river confluences have been actively studied during the past two decades, the complexity of currents at these locations hinders accurate theoretical descriptions. Numerical simulation models currently represent the most sophisticated tools for trying to characterize the complexity of river currents at confluences. However, ongoing studies of shallow mixing layers and free recirculating flows suggest that generalized theoretical models may emerge in the next decade.

Currents at Engineering Structures

Lateral nonuniformity of river currents is associated with natural riverbank protrusions and various engineering structures, among which groynes (spur dykes) are the features most widely used to support navigation and protect banks against erosion. Groynes are usually placed in sequences so that the area between successive groynes is referred to as a groyne field. Flow separates at the tip of the protruding groyne, or groyne head, and forms a rotating current in groyne fields depicted by large-scale vortexes with a vertical axis of rotation called gyres. Since water is forced to recirculate within the groyne field in spiraling trajectories, the local residence time for suspended particulate matter can increase substantially and may be sufficient to maintain local phytoplankton reproduction. Therefore, understanding of these complex currents can have important ecological implications.

Specific patterns of flow in a groyne field are controlled by the geometry of the field: the aspect ratio between the groyne length (L_g) and the streamwise length of the groyne field (L_f). Observations indicate that a two-gyre circulation pattern develops when the aspect ratio is less than a critical value ($L_g/L_f < 0.5$), and a one-gyre circulation forms in groyne fields aspect ratios greater than the critical value ($L_g/L_f > 0.5$) (**Figure 7**).

The distribution of mean velocities within the flow in groyne fields can also be reasonably described by a shallow mixing layer model and hyperbolic tangential equation. A canonical free mixing layer (family of free-turbulent flows) evolves in coflowing liquids of different densities or flows of different mean velocities, and can be described by a simple model

$$\frac{\partial \delta}{\partial x} = \alpha \lambda \qquad [7]$$

$$\lambda = \frac{\Delta U}{U_1 + U_2} = \frac{\Delta U}{2U_c}, \quad \Delta U = U_1 - U_2 \qquad [8]$$

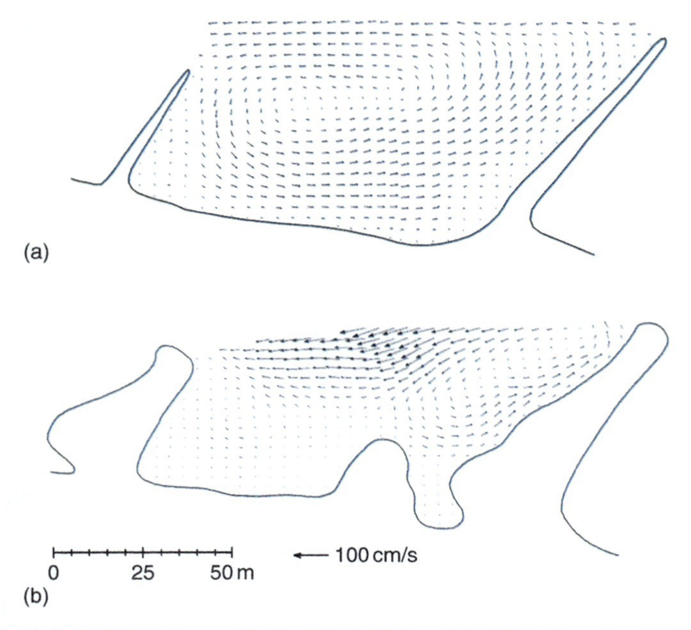

Figure 7 Distribution of depth-averaged time–mean velocity vectors (interpolated from measurements) in the groyne field with aspect ratio (a) 0.6, and (b) 0.4 (the Elbe River, Germany).

where α is the spreading rate (a constant in canonical mixing layer, equals 0.18), δ is the width of mixing layer, x is the downstream coordinate, λ is the velocity ratio, U_1 is the velocity in free part (above) of flow, and U_2 is inside the stand, and U_c is the velocity in the centre of the mixing layer. Mean velocity profiles in mixing layers have been shown to comply with a hyperbolic tangential distribution

$$\frac{U}{U_s} = 1 + \tanh\left(\frac{2y}{L}\right) \quad [9]$$

where $U_s = \Delta U/2$, y is the distance across the layer, and $L = \gamma\delta$ is a characteristic length scale normally proportional to the width of the mixing layer. An example of the distribution of depth-averaged velocity across the groyne field and its interface with the mean flow is shown in **Figure 8**.

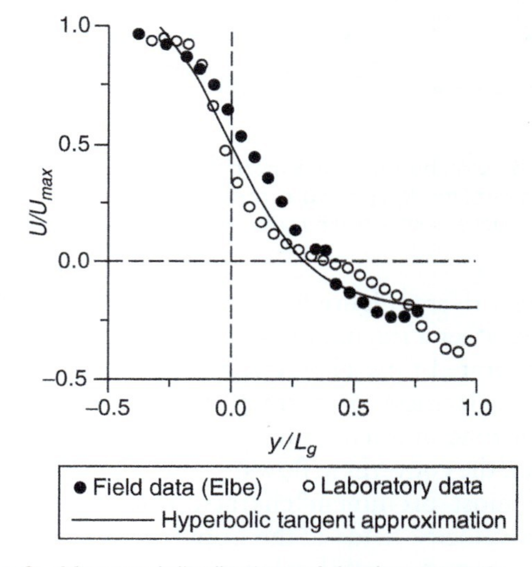

Figure 8 Measured distributions of depth-averaged time–mean velocities (section through the gyre center) in the mixing layer between main flow and groyne field, and their approximation with hyperbolic tangent (eqn [10]).

The specific pattern of recirculation has important implications for distribution of deposited fine sediments within the groyne field. The low-velocity area in the centre of gyres promotes accumulation of relatively fine sediments. The thickness of the layer of deposited fines decreases toward the gyre margins, where flow velocities increase.

Navigation-Induced Currents

Rivers have been always heavily exploited as the inland waterways for commercial and recreational navigation. A vessel moving along a river channel expends energy to overcome resistance of water. The energy is transformed into complex pattern of currents and waves in the lee of the vessel. In width-restricted channels, when commercial tugs towing loaded barges cruise with speed close to the navigational limits, the navigation-induced currents maintain extremely large velocities (**Figure 9**). An analytical

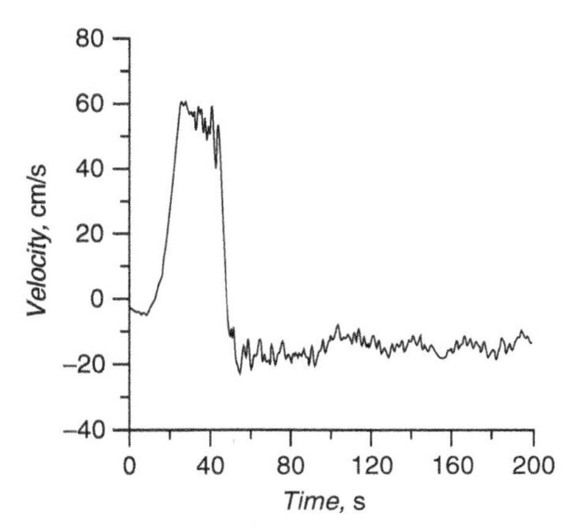

Figure 9 Navigation-induced current from a towing barge measured 15 cm above the riverbed at 2 m distance from the water edge (Oder-Havel Canal, Germany).

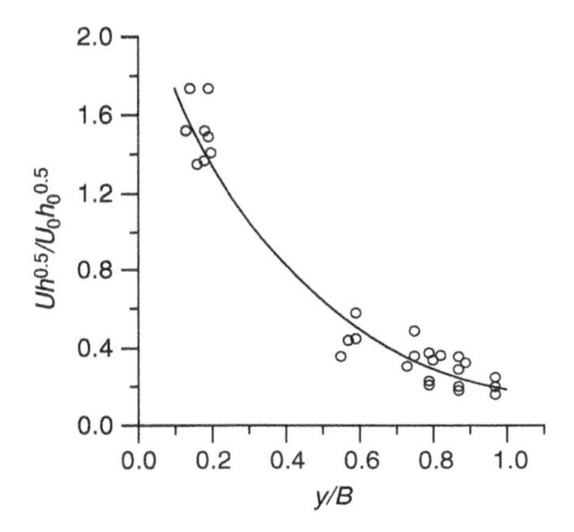

Figure 10 Measured (circles) and predicted (line) values of return currents.

framework for assessment of such currents was deduced from the balance of kinetic energy and represented by a relationship

$$\frac{U}{U_0} = \alpha_0 \sqrt{\frac{h_0}{h}} \exp\left[-\beta \frac{y}{B}\right] \quad [10]$$

where α_0 and β are parameters depending on the characteristics of kinetic energy transfer (dispersion coefficient, wave celerity, and channel width B), U is maximal value of depth-averaged velocity in navigation-induced current, U_0 is the speed of the vessel, h is the flow depth, h_0 is the draught of the vessel, and y is the transverse distance from the vessel towards waterway bank. Performance of the relation [10] is illustrated in **Figure 10**, where the results of field measurements are compared with values predicted by the model [10].

Conclusions

The major factors controlling currents in rivers are the macro- and microscale geometry of the river channel, the joining of streams induced by the structure of fluvial networks, properties of alluvium composing the riverbed material, biological features, and anthropogenic influences. Although quantitative descriptions of currents in rivers are based on well-known theoretical principles, a universal theory of river currents has yet to be developed mainly because of the lack of a universal theory for turbulence aspects of river flows. The problem of quantitative description is presently solved with application of case-specific models and relies substantially on the application of empirical knowledge.

Mutual interactions among flow, the river channel, and biota at different spatial and temporal scales ranging from a sediment grain to the scale of a river reach, and from milliseconds to hundreds of years complicate unambiguous studies of currents. Therefore, available data on important parameters in models of river currents include significant scatter that leads to uncertainties in prediction of magnitudes and patterns of flows. Some of this uncertainty reflects the fact that most theoretical, laboratory, and field studies investigate or assume uniform steady flow, while in nature such flows are always an idealization.

Nomenclature

B	width of a channel (m)
B_0	integration constant
C	Chezy coefficient ($m^{1/2}$ s^{-1})
c_f	friction factor
F	friction force (kg m s^{-2})
Fr	Froude number
F_1, F_2	integral functions
G	gravity force (kg m s^{-2})
g	gravity acceleration (m s^{-2})
h	mean depth (averaged over cross-section) (m)
h_0	draught of a vessel (m)
k_s	height of roughness elements (m)
L	length scale (m)
L_g	length of groyne (m)
L_f	length of groyne field, m
n	Manning–Strickler coefficient ($m^{-1/3}$ s)
r	radial coordinate (m)
S	longitudinal gradient of the water surface
S_r	radial slope
S_θ	tangential slope
U	time–mean velocity (m s^{-1})
U_c	time–mean velocity in the centre of the mixing layer (m s^{-1})

U_0	cruising velocity of a vessel (m s^{-1})
u'	streamwise velocity fluctuations (m s^{-1})
u_r	radial mean velocity (m s^{-1})
u_θ	tangential mean velocity (m s^{-1})
U_*	shear velocity (m s^{-1})
U_s	mean velocity half of velocity difference in mixing layer (m s^{-1})
U_1, U_2	velocity in fast and slow flows (m s^{-1})
w'	vertical velocity fluctuations (m s^{-1})
y	transverse distance (m)
z	vertical coordinate (m)
z_0	hydrodynamic roughness parameter (m)
ΔU	velocity difference (m s^{-1})
α	spreading rate
α_0	parameter
β	parameter
γ	dimensionless coefficient
δ	width of the mixing layer (m)
η	dimensionless distance
θ	tangential coordinate (grad.)
κ	von Karman parameter
λ	velocity ratio
ν_t	turbulent viscosity (m^2 s^{-1})
ρ	density of water (kg m^{-3})
τ	shear stress (kg m^{-1} s^{-2})
τ_0	bottom shear stress (kg m^{-1} s^{-2})

Further Reading

Cunge JA, Holly FM Jr, and Verwey A (1980) *Practical Aspects of Computational River Hydraulics*. London: Pitman.

Best JL and Reid I (1984) Separation zone at open-channel junctions. *Journal of Hydraulic Engineering* 110: 1588–1594.

Ghisalberti M and Nepf H (2002) Mixing layers and coherent structures in vegetated aquatic flows. *Journal of Geophysical Research* 107(C2). doi 10.1029/2001JC00871.

Gordon ND, McMahon TA, and Finlayson BL (1992) *Stream Hydrology: An Introduction for Ecologists*. Chichester: Wiley.

Leopold LB (1994) *A View of the River*. Cambridge: Harvard University Press.

Nezu I and Nakagawa H (1993) *Turbulence in Open-Channel Flow*. Rotterdam: Balkema.

Rhoads BL and Sukhodolov AN (2001) Field investigation of three-dimensional flow structure at stream confluences: 1. Thermal mixing and time-averaged velocities. *Water Resources Research* 37(9): 2393–2410.

Rozovskii IL (1961) *Flow of Water in Bends of Open Channels*. [trans.]. Jerusalem: Israel Program for Scientific Translations.

Schlichting H and Gersten K (2000) *Boundary Layer Theory*. Berlin: Springer.

Sukhodolov AN, Uijttewaal WSJ, and Engelhardt C (2002) On the correspondence between morphological and hydrodynamical patterns of groyne fields. *Earth Surface and Landforms* 27: 289–305.

Sukhodolov AN and Sukhodolova TA (2006) Evolution of mixing layers in turbulent flow over submersible vegetation: field experiments and measurement study. In: Ferreira RML, Alves ECTL, Leal JGAB, and Cardoso AH (eds.) *RiverFlow2006*, pp. 525–534. London: Balkema.

Wolter C, Arlinghaus R, Sukhodolov AN, and Engelhardt C (2004) A model of navigation-induced currents in inland waterways and implications for juvenile fish displacement. *Environmental Management* 35: 656–668.

Yalin MS (1992) *River Mechanics*. Oxford: Pergamon.

Relevant Websites

http://cwaces.geog.uiuc.edu – Center for Water as a Complex Environmental System (CWACES), Urbana-Champaign, Illinois, USA.

http://www.wldelft.nl – WL|Delft Hydraulics, the Netherlands.

http://www.nced.umn.edu – National Center for Earth-surface Dynamics, Minneapolis, USA.

http://www.niwascience.co.nz – National Institute of Water and Atmospheric Research, New Zealand.

http://www.ifh.uni-karlsruhe.de – Institute for Hydromechanics in Karlsruhe, Germany.

http://www.igb-berlin.de – Institute of Freshwater Ecology and Inland Fisheries, Germany.

Hydrodynamical Modeling

B R Hodges, University of Texas at Austin, Austin, TX, USA

Introduction

We model lakes to visualize and quantify fluid flow, mass transport, and thermal structure. Understanding the evolving physical state (e.g., surface elevation, density, temperature, velocity, turbidity) is necessary for modeling fluxes of nutrients, pollutants, or biota in time-varying fields of one, two, or three space dimensions (1D, 2D, or 3D). Hydrodynamic modeling provides insight into spatial–temporal changes in physical processes seen in field data. For example, **Figure 1** shows temperature profiles simultaneously recorded at different stations around Lake Kinneret (Israel). Extracts from model results (**Figure 2**) provides a context for interpreting these data as a coherent tilting of the thermocline. A time series of the thermocline can be animated, showing the principal motion is a counter-clockwise rotation of the thermocline. The complexities of the thermocline motion can be further dissected by using spectral signal processing techniques to separate wave components, illustrating a basin-scale Kelvin wave, a first-mode Poincaré wave, and a second-mode Poincaré wave.

It is said we build by 'measuring with micrometer, marking with chalk, then cutting with an axe'. However, hydrodynamic modeling inverts this process: we take an axe to the real world for our governing equations; we chalk a model grid on our lake, then numerically solve to micrometer precision. Thus, the governing equation approximations, grid selection, and numerical method all affect how a model reflects the physical world. Selecting an appropriate model requires understanding how model construction may affect the model solution.

As a broad definition, hydrodynamic modeling is the art and science of applying conservation equations for momentum, continuity, and transport (**Figure 3**) to represent evolving velocity, density, and scalar fields. The modeling science is founded upon incompressible fluid Newtonian continuum mechanics, which can be reduced to (1) any change in momentum must be the result of applied forces, and (2) the net flux into or out of a control volume must balance the change in the control volume. The modeling art is in selecting approximations, dimensionality, and methods that fit the natural system and provide adequate answers to the question asked.

Tables 1–4 list some of the 1D, 2D, and 3D lake modeling work from the mid-1990s to the present in the refereed literature. Unfortunately, much of the details of model development have been relegated to technical reports that are often either unavailable or difficult to obtain. Similarly, modeling applications conducted by or for government agencies often does not reach refereed publication. However, technical reports are generally detailed and valuable resources for applying and analyzing results; the reader is encouraged to seek out these publications before applying any model.

Dimensionality and Capabilities

Lake and reservoir models range from simple representation of thermocline evolution to multidimensional modeling of transport and water quality. Averaging (or integrating) the governing equations across one or more spatial dimensions provides representation of larger areas with less computational power. Such reduced-dimension models require less boundary condition data but more parameterization/calibration data. The simplest lake models average over horizontal planes (i.e., x and y directions) to obtain a 1D-model of the vertical (z-axis) lake stratification. Narrow reservoirs are modeled in 2D by laterally-averaging across the reservoir, thereby representing both vertical stratification and horizontal gradients from the headwaters to the dam. In shallow lakes, 2D-models in the x-y plane are used to examine depth-averaged circulation (without stratification). These reduced-dimension approaches cannot directly simulate variability in an averaged direction. However, such variability may be parameterized where processes are sufficiently well understood; in contrast, where processes are not understood or cannot be parameterized, the missing variability affects calibration and model results. Modelers must be careful not to simply parameterize or calibrate a poorly understood process by arbitrarily altering model coefficients.

Increasing model dimensionality and complexity reduces the time-scale over which a lake can be modeled. Typically, 1D-models can be applied for decades; 2D-models over multiple years to decades; 3D-models over days/weeks/months (but have been applied for longer in a few cases). This inverse relationship between dimensionality and time is not simply a function of computational power, but is inherent in the effects of stratification, mixing

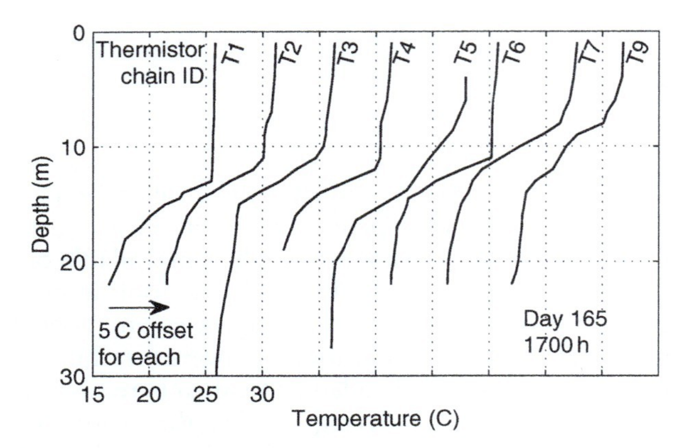

Figure 1 Temperature profiles collected in Lake Kinneret in 1997 (data of J. Imberger, Centre for Water Research, University of Western Australia).

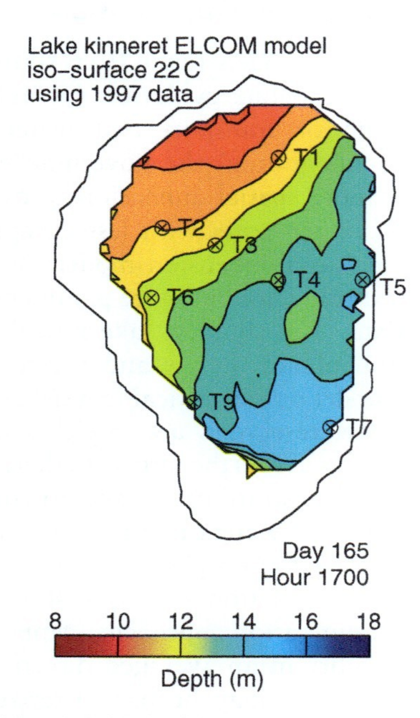

Figure 2 Modeled depth of temperature isosurface in the thermocline. Results from 3D-model at same time as field data collected in **Figure 1**.

parameterizations, and model data requirements. For 1D-models, vertical mixing is readily parameterized or calibrated to maintain sharp thermal stratification. However in 2D- and 3D-models, vertical mixing is caused by both the turbulence model and numerical diffusion of mass (a model transport error). Numerical diffusion always leads to weakening the thermocline, which affects modeled circulation and mixing that further weakens the thermal structure in a positive feedback cycle. Thus, longer-term 2D- and 3D-models require careful setup and analysis or results may be dominated by model error that accumulates as excessive mixing across the thermocline, resulting in poor prediction of residence time, mixing paths, and lake turnover.

For 1D-models (**Table 1**), we distinguish between turbulent mixing models derived from energy principals (e.g., DYRESM) or averaging transport equations (e.g., k-epsilon models) and pure calibration models that simply fit coefficients to nonphysical model equations (e.g., neural networks). Between these extremes are vertical advection/diffusion models (e.g., MINLAKE), which represent hydrodynamics by an advection/diffusion equation that requires site-based calibration. To the extent that more mechanistic models have all physical processes represented and correctly parameterized without site-specific calibration, they can be reasonably used for long-term predictions and readily transferred from lake to lake. Models relying solely on parameter fitting are questionable outside the calibration range, but are often easier to apply when sufficient calibration data is available. Although mechanistic models are arguably preferred, we rarely have sufficient data for a completely calibration-free mechanistic model; thus, in practical application such models are generally calibrated to some extent.

Laterally-averaged 2D-models (**Table 2**) are the workhorse of reservoir hydrodynamic/water quality modeling. For a narrow reservoir, the lateral-averaging paradigm is successful in capturing the bulk thermal/hydrodynamic processes, which are dominated by the large pelagic volume. However, where water quality processes are dependent on concentrations in shallow littoral regions, such models must be used carefully and with some skepticism. For example, a littoral algae bloom may depend on high nutrient concentrations that result from reduced circulation in the shallows. To correctly represent the bloom, a 2D water quality model must distort the biophysical relationships between growth rate and concentration. Furthermore, any scalar (e.g., toxic spill) represented simply by a concentration will automatically be diffused all the way across the reservoir, whether or not there is sufficient physical transport. Thus, a laterally averaged model will represent a toxic spill that mixes as a function of the reservoir width rather than physical processes.

Although 3D-models (**Table 3**) provide good representations of lake physics, they are notoriously complicated to set up and run. Although their operation is becoming easier with more established models, the progression to 'black box' modeling (i.e., where the user is not intimately familiar with the model itself) remains problematic. The interaction of the lake physics with the numerical method, governing equation approximations, time step, grid size, and initial/

Momentum:

$$\underbrace{\frac{\partial u_i}{\partial t}}_{\text{Unsteady velocity}} + \underbrace{\sum_{j=1}^{3} u_j \frac{\partial u_i}{\partial x_j}}_{\text{Nonlinearity}} = \underbrace{-g_i}_{\text{Gravity force}} \underbrace{- \frac{1}{\rho_0}\frac{\partial P_{nh}}{\partial x_i}}_{\text{Non-hydrostatic pressure gradient}} \underbrace{-g\frac{\partial \eta}{\partial x_i}}_{\text{Free-surface or 'barotropic' pressure gradient}} \underbrace{- \frac{g}{\rho_0}\frac{\partial}{\partial x_i}\int_{z}^{\eta} \Delta\rho dx_3}_{\text{Stratification or 'baroclinic' pressure gradient}} + \underbrace{\sum_{j=1}^{3} \nu \frac{\partial^2 u_i}{\partial x_j^2}}_{\text{Viscous force}} \quad : i=1,2,3$$

Continuity:

$$\frac{\partial u_1}{\partial x_1} + \frac{\partial u_2}{\partial x_2} + \frac{\partial u_3}{\partial x_3} = 0$$

Free surface evolution

$$\frac{\partial \eta}{\partial t} + \frac{\partial}{\partial x_1}\int_{B}^{\eta} u_1 dx_3 + \frac{\partial}{\partial x_2}\int_{B}^{\eta} u_2 dx_3 = 0$$

x_i	Cartesian space ($i=1, 2$ are horizontal; $i=3$ is vertical)
u_i	Velocity
η	Free surface elevation
B	Bottom elevation
g	Gravitational acceleration
ν	Kinematic viscosity
P_{nh}	Non-hydrostatic pressure
ρ_0	Reference density
$\Delta\rho$	Difference between local density and reference density

Figure 3 General 3D incompressible flow equations (with the Boussinesq approximation) that are the basis for most hydrodynamic models. Hydrodynamics in lake modeling also requires transport equations for temperature, salinity (if important), and an equation of state for density.

boundary condition data provides a wide scope for model inaccuracies. The effectiveness of a 3D-model presently depends on the user's understanding of the model capabilities and limitations. It is doubtful that we will see scientifically dependable black box models for at least another decade or so. Development of such models depends on development of expert systems that can replace a modeler's insight and experience in diagnosing different error forms.

Boundary and Initial Conditions

A hydrodynamic model is a numerical solution of an initial-and-boundary-value problem of partial differential or integral–differential equations. The model solution is never better than the initial and boundary conditions used for model forcing. Lake boundary conditions include spatially-varying wind field, thermal and mass exchange with the atmosphere, river inflows/outflow, groundwater exchanges, local catchment runoff, and precipitation. Boundary conditions may be poorly known, so understanding the model sensitivity to possible perturbations boundary conditions is necessary for setting the bounds of model believability.

Initial conditions are a snapshot of the system at time $t=0$ (the model start time). Problems arise from our inability to obtain the full data set necessary to initialize a model (a problem that increases with model dimensionality). These problems can be somewhat reduced by providing sufficient model 'spin-up' time so that the boundary forcing dilutes the initial condition error. However, spin-up is only successful when (1) the initial conditions are a reasonable approximation of $t=0$ and (2) the boundary forcing dominates the initial conditions by the end of spin-up. For example, in 3D-models the velocity initial condition is usually zero and the spin-up time is approximated by the 'spin-down' time from typical

Table 1 Examples of 1D hydrodynamic lake models

Model	*Name/source*	*Notes*	*Details*	*Applications*
AQUASIM		tu	15, 31	15, 31, 32
DLM	Dynamic Lake Model	tu	23	7, 23, 24, 27, 33
DYRESM	Dynamic Reservoir Simulation Model	tu	14, 21	1, 5, 6, 14, 16, 17, 18, 22, 26, 35, 36
MINLAKE	Minnesota Lake Model	adv/dif	9, 20, 25	8, 9, 10, 11, 12, 13, 19, 34
Other		adv/diff	4, 30, 37	2, 3, 4, 30, 37
Other		tu	28	28, 29

tu: turbulent transport model; adv/diff: calibrated advection/diffusion transport model.

Sources

1. Balistrieri LS, Tempel RN, Stillings LL, and Shevenell LA (2006) Modeling spatial and temporal variations in temperature and salinity during stratification and overturn in Dexter Pit Lake, Tuscarora, Nevada, USA. *Applied Geochemistry* 21(7): 1184–1203.
2. Bell VA, George DG, Moore RJ, and Parker J (2006) Using a 1-D mixing model to simulate the vertical flux of heat and oxygen in a lake subject to episodic mixing. *Ecological Modelling* 190(1–2): 41–54.
3. Bonnet MP and Poulin M (2004) DyLEM-1D: A 1D physical and biochemical model for planktonic succession, nutrients and dissolved oxygen cycling application to a hyper-eutrophic reservoir. *Ecological Modelling* 180(2–3): 317–344.
4. Bonnet MP, Poulin M, and Devaux J (2000) Numerical modeling of thermal stratification in a lake reservoir. Methodology and case study. *Aquatic Sciences* 62(2): 105–124.
5. Bruce LC, Hamilton D, Imberger J, Gal G, Gophen M, Zohary T, and Hambright KD (2006) A numerical simulation of the role of zooplankton in C, N and P Cycling in Lake Kinneret, Israel. *Ecological Modelling* 193(3–4): 412–436.
6. Campos H, Hamilton DP, Villalobos L, Imberger J, and Javam A (2001) A modelling assessment of potential for eutrophication of Lake Rinihue, Chile. *Archiv Fur Hydrobiologie* 151(1): 101–125.
7. Coats R, Perez-Losada J, Schladow G, Richards R, and Goldman C (2006) The warming of Lake Tahoe. *Climatic Change* 76(1–2): 121–148.
8. Fang X and Stefan H G (1996) Long-term lake water temperature and ice cover simulations/measurements. *Cold Regions Science and Technology* 24(3): 289–304.
9. Fang X and Stefan HG (1997) Development and validation of the water quality model MINLAKE96 with winter data, Project Report No. 390, 33 pp. Minneapolis, MN: St. Anthony Falls Laboratory, University of Minnesota.
10. Fang X and Stefan HG (1998) Temperature variability in lake sediments. *Water Resources Research* 34(4): 717–729.
11. Fang X and Stefan HG (1999) Projections of climate change effects on water temperature characteristics of small lakes in the contiguous US. *Climatic Change* 42(2): 377–412.
12. Fang X, Stefan HG, and Alam SR (1999) Simulation and validation of fish thermal DO habitat in north-central US lakes under different climate scenarios. *Ecological Modelling* 118(2–3): 167–191.
13. Fang X, Stefan HG, Eaton JG, McCormick JH, and Alam SR (2004) Simulation of thermal/dissolved oxygen habitat for fishes in lakes under different climate scenarios – Part 1. Cool-water fish in the contiguous US. *Ecological Modelling* 172(1)
14. Gal GJ, Imberger T Zohary, Antenucci J, Anis A, and Rosenberg T (2003) Simulating the Thermal Dynamics of Lake Kinneret. *Ecological Modelling* 162 (1–2): 69–86.
15. Goudsmit GH, Burchard H, Peeters F, and Wuest A (2002) Application of k-epsilon turbulence models to enclosed basins: The role of internal seiches. *Journal of Geophysical Research-Oceans* 107(C12): 13.
16. Hamblin PF, Stevens CL, and Lawrence GA (1999) Simulation of vertical transport in mining pit lake. *Journal of Hydraulic Engineering-ASCE* 125(10): 1029–1038.
17. Hamilton DP, Hocking GC, and Patterson JC (1997) Criteria for selection of spatial dimension in the application of one- and two-dimensional water quality models. *Mathematics and Computers in Simulation* 43(3–6): 387–393.
18. Han BP, Armengol J, Garcia JC, Comerma M, Roura M, Dolz J, and Straskraba M (2000) The thermal structure of Sau Reservoir (NE: Spain): A simulation approach. *Ecological Modelling* 125(2–3): 109–122.
19. Herb WR and Stefan HG (2005) Dynamics of vertical mixing in a shallow lake with submersed macrophytes. *Water Resources Research* 41(2): 14.
20. Hondzo M and Stefan HG (1993) Lake water temperature simulation model. *Journal of Hydraulic Engineering-ASCE* 119(11): 1251–1273.
21. Imberger J and Patterson JC (1981) A dynamic reservoir simulation model: DYRESM 5. In Fischer HB (ed.) *Transport Models for Inland and Coastal Waters*, pp. 310–361. New York: Academic Press.
22. Kusakabe M, Tanyileke GZ, McCord SA, and Schladow SG (2000) Recent pH and CO_2 profiles at Lakes Nyos and Monoun, Cameroon: Implications for the degassing strategy and its numerical simulation. *Journal of Volcanology and Geothermal Research* 97(1–4): 241–260.
23. McCord SA and SG Schladow (1998) Numerical Simulations of Degassing Scenarios for CO_2-Rich Lake Nyos, Cameroon. *Journal of Geophysical Research-Solid Earth* 103(B6): 12355–12364.
24. McCord SA, Schladow SG, and Miller TG (2000) Modeling Artificial Aeration Kinetics in Ice-Covered Lakes. *Journal of Environmental Engineering-ASCE* 126(1): 21–31.
25. Riley MJ and Stefan HG (1988) MINLAKE – A dynamic lake water-quality simulation-model. *Ecological Modelling* 43(3–4): 155–182.
26. Romero JR, Antenucci JP, and Imberger J (2004) One- and three-dimensional biogeochemical simulations of two differing reservoirs. *Ecological Modelling* 174(1–2): 143–160.
27. Rueda FJ, Fleenor WE, and de Vicente I (2007) Pathways of river nutrients towards the euphotic zone in a deep-reservoir of small size: Uncertainty analysis. *Ecological Modelling* 202(3–4): 345–361.
28. Sahlberg J (2003) Physical modelling of the Akkajaure reservoir. *Hydrology and Earth System Sciences* 7(3): 268–282.
29. Sahlberg J and L Rahm (2005) Light limitation of primary production in high latitude reservoirs. *Hydrology and Earth System Sciences* 9(6): 707–720.
30. Salencon MJ (1997) Study of the thermal dynamics of two dammed lakes (Pareloup and Rochebut, France), using the EOLE model. *Ecological Modelling* 104(1): 15–38.

31. Schmid M, Lorke A, Wuest A, Halbwachs M, and Tanyileke G (2003) Development and sensitivity analysis of a model for assessing stratification and safety of Lake Nyos during artificial degassing. *Ocean Dynamics* 53(3): 288–301.
32. Schmid M, Halbwachs M, and Wuest A (2006) Simulation of CO_2 concentrations, temperature, and stratification in Lake Nyos for different degassing scenarios. *Geochemistry Geophysics Geosystems* 7: 14.
33. Sherman B, Todd CR, Koehn JD, and Ryan T (2007) Modelling the impact and potential mitigation of cold water pollution on Murray cod populations downstream of Hume Dam, Australia. *River Research and Applications* 23(4): 377–389.
34. Stefan HG, Fang X, and Hondzo M (1998) Simulated climate change effects on year-round water temperatures in temperate zone lakes. *Climatic Change* 40(3–4): 547–576.
35. Straskraba M and Hocking G (2002) The effect of theoretical retention time on the hydrodynamics of deep river valley reservoirs. *International Review of Hydrobiology* 87(1): 61–83.
36. Wallace BB and Hamilton DP (2000) Simulation of water-bloom formation in the cyanobacterium *Microcystis aeruginosa*. *Journal of Plankton Research* 22(6): 1127–1138.
37. Wiese BU, Palancar MC, Aragon JM, Sanchez F, and Gil R (2006) Modeling the Entrepenas Reservoir. *Water Environment Research* 78(8): 781–791.

circulation velocities (i.e., the time over which inertia can be expected to keep water moving). However, a lake model with temperature initial condition that does not reflect the initial real-world stratification cannot recover through spin-up.

Calibration

Ideally, hydrodynamic models should not require calibration; i.e., with sufficient data for boundary conditions, initial conditions and turbulence coefficients, a model should adequately represent the physics of the real system. Unfortunately, our data and parameterizations are often inadequate. Calibration may be either through adjusting turbulence coefficients (e.g., the various 'c' values in a k-epsilon model) or by adjusting boundary conditions. Modelers often jump straight into adjusting a turbulence model rather than examining the sensitivity of the model to inaccuracies in the boundary conditions. For example, wind-driven lakes may have unknown spatial gradients of the wind, or the wind sensor may be biased (e.g., in the wind shadow of a building). If the applied wind data under-predicts the actual wind forcing, then calibrating the turbulence model could lead to the 'right' answer for the wrong reason! There should be evidence that the calibrated process is the data mismatch problem (not just the solution). Furthermore, naïve calibration of turbulence coefficients can lead to unphysical values (e.g., an efficiency greater than unity should be a warning sign that something has been missed).

Hydrostatic Approximation

Horizontal length scales are larger than vertical scales in lakes and reservoirs, so the hydrostatic approximation is generally employed. This approximation neglects vertical acceleration ($\partial u_3/\partial t$) and nonhydrostatic pressure gradients ($\partial P_{nh}/\partial x_i$). Note that vertical transport may be reasonably represented in a hydrostatic model, even while vertical acceleration is neglected. Vertical transport has multiple causes: continuity applied to divergence/convergence in the horizontal plane, turbulent mixing, and vertical inertial effects; only the latter is hydrostatically neglected.

Although large-scale free surface motions and the resulting circulations are well-represented by a hydrostatic model, internal seiches are more problematic. Tilting of a pycnocline (e.g., a thermocline) may be relatively steep and ensuing basin-scale waves may evolve in a nonhydrostatic manner. However in a hydrostatic model, the numerical dispersion errors may mimic nonhydrostatic behavior. Thus, physically correct wave dispersion may be serendipitously achieved when numerical dispersion is similar to physical dispersion. Such results must be used with caution as they are highly grid-dependent and the serendipitous confluence of errors may disappear when the model grid is refined. When model results show greater disagreement with the physical world as the model grid is made finer, this type of error may be a suspect.

Applying nonhydrostatic models for large-scale natural systems requires significantly more computational time, model complexity, and modeling expertise than for similar hydrostatic models. An extremely fine model grid and time step is necessary resolve vertical accelerations and nonhydrostatic pressure gradients. Nonhydrostatic lake and ocean models are actively under development, but their widespread application is not imminent.

Model Grid

Overview

Hydrodynamic modeling requires discretizing physical space on a model grid. The size and characteristics

Table 2 Examples of 2D hydrodynamic lake models

Model	*Name/Source*	*Notes*	*Details*	*Applications*
CE-QUAL-W2	U.S. Army Corps of Engineers	fd, la, Ca		1, 2, 3, 7, 8, 9, 10, 17, 19, 20, 21, 24
RMA2	Research Management Associates	fe, da, cu		14, 23
HYDROSIM	Hydrodynamic Simulation Model	fe, da, cu	11	18
others		la	5, 26	5, 12, 13, 26
others		da	4, 16, 22	4, 6, 16, 22

Numerical Method: fd = finite difference; fe = finite element
Horizontal Grid: Ca = Cartesian grid; cu = curvilinear grid
2D form: da = depth-averaged; la = laterally-averaged
Sources

1. Adams WR, Thackston EL, and Speece RE (1997) Modeling CSO impacts from Nashville using EPA's demonstration approach. *Journal of Environmental Engineering-ASCE* 123(2): 126–133.
2. Bartholow J, Hanna RB, Saito L, Lieberman D, and Horn M (2001) Simulated limnological effects of the Shasta Lake temperature control device. *Environmental Management* 27(4): 609–626.
3. Boegman L, Loewen MR, Hamblin PF, and Culver DA (2001) Application of a two-dimensional hydrodynamic reservoir model to Lake Erie. *Canadian Journal of Fisheries and Aquatic Sciences* 58(5): 858–869.
4. Borthwick AGL, Leon SC, and Jozsa J (2001) Adaptive quadtree model of shallow-flow hydrodynamics. *Journal of Hydraulic Research* 39(4): 413–424.
5. Botte V and Kay A (2000) A numerical study of plankton population dynamics in a deep lake during the passage of the Spring thermal bar. *Journal of Marine Systems* 26(3–4): 367–386.
6. Boudreau P, Leclerc M, and Fortin GR (1994) Modelisation Hydrodynamique du lac Saint-Pierre, fleuve Saint-Laurent: l'influence de la vegetation aquatique. *Canadian Journal of Civil Engineering* 21(3): 471–489.
7. Gelda RK and Effler SW (2007) Modeling turbidity in a water supply reservoir: Advancements and issues. *Journal of Environmental Engineering-ASCE* 133(2): 139–148.
8. Gelda RK and Effler SW (2007) Testing and application of a two-dimensional hydrothermal model for a water supply reservoir: implications of sedimentation. *Journal of Environmental Engineering and Science* 6(1): 73–84.
9. Gu RR and Chung SW (2003) A two-dimensional model for simulating the transport and fate of toxic chemicals in a stratified reservoir. *Journal of Environmental Quality* 32(2): 620–632.
10. Gunduz O, Soyupak S, and Yurteri C (1998) Development of water quality management strategies for the proposed Isikli reservoir. *Water Science and Technology* 37(2): 369–376.
11. Heniche M, Secretan Y, Boudreau P, and Leclerc M (2000) A two-dimensional finite element drying-wetting shallow water model for rivers and estuaries. *Advances in Water Resources* 23(4): 359–372.
12. Holland PR, Kay A, and Botte V (2001) A numerical study of the dynamics of the riverine thermal bar in a deep lake. *Environmental Fluid Mechanics* 1: 311–332.
13. Holland PR, Kay A, and Botte V (2003) Numerical modelling of the thermal bar and its ecological consequences in a river-dominated lake. *Journal of Marine Systems* 43(1–2): 61–81.
14. Jennings AA (2003) Modeling sedimentation and scour in small urban lakes. *Environmental Modelling & Software* 18(3): 281–291.
15. Kim Y and Kim B (2006) Application of a 2-dimensional water quality model (CE-QUAL-W2) to the turbidity interflow in a deep reservoir (Lake Soyang, Korea). *Lake and Reservoir Management* 22(3): 213–222.
16. Kramer T and Jozsa J (2007) Solution-adaptivity in modelling complex shallow flows. *Computers & Fluids* 36(3): 562–577.
17. Kuo JT, Lung WS, Yang CP, Liu WC, Yang MD, and Tang TS (2006) Eutrophication modelling of reservoirs in Taiwan. *Environmental Modelling & Software* 21(6): 829–844.
18. Martin C, Frenette JJ, and Morin J (2005) Changes in the spectral and chemical properties of a water mass passing through extensive macrophyte beds in a large fluvial lake (Lake Saint-Pierre, Quebec, Canada). *Aquatic Sciences* 67(2): 196–209.
19. Martin JL (1988) Application of two-dimensional water-quality model. *Journal of Environmental Engineering-ASCE* 114(2): 317–336.
20. Nestler JM, Goodwin RA, Cole TM, Degan D, and Dennerline D (2002) Simulating movement patterns of blueback herring in a stratified southern impoundment. *Transactions of the American Fisheries Society* 131(1): 55–69.
21. Saito L, Johnson BM, Bartholow J, and Hanna RB (2001) Assessing ecosystem effects of reservoir operations using food web-energy transfer and water quality models. *Ecosystems* 4(2): 105–125.
22. Sanmiguel-Rojas E, Ortega-Casanova J, del Pino C, and Fernandez-Feria R (2005) A Cartesian grid finite-difference method for 2D incompressible viscous flows in irregular geometries. *Journal of Computational Physics* 204(1): 302–318.
23. Shrestha PL (1996) An integrated model suite for sediment and pollutant transport in shallow lakes. *Advances in Engineering Software* 27(3): 201–212.
24. Sullivan AB, Jager HI, and Myers R (2003) Modeling white sturgeon movement in a reservoir: The effect of water quality and sturgeon density. *Ecological Modelling* 167(1–2): 97–114.
25. Wu RS, Liu WC, and Hsieh WH (2004) Eutrophication modeling in Shihmen Reservoir, Taiwan. *Journal of Environmental Science and Health Part A-Toxic/Hazardous Substances & Environmental Engineering* 39(6): 1455–1477.
26. Young DL, Lin QH, and Murugesan K (2005) Two-dimensional simulation of a thermally stratified reservoir with high sediment-laden inflow. *Journal of Hydraulic Research* 43(4): 351–365.

Table 3 Examples of 3D hydrodynamic lake models

Model	*Name*	*Notes*	*Details*	*Applications*
CH3D	Curvlinear Hydrodynamics in 3-Dimensions	cfd, cu, zl/ sg, ms	10, 34	16
EFDC	Environmental Fluid Dynamics Code	cfd, cu, zl, ms		9, 12, 13, 14, 15
ELCOM	Estuary and Lake Computer Model	cfd, Ca, zl, si	11	2, 7, 8, 11, 17, 18, 19, 20, 21, 23, 25, 26, 27, 28
GLLVHT	Generalized Longitudinal Lateral Vertical Hydrodynamic and Transport Model	fd, cu, we, si		24, 36
POM; ECOM	Princeton Ocean Model; Estuary and Coastal Ocean Model	cfd, cu, sg/ zl, ms/si	1	1, 3, 4, 5, 33, 35, 37
RMA10	Research Management Associates 10	fe, un, zl		6, 29
SI3D	Semi-Implicit 3D	cfd, cu, zl, si	31	30, 31, 32

Numerical Method: cfd – conservative finite difference; fd – finite difference; fe – finite element.

Horizontal Grid: Ca – Cartesian grid; cu – curvilinear grid; un – unstructured grid.

Time-stepping: ms – mode-splitting; si – semi-implicit.

Vertical Grid: zl – z-level vertical grid; sg – sigma grid.

Sources

1. Ahsan A and Blumberg AF (1999) Three-dimensional hydrothermal model of Onondaga Lake, New York. *Journal of Hydraulic Engineering-ASCE* 125(9): 912–923.
2. Appt J, Imberger J, and Kobus H (2004) Basin-scale motion in stratified upper Lake Constance. *Limnology and Oceanography* 49(4): 919–933.
3. Beletsky D (2001) Modeling wind-driven circulation in Lake Ladoga. *Boreal Environment Research* 6(4): 307–316.
4. Blumberg AF, Khan LA, *et al.* (1999) Three dimensional hydrodynamic model of New York harbor region. *Journal of Hydraulic Engineering-ASCE* 125(8): 799–816.
5. Chen CS, Zhu JR, Kang KY, Liu HD, Ralph E, Green SA, and Budd JW (2002) Cross-frontal transport along the Keweenaw coast in Lake Superior: A Lagrangian model study. *Dynamics of Atmospheres and Oceans* 36(1–3): 83–102.
6. Cook CB, Orlob GT, and Huston DW (2002) Simulation of wind-driven circulation in the Salton Sea: Implications for indigenous ecosystems. *Hydrobiologia* 473(1–3): 59–75.
7. Dallimore CJ, Hodges BR, and Imberger J (2003) Coupling an underflow model to a three-dimensional hydrodynamic model. *Journal of Hydraulic Engineering-ASCE* 129(10): 748–757.
8. Dallimore CJ, Imberger J, and Hodges BR (2004) Modeling a plunging underflow. *Journal of Hydraulic Engineering-ASCE* 130(11): 1068–1076.
9. Elci S, Work PA, and Hayter EJ (2007) Influence of stratification and shoreline erosion on reservoir sedimentation patterns. *Journal of Hydraulic Engineering-ASCE* 133(3): 255–266.
10. Gessler D, Hall B, Spasojevic M, Holly F, Pourtaheri H, and Raphelt N (1999) Application of 3D mobile bed, hydrodynamic model. *Journal of Hydraulic Engineering-ASCE* 125(7): 737–749.
11. Hodges BR, Imberger J, Saggio A, and Winters KB (2000) Modeling basin-scale internal waves in a stratified lake. *Limnology and Oceanography* 45(7): 1603–1620.
12. Jin KR and Ji ZG (2005) Application and validation of three-dimensional model in a shallow lake. *Journal of Waterway Port Coastal and Ocean Engineering-ASCE* 131(5): 213–225.
13. Jin KR, Hamrick JH, and Tisdale T (2000) Application of three-dimensional hydrodynamic model for Lake Okeechobee. *Journal of Hydraulic Engineering-ASCE* 126(10): 758–771.
14. Jin KR, Ji ZG, and Hamrick JH (2002) Modeling winter circulation in Lake Okeechobee, Florida. *Journal of Waterway Port Coastal and Ocean Engineering-ASCE* 128(3): 114–125.
15. Jin KR, Ji ZG, and James RT (2007) Three-dimensional water quality and SAV modeling of a large shallow lake. *Journal of Great Lakes Research* 33(1): 28–45.
16. Kim SC, Cerco CF, and Johnson BH (2006) Three-dimensional management model for Lake Washington, part I: Introduction and hydrodynamic modeling. *Lake and Reservoir Management* 22(2): 103–114.
17. Laval BE, Imberger J, and Findikakis AN (2005) Dynamics of a large tropical lake: Lake Maracaibo. *Aquatic Sciences* 67(3): 337–349.
18. Laval B, Hodges BR, and Imberger J (2003) Reducing numerical diffusion effects with pycnocline filter. *Journal of Hydraulic Engineering-ASCE* 129(3): 215–224.
19. Laval B, Imberger J, and Findikakis AN (2003) Mass transport between a semienclosed basin and the ocean: Maracaibo System. *Journal of Geophysical Research-Oceans* 108(C7).
20. Laval B, Imberger J, Hodges BR, and Stocker R (2003) Modeling circulation in lakes: Spatial and temporal variations. *Limnology and Oceanography* 48(3): 983–994.
21. Leon LF, Lam DC L, Schertzer WM, Swayne DA, and Imberger J (2007) Towards coupling a 3d hydrodynamic lake model with the Canadian Regional climate model: Simulation on Great Slave Lake. *Environmental Modelling & Software* 22(6): 787–796.
22. Leon LK, Imberger J, Smith REH, Hecky RE, Lam DCL, and Schertzer WM (2005) Modeling as a tool for nutrient management in Lake Erie: A hydrodynamics study. *Journal of Great Lakes Research* 31: 309–318.
23. Marti CL and Imberger J (2006) Dynamics of the benthic boundary layer in a strongly forced stratified lake. *Hydrobiologia* 568: 217–233.
24. Na EH and Park SS (2006) A hydrodynamic and water quality modeling study of spatial and temporal patterns of phytoplankton growth in a stratified lake with buoyant incoming flow. *Ecological Modelling* 199(3): 298–314.
25. Okely P and Imberger J (2007) Horizontal transport induced by upwelling in a canyon-shaped reservoir. *Hydrobiologia* 586: 343–355.

26. Romero JR and Imberger J (2003) Effect of a flood underflow on reservoir water quality: Data and three-dimensional modeling. *Archiv Fur Hydrobiologie* 157(1): 1–25.
27. Romero JR, Antenucci JP, and Imberger J (2004) One- and three-dimensional biogeochemical simulations of two differing reservoirs. *Ecological Modelling* 174(1–2): 143–160.
28. Romero JR, Antenucci JP, and Imberger J (2004) One- and three-dimensional biogeochemical simulations of two differing reservoirs. *Ecological Modelling* 174(1–2): 143–160.
29. Rueda FJ and Schladow SG (2002) Quantitative comparison of models for barotropic response of homogeneous basins. *Journal of Hydraulic Engineering-ASCE* 128(2): 201–213.
30. Rueda FJ and Schladow SG (2003) Dynamics of large polymictic lake. Ii numerical simulations. *Journal of Hydraulic Engineering-ASCE* 129(2): 92–101.
31. Rueda FJ, Schladow SG, and Palmarsson SO (2003) Basin-scale internal wave dynamics during a winter cooling period in a large lake. *Journal of Geophysical Research-Oceans* 108(C3), 3097, doi:10.1029/2001JC000942.
32. Rueda FJ, Schladow SG, Monismith SG, and Stacey MT (2005) On the effects of topography on wind and the generation of currents in a large multi-basin lake. *Hydrobiologia* 532: 139–151.
33. Schwab DJ and Beletsky D (2003) Relative effects of wind stress curl, topography, and stratification on large-scale circulation in Lake Michigan. *Journal of Geophysical Research – Oceans* 108(C2).
34. Sheng YP (1990) Evolution of a three-dimensional curvilinear-grid hydrodyamic model for estuaries, lakes and coastal waters: CH3D. In *Estuarine and Coastal Modeling: Proceedings of the Conference*, Newport, Rhode Island, November 15–17, 1989. ASCE.
35. Song Y, Semazzi FHM, Xie L, and Ogallo LJ (2004) A coupled regional climate model for the Lake Victoria basin of East Africa. *International Journal of Climatology* 24(1): 57–75.
36. Wu J, Buchak EM, Edinger JE, and Kolluru VS (2001) Simulation of cooling-water discharges from power plants. *Journal of Environmental Management* 61(1): 77–92.
37. Zhu JR, Chen CS, Ralph E, Green SA, Budd JW, and Zhang FY (2001) Prognostic modeling studies of the Keweenaw current in Lake Superior. Part II: Simulation. *Journal of Physical Oceanography* 31(2): 396–410.

of the grid determine the scales of what a model can and cannot represent. In the horizontal plane, there are three grid systems: Cartesian, curvilinear, and unstructured. For models including a vertical dimension, the vertical grid may be terrain-following (sigma coordinate), Cartesian (z-level) or isopycnal (Lagrangian). Unstructured grids can also be used in the vertical plane, but have not been widely adopted.

Grid Size and Convergence

The local grid size controls the 'resolution' of local processes; e.g., a single grid cell has only a single velocity on a simple Cartesian finite-difference grid. Thus, the grid mesh is a top-level control on the resolvable physics and transport. For example, if only two grid cells are used across a narrow channel the transport may be theoretically either unidirectional or bidirectional; however, two grid cells cannot represent a deep center channel flow with return flows along both shallow banks. A useful exercise is to consider how many grid cells are necessary to represent 1.5 periods of a sine wave: although three cells is clearly the minimum, the resulting discrete pattern will not be particularly sinusoidal. Arguably, 10–15 grid cells should be the minimum resolution for most important flow features. An effective model grid resolves the key physical features at practical computational cost. Grid design should be an iterative process wherein model results at different grid scales are compared to gain insight into model performance. A model grid is 'converged' when further grid refinement does not significantly change model results. Unfortunately, obtaining a converged grid is not always practical; indeed, most large-scale models suffer from insufficient grid resolution. Such models may still have validity, but grid-scale effects may dominate physical processes.

Horizontal Grid Systems

Cartesian grids are obtained with a square or rectangular mesh (**Figure 4(a)**). The mesh structure allows simple model coding since a grid cell's neighbors are easily determined. For multidimensional models, simple Cartesian grids cannot be applied with fine resolution in some regions and coarse resolution in others. These deficiencies can be addressed with 'plaid' structured meshes (i.e., nonuniform Cartesian grid spacing), domain decomposition or nested grid (e.g., quadtree) techniques. To use an efficient rectangular mesh on a sinuous reservoir, the topography may be straightened along the channel centerline before applying the Cartesian mesh.

Curvilinear grids in the horizontal plane are structured meshes (similar to a Cartesian grid) that smoothly distort the quadrilateral elements throughout horizontal space (**Figure 4(b)**). The distortion between physical (x,y) space and curvilinear (ξ,η) space requires transformation of the governing equations. Curvilinear meshes allow fine grid resolution in one area and coarse resolution in another, as long as the mesh changes smoothly between regions. The smoothness and orthogonality of the mesh (as seen in physical space) will affect the model solution. Reasonable rule-of-thumb criteria are (1) adjacent grid

Table 4 Model applications

Lake	*1D*	*2D*	*3D*
Akkajaure Reservoir (Sweden)	28		
Lake Alpnach (Switzerland)	15		
Lake Baldegg (Switzerland)	15		
Lake Balaton (Hungary)		4	
Lake Baikal (Russia)		5, 12	
Bassenthwaite Lake (UK)	2		
Lake Belau (Germany)		22	
Lake Beznar (Spain)	27		
Brenda Pit Lake (Canada)	16		
Brownlee Reservoir (USA)		24	
Lake Burragorang (Australia)	26		26, 27, 28
Cheatham Lake (USA)		1	
Clear Lake (USA)		23	30, 32
Lake Constance (Germany/Switzerland)			2
Cummings Lake (Canada)	24		
Dexter Pit Lake (USA)	1		
East Dollar Lake (Canada)	24		
Entrepenas Reservoir (Spain)	37		
Lake Erie (Canada/USA)		3	22
Flint Creek Lake (USA)			36
Great Slave Lake (Canada)			21
Hartwell Lake (USA)			9
Hume Reservoir (Australia)	33		
Isikli Reservoir (Turkey)		10	
Kamploops Lake (Canada)		13	
Lake Kinneret (Israel)	5, 14		11, 18, 20, 23
Lake Ladoga (Russia)			3
Lake Maracaibo (Venezuela)			17, 19
Lake Michigan (Canada/USA)			33
Lake Monoun (Cameroon)	22		
Mundaring Weir (Australia)			25
Lake Neusiedl		16	
Lake Nyos (Cameroon)	22, 23, 31, 32		
Lake Ogawara (Japan)			7
Lake Okeechobee (USA)			12, 13, 14, 15
Onondaga Lake (USA)			1
Orlik Reservoir (Czech Republic)	35		
Otter Lake (USA)	19		
Lake Paldang (South Korea)			24
Pareloup Reservoir (France)	30		
Pavin Crater Lake (Canada)	16		
Prospect Reservoir (Australia)	26		
Lake Rinihue (Chile)	6		
Rochebut Reservoir (France)	30		
Lake Saint Pierre (Canada)		6, 18	
Salton Sea (USA)			6
Sau Reservoir (Spain)	18		
Schoharie Reservoir (USA)		7, 8	
Lake Shasta (USA)		2, 9, 21	
Lower Shaker Lake (USA)		14	
Shihmen Reservoir (Taiwan)		25	
Slapy Reservoir (Czech Republic)	35		
Lake Soyang (Korea)		15	
J. Strom Thurmond Lake (USA)		20	
Lake Superior (Canada/USA)			5, 37
Lake Tahoe (USA)	7		31
Te Chi Reservoir (Taiwan)		17, 26	
Tseng-Wen Reservoir (Taiwan)		17	
Lake Victoria (Kenya/Tanzania/Uganda)			35
Villerest Reservoir (France)	3, 4		
Lake Washington (USA)			16
Wellington Reservoir (Australia)			8
Lake Yangebup (Australia)	36		

Numbers correspond to notes from **Tables 1, 2,** and **3** for 1D, 2D and 3D models, respectively.

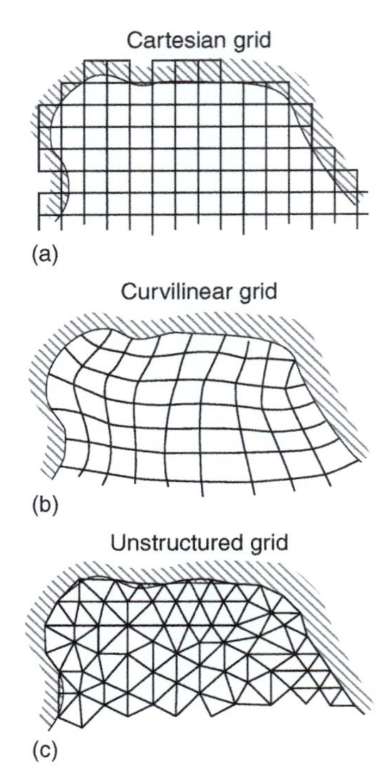

Figure 4 Plan view illustrating different horizontal grid systems.

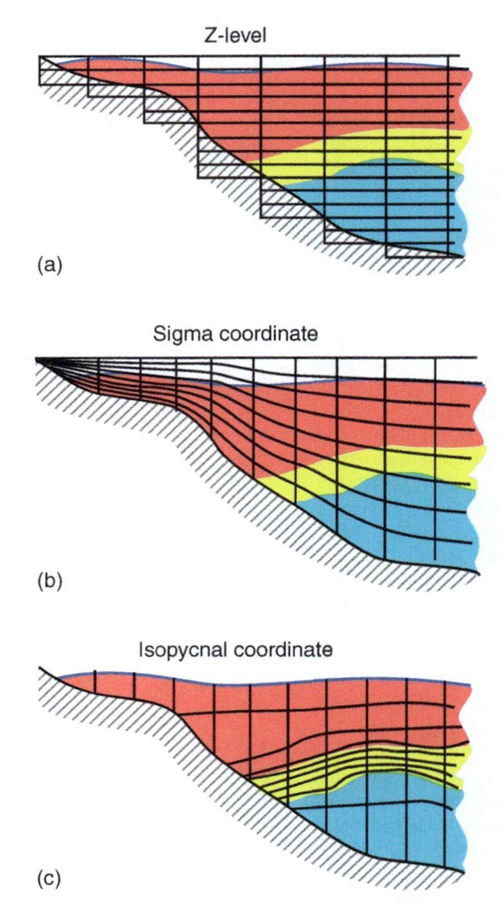

Figure 5 Elevation view illustrating different vertical grid systems relative to a stratified lake with warm (red) surface water, thermocline (yellow) and cooler (blue) hypolimnetic water.

cells should have increased/decreased volume by no more than 10% and (2) off-orthogonal transformation metrics should be an order of magnitude smaller than orthogonal metrics. Smooth curvilinear meshes can be manually designed with simple drafting tools, but stand-alone mesh creation software is generally used. Some models require orthogonality or near-orthogonality for the mesh, which severely constrains mesh creation.

Unstructured grids in the horizontal plane are composed of n-sided polygons (**Figure 4(c)**); triangular and quadrilateral elements are typically preferred or required. An unstructured mesh is the easiest for fitting complicated topography and arguably has the greatest flexibility for providing fine resolution in some areas with coarse resolution in others. However, model solutions are still affected by local gradients of grid cell volume and grid orthogonality. Creating a good unstructured grid is an art, requiring separate grid creation software and a lengthy trial-and-error process. It is often necessary to carefully examine model performance on several different unstructured grids to gain an understanding of how different grid choices affect the solution. Finite difference and finite volume models for unstructured meshes are relatively recent developments, but have not yet seen extensive use in lakes or reservoirs.

Vertical Grid Systems

Z-level grids are the simplest vertical system, using layers with whose thickness is uniform across the horizontal plane (**Figures 5(a)**). Layer thicknesses may vary in the vertical, but should do so smoothly (i.e., no more than about 10% expansion of thickness in adjacent layers). *Z*-level grids are generally preferred for 2D and 3D lake models due to their simplicity. A disadvantage is that steep bottom slopes are represented as discrete stair steps, which distorts along-slope flow. Coupling a 2D- or 3D-model with a benthic boundary layer model can overcome the stair-step problem, albeit by increasing model complexity and introduction of empiricism and ad hoc coupling mechanisms.

Sigma-coordinate (terrain following) vertical grid systems are commonly used in oceanic-scale modeling (e.g., the Great Lakes), but have significant drawbacks for inland waters. Sigma-coordinate systems divide each water column into a fixed number of layers, resulting in thick layers in deep water and thin layers in shallow water (**Figures 5(b)**). For sloping boundaries, the sigma-coordinate system must be truncated or a singularity occurs where the depth goes to zero. Sigma-coordinates are preferred for modeling along-slope processes, but may distort internal wave dynamics along the slope.

Isopycnal coordinate systems require a moving grid that tracks the Lagrangian movement of predefined isopycnals (**Figures 5(c)**). This approach is common for 1D lake models as a means of easily tracking stratification creation and destruction. Multidimensional isopycnal models have been developed for ocean simulations to limit numerical diffusion that otherwise weakens stratification; these models have not seen wide application in lakes or reservoirs.

Time Step

Unsteady models take an initial density/velocity field and advance these forward in time (subject to the boundary conditions of the system). A model that is stable at a large time step is often prized as being more computationally efficient. The model time step is generally limited by a Courant–Lewy–Friedrichs (CFL) condition, defined as $u \Delta t \Delta x^{-1} < C_a$, where u is a velocity (fluid or wave), Δt is the time step, Δx is the grid spacing (in the same direction as u), and C_a is a constant that depends on the numerical method (typically $C_a \leq 1$). Some models also have a viscous limitation controlled by the turbulent vertical eddy viscosity (ν_z) such that $\nu_z \; \Delta t \; \Delta z^{-2} < C_\nu$. It is possible to design stable numerical methods for $C_a > 1$ or $C_\nu > 1$; however stability at large time step does not imply accuracy. For example, a reservoir that is 10 m deep × 10 km length will have a surface seiche period of ~30 min; the physics of this seiche cannot be modeled with a 20 min time step, even if the model is stable. Thus, the model time step should be chosen *both* for model stability and to accurately resolve the time-scale of processes. In particular, a large model time step will mask the cumulative effect of nonlinearities from short-time-scale processes.

Numerical Methods

There are three basic methods for discretizing the governing equations on a model grid; in order of increasing complexity these are: (1) finite difference, (2) finite volume, and (3) finite element. *Finite difference methods* represent spatial derivatives by discrete gradients computed from neighboring grid cells. *Finite volume methods* pose an integral form of the governing equations for conservative cell-face fluxes. Both finite difference and volume methods provide a set of discrete algebraic equations representing the continuous governing equations. For a model with a sufficiently refined grid and time step, the solution of the discrete equations is an approximate solution of the continuous equations. In contrast, *finite element methods* directly approximate the *solution* of the governing equations rather than the governing equations themselves. Finite element methods are often characterized as being appropriate for unstructured grids, whereas finite difference methods are often characterized as appropriate for structured grids; this outdated canard needs to be put to rest. Finite difference and finite volume methods have both been successfully applied on unstructured grids, and finite element methods can also be successfully applied to structured grids. The choice of grid and numerical method are entirely independent in model development. However, most models are designed for only one type of grid.

The finite element method is mathematically appealing but requires considerable computational effort, especially for density-stratified flows. Because temperature gradients are directly coupled to momentum through density and hydrostatic pressure gradients, a pure finite element discretization requires simultaneous solution of momentum, temperature transport, and an equation of state. As a further complication, global and local conservation is not always achieved in finite element methods; i.e., local scalar transport fluxes into and out of an element may only approximately balance the scalar accumulation in the element, and the integrated global scalar content may not be conserved. These effects can create problems for water quality models that are directly coupled to finite element hydrodynamic models as source/sink water quality terms may be dominated by numerical nonconservation. Note that consistent finite element methods may be implemented for global scalar conservation, but many existing codes have not been tested or proven consistent.

Finite difference and finite volume methods are conceptually quite different, but may be very similar in the model code. Finite differences are often described as point-based discretizations, whereas finite-volume methods are described as cell-based. However, most multidimensional hydrodynamic models apply a hybrid approach: momentum is discretized with finite differences, but continuity is discretized on a staggered grid, which is discretely equivalent to a volume integral (i.e., a finite volume approach). Thus, these hybrid or 'conservative finite difference' methods ensure exact volume conservation for fluxes into and out of a grid cell. This exact local and global scalar transport conservation, along with their simplicity, has made these methods the most popular 3D-modeling approaches.

Order of Accuracy

Multidimensional models are often judged by the 'order of accuracy' of their time and space discretizations for the governing equations. This order reflects how the error changes with a smaller time step or smaller grid spacing. For example with 2nd-order spatial discretization, model error reduces by two orders of magnitude for each magnitude reduction in grid size. Higher-order methods are generally preferred, although they are more computationally expensive than low-order methods for the same number of grid cells. There is a trade-off when computational power is limited: a higher-order method may only be possible with a larger time step and/or grid cell size than a lower-order model. It is generally thought that for converged grids the absolute error of a higher-order method on a coarse grid will be less than the absolute error of a lower-order method on a fine grid. However, this idea presupposes that both the model grids provide converged solutions. When the grid cannot be fully converged due to computational constraints (often the case for practical problems), the comparative efficiency of high-order or low-order methods must be determined by experimentation.

As a general rule, 1st-order spatial discretizations (e.g., simple upwind) are too numerically diffusive for good modeling. Spatial discretizations that are 2nd-order (e.g., central difference) often have stability issues, so 3rd-order (e.g., QUICK) is generally preferred. The best 3rd-order spatial methods include some form of flux limiting (e.g., TVD or ULTIMATE) to reduce unphysical oscillations at sharp fronts. Fourth-order and higher spatial discretization methods can be found in the numerical modeling literature, but have not been applied in any common lake models.

For time discretizations, 2nd-order methods (e.g., Crank–Nicolson) are preferred, but many models are only 1st-order due to the complexity of higher-order methods. In general, if one process is modeled with a 1st-order time-advance, then the entire scheme is 1st-order. As a note of caution, some semi-implicit 2nd-order methods may be only 1st-order accurate (albeit stable) for $CFL > 1$.

Model Errors

We separate the idea of 'model error' from 'data error'; the latter is associated with incorrect or unknown boundary/initial conditions, while the former is inherent in the model itself. Model errors are not randomly distributed. Instead, models provide an exact solution of an approximation of the governing equations, so the errors are determined by the discrete approximations. Three different types of fundamental errors will occur in any sufficiently complicated transport field: numerical diffusion of mass, numerical dissipation of energy and numerical dispersion of waves.

Numerical diffusion of mass occurs when advection of a sharp density gradient causes the gradient to weaken (as if mass diffusivity were greater). In a stable model, this error is has a net bias towards weakening sharp gradients and can be a significant problem for representing the evolution of stratification when an active internal wave field is modeled.

Numerical dissipation of energy occurs when momentum is numerically diffused (as if viscosity were greater). This effect is generally referred to as 'numerical viscosity'. It typically occurs near sharp velocity gradients and tends to weaken the gradients. A stable model requires nonnegative numerical dissipation, as negative (or anti-) dissipation leads to positive feedback and the exponential growth of kinetic energy (i.e., the model 'blows up').

Numerical dispersion of waves occurs when a model propagates a wave component (free surface or internal) at the wrong speed. This effect can have interesting consequences for hydrostatic models (as discussed in Hydrostatic Approximation above).

In general, higher-order models have smaller errors, but may lead to antidiffusion (i.e., artificial resharpening of a gradient) or antidissipation (i.e., artificial increase in energy) that can destabilize a model. For any model to be reliable, the numerical diffusion of mass should be an order of magnitude smaller than turbulent mixing, and numerical dissipation of energy should be an order of magnitude less than turbulent dissipation.

Modeling Turbulence and Mixing

The governing equations for lake and reservoir hydrodynamic modeling are generally the Reynolds-Averaged Navier Stokes (RANS) equations, although some Large-Eddy Simulation (LES) methods may be suitable for future applications. With either method, processes smaller than the grid and time scales are empirically-modeled rather than directly simulated. Local values for eddy viscosity and eddy diffusivity are generally used to represent the nonlinear turbulent advection of momentum (viscosity) and scalars (diffusivity). As turbulence varies in both time and space, constant and uniform values of eddy viscosity are rarely appropriate. In particular, the ability of stratification to suppress vertical turbulence and mixing leads to nonuniform profiles with near-zero values at strong stratifications. A wide variety of RANS turbulence models are in use, the most popular being $k-\varepsilon$, $k-l$, and mixed-layer approaches, which must be modified to account for stratification. Performance of turbulence models may be highly dependent on the model grid resolution, so grid selection must be combined with selection of the appropriate turbulence model and settings. A key difficulty is that discretization on a coarse model grid (often required due to computational constraints) leads to high levels of numerical dissipation and diffusion. Indeed, it is not unusual to find that the model error dominates the turbulence model, particularly in the horizontal flow field. The relative scales of numerical dissipation and diffusion may also have an impact. If numerical dissipation is dominant, then internal waves may be damped before they cause significant numerical diffusion of mass. Thus, a 2D- or 3D-model that artificially damps internal waves may provide a 'better' long-term representation of the thermocline, but at the cost of poorly representing the 2D or 3D transport processes!

Similarities and Differences between Lake and River Modeling

Although the focus of this article is on lake models, many of the underlying discussions of model types and errors are equally applicable to river modeling. Such models can also be 1D, 2D, or 3D, may be hydrostatic or nonhydrostatic, and have difficulties with turbulence modeling and grid resolution (especially at finer scales). River models are perhaps easier to validate because there is a single major flux direction (downstream) that quantitatively dominates the hydrodynamics; this directionality is in dramatic contrast to the unsteady oscillatory forcings in a lake that make collecting sufficient validation data a complex and time-consuming task. On the other hand, the higher flow rates typical of rivers lead to bed motion and sediment transport that may strongly affect the flow patterns. At high flows, rivers may be geomorphically active and the use of simple fixed-bed models (appropriate for lakes over shorter time scales) may be entirely unsuitable. Thus, knowledge gained in lake modeling cannot always be transferred directly to rivers or vice versa – each discipline has its own key challenges. For lakes, modeling evolving temperature stratification is the critical requirement; for a river model, the correct representation of the riverbed geometry and its geomorphologic evolution is critical.

Summary and Future Directions

Selecting whether to use a 1D-, 2D-, or 3D-model depends on the water body, available computational power, available field data and the type of answers desired. Applying 1D-models is always fastest and simplest, whereas 3D-models are computationally intense and require the greatest user skill and effort. 2D- and 3D-models need extensive field data to drive spatially-varying model boundary conditions and provide validation. In contrast, 1D-models need less extensive boundary condition data, but may require field studies to parameterize variability in the averaged directions. Whether a 3D-model is 'better' than a 1D-model will depend on the physics of interest. For example, if the physics of internal waves in a lake are unknown, a 3D-model may be needed to understand their effects. However, if the basic internal wave physics are already understood, then a 1D lake model (appropriately parameterized) may be adequate. The ideal conjunction of 1D and 3D lake models has yet to be attempted: the strength of 3D-models lies in quantifying the short-time, space-varying lake response to an event. Theoretically, such a model could be used to develop better parameterizations of 1D-models, increasing our understanding of how short-term events modify longer-term system behavior.

In considering hydrodynamic models coupled to water quality models, the ability to adequately capture bulk transport of hydrodynamic fields (e.g., velocity, temperature) should not be taken as proof of the ability to capture greater complexities in scalar biogeochemical distributions. Modeling the temperature is relatively easy because the problem is bounded and provides negative feedback. That is, lake temperatures are typically between 4 and 35 °C with the warm side facing up, and any attempt to turn the warm side downwards leads to horizontal density gradients and pressure forces that oppose overturning. Similarly, warming of the lake surface leads to increased heat loss to the atmosphere, which tends to moderate and limit errors. Velocity is also subject to large-scale forcing (wind) and is a bounded problem as unphysically large velocities will cause a model to blow up. Furthermore, dissipation is a limiting mechanism that works everywhere and at all times to bring the velocity towards zero. Thus, both velocity and temperature have preferred 'rest' states and model error cannot accumulate indefinitely without the results becoming obviously wrong. In contrast, scalar dispersion is driven by local turbulence and advection, without any global bounds to limit model error accumulation. Thus, even while the large-scale velocity and temperature fields look reasonable, a model may produce localized features that lead to unrealistic transport of scalars. Even simple passive tracer transport leads to complicated model-predicted gradient features as illustrated in **Figures 6** and 7 and associated animations. Although such tracer fields illustrate model-predicted transport, there are relatively few field studies or methods for effective validation. These problems become even more pronounced for water quality models as biogeochemical scalar concentrations (such as phytoplankton biomass) are locally forced by nutrient concentrations, do not have a preferred 'rest' state, and have source/sink behaviors that may be affected by model transport errors. As such, 2D and 3D hydrodynamic/water modeling without validating field data should be considered cartoons that may be informative, but are also speculative and may be simply wrong!

As computers grow more powerful, there is a tendency to throw more grid cells at a system to improve model results. However, as the model grid is made finer, there is some point where neglect of the nonhydrostatic pressure is inconsistent with the grid scale – i.e., the model provides a better solution to the wrong equations. As a reasonable rule of thumb, if the horizontal grid scale is substantially smaller than the local depth of water, then the hydrostatic approximation may be inappropriate. Where internal wave evolution is important, nonhydrostatic pressure gradients should be included in future models. Although nonhydrostatic models presently exist, they have not yet been practically demonstrated for large-scale lake modeling.

Model calibration should be used carefully and in conjunction with sensitivity analyses. Indeed, the difference between an uncalibrated hydrodynamic model and field data may provide greater insight into the physical processes than a calibrated model. A careful modeler will estimate the uncertainty in various boundary conditions and conduct model sensitivity tests to understand how the uncertainty may affect results. Unnecessary calibration can be avoided by gaining a better understanding of the model error characteristics. Before applying any 3D-model to a lake or reservoir, the model should be tested on 2D rectangular domains at similar scales; e.g., simple models of internal waves, river inflows, and wind-driven mixing can provide relatively rapid insight into the relationship between model error, grid scale, time step and physics.

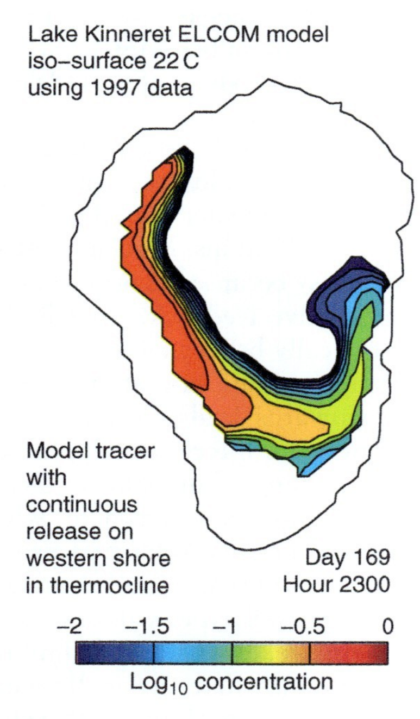

Figure 6 Modeled passive tracer concentrations in the thermocline of Lake Kinneret for a tracer concentration of 1.0 continuously released from the western boundary. This tracer motion is principally due to a basin-scale Kelvin wave.

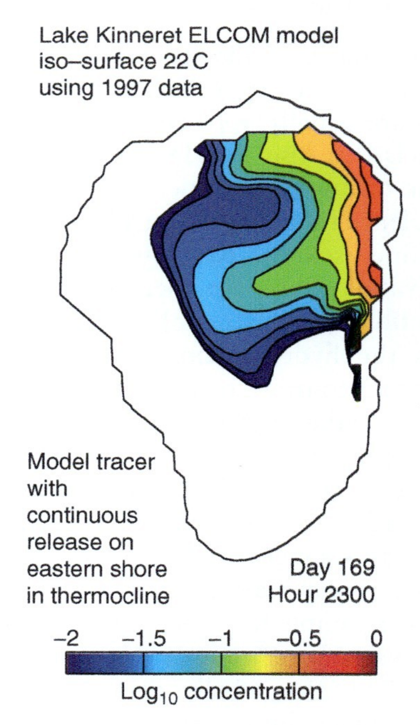

Figure 7 Modeled passive tracer concentrations in the thermocline of Lake Kinneret for a tracer concentration of 1.0 continuously released from the eastern boundary. This tracer motion is due to the combination of a Kelvin wave and a 2nd-mode Poincaré wave.

The horizontal grid for lake models may be Cartesian, curvilinear, or unstructured; these methods have different strengths, weaknesses and complexities, such that the practical choice depends on the system, model availability and the modeler's bias. Where fine grid resolution is needed over a part of a domain (e.g., littoral zones), future developments in automated quadtree meshing of Cartesian grids may be easier to use than either curvilinear or unstructured grids.

Both z-level and sigma-coordinate vertical grids have significant drawbacks that remain unaddressed in the literature. Boundary layer sub-models have attempted to patch these problems, but are relatively complicated to develop and apply. Isopycnal methods may provide some future improvement, but it is not clear that they will be a panacea. Although a few isopycnal simulations have been made in lakes, we presently lack a thorough analysis of how isopycnal models represent internal wave dynamics at lake scales and along sloping boundaries.

There have been significant advances in ocean and estuarine modeling that have not yet appeared in lake models, but one must be careful about generalizing their applicability. Lake modeling faces two key problems: (1) long residence times allows model error to accumulate, unlike error that washes out with the tide in an estuary, and (2) the forcing is inherently unsteady in direction/amplitude, and may have sharp spatial gradients. Thus, methods suitable for a strong tidal exchange or a unidirectional ocean current with a smoothly varying wind field may not be effective for weak, unsteady forcing of a lake in the wind-shadow of a mountain. Indeed, despite our advances there remains significant work ahead before the art of hydrodynamic modeling is replaced by simple engineering.

Further Reading

Chen XJ and Sheng YP (2005) Three-Dimensional Modeling of Sediment and Phosphorus Dynamics in Lake Okeechobee, Florida: Spring 1989 Simulation. *Journal of Environmental Engineering-ASCE* 131(3): 359–374.

Edinger JE (2001) *Waterbody Hydrodynamic and Water Quality Modeling,* 215 pp. Reston, VA: ASCE Press.

Ferziger JH and Peric M (2002) *Computational Methods for Fluid Dynamics,* 423 pp. Berlin: Springer.

Hodges BR, Laval B, and Wadzuk BM (2006) Numerical error assessment and a temporal horizon for internal waves in a hydrostatic model. *Ocean Modelling* 13(1): 44–64.

Hodges BR, Imberger J, Saggio A, and Winters KB (2000) Modeling basin-scale internal waves in a stratified lake. *Limnology and Oceanography* 45(7): 1603–1620.

Kowalik Z and Murty TS (1993) *Numerical Modeling of Ocean Dynamics,* 481 pp. Singapore: World Scientific.

Leon LF, Lam DCL, Schertzer WM, Swayne DA, and Imberger J (2007) Towards coupling a 3d hydrodynamic lake model with the Canadian regional climate model: Simulation on Great Slave Lake. *Environmental Modelling & Software* 22(6): 787–796.

Rueda FJ, Schladow SG, and Palmarsson SO (2003) Basin-scale internal wave dynamics during a winter cooling period in a large lake. *Journal of Geophysical Research-Oceans* 108(C3): 3097, doi:10.1029/2001JC000942.

Schwab DJ and Beletsky D (2003) Relative effects of wind stress curl, topography, and stratification on large-scale circulation in Lake Michigan. *Journal of Geophysical Research-Oceans* 108(C2): 3044, doi:10.1029/2001JC001066.

Floods

J A Hubbart and J R Jones, University of Missouri, Columbia, MO, USA

Definition of Flood

A flood is loosely defined as river discharge exceeding bankfull limitations. It is also considered a temporary rise of the water level, as in a river or lake or along a seacoast or wetland, resulting in its spilling over and out of its natural or artificial confines onto land that is normally dry. Floods are sometimes described according to their statistical occurrence. For example, a 50-year flood is a flood having a magnitude that is reached in a particular location on average once every 50 years. This is often referred to as a return interval (Tr), and is calculated as follows:

$$\mathrm{Tr} = (n+1)/m$$

where n is the total number of events, and m is the specific event number in question. With this calculation, probability (p) of event occurrence is calculated by

$$p = 1/\mathrm{Tr}$$

A helpful way to visualize the difference in flow between a flood and normal runoff is to visualize the flood channel width, which is the floodplain portion that will discharge the 50-year flood (**Figure 1**). In this example, the flood zone is centered over the main channel, an unusual situation in natural systems where the flood zone can be offset or split into several zones depending on the topography.

Forecasting annual flows and the magnitude and frequency of flood events is a challenge. Hydrologic data from unaltered, wildland systems show great variation in annual floods (coefficients of variation ≥ 1.0), making it difficult to predict when the floodplain will be inundated. Human land use changes further complicate flood forecasting because impermeable surfaces reduce infiltration and accelerate runoff. Ultimately, both natural and anthropogenic processes can result in watershed morphological changes that may modify flow. It is noteworthy that even minor changes in flood magnitude, duration or frequency, although statistically undetectable, should not be misinterpreted as ecologically or culturally benign.

Causes and Effects of Floods

All rivers and streams are subject to fluctuations in flow. During a rainstorm, the amount, intensity, duration, area, and path of the storm all influence the runoff reaching the stream. Multiple land form and use factors affect the ability of land to absorb precipitation and therefore affect the rate of runoff. Area and path of the storm relate to the area of the basin receiving rainfall, which in turn, represents the area contributing runoff. Area and the runoff rate determine the volume of water that will pass a given point downstream.

Modification of runoff rates occurs by variations in topographic relief, soil infiltration processes, vegetative cover, and surface retention (e.g., ponding) within a given catchment or watershed. The key physical mechanism controlling runoff and flooding processes is precipitation. River flow is largely determined by the precipitation regime (i.e., amount and type of precipitation). Precipitation type is determined by factors such as elevation (rain, snow) and orographic uplift, and whether it is on the leeward or windward side of mountains. Orographic uplift often results in heavier precipitation on the crests and windward slopes of mountain ranges. It also accounts for much heavier precipitation than in surrounding lowlands. The same process also causes rain shadow effects on leeward sides of mountains. The nature and condition of the drainage basin, and variable climates also affect streamflow, and therefore flood potential. Vegetative cover also affects the rate at which surface water flows to a main channel by slowing and spreading out runoff. The passage of water is similarly ameliorated in basins with natural storage areas, including lakes and wetlands. Consequently, smaller peak flows are produced in basins with dense vegetation and lakes, reservoirs and wetlands than those without.

Snowmelt: In high elevations and northern latitudes, most precipitation is snow. During snowmelt, large quantities of water are released. Snowmelt induced runoff floods are the most common type of flooding in these areas and generally occur in the spring but also occur during sudden winter thaws and as a consequence of rain-on-snow events. Heavy runoff results from rapid melting of snow under the combined effect of sunlight, winds, warmer temperatures, and rain. When there is an above average snow depth, a sudden thaw, or both, the potential for high volumes of runoff and subsequent flooding increases. This process is made more severe when the rising snowmelt runoff is compounded by runoff from heavy rainfall. Because climatic factors influencing the rate of snowmelt are often widespread, snowmelt runoff flooding conditions can exist over vast areas

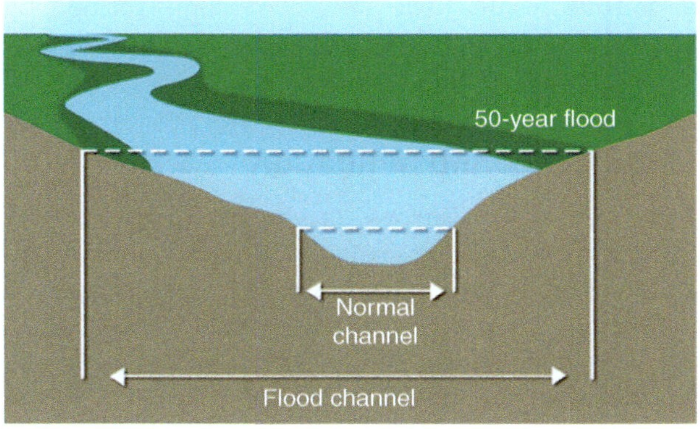

Figure 1 Example 50-year flood event.

and mobilize and transport a great deal of debris and sediment.

Rainfall: Heavy rainfall can result in flash flooding. Although thousands of hectares are frequently flooded as a result of flash flooding (for example, northern Queensland, Australia, and the upper Mississippi River drainage, USA), flash floods usually occur in small watersheds as a result of large rain events and are characterized by peak flow within six hours of the onset of rainfall. Flood conditions develop rapidly because heavy rainfall surpasses the infiltration capacity of the soil, resulting in a very high runoff rate. These types of events are generally locally intense and damage is usually restricted to a limited area. Large rivers generally remain unaffected, while smaller streams can overtop their banks, even in a drought year.

Seasonal and inter-annual variability in rainfall and flooding are often a result of El Niño-Southern Oscillation (ENSO). ENSO is a global system of ocean–atmosphere climate fluctuations arising from warmer ocean currents coupled to higher air temperatures. The result is markedly increased evaporation and large-scale interaction between the ocean and the atmosphere. Effects of ENSO are observed by changes in the distribution of rainfall, causing floods in some areas and drought in others. This process leads to drastic alterations to normal weather patterns, including heavy rains and catastrophic flooding in the United States, Asia, and other parts of South and Central America. For example, large areas of Asia receive more than 80% of annual precipitation during the seasonal monsoon season. Conversely, regions in Australia, Indonesia, and India may undergo severe drought because moisture normally dispersed around the world is evaporating too quickly and staying within the Eastern Pacific Ocean. ENSO is the most prominent known source of inter-annual variability in weather and climate around the world (range: 3–8 years).

Ice jams: Ice jams are a major cause of flooding in northern latitudes. In fact, for most northern rivers, the annual peak water levels are due to ice jams resulting from the accumulation of ice fragments, which build up and restrict the flow of water. A rise in water levels may result from spring snowmelt, or a sudden midwinter thaw accompanied by substantial rainfall, resulting in a rapid increase in water levels and severe ice jams. Ice jams can lead to flooding because of two main features. First, ice jam thickness can be considerable, amounting to several meters. Second, the underside of the ice cover is usually very rough. Therefore, the flow depth has to be much greater than for open water, resulting in relatively high water levels with relatively small discharges. This condition leads to a great deal of water and pressure that when released can lead to substantial flooding.

Outburst floods: Outburst floods are also common in northern latitudes, and have some similarities to ice jams. Lakes dammed by glaciers or moraines suddenly drain and large quantities of water, mud, and debris are released. An outburst flood typically occurs when the water level becomes high enough to actually float the ice or when a small channel forms under the ice and causes rapid melting, thus expanding the channel. Another common mechanism is overtopping of the ice dam and the rapid thermal and mechanical erosion of a channel, leading to sudden large scale drainage. The release of water is often sudden and catastrophic.

Coastal storms: Humans living along the shores of major lakes or along ocean coasts are occasionally subject to flooding as a result of high wind and wave action, or the interaction between high estuarine flows and tides. Shoreline flooding may be caused by storm surges often occurring simultaneously with high waves. Surges are caused by sudden changes in atmospheric pressure and by wind stress accompanying moving storm systems. In certain coastal regions, maximum storm surges are produced by severe tropical hurricanes. Along the coasts, severe storms can produce surges of up to 2 m, but in some areas of the world, for example, Bangladesh, severe storms can produce surges in excess of 8 m. Generally, surges in lakes and reservoirs are less, though they have been recorded as high as 2.5 m on Lake Erie. Specific types of coastal flooding events include tsunamis, cyclones, and hurricanes.

Urban runoff: Urban stormwater runoff can flood local rivers as well as the urban area itself. Urbanization drastically alters the drainage characteristics of natural catchments by increased impervious surface area and thus volume and rate of surface runoff (**Figure 2**). Other effects of increased impervious

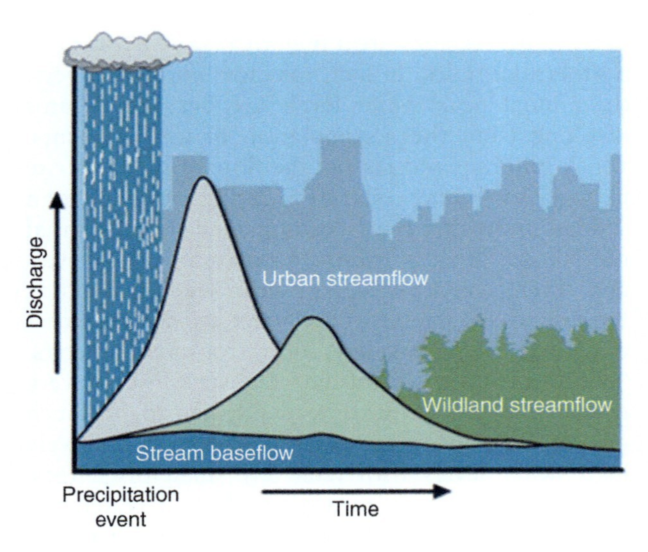

Figure 2 Example of difference between wildland (unaltered) and urban (human modified) event based runoff regimes.

surface area can include decreased water quality, changes to microclimates, habitat degradation and destruction, and diminished stream and landscape aesthetics. Although the impact of urban runoff on major river systems may be minimal, the carrying capacity of small streams may be quickly exceeded, causing flooding and channel erosion problems. Runoff from intense rainfall can exceed the carrying capacity of sewer systems, creating a backup and thus, flooding. In urban settings, streamflow-gauging stations are often used to provide continuous flow records that can be used in the design of new urban infrastructure, including roads, bridges, culverts, channels, and detention structures.

Tropical flooding: Tropical floods are usually caused by cyclones, otherwise known as hurricanes, typhoons, or tropical storms. Cyclones often result in large quantities of rain falling in a short time and can result in a great deal of flooding and sometimes human loss of life. There are two basic types of flood caused by tropical cyclones. Flash flooding occurring in streams and urban areas almost immediately following rainfall and rising water can reach depths of multiple meters. River flooding generally occurs from heavy rains coupled with recent cyclonic activity and can persist for weeks. The impact of tropical floods is locally variable. Water levels that exceed flood stage can constitute minor, moderate, and major flooding over relatively short geographic distances due to topographic variability and ability for the terrain to attenuate flood flows.

Dam failure: Flooding also results from the failure of dams or other hydraulic structures. The suddenness and magnitude of these events often have disastrous results. The failure of dams formed by beaver (*Castor canadensis*) can also result in an outburst flood of impounded water.

Flood Costs and Mitigation

Floods can be disastrous. Often, personal property, businesses, industries, crops, and roads are damaged and human lives can be lost. Floods cost humans many millions of dollars every year in property damage, lost production, lost wages, and lost business. Floods however, are also a natural phenomenon and are often necessary for ecosystem health. The concept of 'environmental flows' was developed in this case to determine how much water needs to be left in a river in order to maintain its ecological health. These flows are critical (especially in dry regions) to provide water for floral and faunal communities, as well as security for human use and socio-economic stability.

Economic development in concert with an increasing population has brought pressure altering the flow regime of surface water systems. Human efforts to constrict the active zone of floodplains that attenuate floods include the construction of dams, dykes, and concrete diversion channels, channel dredging and realignment, and drainage of wetlands. These measures, although perhaps beneficial in the interests of economic development, have resulted in the decline of fish and wildlife habitats, and the disruption of entire ecosystems. Many humans have a false sense of security owing to the size and proclaimed strength of levees protecting their livelihood. Hence, levees may increase the potential for floods by a reduction in flood preparedness and by creating incentives to build structures in areas subject to flooding.

Flood amelioration practices can include materials that are put in place to increase bank stabilization, reduce bank erosion, and stop the meandering of streams and rivers. Common strategies include dikes, loose rock paving, and establishment of plant communities on banks. Often, these practices are successful in reducing bank erosion, but hinder the ability of the channel to widen naturally through flooding, leading to increased flooding by reducing the length of the stream or river, and increasing flow velocity resulting in channel incision. The amount of sediment transported depends mainly on flow volume (**Figure 3**), which also depends on size of the drainage basin, and rate and volume of precipitation. Increases in suspended sediment concentrations above natural levels often have a detrimental impact on fish and invertebrate habitat in streams.

High levels of suspended sediment can reduce the effectiveness of drinking water treatment processes and may increase maintenance costs by clogging or

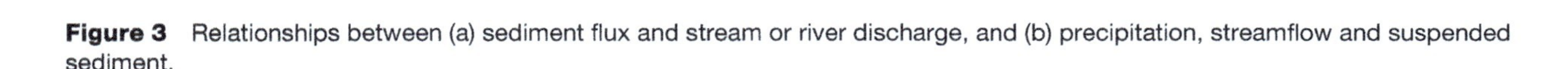

Figure 3 Relationships between (a) sediment flux and stream or river discharge, and (b) precipitation, streamflow and suspended sediment.

reducing the capacity of filtration systems. Suspended sediment and other particulates are aesthetically undesirable for domestic use and can be associated with higher bacterial concentrations. Suspended sediment carried by flood waters can reduce light penetration and temporarily decrease plant productivity in lakes and streams. The same flood waters can also transport nutrients such as phosphorus with the sediment often resulting in an infusion of otherwise nutrient limited waters.

Flood waters are a natural driving force in maintaining the productivity of rivers and floodplain systems. Floods inundate adjacent floodplains and connect river channels with streamside soils and vegetation that result in chemical and faunal exchanges that influence both communities. Peak flows that move or abrade stream substrate can scour attached algae and aquatic macrophytes and reduce or change the community structure of benthic invertebrates and fish populations. Often recovery of these communities to pre-flood conditions is quite rapid. Floods are major contributor to annual inputs of nutrients to lakes and reservoirs. Dissolved nitrogen and phosphorus, delivered with flood inflow, drive aquatic productivity and generally lake systems are more productive during wet years than during periods of drought.

Because of the intricate nature of river ecosystems, water quality sampling and analysis programs are necessary to provide data essential to the understanding and behavior of natural systems and influences of human activity. Thus, simultaneous and continuous sampling of suspended sediment is critical for accurate understanding of pollutant loading of the watershed which is (a) governed by hydrological processes, and (b) closely correlated to chemical pollutant concentrations. These processes are controlled to a large degree by local microclimates.

Ultimately, flooding is a necessary natural process that only has negative effects when humans are involved. Human inhabitation of land that naturally floods causes great losses of life and property. Anthropogenic alterations to these flood-based systems have only worsened the problem in most cases.

See also: Hydrology: Streams.

Further Reading

Junk WJ (2005) Flood pulsing and the linkages between terrestrial, aquatic and wetland systems. *Verh. Internat. Verein. Limnol.* 29: 11–38.

Mays LW (1996) *Water Resources Handbook*. NY: McGraw-Hill.

Novotny V and Olem H (1994) *Water Quality. Prevention, Identification, and Management of Diffuse Pollution*. New York: Van Nostrand Reinhold.

Wanielista M, Kersten R, and Eaglin R (1997) *Hydrology: Water Quantity and Quality Control*. 2nd edn. NJ: Wiley.

Ward AD and Elliot WJ (1995) *Environmental Hydrology*. Boca Raton, FL: CRC.

ECOLOGY OF FLOWING WATERS

Contents

Streams and Rivers as Ecosystems

S G Fisher and R A Sponseller, Arizona State University, Tempe, AZ, USA

Stream Ecosystem Definition

Streams and rivers (hereafter simply referred to as streams) are delightful places in their own right and have attracted over a century of concerted research attention by biologists. Aquatic insects, which dominate the invertebrate fauna of streams and possess intriguing adaptations to current, are favorites of entomologists and naturalists. Popular sport fishes such as cold water trout and salmon thrive in small streams and have justified a vigorous research effort on headwater streams motivated by fisheries science. Until the late 20th century, the stream provided an arena for ecological research – on bugs and fish and their interactions – but these studies did not embrace streams as integrated whole ecosystems. The reasons for this are inherent in the very definition of ecosystems.

Ecosystems are bounded systems containing interacting biotic and abiotic components. It is the definitional element dealing with boundaries that presents the most significant problem in conceptualizing streams as ecosystems. First, streams are long and thin and have a very large surface to volume ratio. As a result they are open to exchanges with the adjacent terrestrial system – not only to exchanges of water, but also of materials dissolved and suspended in water, to particulate materials such as autumn-shed leaves entering from the adjacent riparian zone, and to organisms such as flying insect adults, drifting algae and invertebrates, and migrating fishes. While all ecosystems have boundaries that are relatively closed, streams are among the most open of all ecosystems. Second, stream segments have no natural length, that is, no starting or stopping point and thus have to be dealt with as reaches or segments with indefinite and elusive upstream and downstream boundaries. By necessity, stream ecosystem segments are of arbitrary length. Length influences the ratio of water flow to volume, which is thus correspondingly arbitrary. This is not true of lakes that have distinct boundaries and usually well-defined inlets and outlets. Furthermore, lakes have a longer hydrologic residence time (volume/water flux per time) than do streams that typically turn over (replace water volume) very rapidly. Because of this physical characteristic, fluxes of organic matter and nutrients tend to overwhelm processes of production and transformation that occur within ecosystem boundaries of streams.

Early studies of streams as ecosystems were of a series of Florida springs, most notably Silver Springs, by Dr. Howard T. Odum (1924–2002) in the mid-1950s. Dr. Odum measured inputs and outputs or organic matter to Silver Springs as well as biological productivity and energy use by respiration within the system's boundaries. Boundary designation was facilitated by the fact that these systems boiled up out of the ground at a high rate at a well-defined point – the springhead. Water then flowed in a confined channel at a relatively constant rate until it entered a larger downstream river. Input and output points were anything but arbitrary. Odum was able to construct an organic matter budget for this somewhat atypical running water ecosystem.

Meanwhile, more typical small forest headwater streams were studied from another angle. Autumn-shed leaves enter streams in large amounts and represent an important source of food for a complex suite of consumers, largely invertebrates and particularly immature aquatic insects. Showing how these particulate materials crossed the stream-ecosystem boundary and then were processed by a complex but stereotypical network of organisms and activities, underlined the distinctness of the stream as an ecosystem and the existence of a well-organized community (the biotic component) dependent upon allochthonous materials. These studies revealed streams to be open, to be heterotrophic (i.e., dependent upon imported energy), and to function as a unit – the ecosystem.

The Hubbard Brook Ecosystem Study initiated by Gene Likens and Herb Bormann in the 1960s promoted the use of small watersheds as ecosystems. Watersheds consist of a terrestrial (e.g., the forest) and an aquatic element, the small stream draining the catchment. Small watersheds are hydrologically defined and material input–output budgets of elements, ions, and simple compounds can be constructed by carefully comparing inputs and outputs and inferring intrasystem processes from them. Stuart Fisher and Gene Likens applied the conceptual framework of the small watershed chemical budget to the catchment stream alone and constructed not ion budgets, but energy (organic matter) budgets for the stream ecosystem (**Figure 1**). In this scheme, the stream ecosystem was taken as a linear 1700 m reach. Inputs occurred via leaf fall, by transport in at tributary junctions, and via subsurface groundwater flow. Outputs occurred at the downstream terminus of the reach and, since this was an energetic study, by degradation of organic matter via respiration by consumers – bacteria, insects, fish. Respiration is the biological breakdown of complex organic molecules to inorganic carbon dioxide. This was a landmark study in stream ecosystem science in that it dealt with all inputs and outputs with respect to an explicitly bounded stream ecosystem in space and time – time being an annual cycle in this case.

Stream versus Lake Ecosystems

Streams and lakes together comprise the vast majority of freshwater ecosystems, but they differ from each other in fundamental ways. Streams are lotic ecosystems, characterized by flowing water oriented along an ecosystem axis. Streams and rivers thus have an inexorable spatial orientation and are characterized by material transport in space. Water currents of many types occur in lentic ecosystems (lakes), but none is as dominant as river currents. In lakes, important vertical transport processes mediated by gravity and diffusion occur. Longitudinal movements (e.g., wind-driven currents) occur in lakes but these are not as striking as they are in streams. Stream currents are by definition competent to transport stream bed and bank materials (sediments) and to shape the stream channel but, except for wave action in large lakes, this is not true in lentic ecosystems. By virtue of flow, streams are turbulent and well mixed and have little vertical structure. Lakes, on the other hand, are typically physically and chemically stratified during much of the year. Stratified lakes may experience low oxygen (and other gas) concentrations at depth while streams are better aerated and seldom anoxic, unless polluted. A stream's turbulent air–water

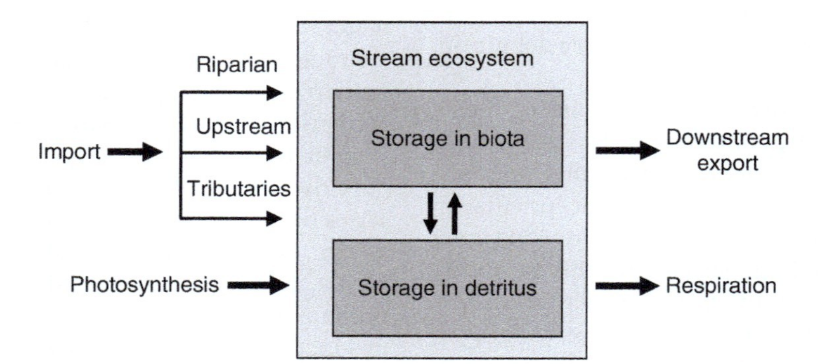

Figure 1 Compartment model of a stream ecosystem showing major input and output fluxes. Conceptual models such as this exhibit no explicit internal heterogeneity and apply to a time period that must be specified (e.g., a year). Sequences of these budgets in space or time can show longitudinal trends and temporal change, but are seldom assembled.

interface facilitates diffusion and mixing of gasses from the atmosphere, into the water column, and to the bottom.

The stream surface is turbulent because water is flowing over the surface of the earth in response to gravity. An obvious but important consequence of this is that streams and rivers slope while the surface of lakes is horizontally level. River ecosystems, if they are long enough, may exhibit longitudinal pattern owing to this elevation change. Lakes never do. One consequence of this is that turbulent mixing and longitudinal transport in streams precludes the development of a true plankton community in all but the slowest (often largest) rivers. Stream communities are benthic (bottom associated). And finally, lakes and streams act differently in the larger landscapes of which they are a part. Lakes are relatively short-lived ecosystems that accumulate and retain materials, ultimately resulting in their eradication by infilling. Streams and rivers erode, transform and deepen. While rivers may ebb and flow and change course constantly, they are among the oldest continuously existing ecosystems on earth. These physical differences provide a template upon which biological ecosystem processes play out and render these two freshwater systems quite different in terms of their internal dynamics and their interactions with adjacent ecosystems of all types.

Ecosystem Boundaries and the Structural Components of Streams

The all too common perception of the stream as a ribbon of water in a sea of land is, unfortunately, oversimplified. Stream ecosystem boundaries in all three spatial dimensions are indistinct and their conceptual resolution is extremely important to understanding not only streams, but the landscapes of which they are a part (**Figure 2**). The ribbon of water analogy depicts streams as extending from bank to bank, surface to bottom, and for some undefined distance downstream. This definition describes what might be called the surface stream or wetted perimeter – an all important component of the stream ecosystem, but just one of many. Concrete-lined canals, such as the irrigation ditches of the American Southwest, consist only of this surface stream component. Other streams are more complex.

We know now that the surface stream flows in and out of the stream sediments, sometimes to several meters depth where benthic particles are coarse and sediments are thus porous and permeable. Even streams with finer sediments exhibit vertical hydrologic exchange. This area of subsurface through-sediment flow is the hyporheic zone. Water enters the hyporheic zone at downwelling areas and returns to the surface stream at upwelling zones. Down and upwelling zones are connected by subsurface flowpaths, which may be centimeters to kilometers in length. Up and downwelling zones and connecting flowpaths pervade the stream; they are not isolated oddities. Just as water is flowing continuously in the surface stream, so is it flowing continuously in the hyporheic zone, albeit more slowly. The hyporheic zone provides a habitat to the hyporheos, a community of invertebrates adapted to this environment. Bacteria and fungi are abundant there, and organisms such as fish may use the environment periodically

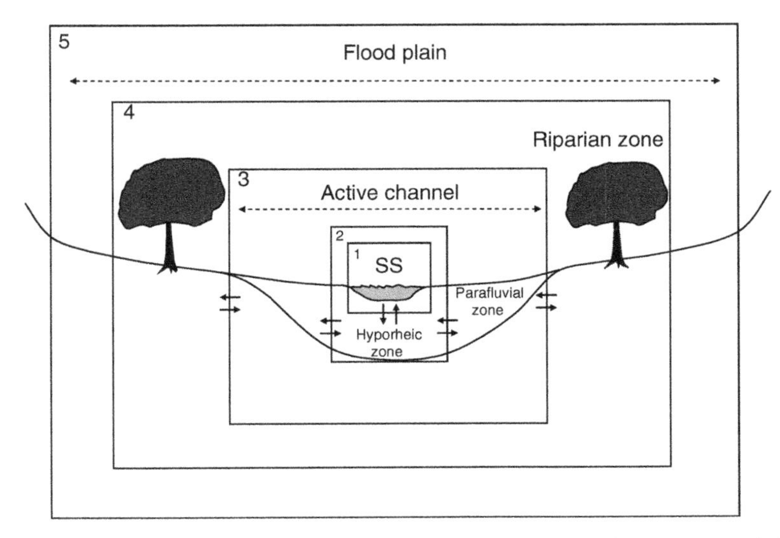

Figure 2 Cross section showing spatial elements of the stream landscape ecosystem. Compartments 1–5 are increasingly inclusive but all represent legitimate ecosystem units. Arrows are examples of flowpath connectivity involving water, materials, or organisms among subsystems.

(aestivating lungfish, developing salmonid eggs and fry). Particulate and dissolved nutrients and gasses move through the hyporheic zone where substantial transformations can occur, and may markedly influence life in the surface stream. In many streams, upwelling water is high in nutrients, and upwelling zones harbor dense communities of algae, unlike downwelling zones. Vertical connections are likely to be important in all but those few streams that flow on bedrock. The lower boundary of the hyporheic zone can be bedrock or a clay lens or it can be indistinct, fading away in influence with depth. The ecosystem concept can accommodate considering the hyporheic zone a separate ecosystem or a subsystem of the larger stream-ecosystem complex. Stream ecologists are increasingly inclusive, and treat the stream as a heterogeneous landscape of interacting patches. This, of course, depends on the scientific question at hand.

The upper vertical limit of the stream is often taken as the water surface. This is easily recognized even though its location may rise and fall in response to floods. Many stream invertebrates are insects that emerge as adults, mate above the stream, and oviposit in it later. During this flying stage, adult insects may move up or downstream. Ovipositing and spent adults become food for fish as they re-enter the surface stream. Because the aerial corridor above the stream is an essential part of the ecosystem, the upper boundary of the stream may properly be located several meters above the water surface – perhaps at the top of the riparian canopy. Incidentally, this is also the zone of first exchange of thermal energy and atmospheric gases. While it is unlikely that gas concentration in the aerial corridor, being part of a well-mixed atmosphere, would either reflect or influence stream metabolic activity, temperature and humidity might do so.

The simplest lateral boundary for the stream ecosystem is the wetted perimeter – the water's edge. Water levels fluctuate however, so even this simple definition requires a model (and measurements) that allows the ecosystem to change size and occasionally disappear (during drought). The edge of the active channel is an alternative boundary. The active channel includes the wetted perimeter, if any, and the parafluvial zone, a subsystem analogous to the hyporeic. The parafluvial zone is connected laterally to the hyporheic and through it to the surface stream, but the parafluvial differs from the hyporheic in having no up- or downwelling zones and no vertical exchange with the surface stream.

Beyond the parafluvial is the riparian zone, often overlain by a grass, shrub, or tree community, which has an underlying saturated zone that is connected hydrologically with the more medial subsystems. The riparian zone differs from the upland vegetation community because of its proximity to the other stream subsystems. This distinctness may be striking in arid lands and subtle elsewhere. A floodplain may be present beyond or overlapping with the riparian zone. Subsurface exchanges connect saturated sediments beneath all stream elements. The direction of flow may be toward or away from the surface stream. In a downstream direction, subsurface water may move toward or away from surface water in gaining or losing reaches respectively, and may have a vertical component as well, recharging regional groundwater aquifers.

Although the ecosystem concept is sufficiently robust to consider any of the described subsystems to be ecosystems in their own right, interaction is intense and all subsystems are involved in the directed movement of water across the landscape. Boundaries between subsystems are relative discontinuities, but they are still quite open, and it is difficult to understand the structure and functioning of any one subsystem divorced from the others. Increasingly, stream ecologists consider the stream ecosystem to incorporate the landscape strip from floodplain edge to floodplain edge. Longitudinal boundaries are ultimately arbitrary, as discussed earlier. A length sufficient to include the spatial heterogeneity representative of a longer reach is usually selected so scaling up can be done reasonably. Downstream changes in ecosystem properties limit the application of this criterion, however.

Ecosystem Metabolism: Energy Flow

Ecosystems are bounded systems containing interacting abiotic and biotic components. We have discussed the unusual boundary attributes of streams. What about interactions? Interactions (between biotic and abiotic components) in the province of ecosystem science include energy flow and nutrient cycling. While there are other interactions involving organisms and populations (e.g., mating behavior, predator-prey responses, territoriality) these activities occur in ecosystems but are more properly in the realm of individual, population or community ecology rather than being attributes of ecosystems per se. Energy flow refers to rates and patterns of production and degradation of organic matter within ecosystem boundaries and transport of organic energy across boundaries – in or out.

Streams vary greatly in latitude, climate, and in the land use and land cover of the catchments they drain. Despite this variety, central paradigms for stream ecosystem metabolism have been developed, largely from studies of small forest streams in North America. While these generalizations are useful, they must be modified to apply to any given stream ecosystem under consideration. Streams are open ecosystems and receive most of their organic energy from autumn-shed leaves of terrestrial trees. These same leaves shade the stream during the warmer seasons, reducing in situ photosynthesis. Thus small streams are dark and dependent on allochthonous organic detritus from adjacent terrestrial ecosystems. Only when the canopy is open is photosynthesis by algae appreciable, thus primary production (P) is low. Despite low production, algae can be an important part of consumer diets because of the fatty acids and other nutrients they provide.

Consumers in streams are organized in food webs based on available inputs, but among invertebrates, detritivores outnumber herbivores greatly. Stream ecologists have found it useful to organize consumers in functional groups based not on what they eat, but how they eat it. Shredders, such as some immature stoneflies and caddisflies, are able to macerate and ingest large particles directly. This is the first step of organic matter processing which involves a sequence of consumers. Shredders produce finer particles available to collector-gatherers and collector-filterers, which differ by whether they gather particles from the stream bottom or from suspension in moving water. Scrapers feed on biofilms, including algae, on substrates. During processing, dissolved organic matter is released to solution and becomes available to bacteria. There is some evidence that dissolved organic matter can also coalesce to form particles by chemical aggregation. In either case, dissolved organic matter is repackaged and is accessible to particle feeders, just as are leaf fragments.

The process by which leaves are made available to consumers is mediated by bacteria and fungi in all stages. These microorganisms help break down leaves and increase their palatability to consumers. They also can extract inorganic nutrients from solution in stream water and thereby enhance the nutritional value of the decomposing organic matrix, especially with respect to protein and nitrogen which are quite low in autumn-shed tree leaves. Consumers of leaves and leaf breakdown products not only benefit from the activities of these microconsumers, but they also consume them in large quantities.

A variety of predators, both invertebrates and vertebrates such as salamanders and fish, fill out the trophic web. In streams, as in many ecosystems, fidelity to one kind of food is rare. Opportunism and omnivory abound. Recent studies have shown that collectors ingest a large fraction of total caloric intake in the form of animals and animal fragments, and exhibit a certain capacity to sort and select particles rather than ingesting in proportion to environmental abundance.

However food webs are organized, an important consequence for ecosystem metabolism is their collective respiration (R); that is, degradation of organic matter to inorganic raw materials including CO_2, and metabolic heat. Organic matter that enters the stream either through in situ photosynthesis (P) or import (I), for example, of dead leaves, represent total ecosystem input. Input that is not respired can be stored or exported downstream (E). In many ecosystems (e.g., lakes and bogs), storage can be high, but in small streams net storage tends to be zero on the time scale of a year or two. These considerations allow construction of an organic matter budget for the stream ecosystem and description of the system holistically, for example calculation of its P/R ratio or its overall efficiency in processing energy as a whole (ecosystem efficiency $= R/(P + I)$).

Material or energy budgets of ecosystems summarize the results of a great complexity of interactions in space and time. The ability to provide this holistic view is a great strength of the ecosystem approach. Budgets such as the one described above are constructed for a finite time period, e.g., one year. An annual budget does not show seasonal dynamics however, thus some resolution is lost by using this approach. There is often significant year to year variation in stream organic matter budgets because fluvial import and export are greatly affected by hydrology (rainfall, runoff, stream flow) which are notoriously variable.

It is also essential to couch budget construction and analysis in terms of an explicit definition of spatial boundaries. The budget discussion above focused on the surface stream as the ecosystem of interest. However, if boundaries were instead extended into the hyporheic zone, respiration would rise because of the added metabolically active volume of sediment and organisms. Photosynthesis would stay the same however because the hyporheic zone is dark and no photosynthesis occurs there. If we were more inclusive and considered the riparian zone to be part of the stream, that would mean that leaf litter is not allochthonous, but is rather in situ productivity. P/R would rise dramatically as a result of this boundary shift and trees would have to be considered stream organisms. Ecosystem science is very flexible in letting the research question dictate where boundaries are drawn in space and time. However, once a decision is made, boundaries must be specified precisely because the consequences of boundary siting for ecosystem properties are enormous.

Longitudinal Patterns

Budgets usually apply to a linear stream segment of a few kilometers in length and thus describe the stream at what is essentially a point in space. In reality, streams may flow for hundreds or thousands of kilometers from their headwaters to the sea and experience remarkable changes as they transition from small brooks to continental rivers. In order to show longitudinal change using a budget approach, a series of linearly arrayed but discrete budgets would be required. This approach is cumbersome, labor intensive, and generally is avoided by stream ecologists. It is well known that streams change greatly in physical and chemical attributes and that the kinds of organisms found in headwaters (e.g., trout) are replaced by large river organisms downstream (bass, carp, catfish). Even at a smaller scale, oxygen sag in response to point inputs of organic matter (e.g., sewage outfall) is a prime example of the need for a continuous model to describe stream ecosystems.

The River Continuum Concept (RCC), developed by a group of collaborating stream ecologists led by R.L. Vannote, describes continuous changes from small headwater streams to large rivers (**Figure 3**). The RCC describes changes in kinds of organisms present and in ecosystem properties (e.g., P/R). These changes are attributable to two major factors: (1) changes in local conditions (e.g., temperature, stream width, slope, light availability at the stream bottom, and amount of leaf litter input), and (2) changes in the nature of transported materials from upstream (e.g., downstream increase in fine particles and a decrease in coarse debris such as leaves).

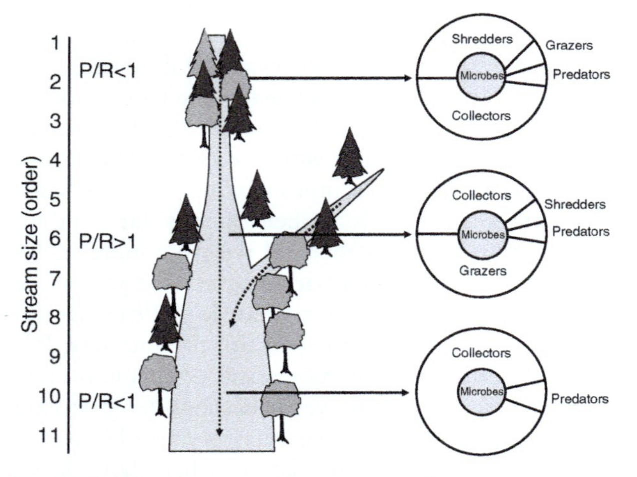

Figure 3 River Continuum Concept showing how ecosystem properties (Photosynthesis/Respiration; P/R) change downstream as the stream widens. Consumer communities, illustrated by pie charts, change as well as available food changes downstream.

A series of elegant predictions about how organisms respond to these changes accompany the RCC and have been found to hold in many temperate North American and European streams. For example, shredders wane in abundance in a downstream direction as whole leaves become less abundant and filtering collectors increase with a rise in suspended particulates. As streams widen, allochthonous inputs decrease and in situ photosynthesis by algae and higher plants increase in the wider, lighter stream. In response to this, the P/R ratio increases in the surface stream. Exceptions to predicted patterns exist where the terrestrial context differs such as in arid lands and grasslands or where streams are dammed or otherwise manipulated by humans. The main point of the RCC holds however: stream ecosystems exhibit continuous change in a downstream direction. Site characteristics and upstream-downstream linkages both contribute to this pattern.

Ecosystem Metabolism: Biogeochemistry

The chemical composition of stream water at any point is a consequence of precipitation input, dissolution of inorganic materials as rainwater finds its way through soils and into stream channels, and changes effected within the stream due to biological uptake and release and to concentration or chemical precipitation of salts (e.g., by evaporation in streams of arid lands). Water chemistry changes continuously in a downstream direction just as do organic matter and the biologic communities using organic matter.

The elements nitrogen and phosphorus can limit the rate biological activity in streams. The bacteria and fungi involved in leaf decomposition for example, can be limited by available nitrogen (nitrate and ammonium). Added nitrogen can accelerate leaf decomposition and, since nitrogen is at such a premium to decomposers, bacteria and fungi can remove nitrogen from solution and cause a downstream decline in transported nutrients as well. Photosynthetic plants in larger, well lighted streams, can also be limited by nutrients. Their uptake of these materials (usually ions of N and P) can cause a downstream decline (or prevent a downstream rise) in these elements. The popularly held belief that streams purify water is based on this kind of process.

Organisms of all ecosystems acquire the elements they need from their environment and release these same elements as metabolic products during life, or as decay products upon death. Such is the essence of nutrient cycling. Streams are no different from other ecosystems except that released elements are subject to downstream transport by virtue of stream

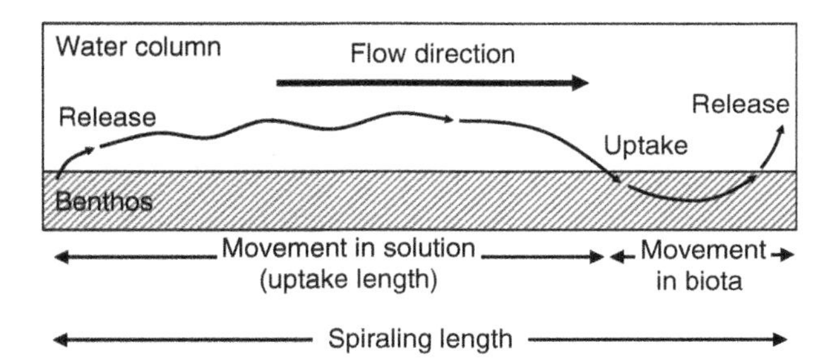

Figure 4 Nutrient Spiraling Concept. Arrow in diagram shows trajectory of a nutrient atom (e.g., nitrogen or phosphorus) as it moves downstream, alternately in solution and in biota. Spiraling length is an index of how tightly nutrients are cycled and how retentive the stream is with respect to transported materials. Short spiraling lengths usually indicate high rate of biologic activity.

flow. Nutrient cycles in streams are thus stretched in space. Instead of cycling, river nutrients spiral (**Figure 4**). In its simplest form, a nutrient spiral consists of two parts: (1) distance moved by the element dissolved in water from release point to point of benthic reuptake, and (2) distance moved while incorporated in the benthic component. The second element of spiraling length is usually quite short compared to the first, thus the first component, termed uptake length, is the critical descriptor of nutrient spiraling in streams. Uptake length is an ecosystem property that varies with stream hydrology, channel morphology, and the biotic status of the system. For example, a limiting element such as nitrogen or phosphorus in a stream choked with algae would likely have a short uptake length (a few meters) while a raging torrent over bedrock would exhibit long uptake lengths (hundreds of meters).

The nutrient spiraling concept is an important conceptual advance for stream ecology. Like the river continuum concept, it transformed a static budget approach derived from terrestrial ecosystem studies, to a more realistic and powerful conceptual approach that is better suited to the unique, spatially distributed stream ecosystem.

Temporal Change

We have seen that spatial boundaries of streams present several conceptual difficulties – what to include as ecosystem components, how to deal with continuous longitudinal change, and how to resolve spatially distributed cycling. Temporal boundaries for stream ecosystems are also problematic. Annual organic matter budgets for small streams in mature forests have been commonly used but these vary from year to year and fail to show details of seasonal responses. Thus, the annual budget is both too long and too short. Furthermore, forests change by succession over a 100–200-year time period, and streams draining these forests are likely to change too. At larger temporal scales, geologists use the erosion cycle to characterize mountain building and degradation that encompass millions of years. While stream ecologists do not normally study ecosystems in that time frame, limnologists do study lake history from formation to extinction (by sedimentation) using paleolimnolgic tools. What legacies over their entire history are still evident in today's rivers and streams? Surely the species present reflect this ancient past and the temporal trajectory of the ecosystem in which they evolved. Do ecosystem properties reflect this long term history as well?

At the other extreme, many streams may respond to events lasting less than a year. A variety of disturbance events such as storm flows, drought-induced drying, chemical spills, and forest fires can have a marked effect on stream ecosystems. Some of these events are of regular occurrence and may be agents of natural selection in species populations of these ecosystems. Others may be rare, one of a kind, or human induced. In any case, stream ecosystems can be devastated by disturbance and bounce back subsequently just as a forest regrows after fire. When disturbance is a significant influence on stream ecosystems, temporal boundaries might be selected to reflect this. Just as the stream was viewed spatially as a string of discrete, segment-specific budgets, so can it be viewed temporally as a string of discrete successional sequences. This approach works well in the desert Southwest of North America where flashfloods and drought alternately affect streams.

Some streams appear to be benign with little exogenous disturbance, while others are regularly disturbed, some by multiple agents of disturbance acting simultaneously. Just like with spatial boundaries, temporal boundaries can be applied flexibly. The stream ecosystem can be defined as a single riffle for a day or as 100 km of a stream-floodplain landscape for a century. Which is best depends on the ecological question at hand.

Conceptual Horizons

Several conceptual advances are in progress in stream ecosystem ecology. Perhaps the most important is to treat streams not as linear systems, but rather as branched convergent networks, which is their true shape. This will require a new set of tools to deal with continuous downstream network change and to resolve material spiraling in dendritic systems.

Another opportunity for ecosystem understanding of rivers is to apply the principles of landscape ecology to streams. We have already acknowledged that streams are heterogeneous and are compopsed of several interacting components (patches). What we are just beginning to learn is how different shapes and configurations of these components affect whole ecosystem function.

Finally, stream ecosystems are spatially oriented systems in structure and functioning owing to hydrologic flow. The path followed by water and its transported load of materials is vital to stream ecosystem functioning. All ecosystems experience flow paths to some degree. Water is often the agent of spatial translation, but other mechanisms are available (gravity, migrations). All ecosystems are involved in processing materials in place and in exchanging materials with adjacent ecosystems. Although streams may be extreme examples of ecosystems dominated by the exchange component, all ecosystems are influenced by flowpaths and thus studies of streams can help us understand ecosystems of all types everywhere.

Glossary

Allochthonous – Organic matter generated outside the system of interest, imported across boundaries, and used inside.

Autochthonous – Material originating within the system of interest.

Catchment = watershed – A hydrologically defined region which receives the precipitation that ultimately drains the region by stream flow.

Detritivore – An organism that eats dead organic matter in varying stages of decay.

Flowpath – A spatial vector that describes the path followed by water and other materials across a landscape.

Functional group – Organisms lumped together based on how they consume food (e.g., shredders, collector-gatherers).

Hyporheic zone – The zone of saturated sediments beneath the stream bottom.

Lentic – Standing water ecosystems such as lakes and ponds.

Lotic – Running water ecosystems such as streams and rivers.

Omnivore – An organism that eats a variety of foods, e.g., animals and plants.

Parafluvial – A stream ecosystem subcomponent located lateral to the stream edge and including the saturated sediments beneath.

Riparian – The stream edge subsystem distal to the parafluvial and usually supporting terrestrial vegetation such as shrubs and trees.

River continuum concept – A synthetic concept describing the longitudinal change is the structure and functioning of lotic ecosystems.

Succession – The sequence of changes occurring in an ecosystem after a disturbance such as fire or flood.

See also: Algae of River Ecosystems; Benthic Invertebrate Fauna, Small Streams; Ecology and Role of Headwater Streams; Hydrology: Streams; Regulators of Biotic Processes in Stream and River Ecosystems; Streams and Rivers as Ecosystems.

Further Reading

Allan JD (1995) *Stream Ecology.* London: Chapman and Hall.

Fisher SG and Likens GE (1973) Energy flow in Bear brook New Hampshire: An integrative approach to stream ecosystem metabolism. *Ecological Monographs* 43: 421–439.

Gibbs RJ (1970) Mechanisms controlling world water chemistry. *Science* 170: 1088–1090.

Hynes HBN (1970) *The Ecology of Running Waters.* Toronto: University of Toronto Press.

Merritt RW and Cummins KW (1984) *An Introduction to the Aquatic Insects of North America.* 2nd edn. Iowa: Kendall/Hunt.

Newbold JD, Elwood JW, O'Neil RV, and Van Winkle W (1981) Measuring nutrient spiraling in streams. *Canadian Journal of Fisheries and Aquatic Sciences* 38: 860–863.

Odum HT (1957) Trophic structure and productivity of Silver Springs. *Ecological Monographs* 27: 55–112.

Vannote RL, Minshall GW, Cummins KW, Sedell JR, and Cushing CE (1980) The river continuum concept. *Canadian Journal of Fisheries and Aquatic Sciences* 37: 130–137.

Algae of River Ecosystems

R J Stevenson, Michigan State University, East Lansing, MI, USA

Introduction

Algae in rivers occupy two distinct habitats, the benthos and water column. Benthic algae are algae attached to or associated with substrates in streams and rivers. (The term benthic algae is used to refer to microscopic and macroscopic algae on or associated with substrates. Thus, periphyton and biofilms (microphytobenthos), macroalgae (microphytobenthos), and metaphyton are benthic algae). Phytoplankton are suspended in the water column. The amount of these algae and their function and biodiversity varies greatly among different types of streams and rivers and with time. River ecosystems, broadly defined, include all flowing water sections within a watershed and may be bordered or interrupted by lakes, reservoirs, and wetlands. Groundwater connections are also important parts of river ecosystems that affect algae. Algal biomass is low on substrates in many shaded mountain headwater streams or when suspended in large rivers during high flow, but biomass can be very high in benthos of open canopy streams draining fertilized lands or as phytoplankton during slow flow, summer conditions. Correspondingly, photosynthesis, respiration, and nutrient transformation by algae increases on an areal or volumetric basis with increasing amounts of algae in the habitat, but cell-specific metabolism often decreases with increasing algal biomass due to competition.

Taxonomic and morphological diversity varies among habitats and seasonally within rivers as well as among rivers with different environmental conditions. Species composition varies in most river ecosystems from diatom-dominated benthos and phytoplankton during spring to a varying diversity of green algae and Cyanobacteria in warmer waters. However, a wider diversity of almost all kinds of algae, euglenoids, reds (Rhodophyta), crytomonads, chrysophytes, xanthophytes, dinoflagellates, and even browns (Phaeophyta) can be found in rivers. Size of these organisms varies from less than 5 μm for small diatoms and chrysophytes in the plankton to meters in length for some filamentous green benthic macroalgae. Morphologically, unicellular, colonial, and filamentous algae occur in both benthic and planktonic habitats. Both unicellular and colonial flagellates use flagella to move vertically in the plankton. Unicellular diatoms may use their raphe, and filamentous cyanobacteria move through sheaths in benthic habitats. Many planktonic species rely simply on the water current or periodic resuspensions to maintain sufficient exposure to light to survive.

Despite the great variation in algae of rivers and the complexity of factors that affect them, we have learned much about the environmental factors that regulate the processes that control algal biomass, diversity, and function. For example, floods and grazing invertebrates can reduce algal biomass, reduce primary production, and shift species composition to flood- or grazer-tolerant organisms. Whether or not floods and grazing invertebrates are important in river ecosystems is determined by the climate, geology, and resulting hydrology of the ecosystem. In this article, I will use a hierarchical conceptual framework for the factors that affect algae in rivers and streams. This framework will help organize the interrelationships among factors that regulate the complex patterns of algal biomass, function, and biodiversity, and thereby, will help predict patterns of spatial variation in these algal attributes among ecosystems and seasonal variation within these ecosystems. Ultimately, this broader conceptual model will help us predict longer-term changes in algae of river ecosystems and how human activities can be managed to solve and prevent environmental problems.

Algae are important elements of river ecosystems and important determinants of the goods and services provided by rivers. Algae are important in the food web for primary production, biogeochemical cycling, and habitat formation and alteration. Algae support human well-being by producing food and cleaning water for drinking. Excess algae in rivers cause many problems by altering physical habitat and depleting dissolved oxygen supplies. This affects biodiversity and productivity of rivers. Algae can also cause taste, odor, and toxicity problems in drinking water supplies and foul the pipes and filters of water users. So understanding algal ecology is important for managing river ecosystems to protect the goods and services that rivers provide.

Factors Affecting Algae in Rivers

Both benthic algae and phytoplankton in rivers live in highly variable habitats. This makes relation of pattern and process difficult. If we focus our attention on algae at a specific location, five fundamental processes affect how much algae occur in that location (**Figure 1**). Immigration of cells or groups of cells

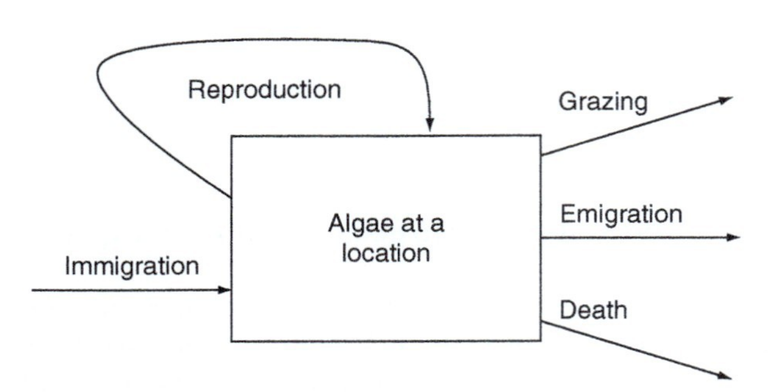

Figure 1 Five fundamental processes determine the biomass and species composition of algae at location.

into the space provides the initial colonists as well as ongoing replacement for cells that are continuously leaving the habitat via loss factors, such as grazing, emigration (drifting and sinking), and death. Reproduction by cells within that location, as well as immigration, affects accrual positively. These fundamental processes determine the structure of algal communities in rivers, the biomass of individual species as well as all algae at a location, and therefore species composition. Many other biological processes, such as photosynthesis, respiration, and nutrient uptake, are considered as functions. Many physical processes, such as eddy mixing and diffusion, affect nutrient availability and sinking rates (equivalent to emigration from plankton and immigration to benthos). Processes such as eddy mixing and diffusion are considered as indirect processes because they regulate fundamental processes, in this case, reproduction and emigration rates. Finally, both abiotic and biotic factors either directly or indirectly regulate the five fundamental processes. These include light intensity and duration, nutrient concentration, and density of grazers, bacteria, and viruses that have direct effects and flood frequency, climate, and geology that have indirect effects. Many patterns are possible because they depend on the relative magnitude of each of the biological processes and their many direct and indirect environmental determinants. Relating patterns in algal biomass and species composition to the biological processes and environmental factors in rivers require, in most cases, application of nonequilibrium models in algal ecology.

Factors that regulate algae in rivers operate at different spatial and temporal scales (**Figure 2**). The processes of accrual and losses of cells at a specific location operate at the cellular or local spatial scale and reflect the biological responses to other ecosystem factors that are regulated at larger spatial and temporal scales. Abiotic factors that directly affect algae include both resources and stressors, such a nutrients and light versus pH, salinity, shear stress, and heavy metals. Biotic factors can also be classified as positive or negative, and include commensalistic and mutualistic interactions as well as competition, predation, disease, and allelopathic interactions. At the habitat scale, riparian canopy, current velocity, and substrate presence and size affect algae, but primarily by mediating the direct biotic interactions, resources, and stressors. At watershed and regional scales, climate and geology ultimately regulate land use, hydrology, and geomorphology of rivers. They also regulate species biogeography and their availability to colonize rivers. The spatial and temporal hierarchy of these factors will regulate the complexity of possible interactions and facilitate prediction of local conditions and algal structure and function in rivers.

Local Abiotic Factors

Resources

Algal reproduction can be regulated by light, nutrients, and space. A variety of metabolic processes have similar functional responses to light intensity and nutrient concentration. Metabolic rates increase relatively rapidly in response to light and nutrient increases at low levels of resource availability, but eventually saturate at high levels of resources and respond little to further increases (**Figure 3**). Light has the possibility of having an additional negative effect at very high intensities because of photoinhibition processes. Light intensity strongly regulates photosynthetic rates and carbohydrate synthesis. Nutrient concentrations regulate their uptake rates and use in protein and lipid synthesis. The differential effects of light and nutrients on algal metabolism have been used to explain the paradox of how algae in thin biofilms can reproduce in very low light as fast as in very high light (10–1000 μm quanta $m^{-2} s^{-1}$), yet photosynthetic rates of algae are known to respond greatly over the same range in light conditions.

Variation in light and nutrients does affect algal biomass, metabolism, and species composition in streams and rivers. High light intensity and nutrient concentrations are required for accrual of high algal biomass on substrata or in the water column. Although reproduction of algal growth saturates at low light and nutrient levels, self-shading and reduced nutrient transport (because eddy mixing and diffusion are less than nutrient uptake rates) limit resource supply at the local scale as algae accumulate on substrates. For example, phosphorus limitation of thin biofilms of diatoms requires concentrations lower than 5 $\mu g\,l^{-1}$ PO_4–P, whereas 30 $\mu g\,l^{-1}$ is required to saturate growth of thicker mats. As algal biomass increases, more light

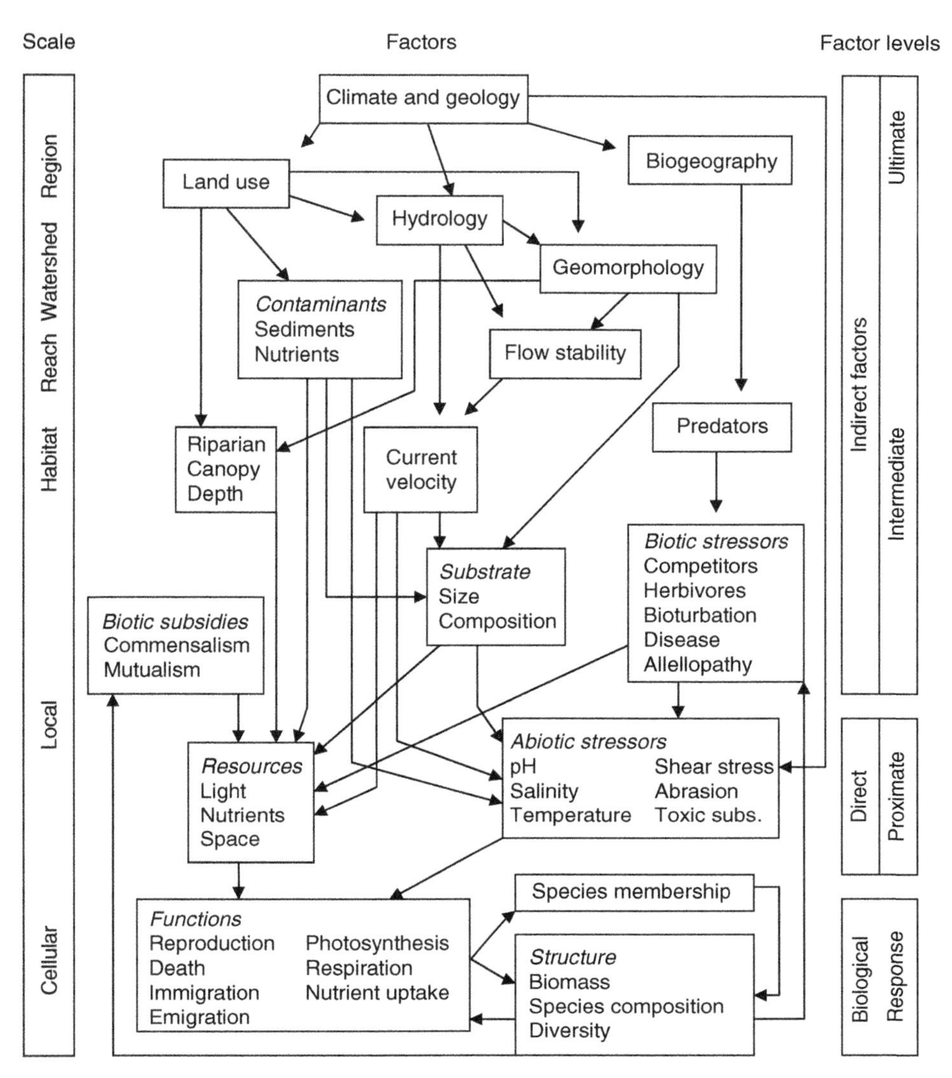

Figure 2 Hierarchical interrelationships among proximate, intermediate, and ultimate determinants of benthic algal structure and function. The relative spatial scale is shown at the left. Modified from Stevenson RJ (1997) Scale-dependent causal frameworks and the consequences of benthic algal heterogeneity. *Journal of the North American Benthological Society* 16: 248–262.

and nutrients are required to produce growth and accrual of high biomasses in relatively short periods of times. However, high algal biomasses can accumulate over longer periods of times at lower light and nutrient concentrations in hydrologically stable habitats, such as some springs, where disturbance does not interrupt accumulation processes. Thus, the process-based approach helps us understand how we can have high algal biomass at intermediate and high nutrient concentrations.

Experiments using nutrient diffusing substrates and dosing in experimental streams show that both nitrogen and phosphorus can limit algal growth in streams. There is also some evidence that micronutrients can limit growth of some algae that need large amounts of an element for enzymes in critical processes, such as Fe availability for nitrogen fixation by cyanobacteria. In several published meta-analyses of research with benthic algae, the trend is for about half of all streams being nutrient limited; and of those streams that are nutrient limited, a quarter are limited by P, a quarter are limited by N, and about half are colimited by both N and P. In streams with high accumulations of diatoms, Si may become limiting for both benthic and planktonic algae. Nutrient ratios have been used successfully to predict which lakes are limited by N or P, but this approach has not been as successful in streams where nutrient concentrations in the water column are highly variable, supply rates may differ with time, groundwater supplies are not often accounted for, and differential leakage of N versus P over time may increase relative N demand.

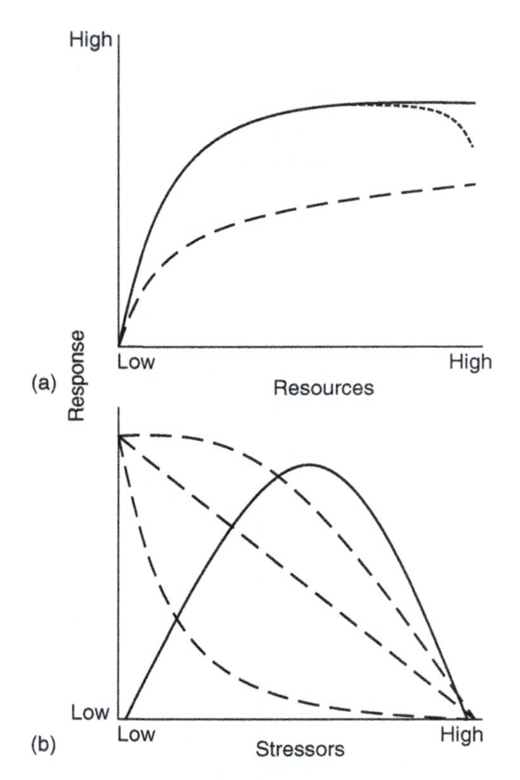

Figure 3 Responses of algae to abiotic factors. (a) Resources have asymptotic responses with rapid increases to saturating levels (solid line). Light can have a negative effect at very high levels (dotted line). Density of algae reduces supply rates and response of algae to given resource levels (dashed line). (b) Stressors negatively affect algal processes either nonmonotonically like temperature (solid line) or monotonically (family of dashed lines), when there is no positive effect at low levels of stressors.

Much less is known about space limitation than nutrients and light. Space limitation has not been evaluated well because it is difficult to isolate space versus density-dependent interactions. Space limitation applies most directly in the case of algae that must attach directly to substrates, such as the diatom *Cocconeis*. Space becomes an indirect factor or resource when algal biomass accumulates on substrata and reduces light and nutrient supply rates to cells in benthic algal mats. Space limitation can be relieved for algae in general when epiphytic microalgae can grow on macroalgae, such as *Cladophora*; but some species may be excluded from this extra space because they can attach to rocks but not well to *Cladophora*.

Species composition as well as biomass and function varies among rivers corresponding to differences in light and nutrient concentrations. Species require different amounts of nutrients and light to survive. Thus, low levels of nutrients and light constrain species membership to those species that can survive in these resource-stressed habitats. High nutrients or light enables species with requirements for high resource levels to colonize a habitat with high nutrient or light. Some experiments, theory, and now, some field evidence suggest that tradeoffs for species do exist between being able to grow fastest in low and high nutrient concentrations, which would help explain the dramatic changes in species composition along nutrient gradients.

Diversity is complexly related to resource availability. As resources increase, more algal species can invade and successfully reproduce in the river based on physiological requirements. Evenness of species abundances change nonmonotonically with increasing nutrients, from low to high to low evenness as nutrient concentrations increase. Evenness is low in low nutrient conditions because only a few species are adapted to grow well in low nutrient concentrations. As nutrients increase, more species can invade the habitat and their growth rates increase faster than species adapted to low nutrients; therefore, evenness of species abundances is highest when evenness of species reproduction rates is highest. At high nutrient concentrations, growth rates and species abundances become uneven again as some species can grow faster than others in high nutrient concentrations. Many models predict that numbers of species (richness) would follow the same pattern along resource gradients. These models predict that richness would increase as the habitat became more available to more species and could decrease in high nutrients conditions because one or more resources became depleted, competitive relationships shift, and some species are competitively excluded. Observations of this pattern have not been satisfactory because estimating species richness of organisms is difficult in habitats where they are so abundant and hard to see.

Stressors

Stressors are environmental factors that can have negative effects on metabolism and other attributes of species performance. Stressors affect reproduction according to one of two groups of response functions (**Figure** 3(**b**)). One group consists of negative linear and nonlinear responses in which there is no or very little positive effect of a stressor on species performance. Fine sediments that bury cells and toxic inorganic and organic substances are examples of factors with monotonic negative effects. Suspended sediment has strong indirect effects on phytoplankton, by reducing light transparency. Temperature and the ionic factors (salinity, pH, and alkalinity) produce nonmonotonic effects on species performance. Algal performance is optimal at some intermediate level of these factors because high and low levels of temperature, salinity, pH, and alkalinity limit species performance for one reason or another. Temperature, for example, stimulates metabolism as temperature increases from relatively low to intermediate levels as kinetic energy increases. However, high temperatures denature enzymes and reduce function. Salinity, pH, and alkalinity probably affect enzyme-mediated processes, and thereby create an intermediate condition which is optimal for species.

Because almost all algal species are negatively affected by sediments and toxic substances, biomass of algae can be negatively affected by these stressors. Similarly, the negative effect of high temperatures (above 30 °C) is 'toxic' for most algae, except for some cyanobacteria that tolerate temperatures as high 55 °C in hot springs. So, high temperatures may reduce algal biomass, as low temperatures may. However, algal species are adapted to an unusually wide range of pH, alkalinity, and salinity, so that biomass is only affected by the ionic factors in extreme conditions. Throughout much of the range of the ionic factors, species composition differs with varying ionic factors, but functional redundancy is able to maintain community-level reproduction and biomass.

Local Biotic Factors

Biotic factors can have positive or negative effects on algae in rivers. We know relatively little about commensalistic and mutualistic interactions, but some examples do exist. One example of commensalism would be the attachment of some benthic diatoms on stalks of other diatoms, which provides an advantage to species that can attach to stalks and has little negative effect on the stalked diatoms. Diatoms with endosymbiotic cyanobacteria provide an example of mutualistic interactions. Of the negative interactions, much more is known about the herbivory and competition versus allellopathic interactions and disease-like effects of fungi, bacteria, and viruses. Many have hypothesized that the latter two biotic stressors should have great effects in dense microbial assemblages like benthic algae. They are known to be important in lake and ocean phytoplankton. Unusual white circles in periphyton with high numbers of bacteria and fungi indicate 'disease,' but little investigation has pursued this line of research.

Competition is probably a more important determinant of algal biomass, function, and species composition for benthic algae than for phytoplankton. Phytoplankton seldom accumulate to sufficiently high densities to deplete nutrient concentrations in rivers because of the relatively short residence times of these organisms in their habitat. However, competition may be important in very slow flowing, lake-like rivers where residence time is sufficiently high to deplete nutrients or light by biological uptake or shading. These processes are thought to be important in benthic algal communities. If species membership is constrained to diatom-dominated communities, peak biomass of communities may be constrained by light and nutrient depletion. We know that light and nutrient availability within benthic algal mats decreases with increasing density; less light penetration and decreasing nutrient transport rates and cell nutrient content has been documented as diatom biomass increases on substrata. Per capita rates of metabolism and reproduction decrease with increasing benthic algal biomass. Species composition changes with increasing biomass of diatoms on substrata. All indicate autogenic changes in environment that are consistent with strong competitive regulation of benthic algal communities.

Herbivory is also an important determinant of algae in rivers. It is more frequently important for benthic algae than for phytoplankton because of the lack of time for zooplankton to accumulate in the water column of rivers. Low disturbance frequency by floods and drought is important for determining whether herbivory is important for benthic algae, too. When river conditions allow herbivores to accumulate to high densities, they can regulate biomass, function, and species composition of benthic algae in rivers. The importance of zebra mussels in some rivers is a good example of herbivores affecting river phytoplankton, but examples of zooplankton regulation are not common. Filter feeding invertebrates like blackflies and net-spinning caddisflies may also be important regulators of suspended algal abundance in streams, but this is not well understood.

Aquatic insects, snails, and some fish consume benthic algae, but aquatic insects and snails are the most important in most situations. Protozoa have also been shown to consume benthic algae, but their importance does not seem to be as great as cased caddisflies, mayflies, and snails that graze algae from substrata. These invertebrates can constrain diatom biomass to very low levels. Many of them can consume filamentous green algae during early stages of growth, but not after they have exceeded the size that can be controlled. They seem to avoid consumption of filamentous Cyanobacteria, but push it back from actively grazed areas or may knock it off the substrate. Invertebrate herbivores can reduce algal biomasses from 10 to 0.5 μg chl *a* cm^{-2}. In addition, they selectively graze overstory versus understory diatoms (stalked and filamentous forms versus tightly adnate and prostrate forms). Although grazers consume algae, not all are killed. Estimates of algae passed alive and viable through guts of aquatic invertebrates often exceed 50%. In a sense, grazing of cells removes cells from the substratum and either causes death or emigration of the cells downstream when they are egested alive. Because of the importance of grazing as a process of removing algae from a location and conceptually within food webs, I have included it as one of the five fundamental processes.

Bioturbation is another process that affects benthic algal biomass. Invertebrates, fish, and terrestrial animals are common sources of disturbance of benthic algae as they move through streams. Diel patterns in algal drift are observed in some stream that correspond to the dawn and dusk activity patterns of aquatic invertebrates. Paths of disturbed benthic algae in shallow riffles can be observed in deeper upstream–downstream channels where fish have moved from pool to pool. Movement of fish in pools disturbs the development of periphyton and clouds the water. Raccoons and larger animals, such as manatees, crocodiles, hippopotamus, and humans, probably have great effects on benthic algae when moving, but these effects have not been quantified.

Habitat-Scale Factors

Riparian canopy, depth, substrate, and current velocity vary among habitats within a reach and have indirect effects on the five fundamental algal processes. Covariation among depth, substrate, and current velocity are regulated by the riffle-pool structure carved in most rivers. Coarser substrates are found in the shallow, high-velocity riffles. Finer substrates are found settling in the deeper, low-velocity pools. Riparian canopy and depth regulate light availability, whereas current velocity and substrate have relatively complex effects that warrant further discussion.

Substrate is very important for algae as a stable surface for attachment and growth and potentially as a source of nutrition. Some algae have special morphological adaptations for attaching to substrata, such as the raphe and mucilaginous stalks and tubes of diatoms. Many filamentous green algae produce specialized cells that attach to a substrate and then form filaments when they divide. Thus, a major ecological division occurs in the algae about which we know relatively little, except for these morphological adaptations. Benthic algae have morphological adaptations that enable their attachment to substrata, and phytoplankton do not. Given the great differences in habitat conditions when suspended in the water versus attached to substrata, such as greater ranges in current velocity and denser packing of cells for benthic algae, great physiological differences must accompany these morphological adaptations.

For benthic algae, substrate size and nutritional value may have great effects. The smallest substrates, silt and organic sediments, are smaller than the smallest algal cells. So organisms that live in this habitat tend to be large and motile, such as the diatoms *Nitzschia* and *Surirella* or the flagellated euglenoids. Sands, slightly larger and inert, are large enough for attachment by small diatoms in streams. During hydrologically stable periods, other algae can colonize sands, but only small diatoms can survive in the crevices of individual sand grains when the sands tumble across the bottom of streams. As rock substrates become larger, they tend to become more resistant to hydrologic disturbance. Almost all pebble and larger substrates can support diverse diatom assemblages. As substrates become larger and more stable, they can support luxurious growths of filamentous green algae that require longer stable periods for colonization by spores and growth of filaments. Woody debris and plants are other common substrates in rivers and both may have nutritional properties that affect algae. Plants leak nutrients that become available to epiphytic algae attached to them. Wood provides nutrition to bacteria, which may actually have a negative effect on benthic algal colonization because of competition with bacteria for nutrients.

Current velocity has watershed- and habitat-scale effects on algae. For phytoplankton, current velocity determines residence time of water and algae in the river and the time for cells to accumulate. For benthic algae, increases in current velocity during rain events may be sufficient to scour algae from stable substrates. Higher velocities can disturb the substrate in riffles and cause severe scouring of algae from substrates. The time between scouring events provides the time for recovery of benthic algal communities. Thus, variations in current velocity at the watershed scale determine the frequency and intensity of disturbance and are important, ultimate determinants of nonequilibrium ecological dynamics of algae in rivers.

The habitat-scale effects of current velocity on benthic algae are relatively well understood compared to their effects on river phytoplankton. Eddy mixing of the water column surely slows sinking of phytoplankton and shear in the water column may reduce nutrient depletion in waters surrounding cells. One of the important premises of current effects on algae is that algae, in still waters, develop nutrient-poor shells of waters around cells. This shell develops because nutrient uptake rates from waters around cells exceed diffusion rates of nutrients into those waters. Water near surfaces, whether cell or substrate surfaces, has different physical properties than water away from these surfaces. Although measurements of nutrient concentrations in these 1–2-μm thick layers around cells has not been practical, many observations between current velocity, metabolic rates, and transport of nutrients through periphyton mats indicate that these shells do exist. Thus, as current velocity around cells increases, the shearing of layers of water around cells increases and disrupts this layer of nutrient-poor (and potential waste-rich) water from around cells.

Habitat-scale effects of current velocity on benthic algae have both positive and negative effects. Although the shear stress of current velocity decreases immigration rates and likely increases emigration rates from habitats with moderate and fast current velocities, the increased physical mixing of water through attached masses of microalgae or macroalgae stimulates metabolic rates. Concentration gradients in micro- and macroalgal communities can cause severe nutrient depletion within this microhabitat. Assuming a common 10–30 cm s^{-1} range in velocities from slow (pool) to fast (riffle) current habitats, profound differences in current effects on immigration and reproduction help relate the patterns of benthic algal colonization after flood disturbance to these processes. Initially, algal biomass and accrual rates will be slower in riffles than in pools because immigration is the dominant process affecting colonization when algal biomass on substrates is low, for example $(5–10) \times 10^3$ cells cm^{-2} (**Table 1**). As more cells accumulate, reproduction becomes more important than daily accrual of cells and the positive effects of current outweigh the negative effects. So we eventually observe higher algal biomasses in riffles than in pools, even though that pattern may not be evident immediately after flood-related disturbances.

Table 1 Illustrating the changing importance of immigration versus reproduction during algal community development on substrates

Day	*Density*	*Immi*	*Repro*
1	1100	1000	100
2	2419	1000	319
3	4001	1000	582
4	5897	1000	896
5	8168	1000	1271
6	10 887	1000	1719
7	14 139	1000	2252
8	18 025	1000	2886
9	22 663	1000	3638
10	28 190	1000	4527
11	34 766	1000	5576
12	42 573	1000	6807
13	51 820	1000	8247
14	62 741	1000	9921
15	75 595	1000	11 854
16	90 663	1000	14 068
17	108 242	1000	16 579
18	128 636	1000	19 394
19	152 140	1000	22 504
20	179 023	1000	25 883
21	209 499	1000	29 476
22	243 699	1000	33 200
23	281 636	1000	36 937
24	323 171	1000	40 535
25	367 984	1000	43 813
26	415 561	1000	46 577
27	465 193	1000	48 632
28	516 004	1000	49 811
29	567 001	1000	49 997
30	617 146	1000	49 145

Accumulation of algal cells (density, cells cm^{-2}) on substrates for 30 days is modeled with a spreadsheet. In this model, only immigration (Immi) and reproduction (Repro) are processes affecting algal cell accumulation. Grazing, death, and emigration are assumed to be 0. Immigration is assumed to stay at 1000 cells cm^{-2} day^{-1} throughout the colonization period. Twenty percent of all cells from the last day and 10% of the immigrating cells reproduce each day. A carrying capacity 1 000 000 cells cm^{-2} was included to slow reproduction as density dependent competition increased on substrates.

Reach to Regional Scale Factors

Although reach or segment scale factors (such as hydrology, channel geomorphology, flow stability, and stream size) interact to shape habitat structure and local abiotic conditions, regional processes associated with climate, geology, and biogeography regulate the more general spatial and temporal patterns of algae in rivers (**Figure 2**). Climate and geology also affect human activities in watersheds, which have profound effects on contaminants in streams, channel geomorphology, and hydrology. Extensive discussion of human effects of algae in rivers is beyond the scope of this article and is covered elsewhere in this encyclopedia. I want to synthesize the review of algal ecology in rivers by describing three common patterns: seasonal and longitudinal patterns of phytoplankton in rivers, algae and the river continuum hypothesis, and the effects of disturbance on benthic algal–grazer interactions.

Phytoplankton in rivers have a distinct longitudinal pattern that varies in magnitude with seasonal variations in discharge. Suspended algal densities are usually low in headwater sections of rivers and largely composed of algae that have emigrated from substrates and are entrained in the water column. As water flows downstream, species better adapted to life in the water column immigrate into a mass of water and stay in it as water moves downstream. Gradually, fewer benthic species occur in the water than do planktonic species. The longer that mass of water resides in the river before reaching a lake, estuary, or ocean, the higher the biomass of algae in the water column will become. More phytoplankton accumulates in rivers of geologic regions where rivers have low gradients versus those with high gradients. Climate determines the time of year and the extent of hydrologically stable periods when algae can accumulate in streams. In climates with periodic storms during the dry season, high discharge events may disrupt the development of phytoplankton communities. During long dry and hot periods, high and often problematic biomasses of algae can develop with sufficient quantities that deoxygenation and accumulation of toxic algae occurs.

The river continuum hypothesis proposes that the ecology of rivers is regulated by upstream–downstream patterns in the hydrogeomorphology and connectivity within rivers. Although originally defined for free-flowing rivers in temperate climates with forested landscapes, the model has been adapted for dammed rivers and many ecological regions. The original model described how narrow streams in forests would be covered by a riparian canopy that would limit light, and therefore, algal production. As streams became wider downstream, light would become more available and benthic algae would be able to accumulate. Correspondingly, the base of the food web was predicted to shift from allochthonous detritus in the headwaters to autochthonous primary production in the mid-reaches of the river. Farther downstream as algae became entrained in the water column and depth increased, autochthonous productivity in rivers would shift from benthic to planktonic algae. Although this upstream–downstream pattern in hydrogeomorphology and connectivity may vary greatly among regions, it tends to vary with climate and geologic conditions of a region, and to some extent with local landscape conditions. Most differences occur in headwater regions where the river may not have accumulated sufficient erosional power to carve channels and broad floodplains. Headwaters vary from springs arising in deserts to high gradient channels in mountains or low gradient channels emerging out of wetlands. Each of these provides a different starting point in downstream regulation of the ecology of rivers and very different environments for algae.

One of the challenges for understanding the ecology of algae in streams has been understanding the relationship between resource availability and disturbance intensity. This challenge underpins: (1) the prediction of top-down or bottom-up regulation of river food-webs, (2) relationships between nutrients and algal biomass for management of these important contaminants of streams, and (3) biodiversity of algae in streams. Increases in nutrient concentrations should result in an increase in algal biomass of streams. Increases in hydrologic stability also should result in increases in algal biomass of streams because algae have, on average, longer periods of time to accumulate between disturbances. Thus, in streams that are relatively hydrologically stable, we should have greater responses of algae to nutrients than in frequently disturbed streams. This is what we see in New Zealand streams where hydrologic stability ranges from great to average on the global scale. However, in the middle of North America, we see the lower end of the disturbance continuum. Low disturbance streams respond relatively little to increases in nutrient concentrations compared to relatively high disturbance streams in this region. Here, accrual of algae is constrained by high grazing pressure in hydrologically stable streams. Flood and drought disturbances in relatively hydrologically variable streams of this region constrain grazer abundances, so grazers no longer regulate algal biomass in the most hydrologically disturbed regions of central North America – which compared to the global range of possibilities, is only intermediate on the hydrologic disturbance scale. Globally, we find complexity arising from the nonlinear and multitrophic level effects of disturbance on algal-resource interactions. Herbivores and disturbance constrain algal response to nutrients in the most hydrologically stable habitats and the most hydrologically variable habitats, respectively; and algae respond most to nutrients at intermediate levels of disturbance when they have sufficient time to recover from disturbance, but not so much time that invertebrates can also recover.

Summary

Interactions between algae and their environment are complex, yet predictable based on accurate linkage of processes, scale of determining factors, and algal attributes. Although much is yet to be understood about algae in rivers, we have learned much by a marriage of three basic scientific approaches: large-scale field observations from environmental monitoring projects, experiments, and process-based models. Study of algae in rivers provides a model for how to understand complex systems that are tightly coupled to human effects on the environment. Future work in rivers should address the importance of biodiversity and ecosystem function, how ecological systems respond to environmental change, and the importance of algae in ecosystem goods and services. The lessons that we have learned in the study of algae in rivers should be applied to studies of algae in other habitats as well, such as wetlands, terrestrial habitats in rainforests, and the arctic tundra.

See also: Biological Interactions in River Ecosystems; Climate and Rivers; Currents in Rivers; Ecology and Role of Headwater Streams; Geomorphology of Streams and Rivers; Hydrology: Streams; Riparian Zones; Streams and Rivers as Ecosystems; Wetlands of Large Rivers: Flood plains.

Further Reading

Biggs BJF (1996) Patterns in benthic algae of streams. In: Stevenson RJ, Bothwell ML, and Lowe RL (eds.) *Algal Ecology: Freshwater Benthic Ecosystems*, pp. 31–56. San Diego, CA: Academic Press.

Biggs BJF, Duncan MJ, Jowett IG, Quinn JM, Hickey CW, Davies-Colley RJ, and Close ME (1990) Ecological characterisation, classification, and modeling of New Zealand rivers: An introduction and synthesis. *New Zealand Journal of Marine and Freshwater Research* 24: 277–304.

Descy JP (1987) Phytoplankton composition and dynamics in the River Meuse (Belgium). *Archiv für Hydrobiologie* 78(Suppl): 225–245.

Hauer FR and Lamberti GA (2006) *Methods in Stream Ecology*, 2nd edn. Amsterdam: Elsevier.

Reynolds CS Descy JP (1996) The production, biomass, and structure of phytoplankton in large rivers. *Archiv für Hydrobiologie.* 113(Suppl): 161–187.

Reynolds CS, Descy JP, and Padisák J (1994) Are phytoplankton dynamics in rivers so different from shallow lakes? *Hydrobiologia* 289: 1–7.

Stevenson RJ (1997) Scale-dependent causal frameworks and the consequences of benthic algal heterogeneity. *Journal of the North American Benthological Society* 16: 248–262.

Stevenson RJ, Bothwell ML, and Lowe RL (1996) *Algal Ecology Freshwater Benthic Ecosystems.* San Diego: Academic Press.

Stevenson RJ, Rier ST, Riseng CM, Schultz RE, and Wiley MJ (2006) Comparing effects of nutrients on algal biomass in streams in 2 regions with different disturbance regimes and with applications for developing nutrient criteria. *Hydrobiologia* 561: 140–165.

Vannote RL, Minshall GW, Cummins KW, Sedell JR, and Cushing CE (1980) The river continuum concept. *Canadian Journal of Fisheries and Aquatic Sciences* 37: 130–137.

Benthic Invertebrate Fauna, Small Streams

J Bruce Wallace, University of Georgia, Athens, GA, USA
S L Eggert, USDA Forest Service, Northern Research Station, Grand Rapids, MN, USA

Introduction

Small streams (first- through third-order streams) make up >98% of the total number of stream segments and >86% of stream length in many drainage networks. Small streams occur over a wide array of climates, geology, and biomes, which influence temperature, hydrologic regimes, water chemistry, light, substrate, stream permanence, a basin's terrestrial plant cover, and food base of a given stream. Small streams are generally most abundant in the upper reaches of a basin, but they can also be found throughout the basin and may enter directly into larger rivers. They have maximum interface with the terrestrial environment, and in most temperate and tropical climates they may receive large inputs of terrestrial, or allochthonous, organic matter (e.g., leaves, wood) from the surrounding plant communities. In locations with open canopies such as grasslands and deserts, autochthonous or primary production in the form of algae, or higher aquatic plants, may serve as the main food base. Hence, headwater streams display a diverse fauna, which is often adapted to physical, chemical, and biotic conditions of the region.

Diversity of Benthic Invertebrates in Small Streams

The benthic invertebrate fauna of small streams is composed primarily of aquatic insects, crustaceans, mollusks, and various other invertebrate taxa. The insect fauna consists primarily of Odonata (dragonflies and damselflies), Ephemeroptera (mayflies), Plecoptera (stoneflies), Megaloptera (alderflies and dobsonflies), Coleoptera (beetles), Trichoptera (caddisflies), occasional Lepidoptera (moths), and Diptera (true flies). Crustaceans (including amphipods, isopods, and crayfish) can also be found in small streams as well as microcrustaceans such as cladocerans, ostracods, and copepods. Other common invertebrates found in small streams include nematodes, oligochaetes, turbellarians, and mollusks such as snails, limpets, and finger-nail clams. Total invertebrate diversity in small streams can be quite high. The Breitenbach, a first-order stream in Germany, contains at least 1004 described invertebrate taxa. At least 293 invertebrate taxa have been found in headwater streams in the southern Appalachian mountains of the United States. Over 182 known invertebrate taxa have been recorded in a mountain stream on Bougainville Island, Papua New Guinea. Incredibly, there are many headwater invertebrate species that remain undescribed in both isolated and populated regions of the world.

With the great diversity of foods available for consumption by invertebrates (i.e., deposited and retained on substrates, or suspended in the water column), it is not surprising that invertebrates have evolved diverse morphobehavioral mechanisms for exploiting food resources. Their diverse feeding behaviors have been lumped into a broad functional classification scheme, which is based on mechanisms used by invertebrates to acquire foods. These functional groups are as follows: scrapers, animals adapted to graze or scrape materials (periphyton, or attached algae, fine particulate organic matter, and its associated microbiota) from mineral and organic substrates; shredders, organisms that comminute large pieces of decomposing vascular plant tissue such as leaf detritus (>1 mm diameter) along with its associated microflora and fauna, or feed directly on living vascular hydrophytes, or gouge decomposing wood; gatherers, animals that feed primarily on deposited fine particulate organic matter (FPOM $\leq$ 1 mm diameter); filterers, animals that have specialized anatomical structures (e.g., setae, mouth brushes, or fans) or silk and silk-like secretions that act as sieves to remove particulate matter from suspension; and predators, those organisms that feed primarily on animal tissue by either engulfing their prey, or piercing prey and sucking body contents.

Functional feeding groups refer primarily to modes of feeding and not type of food per se. For example, many filter-feeding insects of high gradient streams are primarily carnivores, whereas scrapers consume quantities of what must be characterized as epilithon, a matrix of polysaccharide exudates, detritus, microflora, and microfauna associated with stone surfaces, and not solely attached algae. Shredders may select those leaves that have been 'microbially conditioned' by colonizing fungi and bacteria. Shredders also ingest attached algal cells, protozoans, and various other components of the fauna during feeding. Some 'shredders' have been shown to grow by harvesting primarily the epixylic biofilm, the matrix of exudates, detritus, microflora, and microfauna found on wood surfaces. Although it appears valid to separate benthic invertebrates according to these mechanisms used to obtain foods, many questions remain concerning the ultimate sources of protein, carbohydrates, fats, and assimilated energy to each of these functional groups.

Quantitative Measurements of Benthic Invertebrates in Headwater Streams

Invertebrates are often enumerated by abundances or average numbers per unit area of stream bottom. Other measures include average biomass, or weight, per unit area of stream, or more rarely, secondary production per unit area of stream bottom. Each of these will provide a different picture of the invertebrate community. For example, **Figure 1(a)** shows abundances per unit area of moss-covered bedrock outcrop and mixed substrates in three headwater streams ($n = 20$ total stream years) at the Coweeta Hydrologic Laboratory in western North Carolina, USA. Note that abundances are dominated by members of the collector-gatherer (Cg), functional group. In contrast, three groups, predators, shredders, and collectors, represent the majority of the biomass in these small streams (**Figure 1(b)**). Secondary production, which represents the living organic matter or biomass produced by each functional group over a year regardless of its fate, i.e., losses to predation, or other sources of mortality, is fairly evenly distributed between the predator, collector, and shredder functional groups (**Figure 1(c)**). The integration of production, feeding habits, and bioenergetic data can yield a much better understanding of the role of animal populations in ecosystem function than either abundance or biomass.

Distributional patterns for functional feeding group abundance, biomass, and production in small streams may differ among substrate types (**Figure 1(a)–1(c)**). Collectors and predators dominate abundances on both substrates. For biomass, predators > shredders > collectors dominate the mixed substrates, whereas filterers > collectors > predators contribute most to biomass on bedrock outcrop substrates. Most production is attributed to predators > shredders > collectors in mixed substrates compared to filterers > collectors > predators on bedrock out crop substrates. Thus, distinct differences exist in functional feeding group production among different substrates within a stream, which correspond to different available food resources. Filterer production predominates in the bedrock habitats with high current velocities that transport FPOM. Collector production is also enhanced by FPOM trapped in the moss on the bedrock outcrops. Conversely, predator, shredder, and collector production are similar in the retentive mixed substrate habitats, which also have the greatest biomass, abundances, and organic matter retention. Scraper abundance, biomass, and production are low for all

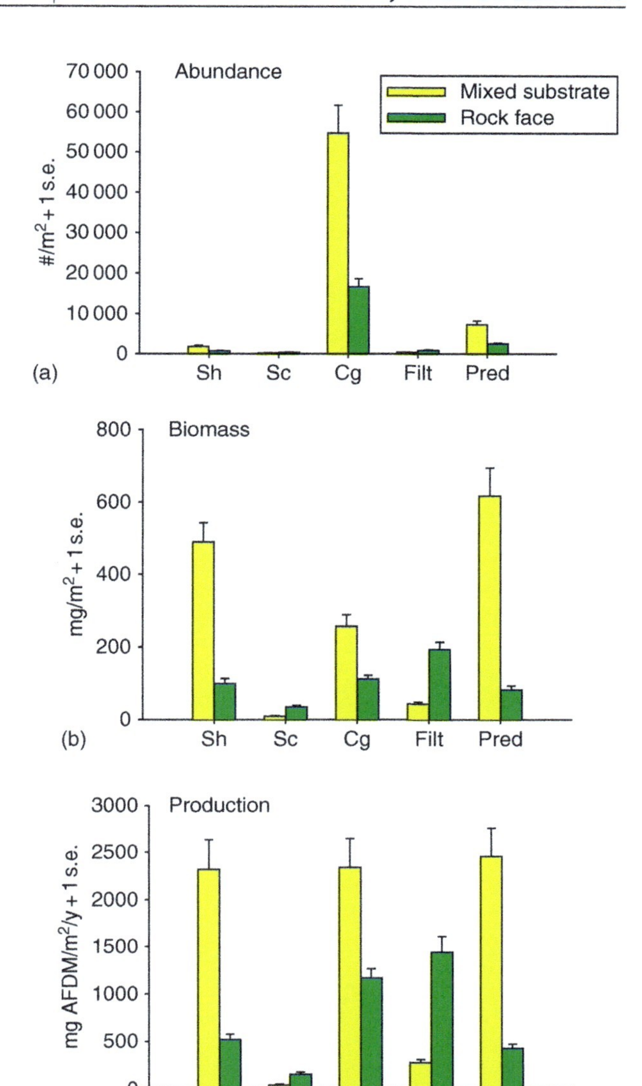

Figure 1 Total abundance, biomass, and production of invertebrates by functional feeding group for mixed substrate and rockface habitats in headwater streams at Coweeta Hydrologic Laboratory in western North Carolina, USA. These data represent 20 total years of abundance, biomass, and production estimates.

habitats as these small streams are heavily shaded year round by dense riparian rhododendron. These data emphasize the influence of local geomorphic processes and riparian linkages on invertebrate productivity in forested headwater streams.

Comparison of Secondary Productivity Measurements from Small Streams

Secondary productivity measures for benthic invertebrates from small streams from various temperate areas around the world are given in **Table 1**. Total annual productivity is quite variable ranging from

Table 1 Estimates of secondary production (g m^{-2} $year^{-1}$) for various functional feeding groups, or primary and secondary consumers, for small streams from various regions of the world

Country	*Biome*	*Stream*	*Scrapers*	*Shredders*	*Collectors*	*Filterers*	*Predators*	*Total production*	*Source*
USA, ID	Cool desert	Douglas Creek*	2.65	0.64	15.28	4.20	0.45	23.22	2
USA, ID	Cool desert	Snively Springs*	0.00	1.32	9.33	3.18	0.33	14.15	2
USA, ID	Cool desert	Rattlesnake Springs*	0.00	0.17	3.62	11.80	0.77	16.36	2
Denmark	Deciduous	Rold Kilde	0.14	5.93	2.58	0.01	0.88	9.54	6
USA, NC	Eastern deciduous	Coweeta Catchment 53	0.09	3.23	5.78	0.56	4.10	13.77	7
USA, NC	Eastern deciduous	Coweeta Catchment 54	0.22	3.53	3.77	0.72	3.17	11.41	7
USA, NC	Eastern deciduous	Coweeta (1985) Catchment 55	0.17	2.51	2.86	0.73	2.52	8.79	7
USA, NC	Eastern deciduous	Coweeta (1986) Catchment 55	0.50	2.75	3.41	0.54	3.37	10.57	7
USA, NC	Eastern deciduous	Upper Ball Ck	0.68	1.67	2.95	0.53	1.68	7.51	5
USA, NC	Eastern deciduous	Bear Pen Ck	0.68	2.02	4.45	2.32	1.66	11.12	12
USA, VA	Eastern deciduous	Buzzard's Branch	0.11	2.84	6.21	0.98	3.78	13.92	10
USA, VA	Eastern deciduous	Collier's Ck	0.11	1.69	0.95	1.28	1.58	5.60	10
USA, NH	Eastern deciduous	Bear Brook	0.74	1.45	0.61	0.43	0.94	4.17	3
USA, ME	Eastern deciduous	Goosefare Brook	0.16	11.78	5.28	3.96	5.20	27.35	13
USA, ME	Eastern deciduous	West Bear Brook	<0.00	0.86	0.21	0.22	0.37	1.66	1
USA, ME	Eastern deciduous	East Bear Brook	<0.00	0.80	0.20	0.27	0.41	1.68	1
USA, KS	Tall grass prairie	Kings Creek	3.80	4.50	6.00	1.70	3.60	19.60	11
Germany	Deciduous forest	Steina (1986)	4.63	5.10	4.53	2.43	2.33	19.02	8
Germany	Deciduous forest	Steina (1987)	7.96	2.38	3.84	3.72	2.93	20.83	8

The following studies calculated production based on primary and secondary consumers rather than functional groups

			Total primary consumer production[a]	Total secondary consumers production[b]	Total production	
USA, MA	Eastern deciduous	Factory Brook*	4.00	0.56	4.56	9
New Zealand	Tussock Grass	Sutton Stream	13.35	2.54	15.89	4

Note that data for those streams marked with an asterisk were in dry mass (DM), whereas others were in ash-free dry mass (AFDM) DM values are ~10–20% greater than AFDM.

[a]Primary consumers includes scraper, shredder, collector, and filterer functional feeding groups.

[b]Secondary consumers = predators.

Sources

1. Chadwick MA and Huryn AD (2005) Response of stream macroinvertebrate production to atmospheric nitrogen deposition and channel drying. *Limnology and Oceanography* 50: 228–236.
2. Gaines WL, Cushing CE, and Smith SD (1992) Secondary production estimates of benthic insects in three cold desert streams. *Great Basin Naturalist* 52: 11–24.
3. Hall RO, Likens GE, and Malcom JM (2001) Trophic basis of invertebrate production in 2 streams at the Hubbard Brook Experimental Forest. *Journal of the North American Benthological Society* 20: 432–447.
4. Huryn AD (1996) An appraisal of the Allen paradox in a New Zealand trout stream. *Limnology and Oceanography* 41: 243–252.
5. Huryn AD and Wallace JB (1987) Local geomorphology as a determinant of macrofaunal production in a mountain stream. *Ecology* 68: 1932–1942.
6. Iversen TM (1988) Secondary production and trophic relationships in a spring invertebrate community. *Limnology and Oceanography* 33: 582–592.
7. Lugthart JG and Wallace JB (1992) Effects of disturbance on benthic functional structure and production in mountain streams. *Journal of the North American Benthological Society* 11: 138–164.
8. Meyer EI and Poepperl R (2003) Secondary production of invertebrates in a Central European mountain stream (Steina, Black Forest, Germany) *Archiv für Hydrobiologie*, 158: 25–42.
9. Neves RJ (1979) Secondary production of epilithic fauna in a woodland stream. *American Midland Naturalist* 102: 209–224.
10. Smock LA, Gladden JE, Riekenberg JL, Smith LC, and Black CR (1992) Lotic macroinvertebrate production in three dimensions: Channel surface, hyporheic, and floodplain environments. *Ecology* 73: 876–886.
11. Stagliano DM and Whiles MR (2001) Macroinvertebrate production and trophic structure in a tallgrass prairie headwater stream. *Journal of the North American Benthological Society* 21: 97–113.
12. Wohl DL, Wallace JB, and Meyer JL (1995) Benthic macroinvertebrate community structure, function and production with respect to habitat type, reach and drainage basin in the southern Appalachians (USA) *Freshwater Biology* 34: 447–464.
13. Woodcock TS and Huryn AD (2007) The response of macroinvertebrate production to a pollution gradient in a headwater stream. *Freshwater Biology* 52: 177–196.

<2 to >27 g m^{-2} $year^{-1}$ for streams in a variety of landscapes (**Figures 2–10**). The following is an indication of the average secondary production (percentage of total) in invertebrates: collectors (31.2%, range 11.9–65.9%) > shredders (avg. = 27.0%, range 1–62.2%) > predators (avg. = 19.0%, range 1.9–31.9%) > filterers (avg. = 15.1%, range 0.1–72.1%) > scrapers (avg. = 7.4%, range = 0.0–38.2%). Scraper production as a percent of total production was greatest in the Steina, Germany (deciduous forest), followed by Kings Creek, a tall grass prairie stream in Kansas, and Bear Brook, NH (deciduous forest). The differences in scraper production in small streams in the eastern deciduous forest are striking. In Bear Brook NH, scraper production comprised >17% of total production, compared with 0.7–9% for the Coweeta streams in NC, which are heavily shaded by dense riparian rhododendron. With the exception of three cold desert streams in southeastern Washington, USA, shredder production was always greater than 10% of the total production. Percent collector–filterer production was lowest in a Danish Spring and highest in a cool desert stream. These data demonstrate that invertebrate production in small streams can be quite variable among various temperate regions. Given the usefulness of this integrative measure for comparing small stream functioning in natural and disturbed environments, additional efforts at quantifying total secondary production in small streams are badly needed.

Figure 2 A steep-gradient headwater stream in the Appalachians of western NC. Note the dense rhododendron understory, which shades the stream. (Photo by S. L. Eggert.)

Factors that Influence Invertebrates in Small Streams

Small stream invertebrates are influenced by physical, chemical, and biological factors (**Table 2**). Physical factors include climate, (e.g., temperature and

Figure 3 Headwater stream draining catchment 53 at the Coweeta Hydrologic Laboratory in western North Carolina during autumn. Note large amounts of leaves and woody debris in the stream channel. (Photo by S. L. Eggert.)

Figure 4 Headwater stream draining watershed 6 in the Hubbard Brook Experimental Forest in central New Hampshire, USA. (Photo by R. O. Hall.)

Figure 5 Intermittent stream in the Huron Mountains of Michigan's upper peninsula, USA. (Photo by S. L. Eggert.)

precipitation), hydrology, and geology. Hydrology and the frequency of flooding or drying can also influence community structure and productivity. For example, the lowest annual productivity shown in **Table 1** occurs in two intermittent streams, East Bear Book and West Bear Brook in Maine, USA. Geology influences both stream substratum and chemistry. Substratum and the proportions of eroding and depositional substratum within a given stream can have an important effect on invertebrate functional distribution and production. The faster flowing erosional reaches in small streams are often dominated by filter-feeders and scrapers, whereas depositional reaches with greater amounts of retained organic matter are often dominated by shredders and collectors. Stream chemistries can be strikingly different from those in nearby streams if they have different underlying geologies. For example, in the southern Appalachians, streams draining limestone regions such as the ridge and valley province have higher nutrients, conductivity, pH, and primary production than those draining the crystalline Appalachians. Geology and climate also influence the vegetation of catchments, including the abundance and type of

Figure 6 Bison grazing in the riparian area of the headwaters of King's Creek, at the Konza Prairie LTER in Kansas, USA. (Photo by Walter Dodd.)

Figure 8 Rattlesnake Creek, a cold desert stream in eastern Washington, USA. (Photo by C. E. Cushing.)

Figure 7 Douglas Creek, a cold desert stream in eastern Washington, USA. (Photo by C. E. Cushing.)

Figure 9 A spring-fed tundra stream in Alaska, USA. (Photo by A. D. Huryn.)

riparian vegetation. Streams that are open, and which receive large solar inputs compared with those draining dense forested catchments, may have a very different food base (autochthonous) compared with those receiving primarily allochthonous inputs from forested catchments. Depending on the food base, small streams may display large differences in functional group abundance, biomass, and production.

Figure 10 A stream draining tall tussock grass on South Island, New Zealand. (Photo by A. D. Huryn.)

Ecological Roles of Invertebrates in Small Streams

Feeding activities of invertebrates in small streams link headwater streams to larger rivers downstream by altering resource quantity, size, and shape (**Table 2**). For example, the shredding of leaf detritus and woody debris by shredders in headwater streams increases the rate of coarse organic matter breakdown to fine organic matter, which is transported by the current to downstream reaches. Scrapers, through their feeding activities and dislodging of epilithon, can enhance the movement of downstream organic particles. Heavy grazing by scrapers results in periphyton mats with adnate, or closely attached forms of diatoms that are less susceptible to scouring during disturbances such as large storms and also promotes nutrient turnover in periphyton communities. Thus, both shredding and grazing activities may result in a consistent, prolonged release of materials to downstream reaches, in contrast to large storms that induce pulsed massive export over short time intervals. The role of gatherers in FPOM transport has been implicated in an Idaho stream, which exhibited continuous deposition and resuspension as particles moved downstream. In montane Puerto Rican streams, feeding activities of atyid shrimp reduce organic matter accrual on benthic substrates. Other invertebrate gatherers such as *Ptychoptera* (Diptera: Ptychopteridae) and sericostomatid (Trichoptera) larvae may transfer fine organic matter buried in depositional areas to substratum surfaces as feces.

Filter-feeding stream invertebrates enhance retention of organic matter and nutrients. Certain invertebrates can transport superficial organic matter to deeper sediments, which reduces downstream transport. However, direct removal of transported material by filter-feeders has received the bulk of attention and has been shown to have variable effects on retention depending on size of stream, abundance of filterers, and taxon-specific differences in feeding, (e.g., feeding on extremely fine organic particles or drifting invertebrates). Studies using radioactive tracers in Alaskan streams have also suggested that particles generated by invertebrate scrapers such as baetid mayflies and chironomid larvae were instrumental in supplying fine particles to downstream black flies. Microfilterers such as the caddisfly family Philopotamidae, and black flies (Diptera: Simuliidae), and bivalve mollusks increase particle sizes by ingesting minute particles and egesting compacted fecal particles larger than those originally consumed. Such microfilterers perform two very important functions in streams. First, they remove FPOM from suspension (which would otherwise pass unused through the stream segment) and second, they defecate larger particles, which are available to deposit-feeding detritivores.

Predators can play numerous roles at scales ranging from individuals to ecosystems and invertebrate predators in small streams are no exception. Predators can influence export and retention of energy and nutrients through their effects on the standing stocks of other functional groups. Other mechanisms include decreasing rates of nutrient cycling by immobilizing nutrients in long-lived predator taxa versus short-lived prey. Besides direct consumption, foraging by invertebrate predators can enhance invertebrate drift and suspended FPOM, which also increases export of nutrients. Invertebrate predators can enhance retention of organic matter by retarding breakdown rates of leaf litter as well as subsequent generation of FPOM. Predaceous stoneflies and caddisflies can significantly decrease the rate of leaf litter processing by reducing shredder populations in leaf packs. Invertebrate predators can also increase the rate of downstream movement of organisms and sediment. Many stream invertebrates exhibit different responses to fish and invertebrate predators, and the local impact of invertebrate predation on benthic prey may exceed the impact of fish predators. In the presence of fish, invertebrate prey often reduces movement and seeks refuge in the substrate. In contrast, invertebrate

Table 2 Factors influencing benthic macroinvertebrates and effects of invertebrates on stream processes in headwater streams

Factor	*Effect on benthic macroinvertebrates and stream processes*	*Source*
Physical		
Water temperature	Lower water temperature resulted in higher total taxa richness and Ephemeroptera, Plecoptera and Trichoptera richness; snail production regulated by thermal regime; water temperature positively related to growth rates of invertebrates	5, 20, 21, 47
Hydrology	Functional feeding group production determined in part by hydrology; current velocity affected invertebrate movement and drift; channel drying altered organic matter standing crops, and invertebrate abundance and production; invertebrate abundance and diversity declined as a result of severe flow diversions	4, 19, 22, 35, 41
Geology	Higher alkalinity resulted in greater benthic invertebrate abundance, biomass, drift biomass, and organic matter standing crop; higher snail biomass and caddisfly production	20, 23, 24, 27
Substrate	Invertebrate taxa show distinct substrate preferences; functional feeding group production varied with factors associated with stream geomorphology; invertebrate diversity and abundance increases with substrate stability and presence of detritus	9, 22, 30
Chemical		
pH	Low pH resulted in decreased taxa richness, loss of sensitive taxa (Mollusca, Crustaceans, Ephemeroptera); increased drift immediately after acidification; decreased emergence; long-term decreased abundance and drift; reduced Ephemeroptera growth; reduced leaf breakdown rates; increased detritus standing crop	2, 11, 13, 16, 17, 39, 45
Conductivity	Loss or reduction of Ephemeroptera at conductivities $>400\ \mu S/cm$; reduced Ephemeroptera, Plecoptera, and Trichoptera diversity with increased conductivity; replacement of sensitive Ephemeropteran taxa with tolerant Dipteran taxa with increasing conductivity	14, 32, 38
Nutrients	Increase in abundance, biomass and production of invertebrates; increased growth rates for short-lived invertebrates; increased growth and abundance of limnephilid caddisflies in the presence of salmon carcasses	7, 8, 51
Biological		
Riparian vegetation	Invertebrate distribution, abundance, biomass, production, diversity, and growth rates, as well as leaf breakdown rates, were strongly related to riparian vegetation composition	1, 15, 29, 34, 42, 46, 52
Competition	Evidence of interspecific competition: between snail and caddisfly grazers, between caddisfly and mayfly grazers, between net-veined midges and blackflies, and between caddisflies and filterers; intra- and interspecific competition between snails; intraspecific competition for Trichopteran and Ephemeropteran shredders. Competition resulted in varied responses with regard to survivorship, growth rates, colonization of habitat, and feeding rates.	3, 6, 12, 18, 25, 26, 33
Predation	Predation on shredders resulted in reduced leaf breakdown and FPOM generation; predation on scrapers resulted in increase periphyton biomass; long-lived predators retain nitrogen; predation causes downstream movement of inorganic material; predators cause prey to drift or seek refuge in the substrate	26, 31, 37, 43, 44, 54

Altered resource size and shape	Grazing caddisflies increase algal turnover; shredders convert large organic matter to FPOM; microfilterers increase particle size	28, 36, 55
Material transport	Gatherers feeding on and egesting FPOM enhance downstream transport; invertebrates transfer FPOM to surface as feces; shrimps reduce organic matter accrual on substrates through feeding	10, 40, 48, 53, 55
Material retention	Incorporation of labeled nitrogen by scrapers and filters downstream; microfilterers retain FPOM; filterers reduce downstream transport of particulate organic matter	37, 49, 50

Sources

1. Aguiar FC, Ferreira MT, and Pinto P (2002) Relative influence of environmental variables on macroinvertebrate assemblages from an Iberian basin. *Journal of the North American Benthological Society* 21: 43–53.
2. Bernard DP, Neill WE, and Rowe L (1990) Impact of mild experimental acidification on short term invertebrate drift in a sensitive British Columbia stream. *Hydrobiologia* 203: 63–72.
3. Boyero L and Pearson RG (2006) Intraspecific interference in a tropical stream shredder guild. *Marine and Freshwater Research* 57: 201–206.
4. Chadwick MA and Huryn AD (2007) Role of habitat in determining macroinvertebrate production in an intermittent-stream system. *Freshwater Biology* 52: 240–251.
5. Collier KJ (1995) Environmental factors affecting the taxonomic composition of aquatic macroinvertebrate communities in lowland waterways of Northland, New Zealand. *New Zealand Journal of Marine and Freshwater Research* 29: 453–465.
6. Cross WF and Benke AC (2002) Intra- and interspecific competition among coexisting lotic snails. *Oikos* 96: 251–264.
7. Cross WF, Johnson BR, Wallace JB, and Rosemond AD (2005) Contrasting response of detritivores to long-term nutrient enrichment. *Limnology and Oceanography* 50: 1730–1739.
8. Cross WF, Wallace JB, Rosemond AD, and Eggert SL (2006) Whole-system nutrient enrichment increases secondary production in a detritus-based ecosystem. *Ecology* 87: 1556–1565.
9. Cummins KW and Lauff GH (1969) The influence of substrate particle size on the microdistribution of stream macrobenthos. *Hydrobiologia* 34: 145–181.
10. Cushing CE, Minshall GW, and Newbold JD (1993) Transport dynamics of fine particulate organic matter in two Idaho steams. *Limnology and Oceanography* 38: 1101–1115.
11. Dangles O, Gessner MO, Guerold F, and Chauvet E (2004) Impacts of stream acidification on litter breakdown: Implications for assessing ecosystem functioning. *Journal of Applied Ecology* 41: 365–378.
12. Dudley TL, D'Antonio CM, and Cooper SD (1990) Mechanisms and consequences of interspecific competition between two stream insects. *Journal of Animal Ecology* 59: 849–866.
13. Fiance SB (1978) Effects of pH on the biology and distribution of *Ephemerella funeralis* (Ephemeroptera) *Oikos* 31: 332–339.
14. Garcia-Críado F, Tomé A, Vega FJ, and Antolín C (1999) Performance of some diversity and biotic indices in rivers affected by coal mining in northwestern Spain. *Hydrobiologia* 394: 209–217.
15. Grubbs SA and Cummins KW (1996) Linkages between riparian forest composition and shredder voltinism. *Archiv für Hydrobiologie* 137: 39–58.
16. Guerold F, Boudot JP, Jacquemin G, Vein D, Merlet D, and Rouiller J (2000) Macroinvertebrate community loss as a result of headwater stream acidification in the Vosges Mountains (N-E France). *Biodiversity and Conservation* 9: 767–783.
17. Hall RJ, Likens GE, Fiance SB, and Hendrey GR (1980) Experimental acidification of a stream in the Hubbard Brook experimental forest, New Hampshire. *Ecology* 61: 976–989.
18. Hill WR, Weber SC, and Stewart AJ (1992) Food limitation of two grazers: Quantity, quality, and size-specificity. *Journal of the North American Benthological Society* 11: 420–432.
19. Hoffman AL, Olden JD, Monroe JB, *et al.* (2006) Current velocity and habitat patchiness shape stream herbivore movement. *Oikos* 115: 358–368.
20. Huryn AD, Benke AC, and Ward GM (1995) Direct and indirect effects on geology on the distribution, biomass, and production of the freshwater snail *Elimia*. *Journal of the North American Benthological Society* 14: 519–534.
21. Huryn AD and Wallace JB (1985) Life history and production of *Goerita semata* Ross (Trichoptera: Limnephilidae) in the southern Appalachian Mountains. *Canadian Journal of Zoology* 63: 2604–2611.
22. Huryn AD and Wallace JB (1987) Local geomorphology as a determinant of macrofaunal production in a mountain stream. *Ecology* 68: 1932–1942.
23. Jin HS and Ward GM (2007) Life history and secondary production of *Glossosoma nigrior* Banks (Trichoptera: Glossosomatidae) in two Alabama streams with different geology. *Hydrobiologia* 575: 245–258.

24. Koetsier P, Minshall GW, and Robinson CT (1996) Benthos and macroinvertebrate drift in six streams differing in alkalinity. *Hydrobiologia* 317: 41–49.
25. Kohler SL (1992) Competition and the structure of a benthic stream community. *Ecological Monographs* 62: 165–188.
26. Kohler SL and Wiley MJ (1997) Pathogen outbreaks reveal large-scale effects of competition in stream communities. *Ecology* 78: 2164–2176.
27. Krueger CC and Waters TF (1983) Annual production of macroinvertebrates in three streams of different water quality. *Ecology* 64: 840–850.
28. Lamberti GA and Resh VH (1983) Stream periphyton and insect herbivores: An experimental study of grazing by a caddisfly population. *Ecology* 64: 1124–1135.
29. Lecerf A, Dobson M, Dang CK, and Chauvet E (2005) Riparian plant species loss alters trophic dynamics in detritus-based stream ecosystems. *Oecologia* 146: 432–442.
30. Mackay RJ and Kalff J (1969) Seasonal variation in standing crop and species diversity of insect communities in a small Quebec stream. *Ecology* 50: 101–109.
31. Malmqvist B (1993) Interactions in stream leaf packs: Effects of a stonefly predator on detritivores and organic matter processing. *Oikos* 66: 454–462.
32. Marqués MJ, Martínez-Conde E, and Rovira JV (2003) Effects of zinc and lead mining on the benthic macroinvertebrates of a fluvial ecosystem. *Water, Air, and Soil Pollution* 148: 363–388.
33. McAuliffe JR (1984) Resource depression by a stream herbivore: Effects on distributions and abundances of other grazers. *Oikos* 42: 327–333.
34. Milner AM and Gloyne-Phillips IT (2005) The role of riparian vegetation and woody debris in the development of macroinvertebrate assemblages in streams. *River Research and Applications* 21: 403–420.
35. Minshall GW and Winger PV (1968) The effect of reduction in stream flow on invertebrate drift. *Ecology* 49: 580–582.
36. Petersen RC and Cummins KW (1974) Leaf processing in a woodland stream. *Freshwater Biology* 4: 343–368.
37. Peterson BJ, Bahr M, and Kling GW (1997) A tracer investigation of nitrogen cycling in a pristine tundra river. *Canadian Journal of Fisheries & Aquatic Sciences* 54: 2361–2367.
38. Pond GJ and McMurray SE (2002) A macroinvertebrate bioassessment index for headwater streams in the eastern coalfield region, Kentucky. Kentucky Department for Environmental Protection, Division of Water, Frankfort, KY.
39. Pretty JL, Gibberson DJ, and Dobson M (2005) Resource dynamics and detritivore production in an acid stream. *Freshwater Biology* 50: 578–591.
40. Pringle CM, Blake GA, Covich AP, Buzby KM, and Finley A (1993) Effects of omnivorous shrimp in a montane tropical stream: Sediment removal, disturbance of sessile invertebrates and enhancement of understory algal biomass. *Oecolgia* 93: 1–11.
41. Rader RB and Belish TA (1999) Influence of mild to severe flow alterations on invertebrates in three mountain streams. *Regulated Rivers: Research & Management* 15: 353–363.
42. Ross HH (1963) Stream communities and terrestrial biomes. *Archiv für Hydrobiologie* 59: 235–242.
43. Sih A and Wooster D (1994) Prey behavior, prey dispersal, and predator impacts on stream prey. *Ecology* 75: 1199–1207.
44. Statzner B, Fuchs U, and Higler LWG. (1996) Sand erosion by mobile predaceous stream insects: Implications for ecology and hydrology. *Water Resources Research* 32: 2279–2287.
45. Sutcliffe DW and Carrick TR (1973) Studies on mountain streams in the English Lake District. I. pH, calcium and the distribution of invertebrates in the River Duddon. *Freshwater Biology* 3: 437–462.
46. Synder CD, Young JA, Lemarié DP, and Smith DR (2002) Influence of eastern hemlock (*Tsuga canadensis*) forests on aquatic invertebrate assemblages in headwater streams. *Canadian Journal of Fisheries & Aquatic Sciences* 59: 262–275.
47. Vannote RL and Sweeney BW (1980) Geographic analysis of thermal equilibria: A conceptual model for evaluating the effect of natural and modified thermal regimes on aquatic insect communities. *The American Naturalist* 115: 667–695.
48. Wagner R (1991) The influence of the diel activity pattern of the larvae of *Sericostoma personatum* (Kirby and Spence) (Trichoptera) on organic matter distribution in stream-bed sediments – A laboratory study. *Hydrobiologia* 224: 65–70.
49. Wallace JB and Webster JR (1996) The role of macroinvertebrates in stream ecosystem function. *Annual Review of Entomology* 41: 115–139.
50. Wallace JB, Webster JR, and Woodall WR (1977) The role of filter-feeders in flowing waters. *Archiv für Hydrobiologie* 79: 506–532.
51. Walter JK, Bilby RE, and Fransen BR (2006) Effects of Pacific salmon spawning and carcass availability on the caddisfly *Ecclisomyia conspersa* (Trichoptera: Limnephilidae) *Freshwater Biology* 51: 1211–1218.
52. Whiles MR and Wallace JB (1997) Leaf litter decomposition and macroinvertebrate communities in headwater streams draining pine and hardwood catchments. *Hydrobiologia* 353: 107–119.
53. Wolf B, Zwick P, and Marxsen J (1997) Feeding ecology of the freshwater detritivore *Ptychoptera paludosa* (Diptera, Nematocera). *Freshwater Biology* 38: 375–386.
54. Wooster D (1994) Predator impacts on stream benthic prey. *Oecologia* 99: 7–15.
55. Wotton RS, Malmqvist B, Muotka T, and Larsson K (1998) Fecal pellets from a dense aggregation of suspension-feeders in a stream: An example of ecosystem engineering. *Limnology and Oceanography* 43: 719–725.

predators have the ability to search in sites similar to those being used by their prey, and the latter may respond by actively entering the water column and drifting downstream. Foraging by invertebrate predators can also influence the downstream movement of inorganic material through their physical activities. Several studies have suggested that the foraging activities increase erosion and downstream transport of sand and fine sediments. Furthermore, some specialized parasites (a subcategory of predators) of scraping *Glossosoma* caddisflies have been shown to influence periphyton biomass in Michigan streams.

Anthropogenic and Natural Disturbances to Invertebrate Productivity in Small Streams

Many unique fauna are found in small streams. Unfortunately, invertebrate fauna in these streams are under assault by anthropogenic and natural disturbances such as invasive species, agriculture, development, logging, mining, recreational activities, global climate change, and wildfires (**Table 3**). Macroinvertebrate communities and productivity can be altered, which can affect higher trophic levels (e.g., fish production) and other stream processes (e.g., organic matter processing).

Invasive Species

Invasive species within riparian habitats can have lasting effects on headwater stream functioning because of the tight linkage between riparian forests and stream processes. Macroinvertebrates abundance and diversity in small streams can be altered by changes in microclimate, energy availability, and habitat that results from loss of tree species within the riparian forest. Outbreaks of terrestrial invaders such as the balsam and hemlock woolly adelgids and the gypsy moth result in losses of some riparian tree species, pulses of slow decaying wood inputs, increases in other tree species, and increases of pesticides used to control invading pests. Indirectly, these changes can affect headwater stream functioning through reductions in the survival, growth, and emergence of macroinvertebrate shredders and detrital processing. The effects of terrestrial invasive species on stream ecosystems are expected to increase in the future. Nonnative scale and fungal diseases such as dogwood anthracnose and beech bark disease have invaded forests of the eastern United States, and the fungus causing butternut canker is beginning to spread rapidly. The fungus causing chestnut blight eliminated the American chestnut from eastern forests and resulted in decreased leaf litter processing, decreased quality of litter inputs, and decreased invertebrate growth rates in headwater streams. The input of dead chestnut logs into streams also facilitated the retention of sediment and served to stabilize stream channels. Few examples of exotic aquatic species invading small streams have been documented in the literature. One species that successfully invaded first- and second-order streams, *Gammarus pulex*, resulted in spatial and temporal reductions in macroinvertebrate diversity.

Agriculture

The filling of former wetlands and headwater streams for agriculture has greatly reduced surface water area worldwide. As an example, 96.6% of the original surface water area of the Kävlinge River catchment in Sweden has been lost due to channelization and drainage of streams for agriculture over a 141-year period. Along with the loss of small streams, intensive agriculture results in excessive nitrate levels in stream water. Overfertilization of agricultural land in low-order sections of river networks affects downstream river reaches. It has been estimated that agricultural sources in Illinois contribute 10–15% of nitrogen and phosphorus loads to the Mississippi River. Nutrient enrichment in small streams can stimulate primary production and higher trophic levels such as scrapers that feed on the abundant periphyton. In detritus-based streams, increased nutrients can lead to increases in microbial production on organic matter, which improves the quality of the food resource for shredder invertebrates. With higher food quality, macroinvertebrate production, particularly those taxa with short life cycles, can increase dramatically in nutrient-enriched streams.

Shifts in the invertebrate community associated with increased sedimentation have been observed in headwater reaches of agriculturally impacted streams. As the percent fine sediment increases, there is usually a shift from clinging and crawling taxa to burrowers. Insecticide runoff from agricultural fields into headwater streams can have more deleterious impacts on macroinvertebrate communities. Pesticides introduced into headwater streams can result in the loss of invertebrate species, cause shifts in functional production of invertebrates, and negatively impact ecosystem processes such as leaf litter breakdown and FPOM export.

Urbanization and Roads

Urban growth scenarios predict substantial increases in population and growth for many regions of the world. The replacement of forested land and riparian habitats with impervious surfaces such as roads,

Table 3 Examples of disturbances and their effects on benthic invertebrates in headwater streams

Disturbance	*Effect on invertebrates and stream function*	*Source*
Invasive species		
Gypsy moth defoliation	Accelerated detritus processing	27
Decline in eastern hemlock forests due to hemlock woolly adelgid	Reduction in alpha and gamma diversity of invertebrates and changes in trophic composition; pesticide inputs caused decline in invertebrate emergence; increased inputs of slow decaying wood; higher hydrologic variability	19, 20, 51
Loss of American chestnut trees as a result of chestnut blight	Decrease in leaf litter processing, quality of litter inputs, and invertebrate growth rates; increase in wood inputs and sediment stabilization	50, 57
Invasion of *G. pulex*	Increased predation on native invertebrates	28, 29
Agriculture		
Filling and tiling of streams	Reduced drainage density of stream network	59
Overapplication of fertilizer	Increased nitrogen alter food resources for invertebrates; increase in abundance, biomass and production of invertebrates; increased growth rates for short-lived invertebrates	13, 14, 16
Sediment runoff	Decline in Ephemeropteran, Plecopteran, and Trichopteran, and Coleopteran taxa; increases in chironomids, oligochaetes and molluscs	2, 6, 7, 24, 39, 44, 53
Insecticide runoff	Decline in invertebrate abundance, biomass, and production; loss of species; shifts in functional structure; decline in organic matter export and leaf breakdown rates	15, 31, 32, 33
Increased water temperature	Decline in Ephemeroptera, Plecoptera, and Trichoptera	24, 44
Urbanization and roads		
Altered hydrology and geomorphology, increased bacterial populations and turbidity, increases of pesticide, herbicide, and fertilizer runoff; decline in habitat	Decline in invertebrate diversity; decline in Ephemeroptera, Plecoptera, and Trichoptera; increase in number of pollution-tolerant taxa; decline in invertebrate production; decline in leaf breakdown rates	4, 9, 30, 39, 40, 48, 58, 60
Increased number of culverts	Reduced adult caddisfly diversity and abundance above culverts	5
Forestry practices		
Increased stream temperature, discharge, nutrients, and primary production; reduced organic matter inputs	Shift from allochthonous to autochthonous energy; increase in abundance, biomass, and production of taxa with short life cycles; leaf litter breakdown altered; significant reduction in invertebrate production with decline in detrital inputs	1, 3, 21, 23, 41, 42, 52, 55, 56
Sediment runoff from logging roads	Decline in total richness and abundance of all invertebrate taxa	22, 54
Mining		
Acid mine drainage and metal uptake	Reductions in abundances of sensitive invertebrate taxa; increase in tolerant taxa; decline in species diversity; increased drift; reduced community respiration; reduced secondary production	8, 11, 12, 18, 34, 45, 49
Mountaintop mining/Valley fill		
Burial of headwater streams; increased sedimentation, conductivity, and metals	Elimination of all biota in buried streams; downstream declines in Ephemeroptera richness, decline in abundances of Ephemeroptera, Odonata, Coleoptera; decline in scraper and shredder abundance	25, 43
Global climate change		
Channel drying	Altered organic matter standing crops, and invertebrate abundance and production; shifts from large-bodied, long-lived taxa to small-bodied, short-lived taxa	10, 17
Increased water temperature	Decline in total invertebrate densities, faster growth rates, reduced size at maturity, and altered sex ratios of some taxa	26

Wildfire		
Intense heating, altered water chemistry, food resources, hydrologic runoff patterns, vegetative cover, sediment transport	Shift in functional feeding groups that parallel changes in food resources; shift toward short-lived, trophic generalists; decline in invertebrate abundance and taxa richness	35, 36, 37, 38, 46, 47
Recreational activities		
Streamside camping, fishing, swimming, rafting, gold mining	Localized decline in abundance of scraper limnephilid caddisfly	61

Sources

1. Baillie BR, Collier KJ, and Nagels J (2005) Effect of forest harvesting and woody debris removal on two Northland streams, New Zealand. *New Zealand Journal of Marine and Freshwater Research* 39: 1–15.
2. Barton DR and Farmer MED (1997) The effects of conservation tillage practice on benthic invertebrate communities in headwater streams in southwestern Ontario, Canada. *Environmental Pollution* 96: 207–215.
3. Benfield EF, Webster JR, Tank JL, and Hutchens JJ (2001) Long-term patterns in leaf breakdown in response to watershed logging. *International Review of Hydrobiology* 86: 467–474.
4. Blakely TJ and Harding JS (2005) Longitudinal patterns in benthic communities in an urban stream under restoration. *New Zealand Journal of Marine and Freshwater Research* 39: 17–28.
5. Blakely TJ, Harding JS, McIntosh AR, and Winterbourn MJ (2005) Barriers to the recovery of aquatic insect communities in urban streams. *Freshwater Biology* 51: 1634–1645.
6. Braccia A and Voshell, JR Jr. (2005) Environmental factors accounting for benthic macroinvertebrate assemblage structure at the sample scale in streams subjected to a gradient of cattle grazing. *Hydrobiologia* 573: 55–73.
7. Braccia A and Voshell JR Jr. (2007). Benthic macroinvertebrate responses to increasing levels of cattle grazing in Blue Ridge Mountain Streams, Virginia, USA. *Environmental Monitoring and Assessment* 131: 185–200.
8. Carlisle DM and Clements WH (2003) Growth and secondary production of aquatic insects along a gradient of Zn contamination in Rocky Mountain streams. *Journal of the North American Benthological Society* 22: 582–597.
9. Chadwick MA, Dobberfuhl DR, Benke AC, Huryn AD, Suberkropp K, and Thiele JE (2006) Urbanization affects stream ecosystem function by altering hydrology, chemistry, and biotic richness. *Ecological Applications* 16: 1796–1807.
10. Chadwick MA and Huryn AD (2007) Role of habitat in determining macroinvertebrate production in an intermittent-stream system. *Freshwater Biology* 52: 240–251.
11. Clements WH (2004) Small-scale experiments support causal relationships between metal contamination and macroinvertebrate community responses. *Ecological Applications* 14: 954–967.
12. Clements WH, Carlisle DM, Lazorchak JM, Johnson PC (2000) Heavy metals structure benthic communities in Colorado mountain streams. *Ecological Applications* 10: 626–638.
13. Cross WF, Johnson BR, Wallace JB, and Rosemond AD (2005) Contrasting response of detritivores to long-term nutrient enrichment. *Limnology and Oceanography* 50: 1730–1739.
14. Cross WF, Wallace JB, Rosemond AD, and Eggert SL (2006) Whole-system nutrient enrichment increases secondary production in a detritus-based ecosystem. *Ecology* 87: 1556–1565.
15. Cuffney TF, Wallace JB, and Lugthart GJ (1990) Experimental evidence quantifying the role of benthic invertebrates in organic matter dynamics of headwater streams. *Freshwater Biology* 23: 281–299.
16. David MB and Gentry LE (2000) Anthropogenic inputs of nitrogen and phosphorus and riverine export for Illinois, USA. *Journal of Environmental Quality* 29: 494–508.
17. Dewson ZS, James ABW, and Death RG (2007) Invertebrate responses to short-term water abstraction in small New Zealand streams. *Freshwater Biology* 52: 357–369.
18. Dills G and Rogers DT (1974) Macroinvertebrate community structure as an indicator of acid mine pollution. *Environmental Pollution* 6: 239–261.
19. Ellison AM, Bank MS, Clinton BD *et al.* (2005) Loss of foundation species: Consequences for the structure and dynamics of forested ecosystems. *Frontiers in Ecology and the Environment* 3: 479–486.
20. Griffith MB, Barrows EM, and Perry SA (1996) Effects of aerial application of diflubenzuron on emergence and flight of adult aquatic insects. *Journal of Economic Entomology* 89: 442–446.
21. Griffith MB and Perry SA (1991) Leaf pack processing in 2 Appalachian Mountain streams draining catchments with different management histories. *Hydrobiologia* 220: 247–254.
22. Growns IO and Davis JA (1994) Effects of forestry activities (clearfelling) on stream macroinvertebrate fauna in south-western Australia. *Australian Journal of Marine and Freshwater Research* 45: 963–975.
23. Gurtz ME and Wallace JB (1984) Substrate-mediated response of stream invertebrates to disturbance. *Ecology* 65: 1556–1569.
24. Harding JS, Young RG, Hayes JW, Shearer KA, and Stark JD (1999) Changes in agricultural intensity and river health along a river continuum. *Freshwater Biology* 42: 345–357.
25. Hartman KJ, Kaller MD, Howell JW, and Sweka JA (2005) How much do valley fills influence headwater streams? *Hydrobiologia* 532: 91–102.
26. Hogg ID and Williams D Dudley (1996) Response of stream invertebrates to a global-warming thermal regime: An ecosystem-level manipulation. *Ecology* 77: 395–407.
27. Hutchens JJ and Benfield EF (2000) Effects of forest defoliation by the gypsy moth on detritus processing in southern Appalachian streams. *American Midland Naturalist* 143: 397–404.
28. Kelly DW and Dick JTA (2005) Effects of environment and an introduced invertebrate species on the structure of benthic macroinvertebrate species at the catchment level. *Archiv für Hydrobiologie* 164: 69–88.
29. Kelly DW, Dick JTA, Montgomery WI, and MacNeil C (2003) Differences in composition of macroinvertebrate communities with invasive and native *Gammarus* spp. (Crustacea: Amphipoda) *Freshwater Biology* 48: 306–315.
30. Kemp SJ and Spotila JR (1997) Effects of urbanization on brown trout (*Salmo trutta*), other fishes and macroinvertebrates in Valley Creek, Valley Forge, Pennsylvania. *American Midland Naturalist* 138: 55–68.
31. Liess M and Schulz R (1999) Linking insecticide contamination and population response in an agricultural stream. *Environmental Toxicology and Chemistry* 18: 1948–1955.
32. Lugthart GJ and Wallace JB (1992) Effects of disturbance on benthic functional structure and production in mountain streams. *Journal of the North American Benthological Society* 11: 138–164.

33. Lugthart GJ, Wallace JB, and Huryn AD (1990) Secondary production of chironomid communities in insecticide-treated and untreated headwater streams. *Freshwater Biology* 24: 417–427.
34. Malmqvist B and Hoffsten P (1999) Influence of drainage from old mine deposits on benthic macroinvertebrate communities in central Swedish streams. *Water Research* 33: 2415–2423.
35. Mihuc TB and Minshall GW (2005) The trophic basis of reference and post-fire stream food webs 10 years after wildfire in Yellowstone National Park. *Aquatic Sciences* 67: 541–548.
36. Minshall GW (2003) Responses of stream benthic macroinvertebrates to fire. *Forest Ecology and Management* 178: 155–161.
37. Minshall GW, Brock JT, Andrews DA, and Robinson CT (2001) Water quality, substratum and biotic responses of five central Idaho (USA) streams during the first year following the Mortar Creek fire. *International Journal of Wildland Fire* 10: 185–199.
38. Minshall GW, Royer TV, and Robinson CT (2001) Response of the Cache Creek macroinvertebrates during the first 10 years following disturbance by the 1988 Yellowstone wildfires. *Canadian Journal of Fisheries & Aquatic Sciences* 58: 1077–1088.
39. Moore AA and Palmer MA (2005) Invertebrate biodiversity in agricultural and urban headwater streams: Implications for conservation and management. *Ecological Applications* 15: 1169–1177.
40. Morse CC, Huryn AD, and Cronan C (2003) Impervious surface area as a predictor of the effects of urbanization on stream insect communities in Maine, USA. *Environmental Monitoring and Assessment* 89: 95–127.
41. Newbold JD, Erman DC, and Roby KB (1980) Effects of logging on macroinvertebrates in streams with and without buffer strips. *Canadian Journal of Fisheries & Aquatic Sciences* 37: 1076–1085.
42. Noel DS, Martin CW, and Federer CA (1986) Effects of forest clearcutting in New England on stream macroinvertebrates and periphyton. *Environmental Management* 10: 661–670.
43. Pond GJ and McMurray SE (2002) A macroinvertebrate bioassessment index for headwater streams in the eastern coalfield region, Kentucky. Kentucky Department for Environmental Protection, Division of Water, Frankfort, KY.
44. Quinn JM, Williamson RB, Smith RK, and Vickers ML (1992) Effects of riparian grazing and channelisation on streams in Southland, New Zealand. 2. Benthic invertebrates. *New Zealand Journal of Marine and Freshwater Research* 26: 259–273.
45. Roback SS and Richardson JW (1969) The effects of acid mine drainage on aquatic insects. *Proceedings of the Academy of Natural Sciences Philadelphia* 121: 81–107.
46. Robinson CT, Uehlinger U, and Minshall GW (2005) Functional characteristics of wilderness streams twenty years following wildfire. *Western North American Naturalist* 65: 1–10.
47. Roby KB and Azuma DL (1995) Changes in a reach of a northern California stream following wildfire. *Environmental Management* 19: 591–600.
48. Roy AH, Rosemond AD, Paul MJ, Leigh DS, and Wallace JB (2003) Stream macroinvertebrate response to catchment urbanisation (Georgia, USA). *Freshwater Biology* 48: 329–346.
49. Short TM, Black JA, and Birge WJ (1990) Effects of acid-mine drainage on the chemical and biological character of an alkaline headwater stream. *Archives of Environmental Contamination and Toxicology* 19: 241–248.
50. Smock LA and MacGregor CM (1988) Impact of the American chestnut blight on aquatic shredding macroinvertebrates. *Journal of the North American Benthological Society* 7: 212–221.
51. Snyder CD, Young JA, Lemarié DP, and Smith DR (2002) Influence of eastern hemlock (*Tsuga canadensis*) forests on aquatic invertebrate assemblages in headwater streams. *Canadian Journal of Fisheries and Aquatic Sciences* 59: 262–275.
52. Stone MK and Wallace JB (1998) Long-term recovery of a mountain stream from clear-cut logging: The effects of forest succession on benthic invertebrate community structure. *Freshwater Biology* 39: 141–169.
53. Stone ML, Whiles MR, Webber JA, Williard KWJ, and Reeve JD (2005) Macroinvertebrate communities in agriculturally impacted southern Illinois streams: Patterns with riparian vegetation, water quality, and in-stream habitat quality. *Journal of Environmental Quality* 34: 907–917.
54. Tebo LB Jr. (1955) Effects of siltation, resulting from improper logging, on the bottom fauna of a small trout stream in the southern Appalachians. *Progressive Fish-Culturist* 17: 64–70.
55. Wallace JB, Eggert SL, Meyer JL, and Webster JR (1999) Effects of resource limitation on a detrital-based ecosystem. *Ecological Monographs* 69: 409–442.
56. Wallace JB and Gurtz ME (1986) Response of *Baetis* mayflies (Ephemeroptera) to catchment logging. *American Midland Naturalist* 115: 25–41.
57. Wallace JB, Webster JR, Eggert SL, Meyer JL, and Siler ER (2001) Large woody debris in a headwater stream: Long-term legacies of forest disturbance. *International Review of Hydrobiology* 86: 501–513.
58. Whiting ER and Clifford HF (1983) Invertebrates and urban runoff in a small northern stream, Edmonton, Alberta, Canada. *Hydrobiologia* 102: 73–80.
59. Wolf P (1956) *Utdikad Civilisation. (Drained Civilization).* Malmo, Sweden: Gleerups.
60. Woodcock TS and Huryn AD (2007) The reponse of macroinvertebrate production to a pollution gradient in a headwater stream. *Freshwater Biology* 52: 177–196.
61. Wright KK and Li JL (1998) Effects of recreational activities on the distribution of *Dicosmoecus gilvipes* in a mountain stream. *Journal of the North American Benthological Society* 17: 535–543.

rooftops, and lawns alters the hydrology and geomorphology of streams. Increases in surface runoff associated with storm flow lead to declines in water quality, increases in bacterial populations and turbidity, and increases of pesticide, herbicide, and fertilizer runoff into nearby streams. Sediment runoff from construction sites and erosion due to downcutting result in habitat loss for aquatic life. Measurable aquatic degradation occurs with 6–10% impervious area and has been long been associated with decreased water quality of nearby streams. Urbanization in watersheds containing small streams usually results in less diverse invertebrate communities consisting of pollution-tolerant species. Culverts have been shown to act as barriers to upstream migration of adult caddisflies. Since many stressors associated with urbanization act synergistically, it is difficult to separate cause and effect of individual stressors on the invertebrate communities in small urban streams.

Forestry Practices

Logging results in changes in stream temperature regimes, increased discharge and altered hydrology, increased nutrient export and increased solar radiation and primary production, increased sediment export, and changes in dissolved organic matter derived from the terrestrial ecosystem. These changes are accompanied by substantial changes in the energy base of headwater streams, with a shift from allochthonous detritus to autochthonous production. The physical and energy base changes can lead to large changes in macroinvertebrate community structure. An experimental long-term reduction of organic matter inputs to a small stream in the southern Appalachians resulted in a significant decline in total invertebrate production. Invertebrate taxa with short life cycles and the ability to exploit increases in primary production greatly increase in abundance, biomass, and productivity. Studies in the central and southern Appalachians show that long-term patterns of leaf litter breakdown can be altered for many years following logging. However, depending upon the extent of terrestrial succession, invertebrate assemblages can revert back toward their prelogged condition.

Mining

Mining has severe consequences for benthic invertebrates in small streams worldwide. Effects of mining on macroinvertebrates in small streams are caused by acid mine drainage, sediments, or burial of the streams themselves. Acid mine drainage and the associated problems of heavy metal contamination usually results in reductions of numbers of sensitive taxa in the orders of Ephemeroptera (particularly those of the family Heptageniidae), Plecoptera, Trichoptera, Megaloptera, Odonata, and Diptera and an overall decline in species diversity. Some studies have shown that functional measures of benthic invertebrates such as drift and community respiration are also negatively affected by mining impacts. In recent years, the practice of mountaintop removal and valley fill mining has resulted in the filling and permanent burial of at least 1450 km of small streams in the Appalachian Mountains. The burial of multiple small streams destroys all aquatic life in these streams and results in declines of sensitive invertebrate taxa immediately below valley fills. The cumulative effects of burying multiple headwater streams on the water quality in downstream rivers should be evaluated.

Recreational Activities

Little information regarding the effects of recreational activities (e.g., horseback riding, cycling, all terrain vehicle use (ATV)) on small streams has been reported in the primary literature. One study suggested that populations of *Dicosmeocus gilvipes*, a scraping limnephilid caddisfly, in a fifth-order stream were affected by localized disturbances associated with multiple recreational activities such as gold mining, streamside camping, swimming, and fishing. With growing public demand for access to undeveloped land harboring networks of small streams for recreational activities such as off-highway vehicle use, there is an urgent need for more research examining the impacts of such use and ways to mitigate potential negative effects.

Global Climate Change

Consequences of global climate change on invertebrates in small streams will vary greatly spatially and temporally, thus making it difficult to predict potential effects. Generally, precipitation and evaporation are expected to become more variable over time. Some regions of the world will become wetter, while others will become drier, affecting runoff patterns. Increased temperatures as a result of global climate change will reduce snow cover and also affect hydrologic patterns in small streams. Shifts in hydrologic patterns (e.g., flooding, drying) will impact transport of nutrients, organic matter, and habitats available for colonization by benthic invertebrates. Changes in riparian vegetation may alter the quality and quantity of detrital inputs to headwater streams, thereby altering ecosystem processes (e.g., production, respiration, organic matter breakdown) within small stream reaches and longitudinally-linked downstream reaches, as well as invertebrate life histories and species composition. There is some evidence in the literature that the timing and duration of small stream channel drying results in altered organic matter standing crops, and invertebrate production. Furthermore, extended channel drying results in shifts from large-bodied, long-lived taxa to small-bodied, short-lived taxa. An experimental manipulation of thermal regime (2.5–3.0 °C increase in water temperature) in a small stream near Ontario, Canada, resulted in a reduction of total invertebrate densities, faster growth rates, reduced size at maturity, and altered sex ratios of some invertebrates.

Wildfire

Invertebrates in small streams are more susceptible to fire disturbance than those in larger streams. Intense heating, severely altered water chemistry, and the smothering of food resources by ash in smaller streams can kill invertebrates directly. Over longer time periods, changes in hydrologic runoff patterns, vegetative cover, channel morphology, and sediment transport also affect invertebrates in fire impacted streams. Changes in food resources over time result in changes in the functional characteristics of the macroinvertebrate community. Initially, scraper densities increase following a fire because of increased primary productivity associated with canopy opening and increased available nutrients. As transportable organic matter levels increase in the stream, abundances of collectors increase. Shredder populations are usually the last to recover since they depend on detrital inputs from the riparian habitat. Recovery of macroinvertebrate communities in intact, normally functioning small streams prior to fire usually occurs quickly (5–10 years) following fire disturbance and parallels the regeneration of the terrestrial vegetation. Short-lived invertebrate taxa that are trophic generalists, and have wide habitat preferences generally recover quicker.

Summary and Knowledge Gaps

The functional contributions of benthic invertebrates to small streams are well known. Hundreds of invertebrate species may be found in a small stream. Since headwater streams make up such a large proportion of total stream length in river networks, total invertebrate production in small stream segments may exceed that in large rivers. Invertebrates also represent an important link between terrestrial and aquatic ecosystems due to the close proximity of the two systems. A variety of environmental factors influence the types and productivity of invertebrates in small streams. Benthic invertebrates are also good indicators of the health of small streams. Human and natural disturbances alter typical macroinvertebrate assemblages in small streams which may have indirect effects on higher trophic levels and small stream processes.

In the last two decades scientists have begun to study macroinvertebrate communities and the ecological processes affected by invertebrates along longitudinal reaches spanning multiple stream orders. However, little information is known about the quantitative and qualitative contribution of headwater benthic fauna to the functioning of downstream ecosystems. In some cases, entire benthic invertebrate communities are being destroyed by burial or stream piping before the true diversity of organisms found in small streams is known. Furthermore, with an increasing number of disturbances that are large scale in magnitude, it is critical that scientists become better able to predict threshold levels of disturbance within headwaters of river networks such that downstream water quality and ecosystem functions are not irrevocably damaged.

See also: Agriculture; Benthic Invertebrate Fauna; Benthic Invertebrate Fauna, Small Streams; Biological Interactions in River Ecosystems; Climate and Rivers; Coarse Woody Debris in Lakes and Streams; Conservation of Aquatic Ecosystems; Deforestation and Nutrient Loading to Fresh Waters; Ecology and Role of Headwater Streams; Floods; Geomorphology of Streams and Rivers; Hydrology: Streams; Regulators of Biotic Processes in Stream and River Ecosystems; Restoration Ecology of Rivers; Riparian Zones; Urban Aquatic Ecosystems.

Further Reading

Allan JD (1995) *Stream Ecology: Structure and Function of Running Waters.* New York, NY: Chapman & Hall.

Cushing CE, Cummins KW, and Minshall GW (2006) *River and Stream Ecosystems of the World.* Berkeley, CA: University of California Press.

Freeman MC, Pringle CM, and Jackson CR (2007) Hydrologic connectivity and the contribution of steam headwaters to ecological integrity at regional scales. *Journal of the American Water Resources Association* 43: 5–14.

Heino J, Muotka T, Mykrä H, Paavola R, Hämäläinen H, and Koskenniemi E (2003) Defining macroinvertebrate assemblage types of headwater streams: Implications for bioassessment and conservation. *Ecological Applications* 13: 842–852.

Huryn AD and Wallace JB (2000) Life history and production of stream insects. *Annual Review of Entomology* 45: 83–110.

Merritt RW, Cummins KW, and Berg MB (2007) *An Introduction to the Aquatic Insects of North America,* 4th edn. Dubuque, IA: Kendall/Hunt.

Merritt RW and Cummins KW (2006) Trophic relationships of macroinvertebrates. In: Hauer FR and Lamberti GA (eds.) *Methods in Stream Ecology*, 2nd edn., pp. 585–609. UK: Elsevier.
Meyer JL, Strayer DL, Wallace JB, Eggert SL, Helfman GS, and Leonard NE (2007) The contribution of headwater streams to biodiversity in river networks. *Journal of the American Water Resources Association* 43: 86–103.
Meyer JL and Wallace JB (2001) Lost linkages and lotic ecology: Rediscovering small streams. In: Press MC, Huntly NJ, and Levin S (eds.) *Ecology: Achievement and Challenge*, pp. 295–317. Oxford, UK: Blackwell Science.
Nadeau TL and Rains MC (2007) Hydrological connectivity between headwater streams and downstream waters: How science can inform policy. *Journal of the American Water Resources Association* 43: 118–133.
Resh VH and Rosenberg DM (1984) *The Ecology of Aquatic Insects*. New York: Praeger.
Thorp JH and Covich AP (eds.) (2001) *Ecology and Classification of North American Freshwater Invertebrates*, 2nd edn. San Diego, CA: Academic Press.
Vannote RL, Minshall GW, Cummins KW, Sedell JR, and Cushing CE (1980) The river continuum concept. *Canadian Journal of Fisheries and Aquatic Sciences* 37: 130–137.
Wallace JB and Merritt RW (1980) Filter-feeding ecology of aquatic insects. *Annual Review of Entomology* 25: 103–132.
Wallace JB and Webster JR (1996) The role of macroinvertebrates in stream ecosystem function. *Annual Review of Entomology* 41: 115–139.
Wipfli MS, Richardson JS, and Naiman RJ (2007) Ecological linkages between headwaters and downstream ecosystems: Transport of organic matter, invertebrates, and wood down headwater channels. *Journal of the American Water Resources Association* 43: 72–85.

Relevant Websites

http://www.americanrivers.org/site/DocServer/WhereRiversAreBorn1.pdf?docID=182 – American Rivers: Where Rivers are Born.
http://www.benthos.org – North American Benthological Society.
http://www.tolweb.org/tree/phylogeny.html – Tree of Life.
http://www.epa.state.oh.us/dsw/wqs/ – Ohio EPA, Division of Surface Water, Ohio Primary Headwater Habitat Streams.
http://www.stroudcenter.org/index.htm – Stroud Water Research Center.

Benthic Invertebrate Fauna, River and Floodplain Ecosystems

M R Whiles, Southern Illinois University, Carbondale, IL, USA
J W Grubaugh, University of Memphis, Memphis, TN, USA

Introduction

River–floodplain systems are considered among the most productive ecosystems in the world. Historically all have supported highly productive fish communities, and today systems that are less impacted by human modifications (e.g., some systems in South America and Africa) still support highly productive fisheries. In addition to fish populations, temperate-region river–floodplain systems serve as critical resources along primary flyways for migratory birds. The fuel for the productivity of river–floodplain systems is organic materials derived from the floodplain, and, to a lesser degree, organic matter transported from tributaries and upstream areas. Abundant and productive invertebrate communities that inhabit these systems are critical links between organic materials and higher trophic levels, as most fishes and many other vertebrates do not directly feed on this material. For example, many migratory bird populations, especially during spring in advance of egg-laying, are highly dependent on aquatic invertebrates in floodplain habitats to provide a protein-rich food resource.

Rivers and their associated floodplains are inherently complex, multidimensional systems that offer a wide diversity of habitats for invertebrates, ranging from swift currents in the main channel, to accumulations of wood (snags) in slower moving water, to ephemeral pools that are disconnected from the channel (**Figure 1**). These distinct habitats represent steep gradients in physicochemical factors and thus harbor different assemblages of invertebrates. Hence, when considered as a whole, river–floodplain systems are some of the most diverse freshwater habitats in the world.

Here we consider channels and floodplains separately because of the distinct habitats associated with each. However, the two are intimately linked through a lateral continuum of material and energy exchanges and many organisms routinely utilize both.

Channel Habitats

Channel habitats are the classical riverine areas of river–floodplain systems. While water levels fluctuate, channel areas are usually inundated and are strongly influenced by unidirectional downstream flow. Channel habitats include the deepest portions of the main channel (the thalweg), areas lateral to the thalweg (channel borders), smaller permanent side channels, natural and anthropogenic structures that impede flow, and the immediate shoreline. These areas provide the invertebrates an array of habitat types that vary in water velocity, substrate particle size, depth, and stability (**Table 1**). Because moving water replenishes dissolved oxygen and entrains living and nonliving organic materials, dissolved oxygen and food resources are less likely to be limiting in most channel habitats than in floodplain areas. However, since channel habitats are permanently inundated, invertebrate communities must cope with vertebrates (primarily fishes and water birds) as both predators and competitors, which can limit populations and the species pool. Channel habitats can be subdivided into two general types: (1) those with stable or hard substrates and (2) those with smaller substrate particle sizes that are unstable.

Stable Substrates

River–floodplain systems are by and large alluvial systems in which geologically stable substrates such as boulders or bedrock are usually a minor geomorphic feature. Naturally occurring stable substrate is primarily large woody debris from uprooted trees (snags) that enters channel habitats through bank sloughing or is imported during flood events from the floodplain or upstream areas. Numerous studies have shown the importance of snags as an invertebrate habitat and as 'hot spots' of productivity. Snags often harbor the highest invertebrate biomass per unit area of any other habitat in the river–floodplain system, with obvious implications for fisheries management (**Figure 2**). Unfortunately, in managed systems, most snags have been removed to facilitate navigation, rendering these systems stable-substrate limited.

Stable substrates in many navigable rivers (where snagging operations have removed most wood from the channel) occur primarily in the form of occasional natural bedrock or boulders and rocks of various sizes that have been added by humans (riprap) and articulated concrete mat revetments (**Figure 3**). Riprap and revetments are artificial habitats designed to maintain navigation channels and stabilize banks. Nevertheless, these can be important to invertebrates living in stable-substrate limited systems. Hard-packed clays often associated with high velocity areas are another type of stable substrate. However, these highly scoured areas are poor habitats and usually support scant invertebrate populations.

Invertebrates associated with stable substrates in channel habitats are characterized as 'clingers,' since they have the ability to maintain their position under high velocities. This is accomplished by morphological features such as the hooked claws of many mayflies (Ephemeroptera), stoneflies (Plecoptera), and caddisflies (Trichoptera) and the dorsoventrally flattened bodies of many taxa (e.g., heptageniid mayflies) that allow them to reside in the low velocity of the boundary layer. Silk attachments are used by groups such as the blackflies (Diptera: Simuliidae), and byssal threads are used by zebra mussels (*Dreissena polymorpha*). Organisms such as *Tortopus* (Ephemeroptera: Polymitarcyidae) remain attached to snags by burrowing into the wood to create a refuge. Invertebrates associated with management structures containing riprap (i.e., wing-dams, closing structures, and hard-points), such as aquatic isopods and midge larvae (Diptera: Chironomidae), use interstitial spaces as refuges from high velocities as well as from vertebrate predators.

Figure 1 An aerial view of Mormon Island on the central Platte River in Nebraska. Main channels of the river are visible on each side of the island. The numerous wetland slough and pond habitats visible on the island vary greatly in hydrology and connectivity to the river. Photograph by G. Lingle.

Most invertebrates on stable substrates in the channel use water flow as an energy subsidy to obtain food resources and dissolved oxygen. As such, these assemblages are dominated by dense populations of relatively sessile, filter-feeding taxa whose substantial abundance, biomass, and productivity are limited only by stable substrate availability. Where stable substrates are abundant, these are some of the most productive invertebrate assemblages in river–floodplain systems, with estimates of secondary production exceeding 600 g dry mass m^{-2} $year^{-1}$. In contrast, similar production estimates for collector-gatherers and others that dominate unstable substrates in channel habitats are generally in the range of 1–5 g dry mass m^{-2} $year^{-1}$.

Adaptations for filter-feeding include appendages with fan-like brushes (e.g., Simuliidae) or comb-like setae (e.g., *Isonychia* [Ephemeroptera: Isonychiidae]), which directly filter particles out of the water column, or silken nets with mesh sizes that vary in accordance with stream velocity, which capture both living and nonliving organic matter (e.g., caddisflies in the families Hydropsychidae, Ecnomidae, Philopotamidae, and Polycentropodidae) (**Figure 4**). Macroinvertebrates (invertebrates that can be seen with the naked eye) that live in the interstitial spaces in riprap or joints between articulated mat revetments tend to be collector-gatherers that utilize deposited organic materials (e.g., Isopoda and Chironomidae) or gastropods and other scrapers that feed on periphyton and biofilms that form on the stable substrates. Predatory invertebrates are also associated with hard substrates, but their abundance and diversity are limited when compared with that of unstable channel substrates and floodplain habitats, perhaps due to

Table 1 Relative comparisons of physical characteristics and macroinvertebrate assemblages associated with primary habitat types across a lateral gradient in the upper Mississippi floodplain–river system

	Main channel stable substrates	*Main channel unstable substrates*	*Channel border & side channel*	*Aquatic macrophyte beds*
Physical Characteristics				
Substrates	Woody debris, rock	Sand	Silt and silt/sand	Silt and silt/clay
Fine particulate organic matter	High: entrained	Low	Moderate	High: deposited
Current velocity	High	Moderate to high	Moderate	Low
Macroinvertebrates				
Diversity	High	Low	Low	High
Density	High (~2000 m^{-2})	Low (~50 m^{-2})	High (~10 000 m^{-2})	Moderate (~500 m^{-2})
Biomass	Moderate to high	Low	High	Low to moderate

Adapted from Anderson RV and Day DM (1986) Predictive quality of macroinvertebrate – habitat associations in lower navigation pools of the Mississippi River. *Hydrobiologia* 136: 101–112.

Figure 2 Biomass of macroinvertebrate groups on benthic substrates and on snag habitats in the main channel of the Ogeechee River in eastern Georgia, USA. Note the much more diverse community and overall higher biomass per unit area found on snag habitats. Data from Benke AC (2001) Importance of flood regime to invertebrate habitat in an unregulated river–floodplain ecosystem. *Journal of the North American Benthological Society* 20: 225–240.

Figure 3 A dike constructed of rock riprap in the Mississippi River. These structures, although artificial, contribute greatly to stable substrate availability for invertebrate communities.

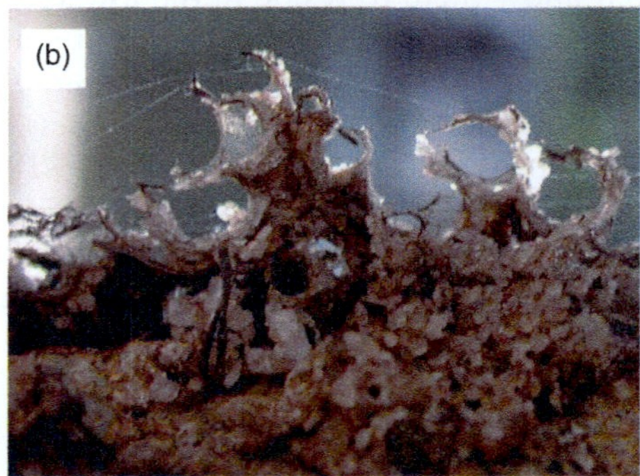

Figure 4 Filtering structures of some freshwater insects typical of river channel habitats: (a) *Neureclepsis*, a polycentropodid caddisfly, on near-shore vegetation in the Mississippi River; (b) retreats and nets of *Hydropsyche* (Trichoptera: Hydropsychidae) on bridge supports in the Satilla River in Georgia. Photograph by Alex Huryn.; (c) an *Isonychia* (Ephemeroptera: Isonychiidae) nymph with an anterior leg removed to show the filtering structure. *Isonychia* orient facing into the current and filter feed with brushes of long setae located on the anterior legs. Of the two caddisflies, *Neureclepsis* live in relatively slow flowing water, feed on relatively small particles, and have a finer mesh net than Hydropsychids, which generally feed in swifter currents in the main channel.

competitive pressure or predation from fishes. Examples of macroinvertebrate predators associated with stable substrates include predatory stoneflies of the families Perlidae and Perlodidae as well as hellgrammites (Megaloptera: Corydalidae).

Unstable Substrates

Channel habitats with unstable substrates of smaller particle sizes (i.e., sand and silt) are usually depositional areas (areas where materials tend to accumulate) associated with diminished water velocities such as along the shoreline or protected areas immediately downstream of flow-impeding hard substrates. However, thalweg areas of many large rivers are characterized by shifting, coarse-sand dunes that are utilized by collector-gatherers (e.g., some Chironomidae and Oligochaeta) as well as some highly specialized suspension-feeding (e.g., *Ametropus* [Ephemeroptera: Ametropodidae]) and predatory (*Pseudiron centralis* [Ephemeroptera: Pseudionidae]) insects that create

Figure 5 Aerial view of the central Platte River in Nebraska showing unstable, shifting sand substrates in the main channel of the river. Photograph by K. Dinan, USFWS.

vortices to remove sand in search of food items (**Figure 5**). Although generally not considered as highly productive habitats, sandy habitats of the main channel can have high invertebrate densities; some estimates from free-flowing portions of the upper Mississippi River indicate densities can exceed 80 000 individuals per square meter, although these assemblages are heavily dominated by small-bodied taxa such as nematodes and small chironomids, and diversity is generally low.

Invertebrate community composition and their life-history characteristics in sand or silt habitats within the channel are greatly influenced by relative stability. In highly unstable sands of the main channel or immediately downstream of wing-dams, communities are dominated by taxa with rapid life cycles (e.g., many Chironomidae) that can quickly colonize and develop in this constantly changing environment. Invertebrates specialized for living in sand are usually burrowers and are primarily collector-gatherers such as chironomids or mobile predators such as *Dolania* (Ephemeroptera: Behningiidae) and *Progomphus* (Odonata: Gomphidae).

In the relatively more stable areas near shore, substrate particle sizes tend to be primarily fine silts or silt/clay mixes. Invertebrates associated with these habitats are primarily burrowers and collector-gatherers or collector-filterers. Unlike the short life cycles of many taxa that live in unstable sand habitats, many silt-dwelling species are semi- or univoltine (e.g., fingernail clams [Sphaeriidae] and burrowing mayflies), and still others, such as the unionid mussels and the exotic *Corbicula fluminea* (Corbiculidae), have life spans that traverse years or even decades. Conditions in these areas are more similar to habitats found in floodplains, and factors such as dissolved oxygen and temperature influence invertebrate community composition more so than in deeper, swifter flowing main channel habitats. Hence, some taxa that are common in floodplain habitats are also found in silt areas within the river.

Floodplain Habitats

The extent of floodplains varies greatly with river size and local geomorphology; floodplains can stretch for kilometers in either direction of the main channel on large river systems such as the Mississippi and Amazon. Although they are important components of streams and rivers of many sizes, floodplains and the communities that inhabit them are most significant to ecosystem structure (e.g., biological diversity) and function (e.g., nutrient cycling, productivity) of higher order, low-gradient rivers. Floodplain habitats harbor invertebrate assemblages that are distinct from those in the main channel, usually dominated by taxa typical of lentic and wetland habitats. Invertebrates can be very abundant in floodplain habitats that hold water for most or all of the year; macroinvertebrate densities exceeding 150 000 individuals per square meter and corresponding biomass values of nearly 10 g dry mass m^{-2} have been reported from floodplain sloughs of the Platte River in Nebraska. Invertebrate diversity is also often higher in floodplain habitats; studies on the Paraguay and Parana River systems found significantly higher diversity in floodplain lakes than in channel habitats, although invertebrate densities were often higher in channel habitats because of the high abundance of the sand-dwelling oligochaete *Narapa bonettoi*.

Although they may appear relatively flat and homogeneous, floodplains contain a mosaic of freshwater habitats, often collectively referred to as off-channel or backwater habitats. These different habitats are home to a variety of invertebrates, including many with special adaptations to cope with the dynamic nature and harsh physicochemical conditions that characterize these areas. Differential connectivity with the main channel, along with subtle differences in elevation, result in hydrologic gradients that, along with other factors discussed below, shape invertebrate assemblages and thus contribute greatly to the biodiversity and productivity of large rivers.

Hydrology

One of the most powerful forces influencing floodplain invertebrate communities, both directly and indirectly, is hydrology, which is generally linked to distance from and connectivity with the main channel of the river. During high flows, the entire floodplain of a river may be inundated. However, most of the time there is a gradient ranging from connected and/or deep, permanently inundated habitats such as oxbow lakes, side channels, and connected sloughs to disconnected, shallower, intermittent habitats such as shallow ponds and pools, and disconnected sloughs in a matrix of terrestrial habitats. These hydrologic gradients, and the associated gradients in water chemistry and physical habitat that accompany them, enhance freshwater invertebrate diversity at the landscape scale. Along with spatial variability in habitats across a floodplain, seasonal changes in habitats result in temporally dynamic assemblages in a given site, and thus also contribute to overall biodiversity. Seasonal changes within a given habitat, such as an off-channel slough, can be striking, often resulting in shifts between aquatic and terrestrial habitats (**Figure 6**), with concomitant shifts in invertebrate assemblages. Overall, habitat characteristics and thus aquatic invertebrate assemblages change greatly moving from the main channel to the floodplain, with communities in the more ephemeral, least connected habitats most different from those of the main channel (**Figure 7**).

Floodplain habitats that dry either frequently or for long periods of time (ephemeral habitats and intermittent habitats with relatively short annual hydroperiods) often have limited freshwater invertebrate diversity, but can harbor unique assemblages of taxa adapted to these harsh habitats. Further, although species diversity in these habitats is relatively low, abundance and biomass, and thus the contribution of these habitats to invertebrate productivity, can be high. As ephemeral and intermittent habitats become increasingly disconnected from the river and begin to dry, their capacity to buffer temperature fluctuations decreases and temperature extremes can surpass the lethal limits of many species. Warming temperatures in turn result in declining dissolved oxygen concentrations, further limiting the pool of species that can persist. As such, communities typical of ephemeral or relatively short hydroperiod intermittent habitats are often dominated by transients (e.g., winged adult beetles and hemipterans), taxa that breathe atmospheric air (e.g., pulmonate snails such as *Physella*, dipterans with respiratory siphons such as mosquitoes [Culicidae] and soldier fly [Stratiomyidae] larvae, and taxa that use transportable air stores such as many hemipterans and adult coleopterans), and those with short developmental times that can complete generations when conditions are favorable (e.g., many small crustaceans [Copepoda and Ostracoda] and midges of the subfamily Orthocladiinae). In particular, larvae and adults of *Berosus*, a hydrophilid beetle (**Figure 8**), can be very abundant in ephemeral and intermittent floodplain habitats in North America, such that the presence of *Berosus* is a reliable indicator of habitats with shorter hydroperiods. Chironomids are ubiquitous in river and floodplain habitats, but show distinct taxonomic shifts

Figure 6 An off-channel slough on the middle Mississippi River floodplain during a period of inundation in late spring 2003 (a) and later in the summer 2003 (b). Vegetation produced during dry periods can represent an important structural habitat and detrital food resource for aquatic invertebrates when it becomes inundated. Photograph by M. Flinn.

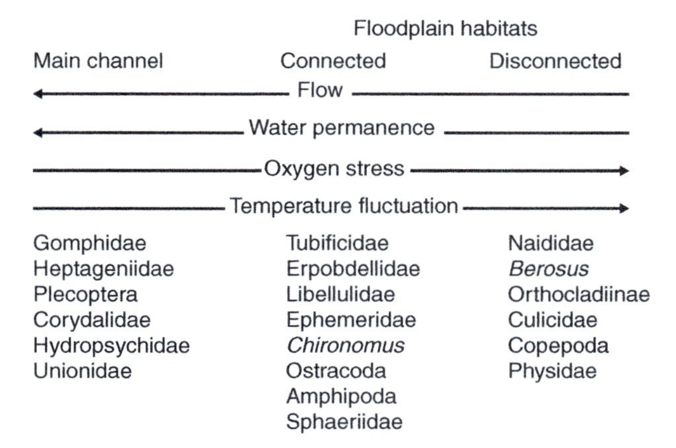

Figure 7 Diagram depicting changes in some important habitat characteristics and some dominant invertebrate taxa from the main channel of a river to connected, wetter (permanent and intermittent with relatively long hydroperiods) and disconnected, drier (ephemeral and intermittent with shorter hydroperiods) floodplain habitats. Arrows indicate the direction of increase for given habitat features. Note that taxa listed for a given habitat type are often dominant in that habitat, but may also be present in other parts of the gradient depicted.

Figure 8 *Berosus* (Coleoptera: Hydrophilidae) larvae (left) and adults (right) collected from intermittent sloughs along the middle Mississippi River. Photograph by S. Peterson.

along hydrologic gradients, with smaller-bodied groups such as the Tanytarsini and Orthocladiinae dominant in ephemeral and shorter-hydroperiod intermittent sites and larger-bodied groups such as the Chironomini more abundant in wetter habitats.

Along with the morphological adaptations and shorter life cycles that are typical of invertebrates inhabiting ephemeral and intermittent floodplain habitats, some taxa also have specialized life histories that allow them to persist. These adaptations can include periods of dormancy (e.g., aestivation or diapause) that generally involve drought resistant stages such as the cysts formed by some oligochaetes and nematodes, drought resistant eggs used by many crustaceans, and the terrestrial pupae of coleopterans and megalopterans. Immature and adult forms of some invertebrates can also avoid desiccation through behavioral adaptations such as movement to more permanent habitats by winged adults, or burrowing and aestivating by groups such as sphaeriid clams, some gastropods, adult amphipods, and some coleopterans. Following inundation, some invertebrate taxa, particularly those that are already present in resting stages, can return quickly. Some nematodes and protists appear within hours, some copepods, cladocerans, and midges appear within a week, and larger taxa such as *Caecidotea* (Isopoda) and ephydrid flies are often present within 10 days.

The Platte River caddisfly (*Ironoquia plattensis* [Trichoptera: Limnephilidae]) is an example of an insect adapted to temporary floodplain habitats. Larval *I. plattensis* live in intermittent sloughs along the central Platte River that dry in the summer. Final instars migrate from the sloughs as they dry in late spring and then aestivate and later pupate in the litter layer of the surrounding mesic prairie. In autumn, when water levels begin to rise because of increased precipitation and decreased evapotranspiration, winged adults emerge from the litter, mate, and lay eggs in the slough channels, where larvae will hatch and develop through the winter–spring hydroperiod. Although nicely adapted to intermittent habitats, the specialized life cycle of *I. plattensis* appears to limit its distribution, as it is rarely encountered in permanent or more ephemeral habitats.

Permanent and intermittent floodplain habitats with relatively long hydroperiods generally harbor higher invertebrate diversity than more ephemeral habitats, and taxa typical of these wetter habitats are often larger-bodied and longer lived. In particular, odonate nymphs, particularly members of the families Coenagrionidae, Lestidae, and Libellulidae, are common predators found on the surface of the substrates or clinging to plants and detritus. One of the most familiar insects of large river systems, burrowing mayflies of the family Ephemeridae, particularly the genus *Hexagenia* (**Figure 9**), can be extremely abundant in permanent floodplain and river margin habitats where they burrow into fine substrates and pump water through their burrows to filter feed. Another mayfly, *Callibaetis* (Baetidae), is typical of floodplain habitats, and being smaller and having a shorter life cycle it can inhabit sites with shorter hydroperiods. Caddisflies are generally not as well represented in floodplain habitats as they are in the main channel of rivers, but some groups, particularly members of the Leptoceridae and

Figure 9 *Hexagenia* (Ephemeroptera: Ephemeridae) adult (a) and nymph (b). Adult photograph by A. Morin, courtesy of the North American Benthological Society image library. Nymph photograph by S. Peterson.

Phryganeidae, can be abundant in some floodplain habitats where they function as detritivores, herbivores, and in some cases (e.g., the leptocerid *Oecetis*), as predators. Midges of the genus *Chironomus* can numerically dominate invertebrate communities in permanent and longer-hydroperiod intermittent habitats. Sometimes called 'bloodworms,' *Chironomus* larvae often appear red because of the presence of hemoglobin in their hemolymph, which aids in respiration in the low oxygen environments of the substrates of floodplain habitats. Mollusks are also generally well represented in floodplain habitats that hold water all or most of the year. In particular, fingernail clams (Sphaeriidae) can be very abundant and thin-shelled varieties of unionid mussels such as *Anodonta* are common in the soft sediments of floodplain habitats.

Along with their contributions to food webs and ecosystem functions in freshwater habitats on the floodplain, abundant invertebrate assemblages in floodplain habitats can contribute to main channel food webs during periods of high flow when they are displaced into channel habitats or predators from channel habitats move into floodplains to forage. Further, for insect taxa, adult emergences, particularly those of some dipterans (e.g., some Chironomidae and Culicidae) and burrowing mayflies can involve spectacular numbers, representing a rich food source for riparian predators such as spiders, amphibians, birds, and bats. Emergence production values in excess of 5 g dry mass m^{-2} $year^{-1}$ have been measured in intermittent floodplain slough habitats of the Platte River.

Hydrology and Vertebrate Predators

Although connected and permanently inundated habitats are more stable, this hydrologic stability and increased connectivity with the main channel result in a predator-permanence gradient, whereby more permanent habitats with greater connectivity to the main channel harbor higher numbers and diversity of predatory fish. Although many studies indicate that invertebrate diversity and abundance often increase with water permanence in floodplain wetlands, they can be suppressed in wetter, more connected habitats because of increased predation. Thus, invertebrate diversity and abundance often peak in intermittent sites that hold water for most of the year but occasionally dry and limit predator populations. Hydrology and the presence of predatory fish are often confounded and thus it can be difficult to assess the relative importance of each. However, one study of systems where the presence of fish and the length of inundation in floodplain wetlands varied independently suggested that the presence of fish was a more important determinant of invertebrate assemblage structure than hydrology. Both are obviously important in determining assemblage structure, and the relative importance of each appears to differ across systems, as results from studies examining these factors vary considerably.

Amphibians may also be present in freshwater habitats on floodplains, particularly those with less connection to the channel and/or shorter hydroperiods, as many amphibians are intolerant of fish predation. In cases where amphibians are present, predatory salamander larvae can shape invertebrate assemblages in the same manner as fish. Tadpoles, which can reach extremely high densities in temporary pools and ponds and are detritivorous, herbivorous, or scavengers, may compete with invertebrates for food and space.

Along with predation and competition with vertebrate groups, competition for resources among invertebrates likely increases in more stable habitats as colonization proceeds and populations grow. Intensifying competitive interactions will, over time, limit diversity as well in sites that hold water for long periods of time.

Vegetation

Aquatic vegetation, which is ultimately influenced by hydrology, also has a pervasive influence on invertebrate communities because both living and dead components contribute to structural habitat diversity, represent a food resource for many invertebrates, and can alter the physicochemical environment through shading, photosynthesis, and respiration. Vegetated habitats can harbor high densities of invertebrates; macroinvertebrate densities of over 50 000 m^{-2} and biomass values approaching 3 g dry mass m^{-2} have been reported from vegetated floodplain habitats of the upper Mississippi River. Particularly during periods of flooding, and especially in tropical systems, flooded vegetation can dominate river floodplain habitats. For example, in the Orinoco River system in South America, flooded forest can represent 80% of inundated area, with much of the rest of the area represented by macrophyte beds.

Vegetation increases surface roughness, thereby lowering velocity during flood events and promoting particle deposition in floodplain habitats. Plants greatly increase horizontal structure, providing emergence sites for many insects such as mayflies and odonates as well as promoting shuttling of collector-gathers between the food-rich/oxygen-poor bottom substrates and oxygen-rich portions of the upper water column. By increasing structural complexity in floodplain habitats, wetted vegetation affords aquatic invertebrates protection from fish predation and attachment sites for eggs. Aerial portions of plants provide resting habitat for winged adults during mating and are also used as oviposition sites by invertebrates such as dobsonflies (Megaloptera: Corydalidae), whose larvae drop from vegetation into the water as they hatch.

Both grazing and detritivorous invertebrates utilize vegetation as a food resource. Examples of herbivores include some aquatic snails (Gastropoda), omnivorous crayfish (Decapoda), and several semi-aquatic insects such as the American lotus borer (*Ostrinia penitalis*) and the water lily leaf beetle (*Pyrrhalta nymphaeae*). In general, herbivory on living plants is uncommon among insects and other invertebrates that are strictly aquatic for substantial portions of their life cycles; rather, these organisms are often part of a highly diverse group of collector-gatherers that utilize plant detritus as a primary food resource. Although detritus can be important for food, biological oxygen demand associated with accumulating detrital materials can depress dissolved oxygen concentrations and result in hypoxic or anoxic conditions, especially at night, when photosynthesis ceases and respiration is high. Vegetated floodplain habitats thus represent challenging habitats for many invertebrate taxa.

The different structural groups of vegetation influence aquatic invertebrate communities in different ways. Woody species – shrubs and trees – are primarily restricted to seasonally dry areas, although some woody species (such as baldcypress [*Taxodium distichum*], blackgum/tupelo [*Nyssa sylvatica*], and buttonbush [*Cephalanthus occidentalis*]) that are common to the middle and lower Mississippi River system are more water-tolerant and once established, can survive in permanently wetted areas. Woody vegetation is important for seasonal inputs of detritus in the form of leaf litter and generation of snags. Exposed roots of water-tolerant species along sloughs and in floodplain lakes also provide a stable and complex habitat for many aquatic invertebrates and are often hot-spots of invertebrate diversity and production. Although relatively poorly studied, there are invertebrate assemblages associated with the submerged roots of floating aquatic plants such as water hyacinth (*Eichhornia crassipes*), with densities of ~1200 individuals per 100 g of wet mass of roots and submerged stems reported from the Orinoco floodplain. Estimates from the Orinoco floodplain indicate that invertebrate assemblages associated with roots are more productive than the benthos or zooplankton. Some large invertebrates, such as shrimps of the genus *Macrobrachium*, are commonly associated with roots of floating plants in tropical systems.

Emergent grasses such as sedges and rushes and a diverse array of dicotyledonous species often dominate shallow-water habitats with shorter hydroperiods and the shorelines of deeper-water areas. Much of this type of vegetation develops while ephemeral and intermittent habitats are dry, and then becomes an important habitat and food source when it is inundated. Some hydric emergent plants are adapted to survive anoxic soil conditions by developing gas-filled aerenchyma tissue (tissues with air passages) that promotes oxygen transport to the roots. Larvae of the beetle *Donacia* (Chrysomelidae) are adapted to take advantage of this and have spiracles mounted on spine-like projections that allow them to tap aerenchyma tissue of emergent vegetation for respiration.

Submersed vegetation such as some pondweeds (*Potamogeton* spp.), eelgrass (*Valisneria americana*), and coontail (*Ceratophyllum demersum*) are readily colonized by clinging invertebrates, including coenagrionid damselflies, libellulids and other dragonflies, and leptocerid caddisflies. Since these plants exchange photosynthetic gases with the water column rather than the atmosphere, they provide an additional benefit to aquatic invertebrates by increasing dissolved oxygen concentrations during photosynthesis. In contrast, floating vegetation such as duckweeds (Lemnaceae) and water hyacinth (*Eichhornia crassipes*) do not release oxygen to the water column and develop as dense stands that limit oxygen diffusion and block light penetration, which depresses dissolved oxygen concentrations in water below them. In shallow, open-water floodplain habitats with floating vegetation, the detrimental effects of low dissolved oxygen may be outweighed by the benefits of greatly reduced diurnal temperature maxima and rich detrital resources. Invertebrates that can tolerate low dissolved oxygen concentrations (e.g., many air-breathing hemipterans and coleopterans, taxa with hemoglobin such as *Chironomus*) are often very abundant and productive under floating vegetation.

Threats to River–Floodplain Invertebrates

Because of their importance to agriculture and industry, and because they drain large expanses of land that are often highly impacted by human activities, river floodplain systems are some of the most imperiled habitats in the world. Major human impacts include the destruction of floodplain habitats, channelization and snag removal, dewatering, and disconnection of floodplain and channel habitats. Impoundments are particularly destructive, as they result in large-scale changes to channel and floodplain habitats and disrupt the longitudinal continuity that characterizes rivers. Massive habitat changes associated with impoundments have predictable, negative impacts on many invertebrates and other consumer groups (**Figure 10**). Over half of the large river systems on the planet are affected by impoundments. In the contiguous United States, only ~40 rivers that are longer than 125 miles remain unimpounded. Impoundments are less numerous on rivers in the tropics, but increasing interest in hydroelectric power threatens unimpounded rivers worldwide.

Rivers are also subjected to myriad pollutants, including nutrients, pesticides, and sediments from point and nonpoint sources, all of which can have negative impacts on invertebrates. Biological pollution, the introduction of exotic species, is a growing problem in river floodplain systems. One of the more notorious invasive species, the zebra mussel (*Dreissena polymorpha*), which was first documented in the United States in 1988, has now spread into many rivers in the eastern United States and is considered a serious threat to freshwater ecosystems in the United States and parts of northern Europe. High densities of zebra mussels alter freshwater food webs by consuming vast quantities of suspended organic particles, which reduces energy available to planktonic food webs, increases light penetration through the water, and increases energy and nutrients in benthic habitats. Zebra mussels also have direct negative impacts on native freshwater mussels and other invertebrates that they attach to (**Figure 11**). Zebra mussels are just one example of the numerous introduced species, including plants, invertebrates, and vertebrates, that negatively impact river– floodplain systems and their inhabitants. Currently, over 800 species of nonnative plants and animals, mostly fishes, have been introduced into freshwater habitats in the United States through a variety of pathways. Given that the bulk of these introductions have occurred in the past 50 years, the effects of many of these species on rivers and other freshwater systems are poorly understood.

As a result of the degradation of river–floodplain systems, freshwater organisms are some of the most imperiled species on the planet, and although they generally receive less attention than vertebrates, invertebrates are no exception. The World Conservation Union lists 1151 endangered freshwater invertebrates globally, many of which live in river floodplain systems. Considering how poorly studied most invertebrate species are, and that thousands have yet to be described, this number is likely a substantial underestimate. Freshwater mussels and crayfish are particularly vulnerable because they have relatively small ranges and poor dispersal capabilities. In the case of freshwater mussels, there are nearly 300 species in the United States, 72% of which are considered

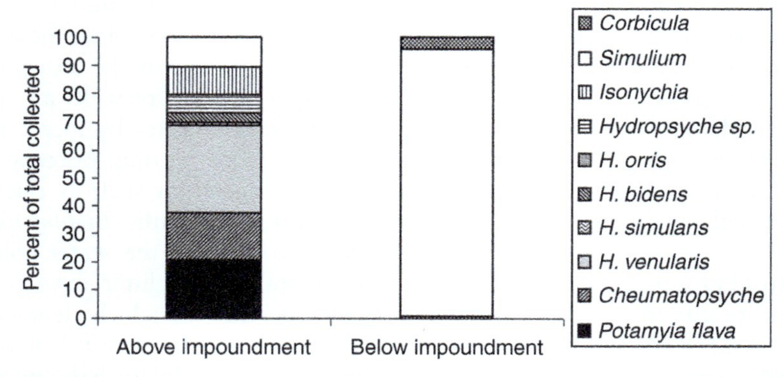

Figure 10 Filter-feeding macroinvertebrate assemblage composition in free-flowing sections of the Big Blue River in northeast Kansas above (~25 km above upstream end of reservoir) and below (~3 km below impoundment) the Tuttle Creek Reservoir Impoundment. Data from Whiles MR and Dodds WK (2002) Relationships between stream size, suspended particles, and filter-feeding macroinvertebrates in a Great Plains drainage network. *Journal of Environmental Quality* 31: 1589–1600.

Figure 11 *Neurocordulia* (Odonata: Corduliidae) nymph from a side channel of the Ohio River with attached Zebra mussels. Photo by S. Peterson.

endangered, threatened, or of special concern, including 21 species presumed extinct. Impoundments and sedimentation are the most widely cited causes of these declines, with exotic species a growing concern.

See also: Africa; Algae of River Ecosystems; Asia – Eastern Asia; Asia – Monsoon Asia; Asia – Northern Asia and Central Asia Endorheic Rivers; Australia (and Papua, New Guinea); Benthic Invertebrate Fauna, River and Floodplain Ecosystems; Benthic Invertebrate Fauna, Small Streams; Benthic Invertebrate Fauna, Tropical Stream Ecosystems; Benthic Invertebrate Fauna; Biological Interactions in River Ecosystems; Climate and Rivers; Coarse Woody Debris in Lakes and Streams; Conservation of Aquatic Ecosystems; Currents in Rivers; Deforestation and Nutrient Loading to Fresh Waters; Ecology and Role of Headwater Streams; European Rivers; Flood Plains; Floods; Geomorphology of Streams and Rivers; Hydrology: Streams; Regulators of Biotic Processes in Stream and River Ecosystems; Restoration Ecology of Rivers; Riparian Zones; South America; Streams and Rivers as Ecosystems; Streams and Rivers of North America: Overview, Eastern and Central Basins; Streams and Rivers of North America: Western, Northern and Mexican Basins; Wetlands of Large Rivers: Flood plains.

Further Reading

Allan JD (1995) *Stream Ecology: Structure and Function of Running Waters.* Boston, Massachusetts: Kluwer.

Batzer DP, Rader RB, and Wissinger SA (1999) *Invertebrates in Freshwater Wetlands of North America: Ecology and Management.* New York: Wiley.

Batzer DP and Wissinger SA (1996) Ecology of insect communities in nontidal wetlands. *Annual Review of Entomology* 41: 75–100.

Benke AC (2001) Importance of flood regime to invertebrate habitat in an unregulated river–floodplain ecosystem. *Journal of the North American Benthological Society* 20: 225–240.

Benke AC and Cushing CE (2005) *Rivers of North America.* Amsterdam: Elsevier.

Corti DS, Kohler L, and Sparks RE (1997) Effects of hydroperiod and predation on a Mississippi River floodplain invertebrate community. *Oecologia* 109: 154–165.

Flinn MB, Whiles MR, Adams SR, and Garvey JE (2005) Macroinvertebrate and zooplankton responses to emergent plant production in upper Mississippi River floodplain wetlands. *Archiv Fur Hydrobiologie* 162: 187–210.

Gladden JE and Smock LA (1990) Macroinvertebrate distribution and production on the floodplains of 2 lowland headwater streams. *Freshwater Biology* 24: 533–545.

Gregory SV, *et al.* (1991) An ecosystem perspective of riparian zones. *BioScience* 41: 540–551.

Harper D, Mekotova J, Hulme S, White J, and Hall J (1997) Habitat heterogeneity and aquatic invertebrate diversity in floodplain forests. *Global Ecology and Biogeography Letters* 6: 275–285.

Junk WJ, Bayley PB, and Sparks RE (1989) The flood pulse concept in river–floodplain systems. In: Dodge DP (ed.) *Proceedings of the International large river symposium Canadian Special publication of Fisheries and Aquatic Sciences,* 106: 110–127.

Lewis WM, Hamilton SK, Rodriguez MA, Saunders JF, and Lasi MA (2001) Food web analysis of the Orinoco floodplain based on production estimates and stable isotope data. *Journal of the North American Benthological Society* 20: 241–254.

Merritt RW and Lawson DL (1992) The role of macroinvertebrates in stream–floodplain dynamics. *Hydrobiologia* 248: 65–77.

Merritt RW, Cummins KW, and Berg MB (2007) *An Introduction to the Aquatic Insects of North America,* 4th edn. Dubuque, IA: Kendall/Hunt.

Sheldon F, Boulton AJ, and Puckridge JT (2002) Conservation value of variable connectivity: Aquatic invertebrate assemblages of channel and floodplain habitats of a central Australian arid-zone river, Cooper Creek. *Biological Conservation* 103: 13–31.

Smock LA, Gladden JE, Riekenberg JL, Smith LC, and Black CR (1992) Lotic macroinvertebrate production in three dimensions: Channel surface, hyporheic, and floodplain environments. *Ecology* 73: 876–886.

Thorp JH and Delong MD (1994) The riverine productivity model: An heuristic view of carbon sources and organic processing in large river ecosystems. *Oikos* 70: 305–308.

Tronstad LM, Tronstad BP, and Benke AC (2005) Invertebrate responses to decreasing water levels in a subtropical river floodplain wetland. *Wetlands* 25: 583–593.

Ward JV and Stanford JA (1979) *The Ecology of Regulated Rivers.* New York: Plenum.

Ward JV and Tockner K (2001) Biodiversity: Towards a unifying theme for river ecology. *Freshwater Biology* 46: 807–819.

Whiles MR and Goldowitz BS (2001) Hydrologic influences on insect emergence production from central Platte River wetlands. *Ecological Applications* 11: 1829–1842.

Whitton BA (ed.) (1975) River ecology. *Studies in Ecology,* vol. 2. Berkeley, CA: University of California Press.

Woodward G and Hildrew AG (2002) Food web structure in riverine landscapes. *Freshwater Biology* 47: 777–798.

Relevant Websites

http://www.benthos.org/index.cfm – North American Benthological Society.

http://www.nature.org/wherewework/greatrivers/resources/ – Great Rivers Partnership.

http://www.sws.org/ – Society of Wetland Scientists.

http://www.entsoc.org/ – Entomological Society of America.

Benthic Invertebrate Fauna, Tropical Stream Ecosystems

M E Benbow and M D McIntosh, University of Dayton, Dayton, OH, USA

Introduction

Tropical regions of the world are located between the Tropic of Cancer and the Tropic of Capricorn between latitudes of about 23°05′ north and south of the equator. The areas just a few degrees north and south of this region are considered the subtropics. Tropical climates are generally mild with temperatures rarely dropping below about 27 °C (80.6 °F) during daylight hours. Most tropical ecosystems are considered humid tropical rainforests characterized by wet and dry seasons, with the number, duration, and magnitude of these seasons highly variable between geographic areas. Rainforests receive a minimum average annual rainfall of >150 cm but this can exceed 10 m. Tropical watersheds reflect local discharge patterns of precipitation that define the frequently flooded habitat conditions experienced by tropical invertebrates (**Figure 1**).

Tropical streams and rivers are found in five of the eight ecozones of the world (**Table 1**). The term ecozone (or biogeographical realm) is a relatively new term that integrates the evolutionary history of both the animal and plant communities of broad geographic areas that have been historically separated, and we use it here to classify tropical stream invertebrate distribution (**Table 1**). The rivers in **Table 1** reflect a myriad of many small tributaries that make up each watershed, and provide a broad basis for the general distribution of tropical invertebrates (**Figures 1–3**). Contrary to terrestrial ecosystems, tropical streams and rivers do not show an increase in diversity compared to similar lotic systems of temperate regions. It has been argued, though, that there has not been enough research to make such general statements, and this argument is especially true of large tropical rivers, so we concentrate on tropical insular and continental montane stream invertebrates (**Figure 4**). David Dudgeon has written extensively on the stream and river communities of tropical East Asia (Indomalaya), including an impressive treatise that has served as a general ecological reference for this article.

Stream Invertebrates

For the purposes of this article, we limit our scope to those invertebrates that have been most extensively studied from two or more ecozones (**Tables 2** and **3**) and typically reflect taxa with body lengths of about 0.5 mm or greater, thus excluding such groups as microcrustaceans (e.g., cladocera and copepods), gastrotricha, tardigrada, and protozoa (e.g., amoeboids and ciliates). Further, the scope of this article is to provide a general introduction to the most common tropical benthic invertebrate groups, thus, we do not provide descriptions of anatomy or evolutionary relationships: this information can be found in other articles of the encyclopedia.

Noninsect Macrofauna

A list of the representative macrofauna common to tropical streams and rivers is given in **Table 2.** Many of the taxa listed in this table have not been intensively studied, and so the generalizations of habitat and occurrence should be taken as such. A brief synopsis is given for each phylum, in addition to more specific information for classes or orders based on the available literature.

Porifera Tropical stream sponges are less diverse compared to their marine relatives, and most can be placed within the family Spongillidae. With perhaps only 100 species (of 5000 total) found in freshwater habitats, tropical sponges are most common in lentic water bodies or backwater, slow-flowing areas of large streams and rivers. There are a few exceptions that can be collected from flowing waters, and these taxa are typically found encrusting stable substrates, probably as a way to avoid dislodgement and scouring during flood events. Freshwater sponges in general are filter-feeders, removing fine particulate matter from the water column using flagellated choanocytes that line the interior body walls of the animal and create a water current through the body. Benthic stream sponges are not thought to play a major ecological role in most tropical stream ecosystems; however, more research is needed.

Cnidaria Freshwater jellyfish are also infrequently collected from freshwaters except for the relatively widespread *Hydra*. The absence of freshwater Cnidaria may not be ecological, but rather a problem of inadequate collection and preservation techniques, so this group of invertebrates may be more commonly distributed than generally thought. All Cnidaria are

Figure 1 Clockwise from top left: the mouth of Halawa Stream, Molokai, Hawaii showing the small estuary of the valley, a representative waterfall barrier (Wailua Falls, Hawaii) common to many tropical streams, another large waterfall from an unnamed stream of Costa Rica, and Rio Sucio (left) flows north from Irazu Volcano in the Braulio Carillo National Park and is joined at the confluence by Rio Patria (right) just north of the park in Costa Rica (Photos by M.E. Benbow).

considered predators of microscopic invertebrates. Although sometimes collected from larger river systems, when freshwater Cnidaria are present, they can be found in backwater and floodplain pools of smaller lotic systems. The full significance of this group to ecosystem function remains unknown.

Polyzoa (Bryozoa) Also known as the phylum Ectoprocta, these colonial animals are represented by about 50 freshwater species worldwide. Much like the sponges, Polyzoa filter suspended matter from the water column, but use numerous ciliated tentacles instead of cell-associated flagella. They are known to encrust various substrates when found in flowing water habitats (even riffles), but are in general found in river backwaters and other lentic systems. Because of the wide variety of habitat types colonized by Polyzoa, this group is generally considered more widespread and probably more abundant in freshwater habitats worldwide than currently documented. Another phylum, the Entoprocta, was once included within the Bryozoa as a class with the Ectoprocta. The two phyla are very similar but phylogenetically distinct. The Entoprocta has only one recorded freshwater genus, *Urnatella*.

Platyhelminthes Many tropical flatworms are parasitic trematodes and cestodes that have great medical importance in developing tropical nations, but those taxa are beyond the scope of this article. The only nonparasitic representatives are within the Turbellaria, best represented by the genus *Dugesia*, and are often considered widespread among tropical streams and rivers of the world. Many times flatworms are lumped into the 'Other' category of benthic stream surveys, or go unnoticed. The general habitat distribution of many turbellaria is considered to be restricted to coldwater streams where most take up a generalist predator life

Table 1 Major tropical rivers watersheds of the world, organized by ecozone

Ecozone	*Major river/watershed*	*Drainage area* (km^2)[a]	*Length (km)*[a]	*Discharge* (m^3 s^{-1})
Afrotropic	Nile	3 349 000	6650	3500
	Niger	1 125 000	4200	6100
	Chari	600 000	950	1270
	Sénégal	441 000	1640	687
	Volta	398 400	1270	1260
	Gambia	77 000	1120	170
Indomalaya	Yangtzc (Ch'ang Chiang)	1 808 500	5980	34 000
	Ganges	1 051 540	2506	18 697
	Mekong	802 900	4360	11 048
	Irrawaddy	429 940	2100	14 079
	Pearl (His Chiang)	425 700	2100	12 500
	Salween	279 720	2400	1493
Neotropic	Amazon	6 000 000	6400	220 000
	Orinoco	950 000	2410	38 000
	Magdalena	270 000	1500	7000
	Coco (Segovia)[b]	6830	780	70
	Motagua[b]	180 034	487	150
	Grande de Matagalpa	15 073	430	580
Australasia	Sepik[b]	77 700	1100	4000
	Mamberamo	79 440	650	4500
	Purari[b]	28 700	470	2700
	Fitzroy[b, c]	150 000	480	168
	Budekin[c]	129 700	710	380
	Normanby[c]	24 408	350	390
Oceania	Kapuas	88 800	1143	na

[a]Most values taken from Cushing *et al.* (2006), except for those as indicated by *c*.

[b]River Discharge Database, Estimates from Center for Sustainability and the Global Environment, Nelson Institute for Environmental Studies, University of Wisconsin-Madison. Drainage areas may only be for area above discharge gages.

[c]From: Brown, John Alexander Henstridge; Australia's Surface Water Resources; Published 1983 by Australian Government Publication Service, Canberra.

style feeding on various small invertebrates or decomposing animals.

Nemertea Ribbon worms are perhaps one of the least studied invertebrate groups of aquatic ecosystems, but at least 12 species have been described from freshwaters including the tropics. They are usually predators, but their functional role in tropical streams is probably marginal.

Nematomorpha Similar to the ribbon worms, hairworms have been little studied in aquatic systems, and their functional role is really unknown, but also thought to be minor. With over 100 freshwater species in the Gordioidae, all known species have a parasitic stage in the life cycle, where they infect terrestrial insects that are feeding along stream banks. When living outside the host, they are probably detritivores.

Nematoda Roundworms are ubiquitous, diverse, and abundant in aquatic ecosystems, and this is true for tropical streams. Many species are parasitic on aquatic insects (e.g., Chironomidae and Ephemeroptera) and other invertebrates, but free-living taxa are found in all aquatic habitats, although their small size (often <10 mm long) makes them difficult to collect and identify. Thus, even though nematodes are thought to play substantial roles in detrital decomposition, difficulty in sampling and identification has hampered quantitative estimates of their role in stream ecosystems. Most genera are considered cosmopolitan, represented mostly by three families (Dorylaimidae, Trilobidae, and Plectridae) with freshwater genera. The biomass of roundworms in aquatic systems is thought to be substantial, but this has been difficult to confirm from field studies.

Annelida The freshwater Annelida are best represented by three major classes: Polychaeta (bristle worms), Oligochaeta (earthworms), and Hirudinea (leeches). A minor class is the Branchiobdellida that are similar in appearance to leeches, but have both parasitic and commensal relationships with a variety of tropical invertebrates (e.g., on the gills of decapod

Figure 2 The near-ocean valley of a tropical oceanic mountain stream (left) showing the high gradient nature of many tropical mountain streams often interrupted by common large waterfalls (right) that prevent migration of many amphidromous stream invertebrates (Photos by M.E. Benbow).

shrimps). As a phylum, these worms are found in most aquatic habitats worldwide; however, the Polychaeta are relatively more abundant and widely dispersed in tropical regions compared to Temperate Zones, which is not the case for the other two classes of aquatic worms.

Polychaeta: Many of the estimated 50 species of freshwater polychaetes are thought to inhabit both fresh and brackish waters, with only about 17 species exclusively restricted to freshwaters. Of polychaetes, the family Nereidae is best represented by freshwater and euryhaline genera, and within this family the free moving (not tube dwelling) species are those found in tropical benthic habitats of streams and rivers worldwide (**Figure** 5). The polychaetes are distinct from the other classes by the presence of parapodial appendages, which allow some species to crawl and swim over substrates. Over half of the 50 euryhaline and freshwater species of Nereidae are located in the tropical and subtropical western Pacific region, whereas only 10 species are found in North America. The nereid polychaetes are free-moving predators that will consume large and small invertebrates using large proboscis jaws; they will also consume large amounts of detritus in the absence of prey items (**Figure** 5). Other families have filter-feeding and deposit-feeding representatives. Although polychaeta have not been well studied, their functional significance in tropical stream communities may be substantial as detritivores in slow-flowing pools and backwater habitats where they can be commonly collected among abundant decaying organic matter.

Oligochaeta: The Oligochaeta are generally more diverse and abundant compared to the Polychaeta and Hirudinea and are common to a wide variety of habitats, ranging from pools to riffles. However, most studies indicate that oligochaetes are most abundant in slow-flowing habitats and floodplains of streams and rivers. Of the estimated 700 species of aquatic oligochaeta, relatively few (probably less than 30%) are found in tropical streams and rivers, with most within the Naididae and Tubificidae families. It is thought that the success of these families in tropical regions is somewhat due to the potential of asexual reproduction in several genera that have temperate zone relatives with only sexual reproduction. Oligochaeta are typical detritus feeders, with some Naididae known to prey on microinvertebrates from the interstitial spaces of the benthic habitat. Densities for various tropical species of oligochaetes have been found to range from <20 to >500 000, providing evidence for their potential functional importance in organic matter recycling.

Figure 3 A representative stream reach of a lowland tropical river (top, Densu River, Ghana, Africa) and of an oceanic mountain stream (bottom, Ngardmau Stream, Republic of Palau) (Photos by M.E. Benbow).

Hirudinea: The Hirudinea are typically thought of as external parasites. Indeed, they commonly feed on blood of other animals including fish, amphibians, birds, and humans, whereas some species are predators on other aquatic invertebrates including worms, snails, clams, and some insects. This phylum is represented by two orders (Rhynchobdellida and Arhynchobdellida) with common taxa found in tropical freshwater benthic communities, most frequently found under stones or among macrophytes in both lentic and lotic habitats. Leeches are not well studied in tropical regions of the world, and their functional significance is not known.

Mollusca Tropical stream Mollusca fall into two classes, the Gastropoda (snails and limpets) and the Bivalvia (clams). The gastropods are further divided into orders Prosobranchia and Pulmonata based primarily on the mode of respiration, where the latter use a lung-like structure formed from the mantle tissue, whereas the former typically use ctenidia, or internal gills.

Prosobranchia: Some of the more common tropical prosobranch families are the Neritidae, Hydrobiidae, Thiaridae, Pomatiopsidae, and Pleuroceridae (but there are others), and they are, in general, more diverse and abundant than pulmonate snails in tropical regions of the world. Several species of Neritidae have unique

Figure 4 Large tropical rivers of South America. Top panel is the Iguazu River, Brazil (Photo by J. Wallace). Bottom panel shows the confluence of the Rio Solimos (brown) and the Rio Negro (black) forming the Amazon River, Brazil (Photo by RW Merritt).

life history strategies that include migrations to and/or from the estuary for reproduction. Some species are known to have planktonic life stages that drift to the ocean for marine development before returning to the freshwater streams as migratory spat (see glossary).

Neritid snails are most often found in flowing habitats of streams and are particularly common among oceanic island streams. Some species (e.g., *Neritina granosa*, endemic to the Hawaiian Islands; **Figure 6**) are found in the highest flow velocities of mountain streams, while a few others (e.g., some Viviparaceae) are more common in the sediments of standing waters such as floodplains. Most tropical stream snails are grazers, using radula to scrap biofilm from stream substrates, and play a major role in nutrient cycling and secondary production in many tropical montane stream ecosystems. Such stream snails serve as prey items for freshwater crabs and prawns, in addition to resources for some terrestrial vertebrates such as birds, rats, and humans.

There are several large species of prosobranch snails common to tropical streams and rivers, most notably those of the Ampullaridae (e.g., *Pila* and *Pomacea*), which can exceed 10 cm, compared to the smaller species of the Bithyniidae and Hydrobiidae with individuals usually <20 mm in height. Several species of tropical prosobranchs have medical importance, as they serve as reservoirs for parasitic trematodes and can play a role in human disease; however, most of these species are restricted to lentic water bodies.

Pulmonata: As a group, the pulmonates are relatively less widespread than prosbranchs in the tropics, and usually have a more omnivorous diet. The pulmonates are most represented by the families Ancylidae, Planorbidae, Lymnaeidae, and Physidae in tropical streams and rivers, some of which are found in the tropics only because of accidental or intentional introduction from Temperate Zone regions. Compared to prosbranchs, pulmonate snails are most

Table 2 Taxa list summary of tropical benthic invertebrate groups, providing common names, representative genera/taxa, and general categories of occurrence and habitat

Phylum	*Sub-phylum*	*Class*	*Order/ suborder*	*Family or infraorder*	*Represenatative common names*	*Represenatative genera*[a]	*Occurrence in tropical streams*[b]	*Stream habitat*[b]
Porifera					Spoages	*Spongilla, Ephydatia*	Occasional	Moderate flow, runs, pools
Cnidaria (= Coelenterata)		Hydrozoa			Jellyfish	*Hydra, Craspedacusta*	Rare	Pools, banks, floodplain ponds
Polyzoa (= Ectoprocta)					Bryozoa – moss-like animals	*Plumatella, Lophopodella*	Occasional	Mainly lentic, but sometimes lotic
Platyhelminthes		Turbellaria			Free living, planaria	*Dugesia, Planaria*	Common	All habitats
		Trematoda			Parasitic flukes		Occasional	Parasites of various invertebrates
		Monogenea			Fish ectoparasites		Common	Parasites of various invertebrates
		Cestoidea			Parasitic tapeworms		Occasional	Parasites of various invertebrates
		Temnocephalidea			Worms on freshwater crustacea		Common	Epizoic on crustacea
Nemertea					Ribbon (or proboscis) worms	*Prostoma*	Occasional	Pools, banks, floodplain ponds
Nematomorpha					Horsehair worms	*Gordius, Baetogordius*	Occasional	Pools, banks, floodplain ponds
Nematoda					Round worms	*Plectus, Hydromermis*	Occasional	All habitats
Annelida		Polychaeta			Bristle worms	*Namalycastis, Lycastis*	Common	Pools, banks, floodplain ponds
		Oligochaeta			Earthworms	*Nais, Tubifex, Enchytraeus*	Common	All habitats, but typically lentic
		Hirudinea (=Hirudinoidea)			Leeches	*Helobdella, Hirudo*	Common	All habitats, but typically lentic
		Branchiobdellida			Ectoparasites of decapod crustaceans		Occasional	Epizoic on crustacea

Mollusca		Gastropoda	Prosobranchia		Snails and limpets	*Neritina, Clithon, Rivularia*	Common	All habitats
			Pulmonata		Snails	*Lymnaea, Ferrissia*	Common	All habitats
		Bivalvia			Clams and mussels	*Corbicula, Musculium*	Common	All habitats, but most lentic
Arthropoda[c]	Crustacea	Malacostraca	Decapoda	Brachyura	Crabs	*Eriocheir, Geosesarma*	Common	All habitats
				Parastacidae	Crayfishes	*Cherax*	Rare	All habitats
				Cambaridae	Crayfishes	*Procambrus*	Rare	All habitats, mostly introduced
				Caridae	Shrimps and prawns	*Caridina, Macrobrachium*	Common	All habitats
		Peracarida	Amphipoda		Scud or side-swimmer	*Gammarus, Hyallela*	Common	Pools, banks, floodplain ponds
			Isopoda		Sow bug or water slater	*Caecidotea, Licerus*	Common	Pools, banks, floodplain ponds
			Mysidacea		Opossum shrimp	*Mysis, Neomysis*	Occasional	Pools, banks, floodplain ponds
		Branchiura			Fish lice	*Argulus*	Occasional	Epizoic on fish
		Ostracoda			Seed or mussel shrimp	*Candona, Cypris*	Common	Pools, banks, floodplain ponds
		Branchiopoda	Anomopoda	Daphniidae	Water fleas	*Daphnia, Ceriodaphnia*	Common	Pools, banks, floodplain ponds
			Anostraca		Fairy shrimp	*Artemia, Branchinecta*	Occasional	Pools, banks, floodplain ponds
			Conchostraca		Clam shrimp	*Lynceus, Leptisthera*	Occasional	Pools, banks, floodplain ponds
			Notostraca		Tadpole or shield shrimp	*Triops, Lepidurus*	Occasional	Pools, banks, floodplain ponds
		Copepoda			Copepods	*Cyclops, Harpacticus*	Common	Pools, banks, floodplain ponds
	Uniramia	Insecta			Insects	various	Common	All habitats
	Chelicerata	Arachnida	Araneae		Spiders	various	Occasional	Pools, banks, floodplain ponds
			Acarina	Hydrachnidae	Water mites	various	Common	All habitats

[a]Many of the representative taxa were found in Dodgeon 1999.

[b]These classifications are based on the general literature, which is incomplete for many of the taxa. Many taxa may be more common in tropical streams than listed here.

[c]Sometimes the arthropods are treated as unrelated groups of phyla that have segmented bodies and jointed appendages. When this is the case, the subphyla are treated as phyla.

Table 3 Taxa list summary of tropical benthic aquatic and semiaquatic insect orders, providing common names, some representative families, life stage in aquatic habitats and general categories of occurrence and habitat

Order	*Representative common names*	*Representative families*	*Life stage in aquatic habitat*	*Occurrence in tropical streams*#	*Stream habitat*#
Collembola	Springtails	Poduridae, Sminthuridae	Larvae, adult	Occasional	Banks, stream margins, water surface
Ephemeroptera	Mayflies	Baetidae, Heptageniidae, Leptophlebiidae	Larvae	Common	All habitats
Odonata	Dragonflies, damselflies	Libellulidae, Gomphidae, Coenagrionldae	Larvae	Common	All habitats
Plecoptera	Stoneflies	Perlidae, Nemouridae	Larvae	Occasional/ common	All habitats
Hemiptera	True bugs	Naucoridae, Belastomatidae, Gerridae	Larvae, adult	Common	All habitats
Blattodea	Cockroaches	Blaberidae, Blattidae	Larvae, adult	Occasional	Banks, stream margins, detritus
Orthoptera	Grasshoppers, locusts	Acrididae, Tetrigidae, Trydactylidae, Gryllidae	Larvae, adult	Occasional	Banks, stream margins
Neuroptera	Spongillaflies	Sisyridae	Larvae	Occasional	On freshwater sponges
Megaloptera	Dobsonflies, alderflies, fishflies	Corydalidae, Sialidae	Larvae	Occasional	All habitats
Lepidoptera	Moths, caterpillars	Pyralidae, Arctiidae	Larvae, pupae	Occasional/ common	Macrophytes (aquatic, semiaquatic), rocks
Coleoptera	Beetles	Elmidae, Psephenidae, Hydrophilidae, Dytiscidae	Larvae, pupae, adult	Common	All habitats
Trichoptera	Caddisflies	Hydropsychidae, Hydroptilidae, Leptoceridae	Larvae, pupae	Common	All habitats
Hymenoptera	Wasps	Agriotypidae, Pompilidae	Larvae, pupae	Occasional/rare	Parasitoids of aquatic/ semiaquatic insects
Diptera	True flies	Chironomidae, Culicidae, Simuliidae	Larvae, pupae	Common	All habitats

*Many of the representative taxa were found in Dudgeon (1999).

#These classifications are based on the general literature, which is incomplete for many of the taxa. Many taxa may be more common in tropical streams than listed here.

often encountered in slower-flowing riverine habitats, or pools and backwaters, but some species (e.g., some Lymnaeidae and Physidae) can be collected from riffles and runs of flowing streams. Similar to prosobranchs, there are several species of tropical lymnaeid and planorbid pulmonates that act as reservoirs for parasitic trematodes that cause human disease.

Bivalvia: Most tropical freshwater bivalves inhabit pools and backwater habitats of floodplains, or bury in the substrate of large, slow-flowing rivers. In smaller streams, species of some families, like the Corbiculidae and Sphaeriidae, are relatively more abundant. Species of *Corbicula* have been distributed over much of the tropics (**Figure** 7). Although introduced from Asia as a human food source into many of these areas, especially tropical oceanic islands, *Corbicula fluminea* can reach extremely high densities in both streams and rivers, effectively displacing indigenous taxa.

Besides the two families already mentioned, there are several Unionidae clams found throughout the tropics, and other taxa represent the families Mytilidae, Margaritiferidae, and Hyriidae. It is not completely known, but the unionid and fingernail clams (Sphaeriidae) are probably the most highly diverse and widespread stream bivalves of the tropics, respectively, and probably play a major functional role in these ecosystems. Both groups have high secondary production in flowing water bodies and act as filterers of suspended matter in these systems. Some

Figure 5 A representative polychaete worm, *Namalycastis sp.* (Annelida: Polychaeta), common to depositional habitats of Hawaiian mountain streams. Polchaetes are typically more common and diverse in tropical compared to Temperate Zones and rivers (Photos by AJ Burky).

unionids have spread from their native regions, perhaps in part because of a life stage that encysts on fish, allowing for long-distance dispersal via fish migrations, or through human fish stocking. Likewise, as the smallest bivalve taxa (rarely >10 mm), fingernail clams are often distributed by waterfowl or through human movement of domestic plants (e.g., taro), when the clams become trapped in feathers or root mats. Because the Sphaeridae brood larvae inside the shell, both the adults and offspring can resist desiccation for relatively long periods.

Arthropoda *Decapod Crustacea:* Physically, the largest freshwater invertebrates that occupy freshwaters are the Decapoda. Nearly 10% of all decapod crustaceans are found in freshwater ecosystems, and in tropical streams they play a large functional role in organic matter processing, habitat structure, and food web linkages that affect both vertebrates and invertebrates. Because of this, the decapods are given more extensive treatment compared to that of other crustaceans. Three families of the Infraorder Caridea have adults that exclusively occupy freshwaters: Atyidae, Alpheidae, Palae-monidae (**Figure 8**). For many species of these families, the larval habitats can

Figure 6 Neritid snails (Gastropoda: Prosobranchia) are very common in tropical streams and rivers. Top panel shows several neritids in a stream of the Republic of Palau, with a pyralid moth larva on the finger (Photo by AJ Burky). Bottom panel shows a closer view of a live *Neritina granosa* in a Hawaiian mountain stream (Photo by M Yamamoto).

Figure 7 A small (about 1 cm) *Corbicula* sp. found in many tropical inland waters of the world (Photo by ME Benbow).

Figure 8 Clockwise from top left: a shrimp (*Xiphocaris elongata*) from a tropical stream of Central America (Photo by JP Benstead), another amphidromous shrimp endemic to Hawaii (*Atyoida bisulcata* Randall; photo by ME Benbow), an amphidromou prawn (*Macrobrachium lar* Holthuis) endemic to Hawaii (Photo by ME Benbow), and *Macrobrachium carcinus* Linneaus from Central America (Photo by JP Benstead).

vary from complete restriction to freshwater to a required period of completely marine development, whereas others require or tolerate brackish exposure before returning to freshwater habitats. The variation of life cycles offers evidence for marine ancestors for most, if not all, of the freshwater shrimps, and this ancestry is exemplified in several species (e.g., *Atyioda bisulcata*, *Atya lanipes*, *Xiphocaris elongata*) with a specialized diadromous life cycle.

The life cycle is called amphidromy and appears to be unique to several decapod crustaceans, neritid snails, and gobiid fishes of Oceania, the Neotropics, and Indomalaya and probably tropical Africa, but this region has been much less studied. There are also known amphidromous taxa from Japan and New Zealand. In the amphidromous life cycle, the adults live and reproduce in freshwater habitats, the larvae or eggs drift to the ocean where they develop for some time before migrating back into the streams as juveniles. The major difference between amphidromy and other forms of diadromy is that the migration is not directly related to immediate reproduction. This life cycle is thought to be an adaptation to life in moving waters, where floods are frequent and unpredictable and have high potential for washing eggs and/or larvae downstream and into the ocean.

Several species of the infraorder Caridae have this amphidromous life cycle; however, there is little definitive knowledge about the life cycle of most shrimp species. The two most diverse and widespread families of freshwater decapods that also have amphidromous species are the Atyidae (shrimps) and the Palaemonidae (prawns). The larval stages of both families can be found in both fresh and brackish water but the adults are usually restricted to freshwater or brackish habitats.

The atyid shrimps are some of the most well-studied tropical decapods, but there is still much debate on the taxonomy of the largest genus, *Caridina* (sometimes synonymous with *Neocaridina*), making identifications difficult in many parts of the world. Genera of the Atyidae are most often large components of tropical mountain streams characterized by high gradient and stochastic flow. They play several functional roles, and often these functions change throughout development. Many species can filter-feed from the water column, act as a scraper, or develop a collector-gatherer mode of food acquisition. The atyid shrimps have been heavily studied in relation to organic matter processing and are known to be very important to tropical stream ecosystem function. The amphidromous life cycle makes populations of many native and indigenous species vulnerable to existing and proposed dams and other stream flow obstructions.

Palaemonid decapods are most often referred to as prawns (even though they are shrimps), mostly being distinguished from atyid shrimps by morphology and overall body size; many species can grow to over 100 mm in carapace length (**Figure 8**). Because of there large size, many palaemonids are cultured as a food source throughout tropical latitudes. With over 175 reported species of *Macrobrachium* throughout the world, the subfamily Palaemoninae has the only freshwater representatives. And although most species of this widely distributed genus are omnivorous, there are also several predatory species that play a large role in benthic community structure. Some *Macrobrachium* species have been introduced into isolated stream ecosystems as a human food source, only to have dire consequences to the native invertebrate taxa. For instance, *Macrobrachium lar* is an introduced prawn to Hawaiian Island streams where it is a generalist predator on several endemic stream invertebrates including a congener brackish water prawn (*Macrobrachium grandimanus*), atyid shrimp, snails, and insects.

The Alpheidae is a small family of freshwater decapods, with only a few species (e.g., *Alpheus cyanoteles* and *Potamalpheops* spp.) reported from tropical freshwaters. Most genera and species of Alpheidae are restricted to the marine environment.

Another group of predominately marine decapods that can only be found in freshwaters of tropical latitudes are freshwater crabs (Brachyura) (**Figure 9**).

Figure 9 An unknown freshwater crab (Brachyura) collected from a stream in Costa Rica (Photo by AJ Burky).

Typically thought of as a major benthic faunal difference between temperate and tropical stream ecosystems, freshwater crabs are sometimes considered the functional replacement of insect shredders, which are arguably reduced in both diversity and quantity in tropical freshwaters. Two major groups of freshwater Brachyura are of the mostly marine and estuarine Grapsidae family, and include species of the genus *Eriocheir* and *Geosesarma*, that differ in life cycle characteristics. Adults of *Eriocheir* migrate to the estuary to breed, whereas *Geosesarma* complete an entirely freshwater life cycle. In general, there is little known about the ecology and life histories of freshwater crabs.

Insect Communities

Insects can be found in aquatic habitats worldwide, with the same major orders commonly found in both temperate and tropical regions. Common orders in tropical streams include the Ephemeroptera, Odonata, Plecoptera, Hemiptera, Trichoptera, Coleoptera, and Diptera (**Figures 10–14**). Although these orders are most common, many other aquatic insect groups, such as the Lepidoptera, Megaloptera, and Neuroptera, also have tropical representatives at the genus and species level (**Table 3**).

Terrestrial insect communities are more diverse in tropical regions compared to temperate regions of the world; however, this general trend does not always hold true for aquatic insect communities. Studies have reported lower, higher, or equal aquatic insect diversity in tropical compared to Temperate Zone streams. At the order level, the stoneflies (Plecoptera) tend to be less diverse in the tropics; whereas, other groups such as riffle beetles (Coleoptera: Elmidae), moths (Lepidoptera; **Figure 10**) damselflies, and dragonflies (Odonata; **Figure 10**) tend to be more diverse in the tropics. However, no general trends are apparent for all tropical streams; this pattern may be due to high variation in geologic history, biogeography, seasonality, hydrologic variability, resource availability, and abiotic/biotic factors between different tropical regions. Possible explanations for lower tropical aquatic insect diversity may be due to insufficient sampling and taxonomy, constant temperatures, and increased disturbance events (e.g., floods), compared to temperate regions. Possible explanations for higher tropical aquatic insect diversity may be from high temperatures that increase mutation rates and lack of historical geologic/climate disturbances (e.g., ice age). The reasons for temperate-tropical taxa richness differences (if any) are still highly debated.

Aquatic insects can be found in all stream habitats (**Table 3**). Insects occupy habitats that provide the best

Figure 10 Clockwise from top left: a damselfly (Odonata) larva from a tropical stream in Costa Rica (Photo by AJ Burky), a stonefly (Plecoptera) from Brazil (Photo by ME Benbow), a mayfly (Ephemeroptera) larva of the family Euthyplocidae collected from Brazil (Photo by ME Benbow), and a moth (Lepidoptera) larva of the family Pyralidae from the Republic of Palau (Photo by ME Benbow).

Figure 11 Two caddisfly (Trichoptera) larvae from the family Hydroptilidae (left) and Hydropsychidae (right) collected from a tropical mountain stream on the Hawaiian Islands (Photo by MD McIntosh).

conditions (e.g., substrate, flow, food availability) for that species. These requirements can vary for the same species during different life stages (e.g., egg, larval, pupae, and adult). In tropical streams, many insect taxa are adapted to fast flowing, erosional habitats, such as torrential cascades and riffles; these groups generally have long tarsal claws, dorsoventrally flattened bodies, use secretions (e.g., silk) or suckers to aid in attachment, and utilize the fast flowing water for food resources (e.g., filter food from the water), dissolved oxygen, and dispersal. Common groups found in these habitats are the Diptera (Chironomidae, Simuliidae, Blephaceridae), Coleoptera (Elmidae), and Trichoptera (Hydropsychidae). Other aquatic insects (Odonata, Hemiptera: Naucoridae, Belastomatidae, **Figure 12**) are adapted for slower-moving depositional habitats, such as pools, using morphological modifications to protect bodies from the accumulation of depositional material such as leaves and silt. For example, some mayflies have an operculate gill, or expanded gill, which shields smaller gills and allowing for respiration. Other taxa cling to or mine into submerged vegetation, while others burrow into the hyporheic zone (area beneath and lateral to the stream bed).

In the tropics, the life cycle of aquatic insects is most influenced by both radiation/temperature and hydrologic variation. Insect growth is dependent on temperature, and although relatively constant temperatures are common in tropical streams there is still some variation in insect growth and life cycles. With consistently warm temperatures, many aquatic

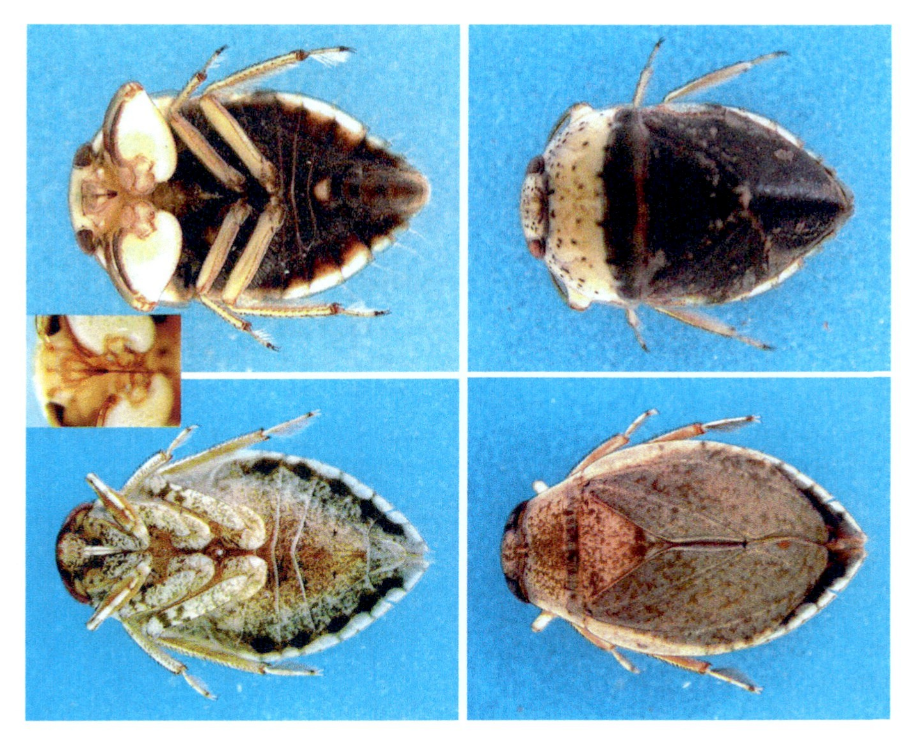

Figure 12 Ventral (left panels) and dorsal (right panels) of adult Hemiptera from the family Naucoridae (top panels) and Belastomatidae (bottom panels) from a tropical river in Ghana, Africa (Photo by T White).

Figure 13 The dorsal (top) and lateral (bottom) views of a beach fly (Diptera) larva from the family Canacidae collected from the Hawaiian Islands (Photo by MD McIntosh).

Figure 14 Life stages of a tropical Diptera: clockwise from left is the larva, adult, and pupa (with yellow eggs in abdomen). This photo is of *Telmatogeton torrenticola* Terry (Diptera: Chironomidae) endemic to the Hawaiian Islands (Photo by AJ Burky).

insects have evolved multivoltine life cycles, with continuous reproduction all year. For example, species of the trichopteran *Cheumatopsyche* (family Hydropsychidae; **Figure 11**) have a multivoltine life cycle in tropical streams of Hawaii, whereas in temperate regions of North America, univoltine life cycles have been most often reported. In general, multivoltine organisms have overlapping generations, shorter generation times, and tend to be smaller in size. Some aquatic insects, despite the warm temperatures, have evolved seasonal life cycles (one to two generations per year); these organisms tend to be larger in size with generations occurring in relation to fluctuations in discharge events. Variable discharge can influence not only aquatic insect life cycles, but also population densities, biomass, and secondary production. Some research suggests that increased natural discharge events (both in magnitude and frequency) are associated with lowered densities and biomass of aquatic insects; however, others studies have not found such a relationship.

Instead of taxonomic classification, aquatic insects can be classified into functional groups based on similar feeding behavior and morphology. In Temperate Zone streams, the functional feeding group of shredders, organisms that breakdown large coarse particulate organic materials (CPOM) (e.g., leaves, wood), have evolved life histories based on the predictable and large input of leaves during the fall season. These shredders play an important role in stream ecosystem functioning by making food resources and nutrients available for other aquatic organisms. However, the absence of insect shredders has been widely reported in tropical stream ecosystems. Several theories have been suggested for this: (1) increased variability in discharge events has reduced the amount of time that CPOM is available in tropical streams; (2) chemical and physical properties of tropical plants prohibit use by shredding organisms; and (3) increased importance of microbial and large crustacean communities on the decomposition of fallen leaf litter. This difference suggests that insect shredders have not evolved in tropical streams due to an absence, or reduction, in resource availability or due to interspecific competition with other macroconsumers such as shrimps or crabs. This pattern is still highly debated among researchers, but future studies should begin to resolve this issue.

Glossary

Amphidromy – A form of diadromy where hatched embryos or larvae spend an obligatory period of larval growth in marine or brackish habitats before migrating back into freshwater habitats; the marine stage is not related to breeding.

Benthos – Community of aquatic organisms generally restricted to the near-substrate area, or bottom, of a water body; animal communities are zoobenthos and plant communities are phytobenthos.

Choanocytes – Also known as collar cells with flagella that line the interior body wall sponges and produce flow currents important for filter feeding.

Commensal – A relationship between organisms, where one organism benefits but the other is neither benefited nor harmed by the other.

Ctenidia – A gill-like structure that usually serves for respiration or filter feeding found in aquatic molluscs.

Ecozone – The largest scale biogeographic division of collective animal and plant communities of the world, with the groups in different ecozones usually being separated from each other over geological time scales.

Epizoic – Living or growing on the surface of an animal.

Hyporheic zone – The region directly underneath and lateral to the benthic substrates of inland waters where there is mixing of surface and ground water.

Insular – Typically referring to islands in biology and ecology.

Lentic – Slow flow or nonmoving aquatic habitats.

Lotic – Flowing water aquatic habitats.

Montane – The area or regions below the tree-line of mountains, typically associated with higher precipitation.

Multivoltine – Referring to life cycles – having more than one generation per year.

Parapodia – Paired and unjointed lobes/appendages of the polychaete worms, often used for locomotion.

Radula – A serrated, or toothed, structure found in the mouths of gastropods, usually used for scraping food from substrates.

Spat – Juvenile stage associated with benthic molluscs.

Univoltine – Referring to life cycles – having only one generation per year.

See also: Africa; Asia – Monsoon Asia; Australia (and Papua, New Guinea); Deforestation and Nutrient Loading to Fresh Waters; Hydrology: Streams; Regulators of Biotic Processes in Stream and River Ecosystems; South America.

Further Reading

Benstead JP, DeRham PH, Gattolliat J-C, *et al.* (2003) Conserving Madagascar's freshwater biodiversity. *BioScience* 53: 1101–1111.

Bright GR (1982) Secondary benthic production in a tropical island stream. *Limnology and Oceanography* 27: 472–480.

Covich AP (1988) Geographical and historical comparisons of neotropical streams: Biotic diversity and detrital processing in highly variable habitats. *Journal of the North American Benthological Society* 7(4): 361–386.

Cheshire K, Boyero L, and Pearson RG (2005) Food webs in tropical Australian streams: Shredders are not scarce. *Freshwater Biology* 50: 748–769.

Crowl TA, McDowell WH, Covich AP, and Johnson SL (2001) Freshwater shrimp effects on detrital processing and nutrients in a tropical headwater stream. *Ecology* 82: 775–783.

Cushing CE, Cummins KW, and Minshall GW (eds.) (2006) River and Stream Ecosystems of the World. Los Angeles, CA: University of California Press.

Dudgeon D (1999) *Tropical Asian Streams: Zoobenthos, Ecology and Conservation.* Hong Kong: Hong Kong University Press.

Dudgeon D (2000) The Ecology of tropical Asian rivers and streams in relation to biodiversity conservation. *Annual Review of Ecology and Systematics* 31: 239–263.

Huryn AD and Wallace JB (2000) Life history and production of stream insects. *Annual Review of Entomology* 45: 83–110.

Jacobsen D, Schultz R, and Encalada A (1997) Structure and diversity of stream invertebrate assemblages: The influence of temperature with altitude and latitude. *Freshwater Biology* 38: 247–261.

Minshall GW (1988) Stream ecosystem theory: A global perspective. *Journal of the North American Benthological Society* 7: 263–288.

McDowall RM (2007) On amphidromy, a distinct form of diadromy of aquatic organisms. *Fish and Fisheries* 8: 1–13.

McDowall RM (1992) Diadromy: Origins and definitions of terminology. *Copeia* 1992(1): 248–251.

Pringle CM, Freeman MC, and Freeman BJ (2000) Regional effects of hydrologic alterations on riverine macrobiota in the New World: Tropical-temperate comparisons. *Bioscience* 50: 807–823.

Wantzen KM and Wagner R (2006) Detritus processing by invertebrate shredders: A neotropical-temperate comparison. *Journal of the North American Benthological Society* 25(1): 216–232.

Relevant Websites

http://www.stri.org/index.php – Smithsonian Tropical Research Institute.

http://www.sage.wisc.edu – Center for Sustainability and the Global Environment, Nelson Institute for Environmental Studies, University of Wisconsin-Madison.

http://rainforests.mongabay.com/0603.htm – Mongabay.com.

http://ites.upr.edu/ – Institute for Tropical Ecosystems Studies – University of Puerto Rico.

http://mit.biology.au.dk/cenTER/ – Center for Tropical Ecosystems Research (center Aarhus).

Benthic Invertebrate Fauna

D T Chaloner, University of Notre Dame, Notre Dame, IN, USA
A E Hershey, University of North Carolina at Greensboro, Greensboro, NC, USA
G A Lamberti, University of Notre Dame, Notre Dame, IN, USA

Introduction

Benthic invertebrates are an abundant and diverse group of aquatic animals (**Figure 1**) that can be found on or in submerged substrates of both flowing (lotic) and standing (lentic) freshwaters. Also referred to as the zoobenthos, benthic invertebrates are an important component of the benthos, a complex community that can also include a diversity of plant and microbial species. Benthic invertebrates have been the subject of considerable research from the perspectives of their diversity, adaptations, community structure, and role in trophic and nutrient dynamics. Currently, more work has been published about lotic than about lentic benthos, which partially reflects the historical emphasis of lotic research on the benthos, whereas lentic research has focused on the overlying water (i.e., the pelagic). However, this distinction may also reflect the higher diversity of lotic benthic habitats and biota, and the more complex management issues associated with flowing waters as compared with more static lakes and ponds.

The Physical and Chemical Context

The contrasting physicochemical challenges of flowing and standing water impose different physiological, morphological, and energetic constraints on benthic organisms. Despite these constraints, a myriad of adaptations has enabled benthic invertebrates to colonize virtually all benthic habitats of inland waters (**Figure 2**), including those of ephemeral streams and wetlands. Major physicochemical factors that influence benthic invertebrates include the characteristics of (1) the substrate, (2) the overlying water, and (3) the water flow. All of these factors are interdependent to varying degrees.

Important substrate characteristics that differentially affect benthic invertebrate taxa include (1) the quantity of organic and inorganic (i.e., mineral) material, (2) the quality of the organic component(s), and (3) the size and composition of substrate particles. Particle size influences physical stability of the substrate and contributes to the food quality of organic components, and therefore is often used to classify substrate materials (**Tables 1** and **2**). Invertebrates can also modify the substrate through feeding, especially their production of fecal pellets, and construction of biogenic structures, such as tubes and cases (**Figure 3**).

Conditions associated with lentic and lotic substrates are influenced by the physicochemical characteristics of the overlying water, including temperature, dissolved oxygen, and dissolved nutrient concentrations. Benthic invertebrates are poikilothermic, so their internal temperature varies with their surrounding environment. Consequently, the growth, metabolism, and movements of benthic invertebrates are all influenced by changes in water temperature, which in turn sets broad limits on their distribution. However, the high specific heat capacity of water means that benthic invertebrates rarely experience temperature extremes; although when they do, large die-offs can occur. One common reason for these die-offs is reduced dissolved oxygen concentrations or complete anoxia. Lotic environments are usually supersaturated with oxygen because of aeration caused by turbulence. Lentic benthos, however, can experience anoxia, especially during thermal stratification, which prevents water and gas exchange with surface waters. Some benthic invertebrates have evolved adaptations to cope with low dissolved oxygen concentrations, such as some chironomid midges (e.g., Chironomini, **Figure 1(i)**), which are red because they produce hemoglobin to increase the efficiency of oxygen uptake, and oligochaetes, (e.g., *Tubifex*, **Figure 1(b)**), which may extend burrows out of anaerobic sediments into the slightly more oxygenated overlying water. High organic content of the substratum can also cause anoxia because of biological oxygen demand associated with decomposition. Inland waters with high dissolved nutrient concentrations typically have substrates with high organic content because dissolved nutrient concentrations usually increase the generation of organic material by primary production. The overall chemical composition of freshwater, including the concentration of various ions, is unique to individual watersheds and can directly influence the presence of certain invertebrates. For example, mollusks require sufficient dissolved calcium to build their shells. The persistence of water in temporary pools and ephemeral streams limits benthic taxa to those that can tolerate desiccation, readily colonize, and complete their life cycle before water disappears.

Water flow strongly influences the lotic benthic environment, including under-surfaces of rocks and deep within the sediment (i.e., the hyporheic). Indeed,

Figure 1 Continued

Figure 1 Continued

Figure 1 Continued

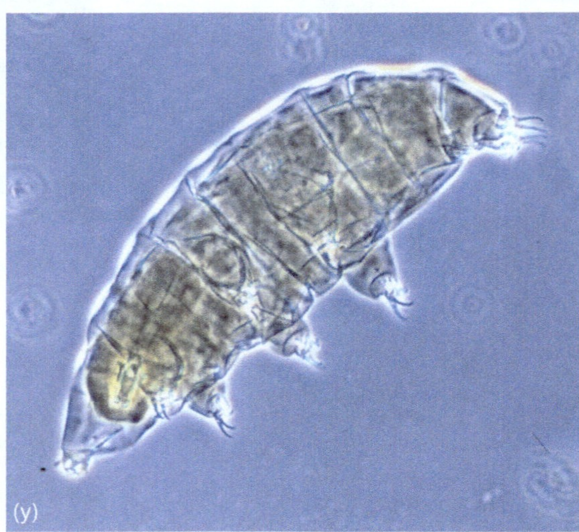

Figure 1 Photographs of major benthic invertebrate taxa, including the leech *Placobdella* (Rhynchobdellida: Glossiphoniidae) (a) (courtesy of L. Serpa), oligochaete *Tubifex* (Haplotaxida: Tubificidae) (b) (courtesy of Peter Chapman), water mite *Torrenticola* (Acariformes: Torrenticolidae) (c) (courtesy of A.J. Radwell), amphipod *Hyalella azteca* (Amphipoda: Hyalellidae) (d) (courtesy of U.S. Department of Agriculture), crayfish *Orconectes propinquus* (Decapoda: Cambaridae) (e) (courtesy of K.A. Crandall), isopod *Asellus aquaticus* (Isopoda: Asellidae) (f) (courtesy of P. Busselen (K.U. Leuven-Campus Kortrijk)), ostracod *Cypris* (Podocopida: Cyprididae) (g) (courtesy of G. Matthews), aquatic beetle larvae *Anchytarsus bicolor* (Coleoptera: Ptilodactylidae) (h) (courtesy of W. Davis), chironomid midge larvae (Diptera: Chironomidae) (i) (courtesy of D. Penrose (NC Division of Water)), blackfly larvae *Simulium* (Diptera: Simuliidae) (j) (courtesy of R.W. Merritt), cranefly larva *Tipula* (Diptera: Tipulidae) (k) (courtesy of R.W. Merritt), mayfly nymph *Baetis* (Ephemeroptera: Baetidae) (l) (courtesy of J. Benda), dobsonfly larva *Corydalus cornutus* (Megaloptera: Corydalidae) (m) (courtesy of S.D. Cooper), dragonfly nymph *Aeshna interrupta* (Odonata: Aeshnidae) (n) (courtesy of Royal BC Museum), stonefly nymph *Isoperla* (Plecoptera: Perlodidae) (o) (courtesy of D. Penrose (NC Division of Water)), caddisfly larva *Hydropsyche* (Trichoptera: Hydropsychidae) (p) (courtesy of A. Elosegi), bryozoan *Lophopodella carteri* (Plumatellida: Lophopodidae) (q) (courtesy of E. Wöss), hydra *Hydra* (Anthoathecatae: Hydridae) (r) (courtesy of J. Benda), mussel *Margaritifera falcata* (Unionoida: Margaritiferidae) (s) (courtesy of U.S. Fish and Wildlife Service), snail *Physella acuta* (Basommatophora: Physidae) (t) (courtesy of N. Yotarou), roundworm (Nematoda) (u) (courtesy of J. Grosse), horsehair worm *Paragordius tricuspidatus* (Gordioidea: Chordodidae) (v) (courtesy of Free Software Foundation), flatworm *Dugesia lugubris* (Tricladida: Planariidae) (w) (courtesy of S.D. Cooper), freshwater sponge *Spongilla lacustris* (Haplosclerida: Spongillidae) (x) (courtesy of H.F. Clifford), and tardigrade *Milnesium tardigradum* (Apochela: Milnesiidae) (y) (courtesy of K. Kendall-Fite).

many aspects of aquatic invertebrate biology, such as body shape, movement, and food acquisition, are influenced by water flow or hydrodynamics. For example, blackflies (**Figure 1**(**j**)) and net-spinning caddisflies (**Figure 1**(**p**)) prefer fast-moving water and have evolved adaptations to trap food items from water flowing to them. In contrast, some chironomid midges (**Figure 1**(**i**)) and cased caddisflies (**Figure 3**(**c**)) prefer slow-moving water, such as in-stream pools, because deposition of organic particles provides food. These deeper pools with finer substrates typically alternate with shallow areas of faster-moving water and coarser substrate to produce characteristic alternating riffle-pool sequences (**Figure 2**(**a**)) that typically support contrasting invertebrate assemblages. Although generally less important in lentic ecosystems, water

Figure 2 Major benthic habitats and associated substratum in flowing (a) and standing (b) waters.

Table 1 Classification of inorganic particles found in the substratum (Wentworth scale)

Size category	*Size range (diameter)*
Boulder	>256 mm
Cobble	64–256 mm
Pebble	16–64 mm
Gravel	2–16 mm
Very coarse sand	1–2 mm
Coarse sand	500 μm–1 mm
Medium sand	250–500 μm
Fine sand	125–250 μm
Very fine sand	62.5–125 μm
Silt	39–62.5 μm
Clay	<39 μm

flow can be important to benthic invertebrates when wind causes sieches (i.e., internal waves) which can generate turbulence within the pelagic, thus reducing anoxia in the profundal zone and bringing nutrients to the surface waters. In the shallow littoral (**Figure 2(b)**), surface waves are also an important source of turbulence, and hence potential disturbance.

Both lotic and lentic ecosystems are characterized by gradients in depth. Depth *per se* probably has little direct influence on benthic invertebrates, but other factors that are correlated with depth have important direct effects. Water depth affects light, temperature, and dissolved oxygen. In the littoral region of lentic ecosystems, light penetrates to the bottom, such that macrophytes and algae can provide substrates and food resources for benthic invertebrates. In contrast, the profundal region is perennially dark, and consequently benthic invertebrates are dependent upon organic material generated in the overlying water rather than by benthic producers. Lotic systems also increase in depth from small headwater tributaries to larger mainstream rivers. With increasing depth of lotic systems, light may become limiting, dissolved oxygen may be reduced, substrates can become finer, and the benthic environment becomes more depositional than erosional, although substrates of large particle size may still be present.

The Biotic Context

Types and Classification of Benthic Invertebrates

Benthic invertebrate fauna includes representatives of nearly all animal phyla, from small, simple organisms such as water bears (<1 mm, **Figure 1(y)**), to large macroinvertebrates such as crayfish (10–150 mm in length, **Figure 1(e)**) (**Table 3**). Aquatic insects are

Table 2 Classification of non-living organic particles found in the substratum (after PS Giller and B Malmqvist (1998) *The Biology of Streams and Rivers*. New York: Oxford University Press

Size category	*Size range (diameter)*	*Source*
Coarse particulate organic material (CPOM)	>64 mm	Large wood
	16–64 mm	Leaf packs
	4–16 mm	Leaf and wood fragments, fruits, buds, and flowers
	1–4 mm	Plant and animal detritus, feces
Fine particulate organic material (FPOM)	0.5 μm–1 mm	Wetting, physical abrasion, microbial colonization, and shredding of CPOM
Ultrafine particulate organic material (UPOM)	0.45–75 μm	Microbes and breakdown of FPOM
Dissolved organic material (DOM)	<0.45 μm	Organic compounds dissolved in water (e.g., humic acids)

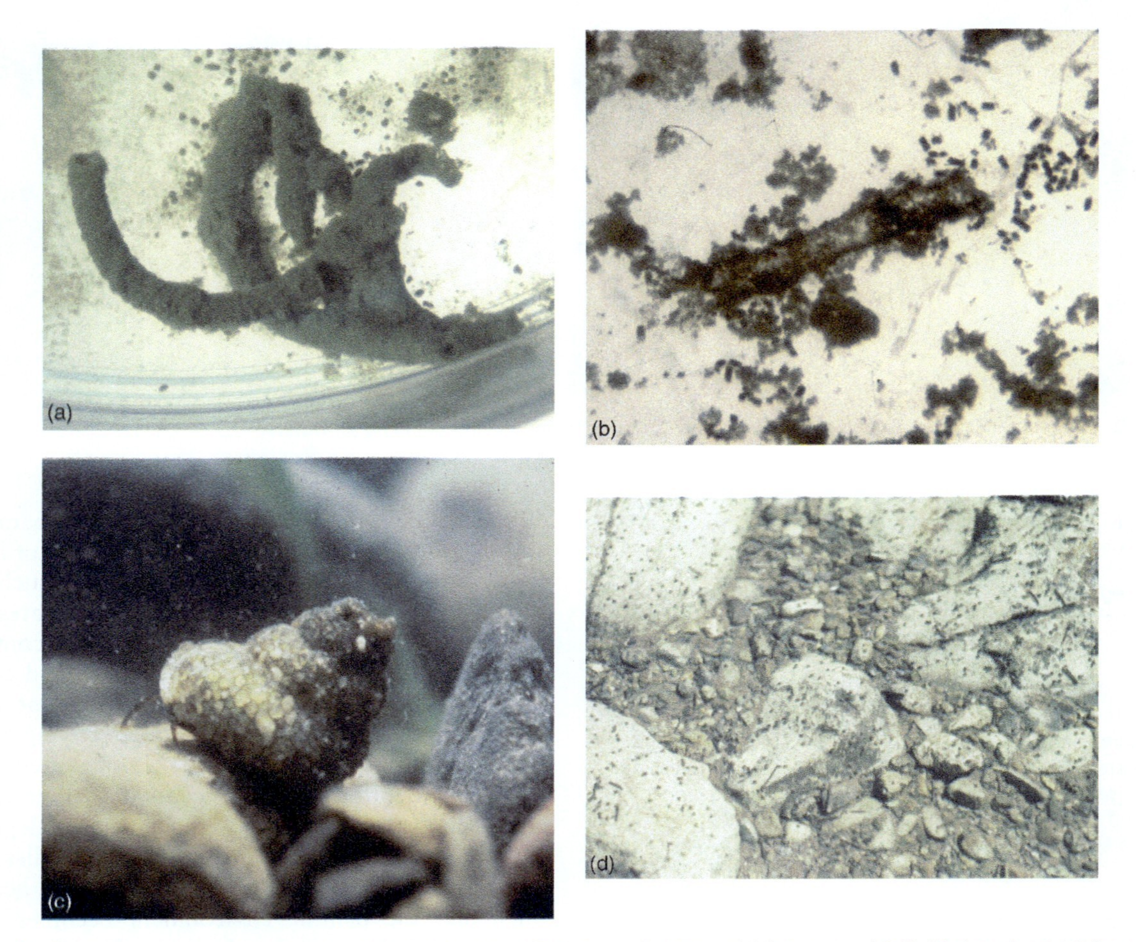

Figure 3 Examples of tubes built by the chironomid midges *Tanytarsus fimbriatus* (a) (courtesy of D.T. Chaloner), and *Cricotopus sylvestris* (b) (courtesy of D.T. Chaloner), and case built by the caddisfly *Helicopsyche borealis* (c) (courtesy of S.D. Cooper), which can be found at high densities covering rocks (d) (courtesy of G.A. Lamberti).

usually the most species-rich and abundant group of benthic invertebrates, having successfully colonized virtually all freshwater as well as many saltwater habitats. The overall diversity of benthic invertebrates, although often high, is variable across both lentic and lotic habitats, with many factors contributing to the complex diversity patterns apparent in both space and time (**Figures 4** and **5**). Such patterns at the regional scale reflect a combination of latitudinal and elevational factors, past glaciation and speciation events, and biogeographical constraints. Patterns at a local scale reflect a combination of biotic interactions and landscape features that influence the physicochemical features already discussed. For example, in the lentic benthos, the shallow littoral zone typically supports a more diverse invertebrate fauna

Table 3 Major groups of benthic invertebrates

Taxonomic grouping	*Common name*	*Size range (mm)*	*Feeding mode(s)*[a]	*Example genus*
Annelida				
Hirudinea	Leeches	5–300	Pr	*Placobdella*
Oligochaeta	Aquatic earthworms	5–400	Cg	*Tubifex*
Arthropoda				
Arachnida	Water mites	<1–8	Pr, Pa	*Torrenticola*
Malacostraca				
Amphipoda	Scuds	2–20	Sh, Cg, Pr	*Hyalella*
Decapoda	Crayfish	10–150	Sc, Sh, Pr	*Orconectes*
Isopoda	Sowbugs	5–20	Sh, Cg	*Asellus*
Ostracoda	Seed shrimp	<1–4	Cf	*Cypris*
Insecta				
Coleoptera	Aquatic beetles	1–80	Sh, Cg, Sc, Pr	*Stenelmis*
Diptera				
Chironomidae	Non-biting midge flies	2–30	Sh, Cf, Cg, Sc, Pr	*Chironomus*
Simuliidae	Blackflies	3–15	Cf	*Simulium*
Tipulidae	Craneflies	10–120	Sh	*Tipula*
Ephemeroptera	Mayflies	1–20	Cf, Cg, Sc, Pr	*Baetis*
Megaloptera	Dobsonflies, alderflies	5–90	Pr	*Corydalus*
Odonata	Damselflies, dragonflies	18–70	Pr	*Aeshna*
Plecoptera	Stoneflies	3–50	Sh, Pr	*Isoperla*
Trichoptera	Caddisflies	2–50	Sh, Cf, Cg, Sc, Pr	*Hydropsyche*
Bryozoa	Moss animals	10–400	Cf	*Lophopodella*
Cnidaria	Hydra	2–20	Pr	*Hydra*
Mollusca				
Bivalvia	Mussels, clams	2–250	Cf, Cg	*Margaritifera*
Gastropoda	Snails	2–70	Sc, Cg	*Physa*
Nematoda	Roundworms	<4	Pr, Cg, Pa	*Rhabditis*
Nematomorpha	Horsehair worms	100–1000	Pa	*Gordius*
Platyhelminthes	Flatworms	1–35	Cg, Pa, Pr	*Dugesia*
Porifera	Sponges	300–600	Cf	*Spongilla*
Rotifera	Rotifers	<1–3	Cf	*Philodina*
Tardigrada	Water bears	<1	Pr	*Milnesium*

Size and feeding mode taken from published literature.
[a]Sh, Shredder; Cf, Collector-filterer; Cg, Collector-gatherer; Sc, Scraper; Pr, Predator; Pa, Parasite.

than the profundal because of the higher habitat heterogeneity, more abundant resources, and generally favorable oxygen conditions compared to the deeper profundal.

Understanding the ecological consequences of benthic invertebrate diversity and abundance requires classification of the invertebrates present into logical groups. The most obvious, and perhaps scientifically defensible, approach is taxonomical grouping. However, identifying large numbers of invertebrates to the generic level or below can be very time-consuming, especially if taxonomic information is limited, and often generates a large amount of difficult-to-interpret data. Alternatively, invertebrates can be categorized into larger groups, based on broader taxonomic groupings, size, preferred habitat, or functional feeding group (FFG). Size and habitat are of limited utility because they both constitute very broad categories that have limited ecological relevance.

The FFG approach is perhaps the most widely used and involves categorizing invertebrates according to their mode of feeding, historically by their morphology, behavior, and gut contents. This approach reduces the number of groups by one to two orders of magnitude compared with the taxonomic approach and allows invertebrates to be studied collectively from the perspective of their ecological functional role. Briefly, FFG categories include (**Table 4**): (1) scrapers and grazers, which remove and consume attached algae and associated material; (2) shredders, which ingest coarse particulate organic material (CPOM) in the form of decomposing leaf litter, living macrophyte tissue, or dead wood; (3) predators, which eat living animals; and (4) collectors, which consume decomposing fine particulate organic material (FPOM). Collectors can be subdivided into (5) gatherers, which collect FPOM from the sediments; and (6) filterers, which trap FPOM from the overlying water. The FFG approach, although important for the development of several paradigms in freshwater ecology, especially the River Continuum Concept (RCC), is not without its limitations. For example, assigning

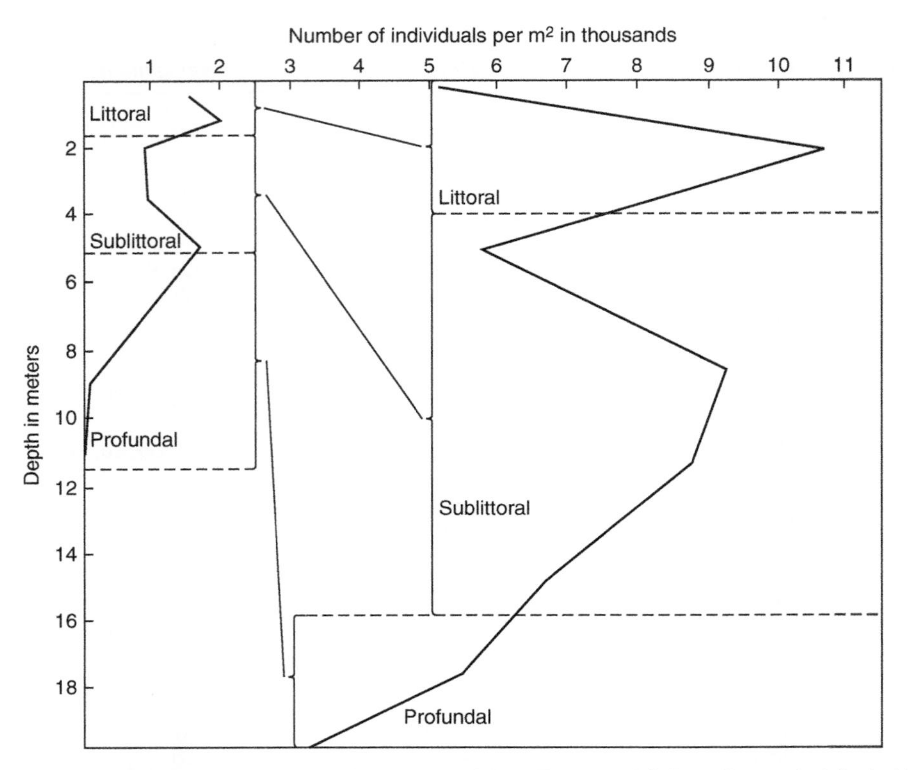

Figure 4 The abundance of benthic fauna at various depths in two lakes, Esrom and Gribsø, Denmark. Adapted from Berg K and Peterson IBC (1956) Studies on the humic acid Lake Gribsø. *Folia Limnologica Scandinavica* 8: 1–273.

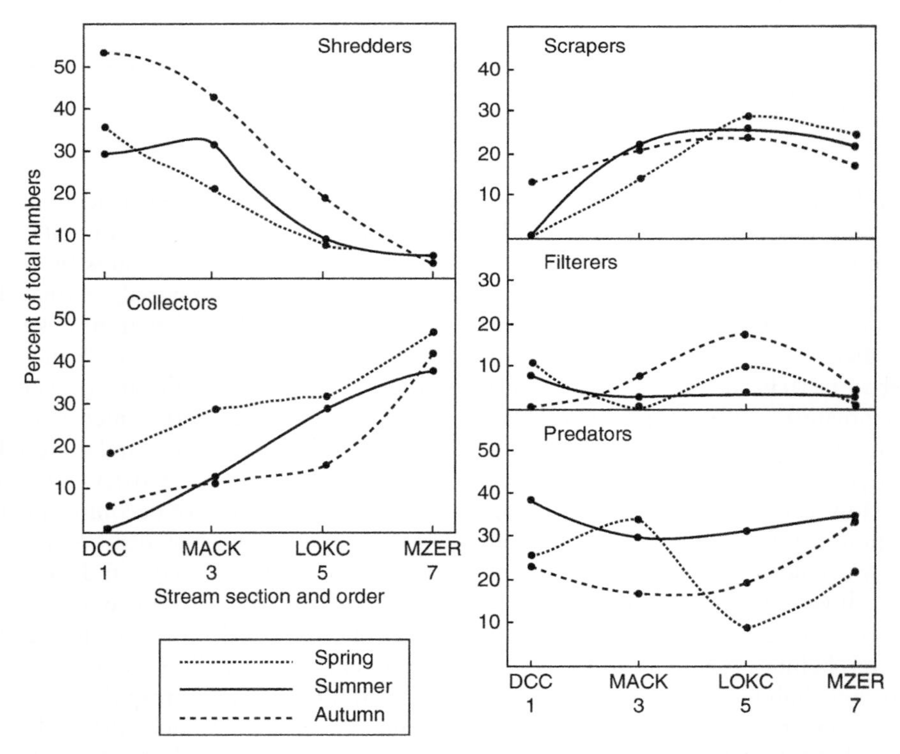

Figure 5 The abundance of FFGs along a gradient of stream order over the course of three seasons. Adapted from Hawkins CP and Sedell JR (1981) Longitudinal and seasonal changes in functional organization of macroinvertebrate communities in four Oregon streams. *Ecology* 63: 387–397.

Table 4 General classification system for benthic invertebrate trophic relationships (after Merritt RW, Cummins KW, and Berg MB (eds.) (2007) *An Introduction to the Aquatic Insects of North America* (4th edition). Dubuque: Kendall/Hunt

Functional feeding group	*Subdivision of functional group*		*Example taxa*	*Food size (mm)*
	Dominant food	*Feeding mechanism*		
Scrapers	Microbiota attached to substrates	Herbivore: scrape, rasp, and browse material from mineral and organic surfaces	Caddisflies, Mayflies, Snails	<1
Shredders	Living vascular plant tissue	Herbivore: chewers and miners of live macrophytes	Aquatic beetles, Caddisflies	>1
	Decomposing vascular plant tissue, wood, and associated microbiota	Detritivore: chewers of coarse particulate material and wood borers	Chironomid midges, Craneflies, Stoneflies	
Collectors	Suspended FPOM and associated microbiota	Detritivore: filterers of material from overlying water	Blackflies, Caddisflies, Mussels	<1
	Deposited FPOM and associated microbiota	Detritivore: gatherers of material from substrate	Aquatic earthworms, Chironomid midges, Mayflies	
Predators	Living animal tissue	Carnivore: pierce tissues and cells, and suck out fluids	True bugs, Water mites, Leeches	>1
		Carnivore: engulf whole or parts of animals	Dobsonflies, Dragonflies, Stoneflies	

organisms to an appropriate FFG can be difficult, especially for the many invertebrates that are opportunistic, omnivorous feeders, or whose feeding changes with habitat, food availability, or age. Recent studies using stable isotope analyses to trace elemental flow through food webs have provided important insights into FFG group assignments, supporting the observation that diet can change with life history or habitat conditions, and sometimes contradicting prior conclusions about diets based on gut analyses and morphological characteristics.

The FFG approach can be augmented by incorporation of broader 'functional traits' into the grouping process for invertebrate taxa. Proponents of this approach argue that 'evolutionarily labile' traits, such as thermal and current preferences, should be used in addition to the FFG approach. Evolutionary labile traits are those that are relatively easily changed in response to modified selection pressures, such as environmental conditions. The addition of functional traits in community-level analyses increases the number of factors used to characterize a particular group of invertebrate taxa, and thus may enhance the resolution for detecting subtle environmental gradients, both natural and anthropogenic.

Ecological Role of Benthic Invertebrates

Benthic invertebrates play an important role in many ecological processes in inland waters. This partly reflects their shear abundance but also the many different relationships that exist between benthic invertebrates, their food resources, competitors, and predators. In addition to being important consumers, benthic invertebrates are an important food resource for most freshwater fishes, many of which are strictly benthivorous. Also, some terrestrial riparian predators, such as spiders and birds, have been shown to be dependant upon emerging adult aquatic insects, representing an important resource transfer from the freshwater benthos to the riparian zone. Benthic invertebrates are also important in benthic–pelagic coupling, both by excreting soluble nutrients back into the water column and by removing material from the overlying water and effectively packaging it as fecal material. Some invertebrate taxa, such as chironomid midges and oligochaete worms, can also dramatically alter the porosity of the substrate through bioturbation or sediment cohesion by tube and burrow construction. Thus, benthic invertebrates provide a critical trophic linkage, and exhibit significant control over energy flow and nutrient cycling in both lotic and lentic ecosystems, and between aquatic and terrestrial ecosystems.

Many benthic invertebrates are vital for decomposition because they recycle organic 'waste' materials, such as leaf and exudates from plants as well as feces from other animals. Shredders consume CPOM and in the process transform CPOM into FPOM. Collectors consume FPOM and dissolved organic material (DOM), and through their feces alter not only the particle-size distribution, but also the quality of the organic matter pool. In lotic ecosystems, the relative importance of the detritivore food chain compared with the grazer food chain is a matter of some debate, but often varies as a function of stream order and riparian conditions. In lentic ecosystems, it is difficult to evaluate the relative importance of primary producers versus detritus in fueling benthic invertebrate food webs. As depth increases benthic primary production declines, and so the detritivore food chain becomes increasingly dependent on the 'rain' of organic material from the pelagic. However, terrestrial DOM, which can be made available to benthic invertebrates via microbial uptake, also provides an important food resource for the profundal detritivore benthos. Decomposition of organic material represents a major ecosystem process that is mediated by both lentic and lotic benthic invertebrates, although the relative importance of various size fractions of particulate organic matter differs considerably among freshwater habitats. In low-order streams and in wetlands, CPOM is often plentiful, and its decomposition by benthic invertebrates has been the subject of considerable research. The relative importance of organic material of allochthonous versus autochthonous origin is variable, and is often influenced by the size of the water body. Consumer resources in small lakes and low-order streams are typically dominated by allochthonous inputs, whereas in large lakes and mid-order streams autochthonous production may dominate. Important detritivores in freshwaters include members of several different dipteran, caddisfly, stonefly, amphipod, and crayfish families (**Figures 1(d)**, **1(e)**, **1(i)**, **1(k)**).

Benthic primary producers, including algae, rooted macrophytes, and bryophytes occur in virtually all shallow inland waters, and are exploited by many invertebrate primary consumers or grazers. Grazing has been extensively studied in lotic systems, and to a lesser extent, in lentic systems. In flowing waters, benthic algae grow on virtually all substrates exposed to light, including on invertebrates and the tubes, shells, and cases they construct. The biomass of benthic algae is typically small relative to detritus accumulations, especially in streams. However, a given biomass of algae can support many times more invertebrate biomass than does detritus because of the high rate at which algae replace themselves. Among benthic invertebrates, caddisflies, mayflies, and snails (**Figures 1(l)**, **1(t)** and **3(c)**) are conspicuous grazers. However, small invertebrates can also be important, such as chironomid midges, because of their high abundance and short generation times. Grazers can exert large effects on benthic algae, controlling algal biomass and production while displacing other benthic invertebrates. Many invertebrates consume living macrophyte tissue, and a number of intimate, evolutionary associations appear to exist between invertebrates and macrophytes.

Compared with other FFGs, there are comparatively few benthic invertebrates that are strictly predaceous. Notable groups of obligate benthic invertebrate predators include dragonflies, damselflies, hemipterans, and dobsonflies (**Figure 1(m)** and **1(n)**), along with some groups of beetles and stoneflies (**Figure 1(o)**). Most invertebrate groups include taxonomic subdivisions that are facultative predators. For example, among the caddisflies, the Rhyacophilidae are free-roaming obligate predators, especially on blackflies, whereas other families include taxa that are facultative predators, such as the sedentary Hydropsychidae (**Figure 1(p)**) that inevitably catch animals in their nets and Limnephilidae that become opportunistic predators as they become bigger. The role of predators in determining the composition of invertebrate assemblages has received considerable attention, demonstrating that predators can often exert significant control over prey distribution and abundance. Predator and prey are often subject to different physical constraints in various lentic and lotic systems, and these physical constraints interact with predation to create different community types. Since predator–prey interactions do not always occur at the same spatial and temporal scales as abiotic factors that affect distribution and abundance of organisms, it is sometimes difficult to extrapolate observations of predator–prey interactions to benthic community dynamics. Often, habitat heterogeneity moderates the effects of fish and invertebrate predation on invertebrate assemblages. Fish can strongly influence specific benthic invertebrate populations, but the specific nature and magnitude of the effect on invertebrate assemblages is variable. Invertebrate predators do not control prey abundance in most lotic ecosystems because these communities are open to immigration and emigration of both predators and prey. However, the indirect effects can be considerable, including altering spatial distribution through predator avoidance and changes in activities such as shredding or movement. Furthermore, predator–prey interactions are under strong selective pressure, which likely has been an important factor contributing to the high diversity of benthic invertebrates.

Secondary Production and Dispersal

Several processes integrate one or more of the ecological relationships among invertebrates, including secondary production and dispersal. Secondary production requires the summation of consumer biomass produced over a given period of time, and thus available to the next trophic level. Secondary production provides an important, albeit time-consuming, method for studying the influence of invertebrate consumers on energy flow, and also provides insights into consumer dynamics not apparent from biomass or abundance measurements alone. For example, many chironomid midge species have a small body size, and thus often constitute small biomass, even at high densities. However, the larvae exhibit rapid growth rates and have short life cycles, often with multiple generations in one season. Thus, their secondary production can be very high, representing a significant conversion of primary production into biomass available to higher trophic levels. Although aquatic insects often dominate secondary production, oligochaetes, crustaceans, and mollusks may also be important. Different benthic habitats can differ substantially in their production. For example, debris dams and snag habitats in streams and macrophyte habitats in lakes have been shown to support especially high benthic invertebrate production compared with other habitats.

Benthic invertebrates are rarely completely sedentary organisms, but rather move to find food resources and to avoid disturbance and predators. Such movements have been shown to take place over small and large distances, and both seasonally and daily. Movements are more apparent when lotic invertebrates enter the 'drift' to be carried by water flow. Drift can be categorized as: (1) constant drift due to accidental dislodgement, (2) catastrophic drift due to disturbance, or (3) active drift due to behavior. Whether drift is predominantly active or passive has been subject to some debate. However, some invertebrates, such as *Baetis* mayflies (**Figure 1(l)**), clearly have a greater propensity to enter the drift, and have evolved adaptations to allow more rapid, controlled return to the substrate. Also, blackfly larvae (**Figure 1(j)**) will spin a silken 'safety line' when moving to a new location to minimize how far they are taken downstream of their original location. In lotic environments, most movements are downstream, but upstream adult flight, and occasionally swimming or crawling of larvae, have also been reported.

Major Paradigms and Concepts

A major ecological paradigm relevant to the ecology of benthic invertebrates is the RCC (**Figure 6**), which describes the longitudinal trends in FFG abundance and has been key to understanding the ecological role of benthic invertebrates in streams and rivers, with parallels being drawn with equivalent situations in lentic systems. The RCC predicts that small, heavily shaded streams will have large inputs of allochthonous detritus from adjacent riparian zones relative to authochthonous material generated within these streams. Furthermore, due to the abundant detritus, shredders will dominate the macroinvertebrate assemblage, while both collector-filterers and collector-gatherers also will be abundant because high-quality FPOM will be produced as CPOM is fragmented (**Figures 5** and **6**). In medium-sized streams, light increases and thus benthic algal production becomes relatively more important. Consequently, scrapers replace shredders, while collectors remain abundant. In large rivers, as benthic algal production and direct riparian inputs decrease, food resources for macroinvertebrates become dominated by suspended and deposited FPOM. As a result, collectors dominate the macroinvertebrate assemblage. In all streams, predators comprise a small but relatively stable proportion of the invertebrate fauna.

The Serial Discontinuity Concept (SDC) is an extension of the RCC, and articulates how natural and artificial impoundments or changes in channel morphology effectively disconnect the upstream to downstream continuum in lotic ecosystems. The SDC predicts lotic responses to changes in flow regulation in the context of recovery downstream of an impoundment or abrupt change in channel morphology. Such changes interrupt the movement of resources downstream as well as alter the physicochemical characteristics of the water, including temperature and chemistry, with important consequences for benthic invertebrate assemblages. Although developed initially to predict the consequences of regulation of streams and rivers, SDC has provided a framework with which to understand lake–stream interactions and broader watershed processes.

The Flood Pulse Concept (FPC) was developed to describe ecological processes in large lowland rivers with extensive floodplains, partly in response to perceived shortcomings of the RCC in these systems. The FPC argues that periodic flooding is critical to the biota in these rivers, because such an increase in water level facilitates the flux of nutrients and organic material between the channel and adjacent floodplain areas. Thus, proponents of the FPC argue that much of the biological production generated, including that of invertebrates, is derived, directly or indirectly, from resources present nearby (i.e., riparian and floodplain areas) rather than elsewhere in the watershed. Such aquatic–terrestrial interfaces are present, to a lesser or greater degree, throughout much of the watershed, not just in lowland rivers and so the FPC is broadly relevant.

An important step in understanding the species composition of benthic invertebrate assemblages was Thienemann's Lake Typology Classification System (LTCS). Originally developed in the 1920s for European lakes, the LTCS linked species composition,

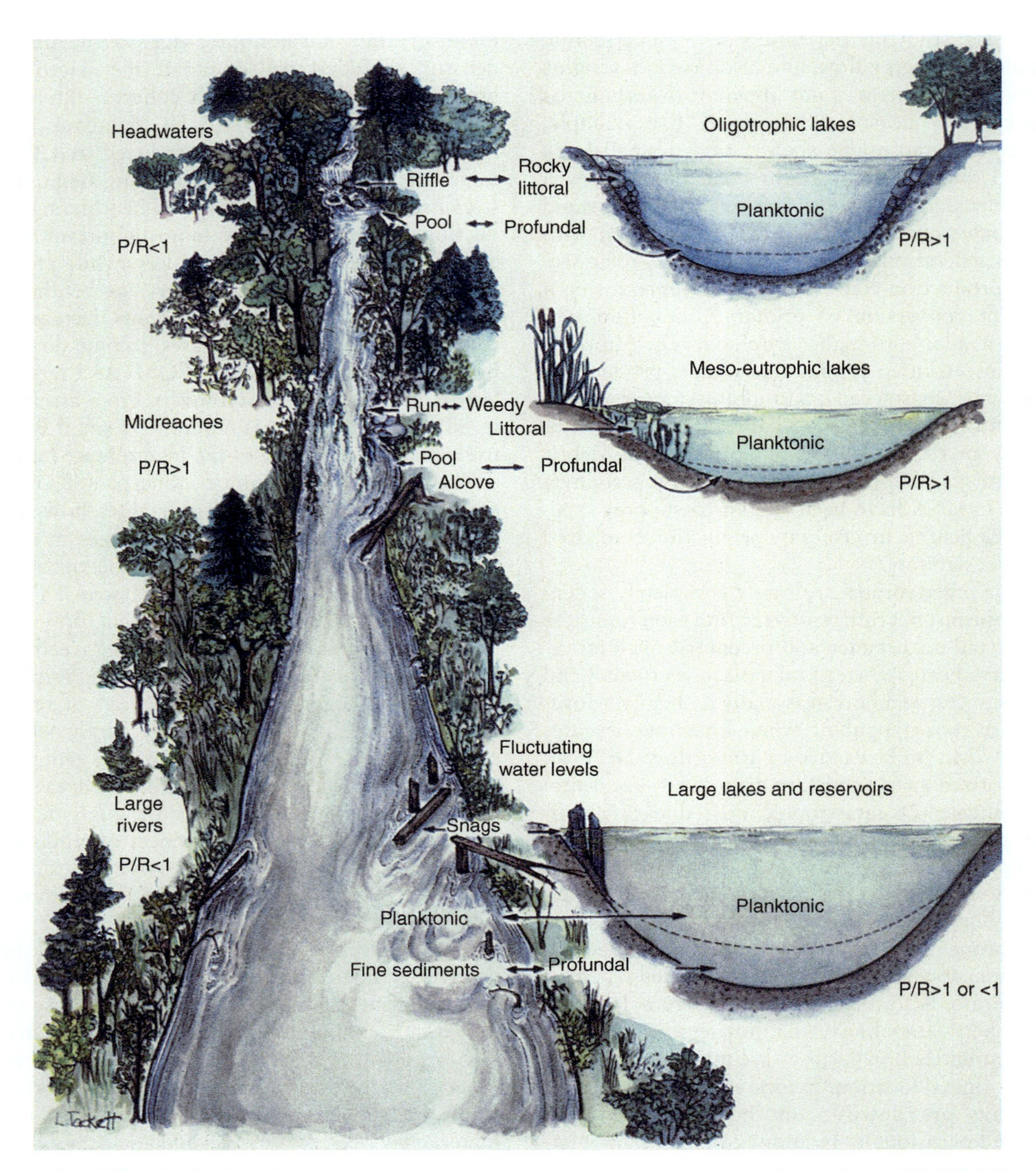

Figure 6 The RCC with extensions to lentic ecosystems. Adapted from Merritt RW, Cummins KW, and Burton TM (1984) The role of aquatic insects in the processing and cycling of nutrients. In: Resh VH and Rosenberg DM (eds.) *The Ecology of Aquatic Insects* pp. 134–163. New York: Praeger Press.

especially chironomid midges, to seasonal changes in hypolimnetic dissolved oxygen. Unfortunately, lakes that did not conform to the established types were soon found, while several other factors, including temperature and depth profiles, were found to also influence the lake type. Consequently, the number of lake types increased and terminology became increasingly complex, especially as more factors were considered, so much so that the LTCS fell out of favor. However, several aspects of modern limnology have their origins in this classification system.

The Permanence–Predator Transition Hypothesis (PPTH) provides a conceptual framework for understanding the diversity and role of predators in structuring benthic invertebrate assemblages in the context of environmental conditions (**Figure 7**). Lentic assemblages exist along an environmental gradient of hydrologic permanence, from ephemeral pools to large

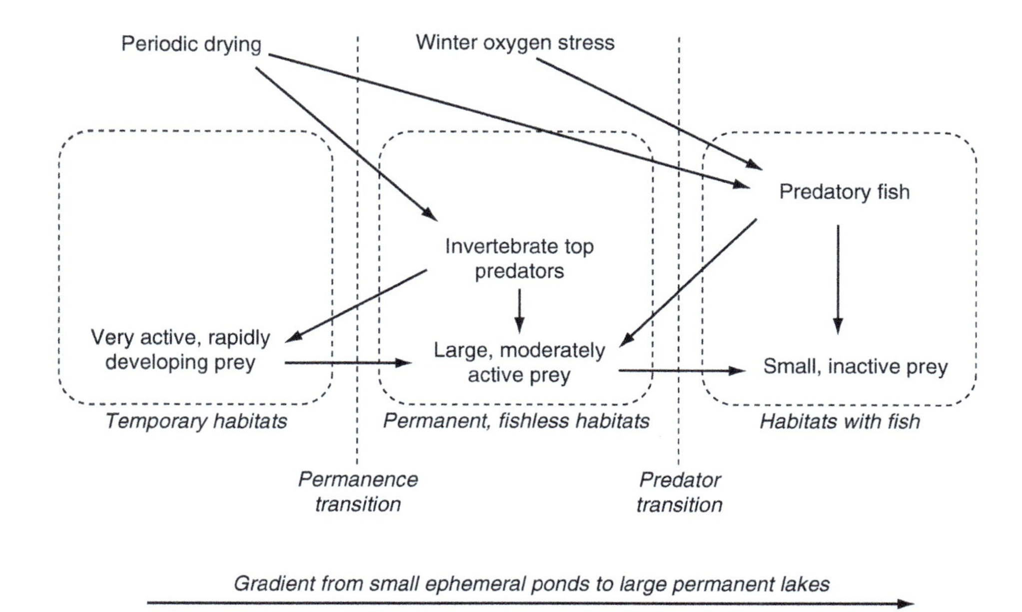

Figure 7 A conceptual diagram illustrating the PPTH. Adapted from Wellborn GA, Skelly DK, and Werner EE (1996) Mechanisms creating community structure across a freshwater habitat gradient. *Annual Review of Ecology and Systematics* 27: 337–367.

lakes. Temporary ponds have simplified communities, supporting few large invertebrate predators. Permanent ponds with no fish represent a transition to more complex benthic invertebrate communities, including a high diversity of predatory species. However, where fish are able to colonize, they exert another threshold effect on community structure by eliminating the larger predatory invertebrate species. Lotic benthic assemblages also experience a gradient of hydrologic permanence, and it has been argued that a similar framework to the PPTH could be used to understand the influence of predators on lotic benthic assemblages.

Lentic benthic food webs are known to be complex because of habitat heterogeneity in the littoral, and constraints imposed by oxygen availability and predatory fish in the profundal. However, benthic–pelagic coupling has not been well integrated into current understanding of lake processes. Since lentic ecosystems are embedded within the landscape, factors associated with the geomorphic setting are likely to indirectly control lake community structure, including invertebrate assemblages. The Geomorphic-Trophic Hypothesis (GTH) integrates lake processes into their geomorphic setting in two ways. First, the stream network constrains or channels fish dispersal, and thus by modifying fish species composition, links dispersal with fish control of food webs. Second, landscape factors determine hydrologic retention, and nutrient, organic matter, and sediment loading in lakes, which affect benthic invertebrates by altering pelagic primary production, oxygen demand, light attenuation, and the ratio of pelagic to benthic primary production. Although originally developed for Arctic lakes, the GTH has been applied elsewhere. However, the specifics of fish dispersal and landscape factors are likely to differ among regions depending upon the local biogeographical and geomorphic setting, including the relative importance of disturbance.

The Intermediate Disturbance Hypothesis (IDH) embraces the role of disturbance in structuring benthic assemblages. Originally developed for intertidal marine communities, the IDH is relevant to benthic invertebrate assemblages that are subject to varying degrees of disturbance. Thus, this concept is especially appropriate to benthos that experience periodic floods, wave action, or drought. The IDH predicts that at the two extremes of a disturbance regime, no disturbance or severe disturbance, assemblages exhibit low species diversity (**Figure 8**). In a benign environment, competitively superior species come to dominate and exclude most other species, whereas in a frequently disturbed environment, most species are eliminated and cannot repopulate prior to the next disturbance. At intermediate levels of disturbance, competitively superior species are kept at sufficiently low densities to permit competitively inferior species to persist, thereby resulting in a higher diversity.

Use of Benthic Invertebrates in Applied Freshwater Biology

Benthic invertebrates are being increasingly used in many different areas of applied biology. For example,

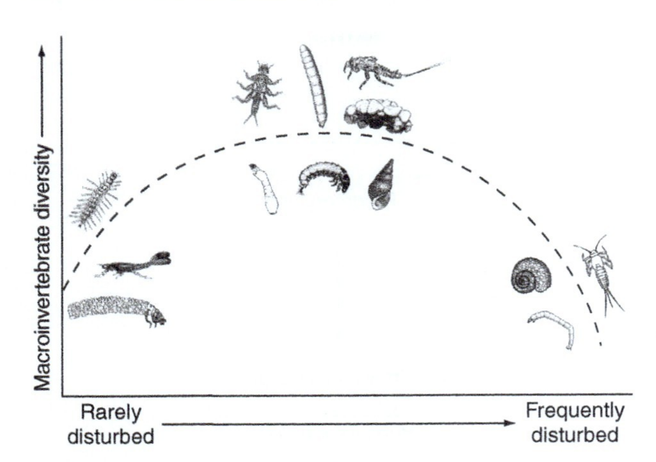

Figure 8 The relationship between disturbance regime and the diversity of benthic invertebrates suggested by the IDH. Adapted from Hershey AE and Lamberti GA (1998) Stream macroinvertebrate communities. In: Naiman RJ and Bilby RE (eds.) *River Ecology and Management: Lessons from the Pacific Coastal Ecoregion*, pp. 169–199. New York: Springer.

in forensic science, freshwater invertebrates found on homicide victims have been used to establish the post-mortem submersion interval, even though there are believed to be few obligate necrophagous invertebrates in freshwaters. In restoration ecology, biologists are now aware that the benthic invertebrate fauna should be considered when designing or assessing restoration schemes. Furthermore, a number of benthic invertebrates have been listed as endangered or threatened, and thus of conservation concern. Examples of threatened benthic invertebrates include many species of crayfish and freshwater mussels (**Figure 1**(**e**) and **1**(**s**)). However, the most common use of benthic invertebrates is in biomonitoring. Biomonitoring is often used to measure response and recovery of ecological communities to human disturbance, and has its origins in Theinemann's LTCS. Biomonitoring studies are used to assess ecosystem health because aquatic organisms can function as sensors of environmental quality in ways that direct measurements of water quality cannot. Many freshwater biomonitoring programs incorporate metrics for many different taxa, but benthic macroinvertebrates are widely used by many agencies to monitor lotic, and less frequently, lentic ecosystems.

Benthic invertebrates, especially insects, offer several advantages in biomonitoring. First, the small size and limited mobility of benthic invertebrates, as compared with vertebrates, means they are relatively easy to sample. Second, invertebrates generally do not require collection permits. Third, a number of taxa are known to be sensitive to pollutants. Finally, invertebrates integrate environmental conditions over a longer time interval than direct measures of water quality. Thus, biomonitoring incorporates and integrates many different aspects of invertebrate biology into a single management and assessment tool. For example, mouthpart deformities in chironomid midges have been used as an index of pollution. Although valuable and widely used, a few disadvantages are evident. Perhaps most critically, the time and training needed to identify invertebrates can be considerable.

Several different metrics of benthic invertebrate communities are commonly used in biomonitoring. Taxa richness is widely used, but is difficult to compare among studies because different levels of resolution are often associated with different investigators and taxonomic groups. Perhaps the most popular metric is the combination of Ephemeroptera, Plecoptera, and Trichoptera richness to generate so-called EPT richness, which takes advantage of relative ease of identification for these taxa and their general pollution-sensitivity. A modification is the EPT to chironomid ratio, since chironomids as a group, unlike EPT taxa, are considered pollution tolerant. Several diversity and similarity indices (e.g., Shannon-Wiener) are readily comparable between sites, and provide easily interpretable scores but are sensitive to the level of taxonomic resolution. Finally, metrics that rely on functional rather than taxonomic information, such FFGs, bridge community- and ecosystem-level approaches to biomonitoring.

Summary

The benthic invertebrate fauna of inland waters are an abundant, diverse, and important group of organisms that exhibit a myriad of adaptations to life in flowing or standing waters. Much is known about benthic invertebrates, especially their central role in several major ecological processes, and this information has been used in the development of important ecological paradigms and concepts. As well as being a subject of much research, benthic invertebrates are used or taken into consideration in many different areas of applied freshwater biology, especially biomonitoring. Despite the wealth of information about benthic invertebrates, more research is needed in several areas, such as the role of benthic invertebrates in lentic processes. Such information will contribute to a more complete understanding of the ecology of inland waters.

Glossary

Allochthonous – Material originating from outside an ecosystem.

Autochthonous – Material generated within an ecosystem.

Benthos – Organisms associated with submerged substrata.

Biomass – Mass of living organisms.

Consumer – Organisms that obtain organic compounds through consumption of other organisms or their parts.

Detritivore – Organisms that consume detritus.

Detritus – Decaying organic material.

Drift – Material, especially invertebrates, that is washed downstream.

Grazer – Organisms that consume primary producers.

Hyporheic – Area of subsurface flow influenced by streams or rivers.

Littoral – Shallow area near the shore where rooted aquatic plants are found.

Lentic – Standing water.

Lotic – Moving water.

Pelagic – Open water.

Profundal – Deep area of sediment where light does not penetrate.

Primary producer – Organisms that generate organic compounds from inorganic nutrients.

Production – Increase in biomass over time within a specific area.

Primary consumer – Organisms that consume primary producers.

See also: Benthic Invertebrate Fauna; Benthic Invertebrate Fauna, River and Floodplain Ecosystems; Benthic Invertebrate Fauna, Small Streams; Benthic Invertebrate Fauna, Tropical Stream Ecosystems.

Further Reading

Allan JD (2007) *Stream Ecology: Structure and Function of Running Waters,* 2nd edn. New York: Chapman & Hall.

Covich AP, Palmer MA, and Crowl TA (1999) The role of benthic invertebrate species in freshwater ecosystems. *BioScience* 49: 119–128.

Hauer FR and Lamberti GA (eds.) (2006) *Methods in Stream Ecology,* 2nd edn. San Diego: Academic Press.

Hershey AE, Gettel G, McDonald ME, *et al.* (1999) A geomorphic-trophic model for landscape control of trophic structure in arctic lakes. *BioScience* 49: 887–897.

Junk WJ, Bayley PB, and Sparks RE (1989) The flood pulse concept in river-floodplain systems. In: Dodge DP (ed.) *Proceedings of the International Large River Symposium,* pp. 110–127. Canadian Special Publication of Fisheries and Aquatic Sciences 106.

Lake PS, Palmer MA, Biro P, *et al.* (2000) Global change and the biodiversity of freshwater ecosystems: Impacts on linkages between above-sediment and sediment biota. *BioScience* 50: 1099–1107.

Merritt RW, Cummins KW, and Berg MB (eds.) (2007) *An Introduction to the Aquatic Insects of North America,* 4th edn. Dubuque: Kendall/Hunt.

Poff NL, Olden JD, Vieira NKM, *et al.* (2006) Functional trait niches of North American lotic insects: Traits-based ecological applications in light of phylogenetic relationships. *Journal of the North American Benthological Society* 25: 730–755.

Rader RB, Batzer DP, and Wissinger SA (eds.) (1999) *Invertebrates in Freshwater Wetlands of North America.* New York: Wiley.

Rosenberg DM and Resh VH (eds.) (1993) *Freshwater Biomonitoring and Benthic Macroinvertebrates.* New York: Chapman & Hall.

Stanford JA and Ward JV (2001) Revisiting the serial discontinuity concept. *Regulated Rivers: Research and Management* 17: 303–310.

Thorp JH and Covich AP (eds.) (2001) *Ecology and Classification of North American Freshwater Invertebrates,* 2nd edn. San Diego: Academic Press.

Wallace JB and Webster JR (1996) The role of macroinvertebrates in stream ecosystem function. *Annual Review of Entomology* 41: 115–139.

Ward JV and Stanford JA (1983) The intermediate disturbance hypothesis: An explanation for biotic diversity patterns in lotic ecosystems. In: Fontaine TD and Bartell SM (eds.) *Dynamics of Lotic Ecosystems,* pp. 347–356. Ann Arbor: Ann Arbor Science.

Wellborn GA, Skelly DK, and Werner EE (1996) Mechanisms creating community structure across a freshwater habitat gradient. *Annual Review of Ecology and Systematics* 27: 337–367.

Wetzel RG (2001) *Limnology,* 3rd edn. San Diego: Elsevier.

Biological Interactions in River Ecosystems

C E Cushing, Streamside Programs, CO, USA
J D Allan, School of Natural Resources and Environment, University of Michigan, Ann Arbor, MI, USA

Introduction

Any discussion of the biological interactions in riverine ecosystems must recognize two points: (1) most of the interactions among the biological components are mediated by a variety of physical and chemical factors that are discussed elsewhere in this publication, and (2) species interactions are very complex, as are their interactions, and it behooves us to begin with a description of the main biological components and the resources that fuel their activities.

Energy Resources in River Ecosystems

Two general types of organisms, in terms of energy production, are found in rivers: autotrophs and heterotrophs. The former produce their own energy from inorganic matter and the latter derive their energy from autotrophs. Autotrophic organisms produce organic matter by the process of photosynthesis, and the material produced is termed primary production because it uses inorganic materials to produce new organic matter. Secondary production is the organic matter created by the consumption and transformation of primary production to new biomass. Primary production originates in two places: instream (autochthonous) and from the terrestrial environment (allochthonous).

Plants are the main instream primary producers, and the three main types are algae, mosses, and large flowering plants called macrophytes. Algae generally fall into two types based on their morphology: (1) the filamentous green algae which may be found as long tendrils several meters in length under favorable conditions, and (2) the diatoms, unicellular brownish/yellowish single-celled organisms that form part of the brown slippery coating on rocks and other objects on the stream bottom. Diatoms occur in uncountable numbers and are one of the most important food resources for stream macroinvertebrates. This slippery film (called a biofilm or periphyton) is composed not only of algae, but also of detritus, microinvertebrates, and bacteria, fungi – all enclosed in a polysaccharide matrix. This microecosystem is a dynamic entity where many microorganisms function in what is called the microbial loop (**Figure 1**).

Algae suspended in the water column are called phytoplankton. These organisms are generally flushed into rivers from lakes or bays, and usually do not proliferate in flowing water conditions. They are a source of food for filter-feeding organisms.

In the colder, headwater reaches of rivers, where the water is well oxygenated, mosses may be found in abundance. These are attached plants with leaves and are also adapted to low light conditions and rapid currents. They can have high rates of growth, but because of their limited distribution (headwater reaches), they are not as important in the overall ecology of a river continuum. Typical genera are *Fontinalis* and *Fisidens*.

Where current slows and fine sediments settle out, conditions are suitable for the growth of the rooted, true flowering plants (angiosperms). Common genera are *Potamogeton*, *Elodea*, and *Ranunculus*. Although these plants can occur sparsely or in dense mats, they are less important as a food resource in their growing state. However, when they die and decompose, the resulting detritus becomes a rich food resource for macroinvertebrates. In their growing state, they provide cover for organisms to avoid predators, oxygen to the water via photosynthesis, and are fed upon by some insects with piercing mouthparts.

Primary production from the terrestrial environment (allochthonous) that reaches the river becomes an important food resource, and can originate as leaves, twigs, grasses, even tree trunks – virtually any organic matter that becomes wetted is included in this category. Early stream ecologists largely ignored material until Noel Hynes presented a paper in 1975 entitled 'The Stream and Its Valley.' In it, he essentially said that until we include the energy inputs from the terrestrial environment into our calculations of energy flow within streams, we would never completely understand their ecology. Organic matter entering the stream is called coarse particulate organic matter (CPOM, >1 mm diameter). In this form, it is not a viable food resource for organisms, but once it enters the water, it is colonized by a wide variety of fungi and bacteria that transform it into a rich and palatable food source. As the CPOM becomes physically broken down by abrasion, microbial action, and shredding, smaller particles are formed that are called fine particulate organic matter (FPOM, <1 mm diameter). FPOM and diatoms are

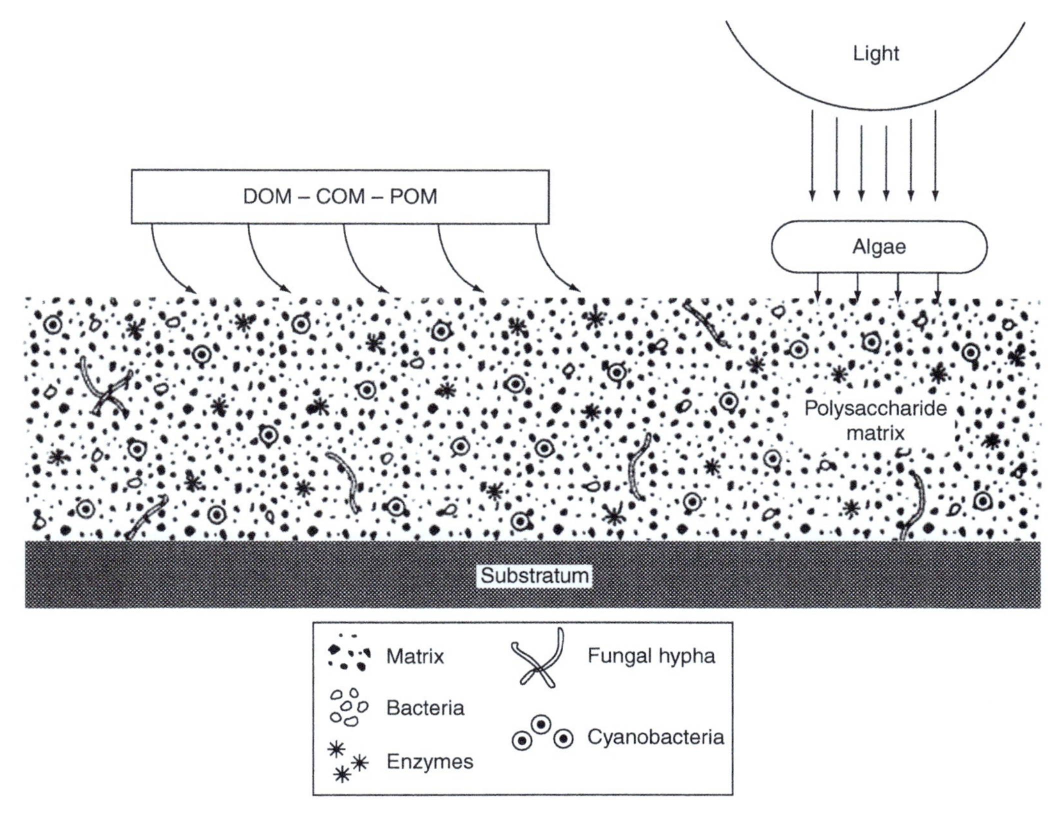

Figure 1 The biofilm found as a surface 'slime' on stones and other submerged objects in streams. A polysaccharide matrix produced by the microbial community binds together bacteria, algae, and fungi and is inhabited by protozoans and micrometazoans, which consume this material. Within the matrix, extracellular release and cell death result in enzymes and other molecular products that are retained due to reduced diffusion rates. From Lock (1981) as modified by Allan (1995), with permission.

the most important food resources in riverine ecosystems.

A third source of energy for lotic ecosystems that does not conveniently fall into either autochthonous or allochthonous because it originates from both sources is dissolved organic matter (DOM). It is a large and complex carbon pool and originates from many sources – decomposition by-products in groundwater, exudates from living plants, microbial action, etc. It is highly variable in quality and its function as an energy source in flowing water is not well understood.

Food Web Dynamics

To live, grow, and reproduce, stream organisms must have sufficient food, in both quality and quantity and at the right time. Moreover, food requirements change over the life cycle; the food requirements of an adult trout are quite different from those of newly hatched fry, or a small yearling. Adults consume larger invertebrates and small fish, while juveniles feed on tiny organisms such as midge larvae until they are large enough to start taking larger food items. Some consumers are quite specialized, others less so; and most are adapted via their mouthparts or other feeding apparatus to capture some food items more readily than others.

What organisms eat and how they obtain their food resources determines their role in riverine food webs. The broad categories are herbivores, organisms that feed on plants; carnivores, organisms that feed on other organisms (i.e., predators); and omnivores, organisms that feed on both plants and animals. A fourth category is sometimes included – detritivores, organisms that feed on organic detritus. The active processes of feeding on these resources are called herbivory, carnivory, omnivory, and detritivory. Classifying organisms, for example, aquatic insects, into these categories describes how they are linked into a particular food web; an example of such a simplified food web is shown in **Figure 2**. The insects are the herbivores and detritivores in this example,

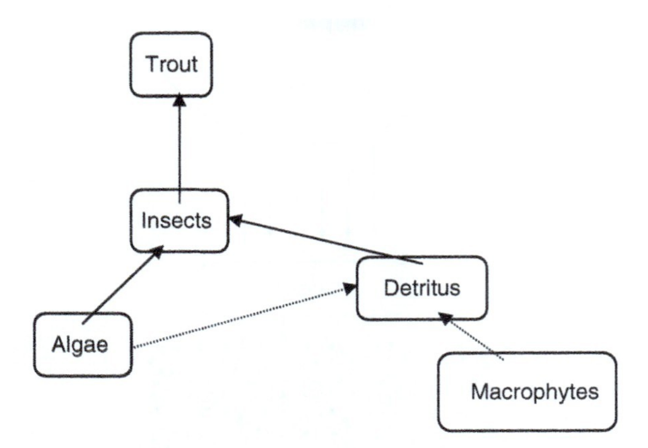

Figure 2 Diagram of food web based on type of food. Solid arrows indicate feeding pathways; dashed arrows indicate decomposition pathways. From Cushing and Allan (2001), with permission.

feeding directly on algae or on both algae and macrophytes after they have been converted to detritus, and the trout are carnivores, feeding on insects.

However, because most aquatic insects are omnivores, food web pathways determined using this classification are not very edifying. We need to know not only what is consumed, but also how animals obtain their food and how this diet might change with prey availability, location, season, etc. This analysis provides more insight at the true pathways of energy flow through the food web.

Ken Cummins proposed a classification scheme designed to augment the herbivore–carnivore–omnivore–detritivore system; it is termed functional feeding groups (FFG). The major categories in this scheme are shredders, grazers (or scrapers), collectors (both gatherers and filterers), and predators. Other categories have been added since inception of this scheme to address specialist organisms, and are described below. This system was mainly developed with insects in mind because they are by far the most numerous organisms present in most rivers and have the greatest diversity of feeding methods.

Shredders are an important functional group, especially in small streams of deciduous forests or streams that receive significant inputs of CPOM. Common shredders include some stoneflies such as *Pteronarcys californica*, many dipteran larvae such as those of the family Tipulidae, and most of the caddisflies in the family Limnephilidae that construct cases of organic material. The eastern caddis species *Pycnopsyche lepida*, however, shows a role reversal during its immature life. Early life stages (instars) have an organic matter case and are shredders; later instars construct sand grain cases and become grazers that exploit the spring periphyton bloom on rocks.

To be useful for shredders, CPOM accumulation sites must remain fairly well oxygenated (this is also a requirement of the fungi that colonize, soften, and nutritionally enrich the leaf). This means that obstructions that trap CPOM in the current are excellent places to find shredders. It also means that usually only the surface layers of CPOM, such as leaf litter, are used in pools and backwaters, because internal layers may lack oxygen. The mouthparts of shredders are adapted for maceration of the CPOM particles, which they tear and shred while feeding. Their feeding results in the initiation of the conversion of CPOM to FPOM by physically breaking up the CPOM and by production of FPOM in the form of fecal pellets. Shredders obtain energy from the leaf itself and from the microbes, primarily fungi that colonize it.

Grazers are an important functional group; they occur most abundantly where light reaches the stream bottom, promoting algal growth, because this is their main food source. Mouthparts of scrapers are adapted to scraping the film of algae, or periphyton, growing on the surfaces of rocks and other large objects; thus, they literally scrape off, or graze, this food source. Some grazers, including some caddis larvae, have mouthparts adapted to scraping or rasping diatoms that lie very tightly adjacent to stone or stick surfaces. Others, including many mayflies, are more adept at 'browsing' the more loosely attached algae. This reminds us that functional groups are useful broad classifications that can be more finely dissected to reveal further differences in feeding strategy. Grazers, too, produce large amounts of FPOM through the production of fecal pellets and dislodgment of algal cells during their feeding.

Several familiar organisms are grazer/scrapers. The genus *Dicosmoecus* is a stone-cased grazer (the general rule is that caddis larvae with cases made from organic matter are shredders, and those with cases made from inorganic matter are scrapers). However, like many large grazers, they may incidentally ingest some larger items such as small midge larvae and CPOM fragments while scraping the algal film. Many common mayflies are grazers, including many members of the family Heptageniidae; this includes such genera as *Epeorus*, *Stenonema*, and *Rhithrogena*. Another common grazer is the caddisfly genus *Glossosoma*, a small caddis that has a tortoise shell-shaped case made of coarse sand grains cemented together.

Collectors are the largest functional feeding group in terms of numbers. This large group is further divided into filtering-collectors and gathering-collectors, the name indicating how they obtain their food. Both groups feed almost exclusively on FPOM. Filtering-collectors obtain their food by filtering FPOM from the water, and although there are a number of refinements and modifications, the two main methods are by filtering with nets and filtering with specially adapted body parts.

Net-weaving filterers are mostly caddisflies, which construct some kind of net in conjunction with their nonportable case or fixed retreat. The larva retreats into the case and lets the current carry FPOM to the net, where the fine meshes of the net catch particles that are then harvested by the larva. Common net-spinners are members of the family Hydropsychidae, one of the most ubiquitous families of caddisflies in the world. Different species occur in relation to such factors as temperature and FPOM size, and the catch nets of different species appear to adapt to function at different current velocities. In the downstream direction, the change in species usually includes a decrease in body size and decrease in net mesh size as CPOM decreases and FPOM increases. Although the occasional microscopic animal may be entrapped and eaten by the filterer, it does not make it a predator in this classification system; how it obtains the food is the important point.

Other filtering-collectors have developed specialized body parts to filter FPOM from the current. The larvae of the caddisfly *Brachycentrus* cement one edge of the opening of the four-sided case to a solid substratum so that the opening faces into the current. The larva has many fine hairs on the middle and hind legs, and with the abdomen inside the case, it extends its legs into the current. These hairs filter FPOM from the current until the larva has enough to comb the material from its legs with its mouthparts. The black fly larva accomplishes its filter-feeding differently. The larva constructs a small silk pad on a rock or stick and anchors its abdomen to this pad with a series of anal hooks encircling the rear of the abdomen. One set of its mouthparts, the labia, are modified to resemble fan-like structures fitted with fine hairs and coated with a sticky mucous, which the larva extends into the current and with which it filters the FPOM. When the filtering fans are full, it collapses them and stuffs them into its mouth, then combs the particles off as the fans are withdrawn for use again. The mayfly *Isonychia* has heavy fringes on the front legs, forming a basket held under the head. The nymph stands up on the middle and back legs allowing water to pass through the basket and under the body. When the basket is full, the nymph raises the front legs and removes the FPOM with its mouthparts.

Gathering-collectors obtain their food, largely FPOM, by simply gathering it from wherever they can find it – under rocks, on the surface of stones, or in deposition zones where the current slackens and allows FPOM to settle from the water column and accumulate on the streambed. Mayflies of the genus *Baetis* are a common example of this functional group. Most gathering-collectors have rather generalized structures and many make-do simply by scurrying around the stream bottom picking up particles wherever they find them.

Predators eat other animals and are the final functional group. They are found throughout the river ecosystem and have many different adaptations to enable them to pursue and capture prey. Most stoneflies are predators, but one family of caddisflies, the Rhyacophylidae, are active predators. Interestingly, this family of caddisflies does not construct cases – their free-moving lifestyle may enable them to better pursue and capture prey. Other common predators are all the species of dragonflies and damselflies of the order Odonata, and the well-known hellgrammites (order Megaloptera).

Ecologists have also given names to some other insects that do not readily fit into the functional feeding groups described above. These additional categories include miners for some larvae that feed on detritus buried in fine sediments, piercers for insects with sucking mouthparts that feed on plant fluids, and gougers for larvae that burrow into large woody debris while feeding on the fungal and bacterial colonies that develop on their surfaces.

Fish fit into the original categories based on the type of food, but are not readily categorized as to FFG. Some fish, such as suckers, are herbivorous, grazing periphyton from rock surfaces; carp are omnivorous; and trout are predaceous. Fishes have also been classified into feeding guilds, depending on where they feed in the water column. Thus, there are top-feeders, midwater-feeders, and bottom-feeders; some obviously fit into more than one of these categories.

Classifying organisms into various categories can be used to describe how they are linked into food webs, and the more diverse the classification system, the more we can learn about the food web. **Figure 2** illustrates a simple food web based on insects being classified as herbivores (feeding on algae) or detritivores (feeding on both algae and macrophytes after decomposition). The trout are carnivores feeding on insects. **Figure 3** shows the interactions of the same food web when the FFG classifications system is used, providing a broader look at the complexity of the food web.

One other aspect that needs mentioning in connection with food webs is that of competition. Competition occurs between and among organisms for food, space, and reproductive partners and is a vital regulatory mechanism that helps determine numbers and location of individuals within the ecosystem.

Structure and Function of Lotic Ecosystems

Consider the cross section of a typical stream (**Figure 4**). It shows the interrelationships among the energy sources produced within the stream and terrestrially derived primary production, and how these energy sources are utilized by different functional groups of organisms within the stream. This is a generalization of what is going on at a particular place in a stream, but not how they change along a continuum from headwaters to mouth. In 1980, Robin Vannote and colleagues published a seminal paper, in which they theorized how a river changes along its length, in terms of size, energy resources, and FFG communities. Known as the River Continuum Concept (RCC), this model generates many useful predictions about patterns that can be seen in any geographical region or biome. Of course, no model is applicable to every situation, but the generalities of this model have been supported by considerable research.

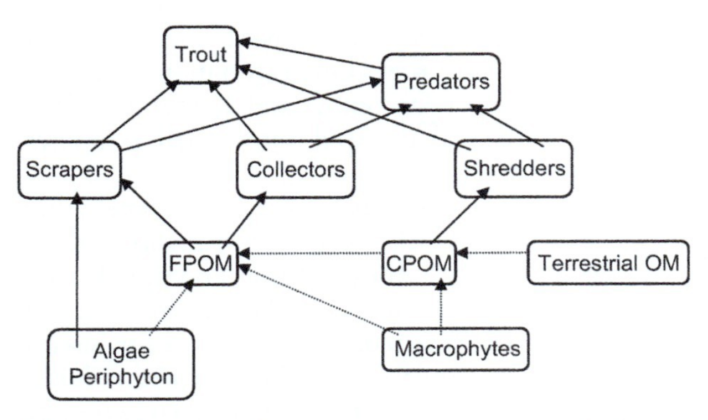

Figure 3 Diagram of food web based on functional feeding groups. Solid arrows indicate feeding pathways; dashed arrows indicate decomposition pathways. From Cushing and Allan (2001), with permission.

The River Continuum Concept: A Model

The following description of the RCC uses a hypothetical stream ecosystem located in a deciduous forest, typical of those found in the eastern part of the United States.

The headwaters of our hypothetical river flows through a heavily shaded forest; it then flows into more open country, and eventually becomes a large, deep, heavily silted river; **Figure 5** is a diagram of this system. We will break the river into three general regions for discussion: the headwaters (orders 1–3), the mid-reaches (orders 4–7), and the lower reaches (orders 8 and above).

In the headwaters, the stream is narrow and generally well shaded by the riparian canopy. The stream bottom may be rocky, sandy, or a combination of the two, depending on the geological characteristics of

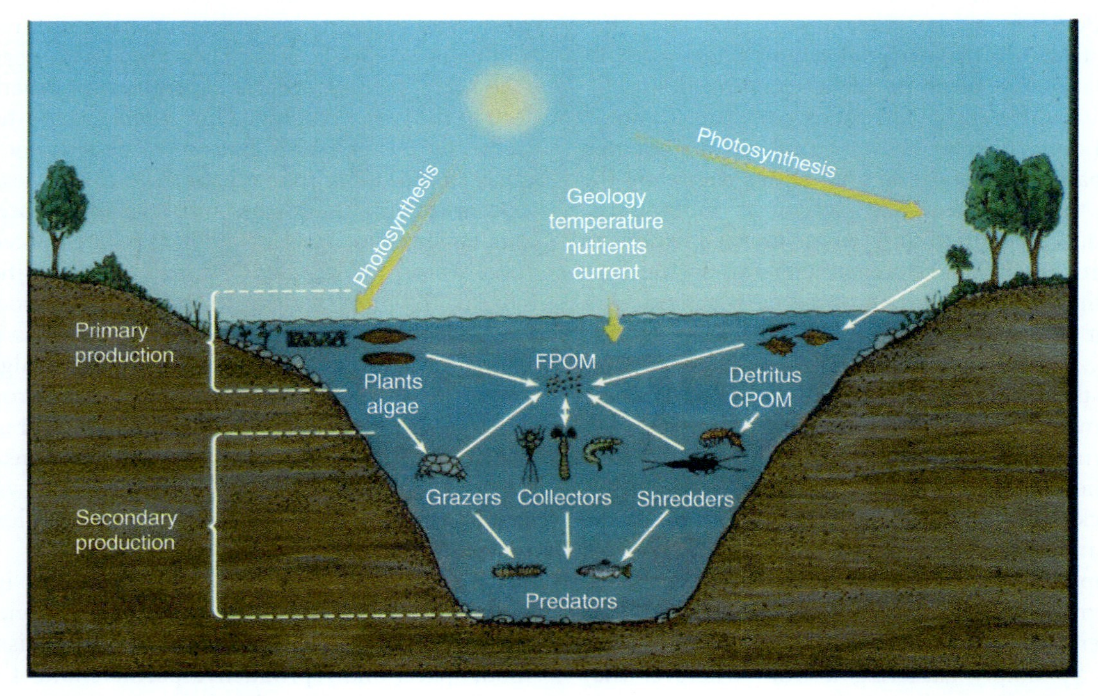

Figure 4 Cross section of a typical stream showing influence of physical and chemical variables and interrelationships among energy resources and functional feeding groups. From Cushing and Allan (2001), with permission.

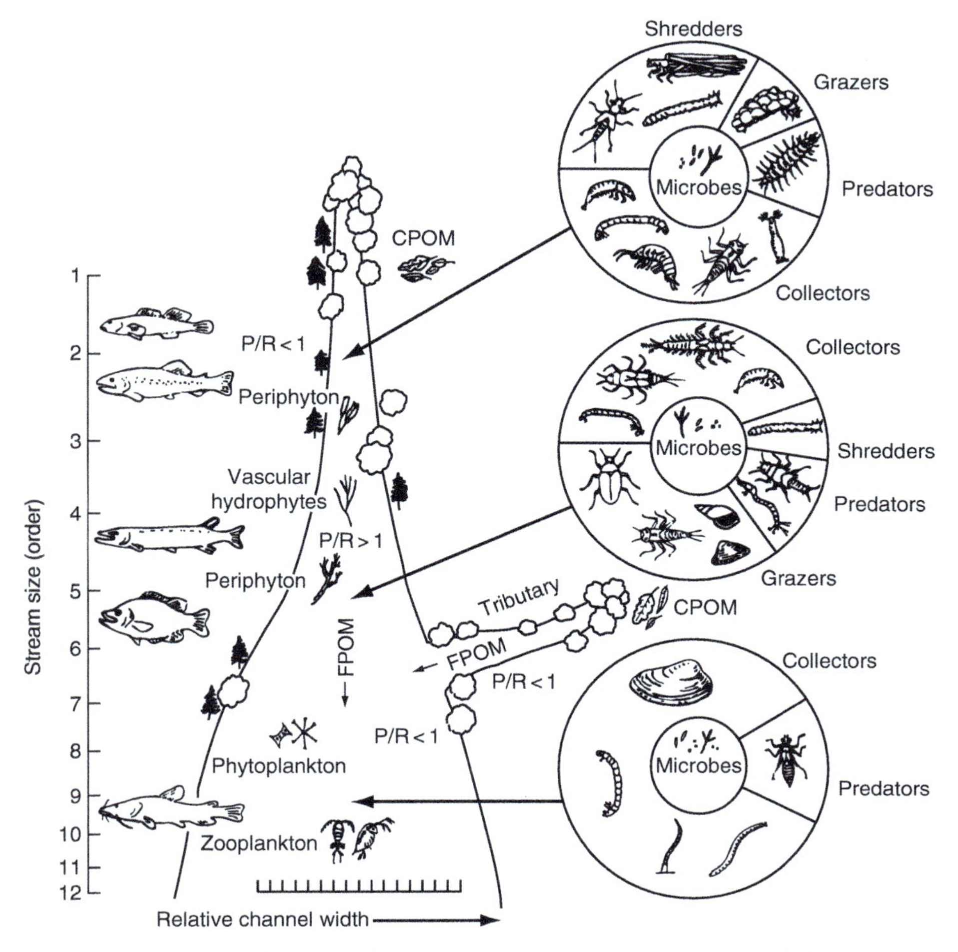

Figure 5 Diagrammatical representation of the River Continuum Concept. From Vannote et al. (1980) as modified by Allan (1995), with permission.

the drainage basin. Insufficient light reaches the stream bed to promote significant algal growth, and the current is too fast, substratum often unsuitable, and nutrients usually too low to allow growth of macrophytes. Mosses are the dominant instream primary producers. Considerable CPOM from the terrestrial environment enters the stream due to its proximity and the narrow stream channel. This may be in the form of leaves, especially during autumn leaf fall, or twigs and branches that fall or reach the streambed by gravity from the usually steep hillsides found in this part of the stream. Comparison of the amount of dissolved oxygen produced by primary production with that used by respiration shows that more oxygen is respired within the stream than is produced; thus, energy from outside the stream is necessary. This energy comes from the CPOM from the terrestrial environment. Ecologists term this a heterotrophic stream reach; it doesn't produce enough of its own food. Analysis of the FFG composition shows that it is dominated by roughly 35%shredders and 45% collectors because of the preponderance of CPOM and FPOM. Grazers (5%) are limited by the lack of algae on the stream bottom, and predators compose the remaining 15%. The percentage of predators present is fairly constant throughout the river continuum; however, different groups of organisms perform this function in the different reaches. This reach of the continuum exports a large amount of FPOM to the mid-reaches. Fish species present in these reaches include various minnows, trout, sculpins, and other typical fish tolerant of seasonal and daily cold temperature regimes.

In the mid-reaches, the streambed has widened, the bottom is well lit by direct sunlight, temperatures have warmed, and nutrient concentrations have increased, all leading to a proliferation of filamentous greens and/or diatoms on the bottom. The stream bottom is usually composed of rubble, rocks, and pebbles, with sand and silt accumulations where the current slackens. Rooted macrophytes will occur in protected places. CPOM inputs have decreased on an areal basis because the stream is wider and the riparian canopy has decreased in extent. Primary production by algae and macrophytes in this reach would exceed respiration by the instream community; hence, this is an autotrophic stream reach, meaning that the stream produces more energy than is needed to support it. This section of the continuum exports excess FPOM to the lower reaches of the river. The FFG composition reveals about 50% collectors, the same as in the headwaters but composed of different species. High export of FPOM from the headwaters and instream production supports the collectors. The major change is that grazers now represent about 30% of the population because of the proliferation of algae, and shredders (5%) are reduced because of the paucity of CPOM. Predators again make up about 15%. Fishes present in the mid-reaches are typical of those species that can tolerate wider fluctuations in daily and seasonal temperatures, although there is considerable overlap with some of the headwater species. Trout are usually present, along with suckers and many minnows. This section of the continuum is highly productive and is the only autotrophic section in the continuum.

In the large, usually slow flowing and deeper lower reaches of our hypothetical river, several changes occur. Increased turbidity and depth prevent sunlight from supporting algal growth on the bottom; this is also adversely affected by the fine grained, shifting nature of the bottom sediments. Nutrients are in high concentrations. Because algae cannot grow on the stream bottom, instream primary production now takes place within the water column where phytoplankton may flourish, and in shallow, littoral areas by macrophytes. Indeed, the presence of phytoplankton and rooted plants hints at the lake-like characteristics of many high order rivers. Terrestrial input of CPOM is negligible. The water column also contains large amounts of suspended FPOM transported from the mid-reaches. Respiration of this material and of senescent phytoplankton exceeds the primary production of the small algal crop and macrophytes, thus making this a heterotrophic reach. The benthic community is composed of about 85% gathering and filtering-collectors utilizing the rich supply of FPOM suspended in the water; most are mud-dwelling molluscs or dipteran larvae. Grazers and shredders are absent, although grazing snails may be found feeding on the surface layers of algae that coat plant stems or other places. Predators, again, comprise about 15% of the population. Fish present in the lower reaches are those found in environments where temperatures fluctuate widely. Species such as suckers, carp, and chubs are typical.

This, then, describes the predictable patterns and biological interactions found in a hypothetical pristine stream continuum from its headwaters to its mouth, and the RCC has proven to be a useful paradigm for lotic ecologists. One other aspect of the patterns described above is the fact that changes in these patterns occur from both natural and human causes; these are called reset mechanisms.

Other Ecosystem Concepts

Other aspects of the river ecosystem are important to mention and influence their ecology. The hyporheic community consists of those organisms occurring in the interstitial spaces between the rocks and stones (hyporheic zone) making up the bottom of the stream. They are usually fairly minute, although the larvae of several stoneflies and midges have been found in this habitat far below the streambed and many meters laterally from the stream shoreline.

It should also be kept in mind that stream habitats and communities exist at differing scales, from those existing on a single grain of sand or rock to those characteristic of the entire stream reach. The study of these habitats is called patch dynamics, and it is important to consider these when describing the animal communities and their interactions. Another important relationship in some river and stream ecosystems is the interaction and exchanges that occur between the stream and its floodplain. This has been termed the flood-pulse concept and describes the exchange of nutrients, organisms, and organic material that occurs when a stream or river floods and then recedes. Nutrient spiraling is an import aspect of the interactions in riverine ecosystems and refers to the fact that nutrients, such as nitrogen and phosphorus, not only cycle within the aquatic community, but also have a downstream component because of the flowing nature of the water. A new concept, the network dynamics hypothesis, has recently been proposed to describe the distribution of FFG communities based on the prevalence of tributary junctions; it has yet to be adequately tested.

Also remember that the dynamic exchange of organisms and energy in riverine ecosystems occurs in several directions – upstream and downstream, laterally and vertically within the hyporheos, and with time. Upstream movement occurs by active movement, either within the stream itself or above it in the case of flying insects. Downstream movement is by either active movement or by drifting of organisms dislodged from the stream bottom. Lateral movement occurs by adult insects dispersing into the terrestrial environment, and both lateral and vertical movements take place within the hyporheos and between the hyporheos and the water column. The fourth dimension, time, describes the seasonal changes that occur among the populations as they grow, mature, and die.

Glossary

Coarse particulate organic matter (CPOM) – All nonliving organic matter within a river that exceeds 1 mm in diameter (e.g., leaves, twigs, grass).

Fine particulate organic matter (FPOM) – Nonliving organic matter within a river that is less than 1 mm in diameter (e.g., detritus, fecal pellets).

Flood plain concept – A concept describing the relationships and exchanges that occur between the flood plain and the water column when a river inundates its flood plain and then recedes to its normal channel. Exchanges include biota, chemicals, and organic matter.

Functional feeding groups – A system that classifies macroinvertebrates according to how they obtain their food, rather that what they eat. Major groups are shredders, grazer/scrapers, collectors (gathering and filtering), and predators.

Microbial loop – The complex interactions of microorganisms, including meiofauna, bacteria, zooplankton, governing food (energy) transfer within the biofilm (= periphyton).

Network dynamics concept – A concept that attempts to explain the distribution of macroinvertebrate communities based on the occurrence and number of tributary junctions along a river continuum.

Nutrient spiraling – The concept that adds a downstream component to the nutrient cycle, indicating that the cycling of nutrients has a spatial component.

Patch dynamics – The study of organisms communities and their distribution on different spatial scales.

River continuum concept – The concept that presents an explanation for the downstream relationships among ecosystem components (energy resources, functional feeding groups, physical attributes) for a river from its source to mouth.

See also: Algae of River Ecosystems; Benthic Invertebrate Fauna; Benthic Invertebrate Fauna, River and Floodplain Ecosystems; Benthic Invertebrate Fauna, Small Streams; Coarse Woody Debris in Lakes and Streams; Currents in Rivers; Hydrology: Rivers; Hydrology: Streams; Regulators of Biotic Processes in Stream and River Ecosystems; Streams and Rivers as Ecosystems.

Further Reading

Allan JD (1995) *Stream Ecology.* Dordrecht, The Netherlands: Kluwer/Academic.

Cummins KW (1973) Trophic relations of aquatic insects. *Annual Review of Entomology* 18: 183–206.

Cushing CE and Allan JD (2001) *Streams: Their Ecology and Life.* San Diego, CA: Elsevier/Academic.

Hynes HBN (1975) The stream and its valley. *Verhandlungen der Internationalen Vereinigung für theoretische und angewandte Limnologie* 19: 1–15.

Junk WJ, Bayley PB, and Sparks RE (1989) The flood pulse concept in river-floodplain systems. In: Dodge DP (ed.) *Proceedings of the International Large River Symposium,* pp. 110–127. *Canadian Journal of Fisheries and Aquatic Sciences No. 106, Ottawa.*

Rice SP, Greenwood MT, and Joyce CB (2001) Tributaries, sediment sources, and the longitudinal organization of macroinvertebrate fauna along river systems. *Canadian Journal of Fisheries and Aquatic Sciences* 58: 828–840.

Townsend CR (1989) The patch dynamics concept of stream community ecology. *Journal of the North American Benthologiccal Society* 8: 35–50.

Vannote RL, Minshall GW, Cummins KW, Sedell JR, and Cushing CE (1980) The river continuum concept. *Canadian Journal of Fisheries and Aquatic Sciences* 37: 130–137.

Webster JR and Patten BC (1979) Effects of watershed perturbation on stream potassium and calcium dynamics. *Ecological Monographs* 49: 51–72.

Regulators of Biotic Processes in Stream and River Ecosystems

M E Power, University of California, Berkeley, CA, USA

Introduction: Downstream Changes and Dynamics in River Environments

River networks sculpt the Earth's landscapes. Along river networks, channel and riparian environments change in partially predictable ways that strongly influence river organisms and their interactions. Width, depth, and average flow velocities of rivers change downstream and also vary through time at a single station according to a set of empirical relationships known as hydraulic geometry. As channels widen downstream, they are less shaded and therefore receive more solar radiation, which increases both water temperature and the potential for primary production (growth of aquatic plants and algae). Habitat structure and disturbance regimes in channels also change along the river gradient. Bed sediments become finer downstream as channel gradients decrease (**Figure 1**). Steep headwaters often have coarse boulder and bedrock substrates, whereas mid-elevation mainstem rivers are likely to have gravel, pebble, and cobble substrates, with boulders and bedrock emerging as habitat islands, and lowland rivers with large floodplains typically have beds of mobile sand and silt. Events that erode or mobilize the river bed are rare at any given site in headwaters and largely result from debris flows that may recur only every 1000–10 000 years. Scouring disturbances can happen several times per year when storm flows mobilize cobble, pebble, or gravel beds in meandering mainstem rivers. In rivers with beds of sand and silt, portions of the substrate (and small organisms attached to them) are in constant motion, except for large debris jams or on floodplains or elevated banks.

In general, ecological gradients in rivers demonstrate the effect of physical environment on food–web interactions. In addition, rivers offer repeated opportunities to study species interactions on varied temporal scales as food webs reassemble after disturbance by drought or flood.

Changes in Communities over Time: Disturbance and Succession

Food webs reassemble after flood or drought disturbances as surviving and colonizing species rebuild their local populations and interact with each other during ecological succession, the period of biotic recovery following disturbance. Disturbances caused by bed movement are rare in headwater streams with gradients steeper than 10%, where channel substrates are bedrock or boulder dominated. Bed mobilization at these sites occurs infrequently (perhaps once or twice in 1000–10 000 years) when debris flows deliver sediments to channels. In lowland floodplain rivers, sand and silt making up the bed sediments are constantly in movement. Here, bed mobilization is not a pulsed event, hence not a disturbance in the ecological sense. Lowland river benthic biota adapt to fine, shifting river sediments by attaching to stable substrate, such as inundated floodplain vegetation or log jams, or by burrowing into channel walls strengthened by clay or roots. Bed mobilization is an important disturbance process at intermediate positions in the drainage network, where gravel, pebble, and cobble bedded rivers offer repeated opportunities to study species interactions as food webs reassemble after flood scour or dewatering.

In Mediterranean climates, which are typical of coastlines, the seasonal timing of flood and drying disturbance is somewhat predictable. Typically, a rainy winter season is followed by a summer drought with little or no rainfall. Variation occurs from year to year, however, in the magnitude and timing of flood and the severity of subsequent drought. In continental (inland) regimes, precipitation heavy enough to cause floods may fall during any month of the year, although where snow melt is important, large spring floods may predictably dominate the annual hydrograph. Under either Mediterranean or continental climate regimes, the timing of disturbance relative to the life histories of organisms will strongly influence the effect of these hydrologic events on abundances of species.

After gravel-dominated river beds are mobilized, rock-bound organisms (attached algae, mosses, or invertebrates with limited mobility) that cannot escape are damaged or removed. Mobile fauna, on the other hand, may escape. Fish can swim above the mobilized bed. When flow velocities become unmanageable, fish can take refuge in the slack water refuges of inundated off-channel habitats, undercut banks, or behind logs, bedrock formations, boulders, or large cobbles. Water-filled pore spaces within coarse bed substrates are critical refuges at high flow for fish and macroinvertebrates. Fish and invertebrates also can seek refuge from drought in groundwater underlying the stream. Fish and invertebrates may reach this

Figure 1 Seasonal cycle of the green macroalga *Cladophora* in the Mediterranean South Fork Eel River (counter clockwise from upper right). Substrates scoured bare by winter floods show deposited silt as the river subsides and clears in early spring. Long turfs of *Cladophora* regrow vegetatively from surviving basal cells on large boulder and bedrock substrates that escaped severe burial or abrasion during the preceding winter. By midsummer, *Cladophora* has detached to form floating mats that accumulate in slack water along river margins; attached turfs persist in mid-channel where higher flow velocities are maintained. By late summer, both turfs and mats have collapsed to stringy, webbed remnants with architecture created by dense infestations of the tuft-weaving midge, *Pseudochironomus*. Food-web interactions contribute to the loss of most *Cladophora* biomass well before the onset of the next winter's scouring floods.

refuge through the burrows of larger organisms (alligators, burrowing catfish). Such refuges from erosion or drying of the riverbed are lost to the community when coarse substrates become embedded with fine sediments. Excessive mobilization of fine sediments is one of the most widespread forms of environmental degradation of river ecosystems; it is caused by land-use practices such as forest clearing, road construction, agriculture, and stock grazing.

Animals that survive drought or flood in refuges, as well as attached algae, mosses, and microbes that are not completely removed from rock surfaces become the seeds for recovery during ecological succession, the process that re-establishes river biota after disturbance. Surviving organisms are joined by colonists that immigrate from other habitats into recently disturbed, sparsely populated areas. For example, aquatic insects rebuild their populations from the 'air force reserve' of winged adult aquatic insects.

Organisms at different trophic positions in river food webs have different mobilities and vulnerabilities to disturbance. Because predators are often more mobile than rooted, drifting, or attached prey, they often survive disturbances that exterminate much of the biota at lower trophic levels. As food webs recover from disturbance during succession, the first prey species to recover or colonize tend to have traits that favor high dispersal and high growth rates, rather than defensive traits such as toxins, armor, or protective attached shells that require organisms to allocate energy and nutrients away from growth. During early stages of succession, food webs tend to have surviving predators that encounter relatively edible, vulnerable prey. Therefore, disturbances in rivers often cause food webs to have food chains that link predators to consumers of plants (herbivores). Energy from primary producers (plants) flows efficiently up these food chains to predators.

The Eel River of Northern California has a Mediterranean hydrologic regime; rainy winters precede the biologically active low flow season of summer. Each spring, the green macroalga *Cladophora glomerata*, which dominates primary producer biomass of summer, initiates growth vegetatively from basal cells that survived winter flood scour on stable boulder and bedrock substrates (**Figure 1**). *Cladophora* and other algae are grazed by invertebrates that vary markedly in their vulnerability to local predators. Mobile, unarmored taxa such as mayflies and chironomids are vulnerable to predatory invertebrates and fish. In contrast, heavily armored taxa (e.g., stone cased caddisflies) and sessile grazers (immobile grazers: some chironomids, caddisflies, and aquatic moth larvae) are less vulnerable. Summer biomass of *Cladophora* is affected by predatory juvenile steelhead trout if flood scour occurred during the preceding winter. During flood years, fish suppress either herbivores or small predators that affect herbivores, thus exerting indirect effects on algae. During drought years without winter flood scour, large armored caddisflies that are invulnerable to fish abound during the subsequent summer, and *Cladophora* grows only when these caddisflies are experimentally removed. Thus, drought years produce shorter food chains in which fish fail to enhance algal biomass because of predator resistant grazers. Flood disturbance, by suppressing these resistant grazers, sets the stage for fish-mediated enhancement of algal growth in this river food web (**Figure 2**).

In a small Montana stream, J. McAuliffe found that interactions among sessile grazers also were influenced by frequency of disturbance. On shallow cobble substrates (15–24 cm deep) that were infrequently disturbed, a sessile and highly territorial caddisfly, *Leucotrichia,* dominated, and excluded other mobile and sessile grazers, whose densities and diversity increased when *Leucotrichia* was experimentally removed. The competitive dominance of *Leucotrichia* also was disrupted when the stream bed dried. After these disturbances removed *Leucotrichia*, short-lived subordinate competitors including the midge *Eukiefferiella* colonized the newly inundated habitats. Thus, changes in river discharge through time determine the interactions of organisms that make up the river food web.

Changes in Interactions across Space: Depth Zonation

In gravel-bed streams of central Panama, armored catfish (Loricariidae) are the dominant grazers of algae. These grazers can outgrow smaller swimming predators in streams. Armored catfish do not, however, graze in shallow water where they are vulnerable to birds and other large predators. Herons, egrets, and kingfisherss fish most frequently and effectively in water <20 cm deep, where there is insufficient warning between the bird's surface splash and its strike for fish to escape. In streams of central Panama, high algal biomass rims the river margin. Rock and wood

Figure 2 The food chains that control *Cladophora* biomass are longer following winters with scouring floods that remove predator resistant grazers, and shorter following drought winters, when invulnerable grazers like the caddisfly *Discomoecus* persist. (Photo credits: mayfly and *Dicosmoecus*: Will Swalling, high flow tributary of Eel River: Bill Trush.)

substrates that have green carpets of attached algae in shallow areas become abruptly barren when deeper than 20 cm, indicating a spatial threshold. Above this threshold, in shallow water, birds protect algae from grazing fish; below the threshold, in deeper water, unimpeded grazers prevent accumulation of algal biomass. Similar patterns occur in pools along prairie-margin streams of Oklahoma (**Figure 3**).

The studies of *Leucotrichia* in a Montana stream and of loricariid catfish in a Panamanian stream both illustrate the difficulty of inferring from casual observation the processes that determine distributions and abundances of river organisms. Without investigating interactions with other organisms, an observer might reasonably assume that attached algae are more abundant in shallow water because they escape flood scour or experienced higher light exposure in shallow water. Likewise, an observer unaware of the competitive effect of *Leucotrichia* might assume that the midge *Eukiefferiella* prefer or differentially settles on shallow stream substrates. In both cases, the simple explanations are incorrect. Ecology is replete with examples in which the importance of a competitor or a consumer on the distribution and abundance of other organisms can be revealed only by an experimental manipulation or a 'natural experiment' in which algal abundance or other food-web characteristics could be compared over space or time in the presence and absence of species that affect them. Because interactions usually are hidden, their role in driving ecological patterns in rivers and streams is easily underestimated.

Changes in Interactions across Space: Position in the Drainage Network

Stream ecologists have long been interested in the effects of systematic downstream changes in environmental factors affecting the distribution, abundance, diversity, and energy sources of stream organisms or functional groups of organisms. We still know very little, however, about changes in species interactions, food-web composition, and ecosystem functions along river drainage networks.

Ecological regimes are sets of conditions and constraints that produce specific outcomes of ecosystem functions or food-web interactions. Ecological regimes may provide answers to questions such as these: Where or when can grazers in streams suppress algal biomass? Where or when can juvenile fish grow and thrive? Where or when can fishing birds remove or exclude fish, or can bats forage effectively over rivers on emerging aquatic insects? Ecological regimes vary across space and time.

A spatial threshold control of algae by grazers occurs in headwater tributaries of the South Fork

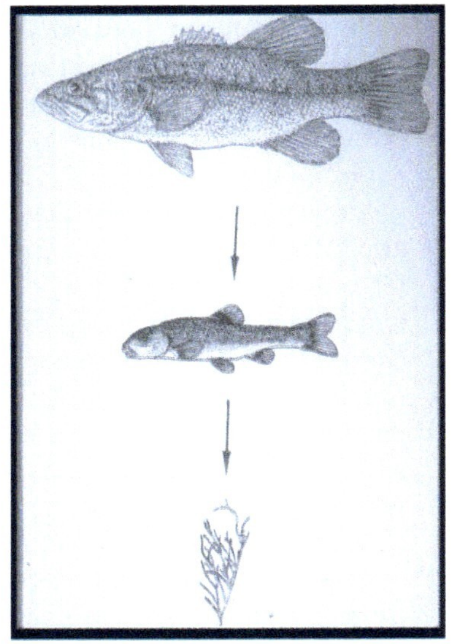

Figure 3 Bathtub rings of green algae persist around the river margin in a barren *Campostoma* pool in Brier Creek, Oklahoma, possibly due to the behavioral avoidance of shallows by grazing fishes in which they are vulnerable to fishing birds and mammals. The shallower algae are more edible (*Rhizoclonium* covered by nutritious diatom epiphytes) and the deeper algae are colonies of the less nutritious *Spirogyra*, which does not acquire epiphytic diatoms. Similar patterns occur due to avoidance of avian and terrestrial predators in small neotropical streams grazed by armored catfish (Loricariidae).

Eel River. Stable isotope analyses have shown that the dominant headwater grazer, larvae of the caddisfly *Glossosoma*, grows upon algal rather than terrestrial carbon (McNeely *et al.*, 2008). Experiments revealed that *Glossosoma* cannot suppress accumulation of algal biomass in very small streams (basins $< 2\ km^2$) but, in streams with watersheds $> 2\ km^2$, *Glossosoma* does reduce the accumulation of attached algae. Removing *Glossosoma* from the larger streams visibly increases turfs of diatoms, and also increases the flow of algal carbon to other insects, such as mayflies, which are more vulnerable to fish than *Glossosoma* is. Understanding how species interactions control algal biomass, and the access of vulnerable grazers to algal carbon that subsequently passes to fish and other predators, shows how the ecosystem might respond to future change. For example, if a parasite eradicated *Glossosoma* from the Eel River, the landscape (drainage area) thresholds at which steelhead trout and other predators could derive energy from herbivores that eat algae might move upstream.

Changes in Biomass Distribution and Elemental Fluxes Mediated by Interactions and Movements of River Organisms

Grazers can track the growth rates of their algal foods. Such tracking has profound effects on the distribution of algal biomass in rivers. Armored catfish in streams of central Panama track algal productivity closely from pool to pool, thus making efficient use of the spatially variable algal productivity, which is affected by heterogeneous shading from forest canopy. The Panamanian studies showed that algae grew up to 16 times faster in sunny than in dark pools, but grazing catfish were 16 times more abundant in sunny than in dark pools (Power, 1984). As a result, algal standing crops (biomass accumulation) were similar in dark, half-shaded, and sunny pools. Pre-reproductive catfish also grew (and survived) at the same rate in sunny, crowded pools and in dark, uncrowded pools. In this Panamanian stream, the grazers suppressed heterogeneity in algal biomass that would otherwise have resulted from pool-to-pool variation in solar radiation reaching individual pools.

The opposite pattern occurred in a stream of similar size in south-central Oklahoma, where light was intense and similar from pool to pool. Algivorous minnows (*Campostoma*) are the dominant grazer in this prairie-margin stream. *Campostoma* is vulnerable to predatory bass. Schools of *Campostoma* can denude stream substrates of algae. Barren pools, however, often are adjacent to pools that are filled with filamentous green algae. The green pools lack *Campostoma* and contain their predators, spotted and largemouth bass. Experimental transfers of bass and *Campostoma* can change pools from green to barren (by removing bass and adding *Campostoma*) or barren to green (adding bass to a *Campostoma* pool) within weeks. Natural floods that rearrange bass and minnows among pools trigger the predicted changes in algal biomass.

If grazers are able to track their resources, they can offset differential algal accumulation despite large spatial heterogeneity in algal growth rates, but when grazers are constrained in their movement by predators, great heterogeneity of algal biomass develops, even when environmental contrasts among habitats in algal growth rates are small. Without experiments, an observer might think that differences in algal biomass distribution among pools in the Oklahoma stream arise because of differential algal resources (e.g., nutrients). Similarly, the uniformity of algal biomass in the Panamanian stream might be interpreted as meaning light is not limiting to algae in this habitat; in fact such an inference that has been made in other rivers without recourse to experiments that could check for the importance of species interactions in controlling algal abundance.

Ecological regimes that determine where algal biomass can or cannot accumulate have important consequences, not only for the animals that feed from or live within algal turfs or mats, but also for fluxes (movements) of nutrients and organic matter through watersheds. Downstream solute fluxes are of societal concern, as nutrients or organic matter not retained high in drainage networks can accumulate downstream, with potentially adverse effects, such as eutrophication of drinking water reservoirs or harmful algal blooms in lowland water bodies or nearshore marine environments.

Stream ecologists have developed models of the spiraling downstream movement of nutrients (e.g., P, N, C), which includes not only downstream movement by water flow, but also periodic interruption of movement caused by nutrient uptake and release, often through uptake by the biota. Flowing water ecosystems that tend to immobilize atoms of nutrients through use and re-use for local biological production are characterized by short spiral lengths. Lateral wetlands greatly increase retention times and spiraling lengths for nutrients in river networks. Less retentive channels, such as those that have been straightened artificially, tend to have longer spiral lengths, less biological production per length of channel, and are less effective at buffering downstream waters from eutrophication. Species interactions influence nutrient spiraling through several paths. Of these, biological backflows caused by salmon migration, predation, and scavenging have recently received considerable attention.

One of the earlier studies of biological flows mediated by fish migration was done by Hall (1972) in New Hope Creek, North Carolina. He found that upstream migrations of large fish such as redhorse suckers (*Moxostoma*) played an important role in maintaining phosphorus reserves in New Hope Creek's headwaters. More recently, other scientists have documented large backflows of marine-derived nutrients on Alaskan rivers by migrating salmon, which are then carried upslope out of river channels to riparian vegetation by bear, eagles, mice, and other birds and mammals that prey on or scavenge these salmon. Nutrients may also be carried upslope across river valleys by emerging river insects and their terrestrial predators. The ecological importance of these backflows relative to nutrient cycles in situ is still largely unknown. Clearly, however, upstream and upslope fluxes mediated by large vertebrates were much larger before humans reduced their populations. Humans have also blocked vertebrate migrations through river habitats with dams, levees, channel simplification, and by eliminating beavers and large trees and logs that increased habitat complexity in rivers by forming deep pools, undercut banks and floodplain water bodies and marginal channels.

Forecasting Ecological Change in River Networks

River networks and their watershed source areas are experiencing accelerating change in climate, land use, and biota (e.g., invasions and extinctions). The pace of global environmental change demands ecological forecasting over time scales of societal importance, e.g., decadal. Insights from local observations and experiments must be expanded in spatial scale ('upscaled') to be useful in predicting effects at the watershed scale. Similarly, effects of regional changes in climate, land use, or biota must be 'downscaled' to local environments. Ecosystem science can reveal the mechanisms that will either foster or destroy the resiliency of watershed ecosystems, and sustain or damage the vital ecosystem services that watersheds provide: clean water supplies, soil stabilization and fertility, and biota with ecological as well as cultural or commercial value. Ecological scaling requires an understanding of the linkages between climate, landscapes, food webs, and ecosystem functions.

Ecological forecasting remains challenging because of three factors: (1) the spatial heterogeneity and temporal fluctuation of environments, (2) the diversity and idiosyncrasies of the biota interacting with these environments, and (3) species–environment interactions that are mediated indirectly through webs of interactions with other species and physicochemical factors. Recent advances in mapping, sensing, and tracing technologies have greatly improved documentation of ecological interactions. Remote sensing technologies (e.g., airborne laser altimetry, multispectral imaging from satellites) can resolve land cover and provide topographic data that allow us to map whole watersheds or even larger regions. Concomitantly, automated wireless sensing networks hold out the promise of monitoring environmental conditions at scales that for the first time have the necessary resolution to capture variation meaningful to individual organisms. Finally, over the past several decades, ecologists have made increasing use of tracer technologies, particularly involving stable isotopes, to study movements of organisms and fluxes of biologically relevant materials. These technological advances will enhance the scope and resolution with which key ecological patterns and processes can be analyzed, and will improve ecological upscaling, downscaling, and forecasting.

A very old and unresolved question in ecology is whether distributions and abundances of organisms are controlled primarily by the physiological constraints and requirements of individual species or by interactions such as competition and predation. The question resurfaces in light of attempts to predict ecological responses to global climate change. Will climate envelopes (specific ranges of climate conditions) based on the physiological requirements and tolerances of individual species suffice to predict their fate under greenhouse warming, or must species interactions also be considered? Indirect feedbacks mediated by species interactions are important in forecasting effects of climate change in several terrestrial systems (e.g., boreal wolf-moose-spruce ecosystems, or California grassland). In rivers, the direct adverse effects of scouring floods on algae can be reversed within a year by the indirect effects of flood scour on the biota that grazes algae, and the direct effect of light environment on algal accumulation can be completely obscured by food-web interactions of algal grazers and their predators. Advanced technologies, combined with careful natural history observations and field experiments (**Figure 4**), can allow us to compare more quantitatively the relative importance of ecophysiological constraints on individual species versus higher-level limitations imposed by ecological interactions. River networks, with their dynamic, heterogeneous, but partially predictable physical environments, and their crucial roles in maintaining health of ecosystems and societies, must remain foremost as an arena for this exploration.

Figure 4 Experiments to determine the effect of fish on food webs in the Eel River are initiated by use of electrofishing clear enclosures of fish, then stocking some with fish and leaving others free of fish.

Glossary

Autotrophs – Organisms that use light or chemical energy to convert inorganic carbon organic matter, thus producing their own biomass.

Climate envelopes – Ranges of climatic conditions (temperature, moisture, radiation intensity, etc.) within which a species can persist, through adequate survival, growth, and reproduction.

Disturbance – A pulsed event that kills or removes organisms, freeing space and resources.

Functional groups – Groups of species or life stages of species that function in similar ways, for example, consuming similar types of resources, or exerting similar effects in biogeochemical cycles. Functional groupings are flexible; they are defined according to the particular process under investigation (e.g., processing of organic matter).

Hydraulic geometry – A set of empirical relationships that predict changes in the depth, width, or velocity of rivers from their discharge, either at a station over changing discharges, or downstream at a discharge of a given recurrence interval.

Hydrograph – A record over time of the stage (water level) or discharge (flow) of water in a river past a specific cross section.

Primary productivity – Rate of conversion inorganic carbon to organic matter by autotrophs.

Sessile – Living in a stationary position. Sessile animals often build and live within protective retreats that they attach to rock, woody debris, or vegetation.

Spiraling (carbon or nutrient) – The downstream movement of atoms through rivers, interspersed with periodic uptake and release (cycling) by biota or abiotic substrates.

Ecological succession – The sequence of processes that re-establishes biota after disturbance.

See also: Algae of River Ecosystems; Benthic Invertebrate Fauna, Small Streams; Hydrology: Streams.

Further Reading

Ben-David M, Hanley TA, and Schell DM (1998) Fertilization of terrestrial vegetation by spawning Pacific salmon: The role of flooding and predator activity. *Oikos* 83: 47–55.

Hall CAS (1972) Migration and metabolism in a temperate stream ecosystem. *Ecology* 53: 585–604.

Leopold LB and Maddock T III (1953) *The hydraulic Geometry of Stream Channels and Some Physiographic Implications.* Washington, DC: US Geological Survey. USGS Prof. Paper 252.

McAuliffe JR (1983) Competition, colonization patterns, and disturbance in stream benthic communities. In: Barnes JR and Minshall GW (eds.) *Stream Ecology*, pp. 137–156. New York: Plenum.

McNeely C, Finlay JF, and Power ME (2008) Grazer traits, competition, and carbon sources to a headwater stream food web. *Ecology* (in press).

Power ME (1984) Habitat quality and the distribution of algae-grazing catfish in a Panamanian stream. *Jounral of Animal Ecology* 53: 357–374.

Power ME, Matthews WJ, and Stewart AJ (1985) Grazing minnows, piscivorous bass and stream algae: Dynamics of a strong interaction. *Ecology* 66: 1448–1456.

Sousa WP (1984) The role of disturbance in natural communities. *Annual Review of Ecology and Systematics* 15: 353–391.

Wootton JT, Parker MS, and Power ME (1996) Effects of disturbance on river food webs. *Science* 273: 1558–1560.

Ecology and Role of Headwater Streams

W H McDowell, University of New Hampshire, Durham, NH, USA

Introduction

Streams are an integral part of the landscape in most biomes, and serve as the most tangible link between aquatic and terrestrial ecosystems. The physical, chemical, and biological characteristics of streams and the drainage networks that they form reflect the terrestrial environment. In addition to serving as important conduits for the delivery of solutes and sediments to downstream ecosystems, streams are important ecosystems in their own right. They contain a variety of aquatic habitats, support distinctive biota, and are the site of biogeochemical transformations that are important in affecting elemental fluxes from various landscapes.

Streams in the Landscape

Linkages to the Watershed

In almost every sense, a stream is the product of its landscape. More than any other inland aquatic systems, the physical, chemical and biological attributes of streams reflect the watershed or drainage basin in which they are located. This is because of their small volume, the large surface-to-volume ratio in the stream channel, and the short residence time of water within the channel. By definition, streams cannot exist as isolated systems. Stream water requires constant replenishment by precipitation, either directly into the stream channel, or indirectly, through groundwater inputs. Indirect input of precipitation through shallow riparian groundwater is the dominant source of stream flow in most streams (see **Hydrology: Streams**), and riparian flow paths provide important connections between the watershed and the stream (see **Riparian Zones**). The residence time of water in stream networks, including shallow groundwater that interacts with the stream channel, is on the order of days to weeks, whereas it can be centuries in large lakes or large groundwater systems.

The intimate connection between a stream and its watershed makes it difficult to provide a simple, uniform definition of a stream that is useful across a range of scientific and management contexts. Streams are typically defined as small bodies of flowing water in perennially wet channels. But this definition is inadequate, as it does not capture the physical reality of a stream in the landscape. Both the longitudinal and lateral extent of the stream are minimized by this definition. Small headwaters may contain channelized, flowing water much of the year, and should be considered an integral part of the stream. Lateral swales and gulleys serve as conduits of water, sediments, and organic matter to the stream during periods of high runoff, and might also be considered a part of the stream. Water in the main flowing portion of the stream, or thalweg, can enter deep into the stream channel, traveling underground through the stream bed (the hyporheic zone) before emerging downstream to join the main channel. Similarly, water from the main channel can pass under the stream banks, and flow through the riparian zone before re-entering the stream channel many meters downstream (**Figure 1**). Thus, the definition of a stream must include the perennially wetted areas, the ephemeral channels that provide significant channelized flow into to the main channel, and all of the stream bed and riparian zone through which channelized flow passes at any point in its travel downstream.

Formation and Importance of Drainage Networks

The assemblage of individual streams into a drainage network is one of the defining features of stream ecosystems. Frequently termed hydrologic connectivity, this linkage between upstream and downstream reaches defines many of the physical, biogeochemical, and biological attributes of a stream. Although other aquatic ecosystems are often connected into a drainage network, the importance of this hydrologic connectivity is paramount for stream ecosystems.

Drainage networks take various forms, depending on the climate, geology, and topography in a region. In more arid regions, much of the drainage network consists of ephemeral streams. In moderately wet and humid regions, the drainage network consists primarily of perennial streams, although the importance of ephemeral channels can be large there as well. Stream networks connected in a random, branching pattern akin to the branching pattern of trees are termed dendritic drainage networks. Geology can influence stream channel geometry by influencing the path that streams take through the landscape. A more rectangular trellis drainage system can result when streams are confined to beds formed in geologic strata that are less resistant to weathering; a radial drainage pattern is found when streams originate from a prominent mountain peak and flow in all directions.

The position of a stream in the drainage network is a fundamental feature of a given stream reach. The

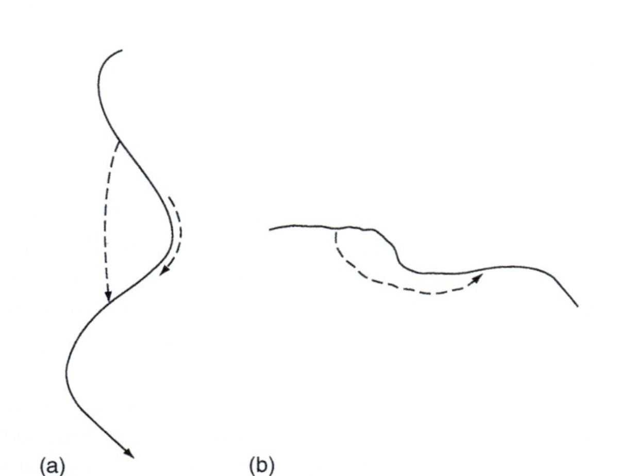

Figure 1 Flow path of water in streams. (a) Plan view of surface waters (solid line) and subsurface waters (dashed line) moving through the stream channel and adjacent riparian flood plain. (b) cross-sectional view of the flow path of water through the stream bed (hyporheic zone) in a gravel- and cobble-filled stream.

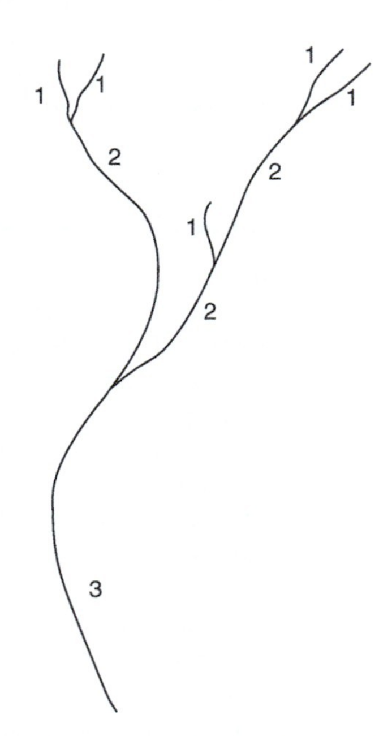

Figure 2 A dendritic stream network showing the Strahler stream order of selected reaches.

Strahler system of stream ordering is used most commonly to describe stream position. Headwater streams are termed first-order reaches, the combination of two first-order reaches produces a second-order reach, two second-order reaches combine to produce a third-order reach, and so on (**Figure 2**). The drainage area of first-order streams is usually a few to a few dozen hectares, and varies as a function of the amount of precipitation in the basin. Second- and third-order streams can be many hundreds of hectares in drainage area. Although the system has been criticized because stream orders tend to vary in width or discharge in different biomes, it is a system of enduring simplicity and utility. The biggest impediment to development of a consistent system of stream ordering is availability of a standard map scale with requisite detail. In the United States, for example, maps of the U.S. Geological Survey 1:24 000 series serve as the benchmark topographic maps, yet they consistently underestimate perennial stream reaches in humid regions. This is apparently due to difficulties in establishing the presence of a perennial stream channel from the aerial photographs that were originally used to establish mapped features. Lack of reliable mapping of stream networks poses an ongoing scientific and management challenge; it is difficult to study or manage streams at a regional scale when their distribution is unknown.

Geomorphology

The primary geomorphic units in streams are pools, riffle and runs. Pools are relatively deep, slow-moving areas; riffles are shallow reaches with fast-moving, turbulent flow, typically occurring over cobble or pebble substrate; runs are fast-moving reaches with relatively smooth or laminar flow, typically occurring over bedrock or other smooth substrates. In steep terrain, these units are often connected in stair-step fashion; in flatter terrain, with open valley floors that are unconstrained by bedrock outcrops or other features that prevent the formation of regular meanders, pools and riffles are typically arranged along the stream channel in a fairly regular sequence.

The bedrock geology of a region can strongly influence the geomorphology of its streams. A striking example of this is provided in the Luquillo Mountains of Puerto Rico, where two primary bedrock types are found. Although the bedrock has similar chemical composition, the rock types are different, resulting in different patterns of weathering and formation of stream bed materials. The most common bedrock type, volcaniclastic andesite and basalt, weathers to produce large boulders and little sand, and stream channels are steep and characterized by numerous pools, riffles, and small waterfalls (**Figure 3(a)**). The second bedrock type, an igneous intrusion of quartz diorite, weathers extremely rapidly and produces large amounts of quartz sand with few boulders (**Figure 3(b)**). It produces a stream channel that is relatively flat, choked with sand, and has many runs and pools but few riffles or small waterfalls (**Figure 3(b)**).

Examination of valley floors over geologic time scales shows that stream channels as well as large rivers are typically in dynamic equilibrium with their valley. Sinuous meanders move back and forth

Figure 3 Effects of bedrock on stream geomorphology. (a) Boulder-strewn stream channel resulting from weathering of basaltic and andesitic bedrock along the Río Sonadora. (b) Sandy-bed, low-gradient stream channel resulting from the rapid weathering of quartz diorite bedrock along the Río Icacos a few kilometers away. Both streams are in the Luquillo Mountains of Puerto Rico and have similar rainfall and temperature in their drainage basins.

across the valley floor, cutting away at banks on the leading edge and depositing material in the trailing edges (see **Hydrology: Streams**). The movement of meanders in streams and small rivers is usually a gradual process accelerated during periods of high flow. An 'avulsion' occurs when the stream forms a completely new channel through the adjacent terrestrial system, creating in a matter of hours or days what typically might take decades to centuries to occur through the process of gradual changes in meandering shape.

Following an avulsion (literally tearing; the creation of an entirely new stream channel), the stream can take years to develop stable geomorphic features and an organized sequence of riffles, runs, and pools. A recent example of the instability in geomorphic features that can result with creation of a new channel is shown by the Suncook River, a small river in southeastern New Hampshire. During a large spring flood in 2006, the river abruptly changed course and carved a new and dynamic channel from former gravel pits and forest land (**Figure 4**). In the first several years following the avulsion, the channel remained highly unstable, and will probably take decades to develop a new channel with an organized series of pools, riffles, and runs.

Figure 4 Newly created stream channel formed by an avulsion on the Suncook River, southeastern New Hampshire in May 2006.

Streams as Habitat

Physical Challenges to Biota

Streams pose a variety of physical challenges to biota, particularly in regions with wide and unpredictable variations in stream flow. Turbulence, shear stress and sediment loads can each affect the ability of aquatic organisms to survive in the stream environment. Species have evolved various adaptations to deal with these habitat challenges, including features such as dorsoventral flattening to resist high velocity (e.g., some mayflies), dorsal fins modified to form suckers that facilitate movement upstream against high velocity (e.g., gobies), and development of nets to capture food in a velocity regime too fast to allow active foraging on specific particles (e.g., caddisflies). Retreat into the stream bed is a common response among benthic invertebrates (those living on the stream bottom) in response to high flows. High concentrations of suspended sediment pose a difficult challenge to many fish owing to the interference of sediment with gill function and gravel-bed spawning. Loss of sensitive species such as trout can occur when sediment loads increase owing to human activities. In very sandy-bed streams, physical features are often transient, reworked by the stream flow after storms, and the stream bed proves to be a difficult place for algae or many invertebrates to inhabit owing to this habitat instability.

Although the heterogeneous physical template in streams can pose challenges to the survival of some species, it also provides others with opportunity. High-flow regimes can help grazing insects avoid predation by macroinvertebrate or fish predators, and vertical drops such as small waterfalls can make grazing on attached algae difficult for most insects or fish. Similarly, this high habitat heterogeneity can provide refugia or nurseries. Small tributaries can provide refuge for fish from acidification, warm or freezing temperatures, and high sediment loads when the main stem is unsuitable habitat. They also provide nurseries for many anadromous fish (fish that spawn in fresh water and spend their adult lives in the ocean) such as Atlantic Salmon, and amphibians such as salamanders.

Connectivity as an Organizing Feature

An important paradigm in stream ecology is the inherent interconnectedness of stream and river ecosystems along the drainage network. Two-way connections of biota, gene pools, nutrients and energy link upstream and downstream ecosystems throughout the drainage network. The physical template provided by the drainage network is an important organizing feature for the biota in streams. A seminal statement describing the importance of hydrologic connectivity is the River Continuum Concept. This concept describes how downstream and upstream communities of invertebrates and fish are inextricably linked, and emphasizes the importance of position in the drainage network as an organizing principle driving community structure. More recently, the topic of hydrologic connectivity has emphasized the many ways in which bidirectional connectivity links upstream and downstream ecosystems, and how lateral connectivity links streams and their associated riparian zones.

One of the best-studied examples of the importance of hydrologic connectivity in maintaining upstream ecosystems is that of bears, salmon, and marine-derived nutrients in North America (see also **Riparian Zones**). Salmon return to spawn in their natal streams, bringing with them nutrients obtained during the marine phase of their life cycle. These nutrients enhance algal primary productivity in the stream, supporting the insects that the juvenile salmon consume during the freshwater phase of their life cycle. Bears contribute to the system as well. When they eat adult salmon returning to spawn, bears defecate in the riparian zone and leave half-eaten carcasses, providing nutrients that enhance the productivity of riparian trees. Leaf litter from these trees provides an important food source to the invertebrates that the juvenile salmon eat. Loss of hydrologic connectivity through dams or water withdrawals begins a negative feedback loop by reducing salmon returns, and decreasing the success of salmon spawning.

Stream Biogeochemistry

Integrators of Watershed Processes

Streams are widely studied in ecosystem science because they integrate the effects of numerous processes occurring within a small watershed into a single convenient sampling point. By incorporating the effects of spatial and temporal variability in processes such as nutrient mineralization by microbes, nutrient uptake by plants and microbes, atmospheric inputs of nutrients, weathering and specific microbial pathways such as denitrification. The stream provides an integrated sample that reflects all the biogeochemical processes occurring in a watershed as well as many human impacts. They are thus powerful tools for estimating the effects of watershed-scale disturbances such as agricultural tillage, human habitation, or forest cutting (see **Deforestation and Nutrient Loading to Fresh Waters**) on elemental losses from watersheds. The integration of watershed function that streams provide also means, however, that the precise controls on elemental losses from watersheds are difficult to infer. Biogeochemical processes that affect stream water chemistry may occur in a distributed fashion throughout the watershed, or they may occur only in small areas of the watershed that have a disproportionate influence on stream chemistry.

Two examples of areas in a watershed that can have a disproportionate effect on stream chemistry are the riparian zone, and the weathering rind of the watershed's bedrock. Processes occurring at these interfaces may mask the effects of biogeochemical reactions occurring in well-drained upper soil horizons, where most biological activity typically occurs, and thus limit the utility of stream chemistry in describing nutrient dynamics in the watershed. The riparian zone is the interface between groundwater and stream water, and it often contains anoxic zones owing to saturation of soil with groundwater and the tendency of organic matter to accumulate at downslope, riparian locations. As groundwater moves through the riparian zone, distinct biogeochemical transformations often occur, such as denitrification (the dissimilatory reduction of nitrate to N_2O and

N_2), and methanogenesis (the production of CH_4 during anaerobic decomposition). These gaseous losses from the riparian zone can reduce the overall amount of carbon and nitrogen in stream water, and thus can alter the perception of nutrient cycling in the watershed (see also **Riparian Zones**). The flow path that water takes through the riparian zone can affect the balance between carbon and nitrogen exported from the site in stream water, and that which is lost as gaseous end products (**Figure 5**). Because flow paths vary with the geomorphology of the riparian zone (which in turn varies with bedrock geology), the extent to which stream export reflects watershed-scale processes may vary with bedrock.

A second example in which processes occurring in a small part of the watershed can have a disproportionate effect on stream chemistry is seen with weathering. Weathering is the physical and chemical breakdown of parent material (bedrock). The weathering rind is the site at which fresh rock surfaces begin the weathering process. It typically occurs at depth in most watersheds, at the interface between watershed bedrock and overlying soils. When the dissolved products of weathering (e.g., SiO_2, Ca^{2+}, PO_4^{3-}) are released to soil water, water moving along the weathering rind can be routed directly to the stream, largely bypassing plant or microbial uptake except in the riparian zone, where depth to the water table is shallow. This can result in a situation where calcium in forest soils is in relatively short supply, for example, and yet calcium concentrations in stream water can be reasonably high owing to bedrock weathering deep below the soil surface.

Pipes, Vents or Ecosystems?

From a geomorphological perspective, 'streams are the gutters down which flow the ruins of continents' (Leopold, LB., Wolman, MG., and Miller, JP. (1964). *Fluvial Processes in Geomorphology*, p. 97. San Francisco, W.H. Freeman and Company). While this is undoubtedly true over geologic time scales for chemical elements with no or limited gaseous phase and little biological demand (e.g., Mg^{2+}, Na^+, Cl^-), it is demonstrably untrue for elements of biological interest such as carbon, nitrogen, and phosphorus. Nevertheless, the role of streams in the landscape has often been characterized as that of a neutral pipe, a conveyance that delivers materials from the landscape smoothly and efficiently to important downstream systems (**Figure 6**). Work on streams as ecosystems that process, transform, and vent carbon and nitrogen to the atmosphere shows that this neutral pipe model should be replaced by an active ecosystem model.

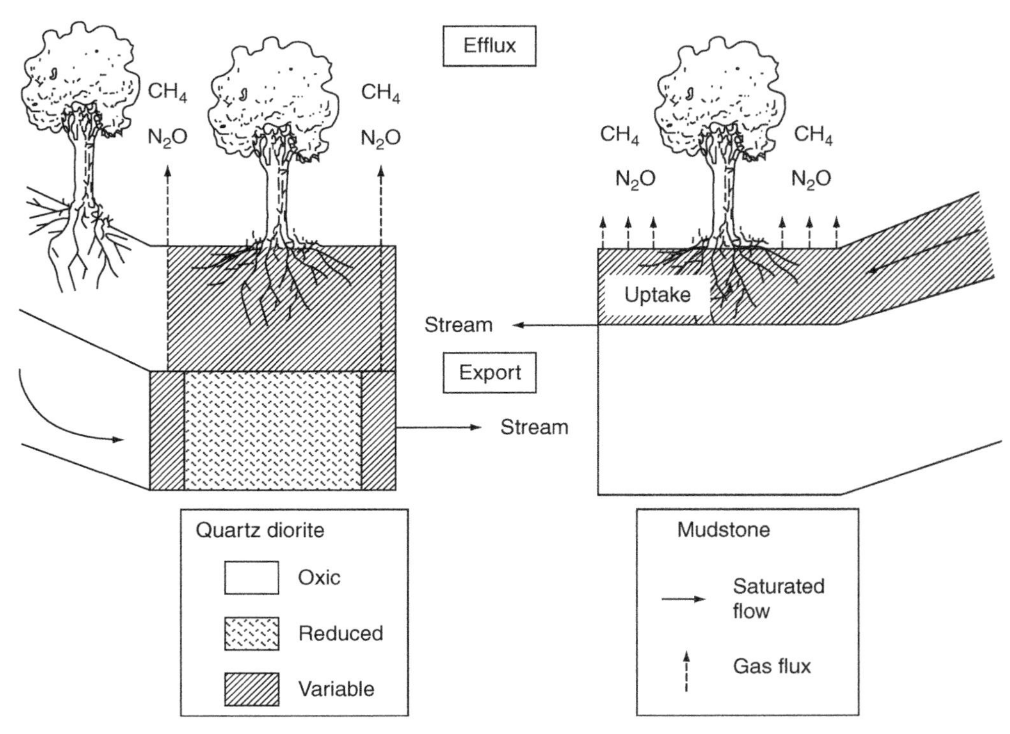

Figure 5 Conceptual model highlighting spatial variability in redox conditions and trace gas flux in riparian zones (basin/stream interface) with different flow paths. The different flow paths result from geomorphic differences associated with two bedrock types, quartz diorite (left) and mudstone (right).

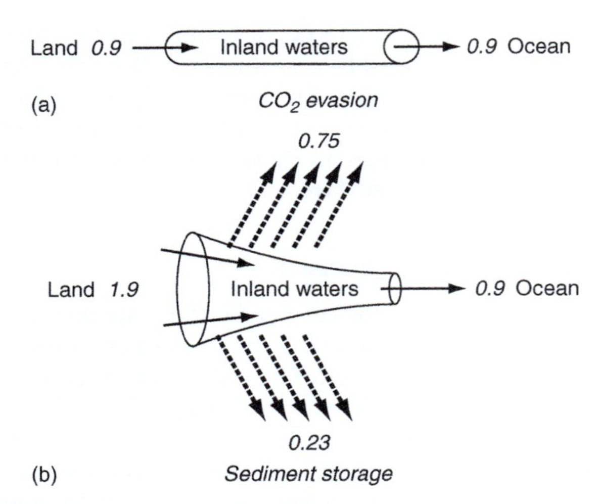

Figure 6 Two views of the role of aquatic ecosystems in terrestrial landscapes. (a) The 'neutral pipe' model, in which aquatic systems are considered to be important solely as pipes that deliver materials unchanged from the land to the sea. (b) The 'reactive ecosystem' model, in which streams and other aquatic ecosystems are important in storing, transforming and releasing to the atmosphere the carbon and nitrogen delivered to them from the landscape. Reprinted with permission from Cole JJ, Prairie YT, Caraco NF *et al.* (2007) Plumbing the global carbon cycle: Integrating inland waters into the terrestrial carbon budget. *Ecosystems* 10: 171–184.

It has been estimated, for example, that the actual delivery of particulate and dissolved organic carbon to water bodies globally is twice that which is exported to the oceans at the mouth of the 'pipe' (**Figure 6**). Some of the carbon lost from the land is stored in reservoirs along river networks, but much of it is vented to the atmosphere, since most aquatic systems are supersaturated with CO_2. The CO_2 vented to the atmosphere by streams is ultimately of terrestrial origin. It either enters the stream directly as CO_2 in groundwater (groundwater is frequently supersaturated with CO_2 owing to microbial decomposition and root respiration), or indirectly as dissolved organic matter in groundwater that is subsequently respired to CO_2 in the stream.

Because of the utility of streams for waste disposal, much is known about the oxygen dynamics of streams following point source inputs. 'Self-purification capacity' is the term most widely used to describe the ability of streams to metabolize organic matter, releasing CO_2 and consuming oxygen in the process. Self-purification has been studied for over a century, first for the oxygen balance and later for the nitrogen balance, and has played an important role in how streams are managed in human-dominated landscapes. It is the basis for modern-day calculations of the allowable load of organic matter and other oxygen-consuming materials (measured as the biochemical oxygen demand, or BOD) that can be discharged to the stream without reducing oxygen concentrations to levels that would harm aquatic biota. The fundamental basis for this self-purification concept is that oxygen depletion that results from microbial respiration is balanced by resupply of oxygen from the atmosphere through diffusion and turbulent mixing. Because the rate of oxygen resupply is proportionally larger in small, high energy streams, they can be important reactors (and not neutral pipes) that process organic matter in treated effluents such as those from sewage treatment plants.

Delivery of nitrogen to downstream water bodies by streams has gained new relevance and urgency with the realization that the near-shore ocean is frequently subjected to 'dead zones'. These are regions with low oxygen levels induced in part by excessive delivery of nutrients to the coast by streams and rivers. Detailed studies of the stream nitrogen cycle across North America have revealed that in-stream transformation of nitrogen can alter the form and total quantity of nitrogen delivered to downstream reaches. Retention of NH_4 is particularly effective in small streams, with the distance that an average molecule travels before being taken up into the stream bed on the order of 15–60 m except when ambient NH_4 concentrations are high. This does not represent 'self-purification' in the way that microbial consumption of organic matter returns that organic matter to the atmosphere in the form of CO_2, however. Much of the ammonium that is taken up (as much as 50%) can be released back to the water column as nitrate. Nitrate is typically removed from the water column much more slowly than ammonium, and its rate of removal declines as ambient NO_3 concentrations increase (**Figure** 7).

Quantifying the rate of denitrification, the only process that returns nitrate to the atmosphere, is critical to understanding the role of small streams in the landscape. By returning NO_3 to the atmosphere, denitrification results in permanent removal of nitrogen and thus protects the coastal zone from eutrophication. Across North America, about half of the uptake of nitrate from the water column is due to denitrification, and rates of denitrification show the same decline in efficiency at higher nitrate concentrations noted earlier for total nitrate uptake (**Figure** 7(**b**)). Modeling of the fate of nitrate in streams suggests that headwater streams (first to third order) are responsible for approximately half of the total nitrogen uptake occurring throughout the drainage network. Despite the short residence time of water in the stream channel, extensive contact between stream water and microbes in the stream bed results in uptake of nitrate in the stream channel and reduced nitrate transport downstream.

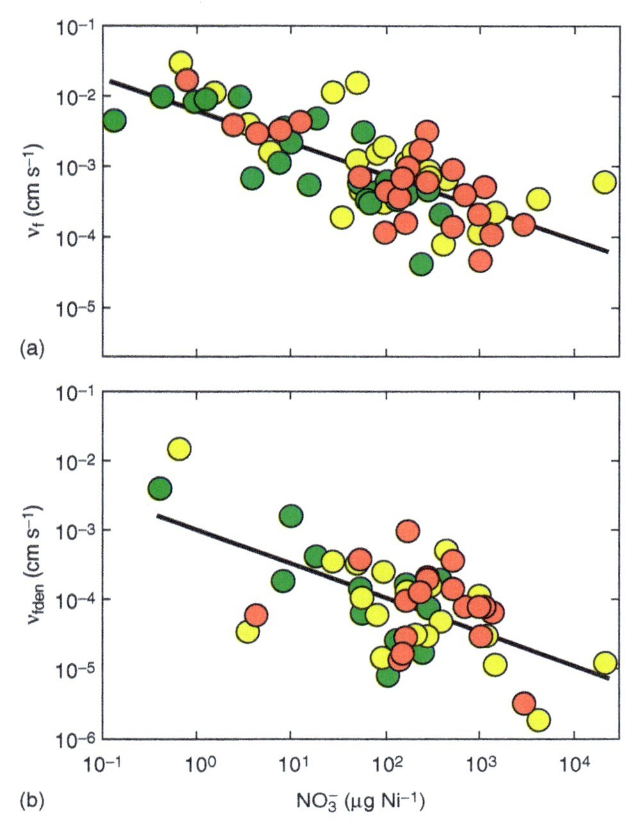

Figure 7 Relationships between NO_3 uptake velocity and NO_3 concentration in streams throughout North America. Nitrate uptake velocity is a measure of the speed at which a nitrate molecule in stream water leaves the water column due to either all uptake processes (v_f) (a) or due to denitrification only (v_{fden}) (b). The color of the symbols denotes data from study streams in watersheds that are largely in native vegetation (green), under agricultural management (yellow), or converted to urban and suburban land uses (red). Note that the rate of nitrate uptake or denitrification varies little with land use alone, but is strongly related to total nitrate concentrations. Reprinted with permission from Mulholland PJ, Helton AM, Poole GC *et al.* (2008) Stream denitrification across biomes and its response to anthropogenic nitrate loading. *Nature* 452: 202–206.

Streams as Indicators of Ecosystem Health

Streams provide an unparalleled opportunity to diagnose conditions in a watershed. Because they integrate conditions occurring throughout the watershed, they can serve as readily sampled sentinels that reflect ecosystem health. Physical, chemical, and biological metrics can be used to assess conditions in a stream. Typically, these metrics focus on geomorphology, nutrient chemistry, and the diversity and composition of aquatic biota. Each of these metrics reflects the health of the stream, and is intimately tied to the health of the landscape.

Figure 8 Attempted bank stabilization in Quebrada Maizales, a Puerto Rican stream with suburban and agricultural land use. Note concrete retaining wall, with bank failure on the upstream (left) side of the retaining wall.

Geomorphology

Urbanization and suburbanization in a watershed result in hydrologic changes that alter the geomorphology of stream channels. One of the most obvious changes is a deepening of the stream bed and undercutting and slope failure of the stream banks. This often leads to attempts by management agencies or individual landowners to stabilize the stream channel. These attempts are frequently unsuccessful (**Figure 8**). Additional changes to the stream often accompany degradation of ecosystem health in the watershed. These include a decrease in sediment grain size due to the combined influence of watershed erosion and the sort of slope failure shown in **Figure 8**. Stream channel widening often occurs, as does the formation of braided channels. Finally, in an attempt to reduce the geomorphic complexity and variability of natural stream channels, they are often hardened with concrete, straightened, and put into drainage pipes. Although the nature of the stream's connection to the landscape has been markedly altered, the chemistry and biology of such urban streams reflects their watersheds all too well. Sediments, nutrients, combustion products, and fecal coliforms are delivered directly from the watershed to the stream, without the beneficial influence of passage through groundwater or riparian zones in reducing these undesirable inputs.

Nutrients

The nutrient content of surface waters is a sensitive indicator of biogeochemical conditions in the watershed. Nutrient loss from a watershed is a function of a variety of physical and biological processes occurring in the watershed (see **Deforestation and Nutrient Loading to Fresh Waters**), which, when disrupted, can result in greater nutrient loads in watershed runoff. Nutrient loading from forestry practices, crop production, animal husbandry, disposal of human wastes, and many other sources can contribute to the nutrient concentrations observed in a stream. Nitrate concentration in drainage waters is one of the most sensitive indicators of ecosystem health. Nitrate concentrations in streams and rivers increase as a function of a wide variety of metrics that measure the human impact on the landscape. Agricultural intensity, human population density, total nitrogen inputs from atmospheric deposition, and other attributes of a watershed have all been shown to be useful predictors of nitrogen concentrations in streams. In contrast to nitrogen, concentrations of phosphate in streams are often less tightly coupled to watershed conditions. This is probably due to the fact that unlike the situation for nitrate, many soils are highly retentive of phosphate as a result of abiotic adsorption. At moderate levels of watershed disturbance, phosphate is retained by the watershed's soils and stream concentrations remain unchanged.

In assessing the sources of nutrients to streams, it is often useful to discriminate between 'point sources' and 'nonpoint sources'. Point sources are those that enter the stream at a specific site, such as a discharge pipe from a sewer system. Nonpoint sources are diffuse, and enter the stream at many points. Erosion-driven discharge of sediments into a stream from an agricultural field is a typical example of a nonpoint source discharge. Point source discharges are more easily regulated and cleaned up than are nonpoint sources. Because of their diffuse nature, reduction of nonpoint sources requires changes in land management practices such as tillage, animal husbandry, and construction of impervious sources throughout a watershed. Point sources, in contrast, are usually reduced through construction of a treatment system that removes contaminants from the effluent stream.

Biota

The biota of a stream provide an important indicator of stream health, and by inference can serve as a proxy for overall watershed health. Two general approaches have been used to assess stream health using stream biota. In the first approach, an assessment is made of the abundance of 'clean water taxa' relative to the entire assemblage of benthic invertebrates. These clean water taxa include the mayflies, stoneflies, and caddisflies (organisms in the orders Ephemeroptera, Plecoptera, and Trichoptera, or 'EPT'). The condition of a stream can thus be judged with a relatively straightforward assessment of the biota by a trained professional or an interested nonprofessional. This approach is widely used in the monitoring and assessment activities carried out by volunteer environmental organizations, as it does not require highly specialized knowledge.

In the second approach, an assessment is made of the biota that are expected to occur in a stream, based on a set of reference criteria from relatively unimpacted streams in the region under study. This approach, often called the Index of Biological Integrity or IBI, is often used with fish and benthic macroinvertebrates in streams. Its advantages are that it provides a much more detailed and nuanced assessment of stream condition than the somewhat simplistic 'EPT' criteria, which do not account for inherent features of the stream that might limit particular taxa. These include stream temperature, channel slope, grain size of the stream bed, and stream bed stability, which can all affect the fauna found in streams of a given region. Furthermore, the biota can be affected by biogeographic constraints that may have prevented colonization of a drainage basin by some groups. The disadvantages of the IBI approach are the difficulty of establishing appropriate baseline data for individual biomes, and the costs of using trained professionals to identify the organisms collected.

See also: Deforestation and Nutrient Loading to Fresh Waters; Geomorphology of Streams and Rivers; Hydrology: Streams; Riparian Zones.

Further Reading

Alexander RB, Smith RA, and Schwarz GE (2000) Effect of stream channel size on the delivery of nitrogen to the Gulf of Mexico. *Nature* 403: 758–761.

Allan JD and Castillo MM (2007) *Stream Ecology: Structure and Function of Running Waters*. 2nd edn. New York: Springer.

Bhatt MP and McDowell WH (2007) Controls on major solutes within the drainage network of a rapidly weathering tropical watershed. *Water Resources Research* 43: W11402, doi:10.1029/2007WR005915.

Battin TJ, Kaplan LA, Findlay S, Hopkinson CS, Marti E, Packman AI, Newbold JD, and Sabater F (2008) Biophysical controls on organic carbon fluxes in fluvial networks. *Nature Geosciences* 1: 95–100.

Caraco NF, Cole JJ, Likens GE, Lovett GM, and Weathers KC (2003) Variation in NO_3 export from flowing waters of vastly different sizes: Does one model fit all? *Ecosystems* 6: 344–352.

Freeman MC, Pringle CM, and Jackson CR (2007) Hydrologic connectivity and the contribution of stream headwaters to ecological integrity at regional scales. *Journal of the American Water Resources Association* 43: 5–14.

Gomi T, Sidle RC, and Richardson JS (2002) Understanding processes and downstream linkages of headwater systems. *BioScience* 52: 905–916.

Hynes HBN (1975) The stream and its valley. *Verhandlungen der Internationale Vereinigung fur Theoretishce und Angewandte Limnologie* 19: 1–15.

Karr JR (1991) Biological integrity – A long-neglected aspect of water resource management. *Ecological Applications* 1: 66–84.

McDowell WH, Bowden WB, and Asbury CE (1992) Riparian nitrogen dynamics in two geomorphologically distinct tropical rain forest watersheds – Subsurface solute patterns. *Biogeochemistry* 18: 53–75.

Meyer JL, Strayer DL, Wallace JB, *et al.* (2007) The contribution of headwater streams to biodiversity in river networks. *Journal of the American Water Resources Association* 43: 86–103.

Mulholland PJ, Helton AM, Poole GC, *et al.* (2008) Stream denitrification across biomes and its response to anthropogenic nitrate loading. *Nature* 452: 202–206.

Paul MJ and Meyer JL (2001) Streams in the urban landscape. *Annual Review of Ecology and Systematics* 32: 333–365.

Peterson BJ, Wollheim W, Mulholland PJ, *et al.* (2001) Stream processes alter the amount and form of nitrogen exported from small watersheds. *Science* 292: 86–90.

Petts GE and Amoros C (eds.) (1996) *Fluvial Hydrosystems.* New York: Chapman and Hall.

Vannote RL, Minshall GW, Cummins KW, Sedell JR, and Cushing CE (1980) The river continuum concept. *Canadian Journal of Fisheries and Aquatic Science* 37: 130–137.

Winter TC (2007) The role of ground water in generating streamflow in headwater areas and in maintaining base flow. *Journal of the American Water Resources Association* 43: 15–25.

Lowe W and Likens GE (2005) Moving headwater streams to the head of the class. *BioScience* 55(3): 196–197.

Riparian Zones

H Décamps, Centre National de la Recherche Scientifique and Université Paul Sabatier, Toulouse, France
R J Naiman, University of Washington, Seattle, WA, USA
M E McClain, Florida International University, Miami, Florida, USA

Introduction

Riparian zones are transitional semiterrestrial/semiaquatic areas regularly influenced by fresh water, usually extending from the edges of water bodies to the edges of upland communities. Because of their spatial position, they integrate interactions between the aquatic and terrestrial components of the landscape. They are dynamic environments characterized by strong energy regimes, substantial habitat heterogeneity, a diversity of ecological processes, and multidimensional gradients. They are often locations of concentrated biodiversity at regional to continental scales.

Riparian zones associated with running waters are three-dimensional biophysical structures set in complex ecological and cultural matrices from headwaters to the sea. Expansion and contraction of riparian zones along rivers occur in relation to precipitation, river flow, and geomorphology. One segment of the channel may be fed largely by upwelling groundwater, whereas at other locations surface runoff may penetrate into bed sediments (alluvium) accumulated over millennia by cut-and-fill alluviation. During flooding, surface flow may recharge groundwater aquifers and spill out over the floodplains, eroding or depositing sediment in accordance with the energy dynamics of water interacting with geomorphic features. During dry periods, flow in the channel may be maintained by alluvial and karstic aquifers. Thus, rivers may be viewed as a collection of dynamic multidimensional pathways along which aquatic–terrestrial linkages vary spatially and temporally. Anthropogenic influences contribute greatly to this variation, as river valleys have been the foci for human settlements and commerce for millennia.

Fires, drought, flooding, mass wasting, wind throw, grazing, and other natural disturbances, coupled with human interventions such as logging, urbanization, farming, and damming, alter vegetative patterns and soil–plant nutrient exchange at a variety of scales. This has direct consequences for ecological processes in riparian zones – such as productivity, biodiversity, sediment transport, and live and dead wood recruitment.

In an increasingly human-dominated world, riparian zones must be viewed in a landscape context – that is, as natural–cultural systems. While surface and subsurface patterns and processes act as key drivers for sustaining riparian goods and services, it is human perceptions and cultural representations of landscapes that shape the dynamic complexity of contemporary riparian zones (**Table 1**).

Here, we focus on the unique ecological functions of riparian zones and how these functions are linked to dynamic biophysical processes and interactions across multiple spatial and temporal scales along natural rivers.

Riparian Zones as Focal Points for Biodiversity

River corridors normally possess highly diverse floral and faunal communities and attendant biological processes. These attributes emerge from the unique spatial organization – a mosaic of habitats continually changing in response to variable water flows above and below ground – and biotic responses to the locally variable topography and climate. The net effect is substantial smallscale heterogeneity. Further, the inherent heterogeneity introduced by local topography affects the frequency and duration of inundation, and the local microclimate influences the ability of individual species to flourish, thereby adding even more physical as well as genetic heterogeneity. Collectively, heterogeneity at local scales offers a great variety of conditions for life, making riparian corridors focal points for diversity. Riparian diversity is further magnified at the catchment scale since riparian corridors extend from the highest to the lowest elevations above sea level, and the higher elevation riparian zones are numerous, which tends to augment regional diversity.

Studies throughout the world indicate that riparian areas generally have high levels of both of plant and animal diversity (**Figure 1**). In the Pacific Northwest of the United States, 74% of all plant species within one catchment were found in the riparian corridor, which corresponds to only ~3–8% of the catchment depending on topography. According to other reports, all periodically flooded forests in the Amazon basin may have about 20% of the 4000–5000 estimated Amazonian tree species, and about 1400 vascular plant species occur along the Adour River riparian corridor in France, representing 30% of the French flora. However, this information may be slightly

Table 1 Summary of the major types of anthropogenic environmental change and their principal effects on riparian zones (Naiman *et al.* (2005)

Environmental change	*Principal effects on riparian zones*
Flow regulation	
Flow regime	Alters community composition and successional processes; loss of life history cues
Dams	Lotic to lentic; inundation above dam; altered flow, nutrient, sediment, and temperature regimes below dam
Withdrawals	Lowers water table; alters flow regime; decreases alluvial aquifer recharge; system simplification
Channelization and dredging	Lowers water table; desiccates riparian forest causing terrestrialization and change in community composition; possible decline in biodiversity
Levees	Isolates river from floodplain, thereby reducing hydraulic connectivity laterally and vertically. Constrains channel migration; alters riparian successional trajectories
Pollution	
Nutrients	Increases productivity; shifts community composition toward tolerant species; organic loading leads to redox changes
Toxic materials and acid rain	Decreases productivity; declining biodiversity; system simplification; shifts community composition toward few tolerant species
Climate	
Precipitation	Modifies entire flow regime, groundwater–surface water exchanges, and channel morphology, and stability; loss of life history cues
Temperature	Spatial patterns and phenology of riparian species are changed
Land use	
Vegetative cover	Modifies albedo and feedbacks to climate; changes local microclimate and successional trajectories
Invasive species	Introgression and hybridization; increased competition for space and resources; may reduce biodiversity
Resource management	Usually alters successional trajectories and community composition

misleading because comparative studies are seldom undertaken in surrounding uplands. It has been suggested that riparian zones increase regional diversity >50% by harboring different species assemblages, rather than more species, when compared with surrounding uplands.

Many riparian corridors vary in species richness along the river's course, but some do not. There appear to be three types of riverine reaches that fundamentally differ in terms of biodiversity: naturally constrained river reaches increase in biodiversity downstream, whereas braided reaches have relatively low diversity and meandering reaches have high biodiversity. In general, longitudinal studies show that plant diversity generally increases downstream, with the peak reached in the piedmont (transition zone between mountain and lowland domains) as the riparian zone widens, or at major tributary deltas.

These catchment-scale patterns suggest a maximum diversity at an intermediate level of natural disturbance, which induces considerable spatial heterogeneity. However, other physical attributes (e.g., channel gradient, lithology, level of confinement) and local climate also modify site-specific species richness and can result in no net trend in longitudinal species richness. Lateral patterns in plant diversity also generally peak at intermediate levels of natural disturbance and moisture availability. Diversity is often low immediately adjacent to the river, increasing with local elevation and geomorphic complexity, and then declining slightly upslope of the riparian–upland interface, although there are variations to this generalization. The factors controlling species expression are also keys to understanding riparian plant diversity: flood frequency, site productivity, and spatial complexity have been related to plant species richness (**Figure 2**). In contrast, human-induced disturbances like deforestation and farming tend to reduce both levels biodiversity and the processes promoting diversification.

Overall, riparian soils appear to have an impressive diversity. There may be as many as 10 000 species of ectomycorrhizal fungi, >3000 species of bacteria, >5000 species of nematodes, and tens of thousands of species of mesofauna (mites, collembolans) and macrofauna (ants, termites, earthworms). Aboveground faunal diversity is better known but it is far from being completely understood. Reasonably comprehensive studies exist for some groups, especially birds, whereas knowledge about other groups is nearly nonexistent (e.g., most invertebrates). Flooding certainly affects invertebrate community dynamics – distribution, abundance, and diversity – with species richness often increasing with elevation above the river even though flood duration will affect groups differentially. Among the invertebrates,

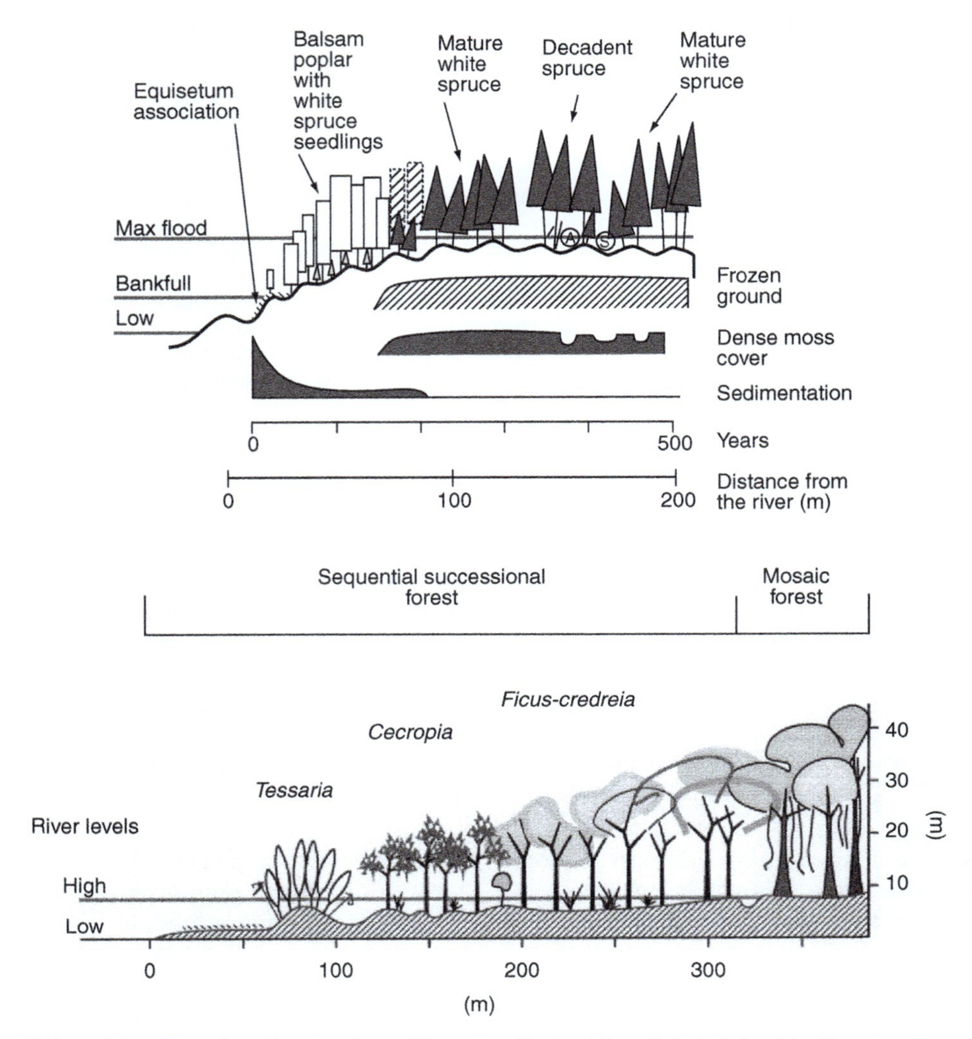

Figure 1 Floodplain profiles of two rivers in natural conditions: the Beaton River, British Columbia, Canada, above (after Nanson and Beach (1977) and the río Manu, Peru in the upper Amazon basin, below (after Salo *et al.* (1986)). Such relationships between landforms and forest succession characterize riparian forests along rivers under natural conditions all over the world (Décamps (1996).

spiders (Arachnida) and beetles (Coleoptera) are the better-investigated groups.

In many areas, a large proportion of vertebrates utilize the riparian zone, but it is not always apparent whether individual groups are more or less diverse there. For example, in western Oregon and Washington (USA), 87% of the 414 resident species of amphibians, reptiles, birds, and mammals use riparian zones or wetlands, but only 10% use specialized habitat within riparian zones. More is known about species depending on the streams for foraging or reproduction than about species using riparian zones only occasionally. Certainly, amphibians are more diverse in riparian zones because of their reproductive needs. For reptiles, birds, and mammals, it depends on the local environment and specific life history needs.

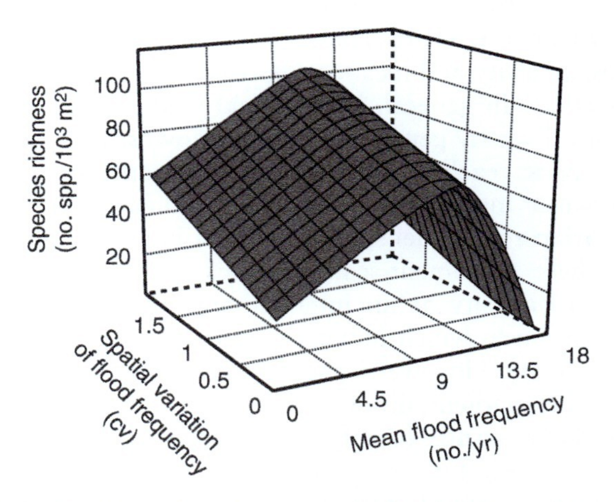

Figure 2 Three-dimensional representation showing relations among species richness, flood frequency, and topographic variation for riparian areas in southeast Alaska (Pollock *et al.* (1998)).

The best information available is for birds. In the drier western United States, bird diversity at the site scale (alpha) is generally high relative to the number of potential species. Between habitats, diversity (beta) varies across a catchment, with riparian assemblages differing most from upland assemblages at the highest and lowest elevations. This pattern is attributed to enhanced avian movements within the riparian corridors. The corridors for bird movements, in turn, facilitate faunal mixing on a broader scale, increasing regional diversity (gamma) within landscapes.

Mammal diversity in riparian zones appears to be great, but this may be deceptive. Mammals are highly mobile, and many, such as moose and elk in winter, make frequent use of riparian zones for feeding and cover. Certainly bats feeding on emerging stream insects are abundant at times, and other medium-sized predators (badgers, foxes, cats) come to feed and drink. Browsers and grazers, as well as rodents and shrews, are often abundant, but most can be found in the uplands too.

Are riparian zones more diverse than the surrounding uplands? The answer is that it depends on the environmental setting and the taxon being considered. Ecological theory would suggest that, in general, riparian areas are highly diverse because of their mosaic structure, presence of refugia, broad ranges of environmental settings, and species assemblages.

Riparian Zones as Buffers against Nutrient Pollution from Upland Runoff

The role of riparian zones as nutrient filters for water flowing from agricultural catchments to streams has led to major government-supported programs in North America and Europe to conserve and restore riparian buffers. The filtering capability of riparian zones is due to their position in the landscape, and to their geomorphic, hydrologic, and biotic processes. Since riparian zones lie at the interface of terrestrial and aquatic ecosystems, virtually all surface and shallow subsurface runoff in catchments must pass through them in order to reach the stream channel.

Riparian zones are generally planar, with lower gradients than surrounding uplands. This, plus the baffling effect of riparian vegetation, dissipates the kinetic energy of surface flows during storms and causes entrained sediments to be deposited. This is an especially effective mechanism for retaining particulate phosphorus and the chemical pollutants associated with sediment particles. The capacity of riparian zones to retain dissolved nutrients such as N, P, Si, Ca, and Mg is controlled by hydrologic characteristics (e.g., water table depth, water residence time, and degree of contact between soil and groundwater) and by biotic processes (e.g., plant uptake and denitrification). The relative influence of these factors depends on soil characteristics, nutrient input rates, and vegetation type.

Elucidating the volume and pathway of water moving through riparia is fundamental for understanding nutrient removal and retention. If local groundwater passes beneath the rooting zone or if extensive piping occurs, the roots of riparian vegetation cannot access the nutrients. Riparian vegetation along small streams normally has good access to the water table, and many studies have shown that in those situations riparian vegetation can act as buffers for nonpoint sources of N and P. However, hydrologic pathways are often more complex along larger rivers, especially where deposits of coarse alluvium are extensive.

Riparian zones are known to be especially effective at protecting surface waters from nitrate runoff. Early studies documented total nitrogen retention by riparian zones ranging from 67–89% of total up-slope inputs. Wetlands and soils are also sites of N retention, but a plethora of studies have shown focused removal of nitrate in groundwater moving laterally through riparian zones. In a large-scale study of N budgets in 16 catchments covering 250 000 km^2 of the northeastern United States, river exports of N accounted for only 25% of the total N inputs to the catchments, implying that most of the N was retained in some manner.

Candidate mechanisms preventing nitrogen flowing through riparia from entering streams include denitrification, assimilation, and retention by the vegetation with the uptake by biota followed by storage in organic detritus. Denitrification is invoked most often as the primary mechanism of nitrate retention and the most important given that N is removed permanently from the system and returned to the atmosphere as N_2 and N_2O. However, the extreme spatial and temporal variability of denitrification rates in riparian zones makes it difficult to determine accurate fluxes and to extrapolate these to wider areas. By difference, denitrification was found to account for 51% of N losses from catchments of the northeastern USA. In individual site studies denitrification rates of 1–295 kg N ha^{-1} $year^{-1}$ have been recorded; the fastest rates occur at the riparian-stream boundary where nitrate-enriched water enters organic surface soil. In addition to nitrate, saturated conditions, available carbon, topography and soil grain size (i.e., water logging potential) are pertinent environmental factors for denitrification.

Plant uptake results in a short-term accumulation of nutrients in nonwoody biomass and a long-term accumulation in woody biomass. Riparian forests are especially important sites for biotic accumulations of nutrients because transpiration may be quite high, increasing the mass flow of nutrient solutes toward root systems, and because morphological and physiological adaptations of many flood-tolerant species facilitate nutrient uptake under low-oxygen conditions. Thus, assimilation by the forest could be the primary mechanism of nitrate removal from groundwater during the growing season. Nitrogen removal efficiency by riparian buffers is mainly positive across a wide range of climate conditions, nitrate inputs, soil characteristics, and vegetation types.

The current consensus that most riparian zones effectively remove nitrate from subsurface water is based largely on studies where groundwater inputs are restricted to shallow subsurface flow paths by impermeable layers that force maximum interaction with riparian soils and vegetation. Limited research suggests that there is less effective nitrate retention in riparian areas connected to large upland aquifers where riparian hydrology is often dominated by surface transport or ground water transport below rooting zones. Additionally, it has been discovered that considerable nitrogen can be released from riparian zones to streams as dissolved organic nitrogen. The importance of this pathway, and the forms and subsequent fates of the organic nitrogen are unknown. Collectively, the evidence assembled so far raises doubts about all riparian zones being efficient filters for nutrients – some may be leaky while others release nutrients after they have been transformed to organic matter.

Riparian Zones as Buffers against High In-Stream Nutrient Levels

In much the same way that riparian zones intercept and retain nutrients from upland overland and subsurface runoff, they may also intercept and retain nutrients flowing in adjacent streams. The same mechanisms of N and P removal are at work, including biotic uptake, physical adsorption, and microbial denitrification, and the effectiveness of riparian buffering depends on the characteristics of riparian soils, vegetation, and connectivity between the river and riparian zone along surface–subsurface exchange pathways. Water in streams exchanges continually with the interstitial waters of bed and bank sediments in what may be viewed as a mosaic of surface–subsurface exchange patches. This zone of mixed surface and subsurface waters is known as the hyporheic zone and may extend from a few centimeters to a few kilometers from the channel margin. The volume of surface water moving along subsurface flow paths can be equal to or greater than that moving in the channel.

Hyporheic flow paths act as both sources and sinks of nutrients to streams and the degree of hyporheic influence on river nutrient levels depends on the extent of hyporheic zones, the fraction of river flow diverted through them, the residence time of individual flow paths, and the specific biological and biogeochemical processes operating along flow paths. In general, streams with greater hyporheic exchange tend to retain and process nutrients more efficiently. Because the amount of surface-hyporheic water exchange relative to channel volume decreases exponentially with increasing channel size, the efficiency of nutrient removal linked to hyporheic exchanges tends to decrease with increasing stream size. The most important sites of nutrient retention therefore lie in the low-order stream networks of any river basin. For example, first order N retention rates in headwater streams of the Mississippi River basin average 0.45 day^{-1}, while the rate of removal in the mainstem river is two orders of magnitude less, or 0.005 day^{-1}.

Plant cover also influences the efficiency of riparian zones in filtering nutrients and pesticides. A riparian zone vegetated with poplar is more effective for winter nitrate retention than one vegetated with grass. Some trees are better than others in filtering nitrate: *Populus x canadensis* effectively removes nitrate from saturated soils with a subsequent accumulation of nitrogen in root biomass. Roots of alder, willow, and poplar seem to favor colonization by proteolytic and ammonifying microorganisms and, particularly for alder roots, to inhibit nitrifying microorganisms. It follows that changing plant cover may affect water quality.

Riparian Zones as Buffers against Suspended Sediments

Riparian vegetation also facilitates the removal of suspended sediments – along with their nutrient, carbon and/or pollutant contents – from overland flow whether from the uplands or from the adjacent river. Sediments and sediment-bound pollutants carried in surface runoff are effectively deposited in mature riparian forests and in streamside grasses. Riparian areas remove 80–90% of the sediments leaving agricultural fields in North Carolina. Sediment deposition may be substantial in the long term, with coarser material deposited within a few meters of the field–forest boundary, and finer material deposited further into the riparian forest. Grassy areas are especially effective as they often transform channelized flows into expanded shallow flows, which are more likely to deposit sediment. However, the performance of grassy vegetation seems to be highly variable and to be of short duration when several floods occur within a limited period.

Removal of fine sediment from runoff by riparian zones occurs as a consequence of the interactive processes of deposition and erosion, infiltration, and dilution. This is important because fine sediments carry higher concentrations of labile nutrients and adsorbed pollutants. In forested catchments, with relatively low nutrient concentrations, fine sediments in riparian zones can be sources or sinks for nutrients, depending on how oxidation–reduction conditions affect absorption/desorption to fine particles. For example, the riparian zone of a small deciduous forest stream in eastern Tennessee (USA) was a net source of inorganic phosphorus when dissolved oxygen concentrations in riparian groundwater were low, but a sink when dissolved oxygen concentrations were high.

In contrast to nitrogen, phosphorus adheres strongly to particles and considerable movement is normally associated with sediment flux. Reductions of 50–85% in total P are observed, with the greatest removal occurring in the first few meters of the riparian zone. Results for soluble P in surface runoff are less consistent due to variability in water flow (i.e., sheet flow, channel flow, or percolation). In general, significant amounts of phosphorus may first accumulate in riparian zones but then be transported to aquatic ecosystems in a different form via shallow groundwater flow, possibly as a result of increased decomposition of organic matter.

Only limited research has been conducted on subsurface P transport in riparian zones. However, the same environmental conditions leading to redox gradients in soils that are responsible for nitrogen removal by denitrification also mediate desorption and release of phosphorus. Although riparian zones may act as effective physical traps (sinks) for incoming particulate phosphorus, they may enrich runoff waters in available soluble phosphorus. Finally, unlike nitrate, phosphorus removal by soil retention and biotic uptake results in accumulation within the system. Consequently, the long-term performance of riparian zones receiving high P inputs remains unclear.

Riparian Zones as Sources of Energy for Adjoining Aquatic Systems

By far the most thoroughly investigated and best understood connection between riparian and stream food webs is via the transfer of riparian plant litter to streams. Riparian organic matter inputs represent allochthonous sources of energy as opposed to the autochthonous organic matter contributed by aquatic primary producers. In low-order streams beneath closed-canopy riparian forests the influx of carbon from riparian plant sources, both surface and subsurface, may amount to 80–95% of total organic carbon influx to streams. Although the area-normalized flux of riparian litter to river systems decreases downstream, the total amount of riparian litter input to the river continues to rise and varies as a function of channel morphology and riparian forest structure and composition.

Once in the stream, riparian organic matter is decomposed by a variety of specially adapted microbial and invertebrate fauna. When litter (mainly leaves and needles) first enters streams there is a brief period (a few days) of rapid leaching in which 25% or more of the initial dry weight can be lost. Biotic decomposition is initiated by hyphomycete fungi that break up the litter's structural integrity by secreting enzymes to hydrolyze cellulose, pectin, chitin, and other difficult-to-digest compounds. Fungal community composition is closely tied to the riparian forest, and fungal species richness has been positively correlated with riparian tree richness. With time, fungi give way to bacteria as the dominant microorganism in the decay process. Decomposition rates are driven by substrate quality, stream nutrient concentrations, redox conditions and temperature. Fungal decomposition alone can fragment leaves into flakes of finer particulate organic matter within weeks. This fragmentation process is critical to energy dispersion in streams and rivers because finer fragments tend to be more mobile and to therefore fuel metabolism in downstream river sections. Microbially colonized litter has higher nutrient concentrations than non-colonized litter, and is therefore the preferred choice of macroinvertebrate consumers – shredders and collectors such as caddisflies and blackflies – that make up the next link in aquatic food webs.

Riparian arthropods are also important energy sources to stream consumers. Arthropods fall into streams from overhanging foliage by accident and the flux is proportional to arthropod abundance in the canopy. Arthropods may also wash into streams and rivers during overland flow events. The normalized flux (per square meter of channel area) is higher in smaller streams flowing beneath a closed riparian canopy but even in larger streams and rivers the flux may remain substantial at the channel margins. Once in the aquatic system, riparian arthropods are consumed by drift foraging fish and may constitute a major proportion of their diet. Riparian-derived arthropods are higher quality food than riparian litter and are directly available to top consumers such as fish. Experimental evidence shows that withholding this energy input from streams has consequences that reverberate through aquatic food webs and ultimately upset the basic composition of the stream community.

Energy flows in both directions across the terrestrial–aquatic interface, and riparian food webs are also subsidized by aquatic resources. The emerging adults of aquatic insects are an important energy source to a variety of riparian arthropods, and this energy subsidy is passed to higher trophic levels by the lizards, bats, shrews, and birds that consume riparian arthropods. Riparian arthropods inhabiting resource-scarce habitats such as exposed gravel bars and desert riparian environments appear to rely almost exclusively on aquatic prey. For example, emerging aquatic insects composed 80–100% of the diet of certain staphylinid and carabid beetles and about 50% of the diet of lycosid spiders inhabiting gravel bars of the Tagliamento River in Northeast Italy.

Reciprocal energy subsidies such as these are especially important over the course of the year in Temperate Zones due to strong seasonal variability in the emergence and abundance of different insects. For example, aquatic arthropod abundance peaks following 'leaf-out' of riparian forests in spring and defoliation in autumn whereas riparian arthropod abundance peaks during the summer when forest productivity is maximal.

Large Animal Influences on Riparian Zones

Large animals influence nutrient and energy flows by consuming and redistributing energy and nutrients within riparian zones as well as across adjacent system boundaries. More importantly, large animals may alter the hydrologic and geomorphic characteristics of riparian zones, causing fundamental changes in energy and nutrient cycles, and altering plant community composition and structure.

Animals that pond water, dig holes, trample plants, or move materials cause fundamental geomorphic changes. For example, hippopotamus (*Hippopotamus amphibious*) increase the ponding of water in African stream networks by creating and maintaining deep pools and forming trails between channels and adjacent terrestrial feeding areas. Beaver (*Castor canadensis* and *C. fiber*) profoundly influence the short- and long-term structure and function of riparian zones of drainage networks in the boreal forests of northern latitudes by cutting wood and building dams. In catchments where beaver are abundant, there may be 2–16 dams per km of stream length, and each dam may retain between 2000 and 6500 m^3 of sediment. Ponds are eventually abandoned as they fill with sediment or as local food resources are depleted and, once abandoned, dams fail and ponds drain to produce nutrient-rich wetland meadows.

Animals browsing riparian and aquatic vegetation strongly influence riparian community structure, soil development, and propagule dispersal. Animals that browse selectively keep preferred plant species from dominating the plant assemblage and thereby provide an advantage to species not browsed. For example, moose (*Alces alces*) prefer willow (*Salix*) and poplar (*Populus*), thus giving a competitive advantage to white spruce (*Picea glauca*), which is not browsed.

The actions of all large herbivores influence the belowground components of riparian systems as well, with consequent effects on interactions that determine long-term ecosystem function. For example, grazing activities that stimulate the growth of early successional species, and therefore retard succession, help to maintain higher belowground productivity.

Pacific Salmon Influences on Riparian Zones

A remarkable example of the consequences of animal-mediated nutrient and energy flows in riparian zones is the migration of salmon (*Oncorhynchus* spp.) from the North Pacific Ocean to spawning areas in fresh water. Migrating Pacific salmon transport marine-derived (MD) carbon and nutrients upstream and, upon death after spawning, hydrologic and animal pathways distribute these elements throughout aquatic and riparian systems. In an important system-scale feedback, fertilization of riparian plant communities with MD nutrients enhances the growth of some riparian plants, positively influencing salmon over the longer term by supplying stream organisms with an increased supply of nutritious litter and by improving salmon habitat via an influx of large diameter riparian-derived wood.

Historically, spawning salmon represented a flux of nearly 7000 Mt of nitrogen and more than 800 Mt of phosphorus to river corridors in California, Idaho, Oregon, and Washington. Although fluxes have been reduced by $>90\%$ during the past century as populations have declined, salmon still are an important source of nutrients to many river and riparian systems of Canada, Alaska, Russia, and Japan.

Conclusions

Riparian zones are highly complex physical and biological systems. Their complexity and distinctive ecological functions are maintained through strong spatial and temporal biophysical connectivity with adjacent riverine and upland systems. Water, sediments, and nutrients enter riparian zones from adjacent uplands and streams, mixing and reacting along dynamic surface and subsurface flow paths. Under normal flow conditions riparian zones retain a significant portion of these materials and generally return chemically purer water to streams and rivers. At the same time riparian zones are important sources of energy to both upland and aquatic systems in the form of plant and insect tissues. Many stream food webs fundamentally depend on these resources and many upland animals depend on them as important subsidies to their diets. At the same time, riparian communities benefit from the enhanced productivity of adjoining ecosystems (especially aquatic systems) through physical and biotic feedbacks that return a portion of that productivity to riparian zones in the form of organic matter and nutrients.

The unique ecological functions of riparian zones are linked to dynamic biophysical processes and interactions across multiple spatial and temporal scales. Maintaining these interactions and the connectivity driving them is a fundamental requirement for maintaining healthy riparian zones and the many services they provide; effective management requires maintaining connectivity, both in the timing and extent of flows as well as in the movements and types of animals.

See also: Floods; Streams and Rivers as Ecosystems.

Further Reading

Dang CK, Chauvet E, and Gessner MO (2005) Magnitude and variability of process rates in fungal diversity litter relationships. *Ecology Letters* 8: 1129–1137.

Décamps H (1996) The renewal of floodplain forests along rivers: a landscape perspective. *Verhandlungen der Internationalen Vereinigung für Theoretische und Angewandte Limnologie* 26: 35–59.

Helfield JM and Naiman RJ (2006) Keystone interactions: Salmon and bear in riparian forests of Alaska. *Ecosystems* 9: 167–180.

Junk WJ, Bayley PB, and Sparks RE (1989) The flood pulse concept in river-floodplain systems. In: Dodge DP (ed.) *Proceedings of the International Large River Symposium. Canadian Special Publication of Fisheries and Aquatic Sciences* 106, pp. 110–127.

Lowrance RR, Altier LS, Newbold JD, *et al.* (1995) *Water Quality Functions of Riparian Forest Buffer Systems in the Chesapeake Bay Watershed.* Annapolis, MD: USEPA. U.S. Environmental Protection Agency Chesapeake Bay Program, E.P.A. Publication 903-R-95-004 CBP/TRS 134/95.

Mahoney JM and Rood SB (1998) Streamflow requirements for cottonwood seedling recruitment: and integrative model. *Wetlands* 18: 634–645.

Malanson GP (1993) *Riparian Landscapes.* Cambridge, UK: Cambridge University Press.

McClain ME, Boyer EW, Dent CL, *et al.* (2003) Biogeochemical hot spots and hot moments at the interface of terrestrial and aquatic ecosystems. *Ecosystems* 6: 301–312.

Naiman RJ, Bunn SE, Nilsson C, Petts GE, Pinay G, and Thompson LC (2002) Legitimizing fluvial ecosystems as users of water: An overview. *Environmental Management* 30: 455–467.

Naiman RJ and Décamps H (1997) The ecology of interfaces: Riparian zones. *Annual Review of Ecology and Systematics* 28: 621–658.

Naiman RJ, Décamps H, and McClain ME (2005) *Riparia: Ecology, Conservation and Management of Streamside Communities.* San Diego, CA: Elsevier/Academic Press.

Naiman RJ, Helfield JM, Bartz KK, Drake DC, and Honea JM (2009) Pacific salmon, marine-derived nutrients and the characteristics of aquatic and riparian ecosystems. In: Haro AJ, Smith KL, Rulifson RA, Moffitt CM, Klauda RJ, Dadswell MJ, Cunjak RA, Cooper JE, Beal KL, and Avery TS (eds.) *Challenges for Diadromous Fishes in a Dynamic Global Environment.* American Fisheries Society Symposium. Bethesda, Maryland.

Nakano S, Miyasaka H, and Kuhara N (1999) Terrestrial-aquatic linkages: Riparian arthropod inputs alter trophic cascades in a stream food web. *Ecology* 80: 2435–2441.

Nanson GC and Beach HF (1977) Forest succession and sedimentation on a meandering-river floodplain, northeast British Columbia, *Canada. Journal of Biogeography* 4: 229–251.

Pollock MM, Naiman RJ, and Hanley TA (1998) Plant species richness in riparian wetlands: A test of biodiversity theory. *Ecology* 79: 94–105.

Rood SB, Gourley C, Ammon EM, *et al.* (2003) Flows for floodplain forests: Successful riparian restoration along the lower Truckee River, Nevada, USA. *BioScience* 7: 647–656.

Sabo JL, Sponseller R, Dixon M, *et al.* (2005) Riparian zones increase regional species diversity by harboring different, not more species. *Ecology* 86: 56–62.

Salo J, Kalliola R, Häkkinen I, *et al.* (1986) River dynamics and the diversity of Amazon lowland forest. *Nature* 322: 254–258.

Stanford JA and Ward JV (1988) The hyporheic habitat of river ecosystems. *Nature* 335: 64–66.

Tockner K, Malard F, and Ward JV (2000) An extension of the flood pulse concept. *Hydrological Processes* 14: 2861–2883.

Van Pelt R, O'Keefe TC, Latterell JJ, and Naiman RJ (2006) Structural development and stand evolution of riparian forests along the Queets River, Washington. *Ecological Monographs* 76: 277–298.

Wallace JB, Eggert SL, Meyer JL, and Webster JR (1997) Multiple trophic levels of a forest stream linked to terrestrial litter inputs. *Science* 277: 102–104.

Flood Plains

S K Hamilton, Michigan State University, Hickory Corners, MI, USA

The term flood plain is defined by the American Geological Institute as 'the surface or strip of relatively smooth land adjacent to a river channel, constructed by the present river in its existing regime and covered with water when the river overflows it banks'. This definition of flood plains includes only seasonally or episodically inundated land, but in fact many flood plains also contain water bodies that are permanent or semipermanent. These water bodies include *floodplain lakes* and channels, as well as shallow wetlands (sometimes called *backswamps*) that are separated from the river by levees. From the standpoint of flood plains as inland waters, these permanently wet areas can be distinguished from land subject to temporary, albeit sometimes prolonged, inundation resulting directly or indirectly from a rise in river level. Seasonal or episodic inundation strongly affects the more permanently wet areas, often completely replacing their surface water and changing the environment for aquatic life. Thus the hydrology and ecology of floodplain water bodies characteristically show strong seasonal dynamics.

The earlier-mentioned definition of a flood plain often does not correspond with how floodplain ecosystems are delineated for ecological studies, which tend to include more distal or slightly elevated lands that originated as flood plains and are contiguous with an active floodplain, but may now rarely or never be inundated by the parent river. Yet these areas tend to share ecological characteristics with active flood plains for several reasons that are discussed later, and the transition between active and relict flood plains may not be clear.

Examples of Floodplain Environments

Remotely sensed images of contrasting types of floodplain environments are depicted in **Figures 1–7**. Images of four large South American flood plains (Amazon, Madre de Dios, Pantanal, and the Llanos de Moxos), as well as the Cooper Creek system in Australia (**Figures 1–5**) depict the diversity of landforms, water bodies, and vegetation in large flood plains with minimal human disturbance. Examples where floodplain hydrology has been strongly altered are shown for the Kalamazoo River (**Figure 6**) and the Mississippi River (**Figure 7**). The ensuing discussion will refer to these images. In addition, photos of many of these sites appear in the online Flood Plain Photo Gallery (cite web site here).

Geomorphological Processes on Flood Plains

Flood plains are built of alluvial fill that normally originates as sediments carried by the parent river and deposited as point bars along the migrating channel, or during overbank flooding. Hydrologists refer to *bankfull discharge* as the river discharge above which the flood plain becomes inundated. The concept of bankfull can be difficult to apply, however, since much water enters and exits the flood plain via low areas or discrete openings in the levees (see later text). Many rivers reach bankfull at least annually, but inundate their flood plains less frequently.

River valleys commonly have one or more abandoned flood plains lying on elevated terraces that are normally above the present-day reach of riverine flooding. Such terraces may still contain wetlands and permanent water bodies that reflect their fluvial origin. Wetlands on such terraces may be sustained by groundwater and surface runoff emanating from adjacent uplands in addition to direct precipitation inputs. Terraces containing palm swamps that retain surface water all year are visible to the north of the Madre de Dios River just above its confluence with the Inambari River (**Figure 2**).

Movements of water and deposition and erosion of sediments sculpt the surface geomorphology of flood plains, particularly during larger flood events. Much attention has been paid to the study of fluvial geomorphology, and the ways in which different hydrogeomorphological regimes produce a myriad of floodplain landforms are well documented. Fluvial deposits can be highly heterogeneous and subject to constant change in active flood plains with high rates of sediment deposition, as for example, in the flood plains close to the Madre de Dios River and on islands in the Inambari River in **Figure 2**. Fluvial landforms tend to become smoothed out over time, as may be reflected by increased distance from or, in the case of terraces, elevation above the parent river. Erosion of elevated features and infilling of depressions contribute to the long-term homogenization of floodplain surfaces. Yet even in humid tropical climates, traces of the geomorphic features produced by fluvial action may remain visible for tens of thousands of years (e.g., Llanos de Moxos in **Figure 4**).

Elevated strips of land known as *levees* often border the river channel and reflect the higher rates of sediment deposition where the river water first decreases in velocity as it exits the channel.

Figure 1 Amazon River (locally known as the Rio Solimões in this reach) and its fringing flood plain above the city of Manaus and the confluence with the Negro River. This image land is about 240 km in width. The Negro River is visible in the upper right. The main river channel contains high inorganic turbidity and clearer waters on the flood plain as well as in the Negro River are dark. The flood plain here shows diverse geomorphic features, including numerous lakes, most conspicuous of which are the dendritic blocked-valley lakes. Image coordinates are 61.20 W, 3.61 S. Image shows Landsat 7 ETM+ data (bands 7, 4, 2) from the Geocover 2000 data set, obtained from NASA World Wind version 1.3.

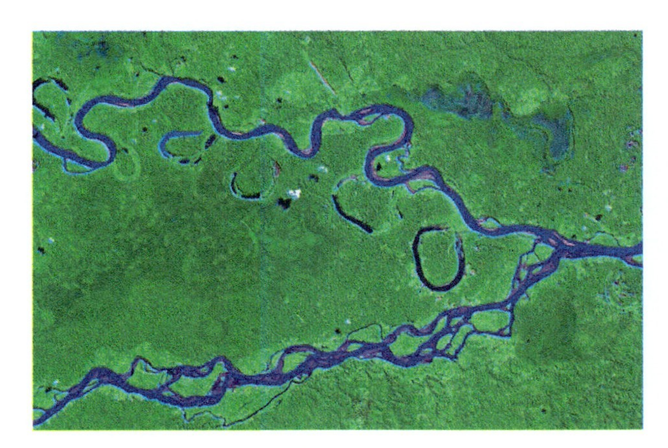

Figure 2 Madre de Dios River (upper) and its tributary the Inambari River (lower) in the Peruvian Amazon basin. This image land is about 29 km in width. The river channels contain high inorganic turbidity while some oxbows contain clearer water and hence appear darker. Lighter areas are vegetation in early successional stages following fluvial disturbance. The flood plain along the Madre de Dios contains oxbow lakes in various stages of separation and infilling. Much of the Inambari flood plain occurs as islands and bars along the highly braided channel. Between the two rivers is an extensive interfluvial backswamp, and in the upper right the darker areas are palm swamps lying on fluvial terraces. Image coordinates are 69.83 W, 12.70 S. Image shows Landsat 7 ETM+ data (bands 7, 4, 2) from the Geocover 2,000 data set, obtained from NASA World Wind version 1.3.

Figure 3 Flood plains of the upper Paraguay River and tributaries in the central Pantanal (Brazil and Bolivia). This image land is about 200 km in width. The flood plains here include the fringing floodplain of the sinuous Paraguay River (running from north to south), large turbid lakes with permanent connectivity to the river, and seasonally flooded savannas of the Taquari River alluvial fan, which covers the right side of the image. Lower lying areas of the fan are subject to backwater effects from the Paraguay River. Image coordinates are 57.14 W, 18.51 S. Image shows Landsat 7 ETM+ data (bands 7, 4, 2) from the Geocover 2000 data set, obtained from NASA World Wind version 1.3.

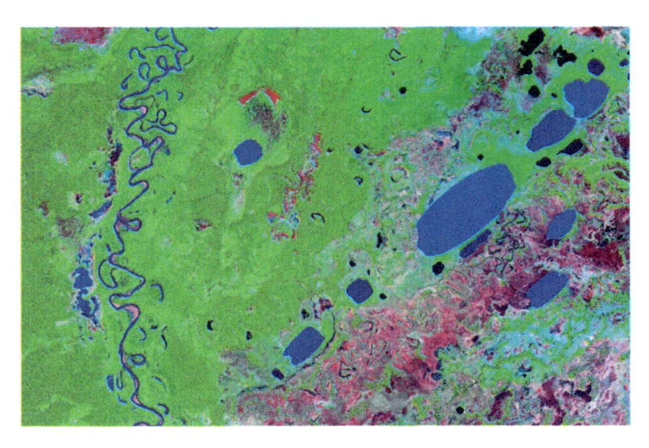

Figure 4 Savanna flood plains of the Llanos de Moxos (also spelled Mojos) in the upper Amazon Basin in Bolivia. This image land is about 108 km in width. The meandering Beni River with numerous oxbow lakes flows from south to north. The flood plain here reveals topographic features created by past fluvial activity. Superimposed on this landscape are strikingly regular depressions filled with shallow turbid water; these lakes have been attributed to subsidence patterns reflecting lineaments in the basement rock far below the alluvium. Image coordinates are 67.20 W, 13.99 S. Image shows Landsat 7 ETM+ data (bands 7, 4, 2) from the Geocover 2000 data set, obtained from NASA World Wind version 1.3.

Figure 5 Flood plains and river channels of Cooper Creek near Windorah (Queensland, Australia). This image land is about 136 km in width. The flood plains were relatively vegetated lands because they recently had been inundated, providing soil moisture for growth of herbaceous plant cover. Surrounding uplands are semiarid, sparsely vegetated shrublands and grasslands. The flood plain here receives river water through a complex system of anastomosed channels that are mostly dry in between the occasional flows. A few deeper channel reaches hold water between flows and are known as waterholes; these are too small to resolve here. Image coordinates are 142.47 E, 25.56 S. Image shows Landsat 7 ETM+ data (bands 7, 4, 2) from the Geocover 2000 data set, obtained from NASA World Wind version 1.3.

Figure 6 Hydrological alterations to the flood plain along the Kalamazoo River (Michigan USA). This image land is about 11 km in width. The reservoir in the lower part of this view was created by impoundment of the main channel for hydroelectric generation. Downstream is a diked area where water levels are managed for wildlife. The remaining flood plain is covered by deciduous forest and retains an approximately natural flood regime because this and upstream reservoirs do not change seasonally in volume. Image coordinates are 85.94 W, 42.57 N. Image source: U.S. Geological Survey digital aerial photography (DOQQ) taken in March 1999, obtained from NASA World Wind version 1.3.

Successively formed, concentric levees and swales may form *meander scrolls*. In some flood plains, these are created by the meandering of smaller channels within the flood plain rather than by the migration of the main channel.

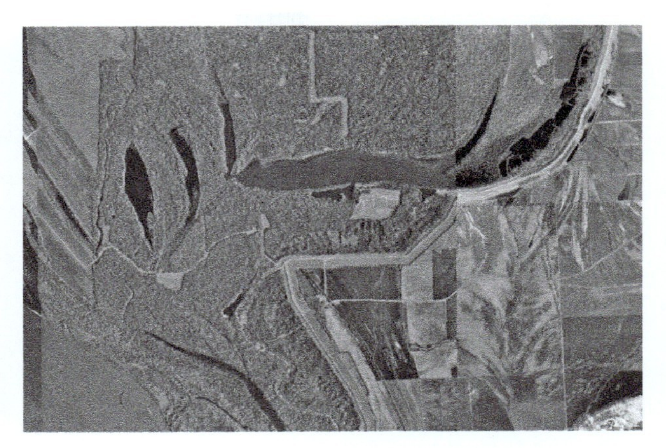

Figure 7 Hydrological alterations to the flood plain along the Mississippi River (northeastern Mississippi east of Helena, Arkansas, USA). This image land is about 9 km in width. Much of the flood plain here has been disconnected from the river by dikes and is used for agriculture; other parts are deforested but not diked. Oxbow lakes are visible in the forested parts; topographic floodplain features can also be seen in cleared and diked areas. Image coordinates are 90.51 W, 34.59 N. Image source is U.S. Geological Survey digital aerial photography (DOQQ) taken in April 2001, obtained from NASA World Wind version 1.3.

Flood plains can be readily classified based on their geomorphology, and several schemes have been proposed. Fewer classification systems deal specifically with the water bodies on flood plains, although various types of floodplain lakes are included in some geomorphological classifications (see later text). Flood plains can be broadly classified as fringing flood plains (i.e., found along the banks of rivers), coastal deltaic flood plains (formed where rivers meet larger rivers, lakes or the ocean), and internal deltaic flood plains (formed internally where a tributary meets a larger river and is subject to backwater effects). Some river channels, particularly those with braided morphology such as the Inambari River in **Figure 2**, have much of their flood plain on mid-channel bars and islands, and smaller channels can exhibit lake-like conditions at lower discharge. Wetlands that resemble riverine flood plains but may not be subject to direct inundation by river water include flood plains on terraces, and poorly drained plains adjacent to rivers that tend to become flooded with local runoff during the wet season. Examples of the latter include much of the land in images of the Pantanal (**Figure 3**) and Llanos de Moxos (**Figure 4**).

Hydrology

By definition, flood plains share the characteristic of being subject to inundation, but key hydrological features of inundation are highly variable. The timing, frequency, and duration of inundation can be collectively considered as the hydroperiod. In many kinds of flood plains, there is a range of hydroperiods depending on the land surface elevation with respect to the river. Subtle differences in topography can result in considerable differences in hydroperiod with corresponding variation in ecological characteristics, particularly when the range of inundation depth is relatively small.

River discharge regimes dictate timing and predictability of inundation, while depth and routing of flow within the flood plain reflect its geomorphology and vegetation. Larger river systems tend to have broader and more predictable flood peaks because their large drainage basins integrate the smaller-scale variability of individual precipitation events. Inundation of their flood plains has been termed the flood pulse (see **Floodplain Wetlands of Large River Systems**). The presence of extensive flood plains or water bodies such as lakes and reservoirs along a large river system also attenuates the flood pulse downstream.

Particularly attenuated and prolonged flood pulses are observed in large, mostly unregulated river systems of South America, where the flood plain typically is inundated once per year, lasting for months and covering much of the flood plain with water to depths up to several meters. **Figure 8** shows examples of daily river stage measurements from three large rivers with extensive flood plains: the Orinoco, Amazon, and Paraguay rivers. In the vicinity of these stage measurement sites, the Orinoco has the least flood plain relative to its discharge, while the Paraguay River has the most extensive flood plain (these data come from the southern Pantanal in Brazil: **Figure 3**). In each case, the inundation lasts for months, with the most protracted inundation in the savanna flood plains of the Pantanal where standing water often persists for more than 6 months of the year. These river systems and their flood plains are so large that the passage of the flood wave through the system takes months, resulting in a time lag between the wet season runoff and the inundation of the flood plain. This time lag reaches 4–6 months in the case of the southern Pantanal.

At the other extreme are flood plains that typically are inundated for only a few days or weeks per year. Examples shown here include the Madre de Dios River (**Figures 2** and **9**), which drains mountain and lowland landscapes in Peru, and the Kalamazoo River (**Figures 6** and **10**), which drains a glacial landscape in southern Michigan, USA. In these river systems, the floodplain inundation is seasonal but its timing is not regular, nor is its duration. In a particular year, inundation can occur several times or not at all. The tropical Madre de Dios responds to rain events in the Andes as well as the lowland plains. The stage record for the Madre de Dios River spans only a few years, which is insufficient to characterize its average behavior. The temperate Kalamazoo River responds to rain events but also to snowmelt and rain on snow, which produce a higher proportion of overland runoff in a landscape where liquid precipitation infiltrates the soils during most of the year. Most floods in the Kalamazoo River occur during cooler months (especially March and April) when biological activity is relatively low, and the mean number of days flooded per year is only eight, with substantial interannual variability.

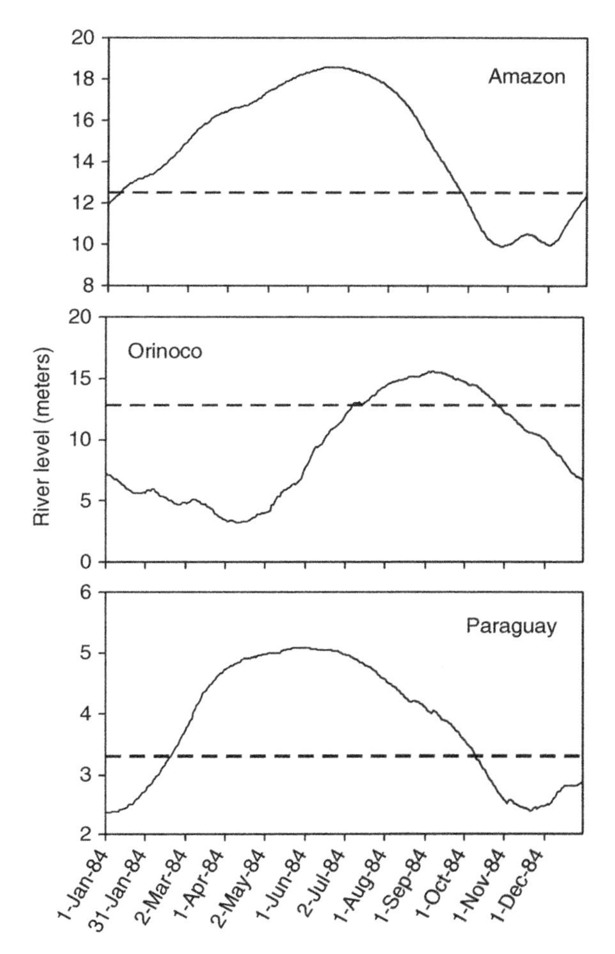

Figure 8 Daily river level measurements over 1984 from the Amazon River (Rio Solimões at Manacapuru, Brazil), the Orinoco River (Ciudad Bolívar, Venezuela), and the Paraguay River (Corumbá, Brazil). Dashed lines show approximate bankfull stages, above which the adjacent flood plains become inundated with river water. These flood regimes are essentially natural with no significant upstream regulation. Data collected by the Brazilian *Agência Nacional de Energia Elétrica* (Amazon River), the Venezuelan *Ministerio del Ambiente y de los Recursos Naturales Renovables* (Orinoco River), and the Brazilian Navy (Paraguay River).

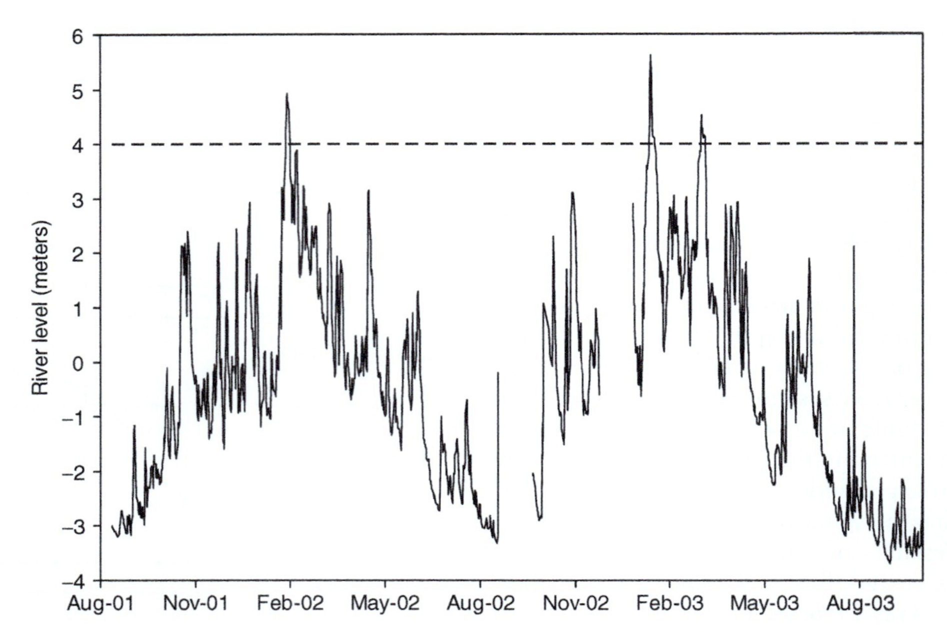

Figure 9 Daily river level measurements over two annual cycles from the Madre de Dios River above the confluence with the Los Amigos River in Peru. This river drains Andean mountain and lowland environments, and has a natural flow regime. Dashed line shows approximate bankfull stage, above which most areas on the adjacent flood plains become inundated with river water. Data collected by the *Centro de Investigación y Capacitación Río de los Amigos* (CICRA), operated by the Amazon Conservation Association.

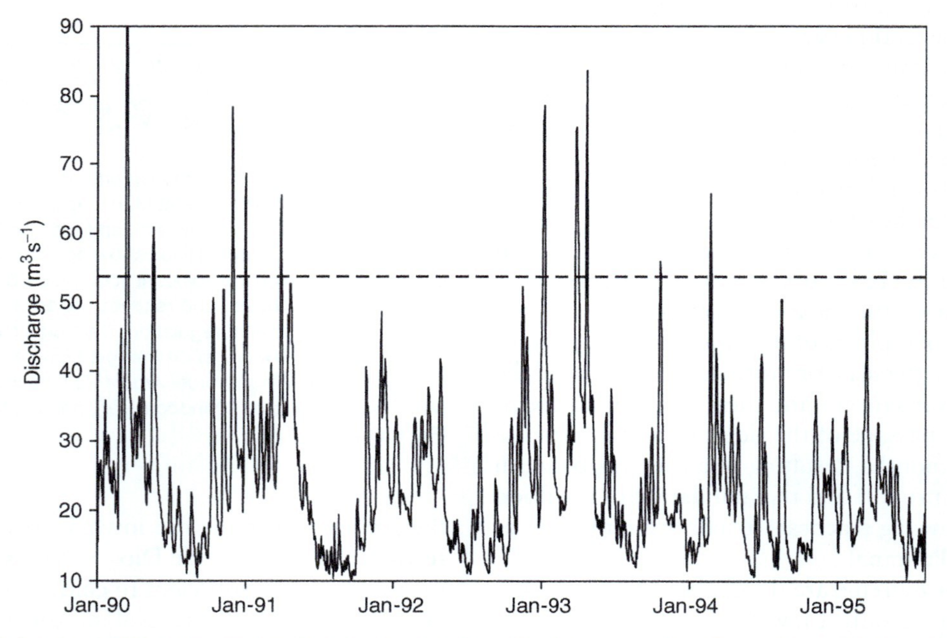

Figure 10 Daily river level measurements over six annual cycles from the Kalamazoo River (Battle Creek, Michigan, USA). Reservoirs and urbanization may have a small effect on the flow regime of this river. Dashed line shows approximate bankfull stage, above which most areas on the adjacent flood plains become inundated with river water. Data collected by the U.S. Geological Survey.

Even brief flooding can be important as a geomorphological force, but from an ecological perspective, brief and unpredictable floods may be viewed as a disturbance that limits the plant and animal life rather than benefits it. In cooler climates, the timing of the inundation is critical as well. Some large boreal/arctic rivers that flow in a northerly direction, such as the Mackenzie River in northwestern Canada and the Ob and Yenisei rivers in Siberia, exhibit spectacular flood pulses as the river breaks through ice in the spring.

The most variable river discharge regimes, and hence the most unpredictable flood pulses, occur in certain dryland rivers such as Cooper Creek and the Diamantina River in the endorheic Lake Eyre basin of interior Australia (**Figure 5**), which tend to flow only when monsoons bring heavy rainfall far into the continent. In these semiarid environments water limits biological activity much of the time, and hence inundation of extensive flood plains and anastomosed channel systems can be important as an ecological process despite its erratic and ephemeral nature. Permanent waters in these systems are known as 'waterholes', which are particularly deep channel reaches of restricted length that hold water long after the river has ceased to flow and most of the channel has dried.

The routing of flood waters across flood plains is complex and often changes over the course of the flood event. Commonly, water first enters the flood plain through low breaks in the levees known as crevasse splays, and this water may follow floodplain channels for some distance before spreading out into backswamps or lake basins. Water may also back up through downstream openings, for example, through the lower end of tributaries or oxbow lakes that are connected with the channel. Thus floodplain inundation commences before the levees become submerged. At the highest river stages, most or all levees may be underwater, and sheet flow proceeds generally in the downriver direction.

The water that inundates flood plains does not necessarily originate from overbank flow of the parent river, even though the parent river may control water levels by backwater effects. Flood plains typically show both spatial and temporal variation in water sources. Locally derived water can enter from lateral tributaries, perhaps becoming impounded temporarily by river flood waters, or traveling down the flood plain as deferred flow before mixing with the mainstem (parent) river. In expansive flood plains, the flood waters can be derived from delayed drainage of precipitation falling directly on the flood plain. Groundwater inputs from adjacent uplands also can be important, especially in smaller flood plains or in floodplain wetlands lying close to the upland boundary. Such inputs may maintain a high water table, and thus create wetland conditions for most or all of the year. These distinct sources of flood waters often differ in chemistry and suspended matter, enhancing the biogeochemical and ecological heterogeneity across flood plains.

Freshwater rivers near their confluences with the sea can have flood plains subject to tidal control of water levels, superimposing a short-term cycle of variability on longer-term, discharge-driven flood pulses. Examples include the Amazon and Orinoco deltas, where freshwater discharge is high enough to prevent seawater intrusion yet tidal cycles can be observed for considerable distances upriver. In contrast, rivers with little or no dry-season flow can experience substantial intrusion of seawater in their lower reaches. In such river systems, seemingly modest changes in relative sea level can alter the zone of seawater influence, producing dramatic implications for floodplain ecosystems. Seawater intrusion as a result of erosion of low ridges, possibly instigated by introduced water buffalo, has been documented in northern Australia east of Darwin, where the salinity caused massive changes in vegetation. Sea level rise associated with climate change increasingly will pose a threat to many low-lying floodplain ecosystems that contain freshwater in close proximity to the coastal zone.

Floodplain Lakes

More or less permanently flooded depressions on the flood plain include the backswamps behind the levees and can also include water bodies that are deep and permanent enough to be called floodplain lakes. There is no universal definition that distinguishes a lake from a wetland and usage of 'lake' or comparable terms varies regionally. Commonly water bodies that are called lakes have an open-water area for most or all of the year, which distinguishes them from vegetated wetlands. Their open-water areas may not fill with aquatic vegetation even though they can be quite shallow at low water; this may be a result of the changing water levels and low light penetration. Interannual variation in the flood regime can produce striking changes in the proportions of open water and emergent vegetation. Although they are often small in area, floodplain lakes are one of the most abundant types of lakes, particularly in the tropics and arctic where large, unregulated floodplain rivers remain. Very large lakes are associated with some flood plains, such as the Grand Lac on the Mekong River, though these may have a distinct geomorphological origin.

Floodplain lakes are formed by a variety of processes, including the isolation of main channel meanders (oxbows), formation of swales between successive levees, subsidence of alluvial fill, and permanent flooding of incised tributary valleys. Examples of each of these are visible in **Figures 1–7**. Floodplain lakes may be permanently or seasonally connected to the parent river, and the connections can be broad or quite restricted. Lakes that become seasonally isolated from the parent river may become perched at higher elevations than the river level as the river falls, and they may accumulate water of local origin during the phase of isolation from the river, often maintaining a stream that drains to the river.

Lakes typically occupy a minority of floodplain area but this depends on the geological history of the flood plain. Sometimes subsidence or backflooding exceed rates of new alluvial deposition (accretion), producing lakes that cover much of the backswamp areas. Neotectonic processes that cause subsidence, tilting, or uplift of basement rock can create areas of permanent flooding on flood plains.

Most kinds of floodplain lakes are no deeper than the parent river, and they can become very shallow during low water. During periods of isolation or at least minimal through-flow of river water, they can be rich in phytoplankton, although in many cases they become shallow enough that sediments are resuspended by wind-induced turbulence and inorganic turbidity greatly restricts underwater light availability (e.g., the large oval lakes in the Llanos de Moxos: **Figure 4**). During inundation they receive through-flowing river water and this may drastically reduce the water residence time to the point where plankton growth is suppressed by flushing.

Remote Sensing of Flood Plains

Delineation of floodplain boundaries by remote sensing can be challenging due to temporal variability in the extent of inundation and the difficulty of detecting standing water beneath vegetation canopies. In some humid tropical flood plains, the nearly perpetual cloud cover can also impede optical remote sensing systems. The difficulty of observing inundation dynamics by remote sensing has led investigators to rely more on vegetation and geomorphological features to delineate flood plains, and these features usually provide a reasonable indication of the overall floodplain extent and the boundaries between flood plains and upland ecosystems. At coarse spatial scales and in remote regions of the world, new remotely sensed, elevation data (e.g., the Shuttle Radar Topography Mission) are proving useful as well because they reveal the extent of relatively level terrain along major rivers.

Remote sensing of flood regimes can provide information for hydrological modeling and ecological and biogeochemical investigations in flood plains. Traditional optical remote sensing, including aerial photography for limited areas and Landsat satellite imagery for extensive areas, can provide snapshots of flood extent, but often lacks temporal resolution and may be inadequate for the reasons mentioned above. Relatively new microwave technologies such as radar can be better for hydrological dynamics because of their all-weather capability (i.e., they are less impeded by cloud cover) and their ability to penetrate vegetation to at least some degree. Several microwave systems are currently deployed on orbiting satellites and data from these sensors have provided new insights into floodplain hydrology. New image analysis approaches that combine multiple types of imagery are especially promising for remote sensing of flood plains.

Functions of Flood plains

The permanent and temporary aquatic environments of flood plains are locations for a number of important ecosystem processes, or functions, which provide values and services to people.

Flood plains tend to be highly productive ecosystems and have long been utilized for production of food and fiber and harvest of wild plants and animals. Perhaps the greatest contrast in productivity between uplands and flood plains occurs in dryland regions, but even in humid climates the flood plain is often desirable for farming and livestock production.

The temporary residence of water on flood plains is an important hydrological function because it delays the passage of flood waters through the fluvial system. This delay tends to attenuate the flood peak downriver, reducing peak water levels, which often is advantageous to riverside communities and agricultural activities. Passage of river water through flood plains can significantly enhance evapotranspirative losses, which in dry regions may be viewed as negative, but also may increase groundwater recharge. Some flood plains overlie extensive alluvial aquifers and these can provide a readily accessible water supply that is recharged by seasonal flood pulses.

Passage of river water through floodplain environments changes its content of dissolved and suspended matter, which can affect the composition of riverine exports. Suspended sediments tend to show net loss by deposition, whereas nutrients may show net retention or transformations. Concentrations of certain pollutants, particularly those associated with particulate material (e.g., trace metals) as well as labile nutrients (e.g., nitrate), are often greatly reduced in water passing through flood plains. High rates of primary production can result in net export of organic matter back to the river, although it is unclear whether flood plains are a net source or sink for riverine organic carbon.

Under certain circumstances, water quality can be diminished by passage through flood plains. Strong oxygen depletion upon initial inundation of some tropical flood plains is known to result in fish kills (e.g., the Pantanal in Brazil). In coastal plains of tropical Australia, initial contact of flood waters with acid sulfate soils can cause marked decreases in pH and result in metal toxicity for fishes. Such leaching effects usually diminish as the flood plain is increasingly flushed by flood waters.

Extensive flood plains can be important as sources of greenhouse gases to the atmosphere. It remains uncertain whether flood plains tend to be net sources of sinks of carbon dioxide, but like other wetlands they certainly tend to be net sources of methane while they are wet, and they may be important sources of nitrous oxide as well, particularly in regions with nitrogen pollution. Tropical flood plains are a globally significant source of methane emission to the atmosphere because of their extensive area and high biological activity. In general methane emission rates are proportional to primary productivity in wetlands, and this presumably extends to flood plains as well.

In rivers with extensive flood plains that are inundated for relatively long periods, much of the primary and secondary production in the overall river–flood plain system can occur in the flood plains. The Flood Pulse Concept articulates how such flood plains serve as the locus of biological production, supporting rich aquatic and terrestrial biodiversity including economically and culturally valuable fisheries (refer to 'see also' section).

Human Modification of Floodplain Hydrology

The age-old proclivity of humans to control river systems, combined with our ever-increasing technological ability to do so, has made flood plains one of the most altered aquatic ecosystems. River regulation, mainly through construction of dams, has strongly impacted flood regimes of rivers across all spatial scales, and even many of the largest rivers of the world have been regulated to some degree.

Impoundments tend to create permanently flooded reservoirs in place of seasonally flooded lands (**Figure 6**), and in many cases, they alter the discharge regime and often the water quality and temperature well downstream. They can trap a large fraction of the suspended sediment load, leading to geomorphological destabilization of the river–flood plain system downriver of the dam. Dams operated for hydroelectric generation may impose highly unnatural, short-term fluctuations in water levels, while those operated primarily for agricultural irrigation tend to change the seasonality of river flow in addition to removing water from the system.

Modification of river channels to facilitate navigation usually impacts flood plains by altering the relation between water levels and discharge. Removal of natural barriers to navigation can entail dredging, channel straightening, and excavation of rock outcrops. All of these measures tend to enhance flow conveyance and diminish the backwater effect that produces overbank flooding. Construction of navigation locks is akin to damming rivers. Low-head navigation dams that allow passage of flood waters, such as those on the upper Mississippi River (USA), are less damaging but still create extensive permanently impounded areas at low water levels.

Flood plains have often been isolated from their parent rivers by construction of dikes, commonly with the goal of farming the land (see Mississippi River example in **Figure 7**). Such land can be highly productive but may require costly measures to remove or control water, and over time land subsidence, loss of fertility, and occasional incursion of flood waters can detract from its sustainability. Nonetheless, agriculture on converted flood plains has played an important role in many societies and continues to be significant throughout the world. In some regions flood plains have been extensively mined for clay, sand or gravel, gold, or diamonds, a provisioning service that is sustainable only if subsequent floods replenish the material that is removed.

Aquatic ecosystems on flood plains are also impacted by urban and agricultural development in the upland watershed, which results in alterations to the flow regime and water quality of the parent rivers. Runoff is intensified by impervious surfaces and stormwater runoff drainage in built areas, and by land clearing and wetland drainage for agriculture. Nutrient loading to rivers and their flood plains increases with development, although the floodplain biota can have a large capacity for nutrient uptake and retention, and the effects of intact flood plains on water quality can be viewed as a valuable ecosystem service.

See also: Floods; Geomorphology of Streams and Rivers; Hydrology: Streams; Riparian Zones; South America; Streams and Rivers as Ecosystems; Wetlands of Large Rivers: Flood plains.

Further Reading

Alsdorf DE, Rodríguez E, and Lettenmaier DP (2007) Measuring surface water from space. *Reviews of Geophysics* 45, RG2002, doi:10.1029/2006RG000197.

Brinson MM (1993) *A Hydrogeomorphic Classification for Wetlands.* Technical report WRP-DE-4. Vicksburg, MS: U.S. Army Engineer Waterways Experiment Station.

Church M (2002) Geomorphic thresholds in riverine landscapes. *Freshwater Biology* 47: 541–557.

Goulding M, Barthem R, and Ferreira E (2003) *The Smithsonian Atlas of the Amazon*. Washington DC: Smithsonian Books.

Hamilton SK (2002) Hydrological controls of ecological structure and function in the Pantanal wetland (Brazil). In: McClain M (ed.) *The Ecohydrology of South American Rivers and Wetlands*, pp. 133–158. Wallingford, UK: International Association of Hydrological Sciences, Special Publication no. 6.

Hamilton SK, Sippel SJ, and Melack JM (2002) Comparison of inundation patterns in South American floodplains. *Journal of Geophysical Research* 107(D20), Art. No. 8038. doi: 10.1029/2000JD000306.

Junk WJ, Bayley PB, and Sparks RE (1989) The flood-pulse concept in river-floodplain systems. In: Dodge DP (ed.) *Proceedings of the International Large River Symposium*, pp. 110–127. Canada Special Publications Fisheries and Aquatic Sciences 106.

Junk WJ (1997) *The Central Amazon Floodplain: Ecology of a Pulsing System* Ecological Studies. Vol. 126. New York: Springer.

Leopold LB (1994) *A View of the River.* Cambridge, MA: Harvard University Press.

Mertes LAK (2000) Inundation hydrology. In: Wohl EE (ed.) *Inland Flood Hazards: Human, Riparian, and Aquatic communities*, pp. 145–166. Cambridge, UK: Cambridge University Press.

Mertes LAK (2002) Remote sensing of riverine landscapes. *Freshwater Biology* 799–816.

Poiani A (ed.) (2006) Floods in an Arid Continent. In: *Advances in Ecological Research 39*, San Diego, CA: Elsevier-Academic Press.

Ward JV, Tockner K, Arscott DB, and Claret C (2002) Riverine landscape diversity. *Freshwater Biology* 47: 517–539.

Welcomme RL (1985) *River Fisheries.* FAO Fisheries Technical Paper 262. Rome: Food and Agriculture Organization of the United Nations.

Winter TC (2001) The concept of hydrologic landscapes. *Journal of the American Water Resources Association* 37: 335–349.

Coarse Woody Debris in Lakes and Streams

G G Sass, Illinois Natural History Survey, Havana, IL, USA

Introduction

Riparian forests fringe many of the world's unperturbed lentic (i.e., lakes) and lotic (i.e., rivers, streams) systems. Aquatic coarse woody debris (or coarse woody habitat (CWH), coarse woody material, coarse woody structure, large woody debris) can be defined as trees and tree fragments (both living and dead) that have fallen into lakes, streams, and rivers from the riparian zone (**Figure 1**). Coarse woody debris is a natural feature of many aquatic ecosystems and may play an important role in many ecosystem processes. For example, many ecosystem processes and organisms are dependent upon or evolutionarily adapted to the presence of coarse woody debris pools (i.e., collections of coarse woody debris) in lakes and streams. Anthropogenic land-use change and habitat degradation can threaten the natural balance and ecological contribution of coarse woody debris to aquatic ecosystems. This article focuses primarily on the less understood role of coarse woody debris in lentic systems, and integrates and synthesizes knowledge from studies of coarse woody debris in rivers and streams to compare and contrast ecological function among aquatic ecosystems.

Sources of Coarse Woody Debris to Aquatic Ecosystems

A fringing riparian forest is generally essential for the presence and sustainability of coarse woody debris pools in aquatic ecosystems. General exceptions may include transport of coarse woody debris from the upstream portions of the watershed in lotic systems, physical transport in lakes, and transport by humans and/or other vectors (e.g., American beaver *Castor canadensis*). Common sources of coarse woody debris include:

1. Senescence
2. Windthrow
3. Beavers
4. Logging activities
5. Habitat additions
6. Fire
7. Flooding/erosion
8. Landslides
9. Ice storms

Senescence and felling of riparian trees where tree or snag height is greater than the distance from the tree/snag to the water of lentic and lotic systems may be a large, yet temporally variable, source of coarse woody debris. Early successional tree species stands (e.g., aspen, birch) may be more frequent contributors to coarse woody debris pools compared with late successional tree species stands (e.g., oak, pine). For example, in a study on one Ontario lake, no mature Eastern white pine (*Pinus strobus*) had fallen into the lake over the last 100 years. After senescence, snags falling toward the waterbody will contribute to the coarse woody debris pool if the tree is taller than the distance to the waterbody. Riparian forests located in the compass direction of the prevailing winds may be disproportionate contributors to coarse woody debris pools. Coarse woody debris was most prevalent on moderate-to-steep slopes, on southwest shorelines, and in areas with low levels of lakeshore residential development in four northern Wisconsin lakes. Congregations of coarse woody debris in lotic systems are generally associated with the edge of the floodplain in areas such as islands, concave banks, and side channels. Windthrow, logging activities, fire, and flooding/erosion are infrequent (>100 year) events, yet may be large contributors of coarse woody debris. Variable flooding conditions on streams and rivers can rearrange or contribute new coarse woody debris on various temporal scales. Beaver are a large and frequent contributor of coarse woody debris to aquatic ecosystems (**Figure 2**). For example, beaver contributed up to 33% of all littoral coarse woody debris examined across 60 lakes in northern Wisconsin, not including wood associated with beaver lodges. Beaver activity was the only source of coarse woody debris input, felling three trees into Little Rock Lake, Wisconsin, from 2002 to 2006 (+0.75 trees year^{-1}), following a whole-lake removal of coarse woody debris in 2002. Net accumulation of coarse woody debris among northern Wisconsin lakes was low and variable (**Table 1**). Wood input rates may be much higher in rivers, yet little wood may be stored in channels. Humans also add coarse woody debris to waterways for fish habitat (e.g., fish cribs), stabilizing structures, and as docks and piers. Docks and piers may provide structure, yet may not provide the same ecological function as coarse woody debris; docks may shade aquatic vegetation, decrease small fish abundances, and decrease benthic macroinvertebrate abundances.

Figure 1 Littoral zone coarse woody debris from Little Rock Lake, Vilas County, Wisconsin. Photograph by Steve Carpenter.

Figure 2 American beaver (*Castor canandensis*) and associated lodges, dams, and ponds can provide major contributions of coarse woody debris to lakes, streams, and rivers. Upper photograph by Steve Carpenter. Lower photograph by the Kansas Department of Wildlife and Parks.

Table 1 Net accumulation of coarse woody debris (CWD) in four northern Wisconsin lakes from 1996 to 2003

Lake	*State*	*Net CWD accumulation ($\pm$ logs km^{-1} $year^{-1}$)*
Allequash	Wisconsin	+1.7
Big Muskellunge	Wisconsin	+1.9
Sparkling	Wisconsin	−1.1
Trout	Wisconsin	+0.5

Reproduced from Marburg AE (2006) *Spatial and Temporal Patterns of Riparian Land Cover, Forests and Littoral Coarse Wood in the Northern Highland Lakes District, Wisconsin, USA*. Ph.D. Thesis, University of Wisconsin-Madison, Madison, WI, 129 pp.

Loss of Coarse Woody Debris from Aquatic Systems

Natural decomposition of aquatic coarse woody debris is a slow process (centuries to millennia) and highly dependent on tree species. Oxygen concentrations in wet pieces of wood may be insufficient to support fungi, which are the major decomposing agent in terrestrial coarse woody debris. The major decomposing agents for coarse woody debris in lakes and streams are bacteria and actinomycetes, which are limited to the surface of the wood. The mean age of Eastern white pine coarse woody debris in an Ontario lake was 443 years. Mean calendar date of all annual rings in Ontario coarse woody debris samples was 1551 and ranged from 1893 to 1982. Although coarse woody debris more than 1000 years old has been found in rivers, most coarse woody debris in streams degrades within 100 years. Coarse woody debris from nonresin producing tree species likely decomposes at faster rates. Decay rates of different species of coarse woody debris in northern Wisconsin and Canadian lakes and streams were variable, ranging from −0.2 to −3.25 g $year^{-1}$ (**Table 2**). Decay rates of coarse woody debris species in northern Wisconsin lakes did not differ among lakes or by sand or muck substrate. The functionality of coarse woody debris can be lost from aquatic ecosystems through decomposition, transport, burial, and physical removal. In lakes, ice and wave action can transport coarse woody debris to the deepest portions of lakes, to the riparian zone, into sediment (burial), and

Table 2 Half-life and decay rate of 12 coarse woody debris tree species in Wisconsin, Ontario, and Alberta lakes and streams

Common name	*Species*	*Location*	*Half life (years)*	*Decay rate (± g year^{-1})*
Eastern hemlock	*Tsuga canadensis*	N. Wisconsin	30.4	−0.75
Red pine	*Pinus resinosa*	N. Wisconsin	29.4	−0.6
Aspen	*Populus* spp.	N. Wisconsin	23.1	−1.2
Sugar maple	*Acer saccharum*	N. Wisconsin	20.9	−1.9
Paper birch	*Betula papyrifera*	N. Wisconsin	17	−2
Eastern white pine	*Pinus strobus*	N. Wisconsin	16.6	−1.25
Tamarack	*Larix laricina*	N. Wisconsin	7.9	−3.25
Eastern white pine	*Pinus strobus*	S. Ontario	106.6	–
Balsam poplar	*Populus balsamea*	Alberta	57.8	–
Northern white cedar	*Thuja occidentalis*	N. Wisconsin	–	−0.25
Spruce	*Picea* spp.	N. Wisconsin	–	−0.2
Red oak	*Quercus rubra*	N. Wisconsin	–	+0.25
Balsam fir	*Abies balsamea*	N. Wisconsin	–	+0.75

Sources
1. Hodkinson ID (1975) Dry weight loss and chemical changes in vascular plant litter of terrestrial origin, occurring in a beaver pond ecosystem. *Journal of Ecology* 63: 131–142.
2. Guyette RP Cole WG Dey, DC, and Muzika R-M (2002) Perspectives on the age and distribution of large wood in riparian carbon pools. *Canadian Journal of Fisheries and Aquatic Sciences* 59: 578–585.
3. Marburg AE (2006) *Spatial and Temporal Patterns of Riparian Land Cover, Forests and Littoral Coarse Wood in the Northern Highland Lakes District, Wisconsin, USA*. Ph.D. Thesis, University of Wisconsin-Madison, Madison, WI, 129 pp.

may also congregate coarse woody debris rafts that accumulate along shorelines. Median distances of in-lake transport of tagged and recaptured coarse woody debris in northern Wisconsin lakes was 24 m from 1996 to 2003 (~3 m year^{-1}). Flooding and water-level fluctuations in lotic systems can redistribute coarse woody debris, permanently or temporarily, into riparian zones with periodic water access. Humans may be the largest, and fastest, physical removers of coarse woody debris from aquatic ecosystems (**Figure** 3). For example, a strong negative relationship exists between coarse woody debris abundances and lakeshore residential development in Wisconsin, Upper Michigan, Washington State, and British Columbia lakes (**Figure** 4). Coarse woody debris is also less prevalent in agricultural and urban rivers and streams. Riparian forests and coarse woody debris pools may be decoupled in lakes that are developed. Coarse woody debris density and riparian forest tree density exhibit a positive relationship in undeveloped lakes and no relationship in developed lakes. Surveys of northern Wisconsin lakeshore homeowners in 1998 found that 55% of those polled either rarely thought about logs in their lakes or did not think about coarse woody debris at all, prior to confronting the issue in the questionnaire. Approximately 25% of respondents had removed at least one log from the water, with 64% of those who had removed wood doing so within one year of the survey. Wood that was a boating or swimming hazard was more likely to be removed. The lakeshore residential development process (including road building) also acts to thin the riparian forest, thus decreasing the source pool of coarse woody debris to lakes and streams. In the past, coarse woody debris was actively removed from Pacific Northwest streams to improve salmonid habitats; that practice has been abolished as recent knowledge suggests that coarse woody debris creates pools in streams that provide critical juvenile salmonid nursery habitats.

Physical and Hydraulic Role of Coarse Woody Debris

Coarse woody debris stabilizes shorelines and riparian zones from erosion, promotes sediment retention and burial, and alters flows of lotic systems. Loss of coarse woody debris from littoral zones of lakes promotes sediment resuspension, increases in turbidity, and loss of sediments to the deepest portions of the lake. Sediment disturbance can resuspend buried contaminants, promote microbial activity, and increase mercury methylation rates, similar to dredging activities. For example, a whole-lake removal of coarse woody debris resulted in a threefold increase in waterborne methyl mercury concentrations in a northern Wisconsin lake. Anthropogenic disturbances to littoral zone coarse woody debris can, therefore, potentially lead to elevated methyl mercury concentrations in fishes. Much the same set of consequences can be expected with coarse woody debris loss and other contaminants that occur in sediments.

Coarse woody debris and congregations of coarse woody debris forming snags in lotic systems alter flow regimes, stabilize banks, and promote undercut and pool formation. The cross-flow field (i.e., a cross-section of water currents around a log) for a single

Figure 3 Lakeshore residential development can thin riparian forests and reduce coarse woody debris abundances in lakes, streams, and rivers. Photograph by Michael Meyer.

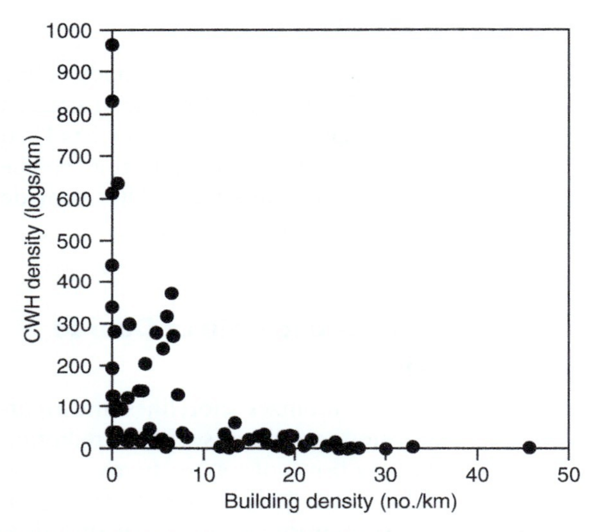

Figure 4 Relationship between CWH abundance and the number of buildings per km of shoreline for several Northern Wisconsin and Upper Michigan lakes. Reproduced from Sass GG, Kitchell JF, Carpenter SR, *et al.* (2006) Fish community and food web responses to a whole-lake removal of coarse woody habitat. *Fisheries* 31: 321–330.

cylindrical log perpendicular to the flow of water is related to the Reynold's number and log diameter. Reynold's numbers for logs in lotic systems range from 100 to 1 million with symmetrical cross-flow patterns resulting in reduced velocities behind the log. Recirculating vortices develop in front and behind logs buried or partially-buried into streambeds. Large coarse woody debris aggregations can be considered solid structures where flow field is determined by bluff surface size and shedding from edges obtuse to the flow direction. Coarse woody debris and associated root wads naturally buffer lakes and streams, thus dampening wave action and preventing erosion and sluffing of shoreline sediment. Stable pieces of coarse woody debris in lotic systems can

1. influence rates of bank erosion;
2. create pools;
3. initiate sediment deposition and bar formation.

Lotic systems with abundant coarse woody debris may retain more sediment, have steeper slopes, and have lower sediment transport rates than coarse woody debris depauperate rivers and streams. Removal of coarse woody debris from a 200 m stretch of a New Hampshire stream resulted in a sevenfold increase in sediment transport and particulate organic matter. Undercuts, pools, and slack water conditions are energetically favorable for salmonids as refuge, feeding, and juvenile nursery and rearing habitats. Coarse woody debris creates fish habitat by increasing the size and depth of stream pools and by providing refuge from predators. Retention of coarse woody debris has been critical in stream salmonid habitat restoration efforts.

Nutrient Properties and Primary Production Associated with Coarse Woody Debris

Coarse woody debris is generally considered to be chemically inert in nitrogen and phosphorus; however, this may be highly dependent and positively related to coarse woody debris age. For example, decayed coarse woody debris may act as a substrate, and likely provides nutrients, to promote algal growth. Periphyton growth is limited by light, but may also be limited by phosphorus when it grows on nonnutrient diffusing substrates such as most coarse woody debris. Epixylic production (i.e., algal growth on wood) (4%) is generally lower than epipelic (i.e., algal growth on sediment) benthic production (50–80%) in lakes. Although the direct contribution of epixylic algae to whole-lake primary productivity is relatively minor, the indirect influences of coarse woody debris, such as increased organic sediment retention, may be important for primary productivity derived from epipelic algae. Removal of coarse woody debris may lead to decreases in lake productivity through loss of organic sediments from the littoral zones of lakes. Organic content of littoral zone sediment and the density of coarse woody debris were negatively related across a lakeshore residential development gradient of Washington State and British Columbia lakes. Littoral sediments of undeveloped lakes ranged from 34 to 77% organic by mass, while developed lakes sediment was 1–3% organic. Accumulations of sedimentary organic matter were highest in littoral zones of lakes with coarse woody debris, and decreased with distance from the shore; the opposite relationship was noted for lakes without coarse woody debris. Organic sediments sequester nutrients. Epixylic algal growth was higher at the sediment–water interface where algae could tap the sediment for nutrients. Epixylic production is likely dependent on nutrient concentrations in surrounding waters, water clarity, proximity to organic sediment, and may be coarse woody debris species-specific. Little information exists on the nutrient leaching properties of certain coarse woody debris species and over time. Loss of coarse woody debris may decouple benthic–pelagic energy and nutrient linkages in lake ecosystems.

Coarse Woody Debris and Secondary Production

Many invertebrates require coarse woody debris for food and habitat. In many streams, the highest values of invertebrate diversity are in areas with accumulations of wood. Zoobenthos secondary production is higher on coarse woody debris than on adjacent habitats, such as sand and muck. Geomorphic properties provided by coarse woody debris serve as rearing sites and habitat for macroinvertebrates. In rivers, 60% of total invertebrate biomass was found on coarse woody debris (4% of habitat availability) compared to sand (80% habitat availability). In Southeastern United States rivers, snags support invertebrate production that is among the highest found in lotic systems. Epixylic algae may attract grazers, thus attracting invertebrate predators (e.g., odonates, crayfish). In streams and rivers, loose streambed wood is generally colonized by shredders (gougers) and stable coarse woody debris is dominated by filterers and gatherers. Zoobenthos secondary production is an important energetic pathway to upper trophic levels in lentic systems and may be underestimated in a continuum of benthic–pelagic coupling in aquatic ecosystems. Production in lentic systems may be dominated by terrestrial and allocthonous sources of carbon, mediated through microbial processes, and enhanced by coarse woody debris presence.

Coarse Woody Debris and Fishes

Coarse woody debris is an essential physical, biological, and chemical attribute of lakes and streams for many fish species. Over 85 species of fish are recognized to rely on coarse woody debris during all or part of their life histories. For fishes, coarse woody debris provides:

1. food
2. refuge
3. spawning substrate
4. nursery and rearing habitat

Coarse woody debris is a direct and indirect source of food to fishes. Directly, several fishes consume decomposing wood and bark, while many fish species are dependent on fruits and seeds provided by felled or flooded coarse woody debris. Indirectly, coarse woody debris forms the base of the food web in many aquatic ecosystems by providing a substrate and nutrients for epixylic algal production to herbivorous fishes and benthic macroinvertebrate grazers. Zoobenthos secondary production is the dominant energetic pathway to upper trophic levels in many aquatic ecosystems, comprising 65% of the total prey consumed by fishes. Coarse woody debris attracts small fishes, and thus serves as a focal point for predator–prey interactions. Coarse woody debris, particularly complex and branchy coarse woody debris, decreases predator foraging success, thus creating refuge habitat for small fishes (**Figure 5**). Complex coarse woody debris causes visual interference for predators and interstices prevent predators from entering complex coarse woody debris arrangements. The presence of coarse woody debris may act to create heterogenous fish distributions with prey and small fishes located in the coarse woody debris refuge and large predatory fishes located on or near the coarse woody debris refuge edge (**Figure 6**). For example, largemouth bass (*Micropterus salmoides*) home range size is negatively correlated with coarse woody debris abundance in

Figure 5 Bluegill (*Lepomis macrochirus*) associated with coarse woody debris refuge in Anderson Lake, Vilas County, Wisconsin. Photograph by Greg Sass.

Figure 6 Largemouth bass (*Micropterus salmoides*) cruising the edge of coarse woody debris refuge in Anderson Lake, Vilas County, Wisconsin. Photograph by Greg Sass.

northern Wisconsin lakes. Loss of coarse woody debris may result in homogenous predator–prey distributions and greater largemouth bass home ranges. Experimental relocation of coarse woody debris in lakes altered largemouth bass movement patterns to focus on redistributed coarse woody debris. Loss of coarse woody debris refuge can cause extirpations of prey species (e.g., yellow perch *Perca flavescens*, cyprinids) and ultimately determine species assemblages in lakes. Persistence of predator and prey populations is enhanced when intense interspecific competition and predation occur between juvenile predators and adult prey fishes located in refuge areas; this is often called a trophic triangle and an example can be found with juvenile largemouth bass and adult yellow perch. Many fish species are dependent upon coarse woody debris for spawning. As one example, yellow perch may use coarse woody debris as a spawning substrate. Experimental removal of coarse woody debris in a northern Wisconsin lake resulted in no perch reproduction in subsequent years. Large- and smallmouth bass (*Micropterus dolomieu*) tend to build nests in association with littoral coarse woody debris. Black basses (*Micropterus* spp.) prefer nesting near physical structure because it increases mating, hatching, and nesting success. A positive, saturating relationship exists between the number of largemouth bass nests $100\,\mathrm{m}^{-1}$ of shoreline and coarse woody debris abundance in northern Wisconsin lakes. Coarse woody debris is indirectly associated with juvenile nursery habitats in streams and rivers. Juvenile survivorship of salmonid smolts is considerably higher in streams with coarse woody debris than those without. Coarse woody debris in streams was associated with increased densities of salmonids such as coho salmon (*Oncorhynchus kisutch*), cutthroat trout (*O. clarkii*), rainbow trout (*O. mykiss*), brook trout (*Salvelinus fontinalis*), and brown trout (*Salmo trutta*) by increasing suitable habitat for both adults and juveniles and by providing ample prey resources. Coarse woody debris-rich sites in lotic systems worldwide exhibit more fish species diversity and numbers of fish compared to sites without wood. Fish species diversity and numbers of fish in rivers ranged from 1.3 to 2 and 1.6 to 50 times higher in sites with and without coarse woody debris, respectively. Loss of coarse woody debris from streams generally results in loss of pool habitat, complexity, and smaller individuals of coldwater and warmwater fish species.

Coarse Woody Debris and Other Organisms

Specific plant assemblages (e.g., leatherleaf *Chamaedaphne calyculata*, *Sphagnum* spp., sundew *Drosera* spp.) are associated with emergent and floating coarse woody debris and rafts in north-temperate lakes. Invertebrates, such as crayfish and some freshwater mussels, are dependent upon coarse woody debris during their life histories. Amphibians, such as frogs and toads, use coarse woody debris as a spawning substrate and for refuge during breeding. A negative relationship exists between turtle abundances and coarse woody debris abundances in lentic and lotic systems. Aquatic and shore birds use coarse woody debris for nesting habitat (e.g., common loon *Gavia immer*) and perching areas (e.g., herons *Ardea* spp., cormorants *Phalacrocorax* spp., kingfishers *Megaceryle* spp., wood ducks *Aix sponsa*). Mammals use coarse woody debris for shelter (beaver) and

Figure 7 Organisms associated with floating coarse woody debris in Little Rock Lake, Vilas County, Wisconsin. Organisms include sundew (*Drosera* spp.), leatherleaf (*Chamaedaphne calyculata*), *Sphagnum* spp., and Eastern painted turtle (*Chrysemys picta picta*). Photograph by Matt Helmus.

for ambush points (e.g., raccoon *Procyon lotor*, river otter *Lutra canadensis*, mink *Mustela vison*). Dams built by beavers can affect groundwater recharge rates and stream discharge, may influence valley floor morphology through sediment retention, and enhance stream habitat quality for fishes. Over 82 species of fish have been known to use beaver ponds. Loss of beavers and removal of dams may accelerate stream incision and promote lowering of groundwater levels and stream drying. Coarse woody debris associated with riparian and littoral areas of lentic and lotic systems provides organisms with proximity to water, unique plant assemblages, and diverse microhabitats that provide habitat structure, shelter, patchiness of habitat, and increased food resources (**Figure** 7).

Riparian Forest/Coarse Woody Debris/ Aquatic Food Web Models

Most models of lotic systems have focused on the amounts and distributions of coarse woody debris in streams and rivers. Fewer models have addressed processes determining patterns such as riparian tree mortality, input, breakage, decomposition, mechanical breakdown, and transport. In most models, forest stand age directly influences the abundance of coarse woody debris in lotic systems and wood dynamics are most sensitive to rates of input and decomposition.

Simulation models that incorporate the riparian forest, coarse woody debris pools, and aquatic food webs show variable effects on fish populations dependent upon lakeshore residential development and harvest rates of the top predator. Prior to development, late successional tree species dominate the riparian forest. During development, both late and early successional tree species decline to low levels and are maintained over time. As a consequence of development, coarse woody debris pools decline and are not replenished over time due to loss of the source pool and active physical removal. Declines in coarse woody debris result in two outcomes in the fish population that are dependent on the harvest rate of the top predator:

1. Without fishing, adult and juvenile top predator biomass increases and dominates and prey fish are extirpated.
2. With high top predator harvest rates, adult and juvenile top predator biomass is suppressed and prey fish biomass dominates.

Ecosystem-Scale Coarse Woody Debris Experiments

Largemouth bass and bluegill (*Lepomis macrochirus*) growth rates are positively correlated with coarse woody debris abundances and negatively correlated with lakeshore housing density in northern Wisconsin and Upper Michigan lakes. Decreases in coarse woody debris alter spatial distribution patterns of fishes in Washington state lakes. Mechanisms for the observed patterns in fish growth and distribution with coarse woody debris are unknown. A whole-lake removal of coarse woody debris was conducted on the treatment (north) basin of Little Rock Lake, Vilas County, Wisconsin, in 2002 (**Figure** 8). The reference (south) basin was left unaltered. Coarse woody debris removal in Little Rock Lake resulted in:

1. a decrease in largemouth bass growth rates;
2. increased reliance by largemouth bass on terrestrial sources of prey, such as insects, reptiles, birds, and mammals;
3. the extirpation of the yellow perch population from the treatment basin;
4. a threefold increase in waterborne methyl mercury concentrations;
5. a decrease in largemouth bass nest density and spawning success;
6. an increase in largemouth bass home range size.

A reciprocal whole-lake coarse woody debris addition was completed on Camp Lake, Vilas County, Wisconsin, in 2004. Over 300 trees (one tree $10\,m^{-1}$ of shoreline) of various species and complexities were added (**Figure** 9). Coarse woody debris addition to Camp Lake resulted in:

1. increased reliance by largemouth bass on fish prey;
2. an increase in proportions of trophy size (>457 mm) largemouth bass;

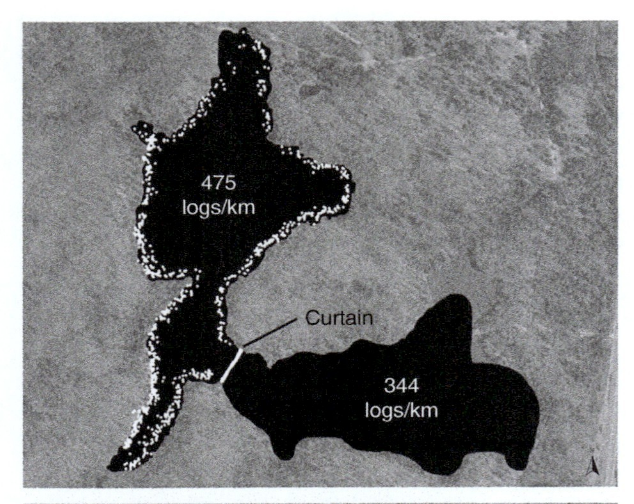

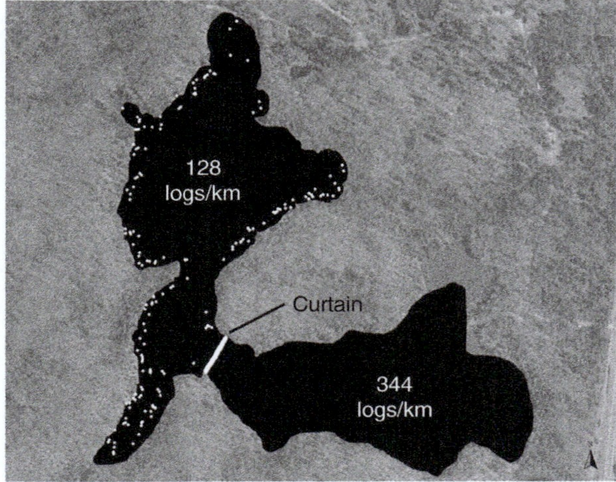

Figure 8 Aerial photographs of Little Rock Lake, Vilas County, Wisconsin with abundances of large (>10 cm diameter) CWH labeled and represented by white dots before and after the CWH removal in the treatment basin (north) in 2002. Reproduced from Sass GG, Kitchell JF, Carpenter SR, *et al.* (2006) Fish community and food web responses to a whole-lake removal of coarse woody habitat. *Fisheries* 31: 321–330.

3. an increase in largemouth bass nest density and nest success;
4. a decrease in largemouth bass home range size;
5. a fourfold increase in usage of complex coarse woody debris by largemouth bass and bluegill;
6. no change in bluegill population dynamics;
7. an increase in yellow perch abundances.

Forty to 70% of the coarse woody debris was removed from three Ontario lakes to determine the link between submerged wood and production of periphyton and invertebrates. Removal of coarse woody debris resulted in:

1. little loss of whole-lake invertebrate productivity, despite greater biomass of invertebrates on wood than in adjacent sediments;
2. highly decayed wood showing higher chlorophyll concentrations and invertebrate biomass and diversity than fresh wood;
3. no measurable effect on whole-lake water chemistry or on residual epixylic periphyton and invertebrate biomass.

Large wood was added to three headwater tributaries of the Jumbo River in the Upper Peninsula of Michigan. Organic matter retention is typically low in these sandy-bottomed, low-gradient systems. Two years after wood addition, there were no significant differences in density or biomass of stream fishes, but selected ecosystem metrics, such as nutrient spiraling and ecosystem metabolism, were higher in treatment reaches. Changes in geomorphology and microbial biofilm dynamics associated with wood addition occurred quickly, but responses by higher trophic levels (e.g., macroinvertebrates and fishes) may lag behind.

Habitat enhancement projects that have added coarse woody debris to streams generally create new habitat and rapidly increase the density of salmonids associated with the habitat. Similarly, coarse woody debris additions to lakes and reservoirs increase the abundance of fishes associated with the structures and subsequently increase angler catch rates.

Temporal Dynamics and Coarse Woody Debris Restoration

Coarse woody debris is a natural attribute to many lakes, rivers, and streams. The fast dynamics of removal rates (days to years; logging, development) and slow rates of natural replacement and decomposition (centuries to millennia) suggest that coarse woody debris removal can have long-lasting or permanent consequences on aquatic ecosystems. The use of the terms CWH or coarse woody structure may be more appropriate than coarse woody debris and may give a positive connotation to wood presence, conservation, and restoration in aquatic ecosystems. Discontinuation of physical removal, logging practices that maintain riparian buffers, and active additions can ameliorate the effects of coarse woody debris loss on aquatic ecosystems. Tree drops (active felling of riparian trees), preservation of intact riparian forests, preservation of pristine systems, and management of riparian forests specifically to accelerate recruitment of coarse woody debris may be effective mitigation tools for restoration efforts. For lotic systems, soft placement of wood to allow movement and transport may be preferable to hard engineering approaches that anchor wood permanently. In addition to reintroduction of coarse woody debris, restoration of natural

Figure 9 Littoral zone CWH addition to Camp Lake, Vilas County, Wisconsin in the spring of 2004. CWH was distributed as one piece 10 m^{-1} of shoreline lake wide. Photograph by Michele Woodford.

hydrologic cycles, riparian vegetation, selected floodplains, and an associated natural disturbance regime should be considered in rehabilitation efforts.

Further Reading

Benke AC and Wallace JB (2003) Influence of wood on invertebrate communities in streams and rivers. In: Gregory SV, Boyer KL, and Gurnell AM (eds.) *The Ecology and Management of Wood in World Rivers*, pp. 149–177. Bethesda: American Fisheries Society.

Christensen DL, Herwig BR, Schindler DE, and Carpenter SR (1996) Impacts of lakeshore development on coarse woody debris in north temperate lakes. *Ecological Applications* 6: 1143–1149.

Francis TB and Schindler DE (2006) Degradation of littoral habitats by residential development: Woody debris in lakes of the Pacific Northwest and Midwest, United States. *Ambio* 35: 274–280.

Gregory SV, Boyer KL, and Gurnell AM (eds.) (2003) *The Ecology and Management of Wood in World Rivers*. Bethesda: American Fisheries Society.

Guyette RP and Cole WG (1999) Age characteristics of coarse woody debris (*Pinus strobus*) in a lake littoral zone. *Canadian Journal of Fisheries and Aquatic Sciences* 56: 496–505.

Magnuson JJ, Kratz TK, and Benson BJ (2006) *Long-Term Dynamics of Lakes in the Landscape: Long-Term Ecological Research on North Temperate Lakes*. New York: Oxford University Press.

Marburg AE, Turner MG, and Kratz TK (2006) Natural and anthropogenic variation in coarse wood among and within lakes. *Journal of Ecology* 94: 558–568.

Newbrey MG, Bozek MA, Jennings MJ, and Cook JE (2005) Branching complexity and morphological characteristics of coarse woody structure as lacustrine fish habitat. *Canadian Journal of Fisheries and Aquatic Sciences* 62: 2110–2123.

Sass GG (2004) *Fish Community and Food Web Responses to a Whole-Lake Removal of Coarse Woody Habitat*. Ph.D. Dissertation, Madison: University of Wisconsin.

Sass GG, Kitchell JF, Carpenter SR, *et al.* (2006) Fish community and food web responses to a whole-lake removal of coarse woody habitat. *Fisheries* 31: 321–330.

Scheuerell MD and Schindler DE (2004) Changes in the spatial distribution of fishes in lakes along a residential development gradient. *Ecosystems* 7: 98–106.

Schindler DE, Geib SI, and Williams MR (2000) Patterns of fish growth along a residential development gradient in north temperate lakes. *Ecosystems* 3: 229–237.

Smokorowski KE, Pratt TC, Cole WG, McEachern LJ, and Mallory EC (2006) Effects on periphyton and macroinvertebrates from removal of submerged wood in three Ontario lakes. *Canadian Journal of Fisheries and Aquatic Sciences* 63: 2038–2049.

Steel EA, Richards WH, and Kelsey KA (2003) Wood and wildlife: Benefits of river wood to terrestrial and aquatic vertebrates. In: Gregory SV, Boyer KL, and Gurnell AM (eds.) *The Ecology and Management of Wood in World Rivers*, pp. 235–247. Bethesda: American Fisheries Society.

Vadeboncoeur Y and Lodge DM (2000) Periphyton production on wood and sediment: Substratum-specific response to laboratory and whole-lake nutrient manipulations. *Journal of the North American Benthological Society* 19: 68–81.

Relevant Websites

http://limnology.wisc.edu – University of Wisconsin-Madison, Center for Limnology, Arthur D. Hasler Laboratory of Limnology.

http://limnosun.limnology.wisc.edu – University of Wisconsin-Madison, Center for Limnology, Trout Lake Station, North Temperate Lakes Long Term Ecological Research.

http://biocomplexity.limnology.wisc.edu – University of Wisconsin-Madison, Center for Limnology, Arthur D. Hasler Laboratory of Limnology, National Science Foundation Biocomplexity Project.

http://www.tlws.ca – Turkey Lakes Watershed (TLW) Study.

http://www.fsl.orst.edu/lter/ – HJ Andrews Experimental Forest and Andrews Long Term Ecological Research.

Wetlands of Large Rivers: Flood Plains

S K Hamilton, Michigan State University, Hickory Corners, MI, USA

Wetlands on Flood Plains

The term flood plain is defined by the American Geological Institute as 'the surface or strip of relatively smooth land adjacent to a river channel, constructed by the present river in its existing regime and covered with water when the river overflows it banks'. Wetlands on flood plains are important for several reasons: (1) their distribution is ubiquitous in association with streams and rivers throughout the world; (2) their total area is a substantial fraction of global wetland area; (3) often they are sites of rich biological diversity and high biological production; and (4) they perform ecosystem services of value to people and historically have been loci for the development of human societies.

The seasonally or episodically inundated land on flood plains qualifies as wetland to the extent that inundation influences the soils, vegetation, and other elements of the biota, and even brief inundation or soil saturation can be a strong ecological influence in flood plains as in other kinds of wetlands. Many flood plains contain water bodies that are permanent or semipermanent, including floodplain lakes and channels, and often have more extensive shallow wetlands that may remain wet throughout the year. From the standpoint of flood plains as wetlands, these permanently wet areas can be distinguished from land subject to temporary, albeit sometimes prolonged, inundation resulting directly or indirectly from a rise in river level. Seasonal or episodic inundation strongly affects the more permanently wet areas, often completely replacing their surface water and changing the environment for aquatic life.

Flood plains subject to riverine inundation often undergo transition into more distal wetland areas that may have been created by fluvial processes but may rarely or never be inundated directly by river water. Nonetheless, these land surfaces are often poorly drained and are generally considered as flood plains in the ecological literature, even though they tend to be distinct in their ecological characteristics from flood plains subject to riverine inundation. There may be an indirect influence of river level on the hydrology of more distal flood plains through backwater effects on surface or subsurface water levels.

The ecological features of floodplain wetlands are highly variable and many wetland classification schemes do not distinguish floodplain wetlands as a separate category, as for example in the widely used system proposed in 1979 by the U.S. Fish and Wildlife Service. Vegetation- and hydrology-based classifications typically identify wetland classes that can be found both on flood plains and in non-flood plain wetlands. For example, within a typical flood plain there may be forested swamps, marshes with emergent herbaceous plants, shallow permanent lakes, and groundwater-fed fens. Seasonal or episodic inundation may be superimposed on hydrologic regimes that are controlled by local water inputs, or the inundation may be the only source that produces wetland conditions. Permanent water bodies, as delineated during periods of isolation from the river, may expand and become merged with inundated floodplain and adjacent lakes during inundation to form a contiguous flooded area. Very narrow flood plains may be considered to be riparian zones, and there is no consistent delineation between the use of these terms in the literature.

Images and Photos of Floodplain Environments

Remotely sensed images of contrasting types of floodplain environments are included in a companion entry on Flood plains (see **Flood Plains**). In addition, photos of many of these sites appear in the online Flood plain Photo Gallery (cite web site here).

Distribution and Extent

Floodplain wetlands tend to scale with the size of the parent river system with which they are associated, and the largest floodplain wetlands are found along the world's largest rivers, although these wetlands may be composites of riverine floodplains inundated by the parent river and its tributaries, and contiguous, poorly drained areas subject to inundation by locally derived rain and runoff. Geomorphological processes can result in exceptionally extensive flood plains relative to the discharge of the river, as for example in the Pantanal along the Paraguay River in Brazil, where neotectonic subsidence appears to have produced a vast sedimentary plain subject to seasonal inundation.

The largest flood plains have attracted academic attention, but flood plains along smaller rivers and streams add up to a sizable area as well. However as streams become smaller and their discharge regime more subject to short term variability, their flood plains are often inundated more irregularly, and as a result their flood plains differ markedly in ecological characteristics. As the duration of inundation decreases the floods become more of a disturbance that limits or excludes some plants and animals, rather than an ecological driver to which specialized elements of the biota can adapt.

Regimes of Flooding and Drying

Ecologists studying flood plains with predictable and protracted inundation have observed that many plants and animals are adapted to cope with and benefit from the seasonal inundation. In 1989 Wolfgang Junk and others synthesized the large body of work on the importance of inundation to articulate the Flood Pulse Concept. The Flood Pulse Concept holds that high species diversity and biological productivity of the overall river-floodplain ecosystem is explained by the seasonal inundation, which maintains a spatially and temporally variable environment with both aquatic and terrestrial characteristics. Much of the biological activity is centered on the flood plains, while the river channels provide critical interconnections among habitats and, at low water, aquatic refugia.

In addition to the frequency, duration, and amplitude of the flood pulse, its timing with respect to climatic seasonality determines its ecological roles. For example, flooding in the north temperate Mississippi River (USA) tends to occur in early Spring before the peak growing season, and flooding in the Mackenzie River delta (Canada) occurs in conjunction with ice breakup because of the northward flow direction of that river. Tropical and subtropical flood plains show the greatest biological responses because inundation occurs at warm temperatures.

Large flood plains can be inundated over vast areas, and in that case the existence and distribution of terrestrial refugia can become a limiting factor for populations of animals that cannot tolerate life in water. Larger terrestrial species of wildlife can abound in flood plains with ample refugia such as tree islands. Presumably these animals take advantage of the abundance of food on flood plains, and the reduced hunting pressure by humans can be important as well.

In some flood plains the characteristics of the isolation phase can be an important ecological driver in addition to the flood pulse of the inundation phase. Relatively small areas of permanent water can host large numbers of aquatic animals as the flooded area contracts. Suitable refugia for aquatic animals such as fishes, either in the river channel or in permanent floodplain water bodies, can enhance their populations. Depending on climate, the soil moisture may become limiting to plant growth. Wildfires are common where vegetation dies during dry periods, as for example in tropical savanna flood plains such as the Pantanal of Brazil and the Orinoco Llanos of Venezuela.

The seasonal alternation between soil saturation or inundation and soil moisture limitation can act to greatly limit the plants and animals that inhabit flood plains, although animals may migrate onto and off of the flood plain in response to changing conditions. Vertical migration of terrestrial invertebrates into forest canopies to escape flood waters has been documented in the Amazon flood plain.

Primary and Secondary Production

Floodplain wetlands commonly support high primary production, particularly in the case of tropical and semitropical floodplains where seasonal inundation often is prolonged and occurs at high temperatures.

Certain aquatic vascular plants (macrophytes) are superbly adapted to the seasonal inundation and variable water levels and attain high rates of primary productivity in spite of the constantly changing environment. Examples include the water hyacinth (*Eichhornia* spp.), water lettuce (*Pistia stratiotes*), and several grasses (e.g., *Paspalum* spp., *Echinochloa polystachya*). Many of these are native to tropical South America and have spread throughout the tropics and subtropics, causing problems in water bodies with artificially regulated water levels such as reservoirs on rivers.

Floodplain lakes can be rich in phytoplankton, particularly during periods of isolation or at least minimal through-flow of river water. In some cases algal blooms are stimulated by nutrient inputs during inundation. However, during inundation floodplain lakes often receive through-flowing river water, and this may reduce the water residence time to the point where plankton growth is suppressed by flushing (i.e., to less than a week or so). Floodplain lakes may become shallow enough over the interval between floods that sediments are resuspended by wind-induced turbulence, and then inorganic turbidity may restrict underwater light availability and limit algal growth.

Algae can be important to ecosystem-level primary production in floodplain water bodies, rivaling that of the more conspicuous floating emergent plants and floodplain forest. Stable isotope studies have shown that algae can contribute disproportionately to the support of aquatic food webs even though their biomass is small compared with that of vascular plants. In waters where flushing limits phytoplankton growth, attached algae may proliferate on submersed plant surfaces, or on sediments if the water is shallow. The importance of algae to aquatic consumers is thought to be explained by the greater nutritional value of algal cells relative to plant material containing more structural biopolymers like cellulose.

The existence of flood plains with natural flood regimes enhances the overall secondary productivity of the river-floodplain system, and this productivity extends to fisheries of cultural and economic importance. Riverine fishes include species that migrate seasonally between river channels and flood plains and others that are largely confined to floodplain waters. Water turbidity can be important in structuring fish species composition because visual feeders are limited to relatively clear waters while tactile and electrosensory feeders do well in turbid waters. When flood plains are no longer inundated in a natural fashion, or are isolated entirely from the river, riverine fish productivity and diversity tends to be diminished.

Biodiversity

Flood plains are often cited as ecosystems that harbor high biological diversity, in spite of the fact that the physical challenges imposed on the biota may well limit the suite of species that can survive and dominate in flood plains. Certainly the existence of floodplains enhances the biodiversity of a river system, and the high spatial heterogeneity typical of flood plains offers a wide range of habitats and niches. Tropical freshwater fishes, particularly in and around the Amazon basin of South America, are an especially diverse group in which many species have direct ties to floodplain environments. Flood plains can also be an important habitat for rare and endangered species that are not floodplain specialists, probably because flood plains often are less accessible to hunting, and because they may not be colonized or developed by people as easily as adjacent upland areas. Furthermore, the constant landscape change that is produced by rivers and flood plains can enhance terrestrial biodiversity by creating new areas for vegetation succession and by leaving a legacy of topographic features, soil formation, and soil drainage.

Biotic Adaptations

Many plants and animals display adaptations to seasonal inundation and, in some cases, to seasonal desiccation of floodplain environments. Most flood plains support some plants and animals that are also found on adjacent upland habitat and can persist in the flood plains despite inhospitable conditions at some times. Other floodplain species are found in wetlands in general, and still others are especially adapted to conditions on flood plains, or move between the flood plain and the river or permanent floodplain lakes for at least part of their life cycles.

Among plants, adaptations to life in floodplain habitats include rapid growth upon the arrival of flood waters, floating emergent growth habits that allow plants to rise and fall with changing water levels, and timing of flowering and seed production to take advantage of flooding for seed dispersal by water or aquatic animals including fishes. Among aquatic animals, adaptations include migration in and out of seasonally inundated areas or along an axis of inundation and drainage, timing of reproduction to match the flood pulse, and dormancy to survive dry periods. Birds, reptiles, and mammals may take advantage of the flood pulse through specific adaptations, but many species can do so opportunistically.

The high biological productivity of tropical flood plains, with much of the photosynthesis conducted by plants whose leaves are above the water surface, produces a high demand for dissolved oxygen beneath the water surface. Decomposition of organic matter and root respiration consume dissolved oxygen, and physical impediment of gas exchange by dense plant canopies reduces reaeration. Consequently, standing waters on vegetated flood plains are often depleted in oxygen to the point where its availability limits the species of aquatic life that can live there. This may be a feature common to other kinds of wetlands with high temperatures, water above the soil surface, and floating or emergent plants (oxygen depletion in water-saturated soils is characteristic of almost all kinds of wetlands).

In tropical flood plains, many aquatic animals including fishes display morphological or physiological adaptations to low oxygen availability. For example, some fish species can breathe air using the physiological equivalent of a lung, while others develop adaptations to facilitate the use of water in the thin surface layer that tends to be more oxygenated. These adaptations allow tropical floodplain fishes to thrive under conditions of low dissolved oxygen that would be lethal to many temperate fish species. However, despite these adaptations, depletion of dissolved oxygen limits the composition of the fish community and can cause fish kills in tropical floodplain waters.

Human Impacts on Floodplain Wetlands

Floodplain wetlands have been strongly altered throughout the world, often through hydrological modifications of either the parent river that inundates the flood plain or the flood plain itself. Control or exclusion of flooding has allowed colonization, agriculture, and sometimes urbanization of flood plains, as, for example, in the Yangtze (China) and lower Mississippi (USA) rivers. In some regions the flood plains are heavily populated by people even though they still undergo seasonal inundation, as, for example, in the delta of the Ganges and Brahmaputra rivers (India and Bangladesh), where much of the flood plain has been converted to rice paddies.

River regulation, mainly through construction of dams, has strongly impacted flood regimes of rivers across all spatial scales, and even many of the largest rivers of the world have been regulated to some degree. Impoundments tend to create permanently flooded reservoirs in place of seasonally flooded lands, and they usually alter the discharge regime and often the water quality and temperature well downstream. They can trap a large fraction of the suspended sediment load, leading to geomorphological destabilization of the river-floodplain system downriver of the dam. Dams operated for hydroelectric generation may impose highly unnatural, short term fluctuations in water levels, while those operated primarily for agricultural irrigation tend to change the seasonality of river flow in addition to removing water from the system.

Modification of river channels to facilitate navigation usually impacts flood plains by altering the relation between water levels and discharge; removal of barriers to navigation can enhance flow conveyance and diminish the backwater effect that produces overbank flooding. This can reduce the extent of wetland subject to inundation as well as the hydroperiod of floodplain inundation. Spoils from dredging the rivers are often placed directly on flood plains.

Flood plains have often been isolated from their parent rivers by construction of dikes, commonly with the goal of farming the land. Such land can be highly productive at first but may require costly measures to remove or control water. Over time, land subsidence, loss of fertility, and occasional incursion of flood waters can detract from agricultural sustainability. Dikes have also been constructed to maintain permanently flooded areas, sometimes for enhancing habitat for wildlife populations (e.g., waterfowl). These diked wetlands or lakes may be managed for variable water levels, but often not following a natural flood regime, and thus they are more akin to reservoirs.

The Florida Everglades (USA) provides a particularly well-studied example of how a strongly modified hydrological regime can produce negative impacts that extend to all levels of the floodplain wetland ecosystem. The hydrology of the region from the Kissimmee River through Lake Okeechobee and south across the Everglades to the southern end of the Florida peninsula originally functioned as a flood plain, even though there was no single parent river channel. During the wet season water would slowly travel from north to south across vast but shallow flooded areas. Throughout the 1900s, massive engineering works to regulate and divert water flow caused untold changes to the ecosystem, yielding a complex system of aqueducts and diked areas in which water levels have been managed for flood control and water supply. The rich soils of drained areas were developed for agriculture but are suffering from land subsidence as the organic soils oxidize and compact. A massive nutrient-enriched zone has been created downstream of the agricultural areas, changing the vegetation to favor dense stands of the cattails (*Typha* spp.). Invasive exotic species have proliferated, while native species such as wading birds that rely on a natural flood pulse have been reduced in numbers. Some coastal marine ecosystems have lost their freshwater inputs (i.e., Florida Bay) while others have experienced an unnaturally high amount of freshwater via diversion channels; both kinds of changes have strongly affected the biota. These problems have been widely acknowledged, and restoration of a more natural hydrological regime is currently being sought.

See also: Floods; Flood Plains; Riparian Zones.

Further Reading

Dudgeon D and Cressa C (eds.) (In press). *Tropical Stream Ecology*. New York: Elsevier-Academic Press.

Goulding M (1980) *The Fishes and the Forest: Explorations in Amazonian Natural History*. Berkeley: University of California Press.

Goulding M, Barthem R, and Ferreira E (2003) *The Smithsonian Atlas of the Amazon*. Washington, DC: Smithsonian Books.

Junk WJ, Bayley PB, and Sparks RE (1989) The flood-pulse concept in river-floodplain systems. In: Dodge DP (ed.) *Proceedings of the International Large River Symposium, Canada Special Publications Fisheries and Aquatic Sciences*, 106: 110–127.

Junk WJ (1997) *The Central Amazon Floodplain: Ecology of a Pulsing System. Ecological Studies*, Vol. 126. New York: Springer.

Lewis WM Jr, Hamilton SK, Lasi MA, Rodríguez M, and Saunders JF III (2000) Ecological determinism on the Orinoco floodplain. *BioScience* 50: 681–692.

Lodge TE (1994) *The Everglades Handbook*. Delray Beach, FL: St. Lucie Press.

Moss B (1998) *Ecology of Fresh Waters: Man and Medium, Past to Future*. 3rd edn. Oxford, UK: Blackwell Science.

Welcomme RL (1985) *River Fisheries*. Food and Agriculture Organization of the United Nations. Rome: FAO Fisheries Technical Paper 262.

Models of Ecological Processes in Riverine Ecosystems

J H Thorp, University of Kansas, Lawrence, KS, USA

Caveat Emptor

The admonition 'let the buyer beware' may seem a strange beginning for an encyclopedia article, but it is important for readers to realize from the outset that ecological models should be viewed as heuristic rather than final truths carved in stone. That is, scientists attempt to build malleable models of the real world which can, with some remolding, aid in our continued accumulation of relevant data and the never-ending search for truth. Sometimes, however, models become cracked or friable and need replacement, requiring scientists to progress to more realistic and accurate models. The problem arises when model creators and the greater scientific community fail to remember their heuristic nature and instead confuse models with reality. This false step can lead the field astray for decades. Keep this warning in mind as you consider the following summaries of a few pivotal models proposed over the last several decades.

With a few exceptions, the following discussion is limited to ecological concepts that are either original to lotic systems or extensively modified from counterparts in other ecosystems. This proviso limits the discussion, unfortunately, because relatively few comprehensive models have been created in stream ecology. This may reflect the nature of stream ecology as a habitat-driven rather than a theory-defined subdiscipline of ecology. All models described are primarily conceptual rather than mechanistic and quantitative, and they were mostly conceived with pristine ecosystems in mind, though some have applications to regulated rivers. The discussion proceeds generally from larger to smaller spatiotemporal scales.

Four Dimensions and Hierarchy Theory

Riverine ecosystems are considered energetically open with an unusually high degree of spatiotemporal variability. A complete understanding of how ecosystem structure and function are regulated requires knowledge of factors operating across four dimensions: longitudinal (upstream-downstream), lateral (main channel to true backwaters, and riverscape to floodscape), vertical (surface, or epigean habitats, downward into the subsurface hyporheic zone), and temporal (microseconds through geological and evolutionary time periods). The temporal dimension is an essential component of all dimensions through effects on expression of many factors, including flow rates, channel bed structure, and variability in material import from watersheds, to name a few. Spatiotemporal scales are integral to each of the four dimensions but to different degrees. Therefore, to understand patterns and processes, one must appreciate scale effects and link the appropriate scale to the phenomenon of interest. As the following discussion will reveal, some riverine models are more adept at this than others.

Hierarchy Theory

Hierarchy theory is especially appropriate when using multiple scientific disciplines to study spatially and temporally complex systems such as rivers. River ecosystems have multiple hierarchies (e.g., geomorphological, hydrological, and ecological), each interacting in disparate ways at different scales (**Figure 1**). A level, or holon, is a discrete unit of the level above and an agglomeration of discrete units from the level below, with higher levels operating on larger/longer spatiotemporal scales than lower levels. Successive levels act like filters or constraints on levels below, particularly the adjacent level. Lower organizational levels can also influence the structure and functioning of those at higher levels. This is influenced by boundary conditions between individual hierarchical levels. Activity rates within and between successive levels alter dynamics of the next higher level in the system via various mechanistic processes. Therefore, different hierarchical levels and scales need to be matched appropriately when seeking answers to specific questions, as illustrated in **Figure 1**. This applies to most ecological models.

Longitudinal Perspectives

The longitudinal complexity of rivers has been recognized for at least a century. Early attempts to cope with this complexity involved the division of riverine ecosystems into specific, longitudinally ordered zones. This model was widely accepted for 80% of the twentieth century but was rapidly discarded in most countries after strong criticisms developed with publication of the River Continuum Concept (RCC) in 1980. The RCC portrays riverine systems as intergrading, linear networks from headwaters to

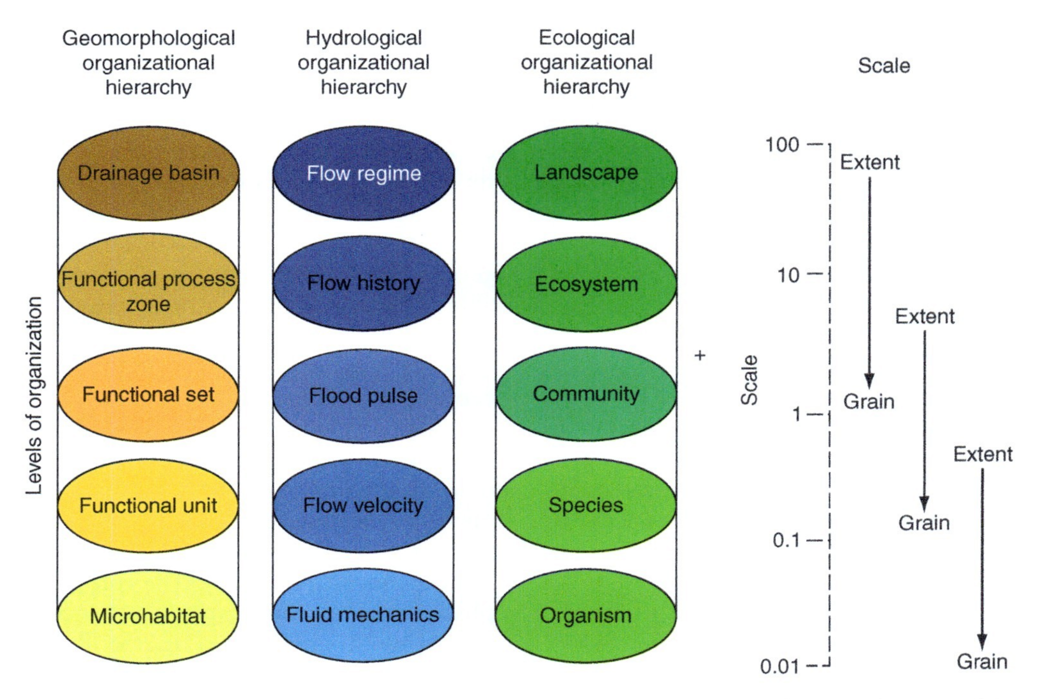

Figure 1 Organizational hierarchies in river science. To use this framework, one must first define the relevant spatiotemporal dimension for the study or question. Scales for each hierarchy are then determined to allow the appropriate levels of organization to be linked. The scale at the right demonstrates that linking levels across the three hierarchies may be vertical depending on the nature of the question. Adapted from Figure 3.2 in Thorp *et al.* (2008).

the mouths of great rivers. The RCC (described in the next section) is still the dominant theory employed intentionally or de facto by riverine ecologists and environmental scientists/managers despite strong criticism over the last decade or two. The Network Dynamics Hypothesis, which emphasizes the ecological importance of geomorphic transitions at tributary junctions, was proposed as a partial replacement for the RCC. A more marked change is represented by a movement of aquatic ecologists and fluvial geomorphologists to replace existing concepts with models based on hydrogeomorphic perspectives on riverine structure and functioning.

River Continuum Concept

Central to the RCC are the linked concepts that: (i) a riverine ecosystem is characterized by a continuous gradient of physical conditions from headwaters to a river's mouth; and (ii) this longitudinal gradient elicits a series of responses within the constituent populations resulting in a continuum of biotic adjustments to mean conditions and consistent patterns of loading, transport, use, and storage of organic matter along the length of a river. For simplicity sake, the model assumes an uninterrupted gradient of physical conditions in natural rivers where physical conditions gradually alter as one moves downstream. The authors noted that regional and local deviations from RCC predictions occur as a result of variations in the influence of watershed climate and geology, riparian conditions, tributaries, location-specific lithology and geomorphology, and floodplain inputs. However, these deviations from the clinal nature of rivers are considered exceptions to the fundamental portrayal of rivers. In more recent models, emphasizing the importance of hydrogeomorphic patches, these exceptions are now considered the rule.

The RCC includes a large array of stimulating ideas, but ecologists have focused only on central themes related to downstream changes in food sources and functional feeding groups. From perspectives on how physical conditions should alter the relative and absolute input of allochthonous carbon and generation of autochthonous organic matter, the RCC postulates a predictable, unidirectional change in functional feeding groups from small streams to large rivers. These categories and their relationship to a predicted continuum were the subject of many research studies in the 1980s and are still widely cited in general biology and ecology textbooks even though their predictions have rarely been confirmed and they ignored lateral components of the riverine landscape.

The greatest value of the RCC is probably that it made scientists think about riverine ecosystems in a different light, which eventually spurred development of many other conceptual models, such as ones related to hyporheic corridors, the importance of flood pulses, and the role of instream primary production. It is, however, a prime example of how the scientific community can become complacent in challenging models, thereby allowing them to gain access to our general ecological literature and even adoption for riverine management without caveats on their heuristic nature.

Serial Discontinuity Concept

Following shortly on the heels of the RCC was the Serial Discontinuity Concept (SDC). The continuous change in physicochemical conditions and biological characteristics stipulated in the RCC is hypothesized in the SDC to be interrupted in predictable ways by the presence of dams. Discontinuity distance (i.e., the amount of disruption in the predicted pattern of the RCC) is defined as the longitudinal shift in a given parameter in a positive (downstream shift), negative (upstream reset), or near zero manner. Somewhat similar types of changes have also been attributed to tributaries.

Network Dynamics Hypothesis

Rivers can also be evaluated using fractals, multifractals, and network approaches. Network theory is a relatively recent development by statistical physicists that has been applied to linkages among tributaries in a downstream progression. It views riverine ecosystems as possessing a scale-free architecture with rivers as nodes and tributary confluences as links. In the Network Dynamics Hypothesis (NDH), deviations from the expected mean state of conditions within a river channel are postulated to occur in response to network geometry, and tributary junctions serve as ecological hotspots. The NDH focuses on branching patterns of rivers and mostly ignores changes to hydrogeomorphic characteristics other than spatially limited increases in habitat complexity.

The use of network theory to riverine ecosystems is in its infancy, but many potential applications having a spatial context may be revealed in the next decade. Although its use for explaining biotic communities and ecosystem processes within a given area is probably limited, it may prove useful in examining processes occurring among river locations (e.g., dispersal studies of exotic species and fish metapopulation dynamics).

Hydrogeomorphic Patches and the Riverine Ecosystem Synthesis

In contrast to the more linear view of rivers embodied in the three previous models (RCC, SDC, and NDH), there is a growing trend to view rivers as composed of large hydrogeomorphic patches. Proponents of these models argue that local conditions are vastly more important in explaining ecological patterns and processes than the simple distance downstream from the headwaters because a disjunct pattern of large patches seems more characteristic of rivers than a gradual physical cline. Consequently, the local hydrogeomorphic nature of the ecosystem is less predictable than clinal models have proposed, especially when comparing among rivers present within and between ecoregions. In the Link Discontinuity Concept, these hydrogeomorphic patterns are thought to resemble a punctuated sawtooth pattern which is highly susceptible to tributary influences. The importance of stream segment structure and tributary catchments to ecological structure and function is also described in a spatiotemporal framework known as the Catchment Hierarchy. The role of local geomorphic conditions, landscape disturbances, and climate variability is emphasized in the Process Domains Concept. Others have described rivers as having a dominant, discontinuous nature characterized by a longitudinal series of alternating stream segments having different geomorphic structures punctuated by tributary confluences. From this perspective, a biotic community in one segment would not necessarily be more similar in diversity and function to an adjacent segment than it would be to assemblages with more similar geomorphic features but located farther upstream or downstream – a view which contrasts sharply with predictions of clinal models like the RCC.

The most dramatic and comprehensive departure from a clinal view of rivers is embodied in the heuristic Riverine Ecosystem Synthesis (RES). As a true synthesis, it contributes new material to the rich tradition of published papers since 1980, and describes the structure and function of riverine landscapes from headwaters to great rivers and from main channels to lateral floodscapes. It contains three primary elements: (i) a hierarchical, physical model describing the longitudinal organization of riverine ecosystems into large, repeatable hydrogeomorphic patches, termed Functional Process Zones (FPZs); (ii) a research framework for studying riverine landscapes based partially on the hierarchical patch dynamics model; and (iii) an expandable set of hypotheses linking the physical model to species distributions though landscape processes. The core physical model portrays rivers as composed of FPZs at the valley-to-reach scale which differ in geomorphic structure and their hydrologic patterns. FPZs can repeat along a longitudinal dimension and are only partially predictable in position (**Figure 2**). Differences in the hydrogeomorphic nature of the

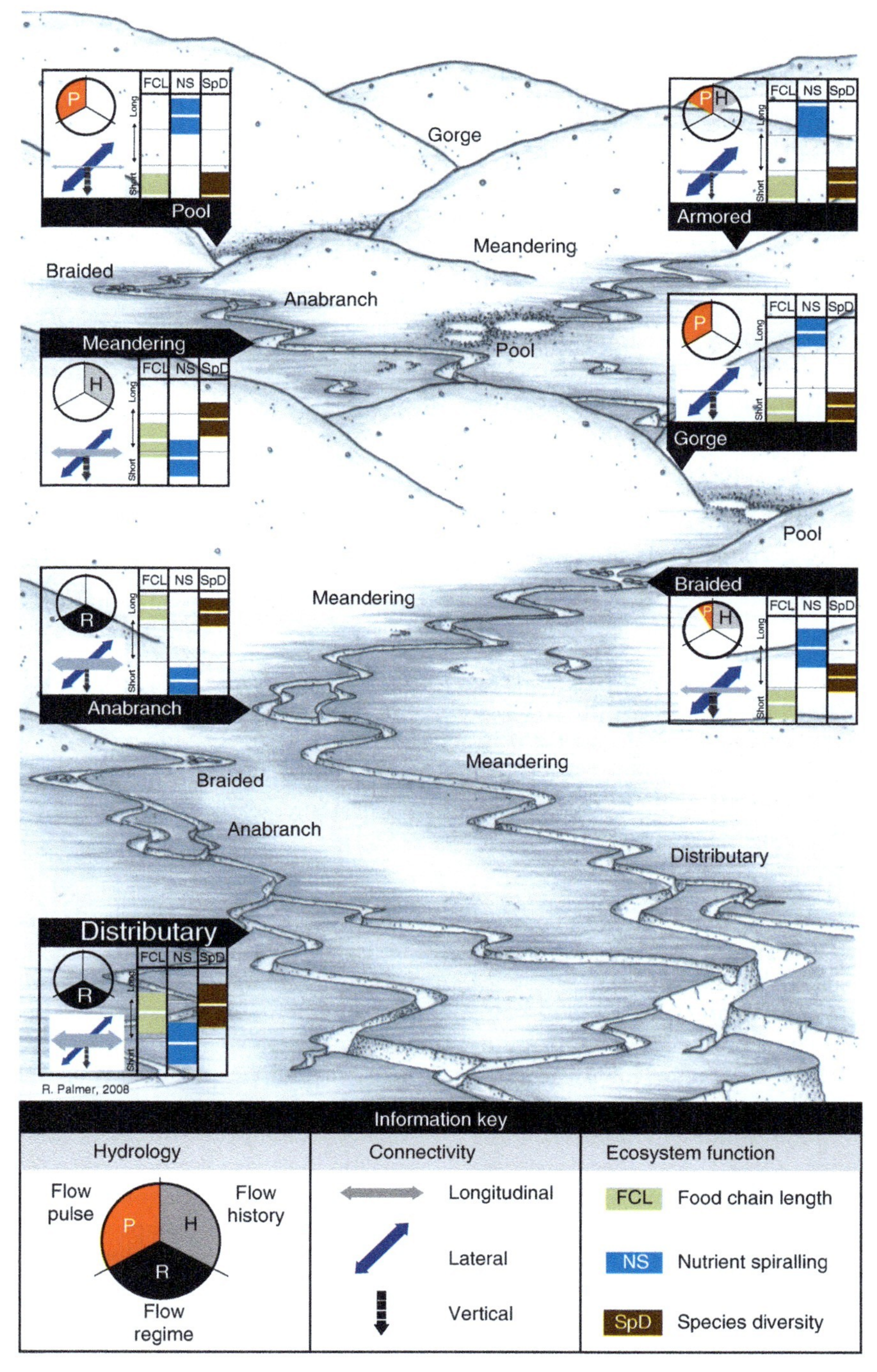

Figure 2 A conceptual riverine landscape is shown depicting various functional process zones (FPZs) and their possible arrangement in the longitudinal dimension. Not all FPZs and their possible spatial arrangements are shown. Note that FPZs are repeatable and only partially predictable in location. Information contained in the boxes next to each FPZ illustrate the following. The relative importance of the hydrological character is depicted in the pie charts in terms of flow regime (R), flow history (H), and flood pulse (P) for that FPZ. The relative strength of connectivity in the longitudinal, lateral, and vertical dimensions are shown by the various arrows: vertical connections by the dashed arrow, lateral connections between the riverscape and floodscape by the gray arrow, and longitudinal connectivity between upstream to downstream locations by the horizontal blue arrow. The strength of the hydrological and material exchange rates for ground and surface water connectivity are based on the size of each arrow for each FPZ. Food chain length (FCL), nutrient spiral length (NS), and species diversity (SpD) are given for each FPZ. These ecological measures are scaled from long to short, with this translated as low to high for species diversity. The light bar within each box is the expected median, with the shading estimating the range of conditions. Adapted from **Figure 1.1** in Thorp *et al.* (2008).

FPZ have profound impacts on community structure and ecosystem functioning (see some examples in **Figure 2**). Unlike many other lotic models, the RES is applicable both to fundamental questions in pristine environments and to environmental challenges related to management, monitoring, and rehabilitation of altered systems.

MacroHabitat Structure

It is generally accepted by ecologists that the principal factor regulating lotic communities is a hierarchically-scaled Habitat Template composed primarily of the habitat structure and water flow patterns. Important features of the habitat structure are its physicochemical nature, diversity of habitats within the riverine landscape along all spatial dimensions, and variability and predictability of habitat structure over time including access to the floodscape during critical life history periods. Another component of habitat is prevailing current velocity and access to habitats of low to moderate flow. The importance of these slackwaters to diversity and productivity of animals and plants is emphasized in the Inshore Retention Concept. The amount and variability of water flow is now considered a major ecosystem driver, as described below.

Natural Flow Regime

Early applied and fundamental models of riverine ecosystems stressed the ecological importance of minimum flows, but with publication of the Flood Pulse Concept, the critical role of floods became apparent. Research on arid ecoregions later revealed the often equally important contribution of droughts. As a partial consequence of these studies, a model on the Natural Flow Regime (NFR) was developed to emphasize that naturally dynamic hydrological patterns are required to maintain the evolved biocomplexity of riverine ecosystems. As explained in the Riverine Ecosystem Synthesis, individual hydrogeomorphic patches within a riverine ecosystem vary in the nature of their typical natural flow pattern according to different time scales. These can be divided principally into flow regime (greater than 100 years), flow history (1–100 years), and flow or flood pulse (less than one year). It is not surprising, therefore, that different ecological processes respond in a diverse manner to the temporal scale of these flow patterns. Observations on the importance of a natural flow regime have begun to make substantial impacts on government policies controlling the timing and amount of water released from dams.

Food Webs

The nature of lotic food webs was a major research initiative stimulated by the River Continuum Concept (RCC). The RCC postulates a predictable, unidirectional change in functional feeding groups (scraping herbivores, predators, etc.) from small streams to large rivers. The model also indicates that the main source of organic matter fueling food webs differs along a longitudinal dimension, with headwaters dependent on allochthonous (in this case from land) coarse POM (particle organic matter; e.g., fallen leaves), small to medium shallow rivers relying on autochthonous production (mostly benthic algae with some contributions from aquatic vascular plants), and large rivers dependent on fine POM leaking from upstream foodwebs. In 1989, this was modified slightly for large floodplain rivers after publication of the Flood Pulse Concept (FPC; see below). The Riverine Productivity Model (RPM; see below) challenged both the RCC and FPC, arguing that autochthonous organic matter is most important to large river food webs. Unfortunately, all three models fail to effectively incorporate the diversity of a river's hydrogeomorphic structure into analyses of lotic food webs except whether rivers are characterized by constricted channels or large floodplains. Only the RES explicitly links organic sources and food web complexity to a river's hydrogeomorphic structure (see below).

Flood Pulse Concept

Publication of the Flood Pulse Concept (FPC) in 1989 caused a dramatic shift in studies of riverine ecosystems from a focus almost strictly on the main channel during nonflood periods to research that better incorporated the breadth of spatial (riverscape and floodscape) and temporal conditions. The model was originally based on tropical floodplain rivers like the Amazon but was extrapolated to temperate rivers like the Mississippi. Two major components of the original FPC are a food web component and a physical model of the importance of flooding to ecological processes. The physical portrayal of the ecological importance of both seasonal and unpredictable floods influenced other important models, such as the Natural Flow Regime (NFR; discussed previously), and has had a very crucial impact on policies for regulation of water discharge from dams. The food web model in the FPC, however, has proved less lasting in importance. It argues that food webs in large rivers are not fueled by upstream leakage of fine POM but are instead supported energetically by the export of recalcitrant organic matter from decaying terrestrial detritus in submerged floodplains and aquatic macrophytes growing in floodscape lakes. It now appears that floodscape food webs derive much of their nutrition from labile phytoplankton and algae attached to aquatic plants. Moreover, the export rate of organic matter from the floodscape is insignificant in a number of major tropical rivers.

Riverine Productivity Model

Unlike the RCC and FPC, the Riverine Productivity Model (RPM) emphasizes the importance of autochthonous organic matter in fueling food webs in large rivers. The RPM was initially developed for constricted channel rivers but was later expanded to the riverscape of all large rivers. The central idea is that phytoplankton and benthic algae comprise the greatest source of organic matter for riverine food webs because their high lability (easily assimilated) and productivity more than compensate for the sometimes greater absolute abundance of recalcitrant (difficult to digest) organic matter entering large rivers from either upstream leakage or import from the floodscape. This concept was challenged based on the fact that large rivers tend to be heterotrophic for major portions of the year; that is, respiration exceeds production, thereby requiring an external input of organic fuel. The answer to this heterotrophy paradox could be the presence of two somewhat distinct food web pathways. An algal-grazer pathway dependent on autochthonous carbon may support most animal production, whereas a microbial–viral loop (or spiral) dependent on both allochthonous and autochthonous carbon may fuel most respiration by bacteria, protists, some rotifers, etc.

Food Web Complexity and Food Chain Length

Models of food web complexity and the related food chain length (FCL) in aquatic systems are almost exclusively based on theories developed for other environments. Those terrestrial models have recently concluded that ecosystem size (a measure of habitat complexity and opportunities for colonization by new species) and possibly disturbance are more important than primary productivity in accounting for differences among systems in FCL. Applying this to riverine ecosystems is challenging because rivers are naturally more variable in spatial and temporal dimensions and technically only one ecosystem is involved, no matter how many tributaries occur. However, emerging evidence suggests that FCL in rivers is related to hydrologic connectivity within the riverscape (peaking at intermediate values) and between the riverscape and floodscape. Furthermore, according to the RES, the hydrogeomorphic complexity of functional process zones significantly influences the complexity of riverine food webs.

Regulation of Community Structure

Equilibrial (Deterministic) vs Non-Equilbrial (Stochastic) States

Most models pertaining to regulation of species diversity and density in rivers are derived with little modification from other ecosystems. Until about the last decade and one-half of the twentieth century, riverine ecosystems were studied as if they were primarily in equilibrium, despite their highly dynamic hydrogeomorphic patterns. This initial perspective promoted numerous descriptive studies and a few experiments on competition and predation. The dominant role of stochastic factors and hierarchical patch dynamics was accepted by many lotic scientists starting in the mid-1980s and continues today. With this shift, studies of competition and predation declined, along with many fundamental studies of lotic communities. As hypothesized in the RES, both deterministic and stochastic factors contribute significantly to ecological regulation of communities, but their relative importance is scale- and habitat-dependent and is affected by the organism's life history characteristics versus time scale and degree of environmental fluctuations. However, stochastic factors are considered more important in general and throughout the riverine ecosystem. The role of hydrogeomorphic fluctuations in controlling community structure and ecosystem processes still needs considerable study.

Competition and Predation Models

Descriptive and experimental research on river ecosystems as equilibrial systems has primarily concerned control of species richness, species diversity, absolute density, and relative abundance by bottom-up (resource limits) and top-down factors (predator- and parasite–prey). Until about the mid-1970s, most research concerned competition, especially interference (aggressive or agonistic) competition for space on rocks and other hard surfaces in small streams where the winner had better access to food (attached algae or drifting prey and other seston). Research on exploitative (resource) competition rarely proved definitive unless an interference component was present. The following decade saw a major shift by freshwater ecologists toward predator–prey studies, following the popular rise in marine predator–prey field experiments. Most research purporting significant predator control was limited to seasonal changes in absolute or relative abundance of prey, with little evidence of multi-year effects. Prominent exceptions dealt with introduction of exotic species. This led to the concept that freshwater predators had two distinct roles. The first element of this model was that abundant predators in a community may suppress prey densities but rarely regulate them (which often requires simpler food webs and some form of switching behavior). The second component was that native predators rarely control species diversity in lotic ecosystems over ecological time periods (shorter than evolutionary periods) unless the predators are recent immigrants, such as invasive or introduced species. An explanation for this lack of control is that many freshwater food webs are highly complex and replete with omnivores, thus complicating the task of predator control. The concept of trophic cascades (a form of alternating, top-down control), which was developed initially for marine communities and lake plankton, has had little impact on river studies because of their food web complexity and the dearth of studies on simpler riverine plankton webs.

Intermediate Disturbance Hypothesis

The Intermediate Disturbance Hypothesis (IDH) has been applied to most ecosystems, including many studies of lotic habitats, but it has rarely been rigorously tested or throughly justified in freshwater systems. The IDH has been used to predict maximum values at intermediate levels of some disturbance for many dependent variables, including species richness, species evenness, and even food chain length. It has been applied to studies of both equilibrial (predator–prey/competition) and stochastic processes (related mostly to flow variability affecting actions like rocks flipping in a stream). A typical curve is hypothesized to be hump-shaped when the dependent variable is on the *y*-axis and disturbance range is on the *x*-axis. A peak value at intermediate disturbance can occur for multiple reasons. For example, some data suggest maximum species richness and food chain length at intermediate values of hydrologic connectivity (and current velocities) within the riverscape. This could occur, for example, because the densities of most species peak at that intermediate value. Alternatively, this may just represent the convergence point of a declining density curve for species more adapted to fast waters and a rising curve for species preferring more static conditions.

Hierarchical Patch Dynamics

The related concepts of patch dynamics and hierarchical patch dynamics (HPD) have received wide theoretical acceptance in riverine ecosystems during the late 1980s but are of rare empirical attention to date. It is recognized that at multiple hierarchical scales a quasi-equilibrial state can exist because patch conditions change at disparate rates in different places within a larger habitat, thereby promoting greater species richness. For example, a species of benthic herbivorous insect may coexist with another invertebrate that is its superior competitor for space because it can more rapidly colonize the bare rocks appearing after a stream flood has overturned some of the stones. Other aspects of this potentially very important model have yet to be exploited in riverine ecosystems to any significant degree, but this may change with a shift toward interdisciplinary research in river science involving, stream ecology, landscape ecology, and fluvial geomorphology. Details of the HPD model and applications to riverine ecosystems are described in a book on the Riverine Ecosystem Synthesis (see Further Reading).

Functional Processes in Riverine Ecosystems

Many functional processes, such as net ecosystem metabolism, are studied extensively in lotic ecosystems from a mechanistic approach, but very few general models have been developed or even modified from terrestrial ecosystems. A notable example pertains to nutrient spiraling. In a simple sense, this process is roughly comparable to the nutrient cycling found in lakes with the addition of a downstream component produced by water currents. As a visual image, this resembles the popular slinky toy. As water velocity increases, the spirals are stretched and the time and distance for a molecule to go between abiotic dissolved and biotic particulate compartments increase. It is not as simple as that, of course, because the spirals are strongly affected by sediment storage (especially if transported to the floodscape) and to a lesser extent by retention within longer lived organisms such as fish. A somewhat similar process has been proposed for the movement of genetic-level information, where information distance measures the influence of upstream genes (e.g., bacterial) on downstream processes such as catabolism of specific organic molecules by the microbial biofilm. The role in ecosystem functional processes played by hydrogeomorphic complexity across the entire riverine landscape has yet to be adequately addressed and could substantially change model predictions.

Glossary

Floodscape – The aquatic and terrestrial components of the riverine landscape that are connected to the riverscape only when the river stage exceeds bankfull (flood stage). These include the terrestrial floodplain (including components of the riparian zone not in the riverscape) and floodplain water bodies, such as floodplain lakes, wetlands, and isolated channels (e.g., oxbows and anabranches).

Functional Process Zone (FPZ) – A fluvial geomorphic unit between a valley and a reach.

Riverine Landscape – The continually or periodically wetted components of a river consisting of the riverscape and floodscape.

Riverscape – The aquatic and ephemeral terrestrial elements of a river located between the most widely separated banks (commonly referred to as the bankfull channel or active channel) that enclose water below floodstage. These include the main channel, various smaller channels, slackwaters, bars, and ephemeral islands.

Further Reading

Allan JD and Castillo MM (2007) *Stream Ecology: Structure and Function of Running Waters*, 2nd edn. New York: Springer.

Benda L, Poff LR, Miller DT, Reeves G, Pollock M, and Pess G (2004) Network dynamics hypothesis: Spatial and temporal organization of physical heterogeneity in rivers. *BioScience* 54: 413–427.

Connell JH (1978) Diversity in tropical rainforests and coral reefs. *Science* 199: 1302–1310.

Junk WJ, Bayley PB, and Sparks RE (1989) The flood-pulse concept in river-floodplain systems. In: Dodge DP (ed.) *Proceedings of The International Large River Symposium (LARS)*, Honey Harbor, Ontario, Canada, September 14–21, 1986, pp. 110–127. Canadian Special Publication in Fisheries and Aquatic Sciences 106.

Montgomery DR (1999) Process domains and the river continuum concept. *Journal of the American Water Resources Association* 35: 397–410.

Poff NL, Allan JD, Bain MB, Karr JR, Prestegaard KL, Richter BD, Sparks RE, and Stromberg JC (1997) The natural flow regime: A paradigm for river conservation and restoration. *BioScience* 47: 769–784.

Poole GC (2002) Fluvial landscape ecology: Addressing uniqueness within the river discontinuum. *Freshwater Biology* 47: 641–666.

Rice SP, Greenwood MT, and Joyce CB (2001) Tributaries, sediment sources, and the longitudinal organization of macroinvertebrate fauna along river systems. *Canadian Journal of Fisheries and Aquatic Sciences* 58: 824–840.

Schiemer F, Keckeis H, Reckendorfer W, and Winkler G (2001) The 'inshore retention concept' and its significance for large rivers. *Archiv für Hydrobiologie* 12(Suppl. 135): 509–516.

Thorp JH (1986) Two distinct roles for predators in freshwater assemblages. *Oikos* 47: 75–82.

Thorp JH, Thoms MC, and Delong MD (2008) *The Riverine Ecosystem Synthesis*. San Diego, CA: Elsevier.

Thorp JH and Delong MD (2002) Dominance of autochthonous autotrophic carbon in food webs of heterotrophic rivers? *Oikos* 96: 543–550.

Townsend CR (1996) Concepts in river ecology: Pattern and process in the catchment hierarchy. *Archiv für Hydrobiologie* 10 (Suppl. 113): 3–21.

Vannote RL, Minshall GW, Cummins KW, Sedell JR, and Cushing CE (1980) The river continuum concept. *Canadian Journal of Fisheries and Aquatic Sciences* 37: 130–137.

Wu J and Loucks OL (1995) From balance of nature to hierarchical patch dynamics: A paradigm shift in ecology. *Quarterly Review of Biology* 70: 439–466.

HUMAN IMPACTS ON STREAMS AND RIVERS

Contents

Deforestation and Nutrient Loading to Fresh Waters

M C Feller, University of British Columbia, Vancouver, BC, Canada

Introduction

Nutrient cycling within forest ecosystems involves nutrient uptake and retention by biota, which retards nutrient movement to fresh waters. Deforestation, or killing of forest vegetation, initially disrupts this uptake and retention resulting in altered nutrient fluxes to fresh waters. These fluxes are in both dissolved and particulate form. Dissolved fluxes dominate for N, S, C, K, Na, Mg, and Ca but particulate fluxes tend to dominate for P, and can also be important for N (**Table 1**). Fresh water nutrient fluxes have decreased or increased to variable extents, or have remained unaffected by deforestation (**Table 2**). Their response to deforestation depends on the nutrient and the effects of deforestation on the factors controlling freshwater chemistry and particulates.

Factors Controlling Nutrient Loading in Fresh Waters

Dissolved Nutrients

The following factors are the most important determinants of chemistry:

1. geological weathering;
2. atmosphere precipitation and climate, including (1) precipitation chemistry, (2) stream discharge, and (3) temperature;
3. terrestrial biological processes, including (1) nutrient uptake, (2) nutrient transformations, and (3) production of soluble chemicals;
4. physical–chemical reactions in the soil; and
5. physical, chemical, and biological processes within aquatic ecosystems, including the physical–chemical processes of (1) ion exchange, (2) oxidation–reduction, (3) evaporation–crystallization processes, (4) pH-induced chemical transformations, and the biological processes of (5) chemical uptake and (6) microbial transformations.

Due to the complexity of these factors, nutrient loading in streams draining nearby, apparently similar watersheds can vary from one watershed to the next as a result of small changes in geology, soil, streambed materials, or stream shading, for example, none of which might be visually obvious. Freshwater nutrient fluxes may also vary temporally from one year to the next as a result of changes in weather or precipitation chemistry, for example. Consequently, accurate determination of the effect of deforestation on freshwater nutrient loading requires sampling of a control watershed which remains undisturbed, as well as a deforested watershed prior to as well as after deforestation, as occurred in the studies used in **Table 2**. Sampling of a watershed only after deforestation has occurred and comparing its nutrient loading to that of an undisturbed watershed or sampling of only a deforested watershed without a control, will not allow a conclusive assessment of the effects of deforestation on freshwater nutrients.

Temporal variations in stream nutrient concentrations, which vary with discharge, require frequent sampling, at least once every 1–2 weeks, to quantify the effects of deforestation on nutrient loading. Such frequent sampling is less necessary for lakes, whose chemistry changes less rapidly than that of streams.

Nutrients in Particulate Form

The most important factors controlling particulates in freshwater are:

1. watershed topography, including (1) slope and (2) roughness or uniformity,

Table 1 Dissolved nutrient load as a percentage of the total nutrient load carried by streams flowing from essentially undisturbed watersheds, and the influence of disturbance on this percentage

Area	*Watershed size (ha)*	*N*	*P*	*S*	*C*	*K*	*Na*	*Mg*	*Ca*	*Source*
A. Undisturbed streams										
North America										
Wilson and Blossom Rivers, SE Alaska	18 100–29 400	–	–	–	>90	–	–	–	–	20
Beaver Creek, Quebec	1 800	82	–	–	–	–	–	–	–	16
Lake Memphremagog area, Quebec	200	–	38	–	–	–	–	–	–	17,18
Experimental Lakes area, Ontario	<1–7 (4 watersheds)	82–88	44–52	–	94–97	–	–	–	–	1
	12–170 (3 watersheds)	84–90	64–72	–	–	–	–	–	–	3
H.J. Andrews Experimental Forest, Oregon	10	78	–	–	–	–	–	–	–	21
	10	–	–	–	–	99	–	99	98	8
	21	71	81	–	–	–	–	–	–	12
Green Lakes Area, Colorado	8	88	–	–	–	–	–	–	–	23
Ward Creek, California	526	–	16–32	–	–	–	–	–	–	9
Nineteen streams in United States	1 800–85 000	75 (range 34–99)	–	–	–	–	–	–	–	11
Hubbard Brook, New Hampshire	13	96	43	99	–	74	96	93	98	5
	22.5	94	26	100	71	67	91	88	96	4
	130	–	11		–	–	–	–	–	15
Coweeta, North Carolina	13–14 (2 watersheds)	–	–	–	–	100	–	–	98	22
Lexington, Tennessee	<1–1 (8 watersheds)	70	31	–	–	–	–	–	–	14
Coffeeville, Mississippi	2–3 (5 watersheds)	53 (range 50–59)	29 (range 24–36)	–	–	–	–	–	–	7,19
Europe										
Rivers throughout Finland	?									
	15 rivers	90	40	–	94	–	–	–	–	13
Rest of World										
Caura R., Venezuela	4 750 000	64	49	–	–	–	–	–	–	10
BuKit Tarek, Malaysia	3	66	80	–	–	86	99	29	36	24
Ballance Stream, New Zealand	9	75	48	–	–	–	–	–	–	2
Purukohukohu basin, New Zealand	2 watersheds	97–100	14–38	–	–	–	–	–	–	6

B. Effects of disturbance										
Hubbard Brook, New Hampshire										
Undisturbed	13.2	96	43	99	–	74	96	93	98	5
After cutting + herbicide		100	10	97		88	87	91	98	
Hubbard Brook, New Hampshire										
Undisturbed	22.5	94	26	100	71	67	91	88	96	4
After debris jam removal		74	5	99	34	17	62	53	80	
Experimental Lakes area, Ontario				–	–	–	–	–	–	3
Undisturbed	12–170	84–87	64–65							
After windstorm	(2 watersheds)	74–87	48–63							

Sources
1. Allan CJ, Roulet NT, and Hill AR (1993) The biogeochemistry of pristine, headwater Precambrian shield watersheds: an analysis of material transport within a heterogeneous landscape. *Biogeochemistry* 22: 37–79.
2. Bargh BJ (1977) Output of water, suspended sediment and phosphorus and nitrogen forms from a small forested catchment. *New Zealand Journal of Forest Science* 7: 162–171.
3. Bayley SE, Schindler DW, Beaty KG, Parker BR, and Stainton MP (1992) Effects of multiple fires on nutrient yields from streams draining boreal forest and fen watersheds: nitrogen and phosphorus. *Canadian Journal of Fisheries and Aquatic Sciences* 49: 584–596.
4. Bilby RE (1981) Role of organic debris jams in regulating the export of dissolved and particulate matter from a forested watershed. *Ecology* 62: 1234–1243.
5. Bormann FH, Likens GE, Siccama TG, Pierce RS, and Eaton JS (1974) The export of nutrients and recovery of stable conditions following deforestation at Hubbard Brook. *Ecological Monographs* 44, 255–277.
6. Cooper AB and Thomsen CE (1988) Nitrogen and phosphorus in streamwaters from adjacent pasture, pine and native forest catchments. *New Zealand Journal of Marine and Freshwater Research* 22: 279–291.
7. Duffy PD, Schreiber JD, McClurkin DC, and McDowell LL (1978) Aqueous- and sediment-phase phosphorus yields from five southern pine watersheds. *Journal of Environmental Quality* 7: 45–50.
8. Fredriksen RL (1971) Comparative chemical water quality – natural and disturbed streams following logging and slashburning. In Krygier JT, and Hall JD (eds.) *Forest Land Uses and Stream Environment*, pp. 125–138. Corvallis: Oregon State University.
9. Leonard RL, Kaplan LA, Elder JF, Coats RN, and Goldman CR (1979) Nutrient transport in surface runoff from a subalpine watershed, Lake Tahoe basin, California, *Ecological Monographs* 49: 28–310.
10. Lewis WM, Jr. (1986) Nitrogen and phosphorus runoff losses from a nutrient poor tropical moist forest. *Ecology* 67: 1275–1282.
11. Lewis WM, Jr. (2002) Yield of nitrogen from minimally disturbed watersheds of the United States. *Biogeochemistry* 57/58: 375–385.
12. Martin CW and Harr RD (1989) Logging of mature Douglas-fir in western Oregon has little effect on nutrient output budgets. *Canadian Journal of Forest Research* 19: 35–43.
13. Mattsson T, Kortelainen P, and Räike A (2005) Export of DOM from boreal catchments: impacts of land use cover and climate. *Biogeochemistry* 76: 373–394.
14. McClurkin DC, Duffy PD, Ursic SJ, and Nelson NS (1985) Water quality effects of clearcutting upper coastal plain loblolly pine plantations. *Journal of Environmental Quality* 14: 329–332.
15. Meyer JL and Likens GE (1979) Transport and transformation of phosphorus in a forest stream ecosystem. *Ecology* 60: 1255–1269.
16. Naiman RJ and Melillo JM (1984) Nitrogen budget of a subarctic stream altered by beaver (*Castor canadensis*). *Oecologia* 62: 150–155.
17. Prairie YT and Kalff J (1988) Dissolved phosphorus dynamics in headwater streams. *Canadian Journal of Fisheries and Aquatic Sciences* 45: 200–209.
18. Prairie YT and Kalff J (1988) Particulate phosphorus dynamics in headwater streams. *Canadian Journal of Fisheries and Aquatic Sciences* 45: 210–215.
19. Schreiber JD, Duffy PD, and McClurkin DC (1980) Aqueous- and sediment-phase nitrogen yields from five southern pine watersheds. *Soil Science Society of America Journal* 44: 401–407.
20. Sugai SF and Burrell DC (1984) Transport of dissolved organic carbon, nutrients, and trace metals from the Wilson and Blossom rivers to Smeaton Bay, southeast Alaska. *Canadian Journal of Fisheries and Aquatic Sciences* 41: 180–190.
21. Triska FJ, Sedell JR, Cromack K, Jr., Gregory SV, and McCorison FM (1984) Nitrogen budget for a small coniferous forest stream. *Ecological Monographs* 54: 119–40.
22. Webster JR and Patten BC (1979) Effects of watershed perturbation on stream potassium and calcium dynamics, *Ecological Monographs* 49: 51–72.
23. Williams MW, Hood E, and Caine N (2001) Role of organic nitrogen in the nitrogen cycle of a high-elevation catchment, Colorado Front Range. *Water Resources Research* 37: 2569–2581.
24. Yusop Z, Douglas I, and Nik AR (2006) Export of dissolved and undissolved nutrients from forested catchments in Peninsula Malaysia. *Forest Ecology and Management* 224: 26–44.

Table 2 Estimated effect of deforestation on the flux (kg ha^{-1} year^{-1}) of nutrients dissolved in streamwater using studies which involve control and deforested watersheds for periods both before and after deforestation

Area	Vegetation[a]	Percent[b] of vegetation removed	Type of disturbance[c]	Watershed size[d] (ha)	Post disturbance period (years)	Annual precipitation (mm)	Soil texture[e]	Bedrock[f]	Org – C	NO_3-N	NH_4-N	DIN	Org – N	TDN	PO_4-P	Org – P	TDP	SO_4-S	K	Na	Mg	Ca	Cl	Source
North America																								
Experimental Lakes	C (H)	50–100	W	12	1	820	S	G	–	0.1	0.0	–	0.3	–	–	–	0.0	–	0.9	–	–	–	–	4
area, Ontario	C (H)	70–100	W	170	1	820			–	0.0	0.0	–	0.2	–	–	–	0.0	–	1.2	–	–	–	–	32
Nashwaak,	C-H	91 (B)	C	391	3	1320		G	–	1.5	–	–	–	–	–	–	–	–	–	–	–	–	–	16
New Brunswick	C-H	91 (B)	C	391	6	1320			–	2.3	0.9	–	–	–	–	–	–	–	5.1	–	0.3	3.3	–	39
Turkey Lakes,	H	100	C	5	3	1230	GSiL	B,G	0.1	18.4	–	–	–	–	–	–	–	–	–	–	–	20.5	–	11
Ontario	H	50	PC	68	3	1230			0.1	4.4	–	–	–	–	–	–	–	–	–	–	–	0.6	–	11
	H	33	PC	24	3	1230			3.8	3.1	–	–	–	–	–	–	–	–	–	–	–	7.9	–	11
UBC Research Forest	C	61	C	23	5	2220	SL	G	–	1.5	0.0	–	–	–	0.0	–	–	–0.1	3.0	–2.4	0.1	–0.7	–2.4	10
British Columbia	C	19	C+S	68	5	2430	SL		–	0.9	0.0	–	–	–	0.0	–	–	–1.5	0.3	–6.9	–2.1	–7.4	–7.5	10
	C	43	H	23	5	2220	SL		–	4.1	0.0	–	–	–	0.0	–	–	–1.7	1.4	1.6	0.6	3.2	2.8	9
Hubbard Brook,	H	100	F+H	16	3[g]	1300	SL	G,M	–	116.5	–	–	–	~107[h]	–	–	<0.1[h]	–3.9	29.0	9.3	12[h]	64.2	–1.7	2,6,15, 19–22,25
New Hampshire					3[g]	1300			–	42.3	–	–	–	–	–	–	–	2.3	15.5	2.7	–	23.8	2.0	2,15, 19–22,25
	H	100 (B)	PC	36	6	1300			–	5.3	–	–	–	–	–	–	–	0.4	2.9	2.0	–	4.7	0.1	2,15, 19–22,25
	H	100	C	12	6	1300			–	11.8	–	–	–		–	–	–	0.3	6.0	0.7	–	7.5	–0.3	
White Mountains, New Hampshire	H (C)	100	C	2–24(9)	4	?	SL	G,M	–	–	–	14.2	–	–	–	–	–	–	3.7	–	–	15.2	–	26
Catskill Mountains, New York	H	23	PC	10	1	1530	Med	?	–	6.3	–0.2	–	–	–	–	–	–	–	0.8	0.7	–0.1	–0.4	–	36
Isaac Creek, North	C	98	C	91	3	1430	L+	?	–	–	–	0.7	1.5	–	0.2	0.1	–	–	–	–	–	–	–	18
Carolina	C	82	C	109	3		Org		–	–	–	0.9	1.2	–	0.3	0.1	–	–	–	–	–	–	–	18
	C	25	C	359	3				–	–	–	0.5	1.6	–	0.0	0.1	–	–	–	–	–	–	–	18
Clemson Experimental Forest, South Carolina	C	100	C	1–2(3)	3	1440	C	G,M	–	0.0	0.0	–	–	–	0.0	–	–	–	0.2	0.1	0.2	0.4	–	35
Grant Memorial Forest, Georgia	C-H	<100(B)	C+M	33	6	1310	?	?	–	0.1	–	–	–	–	–	–	0.6	–	0.9	1.1	2.4	4.1	–	13
Starke, Florida	C	100	C+M	64	2	1400	S	?	–	0.0	–0.1	–	0.9	–	0.0	–	0.0	–	0.4	–	0.6	–	–	30
	C	100	C+M+B	48	2	1400			–	0.1	0.1	–	1.3	–	0.0	–	0.0	–	4.1	–	0.9	–	–	30
Cherokee County,	C-H	<100(B)	C+M+B	3	5	1070	SiL	S	–	0.1	0.0	–	0.2	–	0.0	–	0.1	–	1.6	1.5	0.6	0.2	–	5
Texas	C-H		C+M+B	3	5	1070	above C		–	0.0	0.0	–	0.2	–	0.0	–	0.0	–	0.5	0.4	0.2	–0.1	–	5
Fraser Experimental	C	40	C	41	2	580	GS	M	–	0.0	–	–	–	–	–	–	–	0.5	0.7	0.6	0.5	5.1	–	33
Forest, Colorado	C	30	C	78	2	580			–	0.2	–	–	–	–	–	–	–	1.8	1.5	4.7	9.6	–	–	33
Blue Mountains,	C	50	PC+B	30	2	1430	?	B	–	0.0	–	–	–0.2	–	–	–	<0.1	–	–	–	–	–	–	34
Oregon	C	43	C+B	24	2	1430			–	0.5	–	–	–0.4	–	–	–	<0.1	–	–	–	–	–	–	34
		22	C+B	118	2	1430			–	0.2	–	–	0.1	–	–	–	<0.1	–	–	–	–	–	–	34
Alsea watershed,	C	100	C+B	71	2	2540	GL	S	–	12.1	–	–	–	–	0.0	–	–	–	–	–	–	–	–	7
Oregon	C	25	C+B	303	2	2540			–	2.1	–	–	–	–	0.0	–	–	–	–	–	–	–	–	7
H.J. Andrews	C	100	C+B	13	7–10	2190	SL-SiL	V	–	0.3	–	–	0.2	–	0.0	0.1	–	–	1.8	3.5	3.0	4.4	–	24
Experimental Forest,	C	60	PC+B	15	7–10	2190			–	0.0	–	–	0.0	–	0.0	0.0	–	–	1.6	2.9	2.7	5.9	–	24
Oregon	C	100	C+B	96	1–2	2390	L-SiL		–	1.3	–	–	–	–	0.3	–	–	–	0.8	4.4	3.9	13.6	–	12
Europe																								
Hälsingland, Sweden	C	95	C	40	9	?	SSi	?	–	0.7	0.4	–	1.2	–	–	–	–	6.0	2.7	4.0	1.3	5.1	2.5	31
	C	50	C	200	9				–	0.2	0.1	–	0.7	–	–	–	–	5.4	2.3	2.2	1.1	4.6	1.5	31

Tegernsee Alps, Germany	C-H	40	PC	4	1	1990	C	S	1.3	8.1	2.7	–	–	–	0.0	–	–	5.1	8.6	10.3	28.8	134.6	8.1	3
Kershope Forest, England	C	100	C	2	2	1440	C	?	–	28.0	4.5	–	–	–	–	–	–	–	16.4	–11.8	1.1	1.1	–	1
Beddgelert, Wales	C	62	C	1	5	2600	CL-	S	–	16.8	–	–	–	–	–	–	–	–	12.4	–	–	–	–	29
	C	28	C	6	5	2600	SCL		–	3.2	–	–	–	–	–	–	–	–	3.2	–	–	–	–	29
Plynlimon, Wales	C	47	C	340	4	2400	SiCL-SiL	S	5.2	8.2	0.0	–	–	–	0.0	–	–	9.9	1.3	2.1	3.8	5.0	3.5	8
Karpenissi, Greece	C	15	PC	147	2	1400	L-CL	S	–	0.0	–0.3	–	–	–	–	–	–	–23.4	0.6	0.6	–2.7	7.2	5.0	27
Rest of World																								
Collie River basin,	DE	100	C	94	5	1120	GS	G	–	–	–	–	–	–	–	–	–	–	–	–	–	–	447.2	38
W Australia	DE	54	C	344	5	820	above C		–	–	–	–	–	–	–	–	–	–	–	–	–	–	4.6	38
	DE	38	C	350	5	800			–	–	–	–	–	–	–	–	–	–	–	–	–	–	2.4	38
Cropper Creek, SE Australia	ME	76(B)	C	46	2	1410	L	S	–	–	–	–	–	–	–	–	–	–	2.1	5.0	2.9	2.0	9.0	14
Coranderrk, SE Australia	ME	86(B)	C	53	5	1100	C-	D	–	0.1		–	–	–	–	–	–	–0.7	0.7	12.0	–6.8	3.9	30.2	17
		42(B)	C	65	5	1270	CL		–	0.0		–	–	–	–	–	–	–1.9	0.2	6.7	0.5	2.0	8.0	17
Maimai, New Zealand	HP	100	C+B	5	1	1310	G	?	–	1.0	0.8	–	–	5.7	0.6	–	1.3	–	126.8	31.6	21.7	46.8	–	28
Sabah, Malaysia	TF	100	C+B	6	2.75	3220	C+LS	S	47.2	1.1	0.8	–	4.4	–	0.4	0.3	–	–8.5	30.5	–3.4	–1.9	10.3	0.3	23
	TF	100	C+M	3	2.75	3490	C+LS		72.1	2.5	1.1	–	6.2	–	0.1	0.2	–	3.6	38.5	0.4	2.8	9.1	13.9	23
	TF	100	C+B	10	2.75	3490	C+LS		61.0	4.6	1.2	–	8.7	–	0.2	0.3		4.1	68.9	3.6	5.6	9.7	17.4	23
Central Amazon basin, Brazil	TF	9	C+B	23	1	2750	SCL-LS	?	–	0.2	0.1	–	2.6	–	0.0	0.1	–	0.1	4.4	7.4	0.9	3.1	3.0	37

Fluxes were calculated usually by assuming that the post-deforestation ratio of flux in the deforested stream to that in the control stream would have been the same as the pre-deforestation ratio, had deforestation not occurred.

[a]Vegetation is C, coniferous; DE, dry eucalyptus; C(H), mainly coniferous with some deciduous hardwood; C-H, mixed conifer and deciduous hardwood; H, deciduous hardwood; HP, hardwood-podocarp; ME, moist eucalyptus; TF, tropical forest.

[b]B indicates a buffer strip was left beside the stream.

[c]Type of disturbance is H, herbicide; C, clearcutting; F, tree felling; M, mechanical site preparation; PC, partial cutting; S, slashburning; W, windstorm.

[d]For multiwatershed studies, the number of watersheds is in parentheses.

[e]Soil texture is C, clay; G, gravelly; L, loamy; S, sandy; Si, silty; Org, organic soil; Med, medium.

[f]Bedrock is B, basalt; D, dacite; G, granitic; M, metamorphic (mainly gneiss and schist); S, sedimentary (mainly sandstone, siltstone, mudstone, and shale); V, mixed volcanic.

[g]The first 3-year period was during deforestation; the second 3-year period was the first 3 years after deforestation.

[h]Estimated for a 4-year period during, and for the first year following, deforestation [Bormann, Likens, Siccama, Pierce, and Eaton (1974)].

Sources

1. Adamson JK, Hornung M, Pyatt DG, and Anderson AR (1987) Changes in solute chemistry of drainage waters following the clearfelling of a sitka spruce plantation. *Forestry* 60: 165–177.
2. Bailey SW, Buso DC, and Likens GE (2003) Implications of sodium mass balance for interpreting the calcium cycle of a forested ecosystem. *Ecology* 84: 471–484.
3. Bäumler R and Zech W (1999) Effects of forest thinning on the streamwater chemistry of two forest watersheds in the Bavarian Alps. *Forest Ecology and Management* 116: 119–128.
4. Bayley SE, Schindler DW, Beaty KG, Parker BR, and Stainton MP (1992) Effects of multiple fires on nutrient yields from streams draining boreal forest and fen watersheds: nitrogen and phosphorus. *Canadian Journal of Fisheries and Aquatic Sciences* 49: 584–596.
5. Blackburn WH and Wood JC (1990) Nutrient export in stormflow following forest harvesting and site preparation in east Texas. *Journal of Environmental Quality* 19: 402–408.
6. Bormann FH, Likens GE, Siccama TG, Pierce RS, and Eaton JS (1974) The export of nutrients and recovery of stable conditions following deforestation at Hubbard Brook. *Ecological Monographs* 44: 255–277.
7. Brown GW, Gahler AR, and Marston RB (1973) Nutrient losses after clear-cut logging and slash burning in the Oregon Coast Range. *Water Resources Research* 9: 1450–1453.
8. Durand P, Neal C, Jeffery HA, Ryland GP, and Neal M (1994) Major, minor and trace element budgets in the Plynlimon afforested catchments (Wales): general trends, and effects of felling and climate variations. *Journal of Hydrology* 157: 139–156.
9. Feller MC (1989) Effects of forest herbicide applications on streamwater chemistry in southwestern British Columbia. *Water Resources Bulletin* 25: 607–616.
10. Feller MC and Kimmins JP (1984) Effects of clearcutting and slashburning on streamwater chemistry and watershed nutrient loss in southwestern British Columbia. *Water Resources Research* 20: 29–40.
11. Foster NW, Beall FD, and Kreutzweiser DP (2005) The role of forests in regulating water: the Turkey Lakes watershed case study. *Forestry Chronicle* 81: 142–148.
12. Fredriksen RL (1971) Comparitive chemical water quality – natural and disturbed streams following logging and slashburning. In Krygier JT, and Hall JD (eds.). *Forest Land Uses and Stream Environment*, pp. 125–138. Corvallis: Oregon State University.
13. Hewlett JD, Post HE, and Doss R (1984) Effect of clear-cut silviculture on dissolved ion export and water yield in the Piedmont. *Water Resources Research* 20: 1030–1038.
14. Hopmans P, Flinn DW, and Farrell PW (1987) Nutrient dynamics of forested catchments in southeastern Australia and changes in water quality and nutrient exports following clearing. Forest Ecology and Management 20: 209–231.
15. Hornbeck JW, Martin CW, Pierce RS, Bormann FH, Likens GE, and Eaton JS (1986) Clearcutting northern hardwoods: effects on hydrologic and nutrient ion budgets. *Forest Science* 32: 667–686.
16. Krause HH (1982) Nitrate formation and movement before and after clear-cutting of a monitored watershed in central New Brunswick, Canada. *Canadian Journal of Forest Research* 12: 922–930.

17. Langford KJ, and O'Shaughnessy PJ (1980) *Second progress report. Coranderrk*. Melbourne and Metropolitan Board of Works Report No. MMBW-W-0010. Melbourne, Australia.
18. Lebo ME and Herrmann RB (1998) Harvest impacts on forest outflow in coastal North Carolina. *Journal of Environmental Quality* 27: 1382–1395.
19. Likens GE, Bormann FH, Pierce RS, and Reiners WA (1978) Recovery of a deforested ecosystem. *Science* 199: 492–496.
20. Likens GE, Driscoll CT, Buso DC, Mitchell MF, Lovett GM, Bailey SW, Siccama TG, Reiners WA, and Alewell C (2002) The biogeochemistry of sulfur at Hubbard Brook. *Biogeochemistry* 60: 235–316.
21. Likens GE, Driscoll CT, Buso DC, Siccama TG, Johnson CE, Lovett GM, Ryan DF, Fahey T, and Reiners WA (1994) The biogeochemistry of potassium at Hubbard Brook. *Biogeochemistry* 25: 61–125.
22. Lovett GM, Likens GE, Buso DC, Driscoll CT, and Bailey SW (2005) The biogeochemistry of chlorine at Hubbard Brook, New Hampshire, USA. *Biogeochemistry* 72: 191–232.
23. Malmer A (1996) Hydrological effects and nutrient losses of forest plantation establishment on tropical rainforest land in Sabah, Malaysia. *Journal of Hydrology* 174: 129–148.
24. Martin CW, and Harr RD (1989) Logging of mature Douglas-fir in western Oregon has little effect on nutrient output budgets. *Canadian Journal of Forest Research* 19: 35–43.
25. Martin CW, Driscoll CT, and Fahey TJ (2000) Changes in streamwater chemistry after 20 years from forested watersheds in New Hampshire, USA. *Canadian Journal of Forest Research* 30: 1206–1213.
26. Martin CW, Pierce RS, Likens GE, and Bormann FH (1986) Clearcutting affects stream chemistry in the White Mountains of New Hampshire. USDA Forest Service Research Paper NE-579. Northeastern Forest Experiment Station.
27. Nakos G and Vouzaras A (1988) Budgets of selected cations and anions in two forested experimental watersheds in central Greece. *Forest Ecology and Management* 24: 85–95.
28. Neary DG, Pearce AJ, O'Loughlin CL, and Rowe LK (1978) Management impacts on nutrient fluxes in beech-podocarp-hardwood forests. *New Zealand Journal of Ecology* 1: 19–26.
29. Reynolds B, Stevens PA, Hughes S, Parkinson JA, and Weatherley NS (1995) Stream chemistry impacts of conifer harvesting in Welsh catchments. *Water, Air and Soil Pollution* 79: 147–170.
30. Riekerk H (1983) Impacts of silviculture on Flatwoods runoff, water quality, and nutrient budgets. *Water Resources Bulletin* 19: 73–79.
31. Rosén K, Aronson J-A, and Eriksson HM (1996) Effects of clear-cutting on streamwater quality in forest catchments in central Sweden. *Forest Ecology and Management* 83: 237–244.
32. Schindler DW, Newbury RW, Beaty KG, Prokopowich J, Ruszczynski T, and Dalton JA (1980) Effects of a windstorm and forest fire on chemical losses from forested watersheds and on the quality of receiving streams. *Canadian Journal of Fisheries and Aquatic Sciences* 37: 328–334.
33. Stottlemyer R (1987) Natural and anthropic factors as determinants of long-term stream chemistry. In Troendle CA, Kaufmann MR, Hamre RH, and Winokur RP (tech. cords.) *Management of subalpine forests: building on 50 years of research*. USDA Forest Service General Technical Report RM-149. Rocky Mountain Research Station.
34. Tiedemann AR, Quigley TM, and Anderson TD (1988) Effects of timber harvest on stream chemistry and dissolved nutrient losses in northeast Oregon. *Forest Science* 34: 344–358.
35. Van Lear DH, Douglas JE, Cox SK, and Augspurger MK (1985) Sediment and nutrient export in runoff from burned and harvested pine watersheds in the South Carolina Piedmont. *Journal of Environmental Quality* 14: 169–174.
36. Wang X, Burns DA, Yanai RD, Briggs RD, and Germain RH (2006) Changes in stream chemistry and nutritional export following a partial harvest in the Catskill Mountains, New York, USA. *Forest Ecology and Management* 223: 103–112.
37. Williams MR, and Melack JM (1997) Solute export from forested and partially deforested catchments in the central Amazon. *Biogeochemistry* 38: 67–102.
38. Williamson DR, Stokes RA, and Ruprecht JR (1987) Response of input and output of water and chloride to clearing for agriculture. *Journal of Hydrology* 94: 1–28.
39. Zhu A, Arp PA, Mazumber A, Meng F, Bourque CP-A, and Foster NW (2005) Modeling stream water nutrient concentrations and loadings in response to weather condition and forest harvesting. *Ecological Modeling* 185: 231–243.

2. soil erodibility, as determined by soil physical properties,
3. precipitation characteristics,
4. watershed susceptibility to mass wasting as determined by topography and bedrock,
5. stream channel characteristics, including presence and stability of debris jams, and the nature of materials lining the streambed and streambanks,
6. proximity of vegetation to surface water (which controls the extent to which plant litter is transferred to water),
7. extent of roading, particularly new unsealed roads, in a watershed.

The two concerns noted above for solute sampling also apply to particulate sampling, even more so in the case of sampling frequency, because particulate transport increases greatly while stream discharge is increasing in response to rain or snowmelt.

Effects of Deforestation on the Factors Controlling Freshwater Nutrient Loads

Dissolved Nutrients

Effects on geological weathering By exposing the land surface to greater temperature extremes, and greater amounts of often more acidic water leaching through the systems, deforestation can increase weathering rates. Increased nitrification following forest harvesting can lead to increased acid leaching which increases weathering rates. Quantitative estimates of the impacts of harvesting on weathering rates, however, appear to be generally unavailable, with the exception of the Hubbard Brook Experimental Forest in northeastern United States.

Effects on atmospheric precipitation/climate
Effects on precipitation chemistry and acidity Deforestation has affected the acidity of water passing through the soil as well as the acidity of streamwater (pH decreases of 0.2–0.5 pH units). This occurs due to release of organic acids from decomposing logging debris or to enhanced nitrification which produces HNO_3. Such enhanced acid production has had similar effects to those of acid rain, in terms of flushing nutrient cations from soil into streams. Enhanced weathering release and dissolution of chemicals, such as Ca oxalate, in soil can contribute to a sustained increased loss of Ca to streamwater for 10 years or more.

Deforestation can also increase H^+ production if it results in a greater net biomass increment (excess of biomass gain from growing vegetation over biomass loss from decomposition), which is likely to occur when slow growing or steady state forests are converted to more rapidly growing younger ones. Enhanced acid production within watersheds would appear to be a general response to deforestation, although the intensity of this response varies widely, depending on the extent of nitric acid generation and the degree to which net biomass increment is affected.

The extent of HNO_3 leaching through a watershed after deforestation depends primarily on (1) the extent of N uptake by regrowing vegetation, greater N uptake leaving less N available for nitrification, (2) N immobilization in the soil, greater N immobilization (indicated by higher soil C/N ratios) also leaving less N available for nitrification, (3) lags in nitrification whereby nitrifying organisms are suppressed or exist only in low populations, and (4) soil N availability prior to deforestation, for which the greater the availability, the greater the amount of N for nitrification.

Deforestation generally increases nitrification to a greater extent in the northern hemisphere temperate deciduous hardwood than in conifer forests. Thus, for the studies in **Table 2**, the mean increase in streamwater NO_3-N loading during the first few years following forest harvesting was 13.1 kg ha^{-1} $year^{-1}$ for seven studies in pure hardwood forests (average deforestation of 72%) compared to 3.5 zkg ha^{-1} $year^{-1}$ for 22 studies in pure coniferous forests (average deforestation of 61%). However, considerable variation within both hardwood and conifer forest datasets (coefficients of variation of both mean values are >100%) indicates that deforestation in a hardwood forest will not always increase streamwater NO_3-N loading to a greater extent than in a conifer forest.

Effects on hydrology Deforestation causes an initial increase in streamflow, the extent of which depends on the degree of forest removal. Streamflow decreases to preharvesting levels over a period of years and may decrease below preharvesting levels if the regenerating forest has a higher evapotranspiration or is less able to capture cloud water than the pre-harvesting forest, as is the case for wetter Australian eucalypt forests.

Deforestation can also increase soil water, allowing rises in water tables. If the rising water table contacts nutrient rich soil material, these nutrients can be taken into freshwater bodies, considerably enriching them. Transport of salts from salt-rich soil parent material to streams via rising water tables following deforestation has been notable in southern Australia (**Figure 1**).

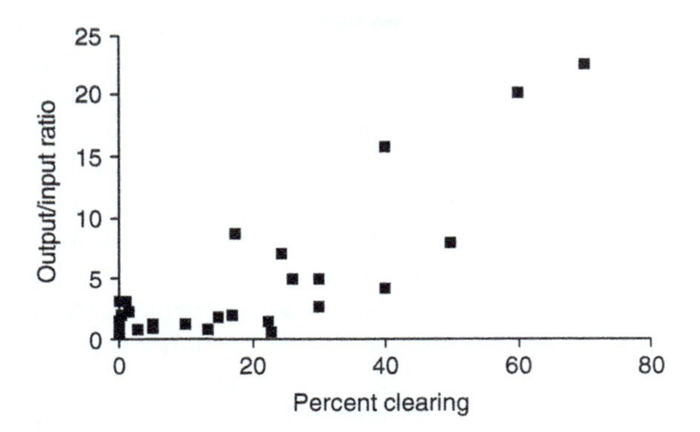

Figure 1 Watershed chloride output/input ratios as a function of the percentage of a watershed deforested by forest harvesting and clearing for agriculture in southwestern Australia. Redrawn from **Figure 5** in Trotman CH (ed.) (1974) *The Influence of Land Use on Stream Salinity in the Manjimup Area, Western Australia*. W.A. Department of Agriculture Technical Bulletin no. 27. Perth, Australia, with the kind permission of the Department of Agriculture and Food of the Government of Western Australia.

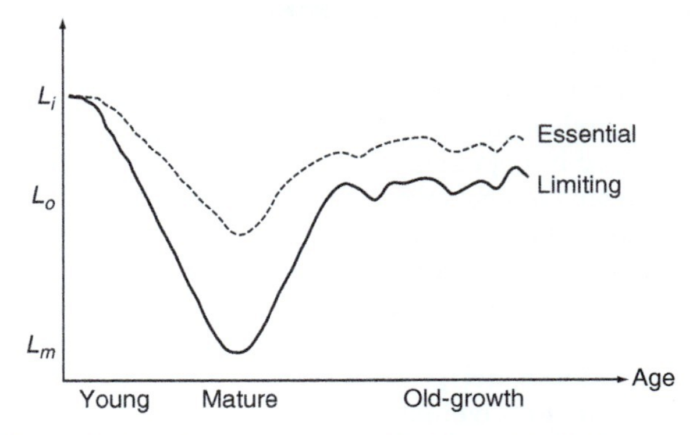

Figure 2 Loading of essential and limiting nutrients in streamwater, expressed as percentage of the maximum, as a function of the age of vegetation in the surrounding watershed. Adapted from Feller MC (2005) Forest harvesting and streamwater inorganic chemistry in western North America: A review. *Journal of the American Water Resources Association* 41: 785–811, and Vitousek PM and Reiners WA (1976) Ecosystem development and the biological control of stream chemistry. In: Nriagu JO (ed.) *Environmental Biogeochemistry*, pp 665–680. Ann Arbor, MI: Ann Arbor Science.

Deforestation has had variable effects on snowpacks, depending on the size and orientation of the deforested area. Greater snowpacks may result in higher streamwater nutrient loadings during the early snowmelt period as a result of preferential elution of ions from the snowpack, but this is unquantified. Thus, deforestation can alter water movement through watersheds, but the influence on nutrient loading of this water increment from deforestation has not been precisely quantified.

Effects on temperature Deforestation has increased summer soil surface temperatures as well as streamwater temperatures, depending primarily on the extent to which the tree canopy is removed. Higher temperatures during summer may accelerate weathering and decomposition as well as chemical and microbial reactions, which may then alter nutrient movement through soil into freshwater, although this, too, has not been quantified.

Effects on terrestrial biological processes

Chemical uptake Deforestation initially reduces nutrient uptake by terrestrial vegetation leading to enhanced nutrient flow from terrestrial to aquatic ecosystems, for periods of usually 1–7 years. If plant uptake was a dominant process influencing freshwater nutrient loading, then this loading would respond to deforestation as shown in **Figure 2**. Thus, essential and growth limiting nutrients would exhibit highest loadings at time 0, shortly after deforestation, then decline as plant growth increased, but limiting nutrients to a greater extent, before increasing again in old-growth forests where mortality reduces uptake. A consequence of this temporal pattern in nutrient loading is that deforestation of mature forests has the potential to cause greater increases in essential and limiting streamwater nutrient concentrations (L_i–L_m for a limiting nutrient), than does deforestation of old growth forests (L_i–L_o for a limiting nutrient) (**Figure 2**).

Dissolved inorganic N can be immobilized in the soil by mycorrhizal assimilation. Deforestation has at least temporarily adversely affected soil mycorrhizae, which might lead to significant initial increases in nitrification and NO_3^- leaching through soil into streamwater. The extent to which deforestation affects mycorrhizae depends on the severity of the disturbance, inherent soil ectomycorrhizal diversity, dryness of the climate, and the abundance of hosts, but the influence of deforestation-induced changes in mycorrhizal populations on NO_3^- leaching is unknown.

Chemical transformations Deforestation usually increases nitrification. This increases NO_3^- flow through soil into streams. This flow may be enhanced if N fixation also increases after deforestation, as it can if the amount of more well-decayed coarse woody debris (CWD) on the ground surface increases and if the pre-existing CWD is not greatly fragmented during deforestation. N fixation may also be increased if the cover of N-fixing plant species, such as *Alnus* spp., increases after deforestation. NO_3^- flow can be retarded if enhanced denitrification offsets enhanced nitrification. Denitrification has been enhanced by deforestation as a result of increased soil moisture (due to decreased evapotranspiration), soil NO_3^- content (due to increased nitrification), and soil dissolved organic C

(DOC) (due to enhanced decomposition of organic material). Nitrification may be decreased, with subsequent reduced freshwater NO_3 loading, if deforestation removes *Alnus* trees. The balance between these different N transformations will help determine the extent of NO_3^- movement to streamwater.

Deforestation has generally lead to declines in streamwater loading more often for SO_4 than for any other nutrient ion (**Table 2**). This decline may be due to (1) decreased activity of S-oxidizing bacteria, perhaps due to toxic effects of high NO_3^- concentrations, (2) increased SO_4^{2-} reduction resulting from increased anaerobic conditions in soil, (3) decreased inputs of dry deposition resulting from loss of foliage area, (4) enhanced microbial immobilization of SO_4^{2-} or reduced organic S mineralization resulting from increased forest floor decomposition, (5) dilution from increased water flow, (6) precipitation of aluminum hydroxy-sulfate minerals, or (7) increased SO_4^{2-} adsorption in the mineral soil as a result of an acid-induced increase in the SO_4^{2-} adsorption capacity of the soil. Studies in the northeastern United States have suggested that soil adsorption (process number 7) was the most important process.

Production of soluble chemicals Deforestation which causes an initial decrease in litter-producing riparian vegetation would result in less organic matter input into streams immediately postharvesting, and an initial decline in litter-originating soluble chemicals in streamwater, although this has not been quantified. Deforestation leaves variable amounts of organic material on the ground surface. This material produces easily soluble chemicals which are leached away as the organic material decomposes (**Figure 3**), although this may not always contribute greatly to enhanced soil N availability postharvesting.

Deforestation has had variable effects on litter decomposition rates, due to differences in (1) the type of material being decomposed, (2) climate, (3) impacts of deforestation on soil organisms, and (4) the degree of mixing of organic matter with soil materials. Thus, deforestation may, or may not, lead to increased organic matter decomposition rates and hence increased production of soluble, easily leached chemicals.

Effects on physical–chemical reactions in the soil Direct effects of deforestation on physical–chemical reactions in the soil have not been reported. Indirect effects, however, include an increase in the anion exchange capacity of a soil as pH decreases. Thus, if deforestation enhances nitrification, increased nitric acid generation can increase the anion exchange capacity of the soil, helping to retard movement of ions through soil into streams, partly explaining the decline in SO_4^{2-} loading in streamwater found

Figure 3 Large amounts of woody debris close to a stream channel following clearcutting of *Eucalyptus delegatensis* forest across a stream in eastern Victoria, Australia, 1987. Photograph by Michael Feller.

after deforestation, as mentioned above. Retardation of P movement through the soil as pH decreases, due to increased fixation by hydrous oxides of Fe and Al, might also help to explain the minimal response of streamwater P loading to deforestation (**Table 2**).

Effects on processes within aquatic ecosystems Deforestation can influence most processes within aquatic ecosystems, but the extent of this influence will depend strongly on the amount of organic debris, fine sediment, solar radiation, and acid reaching the aquatic ecosystem. Ion exchange and chemical redox reactions and microbial transformations all increase with the surface area of the streambed substrate, so these processes are likely to be enhanced if deforestation increases the quantity of colloidal and fine particulate organic material in a stream. Although many studies have investigated nutrient cycling processes within undisturbed forested streams, the direct effects of deforestation on these individual processes have not been studied.

Primary production in a stream ecosystem increases with solar radiation and, to a lesser extent, with temperature, so when these parameters are increased by deforestation, as often happens, enhanced primary production occurs. Increased primary production increases removal of nutrients from water, and hence decreases nutrient loading.

If deforestation increases acid generation within a watershed, and if some of that acid reaches a stream, then ion exchange and pH-dependent chemical redox reactions within the stream will be affected, as will be pH-induced microbial transformations. Additions of acid to streams have resulted in increased streamwater Ca^{2+}, Mg^{2+}, K^{+}, and micronutrient loadings, but not NO_3^-, and NH_4^+ loadings. Increased acid levels in streamwater may also enhance primary production, and hence uptake by aquatic primary producers, as the acid may adversely affect organisms that graze on the primary producers.

The impacts of deforestation on these different processes will affect freshwater nutrient loading in different ways, depending on the nutrient and the process (**Table 3**). Thus, deforestation has the potential to increase, decrease, or have no effect on the loading of every nutrient considered, as has been found. Each arrow in **Table 3** should not be considered to have the same weight for a given nutrient. Thus, Na, Mg, and Ca all have more than twice as many upward as downward arrows but deforestation has sometimes decreased fluxes of these nutrients (**Table 2**).

Other factors contributing to the variable effects of deforestation on nutrient loading Freshwater nutrient loading depends not only on the response of the preceding factors to deforestation, but also on characteristics of the deforestation itself, soil characteristics, and rate of revegetation following deforestation. These additional factors add further complexity so

Table 3 Initial trends in dissolved nutrient chemical loading in freshwaters resulting from deforestation impacts on the different factors controlling freshwater nutrient loading

Factor	*C*	*Ca*	*Cl*	*Fe*	*H*	*K*	*Mg*	*Mn*	*N*	*Na*	*P*	*Si*	*SO_4*	*Trace metals*
Geological weathering	↑	↑	–	↑	↓	↑	↑	↑	↑–	↑	↑	↑	↑	↑
Atmospheric precipitation/climate														
Precipitation chemistry	↑↓	↑–	–	↑–	↑–	↑–	↑–	↑–	↑–	↑–	↑–	↑–	↑–	↑–
Hydrologic influences	↑↓	↑↓	–	↑↓	↑↓	↑↓	↑↓	↑↓	–	↑↓	↑↓	↑↓	↑↓	↑↓
Temperature	↑	↑	–↑	↑	↑↓	↑	↑	↑	↑	↑	↑	↑	↑	↑–
Terrestrial biological processes														
Chemical uptake	–	↑	↑	↑	↑	↑	↑	↑	↑	↑	↑	–	↑	–
Chemical transformations	↑↓	–	–	–	↑↓	–	–	–	↑↓	–	–	–	↑↓	–
Production of soluble chemicals	↑↓	↑–	↑	↑–	↑–	↑–	↑–	↑–	↑–	↑–	↑–	–	↑–	↑–
Physical–chemical reactions in the soil	↓–	–	↓–	–	↓–	–	–	–	↓–	–	↓–	–	↓–	–
Processes within aquatic ecosystems														
Ion exchange reactions	–	↑↓	↑↓	↑↓	↑↓	↑↓	↑↓	↑↓	↑↓	↑↓	↑↓	–	↑↓	↑↓
Chemical redox reactions	–	–	–	↑↓	–	–	–	↑↓	↑↓	–	–	–	↑↓	↑↓
Evaporation–crystallization	↑–	↑–	–	–	–	–	–	–	–	–	–	–	–	–
pH-induced transformations	↑↓	↑↓	↑↓	↑↓	↓↑	↑↓	↑↓	↑↓	↓↑	↑↓	↓↑	↑↓	↑↓	↓↑
Uptake by primary producers	–	↓–	–	↓–	–	↓–	↓–	↓–	↓–	↓–	↓–	↓–	↓–	↓–
Microbial transformations	–	–	–	–	–	–	–	–	↑↓	–	–	–	↑↓	–

↑ Effects lead to an increase in loading.

↓ Effects lead to a decrease in loading.

– Effects have little to no impact on loading.

Adapted from **Table 3** in Feller MC (2005) Forest harvesting and streamwater inorganic chemistry in western North America: A review. *Journal of the American Water Resources Association* 41: 785–811, with the kind permission of the American Water Resources Association.

that deforestation-induced changes in nutrient fluxes can vary widely. These changes have ranged over four orders of magnitude for NO_3-N, for example (**Table 2**). These other factors include:

1. *Extent of the watershed deforested*: In general, the greater the percentage of the watershed deforested the greater will be the impacts on freshwater nutrient loading in that watershed (**Figure 1**). For a wide range of environmental conditions (given in **Table 2**), deforestation of northern hemisphere temperate forests is likely to cause a maximum initial increase in NO_3-N, K^+, and Ca^{2+} loading of approximately 3, 2, and 5 kg ha^{-1} year^{-1}, respectively, for each 10% of a watershed deforested (**Figure 4**).

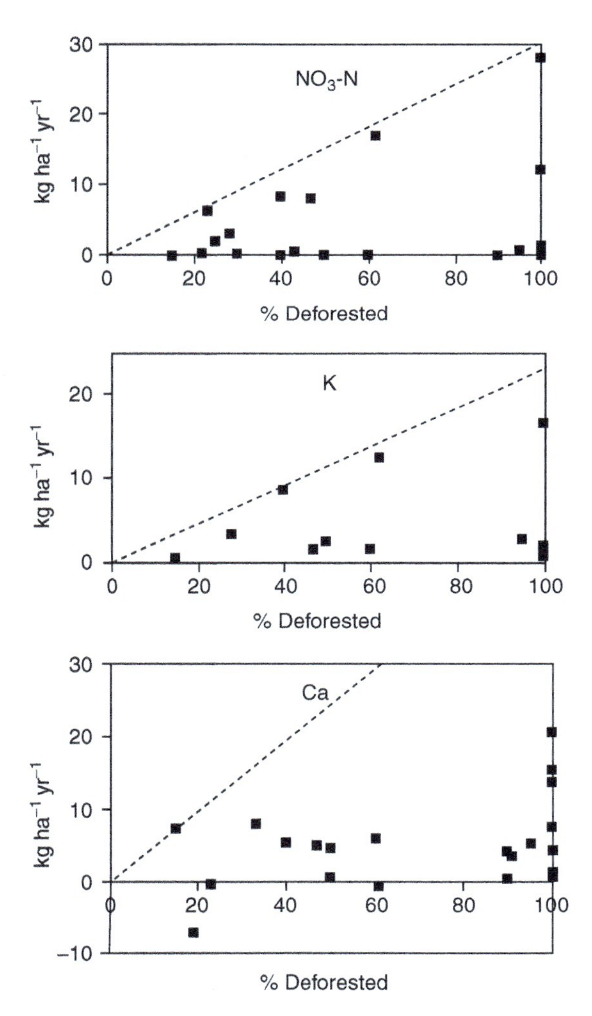

Figure 4 Deforestation-induced changes in streamwater nitrate – nitrogen, potassium, and calcium loads as a function of the extent of deforestation in northern hemisphere temperate forests during the first few years after nonherbicide deforestation. Data were obtained from **Table 2**. Dashed lines represent the maximum likely increase in stream nutrient loading due to deforestation. Calcium data exclude one extremely high outlier.

2. *Presence of buffer strips between freshwater and harvested areas*: Undisturbed buffer strips adjacent to a water body can filter out increased amounts of nutrients flowing from a deforested area to that water body. The efficiency of filtering by buffer strips increases with buffer strip width, with 100% efficiency suggested to occur for widths of approximately 100 m, although the effectiveness of buffer strips also depends on other factors, such as soil properties, slope angle, presence or absence of small ephemeral channels running through the buffer strips to the water body, and the type of vegetation present (**Figure 5**).
3. *Nature of the treatment given to a watershed following deforestation*: Site preparation treatments, such as mechanical scarification, slashburning, and herbicide application, can potentially enhance the effects of deforestation on streamwater chemistry. Thus, clearcutting plus slashburning causes greater increases in streamwater nutrient loading than just clearcutting alone. When deforestation is caused by herbicide application, nutrient loadings in streams appear to increase to a greater extent than from other causes of deforestation (**Figure 6**). This may be explained by greater nitrification following herbicide-induced deposition of lower C/N ratio litter. Removal of woody debris after deforestation can decrease the magnitude of nutrient loading increases, unless this removal causes substantial mechanical disturbance to the soil, in which case nutrient loading increases can be enhanced.
4. *Rate of revegetation following deforestation*: The more rapidly an area revegetates after harvesting, the more rapidly will streamwater nutrient loading return to predeforestation levels. If vigorous revegetation results in greater nutrient immobilization rates than occurred prior to deforestation, nutrient fluxes to freshwater can decline to below predeforestation levels (**Figure 6**).
5. *Nutrient content of the soil (soil fertility) prior to deforestation*: Nutrient loss to freshwater after deforestation increases with soil nutrient content. Consequently, N loading in fresh water following deforestation is likely to be greater as the degree of N saturation of a watershed increases.
6. *Buffering capacity of the soil*: The greater the ability of soil to retain nutrients, the lesser the amount of nutrients that will be washed through soil into a water body, and hence the lesser the increase in freshwater nutrient loading after deforestation.
7. *Abundance of large water bodies in a watershed*: Large nutrient and water storage areas, such as lakes, peatlands, and swamps, can trap nutrients flushed from the land following deforestation.

Figure 5 Small buffer strips of undisturbed vegetation adjacent to major, but not ephemeral, streams in extensively cleared *Eucalyptus* forest prior to planting *Pinus radiata* in the Acheron valley, central Victoria, Australia, 1980. Photograph by Michael Feller.

Consequently, the greater the abundance or the larger the size of such nutrient storages in a watershed, the lesser will be the nutrient loading in the stream leaving the watershed (**Figure 7**). If the flush of nutrients into lakes following deforestation does not exceed their storage capacity, then nutrient loading in outlet streams may be little affected by deforestation.

8. *Timing of deforestation*: Deforestation which occurs before a heavy rainfall period at a time when plant nutrient uptake is relatively low is likely to result in greater movement of nutrients into surface waters than a deforestation that occurs during a period of low rainfall and high plant nutrient uptake.

Nutrients in Particulate Form

Effects on Watershed Topography, Soil Erodibility, Precipitation Characteristics, and Watershed Susceptibility to Mass Wasting

Deforestation generally has little effect on the first four factors controlling particulates in fresh water. Deforestation may increase ground surface roughness by adding coarse woody debris which can prevent materials moved by surface erosion from entering freshwaters. Roads constructed during deforestation activities can concentrate water flows and facilitate movement of runoff into freshwaters (**Figure 8**), but the effects of this on freshwater nutrient loading have not been well quantified.

In general, however, these four factors simply help explain the variable effects of deforestation on freshwater particulate loads. Deforestation is more likely to increase freshwater particulate loads if a watershed (1) has steep, uniform slopes, (2) has erodible soils, such as those with high silt contents, (3) is in an area with intense and heavy precipitation, and (4) is prone to mass wasting by having moderately steep slopes, depressions where soil water can accumulate, or bedrock with fracture planes parallel to the ground surface.

Effects on Stream-Channel Characteristics

Deforestation is most likely to affect stream-channel characteristics if it occurs close to a stream channel. Debris jams tend to trap particulates, decreasing their downstream flux. Creation of stable debris jams is therefore likely to decrease stream particulate fluxes. Destabilization of debris jams will have the opposite effect. If deforestation removes trees immediately adjacent to water bodies, particularly if soil disturbance is involved, bank erosion is likely to enhance particulate loading in the water bodies.

Effects on Proximity of Vegetation to Surface Water

Deforestation which removes vegetation over or close to freshwater surfaces will decrease, at least

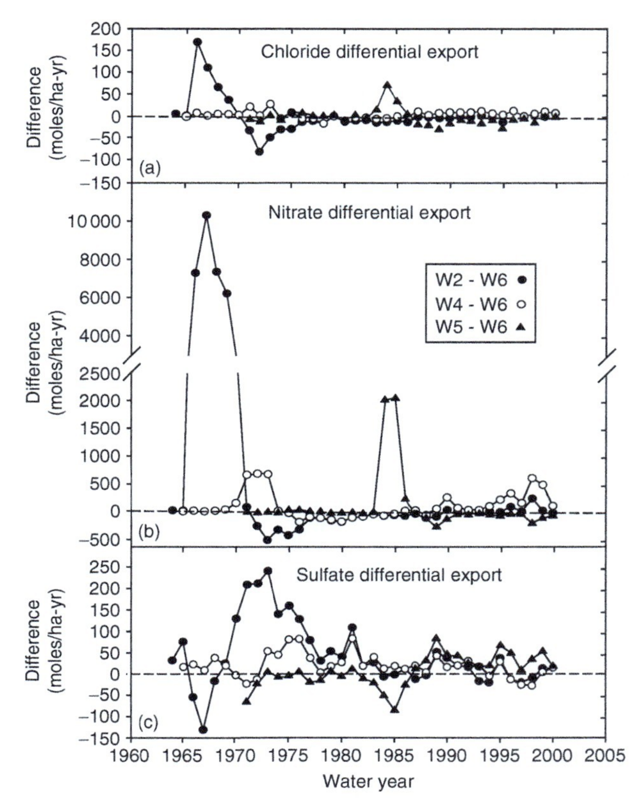

Figure 6 Difference between streamwater ion export in deforested and undisturbed (W6) deciduous hardwood watersheds in the Hubbard Brook Experimental Forest, NE. U.S. A. Deforestation involved tree cutting in 1965–66 followed by herbicide application in 1966–68 for W2, forest harvesting in 25-m wide strips covering one-third of the watershed in each of 1970, 1972, and 1974, for W4, and clearcutting in 1983–84 for W5, leaving a buffer strip adjacent to the stream. Reproduced with the kind permission of Springer Science and Business Media from Figure 19 in Lovett GM, Likens GE, Buso DC, Driscoll CT, and Bailey SW (2005) The biogeochemistry of chlorine at Hubbard Brook, New Hampshire, U.S.A. *Biogeochemistry* 72: 191–232.

temporarily, plant litter input into the water, and hence organic particulate loading. If the deforestation kills vegetation without removing it (e.g., herbicides or wind storms), the opposite occurs and organic particulate loads temporarily increase.

Effects on Extent of Roading

Road construction has often lead to greater increases in freshwater particulates than other human activities, such as tree felling. Consequently, deforestation which involves road construction can increase freshwater particulates to an extent dependent on the care taken with road layout and construction (**Figure 9**).

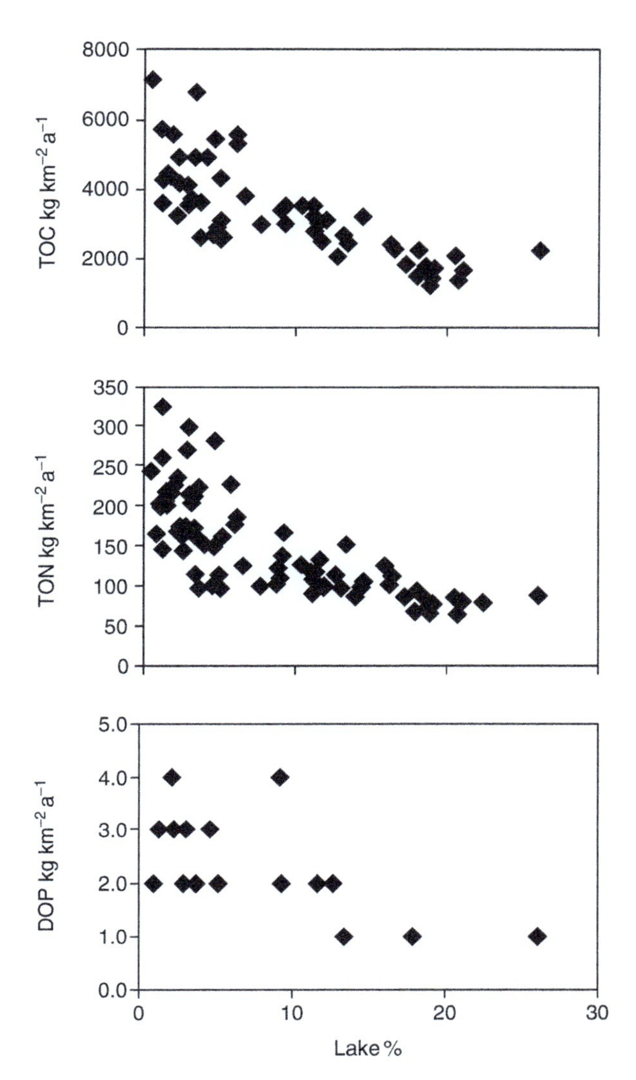

Figure 7 Relationships between export of total organic C (TOC), total organic N (TON), and dissolved organic P (DOP) and the proportion of lakes in coniferous forest watersheds in Finland. Watersheds contained $<$1–44% agricultural land and 0–6.5% urban land. Reproduced with the kind permission of Springer Science and Business Media from **Figure 5** in Mattsson T, Kortelainen P, and Räike A (2005) Export of DOM from boreal catchments: impacts of land use cover and climate. *Biogeochemistry* 76: 373–394.

Deforestation has had variable effects on freshwater particulate loads, depending on the above factors. It has affected the particulate load/dissolved load ratio (**Table 1**) although there have been too few studies to generalize. Many studies have quantified the effect of deforestation on freshwater sediment and particulate loads, but very few have quantified the effect on nutrient loads in particulate form. Increases in sediment and particulate loads, however, will usually mean increases in nutrient loads in particulate form.

Figure 8 Dense network of logging roads on relatively steep slopes in clearcut *Eucalyptus delegatensis* forest in north eastern Victoria, Australia, 1980. Photograph by Michael Feller.

Figure 9 Road construction on moderately steep slopes adjacent to a stream channel, with no buffer strip of undisturbed vegetation in south western British Columbia, Canada, 1999. Photograph by Michael Feller.

Conclusions

Dissolved nutrient loading in freshwater has been explained by five major factors, while particulate nutrient loading has been explained by seven major factors, many of which interact. Thus, nutrient loading is characterized by complexity and variability, both temporal and geographic, and even within one stream at the same time. This presents a challenge to sampling programs and to the development of models which predict nutrient loading.

Variable effects of deforestation on freshwater nutrient loading can be explained by the variability in deforestation effects on the major processes controlling this loading. When this level of variability is added to the variability inherent in undisturbed freshwater bodies, some major knowledge gaps are inevitable. Some of the more important knowledge gaps are as follows.

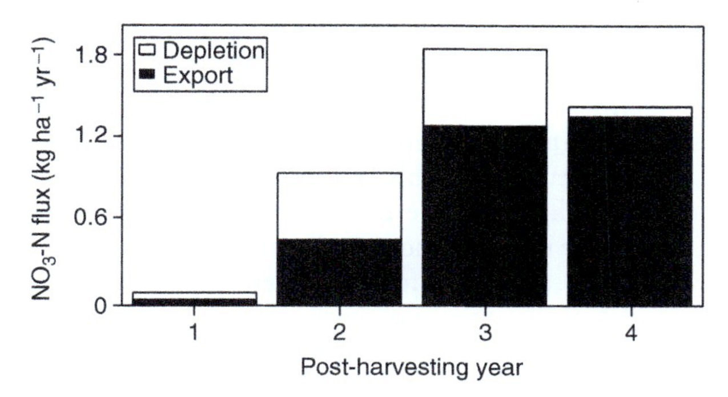

Figure 10 Annual NO_3-N fluxes measured in a stream flowing out of a 59 ha deciduous hardwood-conifer watershed at Coweeta, North Carolina, USA for the first 4 years following forest harvesting. Export is the streamwater dissolved load while depletion is the amount estimated to have been removed from the stream by in-stream processes. Produced from data presented in Swank WT and Caskey WH (1982) Nitrate depletion in a second-order mountain stream. *Journal of Environmental Quality* 11: 581–584, with the kind permission of the American Society of Agronomy.

Knowledge Gaps

1. *Detailed freshwater nutrient budgets are scarce*:
   Very few studies have quantified the different inputs and outputs, whose balance determines freshwater nutrient loading.
2. *Studies of longitudinal variation in stream nutrient loading are scarce*:
   Few studies have quantified nutrient flux trends along extended lengths of streams. Such studies would be important in determining how far downstream the impacts of deforestation might extend.
3. *Studies of deforestation impacts on processes within freshwater ecosystems are scarce*:
   The one published study that could be found on the impacts of deforestation on streamwater nutrient fluxes resulting from within-stream processes suggested that partial deforestation by an ice storm at the Hubbard Brook Experimental Forest increased within-stream retention and transformation of NO_3^-, reducing the potential increase in streamwater NO_3^- loading. Greater quantification of the impacts of deforestation on within-stream nutrient cycling processes is necessary to accurately determine the quantity of nutrients moving into freshwater after deforestation. Nearly all studies which have purported to determine this, have made the tacit assumption, of questionable validity, that deforestation has had no significant effect on processes within the freshwater ecosystem. The true deforestation-induced loss of nutrients to freshwater remains uncertain. Nutrient movement from land to streamwater, for example, is likely to include an amount exported by the stream as well as an amount removed from the stream (**Figure 10**), so that the deforestation-induced loss of nutrients to freshwater may be greater than that inferred from streamwater fluxes alone.
4. *The influence of deforestation on freshwater particulate nutrient loads has been poorly quantified*:
   The effects of deforestation on freshwater sediment and particulate loads has been well studied but not the chemical composition of these particulates or sediment.
5. *The influence of deforestation on geological weathering release of chemicals has been poorly quantified*:
   As geological weathering is a dominant process influencing freshwater nutrient loading, it is desirable to understand its quantitative response to deforestation. This would help in understanding the origin of solutes in freshwater after deforestation and, hence, in determining methods of reducing solute loading, if considered necessary.
6. *The influence of deforestation on the snowmelt flux of nutrients into streams has been poorly quantified*:
   The effect of deforestation on snowpacks and snowmelt has been reasonably well documented but the influence of altered snow hydrology on freshwater nutrient loading during the snowmelt runoff period has not been studied.
7. *The influence on freshwater nutrients of deforestation-induced changes in soil and stream temperatures has been poorly quantified*:
   Changes in soil and stream temperature regimes can affect geological weathering, terrestrial biological processes, physical–chemical reactions in the soil, and processes within aquatic ecosystems. None of these effects seems well quantified.

8. *The effects of deforestation on nutrient loading in drier environments, tropical regions, and the southern hemisphere have been poorly quantified*: Although many studies have quantified deforestation effects on freshwater nutrient concentrations, from which relative trends in loadings can sometimes be inferred, most studies that have more accurately quantified deforestation effects on freshwater nutrient loading have been conducted in areas receiving >1000 mm of precipitation, and in northern hemisphere temperate deciduous hardwood and coniferous forests (**Table 2**).

General Effects of Deforestation on Freshwater Nutrient Loading

Despite the complexity and variability in freshwater nutrient loading, some important generalizations can be made. These are:

1. The usual increase in nitrification following deforestation is one of the most important determinants of not only NO_3 loading in freshwater, but also the loading of other nutrients whose fluxes can be affected by the nitric acid produced by nitrification.
2. The type of vegetation influences its response to deforestation with temperate deciduous hardwood forests likely to exhibit greater nutrient flux changes than temperate coniferous forests in the northern hemisphere.
3. Nutrient flux changes occurring when herbicides cause or accompany deforestation are likely to be greater than those following deforestation without herbicides.
4. Deforestation-caused changes in freshwater nutrient loading are usually short lived, particularly for nitrogen – generally up to 7 years, but usually considerably less. However, base cation changes may occur for more than 10 years.

See also: Chemical Properties of Water; Coarse Woody Debris in Lakes and Streams; Ecology and Role of Headwater Streams; Hydrological Cycle and Water Budgets; Hydrology: Rivers; Riparian Zones; Streams and Rivers as Ecosystems.

Further Reading

Bernhardt ES, Likens GE, Buso DC, and Driscoll CT (2003) In-stream uptake dampens effects of major forest disturbances on watershed nitrogen export. *Proceedings of the National Academy of Sciences* 100(18): 10304–10308.

Bernhardt ES, Likens GE, Hall RO Jr, *et al.* (2005) Can't see the forest for the stream? In-stream processing and terrestrial nitrogen exports. *BioScience* 55: 219–230.

Binkley D and Brown TC (1993) Forest practices as nonpoint sources of pollution in North America. *Water Resources Bulletin* 29: 729–740.

Bormann FH and Likens GE (1979) *Pattern and Process in a Forested Ecosystem.* New York, NY: Springer-Verlag.

Brown GW (1985) *Forestry and Water Quality.* Corvallis, OR: Oregon State University Press.

Feller MC (2005) Forest harvesting and streamwater inorganic chemistry in western North America: A review. *Journal of the American Water Resources Association* 41: 785–811.

Harris GP (2001) Biogeochemistry of nitrogen and phosphorus in Australian catchments, rivers and estuaries: effects of land use and flow regulation and comparisons with global patterns. *Marine and Freshwater Research* 52: 139–149.

Hornbeck JW, Martin CW, Pierce RS, *et al.* (1986) Clearcutting northern hardwoods: effects on hydrologic and nutrient ion budgets. *Forest Science* 32: 667–686.

Malmer A (1993) *Dynamics of Hydrology and Nutrient Losses as Response to Establishment of Forest Plantation. A Case Study on Tropical Rainforest Land in Sabah, Malaysia.* Umeå, Sweden: Dissertation.

Mann LK, Johnson DW, West DC, *et al.* (1988) Effects of whole tree and stem-only clearcutting on postharvest hydrologic losses, nutrient capital, and regrowth. *Forest Science* 34: 412–428.

Martin CW, Noel DS, and Federer CA (1984) Effects of forest clearcutting in New England on stream chemistry. *Journal of Environmental Quality* 13: 204–210.

Naiman RJ, Décamps H, and McClain ME (2005) *Riparia. Ecology, Conservation and Management of Streamside Communities.* Elsevier Academic Press.

Relevant Websites

http://www.hubbardbrook.org – Hubbard Brook.
http://www.fsl.orst.edu/lterhome.html/ – H.J. Andrews.
http://coweeta.ecology.uga.edu/webdocs/1/index.htm – Coweeta.
http://www.umanitoba.ca/institutes/fisheries/ – Experimental Lakes Area.
http://www.tlws.ca/index2.shtml – Turkey Lakes.
http://bangor.ceh.ac.uk/plynlimon – Plynlimon.

Agriculture

J R Jones, University of Missouri, Columbia, MO, USA
J A Downing, Iowa State University, Ames, IA, USA

Introduction

Agriculture accounts for about three-fourths of human use of fresh water and demand will increase with growth in world population. Around 40% of land surface on the planet has been converted to cropland and pastures, which nearly matches current forest cover. Intensified agriculture for humans and livestock relies on maximizing crop production by using nitrogen and phosphorus fertilizers to overcome nutrient constraints on cultivated plants (**Figure 1**). Fertilizers are often applied in excess of minimal crop needs in anticipation of higher than average yields when climatic conditions are favorable during the growing season. This practice has been cost-effective – historically fertilizers have been economical and the value of additional produce has greatly outweighed the extra cost of nutrient application.

Global fertilizer use has increased by 8-fold for nitrogen and 3-fold for phosphorus over the past four decades with measurable increases in crop production. Fertilizer use has highly increased in developing countries to achieve self-sufficiency in food production. Aerial application rates are greatest in parts of China and India but are generally higher in Europe than the United States. Commercial nitrogen fertilizer is produced by an industrial process that converts atmospheric nitrogen to available forms. This product is the largest human-generated contribution to the global nitrogen cycle. Phosphorus is mined from rich mineral deposits, converted to water-soluble forms, and distributed on arable land to increase production. Occasionally, less than half of these nutrient amendments are incorporated into harvested produce with the remainder being stored in soils or leached to surface/groundwater, or in the case of nitrogen – released as gas back to the atmosphere. Nutrient losses from intensified agriculture occur, in part, because fertilizer applications are not always synchronized with periods of peak plant growth or to avoid runoff episodes.

Intensification and specialization in agriculture has resulted in replacement of pasture-based livestock practices by large, confined animal operations. Livestock consolidation results in the transport of crop nutrients from grain- to animal-producing areas, resulting in regional imbalances in nutrient inputs and outputs with surpluses near animal operations. Historically, animal manures were used on the farm but with livestock intensification the large generation of wastes presents a regional disposal problem. Less than a third of the grain produced in North America is currently fed on farms where it is grown. Increasingly this redistribution of nutrients is crossing geo-political boundaries on a global scale. Overall, fertilizer and manure amendments dominate nutrient cycles in agricultural watersheds and substantial losses result in nonpoint pollution, which directly impairs the water quality of surface and groundwater (**Figure 1**). Nitrogen and phosphorus stimulate plant production in streams and lakes much like the terrestrial process. Water quality problems stem from the fact that even minor losses of the total on-farm nutrient budget can cause quantifiable, negative impacts on aquatic ecosystems. Over-enrichment by nutrients (eutrophication) results in algal blooms, oxygen depletion, changes within biotic communities, and reduced utility of the water for most human uses and ecosystem services. These impacts are measurable locally but also in major river drainages, estuaries, and coastal marine areas as a result of cumulative effects.

Nutrient Loss and Fertilization Practices

Nutrient loss from cultivated watersheds that are fertilized is much larger than from forests and grasslands. In agricultural areas, nutrient export varies with vegetation cover, nutrient management, and related agronomic practices, terrain, soil composition, rainfall, and flow paths (hydraulic connections to surface flow), and the extent of buffer vegetation (to take up or trap nutrients). Widely cited export coefficients (**Table 1**) suggest N losses from corn are similar to soybeans, about 3.5-times pastures and about 6-times losses from forested catchments. For P export, losses from corn are about half that found from soybeans and about 20-times that from forest or pasture.

Applications of nitrogen (N) and phosphorus (P) to cropland accumulate in agricultural soils and are subsequently lost to surface and groundwaters by leaching and erosion. N and P have different chemistry and flow pathways. Overall, N is more likely to be transported in runoff than P because of greater mobility of its dissolved forms in water. Estimates are that 20% of N fertilizers leach into surface and groundwaters. Subsurface loss of N is as the mobile form nitrate and agricultural tile drains accelerate export.

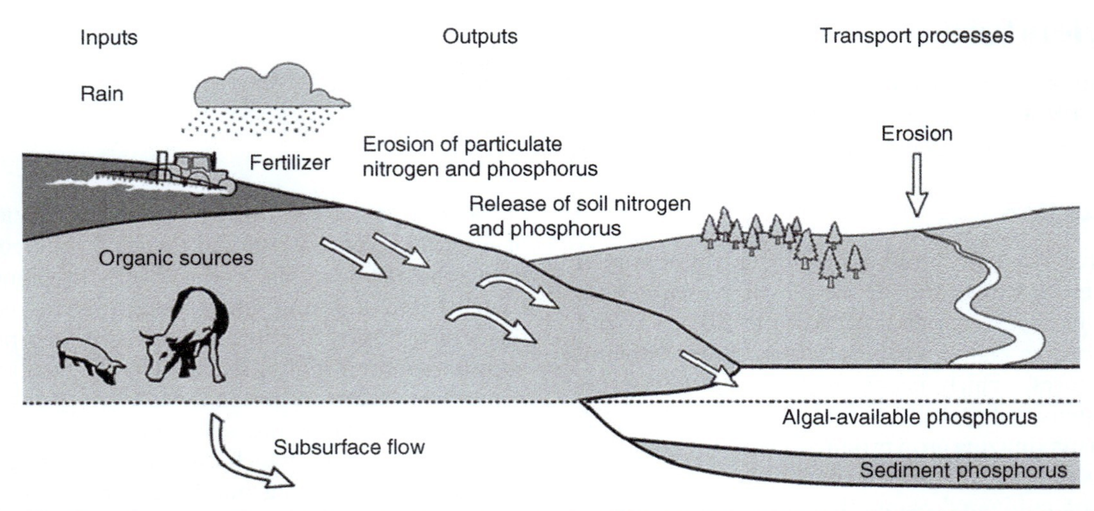

Figure 1 Idealized diagram of nutrient inputs, outputs and transport processes in an agricultural watershed Redrawn from Carpenter *et al.* (1998) Nonpoint pollution of surface waters with phosphorus and nitrogen. *Ecological Applications* 8: 559–568.

Table 1 Export coefficients (kg ha^{-1} year^{-1}) from various agricultural practices

Pollutant budget estimation form

Land use	*Nitrogen export coefficient (kg ha^{-1} year^{-1})*	*Phosphorous export coefficient (kg ha^{-1} year^{-1})*
Forest	1.8	0.11
Corn	11.1	2.0
Cotton	10.0	4.3
Soybeans	12.5	4.6
Small grain	5.3	1.5
Pasture	3.1	0.1
Feedlot dairy	2900	220

Values from Reckhow KH, Beaulac MN, and Simpson JT (1980) Modeling phosphorous loading and lake response under uncertainty: A manual and compilation of export coefficients. Washington, DC: U.S. Environmental Protection Agency. USEPA 440/5-80-011.

The magnitude of volatilized gaseous forms of N from agriculture (from fertilizers and manure) in the United States matches that of nitrate leaching. Most N emissions are redeposited nearby and can enter waterways.

When long term inputs of commercial fertilizer or animal manures exceed the off-take of P in harvested crops, P accumulates in agricultural soils. Inputs of fertilizer P can be greater than double the amount removed as crops causing surplus soil accumulations from an agronomic standpoint. This is common in fields near dairy barns and in one agricultural area >75% of the soils have excess phosphorus. This imbalance increases potential throughput to aquatic ecosystems that could be maintained for decades. Most soil-bound P lost from agricultural watersheds comes in overland flow from hydrologically active source areas during large erosion-causing storms. Losses are mostly particulate P (60–90% sediment and organic-bound), which may not be immediately available for biological uptake and it is the smaller dissolved fraction that is biologically available and can directly impact aquatic ecosystems. Dissolved P dominates runoff from intensive grassland areas, where erosion is modest. Loss of available P in surface runoff is closely correlated with the P content of surface soil, which increases with surplus application. P concentrations can be reduced in water that infiltrates and travels as subsurface flow from agricultural watersheds because of fixation in subsoils. Subsurface losses are regionally important in areas with sandy, well drained soils that lack capacity to retain dissolved P. Losses of P are typically <5% of the soil amendment; these losses considered minor and economically unimportant to farm production but contribute directly to water quality problems. The ultimate sink for this P is the sediments of streams, lakes, and coastal areas.

In developed countries, about 70% of harvested crops are fed to livestock and the resultant manure is comparable to human sewage, but with less stringent regulation. Intensification of the animal industry has resulted in fewer operations supplying a larger share of overall livestock production. In many cases wastes produced from single confined animal operation (housing hogs or poultry) equals the discharge of a medium sized municipality. The economic efficiency of clustering confined feeding operations in rural areas can result in waste output from intensified livestock regions to equal that of major cities.

Like human sewage, animal wastes are rich in organics (high biochemical oxygen demand) and nutrients. Animal wastes from confinement operations are held or partly treated in holding ponds, collection yards, lagoons (dairy and swine facilities), or in litter (poultry operations). During storage, the N content of manure is reduced by volatilization of ammonium and other gases that contribute to the flux of N that can enter surface waters. These materials represent valuable organic matter and nutrient amendments for cultivated fields and pastures but are bulky to move and apply (solids, slurries, or liquids) when compared with commercial fertilizers. Also, there is competition for land suitable for manure application near confined animal operations. Increasingly, manure applications exceed potential crop uptake in these areas increasing the potential for loss to surface waters. Heavy metals, added to feed as micronutrients or bactericides (particularly zinc, arsenic, and copper), can also accumulate in soils. Animal wastes can also enter waterways directly from leakage or overflow of lagoons, and spills have caused fish kills in response to high ammonium (>2 mg l^{-1}) and decreased oxygen.

Ethanol production is projected to increase maize production in North America by about 20%. Intensified crop production and conversion of idle, marginal land into crop production will potentially impact water quality. This is especially serious because much of the idle land is near to waterways and has high slopes and erodible soils. Marked expansion of ethanol production in grain-based biorefineries will also add to regional nutrient imbalances. Distillers grains generated by the production of ethanol are rich in P and protein and are typically used as a feedstock for livestock. Increased use of distiller's grains in feed can increase nutrient content of animal manure particularly on beef and dairy operations.

Manure application is typically aimed at meeting N requirements of crops, which results in a buildup of soil P. Distribution of manure to satisfy crops' P requirements requires vast amounts of land for application because, relative to crop needs, P is higher than N in manure. A United States study concluded livestock waste was the largest source of P contamination in streams and rivers. Manure amendments reduce the ratio of N-to-P in agricultural runoff, which favors cycanobacteria and increases the likelihood of algal toxin production.

Livestock wastes are a source of coliform bacteria (including the toxin producing O157:H7 strain) and protozoan pathogens (such as *Crypotsporidium* and *Giardia*) to surface and groundwaters; disease outbreaks from contaminated municipal water supplies have occurred as a result. Veterinary and nutritional pharmaceuticals, and excreted hormones from livestock enter waterways but there is a lack of information on the widespread impact of these potential pollutants on aquatic communities. About one-third of United States antibiotic use is added to animal feed to prevent infection and promote rapid growth. This practice likely contributes to the increased antibiotic resistance of microbial populations in surface water exposed to livestock wastes. Some of these chemicals affect reproductive endocrine function of wild fishes by mimicking natural hormones.

Nutrients and Sediments in Agricultural Landscapes

Lakes in agricultural landscapes may have naturally greater fertility than lakes in other biomes because of their location in arable soils that have sufficient natural fertility to generate economically viable produce. The inherent tie between soil fertility and lake productivity has been long recognized by ecologists and is nicely demonstrated in the state of Minnesota (USA). Lakes least impacted by human disturbance (reference lakes) in the agricultural plains in southern Minnesota have an order of magnitude larger P values and 5-times the N content of those in the predominately forested northern region. Lakes are intermediate in the central part of Minnesota where land cover is a mosaic of agriculture and forest. Fertilizers and manure amendments in cultivated landscapes increase nutrient export to aquatic ecosystems above background levels and are considered the major water quality impairment.

Data from a suite of Missouri (USA) streams demonstrate how N and P concentrations increase in response to crop cover, a surrogate for nonpoint source nutrient loss from agriculture (**Table 1**), and decrease with forest cover (**Figure 2**). This cross-system pattern is consistent with differences in export coefficients from these two land cover types (**Table 1**) and shows how stream water quality integrates land cover. In Iowa (USA), a region of intensive agricultural production, stream nitrate concentrations increase with watershed density of animal units and row crop and accounted for 85% of the among-stream variation in this nutrient in one study. In the Mississippi River nitrate levels have shown a 2.5-fold increase since 1960 and use of commercial N fertilizers accounts for much of this increase (**Figure 3**). Estimates suggest a $<15\%$ reduction in fertilizer use would achieve a $>30\%$ reduction in riverine nitrate flux with little influence on crop yields.

A strong cross-system pattern between nutrients in Missouri and Iowa reservoirs and cropland demonstrates how lake and reservoir fertility increases directly

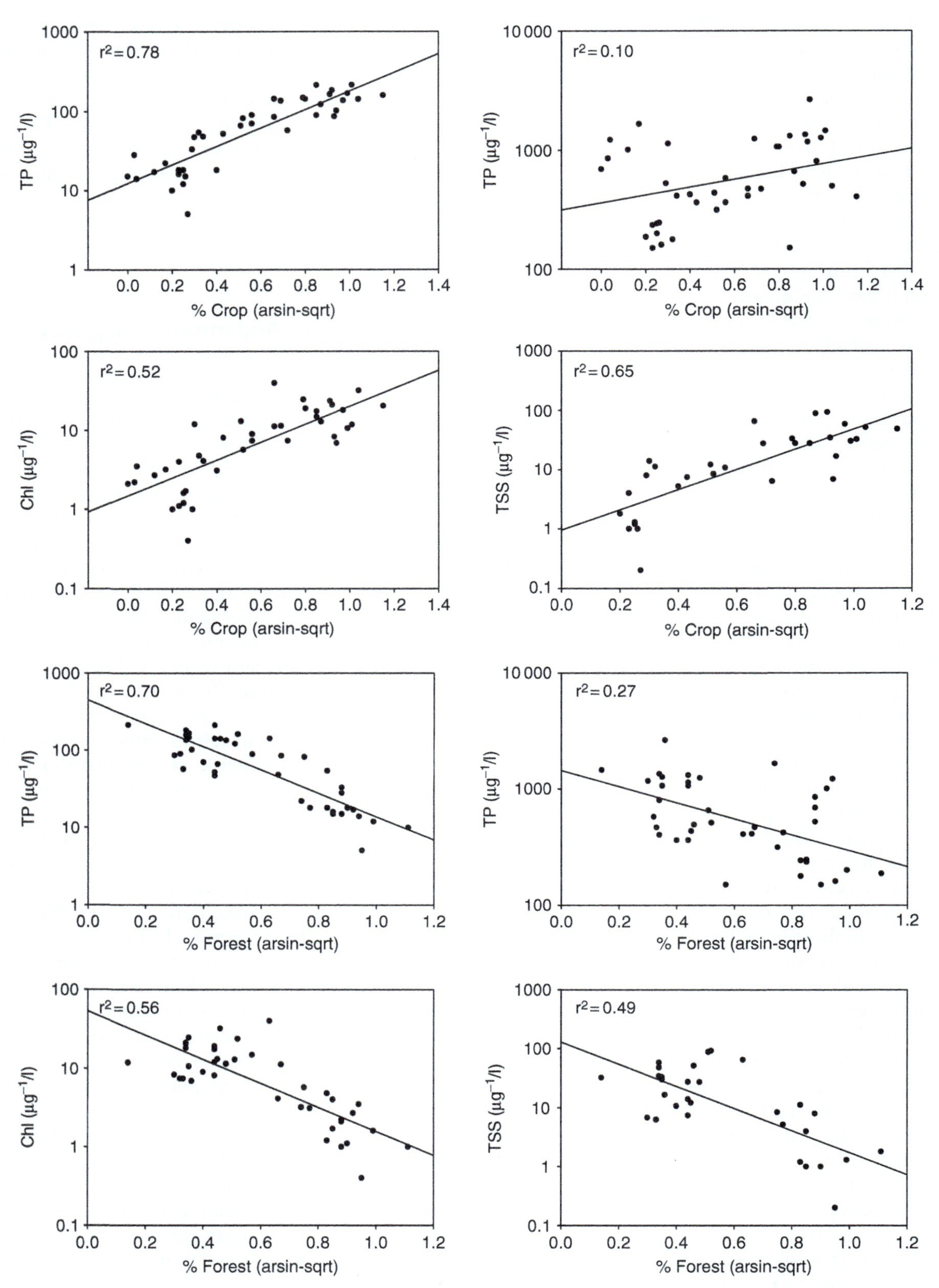

Figure 2 Values of total phosphorus (TP), total nitrogen (TN), algal chlorophyll (Chl) and total suspended solids (TSS) from Missouri Streams (log transformed) plotted against the proportion of crop (upper four panels) and forest (lower four panels) in the watershed (arcsin square root transformed). From Perkins *et al*. (1998) *Verh Internat Verein Limnol* 26: 940–947, with permission.

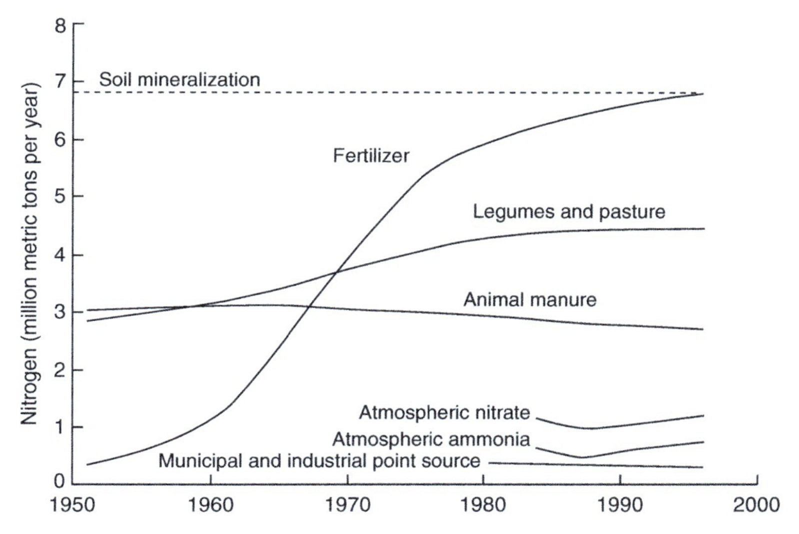

Figure 3 Historic increase in nitrogen in the Mississippi River relative to other sources. Adapted from **Figure 5** in Goolsby and Battaglin (2000) USGS Fact Sheet 135–00.

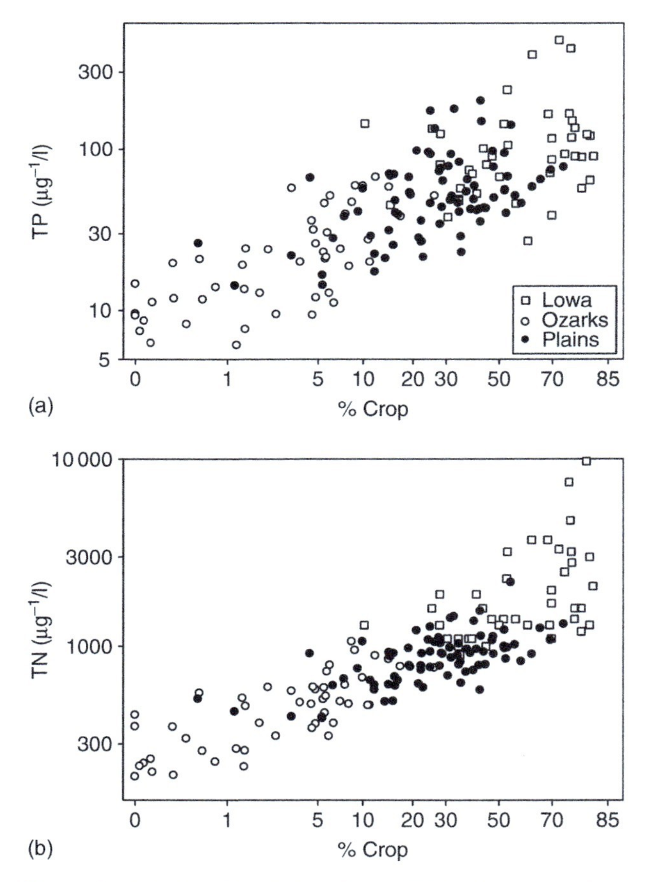

Figure 4 Relation of total phosphorus (TP, log-transformed) and total nitrogen (TN, log-transformed) in reservoirs in the Missouri Ozarks, Missouri Plains and southern Iowa to crop (logit-transformed) in the catchment of each impoundment. Figure from Jones *et al.* (2008) *Lake Reserve Management* 24: 1–9, with permission.

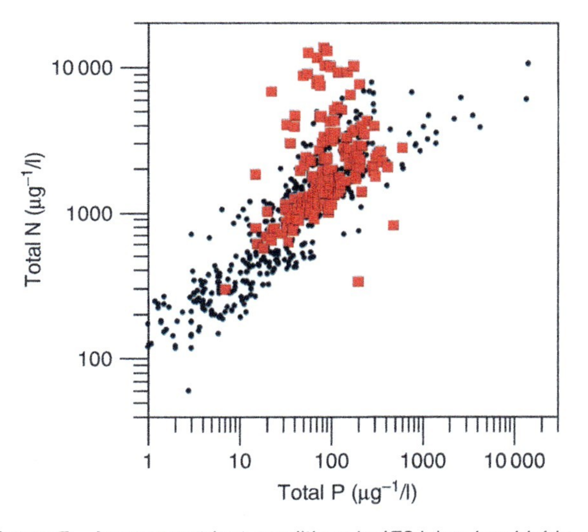

Figure 5 Average nutrient conditions in 172 lakes in a highly agricultural region (Iowa, USA from 2001–2006; red squares) plotted against average conditions in world lakes. Adapted from Downing and McCauley (1992) *Limnology and Oceanography* 37: 936–945. Only a few of the most phosphorus-rich African lakes exceed nutrient levels seen in lakes in agricultural regions.

with agricultural cultivation in the United States mid-continent (**Figure 4**). The increase, which amounts to a 4-fold increase in P and 3-fold increase in N, reflects the increase in cropland agriculture along this geographic axis. Stream nutrients would double or triple across this continuum of increasing crop cultivation. N and P concentrations in lakes receiving run-off from agricultural watersheds in the state of Iowa have some of the highest nutrient concentrations in the world (**Figure 5**). Historical trends in agriculture and water quality in Clear Lake, Iowa (USA) shows the cumulative influence of agriculture on lake water clarity due to land clearing, draining wetlands, and intensification in an individual lake-watershed system (**Figure 6**). Collectively, these studies show extensive nutrient loss from cultivated agricultural land to stream and rivers.

Cultivation increases soil erosion and the total suspended solids load of the Missouri streams show this landscape level disturbance with direct increases with crop cover and decreases with forest (**Figure 2**). Factors that increase erosion and surplus nutrients in soil, as occurs in modern agriculture, increase the potential for nutrient loss to streams. There are no global figures but soil erosion from agriculture is considered responsible for a great deal of the sediment supplied to rivers, lakes estuaries and the oceans. In many agricultural regions of the United States, sediment is considered the greatest water quality problem. Agriculture has modified vegetation structure, riparian areas and hydrological regimes of landscapes which has increased overland flow, stream discharge and bank erosion. Maximum erosion occurs in storm events during planting and after harvest when crop cover is minimal. Soil loss leads to increased turbidity

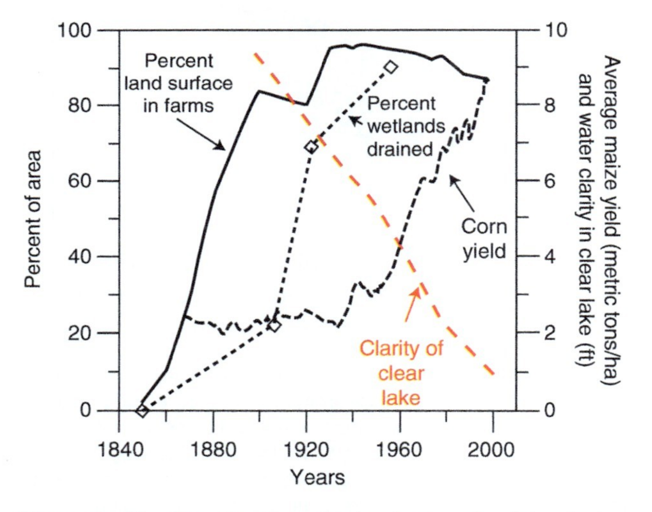

Figure 6 The lines show how the landscape of an intensively agricultural area has been altered by human activities over the last 150 years. This has been achieved by conversion of more than 90% of the land to farmland (1850–1930), drainage and tiling of more than 90% of the wetlands (1900–1960), and four-fold intensification of agricultural yields through fertilization, increased mechaniation and pesticide use (1940 to present). The red line shows the steady decline in the clarity of Clear Lake (shown in feet to fit scale) that has been correlated with changes in the Iowa landscape. Clarity in the late 1800s, allowed aquatic plants to grow to depths of 6 m. Today, water clarity is 0.3–1 m, depending on the weather and season. Data are from the United States Department of Agriculture, the United States Census Bureau and public water quality records.

in receiving waters and light-limitation of photosynthesis. It also causes physical destruction of habitat; deposited sediment disrupts channel hydrology, blankets and intrudes into gravel beds used for fish spawning and by invertebrates, and accelerates the loss of volume in lakes and reservoirs. The clay and silt fraction is a carrier of adsorbed P, pesticides and metals. Conservation tillage, which maintains crop residues, is known to reduce runoff and soil loss.

Coastal Waters and Aquaculture

Nutrients released into lakes and rivers flow downstream leading to coastal eutrophication. Widespread eutrophication of estuaries and coastal areas by nutrient pollution, partly from agriculture, is well documented and expected to increase. In contrast to temperate lakes, eutrophication of coastal waters is frequently controlled by N. The primary symptoms of anthropogenic N enrichment of coastal areas are similar to lakes – increased, and maintained, algal production and reduced transparency. Sustained surface productivity results in greater flux of organic material to bottom sediments where bacteria consume oxygen. Oxygen cannot be renewed from surface waters because of strong stratification during summer (differences in temperature and salinity between surface and bottom layers prevent mixing) and light limitation prevents deep-water photosynthesis. Oxygen consumption results in hypoxia ($<2\ \mathrm{mg\ l^{-1}}$ oxygen) that influences sediment chemistry and the native biota. The condition became more prevalent worldwide in the 1960s as the ability to assimilate additional anthropogenic nutrients from urban and agricultural sources was exceeded and is now common in major bays and estuaries worldwide. Conditions are typically worse near the terminus of major rivers with altered nutrient flux and water quality.

In response to increased loads of inorganic N from the Mississippi drainage (**Figure 3**) hypoxic bottom water in the continental shelf of the Gulf of Mexico has rapidly developed into the largest hypoxic zone in the Western Hemisphere (from the Mississippi Delta to Texas coast). This zone varies annually in response to river flow, nutrient flux and sediment organics but has exceeded 15 000 $\mathrm{km^2}$ most years since the early 1990s. It forms in the area of an important commercial and recreational fishery and when bottom waters are hypoxic, fishing boats do not capture shrimp or bottom dwelling fish. Most aquatic species cannot survive such low oxygen levels and in areas of recurring and severe hypoxia the sea floor community shows reduced abundance, species richness and biomass.

Eutrophication of coastal and estuarine waters creates a favorable nutrient pool for algal blooms and toxic or harmful algal species. Outbreaks along the eastern United States coast and in tributaries of Chesapeake Bay have been linked to nutrient inputs. Increased use of urea fertilizer may also increase the likelihood of dinoflagellates that grow well on this nutrient, including the toxic *Pfiesteria* spp., the phantom dinoflagellate.

Aquaculture ponds and net-pen cages are aquatic livestock operations representing a growing segment of the world food supply. These operations are another source of organic matter, nutrients and antibiotics to rivers, lakes and coastal areas. Water quality impacts of fish culture are typically localized near the aquaculture zone but large intensive operations have altered the tropic state of coastal waters and impoundments; culture operations have been banned from some water supply reservoirs. Discharges from recirculating flow systems to local streams can impact water quality. Culture also increases the risk of disease and parasites. Many aquatic species being cultured are not native to their farm sites or are selectively bred and differ genetically from native strains. Escapes can have irreversible ecological impacts. For example, cultured Asian carp have become established in the Mississippi basin and compete with native fish and in some coastal rivers farmed escapees outnumber wild salmon in spawning rivers.

Changes in Biota

Loss of nutrients and organic matter from agricultural landscapes increase available energy at the base of food web by promoting growth of algae, bacteria and invertebrates with potential impacts on native biota. Increased fertility results in greater overall production and biomass of lake fisheries which is beneficial but with over enrichment there is a concurrent reduction in species richness and community diversity. Changes in the trophic structure of fish communities in agriculturally eutrophic lakes involve a reduction in the proportion of desirable sport fish (primarily piscivores, such as bass) and concurrent increase of less desirable omnivorous fish that feed primarily in bottom sediments (benthivores, such as carp, roach, bream). Most sport fish are sight feeders and depend on clear, oxygenated water. In contrast, fish taxa responding positively to eutrophication tolerate poor water transparency and forage on sediment detritus and/or invertebrates without relying on vision. Feeding and bioturbation of sediments by benthivores increases resuspension of mineral turbidity and results in the translocation of nutrients to the water column which promotes planktonic algal growth and further reduces lake transparency. Omnivores directly compete with early life stages of sport fish by consuming zooplankton causing a shift in the size structure to small zooplankton. The feedback of this results in reduced recruitment of desired fish species and less grazing pressure on planktonic algae. The replacement of picivores by benthic omnivores has been documented in temperate and subtropical lakes located worldwide. In Iowa lakes (USA), with an algal chlorophyll range of 10–100 $\mu g\,l^{-1}$, the catch of sport fish per unit effort decreased by half with increased chlorophyll while the catch of bentivores doubled.

Dramatic alteration of landscapes to accommodate crop and pasture-based agriculture negatively alters stream habitats and biota. Removal of natural vegetation from overland flow pathways and along stream margins alters stream processes. Leaf and organic litter from terrestrial vegetation is an energy subsidy to the natural assemblage of stream organisms. Woody debris provides habitats and shelter for species and so promotes diversity in physical characteristics and biota of the stream channel. Riparian vegetation provides shade, moderates stream temperatures, moderates extremes in streamflow (highs and lows) and reduces diffuse loss of nutrients and sediment from agricultural watersheds. Vegetation removal opens the stream channel to direct sunlight which favors primary production (algal); this is a shift the energy base can cause functional changes in the invertebrate and fish communities that extend downstream. Species richness and composition of stream fish assemblages and overall habit quality declined in a Midwestern watershed (USA) with the extent of agriculture and increased with wetland and forest cover. These ecological processes in streams are well understood but predicting a site-specific or threshold response to nutrient enrichment is complicated by physical and biological interactions. This lack of understanding makes it difficult to quantify the benefits of nutrient management in stream ecosystems.

Pesticides

Pesticides, including insecticides, herbicides, fungicides, and vermicides are used to control insects, undesirable vegetation, microbes, and parasites that can damage plant and animal products in a variety of agricultural, silvicultural, animal husbandry, and domestic activities. Pesticide formulations have changed dramatically over the past four decades to reduce environmental persistence, bioaccumulation and nontarget toxicity. Early organochlorine insecticides such as

DDT were highly toxic to nontarget organisms and bioaccumulated in food chains. These chemicals were linked to developmental and reproductive problems in wildlife and massive pesticide-related fish kills. As a consequence, they were restricted for use in many countries in the 1970s, but are still widely detected in environmental samples. Application of legacy organochlorine insecticides continues in some developing nations. In North America, Europe and elsewhere they have been largely replaced by less persistent compounds such as organophosphates, carbamates, and pyrethroids which do not bio-accumulate. Currently, some organophosphates are being removed from agricultural applications due to high neurotoxicity to birds and mammals. Herbicides account for about half of pesticide use in many areas. Most herbicides exhibit low toxicity to fish and invertebrates; however, some (e.g., atrazine) are extremely water soluble and persistent which has resulted in exceedences of human drinking water criteria. Herbicide loss to surface and groundwater can be minimized by prescription management plans which consider soil type, climate, and tillage practices. Pesticide exposure of aquatic organisms is greatest in regions of multi-crop production where irrigation water is discharged to surface waters; nevertheless, nontarget environmental impacts of pesticides have been reduced in many locations as a result of several decades of research and adaptive management.

Irrigation

Industrialization of agricultural systems to meet demands for grain and protein has moved production from rain-fed cultivation to irrigation. At present 40% of crop production comes from the 16% of irrigated land, and increases in production will require additional river diversions, surface impoundments, and groundwater withdrawal. Crop irrigation increases the potential for soil and water contact and can increase loss of P and other chemicals in return flows. Extensive irrigation withdrawal from major rivers, such as the Colorado, has greatly reduced natural flow to the sea with direct impacts on fish and wildlife. The surface areas of the Aral Sea and Lake Chad have been reduced by diversions for agriculture. Non-sustainable pumping of fossil ground water in aquifers for agriculture occurs in areas with little rainfall and inadequate surface supplies. In the Great Plains (USA), regional depletion of the Ogallala aquifer have driven irrigated agriculture out of production. Salinization is a worldwide problem resulting from irrigation; when irrigation water evaporates from the soil surface or is transpired by vegetation, salts accumulate in the root zone, causing plant stress. Millions of hectares of irrigated land on several continents are damaged by salinity and some are lost to production because of salt accumulation. Most notably in Australia salinization is occurring from ancient salts, deep in the soil profile, rising to the surface in groundwater. This is a result of land cover conversion – removal of deep rooted native vegetation and replacement by shallow rooted crop and pasture plants has reduced evapotranspiration that naturally lowered the water table. Irrigation return flows carry salts, nutrients, and pesticides from crop land to surface waters. Selenium in subsurface irrigation return flows in California (USA) have caused mortality, deformities and reproductive failures in aquatic birds in a wildlife refuge.

Remediation Measures

Blunting the adverse impact of agricultural activities on water quality is complicated because sources are nonpoint and a large number of producers contribute a small individual share to the overall problem. Most localized losses have little short-term impact on farm fertility or economy. By nature, agriculture is leaky – nonpoint pollution from farms is generated over broad areas, losses are subject to weather-related natural variability and site-specific characteristics such as topography, so measuring and controlling impacts are difficult. Regardless, some benefits can be achieved by improving farm management practices and modifying land use/cover in critical source areas where movements from soil to water are the greatest. Controlling nutrient loss from arable land begins with matching fertilizer requirements with crop needs and knowledge of nutrient reserves in soils. N application, timed to crop growth reduces potential losses by runoff and leaching. Conservation tillage increases water infiltration and reduces surface runoff and soil erosion and loss of sediment-bound chemicals, including pesticides. Seasonal cover crops provide some of these same benefits. Manure applications should be calculated to avoid surplus levels of both N and P in soils and incorporation into soil by tillage reduces the potential for losses. Matching P content of feed with the dietary requirements of animals can reduce the amount applied to land. In some locations large volume wastes generated by high-density livestock operations may need to be managed as a point source and treated as municipal wastewater to remove organic matter and nutrients to avoid surplus application to farm land. Other alternatives are pelletization or granulation for easy transport to nutrient deficient areas, composting for the landscape industry or energy generation. Healthy, deep rooted vegetation in the critical source areas where overland flow occurs within the landscape and along the margin of streams (riparian buffer zones) provide a nutrient and sediment sink and promote streambank stabilization while benefiting both wildlife and stream habitat. Wetlands in agricultural landscapes are a sink for N, which is lost as gas by microbial processes. Agriculture has at hand technical support systems that monitor nutrients and provide precision application techniques for fertilizer and pesticides to minimize nonpoint loss. Even with best management practices, however, logistical constraints, site-specific soil properties, inadequate storage facilities for animal wastes, ambient soil moisture, and storm events will influence nonpoint losses to aquatic ecosystems.

Further Reading

Burkholder J, Libra B, Weyer P, Heathcote S, Kolpin D, *et al.* (2007) Impacts of waste from concentrated animal feeding operations on water quality. *Environmental Health Prospectives* 115: 308–312.

Carpenter SR, Caraco NF, Correll DL, Howarth RW, Sharpley AN, and Smith VH (1998) Nonpoint pollution of surface waters with nitrogen and phosphorus. *Ecological Applications* 8: 559–568.

Chadwick DR and Chen S (2002) Manures. In: Haygarth PM and Jarvis SC (eds.) *Agriculture, Hydrology and Water Quality*, pp. 57–82. Wallingford, UK: CABI Publishing.

Edwards AC and Chambers PA (2002) Quantifying nutrient limiting conditons in temperate river systems. In: Haygarth PM and Jarvis SC (eds.) *Agriculture, Hydrology and Water Quality*, pp. 477–494. Wallingford, UK: CABI Publishing.

Hatch D, Goulding K, and Murphy D (2002) Nitrogen. In: Haygarth PM and Jarvis SC (eds.) *Agriculture, Hydrology and Water Quality*, pp. 7–28. Wallingford, UK: CABI Publishing.

Haygarth PM and Jarvis SC (eds.) (2002) *Agriculture, Hydrology and Water Quality*. Wallingford, UK: CABI Publishing.

Howwarth RW, Sharpley A, and Walker D (2002) Sources of nutrient pollution to coastal waters in the United States: Implications for achieving coastal water quality goals. *Estuaries* 25: 656–676.

Leinweber P, Turner BL, and Meissner R (2002) Phosphorus. In: Haygarth PM and Jarvis SC (eds.) *Agriculture, Hydrology and Water Quality*, pp. 29–56. Wallingford, UK: CABI Publishing.

Matson PA, Parton WJ, Power AG, and Swift MJ (1997) Agricultural intensification and ecosystem properties. *Science* 277: 504–509.

Sims JT and Coale FJ (2002) Solutions to nutrient management problems I the Chesapeake Bay watershed. In: Haygarth PM and Jarvis SC (eds.) *Agriculture, Hydrology and Water Quality*, pp. 345–372. Wallingford, UK: CABI Publishing.

Vitousek PM, Aber JD, Howarth RM, Likens GE, Matson PA, *et al.* (1997) Human alteration of the global nitrogen cycle: Sources and consequences. *Ecological Applications* 7: 737–750.

Climate and Rivers

B Finlayson, M Peel, and T McMahon, University of Melbourne, VIC, Australia

Introduction

There is a clear relationship between climate and river flow, driven by the hydrologic water balance. In **Figure 1** we provide an overview of the hydrologic water balance on a global scale. The flow in rivers represents that part of the precipitation not evaporated or transpired (for practical purposes, percolation to groundwater is usually a minor component anyway). This apparently simple proposition is complicated in many ways through issues of scale, topography, climate, geology and vegetation, itself is strongly related to climate. Climate impacts the flow of rivers in terms of quantity, seasonal distribution, persistence, variability, floods, and low (or no) flow. The flow measured at any point in a river represents the integration of all the drivers of discharge in the catchment upstream of that point. Large river basins may cover more than one climate zone so the flow characteristics of the whole basin may reflect a number of different climate influences and a time delay. It is also the case that both the amount and the properties of the flow vary depending on where in the channel network they are measured. For example, small tributaries in the headwaters of rivers in humid climates may have ephemeral flow similar to that of higher order streams in more arid climates.

The nature of river channels, in terms of planform, gradient, network density, and connectivity, is the product of flow characteristics, sediment type, and sediment quantity, and there are complex interrelations with the vegetation cover of the catchment and the riparian zone. All these are, to some extent, determined by climate.

Most major rivers today have experienced substantial human impact and the way in which the human population exploits and regulates rivers is climate related. The need for irrigation is greatest in warm dry climates, where there is relatively little runoff to be harvested. This has led to the construction of large scale interbasin water transfer schemes that started in North America on the Colorado River and spread across the world, a prominent recent example being the south – north transfers in China from the Yangtze River to the drier Yellow River basin.

In this chapter we have tried to isolate the effects of climate from the many and varied human impacts on river flows and we attempt to present an account of the natural system. It is becoming increasingly difficult to do this using measured river flows, as the modification of natural river flows by dams, water diversions into and out of rivers, and riparian extractions for irrigation and water supplies is becoming more widespread. Indeed, it has been argued that there is now so much water stored in dams in the continents that it has produced a measurable, though little, lowering of sea level. In future this may become complicated further by the impacts of climate change, though the magnitude of such changes to the flow totals and patterns of flow in rivers will probably be small when compared with the artificial changes already observed in river systems.

Climate

We have chosen to represent climate in this discussion using the system of climate classification originally proposed in the late nineteenth century by Wladimir Köppen and developed subsequently by him in association with Rudolf Geiger. **Figure 2** is a map of world climates prepared following the Köppen–Geiger system but using a modern data set with interpolations between stations using a 2D thin-plate spline. The precipitation and temperature data that are available for the construction of such a map suffer from limitations similar to the runoff data that are available on which to base a discussion of climate and rivers. Rivers are gauged mainly for the purpose of assessing water resources or for flood warning. It follows that this is more common in areas with significant population concentrations and where there is sufficiently reliable flow to make water resources development a viable proposition. Large areas of the world that are thinly populated, poor, and dry have few data for runoff and often not much more for climate.

The key to the classification used to construct the map in **Figure 2** is shown in **Table 1**. Note that for broad climate zones (as defined by the first letter), the arid zone (B) climates are distinguished from all others on the basis of a relationship between mean annual precipitation and temperature, rather than monthly temperature metrics alone. All other climate zones are distinguished from each other by temperature criteria, and moisture only becomes relevant in the internal subdivisions of those classes. Here we have the basis of the global pattern of streamflow characteristics and river types in a physical sense.

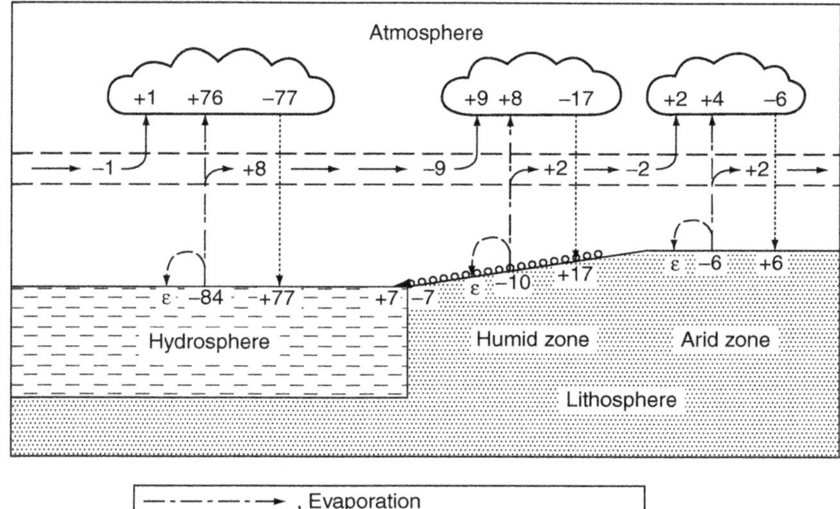

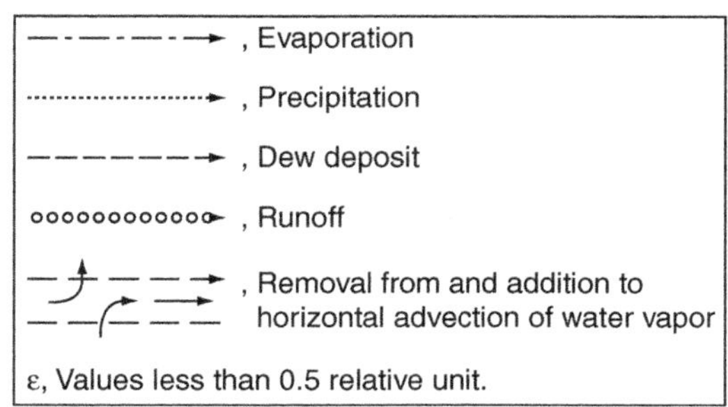

Figure 1 The hydrologic water balance at a global scale. Reproduced from Chow VT (ed.) (1964) *Handbook of Applied Hydrology.* New York: McGraw-Hill, with permission from McGraw-Hill.

Climate and Streamflow

Climate is a major determinant of the amount of flow in a river and of the distribution of that flow through time. Because of the dependence of flow on catchment area (all other things being equal) it is common to compare the flow of rivers using flow per unit area (runoff). This is best achieved by converting the flow to millimeters of depth by dividing flow volume by catchment area. So, flow in thousands of cubic meters per day divided by catchment area in square kilometers gives runoff depth in millimetres per day. This also allows a direct comparison of runoff with rainfall as both are being expressed in the same units (depth per unit time).

Climate as portrayed by the Köppen classification used here represents average conditions over a reasonably long period. There are additional characteristics of climate that this classification does not incorporate but which have important influences on rivers. There are large-scale features of the global circulation system that have the effect of increasing the variability of flow in rivers in the areas that they influence. The best known of these is the El Niño/Southern Oscillation (ENSO) in the equatorial Pacific Ocean, which is the largest of three large-scale ocean–atmosphere fluctuations, the other two being the North Atlantic Oscillation (NAO) and the North Pacific Oscillation (NPO), identified by Sir Gilbert Walker in the 1920s as part of his research into forecasting the Indian monsoon. ENSO type phenomena also influence the occurrence of extreme events, both floods and low flows.

There is also a varying impact around the world of tropical cyclones (hurricanes, typhoons) and this also is not reflected in the classification of climate. Where tropical cyclones occur, they cause extreme flood events in near coastal rivers. When they move over land and become rain depressions, they can generate flooding in rivers at a regional scale. Tropical cyclones do not form within 3° of latitude of the equator and they rarely occur in the tropical oceans adjacent to South America, West Africa, and East Africa north of the Equator. The frequency and seasonal distribution of tropical cyclones are variable in those regions where they occur.

Mean Annual Flow

The effectiveness of precipitation in producing runoff depends on potential evaporation and this is illustrated on a broad scale in **Figure 1.** Frequency distributions of mean annual runoff (MAR) for each

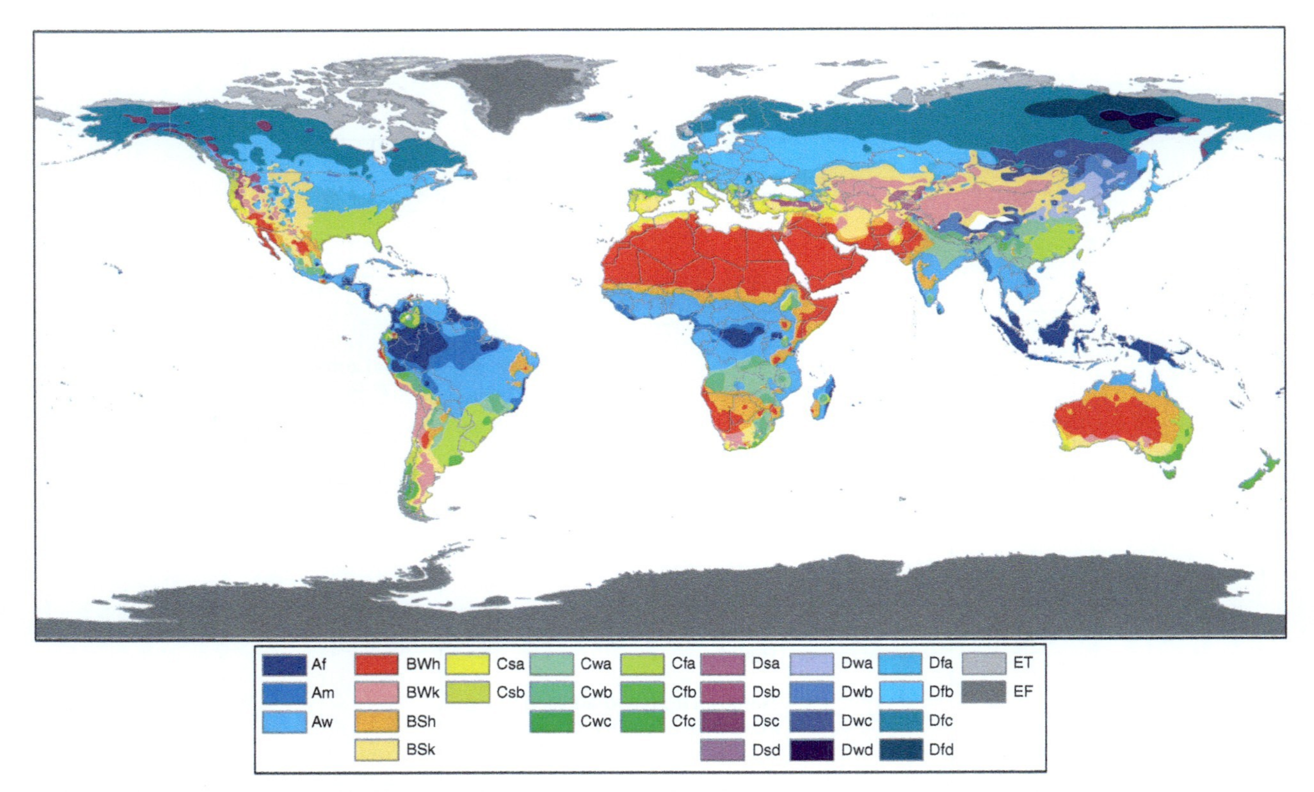

Figure 2 Climate zones of the world classified according to the Köppen–Geiger system. Reproduced from Peel MC *et al.* (2007). Updated world map of the Köppen–Geiger climate classification. *Hydrology and Earth System Sciences* 11: 1633–1644, with permission from European Geosciences Union.

main Köppen climate type are shown in **Figure 3**. Note that the pattern for the whole global data set closely resembles that for the Type C climates. While Type B climates occupy the largest proportion of the land area (30.2%) and the Type C climates cover only 13.4%, the nature of the Type C climates (temperate) and the distribution of the human population, particularly in the highly developed countries, mean that more than half of the river gauging records come from Type C climate zones.

The relationship between climate type and MAR is complicated by seasonality and by topography, which by an orographic effect causes precipitation to be higher (and less variable) than it would otherwise be. There are also the exotic rivers, in which flow is generated in a humid zone and the river runs through an arid zone where there is little or no locally derived addition to the flow. Perhaps the best known example of such a river is the Nile in Egypt where over 80% of the mean annual flow comes from the Blue Nile which has its source in the Ethiopian highlands.

Interannual Variability

The variability of annual flows (measured by the coefficient of variation, C_v) is an important determinant of the beneficial uses to which the flow in a river can be put and a significant factor in the operation of the in-stream ecological system. The relationship between the variability and the amount of annual precipitation is well known; for mean annual precipitation (MAP) below about 600 mm, there is a strong negative correlation between MAP and variability, but there is little trend in the relationship for higher values of MAP. This pattern appears also in the data on the C_v of annual flows as shown in **Figure 4** though the absolute values of C_v for runoff are commonly around twice those of precipitation. The distribution of C_v across the main Köppen climate types is shown in **Figure 5**.

In **Table 2** the average interannual variability of runoff, expressed as the L-C_v for annual runoff for each climate type is listed by continental region. (L-C_v is the L-moment based measure of C_v. L-moments are linear combinations of order statistics and typically L-C_v values are approximately half the magnitude of product-moment C_v values.) Only those climate types that occur in two or more continents and have at least ten values are given. The data in **Table 2** reveal that when intercontinental comparisons are made for the same climate type, there are certain continents that have variability higher than expected. Tests of significance for the 25 pairs available in **Table 2** show that

Table 1 Definitions of classes in the Köppen–Geiger climate classification

1st	*2nd*	*3rd*	*Description*	*Criteria**
A			Tropical	$T_{cold} \geq 18$
	f		Rainforest	$P_{dry} \geq 60$
	m		Monsoon	Not (Af) & $P_{dry} \geq 100 - MAP/25$
	w		Savannah	Not (Af) & $P_{dry} < 100 - MAP/25$
B			Arid	$MAP < 10 \times P_{threshold}$
	W		Desert	$MAP < 5 \times P_{threshold}$
	S		Steppe	$MAP \geq 5 \times P_{threshold}$
		h	Hot	$MAT \geq 18$
		k	Cold	$MAT < 18$
C			Temperate	$T_{hot} > 10$ & $0 < T_{cold} < 18$
	s		Dry Summer	$P_{sdry} < 40$ & $P_{sdry} < P_{wwet}/3$
	w		Dry Winter	$P_{wdry} < P_{swet}/10$
	f		Without dry season	Not (Cs) or (Cw)
		a	Hot Summer	$T_{hot} \geq 22$
		b	Warm Summer	Not (a) & $T_{mon10} \geq 4$
		c	Cold Summer	Not (a or b) & $1 \leq T_{mon10} < 4$
D			Cold	$T_{hot} > 10$ & $T_{cold} \leq 0$
	s		Dry Summer	$P_{sdry} < 40$ & $P_{sdry} < P_{wwet}/3$
	w		Dry Winter	$P_{wdry} < P_{swet}/10$
	f		Without dry season	Not (Ds) or (Dw)
		a	Hot Summer	$T_{hot} \geq 22$
		b	Warm Summer	Not (a) & $T_{mon10} \geq 4$
		c	Cold Summer	Not (a, b or d)
		d	Very cold Winter	Not (a or b) & $T_{cold} < -38$
E			Polar	$T_{hot} < 10$
	T		Tundra	$T_{hot} > 0$
	F		Frost	$T_{hot} \leq 0$

*MAP = mean annual precipitation (mm/yr), MAT = mean annual temperature (°C), T_{hot} = temperature of the hottest month, T_{cold} = temperature of the coldest month, T_{mon10} = number of months where the temperature is above 10°C, P_{dry} = precipitation of the driest month, P_{sdry} = precipitation of the driest month in summer, P_{wdry} = precipitation of the driest month in winter, P_{swet} = precipitation of the wettest month in summer, P_{wwet} = precipitation of the wettest month in winter, $P_{threshold}$ = varies according to the following rules (if 70% of MAP occurs in winter then $P_{threshold} = 2 \times MAT$, if 70% of MAP occurs in summer then $P_{threshold} = 2 \times MAT + 28$, otherwise $P_{threshold} = 2 \times MAT + 14$). Summer (winter) is defined as the warmer (cooler) six month period of ONDJFM and AMJJAS.

Reproduced from Peel MC, Finlayson BL, and McMahon TA (2007) Updated world map of the Köppen–Geiger climate classification. *Hydrology and Earth System Sciences* 11: 1633–1644.

statistically significant differences (at the 5% level) occur in 15 of these, indicated by asterisks (or numbers) in **Table 2**. The average difference in L-C_v for those pairs that show significant difference is 83%, while for the others this difference averages 24%.

The most consistent differences in **Table 2** involve Australia and Southern Africa (defined as Africa south of the equator), when these are compared with other continents. On an average, L-C_v of Australian and Southern African rivers are 2–3 times those of rivers in similar climates on other continents. This is also illustrated in **Figure 4** (for C_v rather than L-C_v) where separate regression lines are fitted to the data for Australia and Southern Africa (combined) and the rest of the world.

The lines of explanation that have been proposed for these differences are as follows. Both Australia and Southern Africa are influenced by ENSO and this increases the annual precipitation C_v compared to other regions. This higher precipitation C_v is then magnified in the conversion of precipitation to runoff. The dominant tree species in the temperate forests of Australia and Southern Africa are evergreen while those of the temperate forests of the northern hemisphere are deciduous. It has been shown that the evergreen forests consistently evapo-transpire more than deciduous forests, leaving a more variable and smaller effective precipitation that becomes more variable runoff. The impact of evergreen vegetation on increasing the annual C_v of runoff declines as annual precipitation increases and is more effective, where the precipitation is winter dominant or nonseasonal.

Seasonal Regimes

The regime of a river is defined as the distribution of flows through the year. Early attempts to develop a global classification of river regimes have been based on the Köppen climate classification, a necessity driven by the lack of adequate global coverage of river flow data. A review of these earlier classifications of river regimes can be found in the chapter by Beckinsale listed below in the further reading. **Figure 6** shows a classification of river regimes derived from a global data set of mean monthly runoff, analyzed using cluster analysis. The global distribution of these regime types is shown in **Figure 7**. Note that this classification does not specifically include ephemeral streams though many of these categories would include periods of cease to flow in the low flow season.

For the most part, the distribution of flow in these regime patterns reflects the distribution of precipitation throughout the year. There are, however, significant exceptions to this. For example, in climatic zones with significant winter snowfall, runoff will be delayed until the onset of warmer conditions causing snowmelt. Across northern Eurasia and North America there is an extensive zone of Df climates – cold climates with precipitation evenly distributed through the year (**Figure 2**). The map of regime types

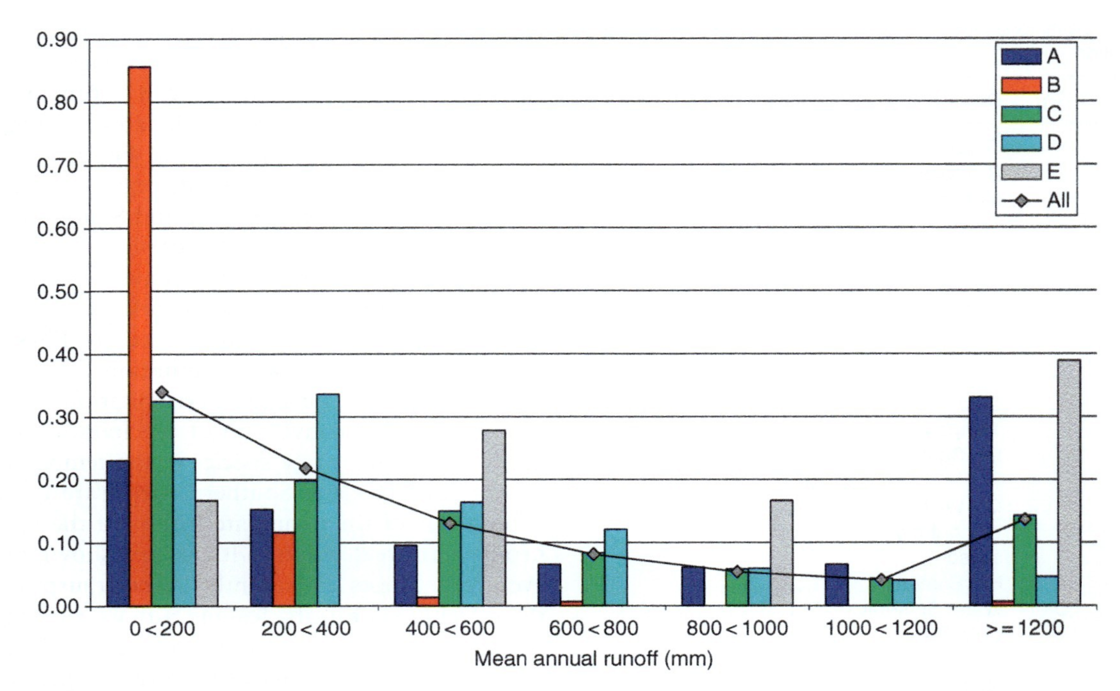

Figure 3 Frequency distributions of MAR for each of the main climate types. The line on the graph is the mean for all stations. Reproduced from Peel MC, McMahon TA, and Finlayson BL (2004) Continental differences in the variability of annual runoff – Update and reassessment. *Journal of Hydrology* 295: 185–197, with permission from Elsevier.

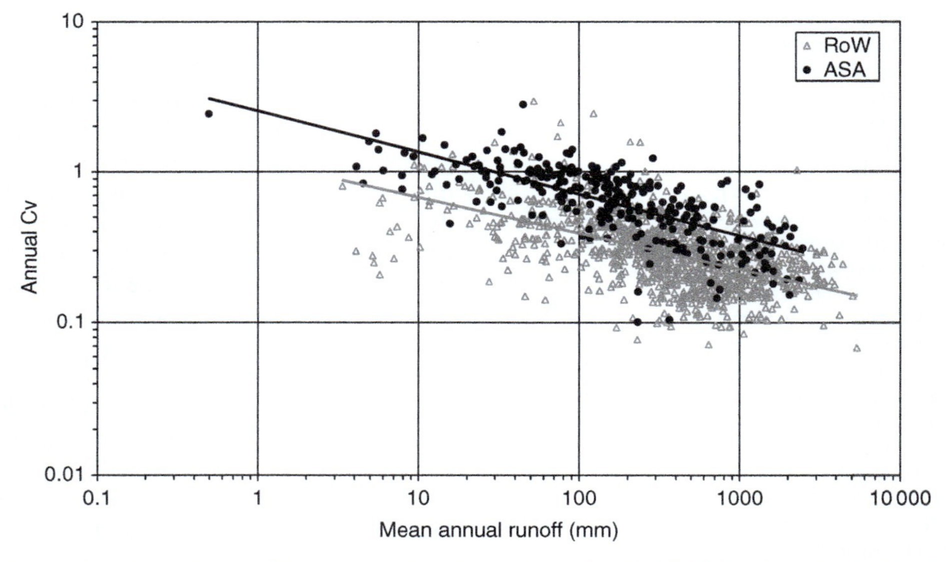

Figure 4 The relationship between the coefficient of variation of annual runoff and MAR. ASA – Australia and Southern Africa; RoW – the rest of the world. Reproduced from Peel MC, McMahon TA, and Finlayson BL (2004) Continental differences in the variability of annual runoff – Update and reassessment. *Journal of Hydrology* 295: 185–197, with permission from Elsevier.

(**Figure 7**) shows extensive regions of regime types 2–5 which have runoff concentrated in the late spring and early summer, driven by snowmelt. Milder C climates at the same latitudes have regimes that reflect the influence of higher evapotranspiration in summer.

The association between Köppen climate zones and these regime patterns is given in **Table 3**, where only flow records for catchments <10 000 km^2 are used in an attempt to ensure that the whole catchment is in a single climate zone. It can be seen from that table that the correlation between regime type and Köppen

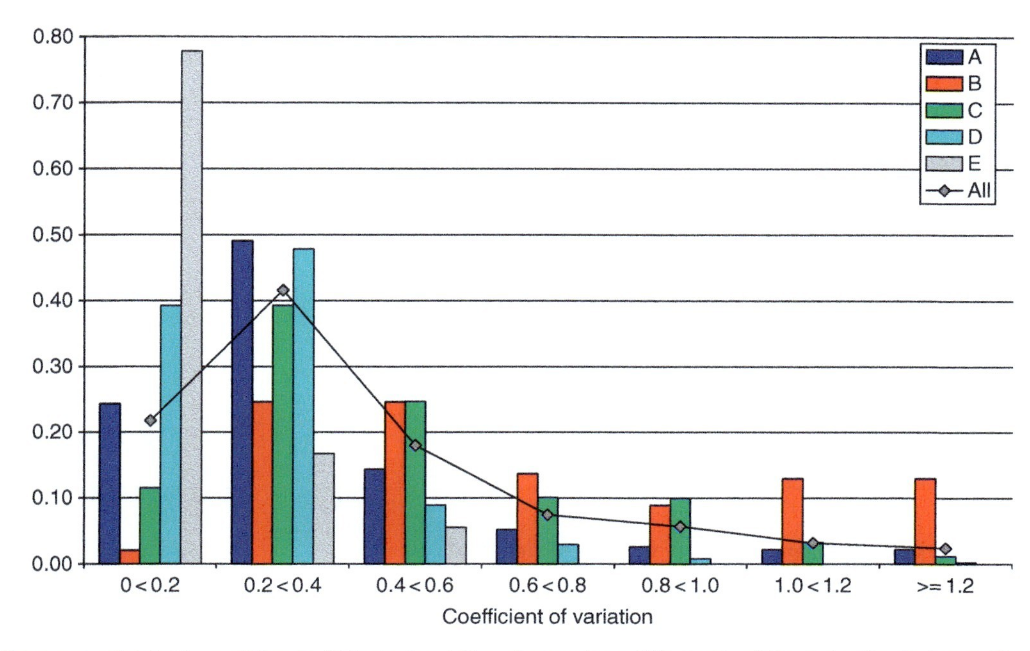

Figure 5 Frequency distributions of the coefficient of variation of annual runoff for each of the main climate types. The line on the graph is the mean for all stations. Reproduced from Peel MC, McMahon TA, and Finlayson BL (2004) Continental differences in the variability of annual runoff – Update and reassessment. *Journal of Hydrology* 295: 185–197, with permission from Elsevier.

Table 2 Interannual variability of runoff (L-C_v) by Köppen climate type and continent

Köppen	*AS*	*AUS*	*EUR*	*NAF*	*NAM*	*SAF*	*SAM*	*SP*
Af	0.15				0.15			
Aw	0.20			0.20^{1}	0.19^{2}		$0.39^{1,2}$	
BSh				0.24*		0.56*		
BSk	0.22*					0.44*	0.31	
Csb		0.34*					0.16*	
Cwa	$0.18^{1,3}$					$0.36^{1,2}$	$0.25^{2,3}$	
Cfa	$0.16^{1,2}$	$0.45^{1,3}$			$0.21^{2,3}$			
Cfb		$0.30^{1,2}$	$0.18^{1,3}$					$0.14^{2,3}$
Dfb			0.17*		0.16*			
Dfc	0.13		0.11		0.14			

Significantly different pairs in each climate type are indicated by asterisks, or numbers when there is more than one pair in a single climate type.
AS – Asia; AUS – Australia; EUR – Europe; NAF – North Africa; NAM – North America; SAF – Southern Africa; SAM – South America; SP – South Pacific (islands).
Source: Peel MC, McMahon TA, and Finlayson BL (2004) Continental differences in the variability of annual runoff - update and reassessment. *Journal of Hydrology* 295: 185–197.

climate zones is not particularly good because of the variety of ways that climate can influence runoff distribution through the year.

Floods

Floods occur as a result of large rainfall events concentrated in a relatively short time and in cold climates they are also generated by the spring thaw. The size of flood produced from any particular rainfall event depends on the interaction between the rainfall event and catchment properties. For example, a given high intensity rainfall event will produce a bigger flood if the catchment is already wet, than if it falls onto a dry catchment. In hydrology, flood magnitude is usually reported as the highest flow reached (the peak) and the standard data set used to describe the floods of a catchment is the annual flood series, consisting of the largest flood in each year of record. Comparison of flood behaviour between catchments

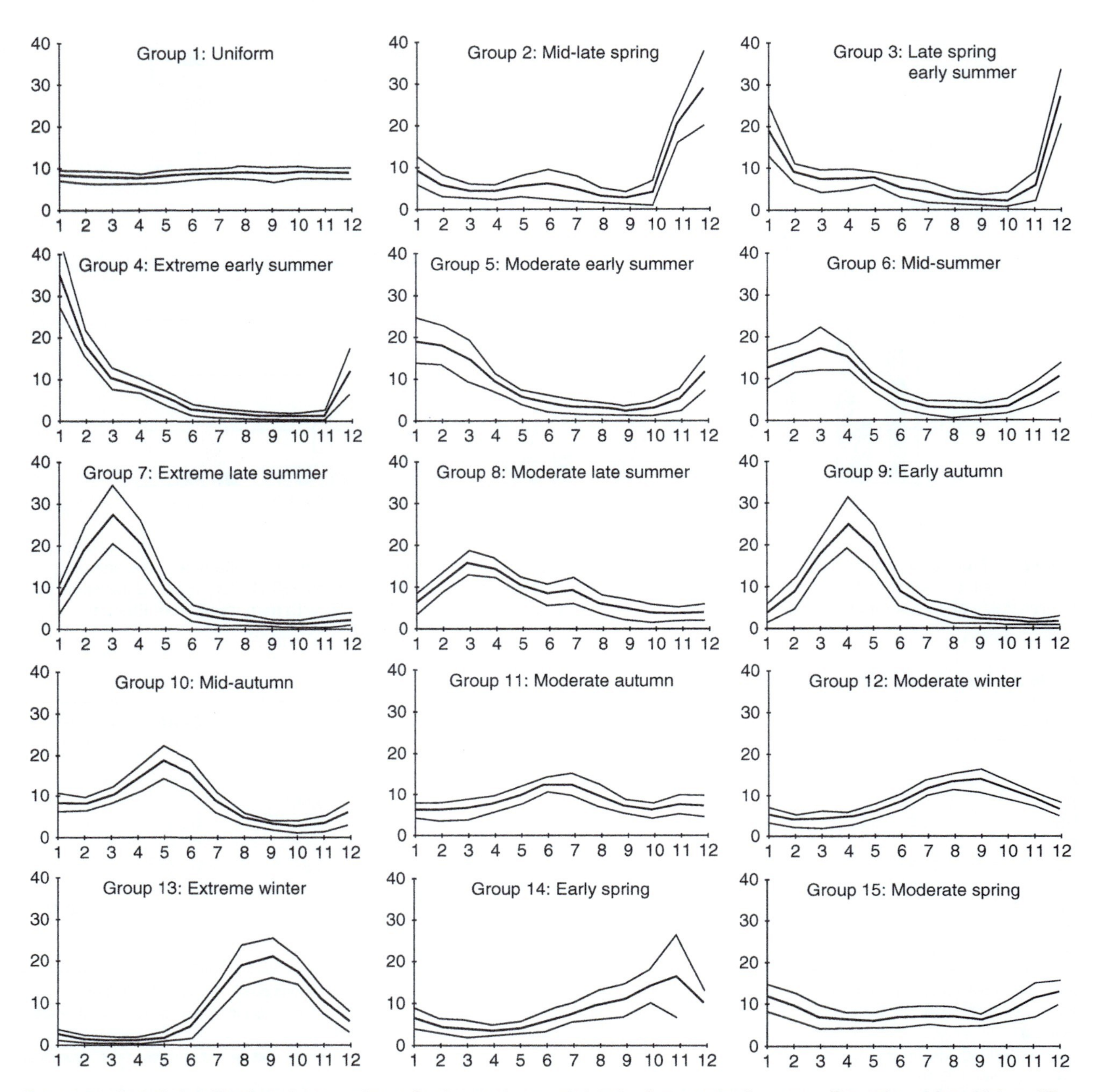

Figure 6 Global classification of river regimes. On the x-axis, month 1 is the first month of summer. Reproduced from Haines AT, Finlayson BL, and McMahon TA (1988) A global classification of river regimes. *Applied Geography* 8: 255–272, with permission from Elsevier.

of different size is made by dividing flood peak discharge by catchment area.

In **Figure 8** the frequency distributions of the specific mean annual flood (SMAF in units of $m^3 s^{-1} km^{-2}$) are plotted for each major Köppen climate type. Here, as in the case of MAR (**Figure 3**) and the C_v of MAR, (**Figure 5**) the world pattern reflects the Type C climate pattern because of the numerical dominance of stream flow records from those climate areas. The largest SMAF tends to occur in the Type A climates, possibly reflecting generally wetter conditions there and large scale weather systems producing rainfall. Type B climates tend to have relatively low values of SMAF possibly as a consequence of generally drier catchment conditions and convective storms that affect only small areas. The low SMAF values for Type D climates are influenced by the temperature controlled snowmelt processes.

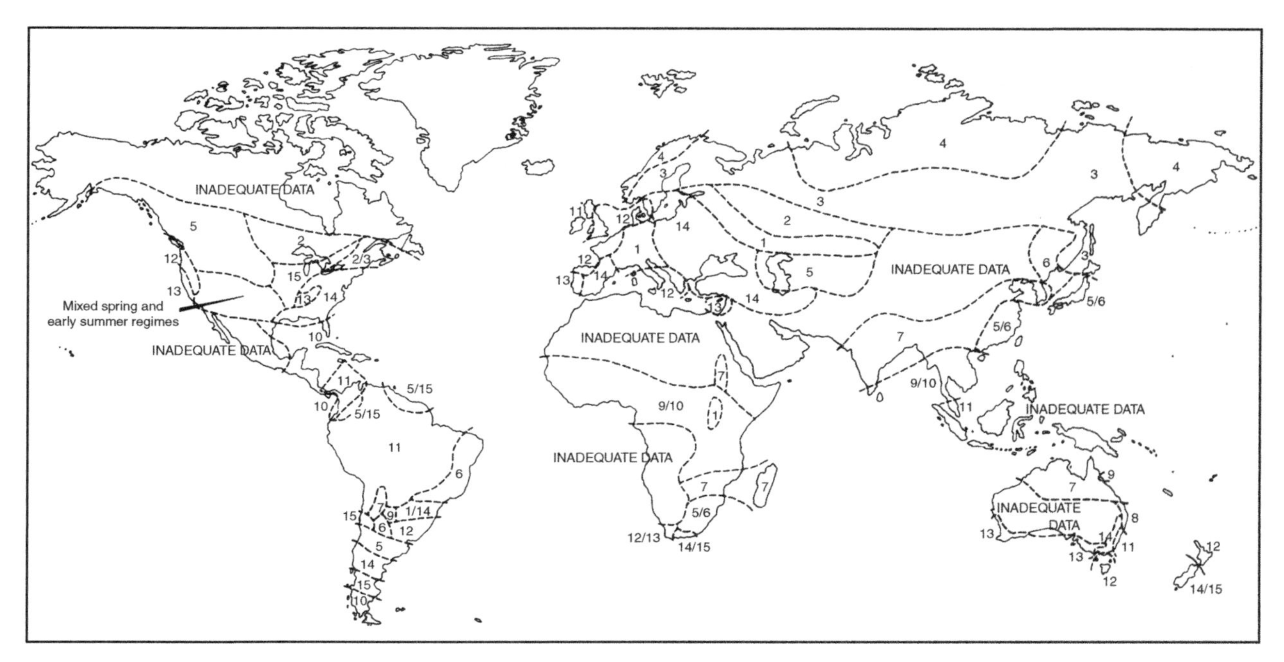

Figure 7 Global distribution of river regimes based on the classification shown in **Figure 6**. Reproduced from Haines AT, Finlayson BL, and McMahon TA (1988) A global classification of river regimes. *Applied Geography* 8: 255–272, with permission from Elsevier.

Table 3 Distribution of river regime types (1–15, see **Figure 6**) and Köppen climate types for rivers with drainage basin areas less that 10 000 km^2

Köppen	*1*	*2*	*3*	*4*	*5*	*6*	*7*	*8*	*9*	*10*	*11*	*12*	*13*	*14*	*15*
Af	4							2		8	17	1			2
Am					2		1	4	4						1
Aw	3		1			3	4	5		41	10	1			
BSh						1	6						1		
BSk		1	1		5	5	2	2		1		1	3	3	2
BWh							1								
BWk												1		2	1
Cfa	3				4	6	4	7	3	7	4	4	18	35	3
Cfb	18		1			5		9			21	87	21	46	9
Cfc			1		1						3	1			1
Csa	2											3	25	4	
Csb												2	12		
Cwa	1					8	24	3	3	1				1	
Cwb					11	23	5	1			1	1			1
Dfa					1									2	
Dfb	1	2	3	1	2									9	8
Dfc		8	26	4	2	1								1	5
Dsa														3	
Dwd			1	1											
E			4	10	3										2

Source: Haines AT, Finlayson BL, and McMahon TA (1988) A global classification of river regimes. *Applied Geography* 8: 255–272.

The variability of flood behaviour is quantified using the index of variability (Iv), also sometimes called the Flash Flood Index. This is calculated as the standard deviation of the logarithms of the annual flood series. Another similar measure of variability is the ratio of the mean annual flood to the flood with an average recurrence interval of 100 years. Average values of this ratio in Europe and North America are in the range 2–4, while in Australia and Southern Africa averages are in the range 5–14, with individual rivers having ratios as high as 40. **Figure 9** presents frequency distributions of I_v for each major Köppen

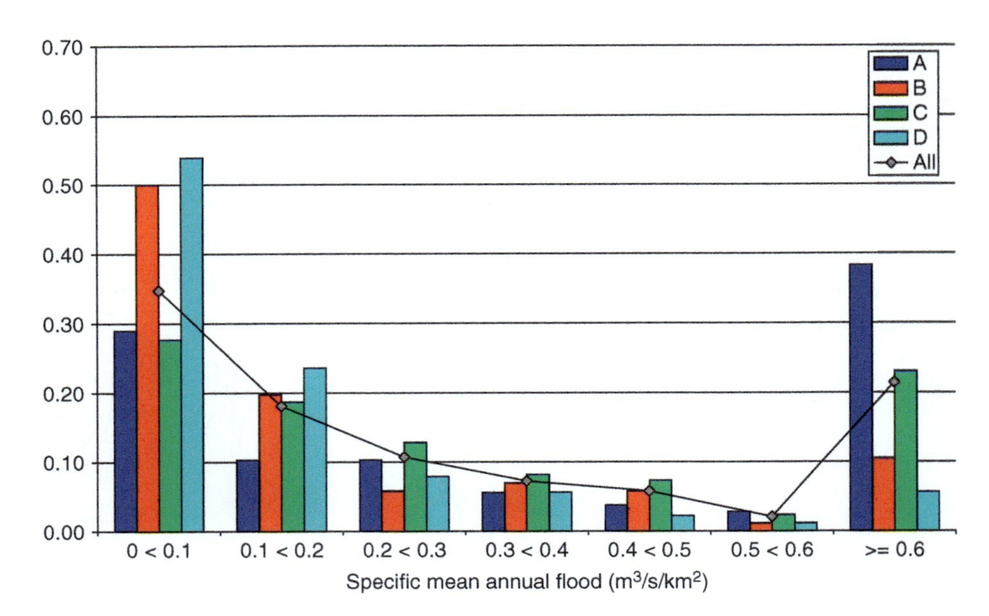

Figure 8 Frequency distributions of the specific mean annual flood for each of the main climate types. The line on the graph is the mean for all stations. Reproduced from Peel MC (1999) *Annual Runoff Variability in a Global Context*. PhD thesis, The University of Melbourne.

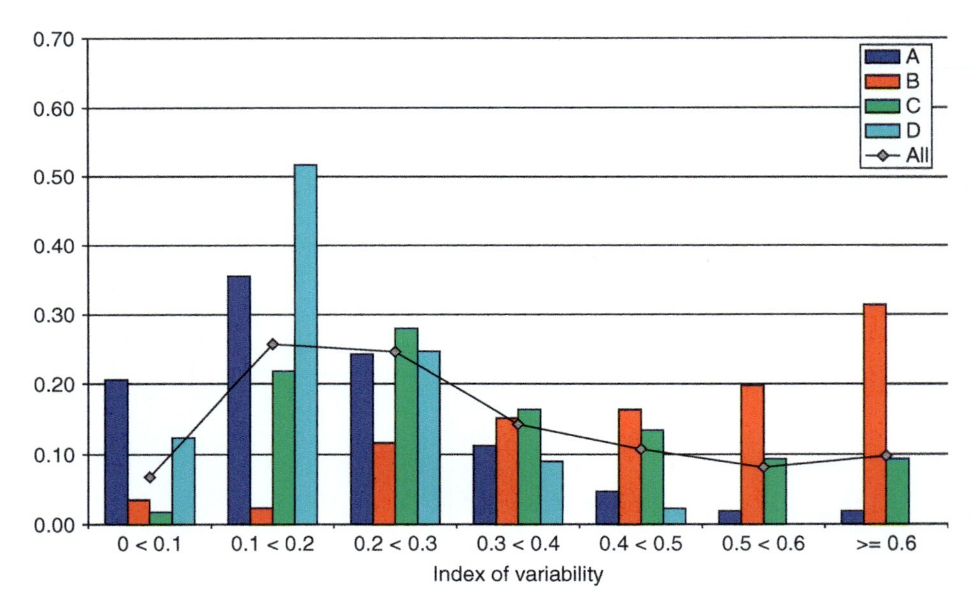

Figure 9 Frequency distributions of the index of variability of the annual flood series for each of the main climate types. The line on the graph is the mean for all stations. Reproduced from Peel MC (1999) *Annual Runoff Variability in a Global Context*. PhD thesis, The University of Melbourne.

climate type. The patterns there are similar to those of the C_v of annual runoff.

In most catchments there is a strong correlation between the largest flood in a year and annual flow. This relationship is strongest in the Type B climates, where the single largest flood of the year commonly produces a large proportion of the total annual flow. In Types D and E climates, the snowmelt peak is generally the largest flood event of the year and also represents a significant proportion of the annual flow. The relationship is weaker, but still persists, in the Type A climates.

With increasing climatic aridity, the peak flows come to dominate stream hydrographs. Runoff occurs following precipitation events, but quickly ceases and there is little baseflow. This is illustrated in **Figure 10**, where daily flows for a typical year are plotted for an arid zone (BWh) and a humid zone (Cfb) stream with similar-sized catchments. This behavior has consequences for stream channel form as will be discussed below.

Low Flows

Low flows are an important characteristic of the flow in rivers that is related to climate. Low flows occur in humid climates on a seasonal basis, depending on the seasonal distribution of precipitation and temperature and these patterns are shown in the regime classes (**Figure 6**). Ephemeral streams are those that cease to flow on a regular basis. Low flows that cease to flow become more common in flow regimes as the climate becomes more arid. One way to express this is with the baseflow index (BFI) which is the ratio of baseflow to total flow. High baseflow ratios are found in humid zone streams and they are also promoted where the catchment bedrock is highly permeable (limestone, for example). The Todd River, an arid zone stream, has a BFI of 0.075, while the Acheron River a humid zone stream has a BFI of 0.65 (**Figure 10**).

River Channels and Climate

The natural channels of rivers are self-formed and self-maintained. In humid zone rivers, where flow is relatively consistent through time, the form and size of the river channel is related to the effective discharge, essentially the range of flows, that because of their magnitude and frequency, are most effective in transporting sediment. Thus, throughout much of the humid zone, river channels can be said to be in equilibrium with the range of flows they carry.

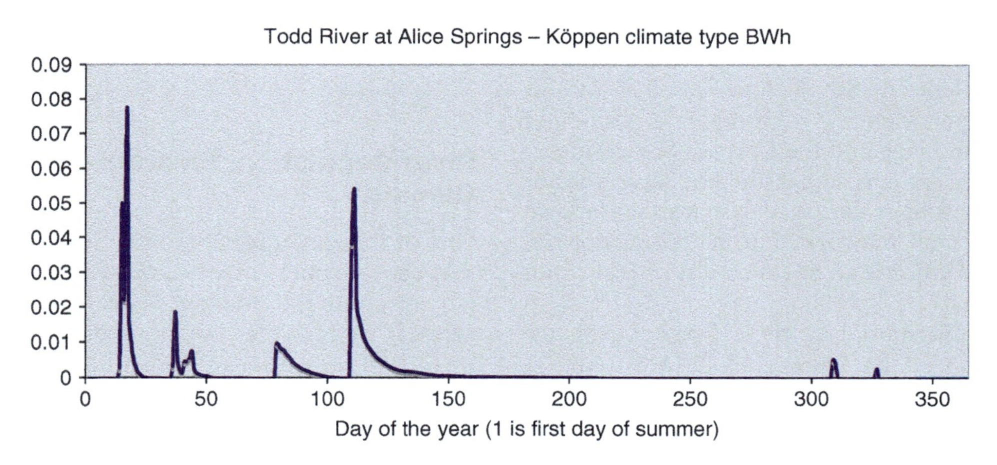

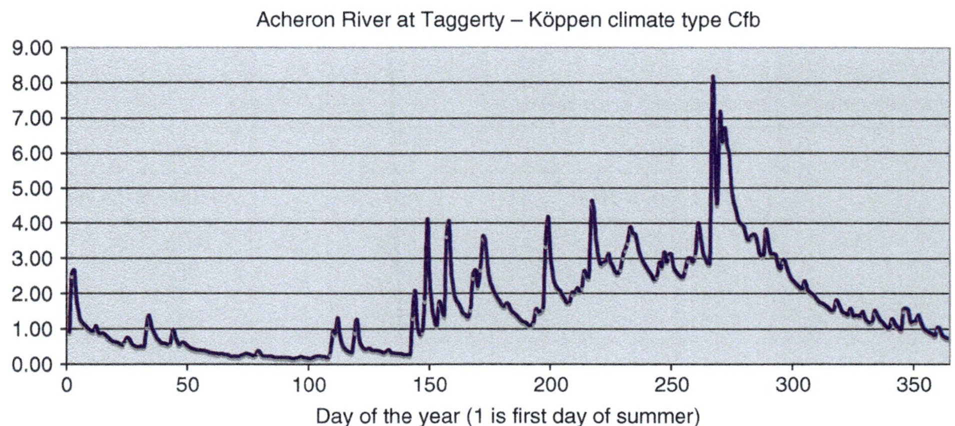

Figure 10 Daily runoff (mm) for a typical year in an arid (Todd River, Central Australia) and a humid zone stream (Acheron River, Southeast Australia). Note the runoff scales on these two graphs differ by two orders of magnitude. These hydrographs are plotted using the same procedure as the regime types in **Figure 6** with the *x*-axis beginning on the first day of summer.

This concept of an equilibrium channel does not apply in semi-arid and arid zone streams. As the hydrograph for the Todd River in **Figure 10** shows, the discharge of streams in semi-arid and arid climates is dominated by flood events and these carry out work in moving sediment, and thereby shaping the form and size of the river channel. The channels and associated features of river morphology in the dry climates are a product of the large flow events and will persist in the landscape, often for decades, until reshaped by a subsequent event of similar or larger magnitude. As climate changes spatially along a gradient, this distinction between arid and humid in terms of the processes driving river morphology also alters along a continuum.

Drainage systems can be classified as exoreic, endoreic, and areic. Exoreic river systems flow to the sea, generally have fully connected channel networks and are most common in humid climate zones. Endoreic rivers flow into an internal basin because of a combination of climatic aridity and continental morphology. Examples of endoreic drainage systems are the Lake Eyre Basin in Australia, the Okavango River system in southern Africa, the Tarim Basin in western China, the Great Basin in the United States, and the Altiplano in South America. The largest endoreic system, found in Central Asia, combines the Aral Sea drainage and the Caspian Sea drainage. Areic drainage systems either do not have surface river channels or have deranged systems that lack an integrated channel network. Further discussion on endoreism and areism can be found elsewhere within this encyclopedia.

The global distribution of these types of drainage systems is mapped by Emmanuel de Martonne in 1927. Generally, endoreic and areic systems are associated with arid climates and/or topographically enclosed basins, though there are some exceptions to this. For example, the Yucatan Peninsula in Mexico is humid but areic, because it is underlain by permeable limestone. **Table 4** lists the areas and percentages of endoreic and areic drainage for the continents. There are significant continental differences in this distribution. Only a small percentage of the areas of North and South America have endoreic and areic drainage due in part to the limited extent of dry climates and also a lack of topographically enclosed basins. Compare this with Asia, Africa, and Australia where the dry lands are extensive and there are significant areas of enclosed basins. Europe is significant as the only continent where the area of areic drainage is small in comparison to the area of endoreic drainage. The main area of endoreic drainage in Europe is the Volga basin, which is climatically humid but flows into the Caspian Sea which is too large for overflow to occur.

As **Table 4** shows, 33% of the surface of the continents is drained by endoreic and areic systems, and the majority (23%) of this is areic – lacking a surface drainage system. The fully connected exoreic drainage networks that form the basis of most textbook descriptions of river systems in fact occupy only two-thirds of the total land area.

River Regulation, Environment, and Climate

One of the consequences of the high flow variability that accompanies climatic aridity is the need to store larger volumes of water to meet the needs of water users. If we take the case of a dam built to supply to users, 80% of the mean annual flow of the river with a reliability of 95%, the size of the storage needed is a function of the square of the C_v of annual flows. **Table 5** lists the size of storages needed for this for

Table 4 Areas and percentages of endoreic and areic drainage

Region	*Endoreic*		*Areic*	
	Area (km^2 × 1000)	%	*Area (km^2 × 1000)*	%
Europe	1753	19	452	5
Asia	4894	11	9935	24
Africa	3452	12	11 771	40
Australia	1611	21	3309	43
North America	1066	5	1070	5
South America	1053	6	1454	8
Northern Hemisphere	10 247	11	22 139	23
Southern Hemisphere	3600	11	5852	17
World	13 847	10	27991	23

Adapted from de Martonne E (1927) Regions of interior basin drainage. *Geographical Review* 17: 397–414.

Table 5 Average storage size needed to supply 80% of the mean annual flow with a reliability of 95%, expressed as a ratio of the mean annual flow

Continental region	*Number of streams*	*Storage size (ratio of mean annual flow)*	*Standard deviation of storage size*
AS	143	0.62	1.45
NAM	189	0.44	0.60
SAM	53	0.44	0.63
EUR	260	0.28	0.36
NAF	23	0.35	0.37
SAF	100	2.07	2.82
AUS	156	1.65	1.57
SP	50	0.22	0.36
WOR	974	0.77	1.43

Continent abbreviations as for **Table 2**.
Adapted from McMahon TA, *et al.*, 1992, *Global Runoff*, Catena Paperback, Cremlingen-Destedt.

the major continental areas, where size is expressed as a ratio of mean annual flow. In Australia and Southern Africa, where, as described above, flow variability is higher than in similar climates in other continents, dams must be large enough to hold multiples of the mean annual flow in order to meet this design draft. This means that water storage is more expensive in those areas and the impact on the flows downstream of the dam is greater than elsewhere, with deleterious consequences for the stream ecology.

In addition to this storage size effect, the impacts of regulation in dry zone rivers will persist much further downstream than is the case for rivers in the humid zone. In the humid zone, there are unregulated tributaries entering the system downstream of the dams that help restore some of the natural flow characteristics to the river. This is not so for dry climates where rivers are exotic and derive little or none of their flow from the areas they traverse downstream. The Nile is a well known example of this effect and the alterations to the patterns of flow caused by the Aswan High Dam persist for over 1000 km downstream to the river's mouth.

Where dams are used to supply water to downstream irrigation systems it is not unusual for the natural flow regime of the river to be substantially changed. Often there is a regime reversal as water is released downstream to irrigators in summer producing high flows at a time when the rivers would naturally carry only low flows. Another effect of regulation is that the low flows in rivers are enhanced by releases from dams, an effect that has been labeled 'antidrought' as it removes periods of low flow that are important to the operation of the in-stream ecological system.

Conclusions

Climate is the main determinant of the flow in rivers through the hydrologic water balance. Rivers are spatially varied systems and small headwater streams in humid climates may have some flow characteristics similar to larger rivers in arid climates. Large river basins invariably include different climate zones, so the relation between climate and flow of the whole basin reflects all these climate effects and a time delay.

The Köppen classification has been used here to characterize climate, but there are additional climate properties not represented in such a classification. Climate extremes of wet and dry are responses to large scale ocean-atmosphere fluctuation and to more localized systems such as tropical storms. The impacts of these are found in flow variability and the flood behavior of rivers.

As a general rule, rivers in drier climates have more variable flows than those in humid climates, reflecting a similar relationship for precipitation. However, there are some areas of the world, notably Australia and Southern Africa, where the variability of annual flows is 2–3 times that of rivers in similar climates in other continents.

The long-term average distribution of flows through the year (regime) reflects the distribution of precipitation in many cases, but where there is significant winter snow and ice storage, the high flow period will be determined by temperature when the snow melts. Temperature also has a strong effect on summer flow when evapotranspiration rates are high.

In relation to climate, floods follow a pattern similar to total flow. Floods (in terms of flow per unit area) tend to be smaller in dry climates and more variable. Floods tend to dominate the stream hydrograph in dry climates and baseflow, as expressed through the Baseflow Index, is reduced as a component of the flow hydrograph.

The importance of floods in dry climate rivers is also reflected in the river channels where channel form is generally a consequence of the most recent large flood. This is in contrast to rivers in the humid zone where channel form is in equilibrium with the range of flows in the river.

Integrated drainage networks flowing to the sea are characteristic of the humid zone and two thirds of the earth's surface is made up of these endoreic drainage systems. Nearly a quarter of the earth's surface lacks a surface drainage system, mainly because of climatic aridity. Ten percent of the earth's land surface has endoreic drainage. This is also mainly a response to dry climates but is also partly determined by the existence of topographic depressions.

The more variable flows in rivers of the dry climates mean that larger volumes of water must be stored to meet demand. Also, in dry climates the effects of river regulation on flow and water quality persist further downstream than they do in humid climates where unregulated tributaries restore some of the natural flow characteristics.

See also: Africa; Asia – Eastern Asia; Asia – Monsoon Asia; Asia – Northern Asia and Central Asia Endorheic Rivers; Australia (and Papua, New Guinea); European Rivers; Floods; Geomorphology of Streams and Rivers; Hydrological Cycle and Water Budgets; Hydrology: Streams; South America; Streams and Rivers of North America: Overview, Eastern and Central Basins.

Further Reading

Reichel A and Reichel E (1975) *The World Water Balance. Mean Annual Global, Continental and Marine Precipitation, Evaporation and Run-off.* Amsterdam: Elsevier.

Beckinsale RP (1969) River regimes. In: Chorley RJ (ed.) *Water, Earth and Man*, pp. 455–471. London: Methuen.

de Martonne E (1927) Regions of interior basin drainage. *Geographical Review* 17: 397–414.

Ghassemi F and White I (2006) *Inter-Basin Water Transfer: Case Studies from Australia, United States, Canada, China and India.* Cambridge: Cambridge University Press.

Haines AT, Finlayson BL, and McMahon TA (1988) A global classification of river regimes. *Applied Geography* 8: 255–272.

Kalinin GP (1971) *Global Hydrology.* Jerusalem: Israel Program for Scientific Translation.

Korzun VI, *et al.* (1978) *World Water Balance and Water Resources of the Earth.* Paris: UNESCO.

McMahon TA, Finlayson BL, Haines AT, and Srikanthan R (1992) *Global Runoff – Continental Comparisons of Annual Flows and Peak Discharges.* Cremlingen: Catena Verlag.

Peel MC, McMahon TA, and Finlayson BL (2004) Continental differences in the variability of annual runoff – Update and reassessment. *Journal of Hydrology* 295: 185–197.

Peel MC, Finlayson BL, and McMahon TA (2007) Updated world map of the Köppen–Geiger climate classification. *Hydrology and Earth System Sciences* 11: 1633–1644.

Trenberth KE (1991) General characteristics of El Nino-Southern oscillation. In: Glantz MH, Katz RW, and Nicholls N (eds.) *Teleconnections Linking Worldwide Climate Anomalies: Scientific Basis and Societal Impact*, pp. 13–42. New York: Cambridge University Press.

Urban Aquatic Ecosystems

J L Meyer, University of Georgia, Athens, GA, USA

Introduction

Early urban centers were established near inland aquatic ecosystems, which provided a reliable source of fresh water or served as important navigation routes. Hence, there is a long history of cities benefiting from and impacting inland aquatic ecosystems (streams, rivers, ponds, lakes, and wetlands). Urban impacts have intensified as urban populations have expanded.

The proportion of the human population living in urban areas has increased considerably in the last century. Only 10% of people were living in cities in 1900; by 1950 the proportion had increased to 29.8%; it was 47.2% in 2000; and projections for 2010 are that 51.5% of people will live in cities. Most of the future increase in urban populations will be occurring in the developing world. For example, urban populations in Europe and North America are projected to increase by only 10% from 1990 to 2010, whereas urban populations in Africa and Latin America are projected to increase by 75%.

There is no single definition of 'urban.' Three characteristics are commonly used: population density (minimum 400–1000 persons per km^2), population size (minimum 1000–5000), and occupation (a maximum of 50–75% employed in agriculture). Urban sprawl is characteristic of urbanization in North America, where the amount of land occupied by urban areas is increasing at a faster rate than is the urban population. This form of urbanization has significant consequences for aquatic ecosystems because of its associated extensive alteration of catchments. Several metrics have been used to quantify the extent of urbanization and relate that to its impacts on aquatic ecosystems (**Table 1**). The simplest is population density, with which all other metrics are correlated; but that metric does not capture the diversity of development patterns and mechanisms of urban influence, which range from the type of infrastructure to socioeconomic conditions.

Cities differ greatly in geographic and climatic setting, population and housing density, types of industry, modes of waste disposal, transportation and water infrastructure, and many other factors that influence aquatic ecosystems. Hence, the term 'urban aquatic ecosystem' encompasses a diverse range of water bodies and impacts. This chapter provides an overview of the diversity of impacts of urbanization on inland, freshwater aquatic ecosystems. It begins with and focuses on urban rivers and streams, then provides a shorter discussion of urban lakes and ponds, and finally urban wetlands. Groundwater ecosystems are not discussed, although urbanization and resultant water withdrawal impact them; in particular, saline intrusion is a result of over-extraction. Withdrawal of groundwater at rates greater than it is recharged leads to declining water tables and drying up of springs and ponds. Urbanization also affects aquatic ecosystems far from cities (e.g., via atmospheric transport of pollutants), but the focus here is on aquatic ecosystems in cities as well as those in the urban fringe. The chapter ends with a summary of services provided to society by urban aquatic ecosystems, general conclusions, and gaps in our knowledge.

Rivers and Streams

Flowing waters in urban areas are impacted by multiple stressors resulting in a characteristic 'urban stream syndrome' (**Figure 1**). Replacing vegetated cover with impervious surfaces (e.g., roads, parking lots, rooftops), altering network structure by burying streams in culverts, encasing them in cement-lined straight channels, and routing storm water through pipes changes water movement across the landscape, which alters hydrologic regime, channel geomorphology, and temperature regime. More sediments, nutrients, pesticides, and contaminants are delivered to streams. These alterations impact aquatic habitats, species assemblages, foodwebs, and ecosystem processes, resulting in a loss of the goods and services benefiting humans that aquatic ecosystems provide.

Geomorphology

Urban development in the catchment impacts the physical features of stream channels even when the channel is not intentionally altered by activities such as straightening and lining with concrete. These changes occur in two stages: an initial increase in sediment loss from the catchment and deposition in the channel as roads and buildings are constructed; this is followed by declining sediment production from the urbanized catchment, but enhanced runoff that often results in channel enlargement. The initial sediment mobilization phase is associated with rates of sediment production many times pre-development rates (**Table 2**), but sediment yields are lower when development is complete. For example, sediment

Table 1 Metrics of urbanization used to evaluate its impact on aquatic ecosystems, components included in each metric, and an example of a study in which the metric was used

Metric	*Measures included*	*Study*
Population density	Humans per km^2	1
Percent urban land use	% of catchment in urban land cover classes: high and low intensity urban, industrial, transportation	1
Percent impervious cover	% of catchment covered by rooftops, roads, parking lots, and other impervious surfaces	2
Effective imperviousness	% of catchment covered by impervious surfaces with a direct hydraulic connection to streams	3
Urban intensity index	Infrastructure (road density, number of point source discharges, number of dams, number of Toxic Release Inventory sites), land use (% urban and % forest + shrublands for entire basin and for 125 m buffer on each side of streams identified on 1:100 000 scale maps of the network), and socioeconomic data (census counts for population, labor, income, and housing variables)	4
Common urban intensity index	% basin in urban landuse, percent of basin in forested or shrubland, % of stream network buffer in developed, % of stream network buffer in forest and shrub lands, and road density	5

1. Meyer JL *et al.* (2005) (see Further Reading).
2. Arnold CL and Gibbons CJ (1996) Impervious surface coverage: the emergence of a key environmental indicator. *American Planners Association Journal* 62: 243–258.
3. Walsh CJ *et al.* (2005) (see Further Reading).
4. McMahon G and Cuffney TF (2000) Quantifying urban intensity in drainage basins for assessing stream ecological conditions. *Journal of the American Water Resources Association* 36: 1247–1261.
5. Tate CM, Cuffney TF, Giddings EM *et al.* (2005) Use of an urban intensity index to assess urban effects on streams in three contrasting environmental settings. *American Fisheries Society Symposium* 47: 291–315.

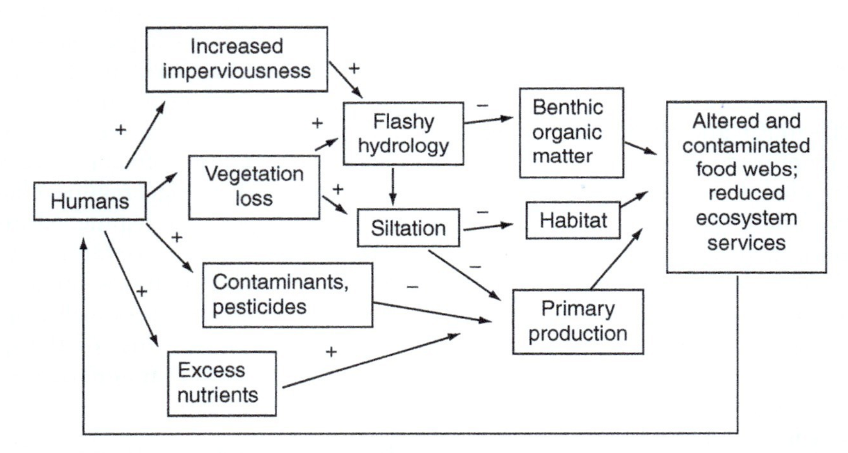

Figure 1 A simplified diagram of the urban stream syndrome showing the pathways of urbanization impacts; + and – signs indicate the direction of change. Human societies cause the changes indicated, and they are also impacted by the changes.

production from developing basins in Maryland averaged 16 times pre-development rates but only 1.7 times pre-development rates in basins where urbanization was complete. The source of sediment differs in the two phases: hillslope erosion is the largest sediment source in the initial aggradation phase, whereas channel and bank erosion is the largest sediment source in the subsequent erosional phase.

Altered delivery of water and sediments from the catchment results in changes in channel form. Data collected around the world indicate larger (i.e., wider and often also deeper) channels in urbanizing rivers, although this generalization has many exceptions. Urban streams also have reduced sinuosity. Where flow has increased and sediment supply has not, bed coarsening is observed; but where accelerated erosion occurs during construction, stream beds are choked with silt and sand. As data are collected from urban streams in different hydroclimatic settings, regional differences in these trends are becoming apparent. For example, reduction in channel capacity because of decreased depth has been observed in humid tropical streams in African and Asian cities; in contrast, British rivers tend to become narrower and deeper with urbanization; and stream channel response to urbanization in arid environments is highly variable spatially with channel widening at some points and incision elsewhere. Differences in channel slope and erodibility of geologic materials as well as bridge and road construction lead to spatial variation in degree of incision.

Urbanization results not only in changes in channel form but also changes in the structure of the entire river network. Headwater channels are filled or buried and encased in pipes, leading to a reduction in drainage density (length of stream/area of catchment). For example, drainage density of natural stream channels was reduced by 58% in an urban Maryland catchment and by 33% in an urban Georgia catchment. When roads and storm drains are included in the river network, drainage density increases by 50–>800%. Hence, small streams, which slow the downstream movement of water, sediments and nutrients, are replaced by an enhanced network of pipes, which are designed to rapidly transport water downstream. This change has profound consequences for the hydrology, chemistry and biology of the river network.

Hydrology

When natural vegetation is replaced by impervious surfaces, the movement of water through the landscape is altered (**Table 3**). Increasing impervious cover results in decreased infiltration and a greater proportion of precipitation leaving as runoff. Not only is the total amount of runoff increased, but its pattern is also altered. Urban streams are characterized as having flashier flows, i.e., floods are more frequent and flows reach peak discharge more rapidly (**Figure 2**). Peak discharges are also higher in urban streams; e.g., discharge during a flood likely to occur every two years in an urban Washington stream is equal to discharge during a flood likely to occur only every ten years in a forested stream. A recent analysis of hydrologic regime in catchments with >15% urban land cover in the southeastern and northwestern United States found increased peak flows, decreased minimum flows, and increased flow variability. Urban peak flows were 3–4 times those in agricultural regions, and annual flood peaks based on daily average discharges were magnified 22–84% in urbanized catchments.

With the decreased infiltration characteristic of elevated imperviousness (**Table 3**), one might expect lower baseflow; but this is not consistently observed because of additional inputs from septic systems, lawn and garden watering, and wastewater treatment plant effluents. Wastewater can constitute a large fraction of urban stream discharge; e.g., effluent is 69% of annual discharge and 100% of discharge during low flow conditions in the Platte River below Denver, Colorado. Effluent-dominated streams are common in cities around the world.

Loss of riparian vegetation, runoff from heated impervious surfaces, direct discharge of heated effluent from power-generating plants, and the urban 'heat island' effect contribute to warmer streams in

Table 2 Rates of sediment production in urbanizing landscapes

Location (% of catchment disturbed)	*Catchment area (km^2)*	*Sediment yield ($t\ km^{-2}\ year^{-1}$)*	*Increase over background*
Maryland, USA (100%)	0.0065	54 056	300×
Maryland, USA (100%)	0.08	30 889	140×
New Jersey, USA (100%)	0.075	1194	47×
Papeete, Tahiti (100%)	0.85	7300	120×
Colorado, USA (53%)	3.7	2913	30×
Maryland, USA (29%)	98.4	236	4×
New South Wales, Australia (23%)	83.8	3829	120×
Maryland, USA (17%)	0.24	9267	30×
Virginia, USA (7%)	24.6	12 549	3×

Data are from studies reviewed by Chin (2006). Increase over background is calculated from pre-development reaches upstream or nearby.

Table 3 Changes in the water budget (fate of precipitation) with increasing impervious cover in urban catchments (modified from Paul and Meyer (2001))

Impervious cover (%)	*Evapotranspiration (%)*	*Shallow infiltration (%)*	*Deep infiltrations (%)*	*Runoff (%)*
<10	40	25	25	10
10–20	38	21	21	20
35–50	35	20	15	30
75–100	30	10	5	55

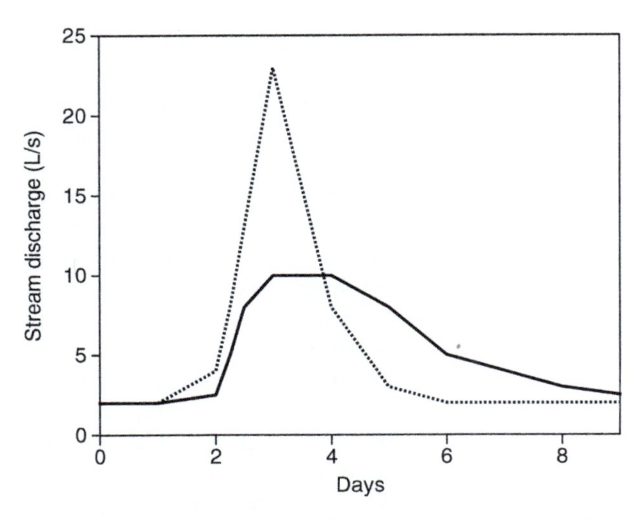

Figure 2 Typical hydrograph after a one-day storm in an urban stream (dashed line) and a forested stream (solid line). Discharge rises and falls faster in the urban stream and reaches a higher maximum.

cities. For example, urban streams have warmer summer temperatures (5–8 °C), cooler winter temperatures (1.5–3 °C), and greater diel change in temperature. These differences are particularly apparent during summer storms, when temperature pulses can be 10–15 °C warmer than forested streams.

Chemistry

Urban streams receive a wide variety of chemical compounds from wastewater treatment plant (WWTP) effluents, industrial discharges, storm sewers, and septic systems, as well as runoff from lawns, gardens, roads, and parking lots. As a consequence, concentrations of both inorganic and organic compounds are usually elevated in urban streams, although the type of chemical pollution varies greatly depending on the nature of human activity in the catchment. Although treatment technologies in WWTP have improved, systems still fail, permitted discharges are exceeded, and many cities still have combined sewer and stormwater pipes so that combined sewer overflows are common during rainstorms. Rivers are the most frequent recipient of effluents from WWTPs; e.g., of 248 urban WWTPs in the United States, 84% discharge into rivers. Non-point sources (e.g., runoff from lawns, roads) are also ubiquitous in urban settings.

High concentrations of phosphorus, nitrogen, and other ions are commonly observed in urban streams. Elevated phosphorus concentrations are observed below WWTPs as well as from fertilizers applied to lawns and gardens. Erosion of soils during construction can carry a considerable load of sediment-bound phosphorus to streams. High concentrations of nitrate and ammonium can extend far downstream of urban centers. Concentrations of other ions such as calcium, sodium, potassium, and chloride, are also commonly elevated in urban streams. For example, runoff from road de-icing in the northeastern U.S. has resulted in chloride concentrations in streams that are elevated throughout the winter, reaching peak concentrations equivalent to 25% sea water. Even during the summer, chloride concentrations remain at levels 100 times those observed in forested watersheds. Elevated electrical conductivity in urban streams is not unique to northern cities, but is a generally useful indicator of urban influence.

Metals such as zinc, copper, lead, chromium, cadmium, and nickel, frequently occur at higher concentrations in urban than in less disturbed streams. Although industrial discharges contribute to these high concentrations, non-point sources such as brake linings (nickel, chromium, lead and copper) and tires (zinc, lead, chromium, copper and nickel) are a greater source. Metal concentrations are generally higher in sediments than in the overlying water, particularly fine-grained sediments with high organic matter content.

Pesticides such as insecticides, herbicides, and fungicides have a high detection frequency in urban streams. Pesticide concentrations in urban stream sediments in the United States and in France frequently exceed those observed in agricultural areas. Pesticides are used on lawns, gardens, and golf courses as well as in homes and industrial or commercial buildings. Urban pesticide use accounts for a third of total use in the United States.

Other organic compounds such as polychlorinated biphenyls (PCBs), polycyclic aromatic hydrocarbons (PAHs), and petroleum-derived hydrocarbons are also found in urban streams. PAHs are largely from organic solvents used in industry and delivered to streams via industrial discharges. In contrast, hydrocarbons from automobiles and trucks enter streams via runoff from impervious surfaces. The amount of hydrocarbons delivered by rivers to the ocean can be considerable; e.g., 48 500 l of oil enters Narragansett Bay via rivers each year.

Pharmaceuticals and compounds from personal-care products (e.g., shampoo, deodorants) are also commonly detected in urban streams. Antibiotics, caffeine, chemotherapeutic drugs, analgesics, narcotics, psychotherapeutic drugs, and contraceptives have been detected, although their impact on aquatic biota and ecosystems is only beginning to be explored. In laboratory experiments where test animals (e.g., fathead minnows, stoneflies) are exposed to water from urban streams, increased mortality rate and altered reproductive characteristics have been observed.

Biology

The organisms in urban streams are impacted by the alterations in geomorphology, hydrology and chemistry described above. Elevated concentrations of coliform bacteria are often recorded, particularly in wet weather and where storm and sanitary sewers are combined. Antibiotic-resistant bacteria have also been observed. Iron bacteria are a common feature of urban streams where anoxic, iron-rich ground water reaches the surface.

Naturally vegetated riparian zones are less common in urban environments. Their elimination alters light and temperature regimes, bank stability, sediment and nutrient delivery, inputs of leaf litter, and habitat availability for plants and animals characteristic of the streamside environment. Even when a forested riparian zone is present, it is often so narrow that it is less effective and vulnerable to extreme events (e.g., wind storms) and bypassed by stormwater pipes so that little removal of sediments, nutrients or contaminants occurs.

Increased light and nutrients in some urban settings can result in elevated algal biomass in urban streams; however, algal biomass may also be lower in urban streams because of the presence of metals and herbicides as well as unstable substrates, variable flow regimes, and high turbidity resulting from excess fine sediments. Pollution-tolerant algal taxa are more abundant in urban streams.

The invertebrate fauna of urban streams is characterized by decreased abundance (and often absence) of sensitive taxa (e.g., Ephemeroptera, Plecoptera, and Trichoptera) and increased abundance of tolerant taxa (chironomids and oligochaetes). Taxa richness declines with increasing urbanization (**Table 4**). Loss of habitat (e.g., excess sedimentation), discharge extremes, elevated water temperature, low oxygen, toxic chemicals, poor food quality, and presence of non-native competitors and predators are some of the many factors responsible for the observed declines. Indices of biotic integrity based on benthic invertebrates reveal threshold effects on invertebrate assemblages when impervious surface cover is 5–18% of the catchment, although linear declines rather than a threshold are also observed (**Table 4**). Effective imperviousness (the area of impervious surfaces with direct hydraulic connection to streams) is a better predictor of urban impact on invertebrates than total impervious cover. Recognition of the importance of effective imperviousness has significant implications for management of urban streams. It suggests that the effects of impervious surfaces on aquatic ecosystems can be reduced by disconnecting impervious surfaces from streams through the installation of features such as rain gardens and infiltration basins rather than stormwater pipes.

The response of stream fish assemblages to urbanization is similar to that described for other taxa: loss of sensitive and native species and increased abundance of tolerant, generalist species, including more non-native species introduced either by accident or for sport fishing. In some cases species richness declines with increasing urbanization, but if urbanization results in invasion of native or exotic generalists, richness may increase (**Table 4**). The numbers of individuals with parasites and lesions often increases with urbanization. These changes are in response to the same kinds of factors causing invertebrate declines as identified in the previous paragraph. Fish-based indices of biotic integrity decline as impervious cover increases, particularly when impervious cover exceeds ~10% or when effective impervious cover is above a threshold of 8–12% (**Table 4**). Urban rivers offer fishing opportunities, although many carry advisories with recommendations for very limited consumption of fish that are caught because of contamination with compounds such as mercury and PCBs.

Declines in other vertebrate taxa have also been observed in urban streams and riparian zones: amphibians, birds, small mammals and marsupials (e.g., platypus). Several factors can lead to amphibian declines in urban streams; these include toxic chemicals, altered conductivity and pH, excess siltation, loss of terrestrial habitat, elevated mortality at road crossings, and introduction of competitors and predators. For example, declines in amphibian richness in southern California streams were related to invasion of an exotic crayfish in streams where flows were perennial because of urban development; amphibians could persist in the intermittent streams characteristic of this climate, but the exotic crayfish could not.

Ecosystem Processes

Changes in hydrology, chemistry and biology result in altered ecosystem processes in urban streams, although functional changes have been studied less than structural changes. Rates of removal of nutrients from stream water are lower in urban streams as a result of reduced storage of benthic organic matter or lower algal biomass. Accelerated rates of leaf breakdown have been observed in urban streams as a consequence of enhanced erosive capacity rather than biological decomposition. Trends in primary productivity and system respiration with urbanization have not been consistent, probably because rates of metabolism reflect a response to several factors (e.g., nutrient and organic matter supply, light availability,

Table 4 Examples of the responses of stream invertebrate (I) and fish (F) assemblages to urbanization

Location of metropolitan area studied	*Nature of response*	*Source*
Maryland, USA	I: ↓diversity with ↑ ISC (1 to 17%) F: ↓diversity at ISC >12–15%; absent at ISC >30–50%	1
Virginia, USA	I: ↓ diversity with ↑ ISC (15–25%)	2
Washington, USA	I: IBI ↓ with ↑ ISC (1 to 6%); no ↓ if riparian intact	3
California, USA	I: ↓ EPT richness and % abundance in EPT taxa with ↑ % urban land cover	4
California, USA	I: all invertebrate metrics lower in concrete-lined streams than in natural or channelized streams	5
Utah, USA	I: ↓ richness metrics and ↑ tolerant taxa with ↑ UII	6
Victoria, Australia	I: ↓ richness metrics with ↑effective ISC	7
Minas Gerais, Brazil	I: depauperate fauna below urban untreated sewage; F: fewer native and more exotic species below city	8
New York, USA	I: ↓ biotic indices with ↑% urban land cover; F: no significant change in indices with % urban land cover but ↓ abundance with ↑road density	9
Maryland, USA	I: metrics go from good to poor at 15% ISC F: ↓diversity when ISC >10–12%	10
Massachusetts, USA	I: ↓ richness metrics and ↑ tolerant taxa with ↑ UII F: ↓ species richness and fluvial specialists with ↑ UII	6, 11
Alabama, USA	I: ↓ richness metrics and ↑ tolerant taxa with ↑ UII F: ↓ species richness and endemic species richness with ↑ UII	6, 11
N. Carolina, USA	F: ↓ IBI with ↑ % urban land cover	12
Georgia, USA	F: ↓ species richness and ↑ relative abundance of centrarchids with ↑ % urban land cover	13
Wisconsin, USA	F: threshold at 8–12% effective ISC; ↓ richness and IBI above threshold	14
Illinois and Wisconsin, USA	F: low IBI when urban land cover >25%	15
Georgia, USA	F: ↓IBI and ↑ fin lesions with ↑ % urban land cover	16
Ontario, Canada	F: ↓IBI at ISC >10%; less impact if riparian intact	17
New York, USA	F: egg and larval density↓to 10% urban land use; absent above that	18

ISC: impervious surface cover; effective ISC as defined in **Table 1**; IBI: Index of Biotic Integrity; EPT: Ephemeroptera, Plectoptera, Trichoptera; UII: urban intensity index as defined in **Table 1**; ↑: increase; ↓: decrease.

Sources

1. Klein (1979) in Paul and Meyer (2001) (see Further Reading).
2. Jones and Clark (1987) in Paul and Meyer (2001) (see Further Reading).
3. Horner RR, Booth DB, Azous A, *et al.* (1997) in Paul and Meyer (2001) (see Further Reading).
4. Carter JL and Fend SV (2005) Setting limits: The development and use of factor-ceiling distributions for an urban assessment using macroinvertebrates. pp. 179–192. In: Brown LR, Gray RH, Hughes RM *et al.* (ed.) (see Further Reading).
5. Burton CA, Brown LR, and Belitz K (2005) Assessing water source and channel type as factors affecting benthic macroinvertebrate and periphyton assemblages in the highly urbanized Santa Ana River Basin, California. pp. 239–262. In: Brown LR, Gray RH, Hughes RM, *et al.* (ed.) (see Further Reading).
6. Cuffney TF, Zappia H, Giddings EM, *et al.* (2005) Effects of urbanization on benthic macroinvertebrate assemblages in contrasting environmental settings: Boston, Massachusetts; Birmingham, Alabama; and Salt Lake City, Utah. pp. 361–408. In: Brown LR, Gray RH, Hughes RM *et al.* (ed.) (see Further Reading).
7. Walsh CJ (2004) Protection of in-stream biota from urban impacts: Minimize catchment imperviousness or improve drainage design? *Marine and Freshwater Research* 55: 317–326.
8. Pompeu PS, Alves CBM, and Callisto M (2005) The effects of urbanization on biodiversity and water quality in the Rio das Velhas Basin, Brazil. pp. 11–22. In: Brown LR, Gray RH, Hughes RM *et al.* (ed.) (see Further Reading).
9. Limburg KE, Stainbrook KM, Erickson JD *et al.* (2005) Urbanization consequences: Case studies in the Hudson River watershed. pp. 23–38. In: Brown LR, Gray RH, Hughes RM *et al.* (ed.) (see Further Reading).
10. Schueler TR and Galli J (1992) in Paul and Meyer (2001) (see Further Reading).
11. Meador MR, Coles JF, and Zappia H (2005) Fish assemblage responses to urban intensity gradient in contrasting metropolitan areas: Birmingham, Alabama and Boston, Massachusetts. pp. 409–423. In: Brown LR, Gray RH, Hughes RM *et al.* (ed.) (see Further Reading).
12. Kennen JG, Chang M, and Tracy BH (2005) Effects of landscape change on fish assemblage structure in a rapidly growing metropolitan area in North Carolina, USA. pp. 39–52. In: Brown LR, Gray RH, Hughes RM, *et al.* (ed.) (see Further Reading).
13. Walters DM, Freeman MC, Leigh DS *et al.* (2005) Urbanization effects on fishes and habitat quality in a southern Piedmont river basin. pp. 69–86. In: Brown LR, Gray RH, Hughes RM *et al.* (ed.) (see Further Reading).
14. Wang L, Lyons J, Kanehl P, *et al.* (2001) Impacts of urbanization on stream habitat and fish across multiple spatial scales. *Environmental Management* 28: 255–266.
15. Fitzpatrick FA, Diebel MW, Harris MA *et al.* (2005) Effects of urbanization on the geomorphology, habitat, hydrology and fish Index of Biotic Integrity of streams in the Chicago area, Illinois and Wisconsin. pp. 87–116. In: Brown LR, Gray RH, Hughes RM *et al.* (ed.) (see Further Reading).
16. Helms BS, Feminella JW, and Pan S (2005) Detection of biotic responses to urbanization using fish assemblages from small streams of western Georgia, USA. *Urban Ecosystems* 8: 39–57.
17. Steedman RJ (1988) in Paul and Meyer (2001) (see Further Reading).
18. Limburg KE and Schmidt RE (1990) in Paul and Meyer (2001) (see Further Reading).

substrate instability, pesticides, turbidity); all of these factors generally increase with urbanization, but some stimulate whereas others decrease primary productivity and ecosystem respiration. The response of invertebrate secondary production to urbanization is unknown because it has not been measured along a gradient of urbanization.

Lakes and Ponds

Many of the impacts of urbanization just described for rivers and streams are also observed in urban lakes and ponds such as altered temperature regimes, elevated concentrations of nutrients and contaminants, reduced total species richness, and a greater proportion of exotic species. Water withdrawals from urban lakes or their tributary streams can significantly reduce lake levels; e.g., lake levels of Lake Chapala in Guadalajara Mexico are currently 7 m below the level of the 1930s, exposing extensive mudflats. Urban lakes and ponds receive of inputs of nutrients and contaminants from both atmosphere and catchment. Alteration of urban catchments and the physical, chemical and biological characteristics of urban streams described above result in enhanced delivery of sediments, nutrients, metals, and organic contaminants from streams and stormwater conduits, as well as direct runoff from impervious surfaces. Air pollution further adds to the contaminant load of urban lakes and ponds.

One of the earliest incidences of urban impacts on aquatic ecosystems is discernible in the sediments of an Italian lake, Lago di Monterossi. Construction of the Via Cassia (a Roman highway) about 2000 years ago resulted in elevated rates of sedimentation, and higher sediment nutrient content. This is an example of cultural eutrophication, commonly resulting from anthropogenic inputs of nutrients. The excess nitrogen (N) and phosphorus (P) usually comes from agricultural or urban sources with higher N:P ratios than in reference settings. Symptoms of cultural eutrophication include increased algal biomass and productivity, a shift from algal species that are palatable to herbivorous zooplankton to inedible cyanobacteria, and an increased incidence of fish kills. Lake Washington in Seattle, Washington, USA is a classic example of this phenomenon, where increased algal blooms and decreased Secchi disc depths were associated with inputs of P from municipal sewage. When these inputs were diverted, algal productivity decreased and Secchi disc depths increased. Simply diverting inputs is not always effective, as was observed in Lake Trummen, Sweden. Because of accumulated P in lake sediments, sediment skimming and elimination of carp (they disturb sediments, thereby releasing P) was also necessary before improvements were observed.

There have been few studies of urban lakes along a gradient of urbanization, but there are many studies of individual lakes in urban settings. Urbanization affects not only lakes within city limits, but also those at the urban-rural fringe, where suburbs are expanding. Although an 'urban lake syndrome' analogous to the 'urban stream syndrome' has not been articulated, its characteristics would include the symptoms of cultural eutrophication described above combined with elevated concentrations of anthropogenic contaminants (e.g., metals and hydrocarbons) and a higher proportion of introduced species. In contrast to reference lakes, where most contaminants are from atmospheric sources, increased inputs of nutrients and contaminants from point (e.g., municipal and industrial effluents) and non-point (e.g., septic systems and stormwater) sources in the catchment alter urban lake chemistry. Concentrations of coliform bacteria can be high, and beach closures occur, especially after storms that result in combined sewer overflows. Many European and older North American cities have conduits that carry both sewage and stormwater; intense rainstorms fill the pipes, overwhelm wastewater treatment plants, and dump untreated wastes directly into receiving waters. In the developing world, untreated municipal and industrial wastes are commonly discharged directly into aquatic ecosystems, resulting in highly degraded urban aquatic ecosystems. The unique biodiversity of Lake Victoria in Africa is threatened by urban development around it because untreated wastes go directly into the lake. Hypoxic bottom waters are common in eutrophic urban lakes, with consequences for biogeochemical cycles as well as benthic biota. The extent to which these symptoms are exhibited in an urban lake depends not only upon urbanization intensity, but also upon lake attributes that result from its geological and biological setting, such as area, volume, depth, water residence time, sediment characteristics, and species present.

Sediments from urban lakes provide a historical record of contamination. Concentrations of metals in sediments from urban lakes are considerably higher than in reference lakes. Stricter discharge limits were enacted in the United States in the 1970s, and concentrations of lead, cadmium, chromium and nickel in urban lake sediments generally declined over the past three decades, whereas there has been no consistent trend for copper and mercury, and increases outnumber decreases for zinc. Concentrations of polycyclic aromatic hydrocarbons (PAHs) and chlordane in sediments from 38 urban and reference lakes across the United States increased as catchment urbanization increased. PAH concentrations increased over the past three decades, whereas chlordane increased in half the lakes and decreased in the other half. Both compounds enter food webs and are the cause of many fish consumption advisories.

Algal diversity is reduced, but biomass and productivity is usually high in urban lakes because of elevated rates of nutrient delivery and altered grazer assemblages. High algal biomass combined with accelerated sediment delivery results in higher turbidity in urban lakes. Water temperature, depth of turbulent mixing, pH, and low N:P ratios in urban lakes can result in seasonal blooms of cyanobacteria that are inedible to zooplankton, unsightly, often create taste and odor problems, and can be toxic to humans, pets, and livestock. In 2007, crews skimmed more than 6000 tons of cyanobacteria from Lake Taihu in China in an attempt to keep them out of the drinking water of the city of Wuxi. Scientists have expressed concern that such blooms will become even more common throughout the world's urban areas because of global warming. Shallow urban lakes often support dense stands of aquatic macrophytes, many of which are invasive weeds (e.g., *Hydrilla*). Housing development along the shoreline can be extensive with resulting loss of riparian forest and reduced input of woody debris to the littoral zone. Those losses represent a loss of nearshore habitat for fishes and other biota. The absence of riparian and littoral vegetation in Wisconsin lakes in urban commercial settings was identified as the factor resulting in fewer zooplankton taxa in those lakes than in lakes in urban residential or forested settings. Regionally common zooplankton taxa were present, but rare taxa were missing from the urban lakes. Artificial lights at night in urban lakes and ponds can interfere with zooplankton migration patterns.

Onondaga Lake near Syracuse, New York, provides a classic example of the impacts of urbanization on a lake. Industrial discharges during the 19th and 20th centuries combined with increasing human population and inputs of sewage effluent resulted in elevated salinity, high concentrations of nutrients and organic contaminants, toxic concentrations of free ammonia, severe oxygen depletion, frequent cyanobacteria blooms, and the loss of native zooplankton and fish species. As industrial and municipal discharges have been reduced and sediment cleanup programs begun, lake water quality has improved, cyanobacteria blooms are less common, and native species of *Daphnia* have returned. Fishing is allowed, but consumption advisories persist.

In addition to naturally occurring lakes, artificially created lakes and ponds are common in urban areas. The European Union has classified 4% of its surface waters as artificial. Artificial lakes and ponds are often fairly shallow and may be purely ornamental, serve as a municipal water source, store storm water, or enhance its infiltration. Reservoirs pooled behind dams provide water for generating electricity. Introduction of non-native plants, invertebrates, and fishes alter food webs and nutrient dynamics in these ecosystems. These introductions occur more frequently in ponds close to roads (a shorter distance to carry an aquarium before dumping). Introductions of bottom-feeding fishes (e.g., goldfish) may enhance cultural eutrophication by accelerating release of phosphorus from the sediments.

Wetlands

Urbanization has resulted in significant wetland loss through draining, dredging, and filling. Even if wetlands are not completely eliminated, urban development fragments them with road crossings and impairs wetland ecosystem function by altering hydrologic regime, increasing input of nutrients and toxins, and introducing exotic species. Wetland species such as turtles and salamanders that spend part of their life on land and part in water are particularly vulnerable to urbanization. Not only does one life history stage have to survive in an altered aquatic environment, but the terrestrial stage has to survive in an often hostile terrestrial environment (e.g., migrating across roads results in high mortality rates). Studies have consistently found anuran abundance and species richness to be negatively correlated with measures of urbanization: % urban land use, road density, % imperviousness, and large inputs of stormwater. Similar findings have been reported for wetland bird species.

Recognition of the impacts of urban stormwater runoff on aquatic ecosystems has resulted in regulations requiring the construction of stormwater retention, detention, or infiltration ponds and wetlands in the United States and the European Union. Wetlands have also been constructed to treat sewage and stormwater. These artificial wetlands can be effective in nutrient removal, but provide habitat that is less desirable than naturally occurring wetlands. Artificial wetlands are characterized by elevated concentrations of contaminants and a high proportion of exotic flora and fauna.

Ecosystem Services

Urban aquatic ecosystems provide a wide range of ecosystem services, which are the goods and services

Table 5 Examples of ecosystem services provided by intact urban aquatic ecosystems. Services are organized according to the framework used in the Millennium Ecosystem Assessment (2005)

Type of ecosystem service	*Services provided by aquatic ecosystems*
Provisioning	Produce food (e.g., fisheries)
	Fresh water for human uses
Regulating	Natural hazard regulation (e.g., flood protection)
	Water purification (e.g., retention of sediments; retention and transformation of nutrients, contaminants and organic matter)
Cultural	Inspiration and aesthetic values; spiritual renewal and a sense of place
	Educational opportunities (e.g., interesting habitats and biota)
	Recreational opportunities (e.g., boating, fishing, swimming, wildlife viewing)

produced by ecosystems that are beneficial to humans (**Table 5**). The impacts of urbanization described in the previous sections have reduced the capacity of aquatic ecosystems to provide these services. For example, increasing impervious cover reduces groundwater recharge, storage of floodwaters and sediments, capacity for nutrient and contaminant removal, all of which impact water quality and aquatic biodiversity. As city dwellers recognize the value of ecosystem services, there is growing interest in preservation (e.g., greenways) and rehabilitation of urban aquatic ecosystems (e.g., riparian planting, daylighting streams previously encased in culverts). Ecologically sensitive development of urban waterfronts and trails along waterways can provide both economic and ecological benefits to city dwellers. As the proportion of the human population living in urban areas continues to increase, aquatic ecosystems in the city offer places for spiritual renewal as well as valuable opportunities to enjoy and learn about the natural world.

Conclusions

An ever-increasing proportion of the growing human population lives in cities and both impacts and depends upon the ecosystem services provided by streams, rivers, lakes, ponds, and wetlands. Urbanization impacts the physical, chemical, and biological characteristics of these ecosystems. Alterations include increased frequency and magnitude of floods; greater range in water temperature; increased sedimentation; altered structure of stream channels and river networks; increased concentration of ions (salinization), nutrients (cultural eutrophication), and contaminants (metals, pesticides, hydrocarbons, pharmaceuticals); reduced capacity for nutrient removal; increased algal biomass with more frequent nuisance algal blooms; a greater proportion of tolerant species of algae, invertebrates, and fishes; fewer amphibian and wetland bird species; and increased prevalence of non-native species. As a consequence of these changes, the ability of urban aquatic ecosystems to provide services benefiting humans has been degraded.

Knowledge Gaps

Scientific understanding of urban aquatic ecosystems has advanced considerably in the past decade, but the complexity of interactions between human infrastructure, institutions, and aquatic ecosystems has only begun to be explored. Effective management and rehabilitation of urban aquatic ecosystems requires improved scientific understanding in the following areas:

- *Measures of aquatic ecosystem processes in a diverse array of cities.* Rates and patterns of nutrient cycling, ecosystem metabolism, and secondary productivity along gradients of urbanization in different geographical and cultural settings are largely unknown; yet these are the processes providing valued ecosystem services. Generalizations about urban impacts are primarily derived from Temperate Zone cities in the developed world, whereas most of the growth in urban populations is occurring in tropical cities in the developing world.
- *Influence of type and pattern of development.* Alternative building designs and development patterns (e.g., clustered housing) are being proposed to reduce urban impacts on aquatic ecosystems; these should be viewed as catchment experiments to explore their impact on physical, chemical, and biotic characteristics of urban waters.
- *Effectiveness of rehabilitation practices.* Many cities have invested heavily in projects to improve conditions in aquatic ecosystems; yet there has been relatively little evaluation of the effectiveness of different practices. For example, given the findings on importance of effective imperviousness, will reducing the hydraulic connectivity between impervious surfaces and streams result in improved ecological conditions?

- *Link ecological, engineering, and socio-economic analyses.* Urban aquatic ecosystems are impacted by human actions and institutions. Better understanding and management of these ecosystems requires collaborative interdisciplinary studies and models. For example, how can economic values be assigned to the ecosystem services provided by urban aquatic ecosystems? Urban systems include processes and pathways that are not found in unmanaged systems and that are influenced by factors not traditionally considered by ecologists such as economic conditions and human decisions on lawn and garden design. Studies of urban aquatic ecosystems are part of the broader discipline of urban ecology, which recognizes that collaboration outside the natural sciences is essential to advance understanding of urban systems.

Glossary

Catchment – Area of the land that is drained by a stream network or land area from which water flows into a lake.

Cyanobacteria – Photosynthetic bacteria (formerly called blue-green algae) that can form dense blooms; some produce toxins and some are able to fix atmospheric nitrogen.

Eutrophic – Very productive.

Hypoxic – Low concentration of dissolved oxygen.

Imperviousness – The extent to which a catchment is covered by surfaces (e.g., roofs, paved roads, and parking lots) that do not allow water to penetrate into the ground.

Infiltration – Gradual movement of water into soil.

Secchi depth – A measure of turbidity in water; the vertical distance that a Secchi disk (black and white disc about the size of a small dinner plate) can be lowered into the water before it disappears from an observer's view; clear water has a large Secchi depth whereas it is small in turbid water.

Urban sprawl – Low-density development on the edges of urban areas usually characterized by single-family homes whose residents are dependent on personal automobiles for transportation.

See also: Restoration Ecology of Rivers.

Further Reading

Azous AL and Horner RR (2001) *Wetlands and Urbanization.* Boca Raton: Lewis Publishers.

Booth DB and Jackson CR (1997) Urbanization of aquatic systems: degradation thresholds, stormwater detection, and the limits of mitigation. *Journal of the American Water Resources Association* 33: 1077–1090.

Brown LR, Gray RH, Hughes RM, *et al.* (eds.) (2005) *American Fisheries Society Symposium 47: Effects of Urbanization on Stream Ecosystems.* Bethesda, Maryland: American Fisheries Society.

Chin A (2006) Urban transformation of river landscapes in a global context. *Geomorphology* 79: 460–487.

Dodds WK (2002) *Freshwater Ecology: Concepts and Environmental Applications.* Academic Press.

Dodson SI, Lillie RA, and Will-Wolf S (2005) Land use, water chemistry, aquatic vegetation, and zooplankton community structure of shallow lakes. *Ecological Applications* 15: 1191–1198.

Effler SW (1996) *Limnological and Engineering Analysis of a Polluted Urban Lake: Prelude to Environmental Management of Onondaga Lake.* New York: Springer.

Grimm NB, Faeth SH, Golubeiwski NE, *et al.* (2008) Global change and the ecology of cities. *Science* 319: 756–758.

Kaushal SS, Groffman PM, Likens GE, *et al.* (2006) Increased salinization of fresh water in the northeastern United States. *Proceedings of the National Academy of Sciences* 102: 13517–13520.

Knutson MG, Sauer JR, Olsen DA, *et al.* (1999) Effects of landscape composition and wetland fragmentation on frog and toad abundance and species richness in Iowa and Wisconsin, U.S.A. *Conservation Biology* 13: 1437–1446.

Mahler BJ, Van Metre PC, and Callender E (2006) Trends in metals in urban and reference lake sediments across the United States, 1970 to 2001. *Environmental Toxicology and Chemistry* 25: 1698–1709.

Meyer JL, Paul MJ, and Taulbee WK (2005) Stream ecosystem function in urbanizing landscapes. *Journal of the North American Benthological Society* 24: 602–612.

Meyer JL and Wallace JB (2001) Lost linkages in lotic ecology: rediscovering small streams. In: Press M, Huntly N, and Levin S (eds.) *Ecology: Achievement and Challenge*, pp. 295–317. Oxford, UK: Blackwell.

Millennium Ecosystem Assessment (2005) *Ecosystems and Human Well-Being. Current State and Trends,* Vol. 1. Washington D.C.: Island Press.

Paul MJ and Meyer JL (2001) Streams in the urban landscape. *Annual Review of Ecology and Systematics* 32: 333–365.

Poff NL, Bledsoe BP, and Cuhaciyan CO (2006) Hydrologic variation with land use across the contiguous United States: Geomorphic and ecological consequences for stream ecosystems. *Geomorphology* 79: 264–285.

Riley SPD, Busteed GT, Kats LB, *et al.* (2005) Effects of urbanization on the distribution and abundance of amphibians and invasive species in southern California streams. *Conservation Biology* 19: 1894–1907.

Scott MC and Helfman GS (2001) Native invasions, homogenization, and the mismeasure of integrity of fish assemblages. *Fisheries* 26: 6–15.

Van Metre PC and Mahler BJ (2005) Trends in hydrophobic organic contaminants in urban and reference lake sediments across the United States, 1970–2001. *Environmental Science and Technology* 39: 5567–5574.

Walsh CJ, Roy AH, Feminella JW, *et al.* (2005) The urban stream syndrome: current knowledge and the search for a cure. *Journal of the North American Benthological Society* 24: 708–723.

Wang L, Lyons J, Kanehl P, *et al.* (2000) Watershed urbanization and changes in fish communities in southeastern Wisconsin streams. *Journal of the American Water Resources Association* 36: 1173–1189.

Relevant Websites

http://beslter.org/ – Baltimore Ecosystem Study Long-term Ecological Research site.

http://caplter.asu.edu/CentralArizona – Phoenix Long-term Ecological Research site.

http://ec.europa.eu/environment/water/water-framework/ – Water Framework Directive for the European Union, which describes the water information system for Europe.

http://water.usgs.gov/nawqa/ – National Water Quality Assessment Program, with projects that include the impact of urbanization on aquatic ecosystems.

http://www.maweb.org/ – Millennium Ecosystem Assessment with information on global trends in urbanization.

http://www.unhabitat.org/ – Provides information on the World Urban Forum and on aquatic ecosystems in the developing world.

Restoration Ecology of Rivers

B G Laub, University of Maryland, College Park, MD, USA
M A Palmer, University of Maryland Center for Environmental Sciences, Solomons, MD, USA

Introduction

Worldwide, the health of rivers is in decline. In the United States, 33% of rivers are impaired or polluted in some way. In Europe, it is around 80–90%. Rivers flow through low-lying areas, making them susceptible to the integrated effects of human activities throughout the landscape. Anthropogenic impacts have impaired the ability of many river ecosystems to provides goods and services that society depends on. Water has frequently become undrinkable and incapable of supporting healthy aquatic communities. When rivers are degraded, they often lose their aesthetic and recreational value that they provide to the people who live near them.

When river ecosystems are degraded to such an extent that natural processes are hampered and social value is lost, people often turn to river restoration. River restoration can be broadly defined as the human-assisted improvement of river integrity through recovery of natural hydrologic, geomorphic, and ecological processes. Thus, river restoration ecology is the branch of science concerned with developing and implementing ecologically effective river restoration. Although the goals of a restoration project are not often ecologically focused, in this chapter, the discussion is restricted to ecological restoration. Before discussing restoration ecology as a science, it is necessary to understand the motivation for pursuing ecological restoration, which is primarily driven by a desire to remediate human impacts on rivers.

Present State of the World's Rivers

The impacts that threaten rivers in different nations depend, in large part, on the history of human use of rivers in that nation. In Europe, humans began manipulating rivers for their needs at least 6000 years ago. Agricultural development of the Nile River floodplain began in about 2000 BC. In the United States, the Industrial Revolution in the late 1800s allowed an increase in heavy engineering projects, including the construction of dams and levees. Some developing nations have recently begun to manipulate rivers using large engineering projects; The Three Gorges Dam on the Yangtze River in China, which was closed in 2003, is one example.

Anthropogenic impacts, though often damaging to river ecosystems, are the result of some structure or development that benefits society in some way. This is an important point, because, as will be discussed later, restoration is only an option when some segment of society decides that the environmental degradation caused by some human system outweighs the benefits derived. The discussion in the upcoming sections focuses on some of the common causes of worldwide river degradation, but the list is not exhaustive. Rivers in each region may be impacted by all or none of the impacts described.

Dams

Dams are built for electricity production, flood protection, and storage of water for municipal and agricultural use. Dams and their reservoirs impact rivers and aquatic organisms in two fundamental ways. First, dams fragment the upstream to downstream connectivity of river ecosystems. Dams can block migration of fish and other aquatic organisms, and reservoirs (large body of standing water) may be uninhabitable to organisms that live in running-water habitats. Water released from some dams is much colder and carries far fewer sediments than the incoming river water. As a result, the stretch of river below the dam may have a thermal regime that favors cold-water exotic species, and it may also erode vital habitats, such as sand or gravel bars, as it picks up sediment to replace the lost sediment load. The combination of these impacts may make the river inhospitable to native species far downstream of the dam, further fragmenting the river system. The second major impact of dams is that they alter the natural variation in flow levels to which native organisms are adapted. Typically, both peak flood magnitudes and daily variability in flow level are reduced (**Figure 1**).

Water Abstraction

In arid regions of the world, the demand for water often outweighs the supply provided by rivers. In some places, the demand is so high that once perennial rivers are now left completely dry for all or part of the year. The high demand may also be met through interbasin water transfers, in which water from one river is piped over drainage, divides into a

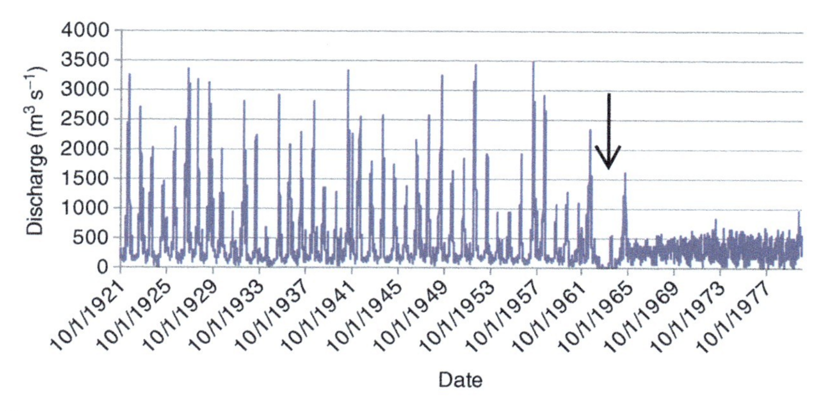

Figure 1 Mean daily discharge of the Colorado River, USA between 1 October 1921 and 1 October 1980 at Lee's Ferry, Arizona. The arrow indicates the closure of Glen Canyon Dam 24 km upstream from the gage station. Note the reduction in peak flows and overall flow variability following dam closure. Data from United States Geological Survey.

different river. The consequences of interbasin water transfers include a reduced flow of multiple rivers and an increased ability of invasive species to spread across watershed divides.

Agriculture

Agricultural land use impacts rivers by creating runoff containing high levels of nutrients and sediments. Nutrients, such as nitrate and phosphate, are added to crop fields during fertilizer application and are transported to rivers through overland flow or through the groundwater. The input of nutrients can substantially increase productivity and lead to eutrophication in rivers and downstream receiving waters such as lakes and estuaries. High inputs of sediments to rivers can be detrimental to aquatic organisms, including fish, whose eggs may become buried. The development of land for agricultural use often involves deforestation, including cutting down of trees within the river floodplains. Floodplain forests can filter nutrients and impurities from the groundwater before that water reaches the river. The ability of a vegetation to perform this function is compromised when a floodplain is developed, and water quality often declines as a result.

Urbanization

Urban development of land can be divided into two stages: the construction stage and the developed stage. Each stage has a different impact on rivers. During the construction stage, vegetation is removed over a large area and the delivery of sediment to rivers is increased as much as 100-fold over preconstruction levels. Sediment inputs are much lower in the developed stage than in the construction stage, but the amount of paved area and other impervious surfaces is greatly increased. The replacement of natural vegetation with impervious surfaces routes rainwater directly to rivers, whereas rainwater would naturally seep slowly to rivers through groundwater flow paths. As a result, the natural hydrograph is substantially altered, with floods occurring more rapidly and with greater intensity following rain events. The increased flashiness of flood flows can cause channel incision and erosion. In addition, when rainwater falls on roads and other paved surfaces it can become heated and may pick up metals, oils, and other pollutants, all of which are delivered to the receiving river (**Figure 2**).

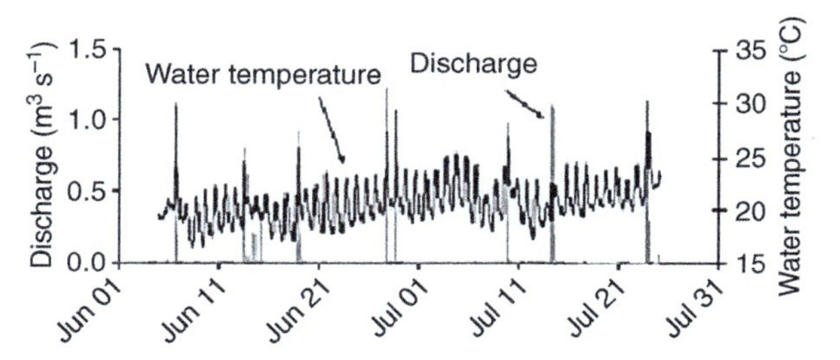

Figure 2 Discharge (light gray line) and temperature (black line) recorded in 30-min intervals on Paint Branch stream, Maryland, USA from 1 June 2001 to 1 August 2001. Paint Branch is an urban stream with 58% impervious surface cover in the watershed. Note the rapid rise and fall of discharge, which is typical of flashy flood flows often observed in urban streams. Also note that many flood flows are accompanied by a spike in water temperatures, sometimes by as much as 10 °C. Both flashy flows and temperature spikes can stress aquatic organisms. Figure from Nelson and Palmer (2007).

The Recommendations of River Restoration Ecology

River restoration ecology as a science can help inform the decisions of people undertaking restoration, but the decision to undertake restoration must be made by people who value rivers, because the perceived degradation of the river ecosystem must outweigh the benefits provided by the systems contributing to degradation. Thus, the motive to restore is ultimately a judgment in which the values placed on river ecosystems outweigh those values derived from the sources of degradation. In some cases, national or international laws mandate restoration, but the laws would not be in place unless people valued river ecosystems or ecosystem services enough to enact laws to protect them. Even if the decision to undertake river restoration is made, restoration in human-dominated landscapes requires a compromise between restoring rivers for purely ecological reasons and preserving human systems that use river ecosystems for some societal benefit. However, it is helpful to understand recommendations under ecologically idealized conditions, not because the conditions are often met – they are rarely if ever met – but because it demonstrates how human systems constrain the river restoration process.

Restoration without Constraints

Ecologically, in the most idealized situation, restoration would remove all human impacts on the river ecosystem and move the river towards a former natural state that existed before human alteration. In their natural state, rivers are dynamic, with channel shape and position on the landscape constantly changing. The dynamic state of rivers is created by natural processes, including flooding and movement of sediments within the river channel and onto the floodplain, and by spatial and temporal variability in process rates. Recovering the river processes and the variability in process rates would be essential in restoring a completely natural river, because river processes support aquatic and floodplain biological communities, many species of which are adapted to thrive in a highly dynamic ecosystem. Recovering processes and process variability would also ensure river self-sustainability. Self-sustainable rivers are able to recover from disturbances and respond to changes on the landscape while maintaining ecosystem services.

Restoration with Constraints

In the idealized situation, all human impacts to a river are removed during restoration. Such a situation will probably never occur, because complete restoration of natural river ecosystems in many parts of the world may be technically infeasible given the long history of human manipulation of rivers. Moreover, the perceived value of a completely natural river ecosystem would have to outweigh the perceived value of all human systems contributing to river degradation. Society is unlikely to place such a high value on a natural river, but is instead likely to impose constraints on the ecologically idealized restoration scenario. The constraints, whether financial, political, social, natural, scientific, or a combination of any of these (see **Table 1** for examples), will restrict the process of ecological restoration. For example, if there are homes near a river that is slated for restoration, people are unlikely to support a restoration project that attempts to recover the process of river flooding. To implement ecologically effective restoration on rivers in human-dominated landscapes, compromises must be reached between people with a vested interest in the river, and restoration must be designed and executed within the constraints imposed. In the idealized situation, restoration would be conceptually simple and success relatively sure: remove all human impacts and allow the river to return to a natural state. In human-dominated systems the river restoration process must involve three steps: (1) planning within the geographical setting and constraints, (2) implementation, and (3) monitoring to determine if restoration was successful.

The River Restoration Process

Achieving ecologically effective river restoration within the constraints imposed by human systems requires that principles from restoration ecology be incorporated throughout the restoration process (**Figure 3**). However, even if restoration ecologists are involved, restoration of rivers with the guarantee of ecological success is as untenable an idea as guaranteeing that medical treatment of an ill patient will lead to a long and healthy life. Like human bodies, river systems are complex and are influenced by their surroundings, their history, and what is put into them. Nevertheless, as will be shown, if restoration is planned for, executed, and monitored properly, each restoration project can become a learning experience and contribute to the progression of restoration ecology as a science. This can happen even if a restoration project is deemed a failure.

The Planning Stage

This is the most important stage in the river restoration process, because it is in this stage that the goals and objectives of the restoration project are stated. The goals should address the problems with

Table 1 A list of common goals in river restoration, associated techniques for accomplishing each goal, and constraints that may limit utility of techniques

Goal	*Number of project records*	*Common techniques*	*Example of a constraint*
Water quality management	11 981	• Planting riparian vegetation • Soil conservation practices such as no-till farming and cover cropping • Controlling point-source pollution	*Political/financial* – If laws protecting water quality are lacking, businesses may be unwilling to pay to remediate pollution or modify production activities
Riparian vegetation management	11 835	• Livestock exclusion • Planting riparian vegetation	*Natural* – Bank erosion may be worsened by debris dams that direct water toward the banks
Instream habitat improvement	5750	• Pool and riffle construction • Boulder and wood addition	*Scientific* – A target species' habitat needs may be unknown
Fish passage	4881	• Fish ladder installation • Culvert redesign • Fish weirs on irrigation canals	*Natural* – Natural mortality factors may keep populations low even if passage is achieved
Bank stabilization	3163	• Planting riparian vegetation • Bank grading • Riprap installation	*Financial/natural* – Disruptive techniques are expensive and may harm biotic communities
Flow modification	1343	• Purchasing water rights • Promoting water conservation • Controlled dam releases	*Political* – Existing policy may prioritize water for industrial, agricultural and municipal use
Aesthetics/recreation/education	1116	• Removing trash • Building footpaths • Placing signs	*Social* – If restoration is not a perceived success, citizens may not support future restoration
Channel reconfiguration	1045	• Channel realignment • Daylighting • Bank grading and reshaping	*Scientific* – Sediment input and flow variability data are needed to design an appropriate channel
Dam removal/retrofit	764	• Dam breaching • Revegetation and sediment removal after dam removal • Multilevel offtake towers	*Financial* – Dam removal and retrofit are both expensive. Costs may be prohibitive
Stormwater management	544	• Pond and wetland construction • Pipe outflow protection or burial	*Political/Financial* – Developers may not pay for stormwater structures if laws are lacking
Reconnecting floodplains	535	• Bank grading • Channel reshaping and elevation	*Social* – Floodplains may be developed
Instream species management	358	• Stocking native species • Exotic species control	*Social* – Exotic species, such as game fish, may be preferred

Data include all river restoration projects in U.S. national databases as of July 2004. Modified from Bernhardt ES, Palmer MA, Allan JD, *et al.* (2005) Synthesizing U.S. river restoration efforts. *Science* 308: 636–637.

overall integrity of the river ecosystem as identified by scientists or managers or brought forth by concerned citizens. The goals are derived by forming a guiding image of what the river ecosystem could potentially be restored to, which may be the river in its original state before human influence, and then restricting the focus to what can potentially be accomplished given the existing constraints. The specific restoration techniques and the monitoring program are designed based on the established goals, and therefore, the goals should be sufficiently broad. Examples of broad goals include recovery of populations of native fish species and reduction of both nutrient and sediment loads. Once the goals are established, measurable objectives with a specific timeline should be defined. One example would be reduction of nitrate by 30% and phosphate by 20% from their prerestoration levels within 5 years following the completion of restoration. Setting specific, measurable objectives helps define the needed restoration techniques and an appropriate monitoring scheme.

The goals and objectives for a restoration project must be set within the spatial and temporal context of

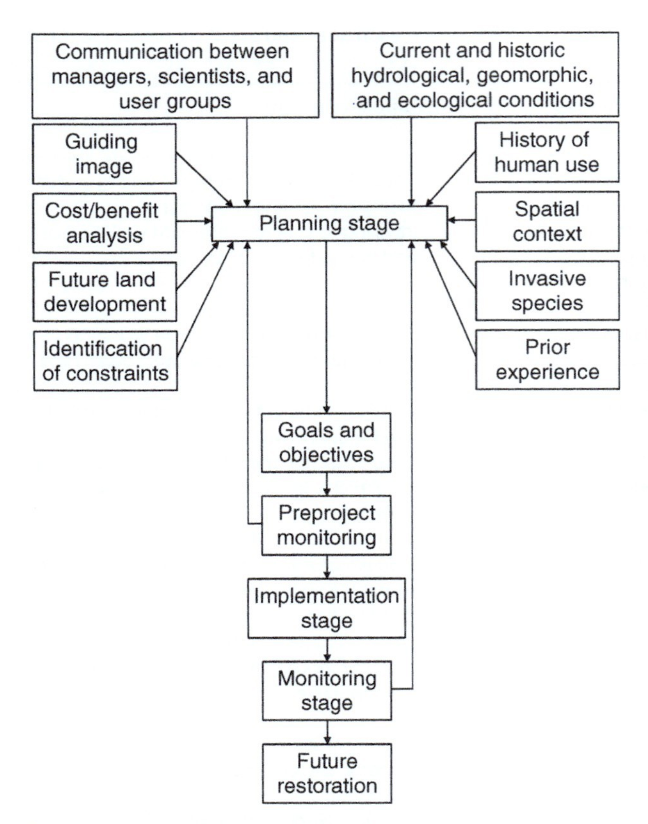

Figure 3 Flowchart illustrating the major stages of the restoration process and examples of inputs during the planning stage. Restoration is an adaptive management process, in which data obtained from pre- and postproject monitoring may prompt managers to change or update the goals and objectives for river restoration.

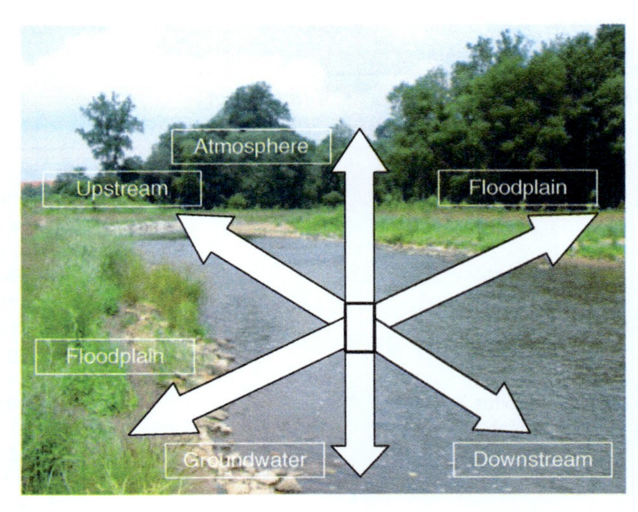

Figure 4 Rivers are connected longitudinally to upstream and downstream rivers, laterally to floodplains, and vertically to the groundwater and atmosphere, and the strength of these connections can vary over time. Such connections are important to consider during restoration, because one section of river is influenced by impacts throughout the surrounding watershed and beyond.

the river ecosystem. Rivers receive runoff from all land in the surrounding watershed, are connected to floodplains and upstream tributaries, and interact with the groundwater (**Figure 4**). In addition, each river has a unique geologic history and a unique history of human development, which will impose unique constraints on any restoration project. Therefore, in planning a restoration project, data must be gathered on the past and current hydrological, geomorphic, and ecological conditions of the subject river. Data from past surveys, old maps, and old aerial photographs can provide information on historic conditions. The knowledge and experience of river managers, scientists, engineers, and other professionals can also be helpful. Regardless of the availability of historical information, a period of data gathering prior to the start of restoration can prove helpful in assessing the success of the restoration project once monitoring is completed, because this allows comparison between prerestoration and postrestoration conditions.

Considering the spatial context of rivers is important when designing river restoration, because the potentially positive benefits of a river restoration project may be overridden by development on the landscape and other upstream or downstream impacts. For example, a fish population is unlikely to respond to a project that adds physical habitat structures to a river if there is an urban area upstream that delivers polluted water and causes frequent and severe flooding. Furthermore, invasive species often thrive in impacted river systems, and a restoration project aimed at recovering habitat for native species can easily provide habitat for invasive species instead. This is especially true if the nearest populations of desired species are not connected to the restoration area through migration pathways. Even if desired species are introduced during restoration, invasive species may quickly colonize a restored area and reduce populations of the desired species. Identifying potential sources of invasive species throughout the landscape should be included in the prerestoration data.

Whatever the situation, people planning the restoration must make a critical decision as to whether the restoration actions will achieve the desired goals. If the restoration involves a localized effort within a large watershed dominated by human impacts, the effects of the restoration project are likely to be minimal. Project planners need to incorporate a cost-benefit analysis, because small, localized projects may have a relatively small cost, but may also provide little benefit. Increasingly, restoration of entire watersheds, in which all impacts on a river system throughout the watershed are identified and prioritized for remediation activities, is becoming accepted as the only viable approach for successful river restoration (**Figure 5**).

In ecological restoration, the aim is to recover natural river processes and river self-sustainability. Planning for how to accomplish this goal within the imposed constraints requires the involvement of multiple groups, including scientists, river managers, people that live near the river, businesses that require river resources for their livelihood, and people that use the river for recreation or cultural purposes.

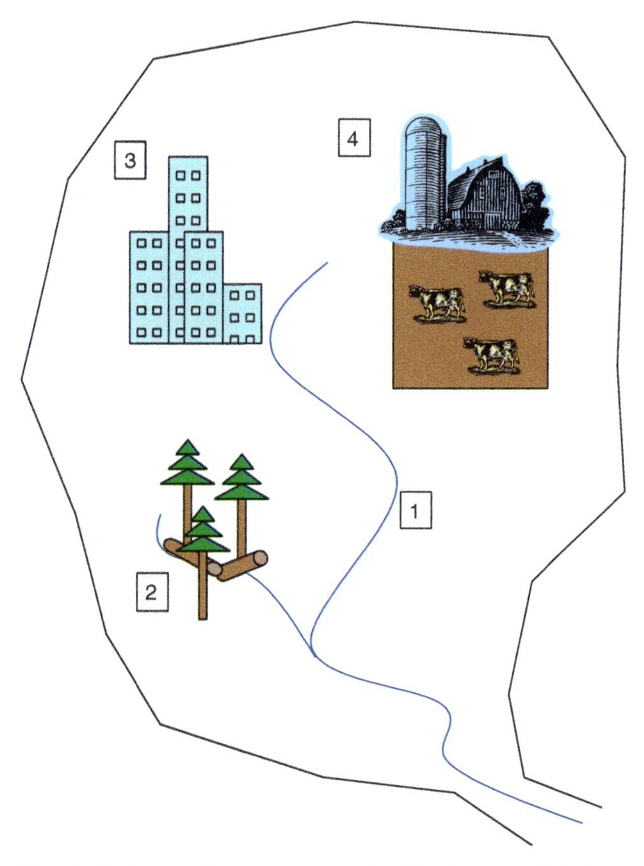

Figure 5 In watershed restoration, impacts to a reach of river are identified and prioritized. In this hypothetical watershed, the degraded river section (1) has much less riparian vegetation and instream habitat compared with a reference river (2). However, these problems would be given a lower priority for restoration than upstream impacts from urban (3) and agricultural areas (4). The urban area is given higher priority, because riparian vegetation and instream habitat structures could be washed away by the intense floods generated from urban runoff. Similarly, high inputs of sediment from agricultural areas could bury riparian vegetation and instream habitat structures. Thus, in this watershed, restoration projects would target the urban and agricultural impacts before incorporating riparian vegetation planting and instream habitat improvement.

In planning restoration, communication between these groups is essential. In fact, lack of effective communication may pose a barrier to effective completion of a restoration project. Ensuring effective communication among all interested parties is one of the greatest challenges faced by those planning restoration projects.

The future must also be considered during the planning stage, primarily because conditions in the river watershed are likely to change over time. For example, population growth in a watershed may lead to increased demand for water, which could lead to reduced river flows. Unless such changes are addressed during the planning process, self-sustainability of restoration projects will be threatened and ecological success in the future will be unsure. Increasingly, restoration planning involves both an evaluation of historical conditions and projected changes in conditions due to future development and shifts in climatic patterns.

The Implementation Stage

In the implementation stage, the restoration project is carried out, using the techniques decided upon during the planning stage. A list of some common goals in river restoration practice and their associated techniques is presented in **Table 1**. Accomplishing river restoration goals often requires techniques that extend beyond direct manipulations of the river channel (**Table 1**). Such off-channel techniques have historically not been labeled as river restoration projects, but they may often be more effective in improving the quality of river systems than within channel manipulations. Whatever techniques are used, restoration ecologists have advocated that no net harm be done to a river in the process of restoring it.

The Monitoring Stage

During the monitoring stage, data is collected and analyzed to determine if the restoration project achieved the goals and objectives. Although conceptually simple, evaluation of ecological success in river restoration may be difficult in practice and is one topic at the forefront of research in river restoration ecology. The topic is of particular interest, because with ever increasing needs for restoration and limited monetary resources, it is important to know if restoration is improving ecological conditions, so that money can be spent efficiently.

One difficulty in evaluating ecological success is that data collected on the restored section of a river must be compared with data collected on the river prior to restoration, an adjacent section of the river upstream of the restoration reach, or that of a set of reference rivers. The most preferable situation is comparing postrestoration data with prerestoration data to determine if the restoration project changed conditions in the river. If reference rivers are used, these must be chosen carefully, because hydrologic and geomorphic conditions and land use are likely to vary substantially between river watersheds. In fact, recent research suggests that the choice of reference rivers may alter the conclusions drawn about restoration success.

Another difficulty in evaluating ecological success is that river processes and variability in process rates must be measured. Evaluating river processes provides a good indication of the likely long term success of a project, because persistence of biological communities depends on river processes. The difficulty for monitoring is that the best methods for evaluating processes are not known or not agreed upon. Research is ongoing to better understand river processes, their variability, and which method to use to evaluate their recovery.

A third difficulty in evaluating ecological success is that restoration should ideally be evaluated on the watershed scale. The focus should be on whether the individual project contributed to overall improvement of ecological conditions locally and downstream. An individual restoration project may appear at first to be an ecological failure, but if the project is coupled with restoration projects on other river reaches and improvement of conditions on the surrounding landscape, the individual project may contribute to ecologically successful restoration of the watershed. Recovery on the watershed scale may take several years or even decades, requiring long term monitoring of conditions over relatively large spatial scales to properly evaluate river restoration success. Even if monitoring is conducted on a small, localized project, monitoring must be conducted for many years to allow for recovery of the restored river reach and to properly assess variability in system attributes over time (**Figure 6**).

Restoration success could also be judged based on nonecological properties such as aesthetics or recreational improvement. Determining success in such areas will still require monitoring, although the monitoring may be aimed more at people's attitudes toward restoration than at ecosystem attributes. However, ecological success and social success must be evaluated separately, because success in one area

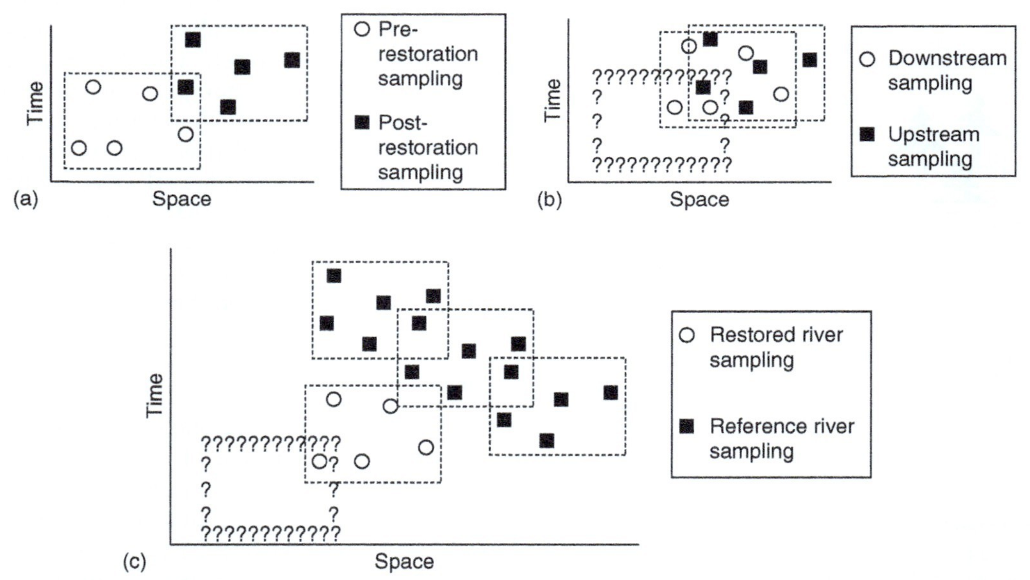

Figure 6 Three methods for monitoring the effects of restoration on river ecosystems. In each method, circles and squares represent one sampling of an ecosystem state or process variable, for example the size of a fish population or the uptake rate of nutrients by river biota. Samples collected at multiple locations and at different times provide a window of variability for the given variable, indicated by the dashed boxes. (a) Pre- and post-restoration monitoring. Sampling is conducted on the river reach to be restored and on the same river reach after restoration is complete. This method allows direct assessment of whether restoration altered the measured variable. (b) Upstream and downstream monitoring. Sampling is conducted after restoration is complete on the restored river reach and on an upstream reach, which is presumably less degraded. The restored reach can thus be compared with the upstream reach, but since the condition of the restored reach prior to restoration is unknown (box outlined by ?'s), the effects of restoration are uncertain.
(c) Reference river monitoring. Sampling is conducted on the restored river and on several reference rivers, which are presumably less degraded. Conditions on the restored river are compared with conditions on the reference river, but the effects of restoration are again uncertain, because the restored river was not sampled prior to restoration.

does not necessarily imply success in the other. Moreover, if monitoring for ecological success is conducted, a project that is deemed an ecological failure can still provide information to project managers that may help improve restoration in the future. In such cases, the restoration project is called a learning success. The large number of river restoration projects being implemented affords an excellent opportunity to evaluate ecological effectiveness. Unfortunately, most restoration projects are not monitored intensively enough to determine ecological outcome. For the few projects monitored there has been a lack of consistency in methods or criteria used to judge restoration success. Incorporating monitoring with consistent criteria on more restoration projects will contribute greatly to progress in developing the science of restoration ecology.

Recommendations for the Future

Many instances can be found throughout the world where river restoration has proceeded even within the most restrictive constraints. As one example, restoration of the Cehonggyecheon River in Seoul, South Korea involved complete removal of a 6-lane elevated highway. However, a critical question to ask of any restoration project is what has been achieved ecologically? Improving aesthetics along urban rivers is a considerable achievement, but does not guarantee that ecological conditions will improve. Often, recovery of key components that contribute to natural ecosystem processes, such as development of a floodplain through flood peak and recession, is simply unachievable. This is a common reality that restoration ecologists are now facing, and will increasingly face in the future. With the long history of development and growing human populations in many areas of the world, restoration will be forced to proceed within the constraints imposed by human systems. A key challenge for restoration ecologists in the future will be to provide guidance during the restoration process on how best to recover ecological integrity within the constraints.

The burgeoning discipline of ecological engineering may help in providing solutions for ecological restoration in human-dominated systems. Ecological engineering aims to create environments that mimic natural systems and perform similar functions as natural systems, but that fit within constraints imposed by society. Ecological engineering requires the integration of engineering and landscape architecture with ecology. Restoration ecology is already a multidisciplinary science, because hydrology and geomorphology must be combined with ecology to understand the effects of flow patterns and sediment dynamics on river biota. Sociology is also essential for resolving disputes when multiple stakeholders are involved in river restoration. The continued integration of fields such as engineering and landscape architecture into restoration ecology will help ensure that ecological improvement can be achieved in human-dominated systems.

Increasing the perceived value of river ecosystems in the eyes of community members is a crucial part of expanding the possibilities for ecological restoration in the future. Public support is needed for any restoration project, and a public that places a high value on river ecosystems may demand restoration. Altering the perceived values of river ecosystems requires that restoration ecologists educate citizenry about river systems, the services they provide, and the process of ecological restoration. More importantly, the public must be involved during every stage of a restoration project (**Figure** 7). Public involvement allows people to learn what problems contribute to river degradation, what important river processes and ecosystem functions are lost as a result of degradation, how it might be possible to recover such processes, what constrains that recovery, whether a restoration project succeeded, and why a project failed or succeeded. People in a given society who are involved in restoration are more likely to feel a sense of responsibility for managing river ecosystems and will be more likely to support future ecological restoration.

Ensuring ecological success in river restoration in the future also requires broadening of the scale on which restoration is planned. Local projects in human-dominated systems are unlikely to solve the

Figure 7 In the foreground, local community members help plant riparian vegetation. In the background, a bridge provides access to the river and floodplain. Local community members learn about restoration by becoming involved and may gain an appreciation for the benefits of restoration by having access to the river and the restored area. Photograph by Margaret Palmer.

underlying problems contributing to river degradation. Increasingly, restoration is being incorporated as a part of ecological management plans for entire river basins. Basin-scale management should be an effective approach, because it provides a holistic view of the river and its watershed. Under a basin-scale management plan, tributaries that are relatively undisturbed may be preserved, or in the much more common situation, the impacts of further development can be minimized. When restoration is deemed necessary, the restoration activities can target the underlying causes of degradation and will not be restricted to a specific reach of the river. Thus, when incorporated into basin-scale management, restoration can much more effectively target the causes of degradation.

Restoration in the future will require collaboration locally, nationally, and internationally. International collaboration will become increasingly necessary as basin-scale management is adopted and restoration is planned throughout the entire course of a river, from headwaters to delta. International collaboration will also be necessary in order for people involved in restoration to share experiences and knowledge, and in doing so, advance the science of restoration ecology.

Glossary

Ecosystem Services – Functions provided by the river and its biota that are valued by humans, such as filtering of pollutants from the water by vegetation.

Eutrophication – A process in which the productivity of a water body greatly increases. Increased productivity is often driven by increased nutrient runoff from land in the surrounding watershed. The process can be harmful to aquatic organisms, because water clarity and oxygen concentration usually decline.

Geomorphology – The study of the distribution and movement of sediments over the Earth's surface. In rivers, fluvial geomorphology is the study of the movement of sediments within river channels and onto floodplains.

Reference rivers – Relatively pristine, undisturbed rivers that are used to compare with restored rivers to evaluate whether restoration achieved the desired goals. A monitoring scheme using restored rivers is less preferable than other types of monitoring designs. In many areas of the world, reference rivers may not exist owing to extensive human development.

River self-sustainability – The ability of a river to change and adjust to changing inputs of water and sediment and to human disturbance on the landscape, without compromising natural river processes and without human intervention.

Stormwater management – As a management practice for controlling storm runoff from urban areas, it involves the construction of ponds, wetlands, and seepage areas around parking lots and other developments that capture runoff from rainfall and allow the water to slowly filter into the soil before it moves to streams. Stormwater systems are designed to reduce the intensity of floods in urban rivers and allow water to be purified naturally by filtering through the soil and groundwater.

See also: Agriculture; Coarse Woody Debris in Lakes and Streams; Conservation of Aquatic Ecosystems; Currents in Rivers; Deforestation and Nutrient Loading to Fresh Waters; Ecology and Role of Headwater Streams; Flood Plains; Floods; Geomorphology of Streams and Rivers; Hydrology: Rivers; Hydrology: Streams; Regulators of Biotic Processes in Stream and River Ecosystems; Riparian Zones; Streams and Rivers as Ecosystems; Urban Aquatic Ecosystems; Wetlands of Large Rivers: Flood plains.

Further Reading

Bernhardt ES, Palmer MA, Allan JD, *et al.* (2005) Synthesizing U.S. river restoration efforts. *Science* 308: 636–637.

DeWaal LC, Large ARG, and Wade PM (eds.) (1998) *Rehabilitation of rivers: principles and implementation.* Chichester, UK: Wiley.

Dynesius M and Nilsson C (1994) Fragmentation and flow regulation of river systems in the northern third of the world. *Science* 266: 753–762.

Harper DM and Ferguson JD (eds.) (1995) *The Ecological Basis for River Management.* Chichester, UK: Wiley.

Kern K (1992) Restoration of lowland rivers: The German experience. In: Carling PA and Petts GE (eds.) *Lowland Floodplain Rivers: Geomorphological Perspectives*, pp. 279–297. Chichester, UK: Wiley.

Kondolf GM and Micheli ER (1995) Evaluating stream restoration projects. *Environmental Management* 19: 1–15.

Kondolf GM, Boulton AJ, O'Daniel S, *et al.* (2006) Process-based ecological river restoration: Visualizing three-dimensional connectivity and dynamic vectors to recover lost linkages. *Ecology and Society* 11: 5. Available online at. http://www.ecologyandsociety.org/vol11/iss2/art5/.

Leuven RSEW, Ragas AMJ, Smits AJM, and Van der Velde G (2006) Living rivers: Trends and challenges in science and management. *Developments in Hydrobiology* 187: 1–371.

Nelson K and Palmer MA (2007) Predicting stream temperature under urbanization and climate change: implications for stream biota. *Journal of the American Water Resources Association* 43: 440–452.

NRC (NationalResearch Council) (1992) *Restoration of Aquatic Ecosystems.* Washington, DC: National Academy Press.

Palmer MA, Bernhardt ES, Allan JD, *et al.* (2005) Standards for ecologically successful river restoration. *Journal of Applied Ecology* 42: 208–217.

Roni P (ed.) (2005) *Monitoring Stream and Watershed Restoration.* Bethesda, MD: American Fisheries Society.

Sear DA (1994) River restoration and geomorphology. *Aquatic Conservation: Marine and Freshwater Ecosystems* 4: 169–177.

Williams JE, Wood CA, and Dombeck MP (eds.) (1997) *Watershed Restoration: Principles and Practices*. Bethesda, MD: American Fisheries Society.

Wissmar RC and Bisson PA (eds.) (2003) *Strategies for Restoring River Ecosystems: Sources of Variability and Uncertainty in Natural and Managed Systems*. Bethesda, MD: American Fisheries Society.

Wohl E, Angermeier PL, Bledsoe B, *et al.* (2005) River restoration. *Water Resources Research* 41(W10301) doi:10.1029/2005WR003985.

Relevant Websites

http://www.restoringrivers.org – National River Restoration Science Synthesis.

http://www.geog.soton.ac.uk/users/WheatonJ/RestorationSurvey_Cover.asp – International River Restoration Survey.

http://www.therrc.co.uk/manual.php – River Restoration Centre.

http://www.rivers.gov.au – Land and Water Australia.

http://www.epa.gov/owow/restore – EPA Office of Wetlands, Oceans, and Watersheds.

Conservation of Aquatic Ecosystems

R Abell, WWF-United States, Washington, DC, USA
S Blanch, WWF-Australia, Darwin, NT, Australia
C Revenga, The Nature Conservancy, Arlington, VA, USA
M Thieme, WWF-United States, Washington, DC, USA

Introduction

Freshwater species and their habitats are on average among the most imperiled worldwide. Because they drain surface runoff from the landscape, freshwater ecosystems – also called inland aquatic systems or wetlands – are subject to impacts from land-based activities in addition to threats like direct habitat alteration and invasive species. Although limnology and related scientific disciplines are arguably well-developed, the field of freshwater biodiversity conservation lags behind that of the terrestrial and marine realms. This article details the state of freshwater biodiversity and habitats, summarizes major threats to freshwater systems, discusses conservation challenges for freshwaters, and provides an overview of more common conservation tools and strategies.

Recent studies show that freshwater species are on average more threatened than those in the terrestrial and marine realms. This is not surprising, as proximity to water bodies has been a preference for the establishment of human settlements for millennia. Society has used rivers for transport and navigation, water supply, waste disposal, and as a source of food. As a consequence we have heavily altered waterways to fit our needs by building dams, levies, canals, and water transfers and by heavily polluting our rivers, lakes, and streams with fertilizers and pesticides, industrial discharges, and municipal waste. And while freshwater ecosystems are very resilient, with examples of species refugia found in highly altered river systems, this resiliency is finite. We know there are thresholds that, once crossed, can put entire ecosystems at risk, with severe consequences for human well-being and biodiversity.

Given the importance of freshwater ecosystems in sustaining human well-being, it is surprising how little we know about their changing condition, their dependent species, or the roles that these species play in sustaining ecological functions. Knowledge is particularly poor for lower taxonomic groups (freshwater plants and invertebrates), especially in tropical regions. Here is a summary of the current status of freshwater biodiversity, given these gaps in our knowledge.

Status of Freshwater Biodiversity

Data on the condition and trends of freshwater species are for the most part poor at the global level, although some countries (e.g., Australia, Canada, New Zealand, South Africa, and the United States) have better inventories and indicators of change of freshwater species. Much of the problem originates from the fact that large numbers of species have never been catalogued and baselines on population status rarely exist, with the exception of a few highly threatened species (e.g., river dolphins) or species of commercial value (e.g., Pacific salmon in the United States).

The leading global effort to monitor the conservation status of species, the World Conservation Union (IUCN) Red List, has limited coverage of freshwater species, although a large effort is ongoing to fill this gap. Because of its harmonized category and criteria classification (i.e., all contributing experts follow the same methodology and guidelines), the IUCN Red List is the best source of information, at the global level, on the conservation status of plants and animals. This system is designed to determine the relative risk of extinction, with the main purpose of cataloguing and highlighting those taxa that are facing a higher risk of extinction globally (i.e., those listed as Critically Endangered, Endangered, and Vulnerable).

The 2006 Red List highlighted that freshwater species have suffered some of the most marked declines. For instance, of the 252 endemic freshwater Mediterranean fish species, 56% are threatened with extinction, and seven species are now extinct. This represents the highest proportion of imperiled species of any regional freshwater fish assessment that IUCN has conducted so far. Similarly, in East Africa, one in four freshwater fish is threatened with extinction. Odonates, another taxonomic group assessed by IUCN, also show high levels of imperilment, with almost one-third of the 564 species assessed being listed as threatened.

In 2004, IUCN completed the first global assessment of more than 5500 amphibian species, which was updated in 2006 to include 5918 species. This assessment considerably improved our overall knowledge of the condition of freshwater species, though its scope and representativeness are limited by lack of information, with 107 species still listed as Data Deficient and therefore unassigned a threat category. The Global Amphibian Assessment serves to reinforce the reality of the imperiled status of freshwater species, with close to a quarter of all assessed species listed as threatened, 34 as extinct, and as many as 165 species described as probably extinct. Overall, 43% of all amphibian species are declining in population, indicating that the number of threatened species can be expected to rise in the future.

Even large freshwater mammals are at increasing risk. For instance, the common hippopotamus, which until recently was not thought to be endangered, was listed in 2006 as threatened because of drastic and rapid declines in its population figures, with recorded reductions of up to 95% in the populations of the Democratic Republic of Congo, because of illegal hunting for meat and ivory. Overall, 41 species of freshwater mammals, including many otter species, freshwater dolphins, two freshwater feline species, as well as freshwater ungulates and rodents are threatened with extinction.

Data on freshwater reptiles, namely freshwater turtles and crocodilians (i.e., crocodiles, caimans, and gharials) also show declining trends. According to the IUCN/SSC (Species Survival Commission) Tortoise and Freshwater Turtle Specialist Group and the Asian Turtle Trade Working Group, of the 90 species of Asian freshwater turtles and tortoises, 74% are considered threatened, including 18 critically endangered species, and 1 that is already extinct: the Yunnan box turtle. The number of critically endangered freshwater turtles has more than doubled since the late 1990s. Much of the threat has come from overexploitation and illegal trade in Asia. The status of crocodilians presents a similar pattern, particularly in Asia. Of the 17 freshwater-restricted crocodilian species, as of 2007, 4 are listed by IUCN as critically endangered (3 of which are in Asia), 3 as endangered, and 3 as vulnerable. The most critically endangered is the Chinese alligator. The major threats to crocodilians worldwide are habitat loss and degradation caused by pollution, drainage and conversion of wetlands, deforestation, and overexploitation.

While information on freshwater plants and invertebrates are not readily available to portray population trends, available data give insight into the condition of freshwater ecosystems and species. In terms of freshwater plants, while many macrophytic species are probably not threatened at a global or continental scale, many bryophytes with restricted distributions are rare and threatened. In the United States, one of the few countries to assess more comprehensively the conservation status of freshwater molluscs and crustaceans, The Nature Conservancy has assessed that one-half of the known crayfish species and two-thirds of freshwater molluscs are at risk of extinction, with severe declines in their populations in recent years. Furthermore, of the freshwater molluscs, at least 1 in 10 is likely to have already become extinct (Master *et al.*, 1998).

While the Red List focuses only on threatened species and therefore does not look at population trends of nonthreatened species, it does provide a good measure of progress in attenuating species loss. Other measures of the change in vertebrate species populations, such as WWF's Living Planet Index (LPI), show a similar downward trend.

As these indices and examples show, freshwater species are in serious decline all over the world. However, available data and information are predominantly from temperate and developed regions. Some progress is being made to collect and compile information elsewhere, but progress is slow and resources needed are high, particularly in developing countries where capacity is limited.

Major Threats

Threats to freshwater systems and species are numerous, overlapping, and operate over a range of scales. The embeddedness of freshwaters within the larger landscape, coupled with the fact that human communities require freshwater resources to survive, means that few freshwaters around the world remain pristine. Most freshwaters are subject to multiple anthropogenic stresses, and this multiplicity can complicate the identification of threat pathways and appropriate conservation levers. Threats can be variously classified, but here we recognize habitat degradation, water pollution, flow modification, species invasion, overexploitation, and climate change as major, often overlapping categories. These threats can be further described in terms of their origins (**Table 1**).

Habitat degradation encompasses habitat alteration, outright habitat destruction, and loss of access due to fragmentation, all of which are described briefly here. Virtually any modification to natural land cover within a catchment has the potential to alter downstream freshwater habitats, including floodplains. Land cover conversion for agriculture, urbanization, forestry, road-building, or other activities can result in changes in flow, sediment regimes, riparian and aquatic vegetation, water chemistry, and other parameters that together define freshwater habitats (**Figure 1**). Direct modifications like streambank mining may also make freshwaters inhospitable for some native species

Table 1 Major threats to freshwater species and habitats

Major threats to freshwater ecosystems	*Description*	*Origin*		
		Local	*Catchment*	*Extra-catchment*
Habitat degradation	Degradation and loss	X	X	
	Fragmentation by dams and inhospitable habitat segments	X		
Flow modification	Alteration by dams	X	X	
	Alteration by land-use change		X	
	Alteration by water abstraction	X	X	
Overexploitation	Commercial, subsistence, recreational, poaching	X	X	
Water pollution	Agricultural runoff (nutrients, sediments, pesticides)		X	
	Toxic chemicals including metals, organic compounds, endocrine disruptors	X	X	
	Acidification due to atmospheric deposition and mining			X
Species invasion	Altered species interactions and habitat conditions resulting from accidental and purposeful introductions	X	X	
Climate change	Results in changes to hydrologic cycle and adjacent vegetation, affects species ranges and system productivity			X

Note that, in nearly all cases where both local and catchment origins are listed, local stresses are transferred downstream to become catchment impacts elsewhere. Introduced species originate outside a catchment but introductions occur at individual locations and can spread both up- and downstream. Modified with permission from Abell R, Allan JD, and Lehner B (2007) Unlocking the potential of protected areas for freshwaters. *Biological Conservation* 134: 48–63, with permission from Elsevier; major categories from Dudgeon D, Arthington AH, Gessner MO, Kawabata Z, Knowler DJ, Lévêque C, Naiman RJ, Prieur-Richard A, Soto D, Stiassny MLJ, and Sullivan CA (2006) Freshwater biodiversity: Importance, threats, status and conservation challenges. *Biological Reviews* 81: 163–182.

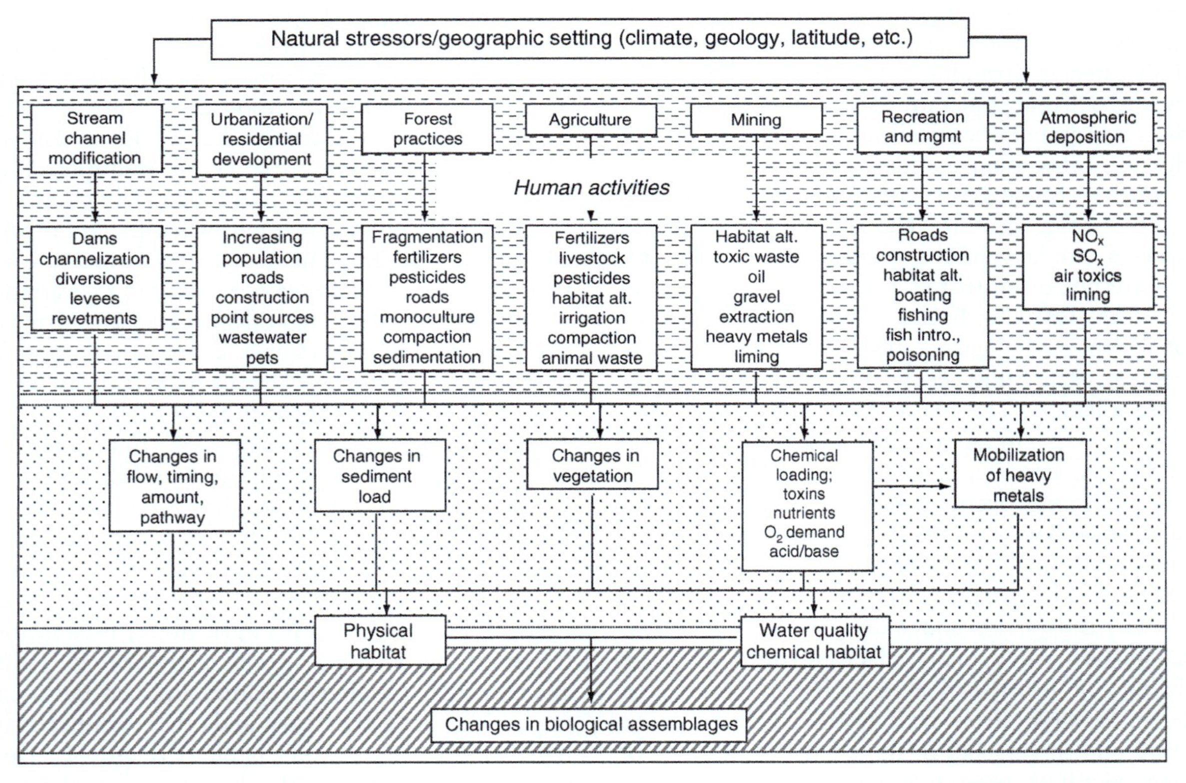

Figure 1 Threats and threat pathways in freshwater ecosystems. Modified from Bryce SA, Larsen DP, Hughes RM, and Kaufmann PR (1999) Assessing relative risks to aquatic ecosystems: A mid-Appalachian case study. *Journal of the American Water Resources Association* 35: 23–36, with permission from Blackwell.

without destroying habitats entirely. Habitat loss can take a variety of forms, such as through wetland draining, dewatering of a river system, disconnecting a river from its floodplain, or conversion of lotic to lentic habitat through reservoir construction. Freshwater species may lose access to habitat when their dispersal or migratory routes are impeded, either by constructed barriers like dams or virtual barriers like highly degraded, and therefore impassable, river reaches. Assessments of the extent of habitat alteration, loss, and fragmentation are notoriously difficult to undertake at broad scales (**Table 2**).

Water pollution is related to habitat degradation and is typically classified as either point or nonpoint source. Point source pollution can be traced to an identifiable, single source like a pipe draining directly into a freshwater. Nonpoint source pollution, like runoff containing fertilizers from agricultural activities or oil from urban centers, comes from multiple diffuse sources and can be far more difficult to mitigate. Acid deposition and other toxic substances transported by air from outside a drainage basin are a special kind of nonpoint source pollution. Many pollutants are chemicals, such as pesticides and endocrine disruptors, but sediments, nutrients, and other 'natural' materials can also act as pollutants when present at abnormal levels. Even temperature can serve as a pollutant, such as when discharge from a power plant is hotter than normal river water or that from a deep reservoir is colder.

Like habitat degradation, flow modification can also result from either landscape activities or direct modifications to freshwaters, and often both simultaneously. Any landscape activity that alters infiltration and associated runoff, or even precipitation in the case of broad-scale climatic impacts, can change a freshwater system's hydrograph or hydroperiod, in the case of flowing and still waters, respectively. The effect of urbanization on reducing infiltration opportunities is well-documented, and soil compaction from activities like forestry can have similar consequences. River impoundments designed for hydropower generation, irrigation, flood control, navigation, or other uses generally alter the timing and volume of flows, as well as sediment and thermal regimes. Water withdrawals as well as returns and interbasin transfers alter flow regimes as well, even if the total volume of water over time may be relatively unchanged. In general, any modification to the natural flow regime, defined by flow magnitude, timing, duration, frequency, and rate of change, has the potential to affect native species adapted to it.

Overexploitation and species invasion can both affect species populations and communities directly. Overexploitation, or the unsustainable removal of individual animals or plants for commercial or other purposes has primarily affected some species of larger fish, some reptiles, as well as mussels and other large macroinvertebrates. Overexploitation has only rarely been implicated as the single cause in the extinction of individual species, but it has likely been a contributing factor in the decline of many. Species invasion, through accidental or intentional introductions of non-native species, including through the opening up of previously inaccessible habitats, has had severe consequences for freshwater species in some instances. Impacts can include direct competition with or predation on native species, hybridization, habitat modification, and the introduction of disease and parasites. Species living in closed habitats like lakes appear to be

Table 2 Alteration of freshwater systems worldwide

Alteration	*Pre-1900*	*1900*	*1950–1960*	*1985*	*1996–1998*
Waterways altered for navigation (km)	3125	8750	–	>500 000	–
Canals (km)	8750	21 250	–	63 125	–
Large reservoirs[a]					
Number	41	581	1105	2768	2836
Volume (km^3)	14	533	1686	5879	6385
Large dams (>15 m high)	–	–	5749	–	41 413
Installed hydrocapacity (MW)	–	–	<290 000	542 000	~660 000
Hydrocapacity under construction (MW)	–	–	–	–	~126 000
Water withdrawals (km^3/year)	–	578	1984	~3200	~3800
Wetlands drainage (km^2)[b]	–	–	–	160 000	–

[a]Large reservoirs are those with a total volume of 0.1 km^3 or more. This is only a subset of the world's reservoirs.

[b]Includes available information for drainage of natural bogs and low-lying grasslands as well as disposal of excess water from irrigated fields. There is no comprehensive data for wetland loss for the world.

Reproduced with permission from Revenga C, Brunner J, Henninger N, Kassem K, and Payne R (2000) *Pilot analysis of global ecosystems: Freshwater systems*. Washington, DC: World Resources Institute.

particularly vulnerable to impacts from species invasion.

Climate change is a final major category of threats to freshwaters, overlapping with habitat degradation, flow modification, and species invasion. Changes in global surface temperature and precipitation patterns will translate to changes in water temperature, water quantity, and water quality in the world's rivers, lakes, and other wetlands. Freshwater biodiversity will be affected indirectly through habitat alteration, and directly where species' life histories are tightly adapted to particular temperature or flow regimes. Dispersal opportunities to more hospitable habitats may be highly limited, especially in systems already fragmented or otherwise modified.

Conservation Challenges

Although recognition of the looming freshwater crisis is growing, freshwater systems and their inhabitants are often still forgotten in local, national, regional, and international processes and plans. In part, this is due to the hidden nature of many freshwater species – they are literally 'out of sight and out of mind' underneath the water's surface. Additionally, many freshwater species are indistinct and small and thus do not engender the same emotional response as the large, colorful, charismatic species found in terrestrial and marine environments. Knowledge about freshwater species and habitats also lags behind that of their terrestrial counterparts. For example, about 3000 freshwater fish species are currently known in the Amazon Basin, but experts estimate that up to 5000 species will be discovered once the basin has been fully explored. A low profile and a lack of knowledge about freshwater systems' biology and ecology make the need for increased awareness from local to international levels even more critical for their conservation.

An even greater challenge to the sustainability of freshwater ecosystems is the direct competition that they are under with human societies for water resources. As the global human population increases, societal needs for water for agriculture, industry, energy generation, and human consumption will continue to grow and put freshwater ecosystems under mounting pressure. Demand is expected to grow fastest in developing countries and agriculture is expected to continue to be the largest consumer of water withdrawals (**Table 3**).

Conservation of freshwater ecosystems requires a paradigm different from that which guides terrestrial conservation activities. Traditional terrestrial approaches to biodiversity conservation center on setting aside areas of high conservation value as networks of protected areas. The inherent connectivity of freshwater systems limits the effectiveness of this approach within the freshwater realm. For example, water withdrawals or a dam upstream of a protected wetland can significantly alter that wetland's hydrology, thus undercutting the conservation effort. Basin-wide processes and interconnectivity must be a central objective in any effective freshwater conservation plan.

Conservation Strategies

Effective freshwater conservation often requires the use of multiple complementary strategies. The most appropriate mix of strategies may depend on the scale of conservation significance of the ecosystem,

Table 3 Water withdrawals for world regions, by sector

Region	*Total (million m^3) 2000*	*Per capita (m^3 per person) 2000*	*Sector Withdrawals (%), 2000[a]*		
			Agriculture	*Industry*	*Domestic*
Asia (excluding Middle East)	2 147 506	631	81	12	7
Europe	400 266	581	33	52	15
Middle East and North Africa	324 646	807	86	6	8
Sub-Saharan Africa	113 361	173	88	4	9
North America	525 267	1663	38	48	14
Central America and Caribbean	100 657	603	75	6	18
South America	164 429	474	68	12	19
Oceania	26 187	900	72	10	18
Developed	1 221 192.0	956	46	40	14
Developing	2 583 916.4	545	81	11	8
Global	3 802 320	633	70	20	10

[a]Sectoral withdrawal data may not sum to one hundred because of rounding.
Source: World Resources Institute, EarthTrends Freshwater Resources 2005.

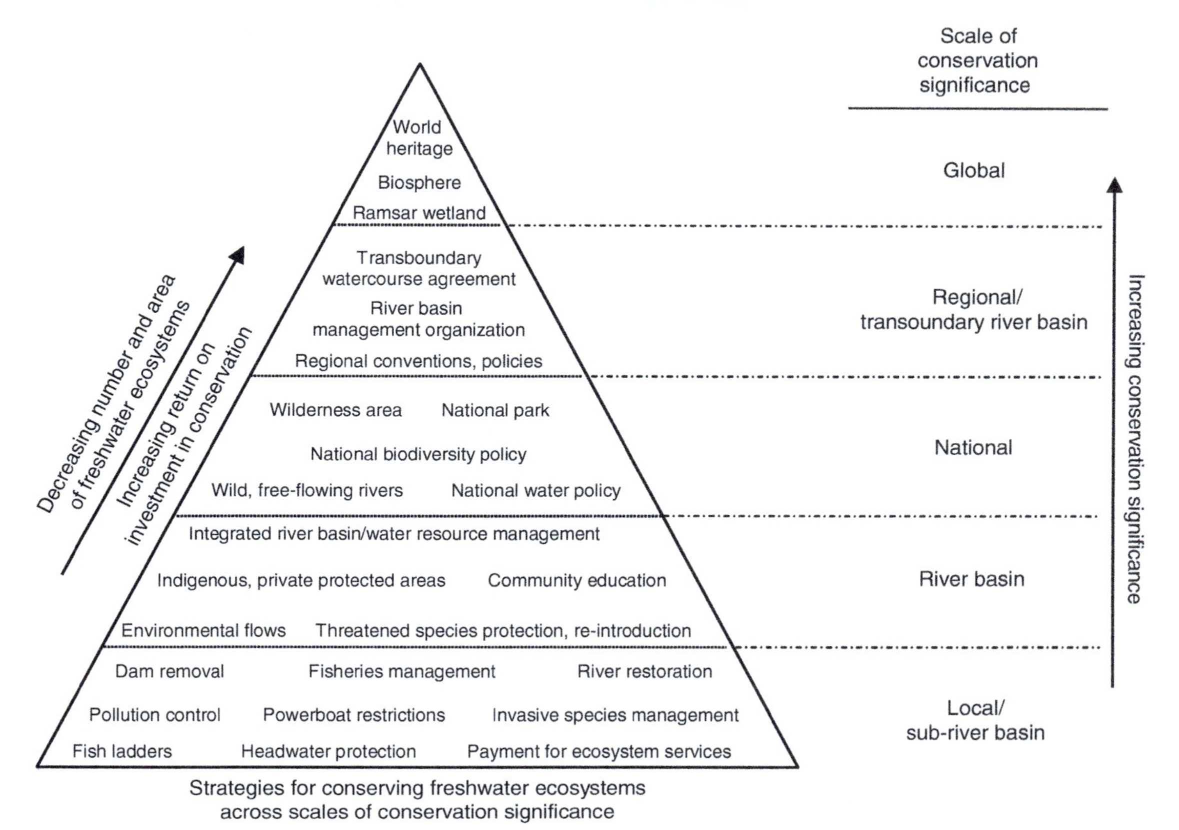

Figure 2 Strategies for conserving freshwater ecosystems. Modified with permission from Blanch SJ (2006) Securing Australia's natural water infrastructure assets. Solutions for protecting high conservation value aquatic ecosystems. A proposal. Sydney: WWF-Australia.

as shown in **Figure 2**. Here we detail a subset of possible strategies, focusing on several with direct and more frequent applications to conserving freshwater biodiversity. Many additional strategies found in **Figure 2** are addressed elsewhere.

Integrated Water Resources Management and Integrated River Basin Management

The concept of integrated water resources management (IWRM) is based on the interconnected nature of water bodies across landscapes, as well as along the river corridor from headwaters to the coast. IWRM promotes the need for participatory planning and implementation processes that bring stakeholders together to determine how to meet society's long-term needs for water while maintaining essential ecological services and economic benefits. A particular adaptation of the principles of IWRM to a river or lake basin is known as integrated river basin management (IRBM). The IRBM approach advocates managing a river and its entire catchment as a single system, and coordinating all the user group activities that take place within this geographic unit.

One of the key elements of IRBM is that it follows the principles of the ecosystem approach. The ecosystem approach framework is based on the central concept of managing water resources as integral parts of the ecosystem, rather than just as a resource to be exploited without regard to the system that nurtures it. Under this approach, water managers must do more than just satisfy one or two key users, but instead accommodate the wide array of economic and social benefits that people derive from aquatic environments, such as recreation, transportation, local livelihoods, cultural identity, and so on. The practical effect of this is that it widens the group of users who have a legitimate say in how the resource is managed.

Applying the ecosystem approach to managing water would ensure, at least in theory, that all goods and services derived from ecosystems, including inherent ecological functions, are taken into account when assessing development plans for a given river or lake. But despite the commitment by many countries to implement IWRM and IRBM approaches, such plans are still in their infancy. In most river basins around the world, allocation of water for irrigation and hydropower continues to take precedence over other water uses, as countries prioritize food and electricity production. Part of the problem is that implementing an IRBM approach requires different legal and institutional frameworks that go beyond our existing national government agencies. It requires cross-sectoral collaboration, and at times, new institutions, such as river basin organizations (RBOs), which link adjacent states along the river corridor in a legal framework that allows for the cooperative management of water resources within a single basin. RBOs can provide a forum for dialogue where the wide array of stakeholders can participate. As a result, development plans and water-use strategies can become more balanced, minimizing environmental and social impacts. For RBOs to be effective, however, they need to be given the authority, funding, and legal mandate to implement long-term water management policies – something that to date has been the exception rather than the norm. They must also have the wide participation of riparian states. The success of such approaches, however, ultimately depends heavily on cooperative governance and political commitment, which unfortunately are still lacking in many parts of the world.

Environmental Flows

Water management laws share water among users, such as towns, agriculture, industry, and the environment. Changing the shares among different user groups is often contentious, particularly in arid areas or where existing water rights are infringed upon, but presents a key conservation opportunity by reserving water for ecosystems. Termed 'environmental flows,' water allocated for ecosystems protects or reinstates key aspects of a river's flow. For example, pumping water may be restricted or prevented when river flows are low to allow aquatic organisms to move along a stream and allow wildlife to drink. Environmental flows do not just benefit the river channel but also floodplains, wetlands, estuaries, and coastal environments. These environments rely upon freshwater, sediments, nutrients, and carbon to be delivered from river channels. Small- and medium-sized floods may be protected from overextraction to ensure lateral hydrological connectivity between a river and its floodplain to allow wetlands to be refilled and fish species to migrate.

Groundwater-dependent ecosystems may be particularly dependent upon environmental flows, even if they occur at long distances from the river channel. Laws exist in some countries to legally protect water from being extracted, or to protect water that is specifically released from dams to maintain or recreate pulses and small floods. For example, environmental water reserves are guaranteed in water statutes in South Africa and the Australian state of Victoria to sustain river ecological functions.

Dams: Operation, Design, Removal

Building, operating, and removing dams, weirs, and barrages (collectively 'dams' for short) can affect river ecosystem function more than nearly any other set of activities. There are an estimated 45 000 large dams worldwide, and millions of small ones. In many parts of the world, new large dams are under construction or planned, especially for hydropower generation, whereas in a few countries like the United States, select dams are now being removed.

River ecology considerations should be fully incorporated into decision making about whether or not to dam a river, as recommended by the World Commission on Dams. For example, leaving the mainstem or large tributary of a river undammed will retain significant ecosystem benefits for river communities. Where dams are in place or under development, trying to mimic natural flow patterns as much as possible can recreate aspects of the flow regime that have been lost. Mitigating unnaturally warm or cold water discharges from dams is often needed to reestablish temperature regimes that trigger fish spawning and allow growth. For example, average water temperatures immediately below deep bottom-release dams in Australia's Murray-Darling Basin may be 5–10 °C cooler than natural. Building a multiple-level off-take at the dam wall allows dam managers to selectively draw surface water from the warmer epilimnion as dam levels rise and fall, thus warming river water below the dam, and hence increasing growth and survival of native warm-water fish.

Old, unsafe, or unnecessary dams can be removed to reinstate more natural flow patterns and permit freshwater organisms to move up and down a stream. Dam removal also reestablishes sediment transport regimes that continually erode and deposit sediments, thus continually creating new habitats. If removal is not feasible, a fish ladder or fish lift may be built to enable some fish and other freshwater animals to move upstream past dams, particularly during spawning migrations. Fish ladders and fish lifts are often poor alternatives to removal, however. They may not allow all fish species to use them, their effectiveness can be reduced by becoming blocked, and they require ongoing expensive maintenance.

Protected Areas

Protected areas – defined as 'areas of land and/or sea especially dedicated to the protection and maintenance of biological diversity, and of natural and associated cultural resources, and managed through legal or other effective means' – have received far less attention as tools for conserving freshwater species and habitats than they have for terrestrial and more recently marine features. Where protected areas have been used for freshwaters it has most frequently been through the establishment of Ramsar sites, identified under the Convention on Wetlands. Traditional protected areas have often been dismissed as ineffective for conserving freshwaters because of the connected and often linear nature of the systems. However, nontraditional protected areas, embedded within basin-wide integrated management efforts, are receiving increased attention. For example, riparian buffer zones can protect critical stream- or lake-side vegetation that filters pollutants, contributes organic material, moderates water temperatures, and provides woody debris for instream habitat. Floodplain reserves are in effect a special kind of riparian buffer zone, typically much wider and designed to protect the highly productive transitional areas that provide habitat to large numbers of both freshwater and terrestrial species. Fishery, or harvest, reserves are designed to provide spatial or temporal refuges for exploited freshwater species, so that populations can be sustainably fished over the long term; broader biodiversity conservation may be a secondary benefit. And protecting rivers as free-flowing, as described later, is a potentially powerful conservation tool gaining traction in certain countries.

Retaining Wild or Free-Flowing Rivers

Maintaining rivers without dams is arguably the single most effective river conservation strategy. Free-flowing rivers have been defined elsewhere as any river that flows undisturbed from its source to its mouth, either at the coast, an inland sea, or at the confluence with a larger river, without encountering any dams, weirs, or barrages and without being hemmed in by dykes or levees. Wild or free-flowing rivers allow water, sediment, nutrients, and biota to move longitudinally from the headwaters to the sea, terminal wetland or lake. Options for designating wild rivers are significantly reduced in most regions (**Table 4**), as only one-third of the world's large rivers remain free-flowing. But, some of the world's greatest rivers remain free-flowing along their mainstem, and sometimes on major tributaries too, including the Amazon, Okavango, Irawaddy, Sepik, and Mackenzie Rivers. No global free-flowing river conservation framework currently exists, although various countries have laws and programs to legally recognize and protect unimpounded rivers. The United States' Wild and Scenic Rivers program protects reaches of over 150 rivers covering 11 000 miles from new dams and some other types of development, but other damaging catchment uses may not be regulated. Wild and Heritage rivers programs and laws also exist in Canada and in the Australian States of Queensland and Victoria.

Restoration

Freshwater restoration involves recreating key features of a stream's ecological processes that have been impaired or lost, and potentially reintroducing species that have become locally extinct. The profession of stream restoration has developed significantly in recent decades, with investments in restoration by communities, industries, and governments totaling billions of dollars annually. Restoration techniques are many and varied. For example, fencing out stock and revegetating riparian zones with indigenous plants helps filter nutrients, reduce erosion, improve habitat, and shade the water. Reintroducing rare and endangered species to streams from which they have disappeared may achieve high community support and serve to raise the profile of stream restoration more broadly, but will only be effective if the threatening processes that drove the species extinct in the first place have been mitigated or stopped. Relatively drastic restoration techniques are sometimes the only realistic option for highly degraded streams. Bulldozers may be required to remove contaminated

Table 4 Regional distribution of rivers longer than 1000 km and percentage of rivers remaining free-flowing

Region	*Number of large rivers*	*Percent free-flowing*	*Example of free-flowing rivers*
Australia/Pacific	7	43	Cooper Creek, Sepik, Fly
Europe (west of Urals)	18	28	Oka, Pechora, Vychegda
Africa	23	35	Chari, Rufiji, Okavango
North America	33	18	Fraser, Mackenzie, Liard
South America	37	54	Amazon, Orinoco, Beni
Asia	59	37	Lena, Amur, Brahmaputra

Modified with permission from WWF (2006) *Free-Flowing Rivers. Economic Luxury or Ecological Necessity?* Zeist: WWF.

sediments that pose an ongoing risk to human health or ecosystems. Formerly channelized streams with low habitat and biodiversity value may have meander bends and rock bars added back. Similarly, logs may be put back into streams after decades of desnagging for navigation to reproduce the eddies and submerged habitats that fish and invertebrates need.

Convention Programs of Work

No single comprehensive international convention currently exists for the conservation and sustainable use of freshwater ecosystems. Rather, global efforts are primarily underpinned by two global conventions, namely the Convention on Wetlands (Ramsar) and the Convention on Biological Diversity, plus a host of transboundary watercourse agreements and widely espoused best management principles. The Convention on Wetlands commits signatory nations to the wise use of all wetlands, the designation of wetlands of international importance, and international cooperation. The Convention's definition of a wetland is very broad and provides a basis for providing some form of conservation for all forms of freshwater and coastal ecosystems. Ramsar sites arguably form the world's largest network of conserved aquatic ecosystems with 1650 sites covering 150 million ha of freshwater and some coastal systems globally as of April 2007, although effective legal protection and on-ground management is lacking for many.

The Convention on Biological Diversity's Programme of Works on Inland Waters and Programme of Work on Protected Areas have goals and actions regarding protecting representative types of freshwater ecosystems within IRBM. Formal cooperation exists between these two Conventions to harmonize global efforts on the conservation of freshwater ecosystems. The United Nations Convention on the Law of the Non-Navigational Uses of International Watercourses contains commitments relevant to freshwater conservation, such as protecting ecosystems, but has not been ratified by enough countries to come into force legally. Notwithstanding, transboundary watercourse agreements have been negotiated for many river basins and provide a framework, at least on paper, for freshwater ecosystem conservation.

Conclusions

The imperilment of freshwater species and habitats around the world is of urgent concern. Not only is a large fraction of the Earth's biodiversity threatened, but essential ecosystem services upon which human communities depend are at risk as well. Conservation strategies for addressing degraded ecosystems will have to be developed and applied within the context of integrated basin management to ensure that threats are mitigated and critical ecosystem processes, often linked to hydrology, can function within natural ranges of variation. Many of the world's freshwaters are already irreparably damaged, but there is time to secure protection for those that remain relatively intact if the political will for such protection can be generated and sustained.

See also: Agriculture; Deforestation and Nutrient Loading to Fresh Waters; Floods; Restoration Ecology of Rivers; Streams and Rivers as Ecosystems; Urban Aquatic Ecosystems.

Further Reading

Abell R, Allan JD, and Lehner B (2007) Unlocking the potential of protected areas for freshwaters. *Biological Conservation* 134: 48–63.

Allan JD (2004) Landscapes and riverscapes: The influence of land use on stream ecosystems. *Annual Review of Ecology Evolution and Systematics* 35: 257–284.

Allan JD, Abell RA, Hogan Z, Revenga C, Taylor BW, Welcomme RL, and Winemiller K (2005) Overfishing of inland waters. *BioScience* 55: 1041–1051.

Allan JD and Flecker AS (1993) Biodiversity conservation in running waters. *Bioscience* 43: 32–43.

Dudgeon D, Arthington AH, Gessner MO, *et al.* (2006) Freshwater biodiversity: Importance, threats, status and conservation challenges. *Biological Reviews* 81: 163–182.

Harrison IJ and Stiassny MLJ (1999) The quiet crisis: A preliminary listing of the freshwater fishes of the world that are extinct or 'missing in action.' In: MacPhee RDE and Sues HD (eds.) *Extinctions in Near Time: Causes, Contexts, and Consequences*, pp. 271–332. New York: Kluwer Academic/Plenum.

Master LL, Flack SR, and Flack BA (eds.) (1998) Rivers of Life: Critical Watersheds for Protecting Freshwater Biodiversity. Arlington: The Nature Conservancy.

Poff NL, Allan JD, Bain MB, *et al.* (1997) The natural flow regime: A paradigm for river conservation and restoration. *Bioscience* 47: 769–784.

Revenga C and Kura Y (2003) Status and trends of biodiversity of inland water ecosystems. Technical Series No. 11. Montreal: Secretary of the Convention on Biological Diversity.

Revenga C, Brunner J, Henninger N, Kassem K, and Payne R (2000) Pilot Analysis of Global Ecosystems: Freshwater Systems. Washington, DC: World Resources Institute.

Silk N and Ciruna K (eds.) (2004) A Practitioner's Guide to Freshwater Biodiversity Conservation. Boulder: The Nature Conservancy.

WWF (2006) Free-Flowing Rivers. Economic Luxury or Ecological Necessity? Zeist: WWF.

Relevant Websites

http://www.dams.org – World Commission on Dams.
http://www.biodiv.org – Convention on Biodiversity (CBD).
http://www.earthtrends.org – Watersheds of the World, and Freshwater Resources 2005.

http://www.fishbase.org – FishBase.
http://www.gemswater.org – UNEP Global Environment Monitoring System (GEMS), Water.
http://www.giwa.net/ – Global International Waters Assessment (GIWA).
http://www.globalamphibians.org – Global Amphibian Assessment.
http://www.issg.org – Global Invasive Species Database.
http://www.iucn.org – Red List of threatened species.
http://www.millenniumassessment.org – Millennium Ecosystem Assessment.
http://www.natureserve.org – NatureServe.
http://www.panda.org – Living Planet Report.
http://www.ramsar.org – Ramsar.

RIVERS OF THE WORLD

Contents

Africa

M Meybeck, Université Pierre et Marie Curie, Paris, France

General Features of the African Continent

The relief, climate, and lithological features of the continent are summarized (**Table 1**) for 21 aggregated coastal catchments that link river networks to the coasts of the Mediterranean Sea and of the Indian, South Atlantic and North Atlantic Oceans, and for few African regions presently not connected with oceans (endorheic rivers).

Relief

Africa has the least rugged relief of all continents together with Australia. Relief categories are defined on the basis of mean altitude and relief rugosity at the 0.5° space resolution. They are aggregated herein three clusters (**Table 1**): low relief (<200 m for the cell average), medium relief (200–2000 m) and high relief (>2000 m).

There are very few regions in Africa with developed high relief: the Ethiopian Plateau, which is drained by the Blue Nile and the Atbara Rivers, both tributaries of the Nile, by the Omo River and by the Awash River, and the North African ranges, Atlas and Aures drained by Moroccan rivers (Sebou, Oum Er Bia, Moulouya), Algerian and Tunisian rivers (Medjerda). Local volcanoes such as Kilimandjaro, Ruzizi, Mount Kenya, and Mount Cameroon may exceed 4000 m; however they do not correspond to important river basins.

Other relief features include the Fouta Djalon, shared by Senegal and Guinea, headwater of Senegal and Niger rivers, the Drakensberg in South Africa and Lesotho mountains (3482 m), headwater of the Orange River, the Katanga highlands, headwater of Zambezi and Congo rivers, the Central Africa highlands, headwater of the Oubangui River, the major tributary of Congo and of the Chari River. In the Sahara desert, the Ahaggar, Tibesti, and Darfour mountains, which reach an altitude of 3000 m, are only fed by very rare rain events.

The Rift Valley, which divides the East African Plateau from the Red Sea to Lake Malawi, is a unique and relatively young feature of the continent. It shapes hydrological networks and has generated two sets of lakes; some of them are among the world's largest and deepest: in the Western Rift lakes Albert, Edward, Kivu, Tanganyika, and Malawi (former Nyassa), in the Eastern Rift lakes Abbe, Turkana. Under the present tectonic and climate conditions the Kivu/Tanganyika lake system is linked to the Congo basin by the Lukuga River and Lake Malawi overflows to the Zambezi by the Shire River. Lake Victoria is a result of regional tectonic uplift and is relatively shallow regarding its area. It is the headwater of the Victoria Nile, which flows to Lake Albert, itself connected to lake Edwards.

Lithology

The continent is largely dominated by shields, plutonic, and metamorphic rocks (**Table 1**) Recent volcanic rocks are abundant in the Rift Valley region and in Cameroon, where numerous small crater lakes as the ill-famous Lake Nyos are found. Limestone regions, which result in hard water rivers, are essentially found in the NE part of the Sahara, i.e., they are not presently drained by rivers, and in the African Horn (**Table 1**, coastal catchments # 05 and 06).

Table 1 General characteristics of African coastal catchments (Meybeck *et al*. (2006))

Sea basin name	*Code (1)*	*Principal rivers*	*Sea basin area*	*Runoff*	*Population density/ sea basin*	*Sediment yield*	*Relief*			*Climate*					*Geology*			
			Mkm²	*mm year⁻¹*	*people per km²*	*t km⁻² year⁻¹*	*% Low*	*% Mid*	*% High*	*% Temperate*	*% Dry < 3 mm (2)*	*% Dry ≥ 3 mm (2)*	*% Tropical < 680 mm (3)*	*% Tropical ≥ 680 mm (3)*	*% Shield plutonic*	*% Volcanic*	*% Carbon*	*% Others*
Algerian Basin	1.0	Moulouya, Cheliff, Medjerda	0.25	86	104	53	1.2	98.8	0.0	78.0	21.0	1.0	0.0	0.0	0.0	1.0	47.5	51.5
South Ionian Sea	2.0	Irharhar, Araye (4)	2.26	0	8	1	53.3	46.7	0.0	1.7	97.7	0.6	0.0	0.0	2.8	4.2	37.6	55.5
East Mediterranean	3.0	Nile, Qattara	4.52	18	28	26	46.9	51.3	1.8	6.9	59.9	8.2	24.8	0.2	28.8	5.4	10.1	55.7
West Red Sea	4.0	No important rivers	0.33	14	37	12	3.5	95.6	0.9	0.0	83.7	12.8	3.5	0.0	61.8	20.2	0.0	18.0
South Aden Gulf	5.0	No important rivers	0.10	0	25	4	4.3	95.7	0.0	0.0	100.0	0.0	0.0	0.0	26.5	5.9	50.0	17.6
Somali Coast	6.0	Jubba, Tana (Kenya)	1.36	14	18	96	49.9	47.6	2.5	7.7	73.8	7.3	8.0	0.0	8.5	11.5	49.4	30.6
Zanzibar Coast	7.0	Rufiji, Rovuma, Galana, Pangani, Lurio, Wami	0.85	187	40	132	22.9	75.3	1.8	13.0	11.8	10.2	64.6	0.4	57.4	9.0	5.0	28.6
North Madagascar Coast	8.0	Ikopa, Sofia	0.17	676	31	69	26.7	73.3	0.0	44.6	0.0	0.0	39.3	16.1	45.4	7.2	21.9	25.5
Mascarenes-Madagascar Basin	9.0	Mananara	0.21	913	31	78	13.4	86.6	0.0	41.7	0.0	0.0	11.5	42.0	71.9	19.7	1.4	7.0
South West Madagascar Coast	10.0	Mangoky, Tsiribihina, Onilahy	0.24	291	20	257	28.4	71.6	0.0	45.3	35.0	2.4	17.3	0.0	40.6	2.5	21.7	35.2
Mozambique Coast	11.0	Zambezi, Save, Pungoe, Licungo, Buzi	1.73	251	20	33	13.7	86.3	0.0	59.7	5.6	7.8	26.1	0.7	42.0	1.2	8.3	48.4
Agulhas Basin	12.0	Limpopo, Gourits, Incomati, Gamtoos, Great Fish, Great Kei	0.88	43	39	182	20.8	78.9	0.3	36.1	54.9	3.8	5.1	0.0	26.8	9.9	8.2	55.1

Cape Basin	13.0	Orange, Olifants	1.29	9	14	28	0.4	97.7	1.9	14.8	83.3	1.9	0.0	0.0	24.1	6.9	12.7	56.3
Angola Basin	14.0	Congo Zaire, Cuanza, Cunene, Kouilou	4.38	344	18	13	27.4	72.4	0.1	13.3	1.4	1.0	82.7	1.5	38.9	0.5	11.6	48.9
Sao Tome-Principe Basin	15.0	Ogooue, Sanaga, Cross, Ntem, Nyong	0.53	793	25	34	36.6	63.4	0.0	0.6	0.0	0.0	51.2	48.2	76.8	5.3	6.6	11.3
Niger Delta Cone	16.0	Niger, Oueme	2.46	172	47	17	78.7	21.3	0.0	0.2	45.4	10.5	38.8	5.0	39.4	1.0	8.8	50.8
Guinea Basin	17.0	Volta, Bandama, Comoe, Sassandra, Pra	0.78	177	51	19	94.1	5.9	0.0	1.2	1.6	4.7	91.4	1.2	73.7	0.0	0.0	26.3
Sierra Leone Basin	18.0	Cavally	0.26	1431	35	189	68.4	31.6	0.0	0.0	0.0	0.0	6.0	94.0	81.6	0.0	0.0	18.4
Cape Verde Basin	19.0	Senegal, Gambia	1.11	111	15	27	95.3	4.7	0.0	0.0	58.9	16.9	20.9	3.3	3.8	4.8	3.8	87.6
South Canary Basin	20.0	Tamanrasett	2.17	0	3	0	76.3	23.6	0.1	0.0	99.9	0.1	0.0	0.0	23.8	2.6	13.2	60.4
North Canary/ Madeira Basin	21.0	Sebou, Oum Er Rbia, Tensift, Sous	0.36	26	52	55	29.4	68.2	2.4	24.3	72.1	3.6	0.0	0.0	8.1	11.4	30.9	49.6
Tenere	A	None	0.15	0	0	0	0.0	100.0	0.0	0.0	100.0	0.0	0.0	0.0	40.9	1.9	0.0	57.1
North Chad, Batha, Shari/ Logone	A	Chari, Logone	1.55	52	21	12	65.0	35.0	0.0	0.0	42.3	22.9	34.6	0.2	25.5	1.2	1.6	71.7
Bodele	*		0.76	18	0	0	55.5	44.5	0.0	0.0	96.1	3.9	0.0	0.0	9.0	1.5	9.6	79.8
Danakil	B		0.02	0	35	1	12.5	87.5	0.0	0.0	100.0	0.0	0.0	0.0	0.0	100.0	0.0	0.0
Awash/Ethiopia	B	Awash	0.14	200	80	81	2.2	67.4	30.5	28.3	26.0	28.3	13.0	0.0	0.0	76.6	6.4	17.0
Turkana	B	Omo	0.23	200	48	133	4.1	82.5	13.5	31.0	29.8	8.1	28.4	1.3	18.7	58.6	0.0	22.7
Ngomba (Chilwa)	*		0.07	155	24	8	0.0	100.0	0.0	36.3	0.0	4.6	59.1	0.0	82.6	0.0	0.0	17.4
Okawango-Kalahari	C	Okawango	0.76	72	2	11	0.0	100.0	0.0	17.4	76.0	6.6	0.0	0.0	7.4	3.4	3.4	85.7
Etocha	C		0.11	6	2	4	0.0	100.0	0.0	14.4	85.6	0.0	0.0	0.0	5.6	0.0	13.8	80.6

(1) See **Figure 1** for location of coastal basins. (2) Annual runoff. (3) Annual precipitation. (4) Presently not flowing.
*Smaller endorheic basins.** Presently dry; carbon = carbonated rocks.

Climate

African climate and hydrology have been synthesized by Korzun. Four major types of seasonal precipitation (P) regimes, the major drivers of Africa river flow regimes, have been identified:

- subtropical regime North of the Sahara (P between 250 and 800 mm year^{-1}) and south of Orange River (P 600 mm year^{-1}) characterized by a winter maximum precipitation and a summer minimum,
- the Saharan regime (Sahara, Namibian desert) very dry with indefinite seasonal and inter annual P variations and precipitation less than 100 mm year^{-1},
- the tropical regime (Sahelian belt from Guinea to Ethiopia and south of 5° S (excepted the southern top of the continent and Namibia) with a summer maximum and a winter minimum,
- the equatorial regime, roughly between 5° S and 8° N, across the continent.

In addition to the seasonal distribution of rainfall, the total amount of precipitation should be considered. It ranges in the Northern tropics from 200 mm in the Sahelian region (e.g., 208 mm at Toumbouctou, Mali) to 1250 mm year^{-1} at Addis Adeba, Ethiopia, and reaches 3600 mm year^{-1} in Sierra Leone; for the Southern tropics it ranges from 200 mm year^{-1} in Lesotho to 3400 mm year^{-1} in places of Madagascar. In the equatorial belt, annual precipitations range from 800 to 1600 mm year^{-1} with local maximums exceeding 4000 mm year^{-1} as on Mount Cameroon (9600 mm year^{-1}).

River Network Organisation and River Regimes

Exorheism, Endorheism, and Arheism are Continuously Changing

Geographers make a clear distinction between rivers connected to oceans (exorheism) and those connected to internal regions (endoreism). They also use the term arheism to describe the permanent absence of flow as in deserts (refer to 'see also' section), which can be conventionally limited at 3 mm year^{-1} runoff corresponding to about one flood event every decade. In Africa, endorheic basins are important and numerous (**Figure 1**) including Lake Chad basin, South of the Sahara, the biggest endorheic basin (1 million km^2) in the middle of the Sub Sahelian region, and the Okawango Basin in Southern Africa. Both are located between the desertic area and the tropical regions. The Rift Valley is also characterized by numerous smaller endorheic basins as the Awash River-Lake Abbe system in Ethiopia, and the Omo River-Lake Turkana system (Ethiopia/Kenya).

These endorheic systems are not stable and are very sensitive to climate variation and/or to tectonic: volcanic movements: (i) the Logone River (Chad Basin) may overflow during very high water periods to the Upper Benue Basin, a main tributary of the Niger River, (ii) there is also evidence of former high water stages of Lake Chad which could exceed 100 000 km^2 in area, (iii) the Okawango swamp may overflow during the wettest years to the Zambezi River basin through the Selinda spillway, (iv) Lake Victoria has once been cut-off from the White Nile system due to tectonic tilting, (v) Lake Turkana system has once been overflowing to the White Nile during Holocene wetter periods, (vi) the Lukuga River is very sensitive to the climate variations; during dryer periods Lake Tanganyika has partially evaporated over hundreds of meters, and (vii) during the wetter Sahara event some 6000 year BP, river networks fed by the Ahaggar, Tibesti and Darfour rains were extended and the lowest tributaries of the Nile basin, now completely arheic, were activated. Similar changes are likely to have occurred in Southern Africa in the Etosha Pan and in the Orange River basin during wetter periods.

The exact timing and mapping of these hydrological changes, which can be very important for the present aquatic biodiversity, remains to be established for this continent.

River Regimes in Africa

River regimes of small and medium basins are essentially controlled by the seasonal rainfall regimes since the potential evapotranspiration (Eo) does not present important seasonal variations and is very high throughout the whole continent; from 1200 to 2300 mm year^{-1}. Also, the snowmelt regime is restricted to the headwaters of some North African rivers. Many large river basins exceeding 500 000 km^2 are flowing across different climate belts and have complex regimes which should be covered separately (e.g., Nile and Niger).

Natural regimes are illustrated in **Table 2** for rivers with minimal impoundment impact based on the first Unesco discharges register (1969) and on a selection of the 1996 register. They are defined on the long-term mean yearly and monthly specific discharges ($\overline{q}$ and $\overline{q}_{min}$ in l s^{-1} km^{-2}) and are presented according to their mean latitudes from Morocco to Madagascar (**Table 2**).

The Sebou River (Morocco) drains the Atlas Mountains and is one of the rare African rivers that is also influenced by snowmelt. The Dra River, also in Morocco, originating from the SE side of the Atlas is facing the Sahara desert and is a very dry basin with a

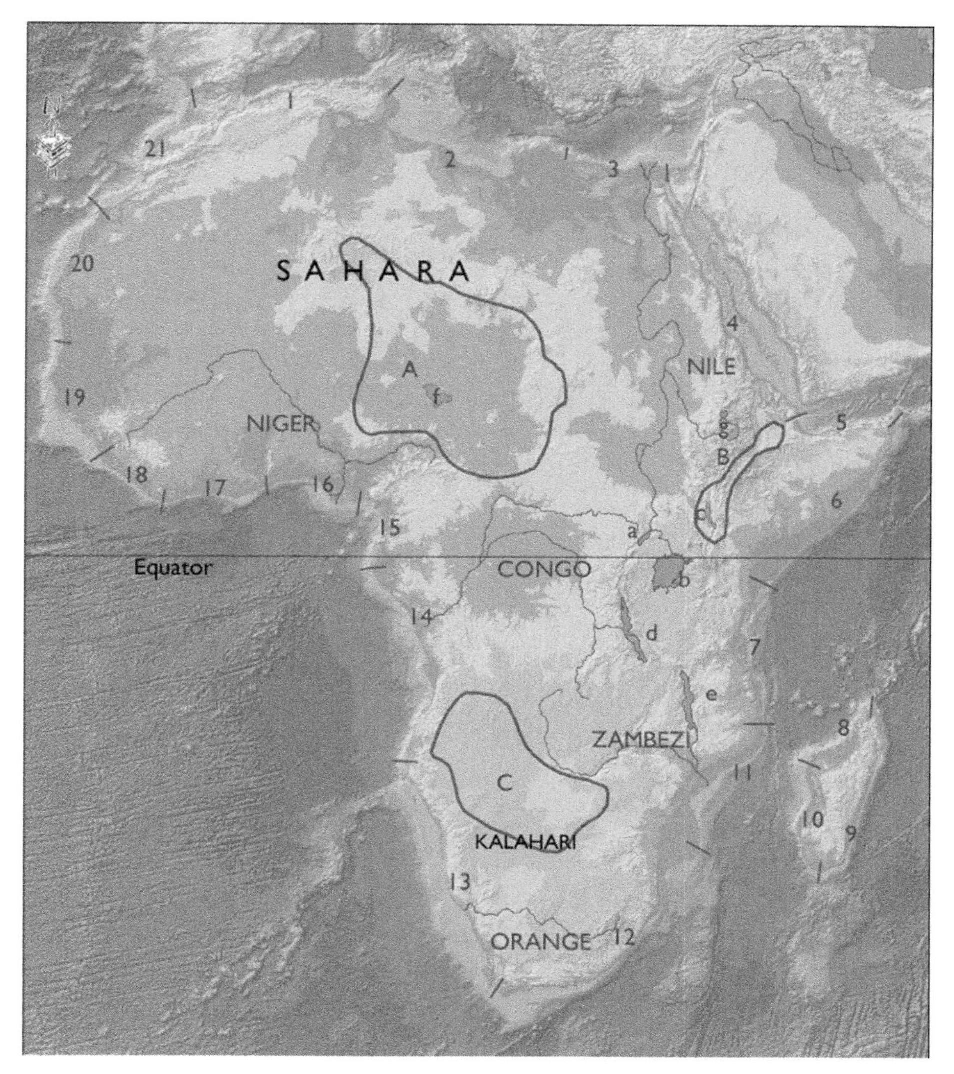

Figure 1 African rivers coastal catchments (Numbers 1–21) and main endorheic regions (A–C).

complete dry-out during 5 months, typical of such type of rivers termed ephemeral rivers or waddis. Waddis are also found South of the Sahara as in the Eastern part of the Niger River basin but most of them do not reach the Niger main course anymore (e.g., Azaouak River). A similar hydrological feature is observed in the Darfour region located between the Lake Chad and Nile River basins where rivers do not generally reach the Nile main course. In the Southern Hemisphere, the Fish River, a large basin in Namibia, is not always flowing to the Orange River.

The tropical regime of the Northern Hemisphere is here illustrated by the Senegal, the Upper Niger at Koulikouro (Mali), the Volta, the Comoe (Ivory Coast), the Benue (N. Cameroon) (**Table 2**), characterized by one maximum discharge occurring during the rainy season between August and September. The Atbara River basin, the ultimate tributary to the Nile from the Ethiopian Plateau has a similar summer high water period but completely dries out from February to April. The other hydrological extreme of the tropical regime is illustrated by the Moa River (Sierra Leone), a small basin where $\overline{q}$ exceeds $60\,l\,s^{-1}\,km^{-2}$ annually, a figure typical of a very wet regime where the high water season extends from June to November.

The equatorial regime is characterized by a typical double peak of discharge, i.e., such rivers do not really have a low water stage. Here again regional differences in annual runoff are important: yearly runoff ($\overline{q}$) is very variable ranging from only $3.6\,l\,s^{-1}\,km^{-2}$ from the Tana River (Kenya) to $80.9\,l\,s^{-1}\,km^{-2}$ for the Cross River (Nigeria/South Cameroon). Despite its position near the Equator, the Oubangui River (3–8° N), a major Congo tributary, has a single q maximum while the Ogooue (Gabon) (3° S–3° N), has two maximums (May and November). The

Table 2 Flow regimes of African rivers before damming (Unesco, 1969 and 1996) – presented from North to South, endorheic rivers and Nile stations in the end

River	*Station*	*Country*	*Jan* $l.\ s^{-1}\ km^{-2}$	*Feb*	*Mar*	*Apr*	*May*	*Jun*	*Jul*	*Aug*	*Sep*	*Oct*	*Nov*	*Dec*	*Year*	*A* $10^3\ km^2$	*Period*
Nile (Main)	Kajnarty	Sudan	0.51	0.39	0.33	0.30	0.27	0.30	0.76	2.90	3.37	2.19	1.10	0.67	1.09	2500	1931/33
Sebou	Azil el Soltane	Morocco	11.6	11.1	10.3	7.7	4.0	2.9	1.6	1.3	0.94	1.25	3.15	7.2	5.25	17.25	1959/64
Dra	Zagora	Morocco	0.18	0.21	0.015	0.25	0.06	0.015	0.01	0.01	0.00	0.00	0.00	0.15	0.075	20.13	1963/64
Senegal	Bakel	Senegal	0.59	0.35	0.21	0.10	0.45	0.56	2.63	10.8	15.9	7.8	2.55	1.05	3.5	218	1901/66
Niger	Koulikoro	Mali	3.35	1.6	0.85	0.56	0.82	3.0	10.4	26.9	44.8	39	17.8	7.3	13.0	120	1907/66
Moa	Moa Bridge	Sierra Leone	13.4	7.7	4.2	21.7	21.2	33.9	35.2	55.2	65.1	59.8	52.9	25.1	31.2	17.15	1976/77
Volta	Senchi	Ghana	0.25	0.14	0.11	0.12	0.28	0.96	2.35	4.75	12.5	13.0	3.3	0.61	3.2	394	1936/63
Comoe	Aniassé	Ivory Coast	0.43	0.21	0.28	0.36	0.50	0.93	2.4	5.85	12.8	12.0	4.55	1.57	3.5		1953/62
Benoue	Garoua	Cameroon	0.40	0.19	0.08	0.03	0.28	1.23	5.1	16.9	29.8	13.4	2.7	0.97	5.95	64	1930/60
Cross	Manfe	Cameroon	10.0	9.1	12.6	19.8	40.6	77.6	136.6	220	216.5	139.4	51.9	20.4	80.9	6.81	1967/79
Sanaga	Edea	Cameroon	6.9	4.5	4.1	5.4	8.0	11.1	16.3	20.1	32.3	44.9	27.9	11.9	16.3	131.5	1950/63
Tana	Garissa	Kenya	2.5	1.6	1.65	5.2	8.8	4.35	2.4	1.95	1.67	2.14	5.7	4.8	3.6	42.2	1933/65
Oubangui	Bangui	Rep. C. Afr	4.5	2.6	2.0	2.4	3.65	5.75	8.4	12.3	16.8	19.1	16.8	8.85	8.64	500	1935/60
Ogooue	Lambarene	Gabon	23.7	21.6	23.6	28.8	31.5	23.1	13.8	9.6	9.5	20.4	35.7	33.8	22.9	205	1929/61
Congo	Brazzaville	Congo	13.9	11.3	10.3	10.8	11.3	10.7	9.3	9.2	10.9	12.8	15.2	16.2	11.8	3475	1971/83
Rufiji	Stieglers	Tanzania	7.26	9.3	12.1	16.8	11.4	5.2	3.1	2.2	1.6	1.2	1.15	2.45	6.15	138.2	1959/65
Shire	Liwonde	Malawi	3.4	3.8	4.1	4.3	4.5	4.3	3.9	3.4	3.1	2.8	2.8	3.0	3.6	130.2	1965/84
Ikopa	Antsatrama	Madagascar	48.5	58.1	48.0	27.8	15.6	11.9	10.1	8.6	6.6	6.8	16.5	33.5	24.1	18.55	1948/66
Mangoki	Banian	Madagascar	28.2	22.0	18.2	7.6	3.65	3.15	2.7	2.35	1.8	1.6	3.5	15.3	9.2	50	1954/65
Awash*	Awash	Ethiopia	2.1	2.1	2.0	1.7	1.7	1.4	2.8	4.6	4.3	2.8	2.3	1.9	2.5	18.7	1976/79
Chari*	Njamena	Chad	1.1	0.6	0.4	0.3	0.2	0.4	0.8	1.9	3.6	4.9	4.6	2.4	1.8	600	1933/91
Victoria Nile	Namasagali	Uganda	2.4	2.4	2.4	2.55	2.7	2.75	2.65	2.55	2.5	2.45	2.4	2.5	2.53	262	1939/62
Blue Nile	Khartoum	Sudan	0.87	0.58	0.48	0.42	0.56	1.42	6.4	18.3	17.4	9.35	3.17	1.53	5.0	325	1912/62
Atbara	Kilo 3	Sudan	0.11	0.03	0.00	0.00	0.014	0.51	9.3	30.4	20.6	4.9	1.14	0.36	5.63	69	1912/62
Nile	El Ekhasse	Egypt	0.53	0.43	0.45	0.46	0.47	0.61	0.70	0.62	0.48	0.46	0.44	0.48	0.51	(2500)	1976/84

River regimes after damming.

*Endorheic basins; Rufiji to Ikopa: Southern Hemisphere basins.

Congo river, the world's second in drainage area (near 4 million km^2 at mouth) and in discharge (1200 km^3 $year^{-1}$) after the Amazon, lies across the equator and has one of the most naturally stable discharge in the world: the range between minimum and maximum daily discharges is only a factor of 2 and the interannual variation is also very limited.

The tropical regime of the Southern Hemisphere, less documented in Unesco data sets, is illustrated here by the Rufiji River (Tanzania), which peaks in April and by two Madagascar rivers, the Mangoki, relatively dry, and the Ikopa, located in a wetter region.

It must be noted that, within a given country as for Cameroon or Madagascar, extreme differences in river regimes can be observed. The complexity of African regimes in large catchments is further illustrated by the Nile (6670 km long) and Niger River (4160 km long), which are both crossing multiple climate belts.

The Victoria Nile discharge is constant throughout the year due to the position of this basin fed by equatorial rainfall and outlet of one of the world's greatest lakes. Downstream of Lake Albert, the White Nile flows North to the Mediterranean Sea some 5000 km away and losses much of its waters by evaporation in the Bahr el Gazal Swamp in Sudan. A project to shortcut this huge wetland area by the Jonglei navigation canal has been postponed due to the conflicts in this region. The Blue Nile originating from Lake Tana (Ethiopia) at an altitude of 1830 m and the Atbara River are actually the major providers of the annual Nile flood when it enters Egypt.

The Niger River regime is also complex. It is fed by the Fouta Djalon mountains (1425 m) and receives waters from the Bani, its main upper course tributary, then losses much of it in the Delta Central wetlands (Mali). From there, it is not fed much until it reaches the waters from the Benue in Nigeria and other very wet local tributaries. The Niger at mouth is mostly fed by the lowest portion of its basin. As for the Nile, large portions of the Niger basin are presently without regular runoff.

The Congo River basin is much simpler: this huge (3.8 million km^2) and flat basin lies across the equator and is fed by multiple tributaries, some of them (Kasai and Oubangui) are comparable in size and discharge to the world's greatest basins.

Several world class wetlands are found in Africa in connection with river courses: the Bahr el Gazal (White Nile) and the Delta Central (Niger), the Central Congo swamps, the Benue-Logone wetlands, and the inland Okawango Delta. Extended delta swamps of Niger and Nile are now very impacted and/or reduced, respectively, by oil and gas extraction and by irrigation, drainage, and urbanization.

Suspended Load

Suspended particulate matter (SPM) contents in African rivers are generally low due to (i) limited sources of sediments in many basins (low relief headwaters and gentle slope), (ii) presence of large lakes, and (iii) development of swamps. Typical yearly average SPM (discharge-weighted figures) are between 20 mg l^{-1} (Congo) and 250 mg l^{-1} (Niger) in natural conditions (**Table 3**). Higher levels have been recorded in East Africa (6700 mg l^{-1} for the Tana and 1270 mg l^{-1} for the Limpopo rivers) and in Northern Africa where average SPM exceeding 1000 mg l^{-1} are common in natural conditions.

In many African rivers, such as the Senegal, Gambia, Upper Niger, Chari, and the Nile, prior to the construction of the High Asswan Dam, the relationship between SPM and discharge is not univocal, but presents a large hysteresis loop with the highest concentration at the very beginning of the high water period, much before the maximum discharge is reached. Owing to the construction of many reservoirs, the natural SPM in many impounded rivers have now been lowered by one order of magnitude.

Desert waddis have a catastrophic type of sediment transport: they only flow after rare and severe rainstorms during which the surficial runoff can be intense. The Oued (waddi) Zeroud, an endorheic basin in Tunisia, has been studied during an exceptional flood event that lasted 2 months: (i) peak specific discharge may exceed 1000 l s^{-1} k^{-2} (reached 17 000 m^3 s^{-1} for only 8950 km^2 during this event), (ii) SPM concentrations exceed 10 g l^{-1}, (iii) river beds in narrow reaches can be scoured over several meters during the rising stage then filled in again at the receding stage, (iv) in the floodplain fresh alluvial deposits exceed 1 m, and (v) the Waddi Zeroud made a connection to the Mediterranean coast, discharging about 240 million tons of sediment, i.e., twice the annual load of the Nile before the construction of Asswan High Dam.

Riverine Chemistry

Riverine chemistry of African rivers is characterized by low to very low ionic contents and by the limited calcium and bicarbonate concentrations (refer to 'see also' section). This is due to the dominance of crystalline rocks from shields and metamorphic rocks, to the limited occurrence of carbonated rocks outside desertic regions (**Table 1**) and to the exceptional development of weathered soil layers, up to dozens of meters in many regions, leaving only quartz, kaolinite and lateritic crust, which are poorly soluble.

Typical African rivers have total cation sum between 150 and 1200 μeq l^{-1} (**Table 3**), i.e., total

Table 3 Water chemistry and suspended solids prior to damming for African rivers (references in Meybeck and Ragu (1996))

River	*Country*	*L*	*A*	*Q*	*SPM*	*SiO_2*	*Ca^{2+}*	*Mg^{2+}*	*Na^+*	*K^+*	*Cl^-*	*SO_4^{2-}*	*HCO_3^-*	*TZ^+*	*DOC*	*$N\text{-}NO_3^+$*	*$N\text{-}NH_4^+$*	*$P\text{-}PO_4^{3-}$*
		km	*$10^3\,km^2$*	*$km^3\,year^{-1}$*	*$mg\,1^{-1}$*	*$mg\,l^{-1}$*	*$mg\,l^{-1}$*	*$mg\,l^{-1}$*	*$mg\,l^{-1}$*	*$mg\,l^{-1}$*	*$mg\,l^{-1}$*	*$mg\,l^{-1}$*	*$mg\,l^{-1}$*	*$\mu eq\,l^{-1}$*	*$mg\,l^{-1}$*	*$mg\,l^{-1}$*	*$mg\,l^{-1}$*	*$mg\,l^{-1}$*
Bandama	Ivory Coast	1050	105	11.5	103	19.9	4.13	2.5		2.2	2.52	2.11	35.5	468				
Congo	Rep.Dem. Congo	4370	3698	1200	19	9.4	2.22	1.43	2.2	1.68	1.31	1.45	15.7	367	8.5	0.09	0.007	0.024
Cunene	Angola	830	106.5	6.8			1.7	2.1	10.3		7.1		30.5	706				
Gambia	Gambia	1200	42	4.9	40.8	10.8	4.25	2.0	3.2	1.1	0.42	0.53	23	544	2.3			0.015
Jubba	Somalia	1600	750	17.2			25	20	25	3	66	75	116	4058				
Konkoure	Guinea	365	16	21.5		5.6	1.06	0.50	1.2	0.62	1.21		7.75	162				
Limpopo	Mozambique	1600	440	26	1270	17.7	19.3	12.3	20.6	4.6	14.2	5.2	144	2990				
Niger	Nigeria	4160	1200	154	259	14	5.5	1.9	1.8	1.1	0.9	0.5	33.5	537				
Nile	Egypt	6670	2870	83		12.8	31	14	52	7.8	44	52	174	5161				
Orange	S. Africa	1860	1000	11.3		16.9	18.1	7.84	13.4	2.25	10.6	7.15	107	2189	2.3	0.72		
Rufiji	Tanzania	1237	178	35.2	483		8.0	4.05	10.6	3.3	8.0		50.0	1280		1.3	0.095	0.01
Sanaga	Ivory Coast	860	119	55	50.9		3.25	1.4	2.41	1.88	0.75	0.91	20.85	430				
Sassandra	Ivory Coast	840	75	18.1		20.8	4.2	2.3	5.1	2.4	2.6		34.1	682				
Senegal	Senegal	1480	440	24.4	78	7.65	3.9	2.9	2.2	1.2	1.0	2.4	29.5	560				
Tana	Kenya	720	42	4.75	6700	20	5.0	2.13	5.5	0.9	8.3			687		0.04	1.6	0.04
Volta	Ghana	1600	394	36.8	516		4.4	3.6			10		38			0.15		
Zambezi	Mozambique	2660	1330	106	190	16.8	10.6	4.1	5.4	1.9	6.5	3.0	32	1150		0.13	0.04	0.01

A, basin area; Q, mean annual discharge; SPM, discharge weighted suspended particulate matter; TZ+, sum of cations; L, river length.

dissolved solids (TDS) from 20 to 100 mg l^{-1} as for the Konkoure (Guinea), Congo, Sanaga (Cameroon), Bandama (Ivory Coast), Gambia, Niger, Senegal, Tana (Kenya), Sassandra (Ivory Coast), Cunene (Angola), Zambezi, and Rufiji (Tanzania), in increasing order of TDS. As in most world rivers, the calcium/bicarbonate type is dominant but the calcium/magnesium ratio is not elevated and can be below 1.0 in some Rift Valley basins; chloride and sulfates anions are generally very low. Higher ionic concentrations are generally associated with volcanic rocks and limited vegetation cover, or both as for the Blue Nile and the Jubba (Somali). Rare ionic contents proportions are found at the outlet of Lakes Kivu (Ruzizi river), Lake Tanganyika (Lukuga River) and Lake Edwards (Semliki River) with dominance of magnesium and potassium over calcium and sodium, mostly due hydrothermal inputs in the Rift Valley.

Compared to other continents, there is still a deficit of information on African river chemistry and water quality (see the Gems water programme register). The sparse information on nutrients in large rivers (**Table 3**) shows very low concentrations on a global scale. Owing to the low suspended load concentrations, the content of particulate organic carbon in SPM is relatively high. The silica concentration of African rivers ranges from 7.65 (Senegal) to 20.8 mg l^{-1} (Sassandra). The latter is about twice the world average concentration and due to crystalline rock weathering under hot climate and, possibly, to the dissolution of soil plant debris rich in amorphous silica (phytoliths).

Human Impacts and Uncertain Future of African Rivers

Our information on African rivers is still not developed, with the noted exception of South Africa. Actually, the gauging efforts have decreased in some regions and the water quality is poorly surveyed unlike in other continents. It is therefore difficult to assess the present impacts on water resources and their future.

Damming

Impoundments are now built for irrigation, water resources security, hydropower and for flood regulations on most major African rivers, except the Congo (Nile, Orange, Volta, Niger, Zambezi, Senegal) and on a great proportion of small and medium rivers (e.g., Sebou, Moulouya, Medjerda, Bandama, Faleme). They have greatly modified their natural flow regimes and have clarified river waters by SPM settling. As low flows are sustained and floods are attenuated, regulated flows downstream of reservoirs present very smooth seasonal hydrogrammes, very different from natural regimes. An example is provided for the Nile at El Ekhasse (31°16′N, Egypt) downstream of Asswan, which presents discharges ranging only from 1072 to 1755 $m^3\ s^{-1}$ (1976/1984) for ca. 2.5 million km^2 i.e., very different from those of the Nile at Kajnarty (21°27′N), which varied from 698 to 8180 $m^3\ s^{-1}$ prior any damming (1912/1962). The comparison between the Atbara River (North Ethiopia) in natural conditions and the Awash River (East Ethiopia) is another example of such regulation (**Table 1**): in the first river, the maximum/minimum monthly discharge ratio exceeds 100 (actually a dry out is observed) while for the second it is regulated at 3.

Another issue related to the construction of reservoirs is the retention of nutrients, particularly of silica, essential for diatoms growth, already well described in other continents where it is linked to coastal dystrophy (refer to 'see also' section), but not yet evident in Africa for lack of regular surveys (silica is generally not put on the list of water quality parameters). Finally, the increasing river fragmentation generates a loss of longitudinal connectivity for aquatic species such as fishes.

Pollution

Information on African rivers pollution is still very sparse compared with other continents. The water quality station density is very low except in Egypt and in South Africa and pesticides and toxic metals analyses are rarely collected. Several types of water quality and aquatic habitat degradation are likely due to: (i) fast growing urban centers without appropriate waste water collection and treatment located on small to medium basins (e.g., Fes, Marrakech, Meknes, Bamako, Niamey, Ouagadougou, Nairobi, Johannesburg) that may be linked to important organic pollution (high BOD_5, COD, NH_4^+) and bacterial contamination, (ii) important mining districts as in the Upper Congo basin, in Namibia, South Africa, that may generate metal contamination (Cu, Zn, Cr, Ni) and in Morocco (cadmium-rich phosphate mines), (iii) industrial agriculture (cotton, oil palm, sugar cane, etc.), which can be extremely polluting with BOD_5 and COD and some pesticides, and (iv) intensive oil and gas exploitation as in the Niger Delta generating hydrocarbons and metal contamination.

The impacts of Industrial development are probably limited in most countries to the main settlements, excepted in Nigeria, Egypt, and in South Africa. When documented, polluted sites reveal contamination levels matching world's records as for some metals in the Nile Delta for Lake Manzalah, which receives Cairo megacity (20 million people) and industrial waste waters.

Decreasing River Flows

Future surface water resources and river flow will depend on climate change, natural climate variability, land use evolution, and water uses. Climate variability is a real threat on many African rivers, which can be very sensitive to regional variations of rainfall. A 100 mm year^{-1} reduction of rainfall may have very different consequences depending on regions: in the Sahel, it would result in complete dryness while in wetter regions it would reduce the river runoff by 10–20% only. Detailed analysis on a medium-sized basin such as the Mgeni in South Africa has shown that the hydrological response to climate change may also vary greatly within one catchment depending on land use and land cover, relief, etc.

It is yet, still difficult to separate natural climate variability and impacts of land use and land cover changes from the climate change and from direct consumptive water use by irrigation, as for the Sahel region: The Niger River in Niamey (Niger) runoff has declined by a factor of two in the last 50 years, compared with 100 m^3 s^{-1} on average for the 1929/1991 period (Unesco, 1996). The Niger also started to develop in the 1990s a complete dryness downstream of the Delta Central in its middle course in June/July. The Chari/Logone discharges to Lake Chad have also been much reduced, thus resulting in a spectacular 20-fold shrinking of the lake area in the last 40 years, although historical recession of the lake is also documented.

The Nile is now completely regulated by a system of large reservoirs on the Blue Nile, the White Nile, and on the Atbara. The Asswan High Dam, constructed in the 1960s, has created the Lake Nasser, one of the world's greatest reservoirs (average water residence time 2 years). Each reservoir is associated with extended irrigation schemes. In Egypt, the River Nile is the only water resource that has been extensively used since antiquity. As the water demand for irrigation has been multiplied by orders of magnitudes, particularly for cotton crops, the actual Nile discharge to the Mediterranean Sea has been reduced by more than tenfold. The sediment flux to the Mediterranean Sea of the Nile, 120 million tons per year before the construction of reservoirs particularly of Lake Nasser, has now dropped to a few million tons per year, generating a long-term erosion of the Nile Delta, which could be enhanced by sea level rise.

Perspectives

As for other continents, there is no typical African river. More than half a dozen types should be considered to describe North Africa mountain rivers, desert waddis, dry tropical rivers, equatorial rivers, humid tropical rain forest rivers, Great lakes outlets, and wetland-dominated rivers courses. As a whole, the ionic and particulate matter concentrations in African rivers are among the world's lowest. On the other side, dissolved silica is about twice the world's average, although likely to be modified by reservoir retention in some highly impounded basins.

Owing to the massive construction of reservoirs (Congo basin excepted), African rivers are now highly regulated and used for irrigation. Other human impacts such as pollution are less documented although likely to occur downstream of big cities, mining districts, and irrigated areas. Unlike for the other continents, the aquatic habitats of African rivers have not yet been massively modified and artificialized as in Europe or in Eastern Asia, where human settlements, diking for flood protection, dredging for navigation have taken place for a long time. Impoundments and river flow decrease owing to irrigation are probably the major threat on aquatic wild life in this continent. Flow decrease could also be amplified by the rainfall reduction. In half of Africa, the climate change, if dryer, could result in river network fragmentation in the Sahel regions from Senegal to Sudan (e.g., central part of Niger and lower White Nile rivers) and in Southern Africa. Despite this uncertain future and the fastest water demand increase in the last 30 years, compared with other continents, our knowledge of African river resources is still lagging behind the water demand, except in a few regions (e.g., Egypt, South Africa).

See also: Asia – Northern Asia and Central Asia Endorheic Rivers; Australia (and Papua, New Guinea).

Further Reading

Colombani J and Olivry JC (1984) Phénomènes exceptionnels d'érosion et de transport solide en Afrique aride et semi-aride. *International Association of Hydrological Sciences Publication* 144: 295–300.

Coynel A, Seyler P, Etcheber H, Meybeck M, and Orange D (2005) Spatial and seasonal behaviour of organic carbon and total suspended solids by the Zaire:Congo River. *Global Biogeogeochemical Cycles.* doi:1029:2004 GB02335.

Korzun VI (1978) *World Water Balance and Water Resources of the World.* In: Studies and Report in Hydrology, no. 25, 663 pp. Paris: UNESCO.

Lévêque C and Paugy D (eds.) (2006) *Les poissons dans les eaux continentales africaines, diversité, écologie, utlisation par l'homme,* 564 pp. Paris: IRD.

Lesack LFW, Hecky RE, and Melack JM (1984) Transport of carbon, nitrogen, phosphorus and major solutes in the Gambia River, West Africa. *Limnology and Oceanography* 29(4): 824–834.

Meybeck M, Dürr HH, and Vörösmarty CJ (2006) Global coastal segmentation and its river catchment contributors: a new look at

land-ocean linkage. *Global Biogeochemical Cycles*, 20, GB IS 90, doi 10.1029/2005 GB 002540.
Meybeck M and Ragu A (1996) *River Discharges to the Oceans. An Assessment of Suspended Solids, Major Ions and Nutrients.* Environnement Information and assessment Rpt. UNEP, Nairobi, 250 nd rologie from Gems Water. http://www.gemsstat.org/descstats.aspx.
Nixon SW (2003) Replacing the Nile: are anthropogenic nutrients providing the fertility once brought to the Mediterranean by a great river? *Ambio* 32: 30–39.
Rodier JA (1983) Aspects scientifiques et techniques de l'hydrologie des zones humides de l'Afrique Centrale. *International Association of Hydrological Sciences Publication* 140: 105–126.
Salomons W, Kremer HH, and Turner RK (2006) The catchment to coast continuum. In: Crossland CJ, *et al.* (eds.) *Coastal Fluxes In The Anthropocene*, pp. 145–200. New York: Springer.
Schulze R, Lorentz S, Kienzle S, and Perks L (2004) Modelling the impacts of land-use and climate change on hydrological responses in the mixed undeveloped/developed Mgeni catchment, South Africa. In: Kabat P, *et al.* (eds.) *Vegetation, Water, Humans and the Climate*, pp. 441–451. New York: Springer.
Siegel F, Slaboda ML, and Stanley DJ (1994) Metal pollution loadings, Manzalah Lagoon, Nile, Egypt, implication for aquaculture. *Environmental Geology* 23: 89–98.
Unesco (1969) Discharges of selected rivers of the world. *Studies and Reports in Hydrology*, no. 5, 70 pp. Paris: UNESCO.
Unesco (1996) *Global river discharge database (Riv Dis). Technical Documents in Hydrology*, 41 pp. Paris: UNESCO.
Walling D (1984) The sediment yield of African rivers. *International Association of Hydrological Sciences Publication* 144: 265–283.

Relevant Website

http://www.gemsstat.org/descstats.aspx – Gems Water program UNEP, global river water quality data.

Asia – Eastern Asia

F Wang, University of Manitoba, Winnipeg, MB, Canada
J Chen, Peking University, Beijing, P. R. China

Introduction

Eastern Asia usually refers to China, Japan, and North and South Korea. It covers an area of about 10.2×10^6 km^2, or 7% of the Earth's total land area. As of 2008, the region is inhabited by more than 1.56 billion of people, almost one fourth of the global population. China is the largest country in Eastern Asia, accounting for 94% of its landmass and 86% of the population.

With an average elevation of over 4000 m asl, the Qinghai-Tibetan Plateau in western China is the roof of the world and the birthplace for some of the largest rivers in the world, including the Changjiang (Yangtze River) and the Huanghe (Yellow River) that flow entirely in China, and the Yarlung Zangbo-Brahmaputra, the Lancang-Mekong, and the Nujiang-Salween whose headwaters are located in China with river mouths in Bangladesh, Vietnam, and Myanmar (Burma), respectively. The latter three rivers empty into the ocean from Southern Asia and are included in the Monsoon Asia chapter. Also not included in this chapter is the Ertix River (Irtysh) which originates in western China and flows northward joining the Ob River before emptying into the Arctic Ocean. Hereafter, the rivers of Eastern Asia shall refer to all other rivers in China, Japan, and North and South Korea (**Figure 1**).

General Features

Table 1 summarizes the major rivers in Eastern Asia and their hydrological and geochemical features. All these rivers empty into the Pacific Ocean. Among the continental rivers (in contrast to island rivers), the Heilongjiang-Amur flows through Russia, Mongolia, China, and North Korea, and empties into the Okhotsk Sea. The Yalu (Amnok) and the Tumen (Duman) Rivers originate from the Changbai Mountains in northeastern China. The Yalu flows southwest into the Yellow Sea draining water from China and North Korea, and the Tumen flows northeast into the Sea of Japan draining water from China, North Korea, and Russia. A small percentage ($\sim$2%) of the Zhujiang (Pearl River) drainage basin is located in northeastern Vietnam. All other major continental rivers are entirely located in continental China.

Most continental rivers of China have their origins in one of the three massive topographic 'stairs' in China (**Figure 1**). Slopping down from west to east, the first and the highest topographic stair is the Qinghai-Tibetan Plateau (>4000 m asl) from where the Changjiang and the Huanghe originate. The second, intermediate stair is formed, from northeast to southwest, by the Greater Hinggan Mountains–Inner Mongolian Plateau–Loess Plateau–Yunnan-Guizhou Plateau series (1000–2000 m asl), and is the source water of the Heilongjiang-Amur, the Liaohe, the Huaihe, and the Zhujiang. The third and the lowest stair includes the Changbai Mountains–Shangdong Peninsula–Southeast Coast Mountains ($\sim$500 m asl) in the east. Examples of rivers originating from the third stair include the Yalu, the Tumen, the Minjiang, and the Qiantangjiang.

Rivers in Eastern Asian islands (e.g., Hainan Island, Taiwan Island, and Japan Islands) and Korean Peninsula are characterized by their relatively short lengths and small drainage areas. However, due to their steep elevation gradients, abundant precipitations, and frequent typhoons, these small island rivers, particularly those in Taiwan Island, can be associated with very high runoff and TSS contents.

In addition to the oceanic rivers, there are also a great number of inland rivers in China that discharge to internal regions and evaporate in salt lakes, a process referred as endorheism. They drain a total area of more than 3×10^6 km^2, and are located mostly on the Qinghai-Tibetan Plateau, and in arid and semiarid northwest China. Among the largest inland rivers are the Tarim River with an active drainage basin of 0.20×10^6 km^2, and the Ili River which has a drainage basin of 0.057×10^6 km^2 with an annual water discharge of $\sim$12 km^3.

Water Discharge

The Changjiang, the Heilongjiang-Amur, and the Zhujiang are among the largest rivers in the world in terms of water discharge. At Datong, the most downstream main-channel hydrological station without tidal influence, the annual water discharge of the Changjiang averages 900 km^3 year^{-1} (range: 680–1360 km^3) during the period 1950–2006 (**Table 1**), ranking it the fourth largest river in the world (after the Amazon, Zaire, and Orinoco). The Heilongjiang-Amur has a mean annual water discharge of 355 km^3 year^{-1},

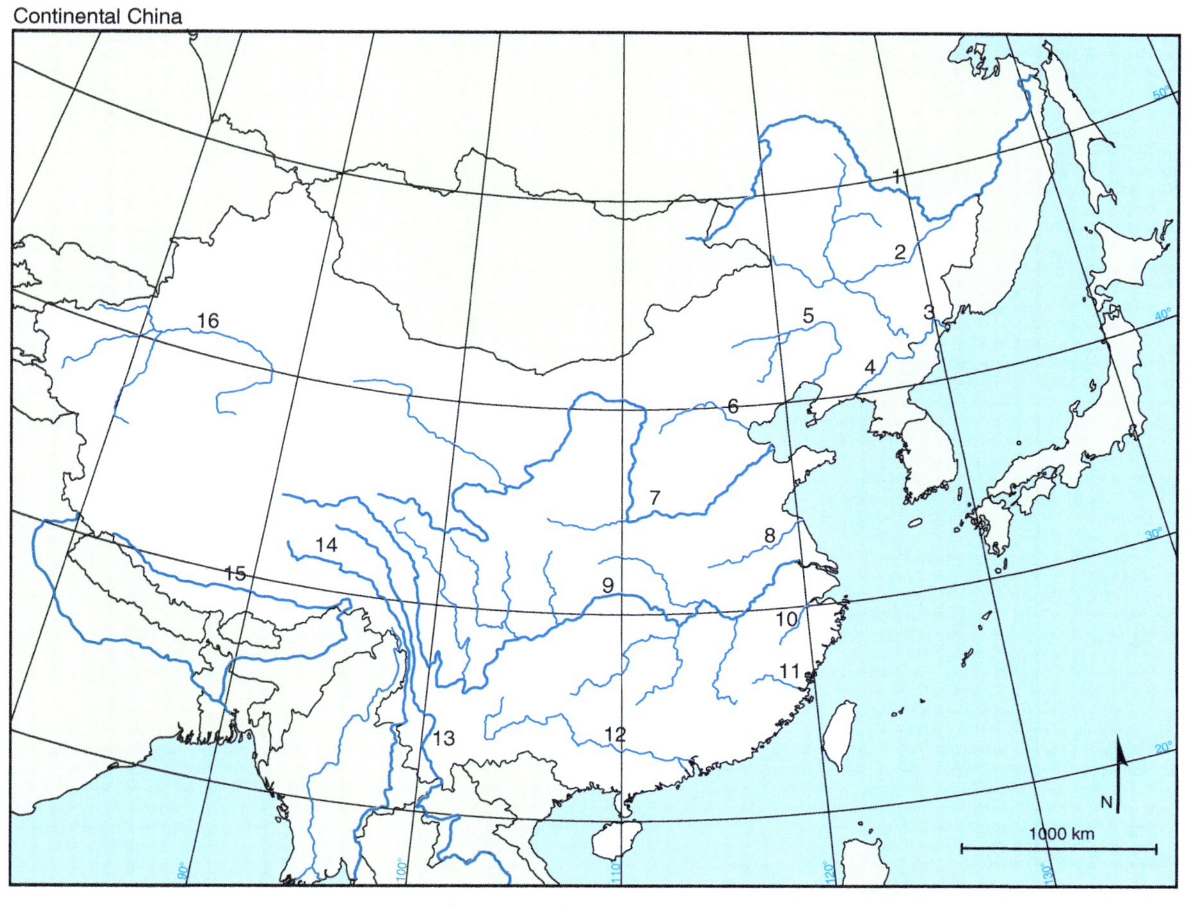

Continental China River Names

1. Heilongjiang (Amur)
2. Songhuajiang
3. Tumen (Duman)
4. Yalu (Amnok)
5. Liaohe
6. Haihe
7. Huanghe (Yellow)
8. Huaihe
9. Changjiang (Yangtze)
10. Qiantangjiang
11. Minjiang
12. Zhujiang (Pearl)
13. Lancangjiang (Mekong)
14. Nujiang (Salween)
15. Yarlung Zangbo (Brahmaputra)
16. Tarim

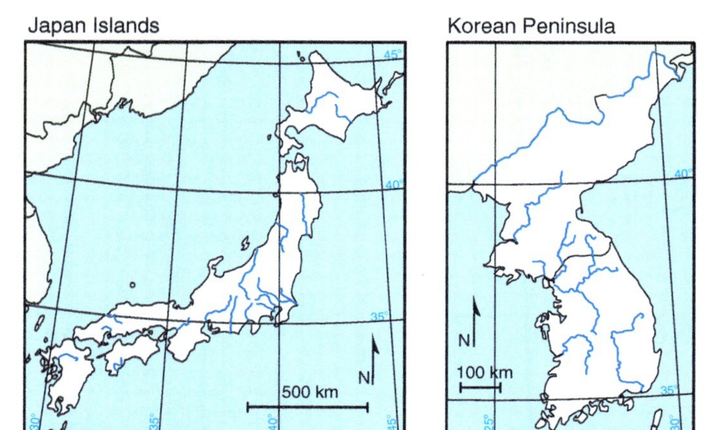

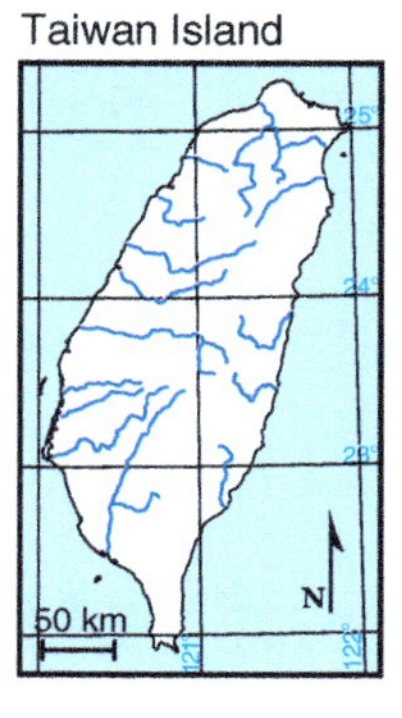

Cartography by Douglas Fast

Figure 1 A map showing major rivers in Eastern Asia.

Table 1 Major rivers in East Asia and their hydrological and geochemical features

River	*Length (km)*	*Drainage area (10^6 km^2)*	*Monitoring station (% of the drainage area)*	*Data period*	*Water discharge (km^3 $year^{-1}$)*	*Runoff (mm $year^{-1}$)*	*TDS (mg l^{-1})*	*TSS (mg l^{-1})*	*TDS flux (10^6 tons $year^{-1}$)*	*TSS flux (10^6 tons $year^{-1}$)*	*TDS yield*[b] *(tons km^{-21} $year^{-1}$)*	*TSS yield (tons km^{-21} $year^{-1}$)*
Continental Rivers[a]												
Exorheic												
Changjiang (Yangtze)	6300	1.81	Datong (94.5%)	1950–2006	900	526	178[b]	458	160[c]	408	93.5[c]	239
Heilongjiang-Amur	4440	1.92		1976–1990	355[d]	185	73[d]	146[d]	25.8[d]	52[d]	13.4[d]	27.1[d]
			Jiamusi[e] (27.6%)	1955–2006	64.9	123	150[f]	197	9.9[f]	12.8	18.7[f]	24.1
Zhujiang (Pearl)	2210	0.45	Gaoyao/Shijiao/Boluo[g] (91.1%)	1954–2006	285	695	166[h]	264	47.5[h]	75.3	116[h]	184
Minjiang	541	0.06	Zhuqi (90.8%)	1951–2006	53.8	988		112		6.0		110
Huanghe (Yellow)	5460	0.75	Lijin (99.9%)	1950–2006	31.9	425	514[i]	23 900	16.5[i]	787	22.0[i]	1049
Yalu (Amnok)	790	0.06			29	483				4.0		
Huaihe	1000	0.19	Bengbu (63.1%)	1950–2006	26.9	224		346		9.3		77.6
Qiantangjiang	668	0.06	Lanxi (32.5%)	1977–2006	16.5	907		121		2.0		110
Tumen (Duman)	520	0.03			8	267				3.0		
Liaohe	1390	0.23	Liujianfang (59.3%)	1987–2006	2.9	215		1565		4.6		33.7
Endorheic												
Tarim	2179	0.20	Aral	1964–2006	2.2			5028		11		
Heihe (Black)	821	0.12	Zhengyixia	1963–2006	1.0			1500		1.5		
Island Rivers												
29 rivers in Taiwan Island		0.022[j]		1970–1999	41.2[j]	1870[j]		9320[j]		384[j]		17 500[j]
Rivers in Hainan Island		0.034[k]		1957–1982	50.4[k]	1210[k]	108[k]	94.4[k]	5.44[k]	4.76[k]	160[k]	140[k]
15 major rivers in Japan		0.12[l]		1938–1992	114[l]	343[l]						

[a]Unless otherwise specified, all the data were compiled from the following sources: The Ministry of Water Resources of the People's Republic of China. River Sediment Bulletin, 2000–2006.

[b]Not corrected for sea salts.

[c]Mean annual average for 1960–1990; Chen J, Wang F, Xia X, and Zhang L (2002) Major element chemistry of the Changjiang (Yangtze River). *Chemical Geology* 187: 231–255.

[d]Fraser AS, Meybeck M, and Ongley ED (1995) Water Quality of World River Basins. United Nations Environment Programme (UNEP) Library No 14. Nairobi: UNEP.

[e]Jiamusi is on the Songhuajiang, the major Chinese tributary of the Heilong-Amur River.

[f]Chen J, Xia X, Zhang L, and Li H (1999) Relationship between water quality changes in the Yangtze, Yellow and Songhua Rivers and the economic development in the river basins. *Acta Scientiae Circumstantiae* 19: 500–505 (In Chinese).

[g]The sum of Gaoyao on the Xijiang, Shijiao on the Beijiang, and Boluo on the Dongjiang.

[h]Chen J and He D (1999) Chemical characteristics and genesis of major ions in the Pearl River Basin. *Acta Sci. Natur. Univ. Pekin.* 35, 786–793 (In Chinese).

[i]Chen J, Wang F, Meybeck M, He D, Xia X, and Zhang L (2005) Spatial and temporal analysis of water chemistry records (1958–2000) in the Huanghe (Yellow River) basin. *Global Biogeochemical Cycles* 19, GB3016. doi:10.1029/2004GB002325.

[j]Compiled from Dadson SJ, Hovius N, Chen HG, Dade WB, Hsieh M-L, Willett SD, Hu J-C, Horng M-J, Chen M-C, Stark CP, Lague D, Lin J-C (2003) Links between erosion, runoff variability and seismicity in the Taiwan orogen. *Nature* 426: 648–651.

[k]Chen, J., Xie, G., and Li, Y.-H. (1991) Denudation rate of Hainan Island and its comparison with Tainwan Island and Hawaiian Islands. Quaternary Sci. (4), 289–299 (In Chinese).

[l]MLIT (Ministry of Land, Infrastructure and Transport, Japan). Major Rivers in Japan. http://www.mlit.go.jp/river/basic_info/english/table.html (visited on June 19, 2008).

TDS: total dissolved solids; TSS: total suspended solids.

followed by the Zhujiang with 285 km^3, similar to that of the Mackenzie and the St. Lawrence Rivers in North America, but larger than the Magdalene River in South America.

Water discharge of most rivers in Eastern Asia varies greatly within a year. Most rivers (e.g., the Changjiang, the Minjiang, and the Huaihe) peak in the flood season from May to September due to the influence of the East Asia monsoon. Rivers in northern China (e.g., the Heilongjiang-Amur, the Yalu, and the Tumen), Korea, and northern Japan also experience high flows during the spring freshet from March to April following snow melting. Rivers in southeast coastal areas (e.g., the Zhujiang, the Minjiang, and rivers in Hainan and Taiwan islands) are subject to the influence of frequent typhoons, and may not see the highest water discharge until later in September or October. The highest degree of seasonal variations occurs with small rivers in semiarid and arid areas (e.g., some tributaries of the Huanghe on the Loess Plateau), where the entire water discharge in a year may be due to a few storm events only.

Water discharge also fluctuates from year to year. While some of the fluctuations are caused by natural and climatic processes, there is evidence that anthropogenic influence has been playing an increasingly important role. Extensive and continuous hydrological monitoring has been in place for 30–100 years for most rivers in continental China and Hainan and Taiwan Islands. These long-term databases make it possible to detect some persistent changes in water discharge from natural fluctuations.

The most notable and alarming change is the dramatic decrease in water discharge of the Huanghe since ~1969. As shown in **Figure 2a**, the mean annual water discharge of the Huanghe has decreased from 49.1 km^3 during 1950–1969 to 14.1 km^3 in the 1990s and 13.3 km^3 in the 2000s (2000–2006). Indeed, the lower reaches of the Huanghe have been experiencing frequent dry-ups since the early 1970s. The most severe dry-ups occurred in 1997, when Station Lijin (the most downstream main-channel station without tidal influence) remained dry for a total of 226 days; the length of the river section dried up extended to near the City of Kaifeng, some 700 km inland from the river mouth. In addition to climatic reasons – the basin seems to have been experiencing a drought in recent years, the significant reduction in water discharge is certainly accelerated by the increasing water withdrawal for agricultural irrigation and cross-basin water diversion to Northern China, and by the evaporation loss from the surfaces

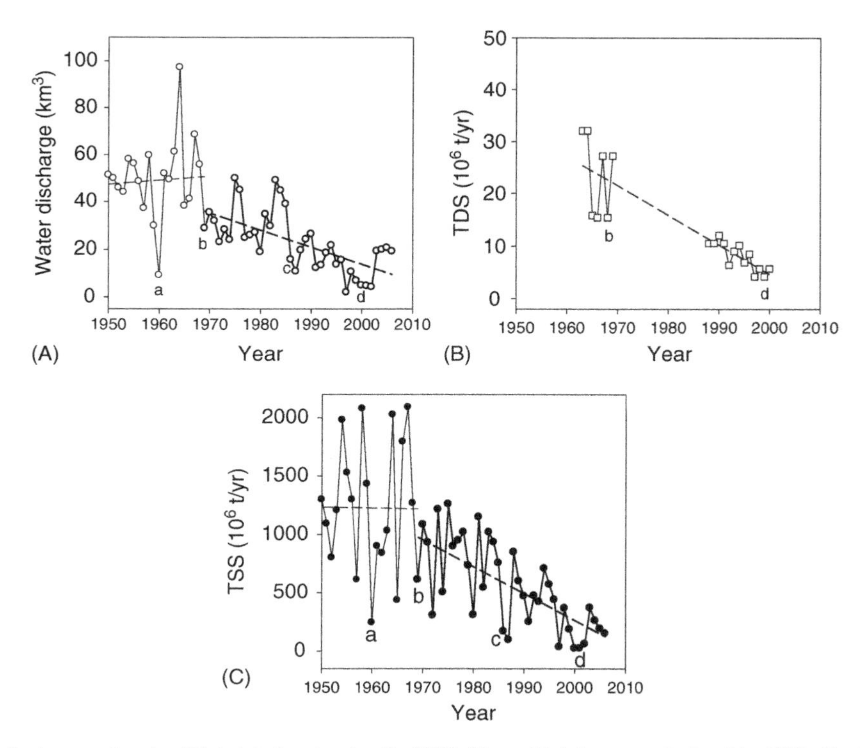

Figure 2 Seaward discharge of water (A), total dissolved salts (TDS; B), and total suspended solids (TSS; C) of the Huanghe at Hydrological Station Lijin, the most downstream station without tidal influence. The dashed lines are linear regression lines. (a)–(d) indicates the years when various large reservoirs started to store water: (a) – Sanmenxia Reservoir (Sept. 1960; designed storage capacity: 3.1 km^3), (b) – Liujiaxia Reservoir (Oct. 1968; designed storage capacity: 5.7 km^3), (c) – Longyangxia Reservoir (Oct. 1986; designed storage capacity: 24.7 km^3), and (d) – Xiaolangdi Reservoir (Oct. 1999; designed storage capacity: 5.1 km^3).

of extensive reservoirs in the basin. The sharp decrease in water discharge around 1969 coincided with the operation of the Liujiaxia Reservoir ('b' in **Figure 2**; designed water storage capacity: 5.7 km^3), the first large reservoir to be built in the upper reaches of the Huanghe, from where the majority of its water is collected.

No significant trend has been observed in the water discharge of the Changjiang, the Zhujiang, and the Songhuajiang (the largest Chinese tributary of the Heilongjiang-Amur). The construction of reservoirs does not seem to have caused major changes in the water discharge of the Changjiang (**Figure 3(a)**), likely buffered by its much larger water discharge and above-normal precipitation in recent years.

Despite their short lengths and small drainage areas, the rivers originating from the third topographic 'stair' in China are abundant in water runoff. The runoffs of the Minjiang and Qiantangjiang, for example, are almost double that of the Changjiang. In particular, rivers from Taiwan and Hainan Islands typically have an annual runoff of more than 1000 mm. With a small drainage basin of 122 km^2, the Shuang River in Taiwan Island, for example, has a mean runoff of 4500 mm $year^{-1}$ during the period 1970–1999, i.e., about 12 times the world's average runoff.

Total Suspended Sediment (TSS)

Eastern Asia is home to some of the most TSS-generating rivers in the world. The Huanghe is the most turbid large river in the world. The TSS concentration averaged 23.9 $g\,l^{-1}$ (range: 4.8–48 $g\,l^{-1}$) during the period 1950–2006 at the most-downstream station Lijin. In 6 out of the past 56 years, the annual average TSS concentration at Lijin exceeded 40 $g\,l^{-1}$, a criterion above which very dense (hyperpycnal) flows would develop in the river mouth due to its higher density than seawater. Many smaller tributaries of the Huanghe have even higher TSS concentrations.

Until very recently the Huanghe was the largest river in the world in terms of the TSS flux to the sea. Over its 0.15 My history, the Huanghe is estimated to have discharged a total of 7.0×10^{12} tons of TSS, the

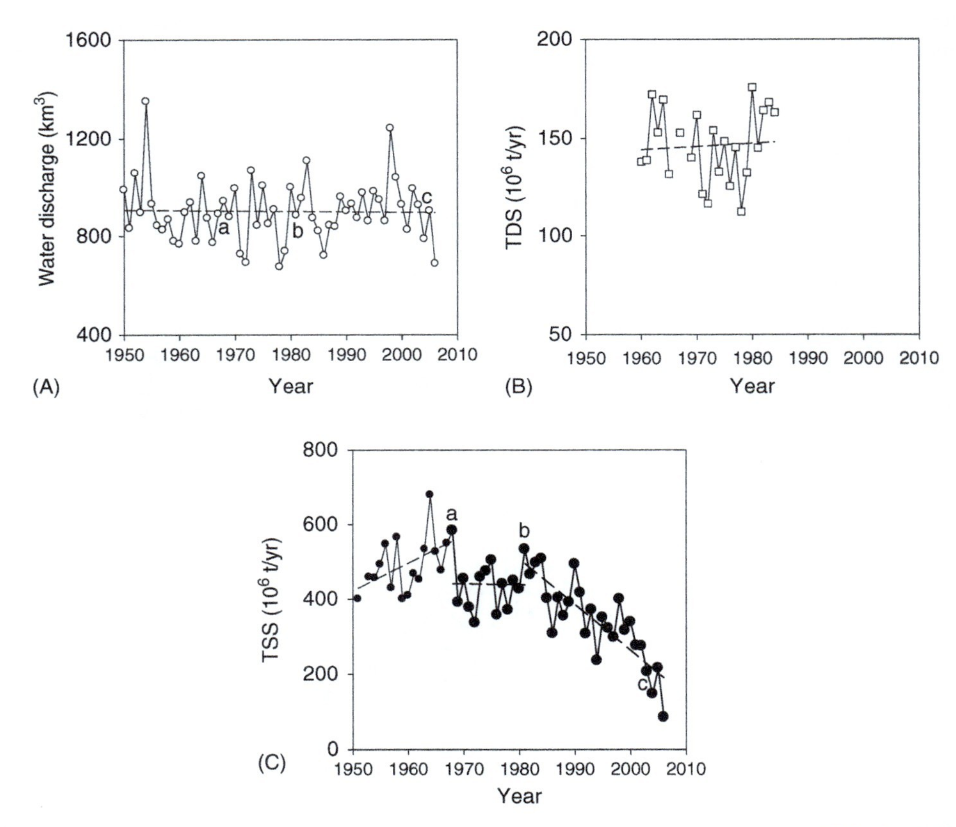

Figure 3 Seaward discharge of water (A), total dissolved salts (TDS; B), and total suspended solids (TSS; C) of the Changjiang at Hydrological Station Datong, the most downstream station without tidal influence. The dashed lines are linear regression lines. (a)–(c) indicates the years when various large reservoirs started to store water: (a) – Danjiangkou Reservoir (July 1967; designed storage capacity: 21.0 km^3), (b) – Gezhouba Reservoir (June 1981), and (c) – Three Gorges Reservoir (June 2003; designed storage capacity: 39.3 km^3).

majority of which was deposited in the lower reaches of the basin to create the North China Plain and the continental delta, but as much as 1.8×10^{12} tons of TSS made its way to the Bohai Sea of the Pacific Ocean. The TSS load of the river is so high that its accumulated deposition in the lower reaches has raised the river bed up to 10 m above the surrounding areas in the North China Plain, making the lower reaches of the Huanghe essentially a 'suspended river.' This has become a major threat to its floodplain which has been inundated regularly since historical periods. The last major flooding event occurred in the 1930s, causing tens of thousands of casualties.

However, the seaward TSS flux of the Huanghe has dramatically decreased in the past decades (**Figure 2(C)**), especially since the 1970s, dropping from 1230×10^6 tons year^{-1} during 1950–1969 to 390×10^6 tons year^{-1} in the 1990s, and 154×10^6 tons year^{-1} in recent years, with a 57-year mean annual seaward TSS flux of 790×10^6 tons year^{-1} during 1950–2006. Note that the widely quoted number of 1600×10^6 tons year^{-1} cannot be regarded as the seaward TSS flux of the Huanghe, as it was based on old data obtained at the station Sanmenxia, some 1060 km upstream inland from the river mouth. The lowest ever TSS flux was observed during 2000–2001 with a mere average of only 22×10^6 tons year^{-1}, less than 2% of its pre-1969 average. Such a sharp decrease in the TSS flux is not indicative of water becoming less turbid; rather, it is mainly caused by the sharp decrease in water discharge (**Figure 2(A)**) and by sediment trapping in reservoirs (e.g., the Sanmenxia Reservoir). Similar to the trend in water discharge, the dramatic decrease in TSS flux of the Huanghe also started at the same time when the Liujiaxia Reservoir started to regulate the water. The reduction in the TSS load may also be attributed, in part, to the extensive soil and water conservation programs launched in the basin some decades ago to improve local agriculture and reduce soil erosion.

The TSS flux of the Changjiang averaged 410×10^6 tons year^{-1} ($85–680 \times 10^6$ tons year^{-1}) at the station Datong during the period 1950–2006, surpassed that of the Huanghe since the 1990s. As shown in **Figure 3(C)**, the TSS load of the Changjiang increased from 400×10^6 tons year^{-1} in the 1950s to around 600×10^6 tons year^{-1} in the later 1960s, then dropped back to $\sim 440 \times 10^6$ tons year^{-1} during 1968–1980. Since the 1980s, a persistent and sharp decreasing trend has been observed in the TSS load with an average value of 150×10^6 tons year^{-1} during 2000–2006. Different from the Huanghe, in the same period there was no significant trend in the water discharge of the Changjiang (**Figure 3(A)**). Sediment retention by large-scale reservoirs and reduced soil erosion by extensive soil and water conservation programs in the basin were the most likely causes. For instance, the reduction in the TSS flux in the 1970s coincided with the operation of the Danjiangkou Reservoir (on the Hanjiang, a major tributary of the Changjiang) since 1967. About 50×10^6 tons year^{-1} of TSS was estimated to have been stored in the reservoir during the first decade of its operation. The further decrease since the 1980s is likely due to the Gezhouba Dam on the main channel of the Changjiang which became operational in June 1981, and more recently, the construction of the Three Gorges Dam. The Three Gorges Dam started to store water in June 2003, and in 2004 the TSS load at Datong dropped to 150×10^6 tons year^{-1}. In 2005, a total of 150×10^6 tons of TSS was estimated to have been stored already in the Three Gorges Dam. The seaward TSS of the Changjiang decreased to 85×10^6 tons year^{-1} in 2006, the lowest in the past 57 years. Further reduction in the downstream and seaward TSS load is expected upon the completion of the Three Gorges Dam in 2009.

In addition to their high runoffs, the rivers in Taiwan Island have the highest TSS yields in the world. With a total drainage area of 0.022×10^6 km^2, the 29 major coastal rivers in Taiwan Island collectively supplied 380×10^6 tons year^{-1} of TSS to the ocean during the period 1970–1999. The mean TSS yield is about 17 500 tons km^{-2} year^{-1}, more than 16 times that of the Huanghe (at Lijin), and can only be matched by a few mountainous rivers in New Zealand. The Pei-Nan River in Taiwan Island has a 30-year average TSS yield of 55 400 tons km^{-2} year^{-1}, which is about 250 times the world's average. Many rivers in Taiwan Island are hyperpycnal (as for the Huanghe), with TSS concentrations frequently exceeding 40 g l^{-1}, particularly during typhoon-related floods. The extremely high TSS yields are due to readily and rapid erosion of poorly consolidated sediments under the subtropical climate (mean annual precipitation of 2500 mm year^{-1}) with high frequency of typhoons and in a tectonically active island (frequent earthquakes with related landslides).

The rivers in Taiwan and Hainan Islands serve as a classic example of how geology and precipitation affect the TSS yield. Both islands are located in subtropical to tropical climate zones. They are of similar size and in close proximity (Hainan Island is about 5° latitude south and 10° longitude west of Taiwan Island). However, the TSS yields of the rivers in Hainan Island average only 140 tons km^{-2} year^{-1} (**Table 1**), less than 1% of that of the rivers in Taiwan Island. Hainan Island was formed long before and is geologically much more stable than Taiwan Island. The lithology of Hainan Island is dominated by less erodible igneous rocks. In addition, the amount of precipitation and the annual runoff of the rivers are also much smaller in Hainan Island when compared to Taiwan Island.

Chemical Composition of TDS

Major solute chemistry of the rivers in China has been surveyed very early and synthesized already in the 1960s, although the data have only become available in the Western literature since the 1980s. Long-term trends in major solute compositions of large rivers have been studied since the late 1990s. **Table 2** summarizes the major solute chemistry of selected rivers in Eastern Asia.

Major Solute Composition of TDS

Much of the drainage basins of the Changjiang and the Zhujiang are dominated by carbonate rocks and are located in humid climatic zones. The southeast basin of the Changjiang and the south basin of the Zhujiang are famous for the well developed karst formations (e.g., near Guilin). As a result, the major solute chemistry of both the Changjiang and the Zhujiang is controlled primarily by the weathering of carbonate rocks, with HCO_3^- being the dominant anion and Ca^{2+} the dominant cation. As the Zhujiang basin is subjected to more intensive weathering and leaching, the TDS concentration of the Zhujiang averaged $189\,mg\,l^{-1}$, slightly lower than that of the Changjiang ($206\,mg\,l^{-1}$); both are well above the world spatial mean (WSM) value of $126\,mg\,l^{-1}$. The Changjiang indeed has the highest relative abundance of HCO_3^- among the major rivers of the world. The Cl^- and SO_4^{2-} in the Zhujiang originate mainly from the sea salt spray due to its close proximity to the ocean, whereas those in the Changjiang basin are mainly derived from the weathering of evaporates in the upper reaches of the basin, as well as from acid deposition.

Much higher TDS concentrations are found in the Huanghe, with a basin-wide average of $450\,mg\,l^{-1}$, about 3.5 times the WSM value. In particular, the concentrations of $Na^+ + K^+$, SO_4^{2-}, and Cl^- are 10–20 times higher than in the other major rivers in the world. With much of its basin underlain by loess and clastic rocks under arid and semiarid climatic conditions, the dissolved salts carried by the Huanghe are predominantly controlled by evaporation and fractional crystallization, and chemical weathering. The evaporation and fractional crystallization are further promoted by intensive irrigation and reservoir constructions.

Among the four major rivers in Eastern China, the Heilongjiang-Amur has the lowest TDS concentration, averaging $73\,mg\,l^{-1}$ for the drainage basin and $150\,mg\,l^{-1}$ for the Chinese tributary the Songhuajiang, due to its much lower weathering intensity, under a cold to temperate climate, of silicate and aluminosilicate rocks that dominate in this basin.

Within the same river system, significant variations in the TDS composition and concentration occur spatially. For instance, the Zhujiang is composed of three tributaries, the Xijiang in the west, the Beijiang in the north, and the Dongjiang in the east, joining at Sanshui just before entering the South China Sea of the Pacific Ocean. The subbasin of each tributary is dominated by very different rocks: the Xijiang by limestone, the Beijiang by red sandstone, and the Dongjiang by granite and shales. Reflecting the chemical weatherability of these different rocks under a humic subtropical climate, the TDS concentration in the Xijiang is the highest ($202\,mg\,l^{-1}$), followed by the Beijiang ($121\,mg\,l^{-1}$) and the Dongjiang ($66\,mg\,l^{-1}$). However, much larger spatial variation is found in the Huanghe basin. At Guochengyi on the Zulihe, where evaporation and fractional crystallization dominates, the TDS concentration is as high as $8500\,mg\,l^{-1}$, i.e., 60 times that at Yimenzhen of the Weihe where chemical weathering of granite dominates the process.

Long-Term Trend in Major Solute Concentrations

Since the major solute concentrations at most of the stations along the rivers in China have been monitored continuously or semicontinuously for more than 30 years, it is possible to statistically detect the changes in the major solute composition of the river water. Three distinctive trends have been reported: (i) acidification; (ii) salinization; and (iii) alkalinization.

Acidification in the Changjiang A significant increase trend has been found in the concentrations of SO_4^{2-} and, to a lesser extent, Cl^- in major tributaries in the Chongqing–Guiyang area and at all downstream main-channel stations of the Changjiang. The rate of SO_4^{2-} increase was the highest in the tributary Tuojiang with an average rate of $1.50\,mg\,l^{-1}\,year^{-1}$ at Lijiawan, just before it joins the Changjiang. Even at the most downstream main-channel station Datong (some 1800 km downstream from Chongqing), the increase was still significant at a rate of $0.22\,mg\,l^{-1}\,year^{-1}$ since 1960. The increasing trend is attributed to acid deposition in the Chongqing–Guiyang area, one of the most severe acid deposition centers in the world. The mountainous Chongqing–Guiyang region has become one of the largest bases of the heavy industry in China since the later 1950s, fuelled by sulfur-rich coals. As a result, acid rain has been reported in the region

Table 2 Major ions and dissolved SiO_2 composition of some Eastern Asian rivers

River	Station	Data period	Ca^{2+} ($mg\ l^{-1}$)	Mg^{2+} ($mg\ l^{-1}$)	$Na^{+}+K^{+}$ ($mg\ l^{-1}$)	HCO_3^- ($mg\ l^{-1}$)	SO_4^{2-} ($mg\ l^{-1}$)	Cl^- ($mg\ l^{-1}$)	SiO_2 ($mg\ l^{-1}$)	TDS	Source
Continental rivers											
Changjiang	Datong[a]	1960–1984	30.1	6.3	5.0	113.2	11.9	4.2	7.0	171.3	1
	191 stations in the basin[b]	1960–1990	34.1 (7.9–53.1)	7.6 (2.0–17.5)	8.2 (3.7–23.5)	134 (41.4–215)	11.7 (2.90–41.4)	2.9 (1.1–17.1)	6.2 (1.6–9.1)	206 (69.2–342)	1
Huanghe	Lijina[a]	1964–1998	51.2	25.2	65.2	197.7	101.5	65.4		508.0	2
	100 stations in the basin[b]	1960–1998	51.0 (27.5–328.6)	18.7 (6.2–256.8)	48.2 (15.6–1150)	206 (115–297)	74.1 (16.2–2020)	30.7 (7.0–1205)	6.0 (3.3–9.1)	452 (221–5258)	2
Zhujiang	Gaoyao/Shijiao/Boluo[a,c]	1959–1984	26.2	4.4	11.2	112.0	8.8	3.5		166.0	3
	96 stations in the basin[b]	1959–1984	32.9 (6.1–57.8)	5.0 (1.8–16.4)	5.5 (2.2–17.0)	130 (35.9–225)	6.6 (2.2–31.7)	1.75 (0.51–6.95)	5.9 (3.8–11.5)	189 (57.0–330)	3
Songhuajiang	Jiamusi[a]	1960–1984	16.3	4.7	18.9	89.8	11.8	8.2	5.0	150.0	4
	68 stations in the basin[b]	1960–1984	14.5 (7.6–43.7)	4.2 (2.1–11.0)	12.5 (6.6–46.4)	78.1 (40.8–250)	6.9 (2.5–19.7)	5.6 (2.6–16.6)	7.9 (4.1–20.8)	126 (65.9–430)	4
Island rivers											
7 rivers in Hainan Island		1959–1988	6.65	1.51	13.46	54.1	7.68	3.63	15.6	79.6	5

[a]Arithmetic mean (not discharge weighted) of the data during the data period.
[b]Median (5%–95% percentiles) of the data for all the stations in the Huanghe basin during the data period.
[c]Discharge-averaged value of the three stations.

1. Chen J, Wang F, Xia X, and Zhang L (2002) Major element chemistry of the Changjiang (Yangtze River). *Chemical Geology* 187: 231–255.
2. Chen J, Wang F, Meybeck M, He D, Xia X, and Zhang L (2005) Spatial and temporal analysis of water chemistry records (1958–2000) in the Huanghe (Yellow River) basin. *Global Biogeochemical Cycles* 19: GB3016. doi:10.1029/2004GB002325.
3. Chen J and He D (1999) Chemical characteristics and genesis of major ions in the Pearl River Basin. *Acta Sci. Natur. Univ. Pekin.* 35: 786–793 (In Chinese).
4. Chen J, Xia X, Zhang L, and Li H (1999) Relationship between water quality changes in the Yangtze, Yellow and Songhua Rivers and the economic development in the river basins. *Acta Scientiae Circumstantiae* 19: 500–505 (In Chinese).
5. Chen J, Xie G, and Li Y-H (1991) Denudation rate of Hainan Island and its comparison with Tainwan Island and Hawaiian Islands. *Quaternary Sciences* 4: 289–299 (In Chinese).

since the 1970s with rainwater frequently having a sulfate concentration of more than 20 mg l^{-1} and a pH lower than 4.1. A strong correlation has been found between the SO_4^{2-} in the Changjiang near Chongqing and the coal consumption in the Sichuan Province. Although no significant trend was observed in total alkalinity and pH along the main channel of the Changjiang due to the abundance of buffering carbonates, the ratio of hardness to alkalinity did increase significantly in some large tributaries (e.g., the Tuojiang and the Wujiang), showing early signs of localized acidification in the river basin.

Salinization in the Huanghe A more profound increase trend is observed in the Huanghe. Concentrations of TDS and all the major ions except for HCO_3^- at all main-channel stations except for the uppermost Lanzhou have been steadily increasing since the mid1970s. The rate of TDS increase was the highest in the middle reaches (10.5 mg l^{-1} $year^{-1}$ at Toudaoguai) and remained high in the lower reaches (5.5 mg l^{-1} $year^{-1}$ at Stations Luokou and Lijin). The increase trend agreed well with the sharp decrease in the water discharge, suggesting primarily a concentrating effect, although the impact of other processes (e.g., industrial discharges) cannot be ruled out.

Alkalinization in the Songhuajiang Among 24 long-term monitoring stations in the Songhuajiang basin, 16 stations showed a distinctive trend of alkalinization since the 1960s, evidenced by an increase in the concentration of $Na^+ + K^+$, HCO_3^-, and TDS, and a decrease in the ratio of hardness to alkalinity. At a few stations the pH value also increased slightly. The alkalinization trend of the Songhuajiang can be attributed primarily to the pulp and paper industry and the practice of groundwater irrigation. The production of pulp and paper in the Heilongjiang province, for example, increased fourfold from <10 000 tons $year^{-1}$ in 1961 to more than 40 000 tons $year^{-1}$ in 1984. As NaOH and Na_2S are commonly used in chemical pulping, an increasing amount of Na^+ may be directly or indirectly discharged into the river. The degradation of high organic pulp and paper wastewater also produces CO_2 and increases the HCO_3^- concentration. Agricultural irrigation with Na^+-enriched groundwater in the region further increases the Na^+ concentration in the return water due to evaporation.

Seaward TDS Flux

Based on long-term monitoring data at the most downstream stations, the Changjiang, the Zhujiang, the Heilongjiang-Amur, and the Huanghe transport a total of 250×10^6 tons $year^{-1}$ TDS (not corrected for sea salts) to the ocean (**Table 1**), accounting for more than 10% of the global total seaward TDS flux. The TDS flux of the Changjiang alone amounts to 160×10^6 tons $year^{-1}$, second only to the Amazon. The Zhujiang transports a total of 47.5 tons $year^{-1}$ TDS, similar to that of the Yukon and Rhine. The seaward TDS flux of the Heilongjiang-Amur is about 25.8×10^6 tons $year^{-1}$, 38% of which is derived from the Songhuajiang. Despite of its high TSS flux, the Huanghe transported only 16.5×10^6 tons $year^{-1}$ of TDS to the sea during the period 1950–2005.

Due to the increasing trend in the SO_4^{2-} concentration of the Changjiang at Datong as a result of acid deposition, the TDS flux by this river is likely to increase further. The rate of increase is rather small at present (0.25 mg l^{-1} $year^{-1}$), but with its massive water discharge it results in an annual increase of 0.23×10^6 tons of SO_4^{2-} to the ocean. In contrary, despite the increasing trend in the TDS concentrations, the seaward TDS flux by the Huanghe, as measured at Luokou and Lijin, has decreased by more than 50% from over 20×10^6 tons $year^{-1}$ in the 1960s to less than 10×10^6 tons $year^{-1}$ at present (**Figure 1(b)**), due to the significant decrease in water discharge.

Elemental Composition of TSS

Elemental composition of riverine TSS in Eastern Asia was not included in many earlier studies on global chemical composition of riverine TSS, due to the lack of relevant data. Such data have become available since the 1990s and are summarized in **Table 3**.

The TSS from Eastern Asian rivers shows a wide variation in elemental composition. The TSS from the northern rivers (e.g., the Heilongjiang, Tumen, and Yalu) in the cold and temperate zones contains higher contents of K and Na, whereas that from the southern subtropical zones is richer in Al, which is in general agreement with the global picture and reflects the mobility and fate of elements under different climates.

The Ca concentration in TSS varies significantly among the rivers. Despite the fact that limestone is the dominant rock type in most southern rivers, especially in the upper and middle reaches of the Changjiang and the Zhujiang, the Ca concentrations are only near the global average of 3.1%, due to the strong leaching by weathering in humid subtropical climates. The highest Ca concentration (5.8%) almost double the global average, is found in TSS from the Huanghe, and is among the highest in the world's large rivers. This is resulted from the relatively weak weathering of the high Ca-containing

Table 3 Elemental composition and flux of riverine TSS in Eastern Asia

River	*Si*	*Al*	*Ca*	*Fe*	*K*	*Mg*	*Na*	*Ti*	*Mn*	*Zn*	*V*	*Cr*	*Cu*	*Ni*	*Pb*	*Co*	*Cd*
Concentrations (mg/kg)[a]																	
Changjiang	297 000	81 800	28 900	48 200	20 400	18 300	7050	5770	892	164	137	76.4	49.2	50.1	38.6	24.1	0.33
Heilongjiang		62 700	19 300	48 000	24 900	14 700	9590	6430	662	155	77.3	52.6	25.4	37.0	34.0	18.6	0.15
Zhujiang	246 000	111 000	13 100	60 900	18 400	10 100	3460	7570	1030	236	118	91.9	58.2	46.7	49.9	24.8	0.78
Minjiang		97 700	6470	50 200	16 200	5460	4900	5130	1072	291	106	55.9	45.0	37.9	76.3	21.4	0.62
Huanghe	304 000	72 500	58 100	35 600	18 400	17 700	10 600	3670	694	120	87.5	69.1	27.4	42.1	20.9	13.7	0.19
Huaihe		54 800	20 600	41 200		10 300		4740	814	80.0	88.3	84.8	23.1	44.1	23.1	23.8	0.27
Qiantangjiang		74 800	7280	33 000	19 200	11 700	5100	4630	938	239	98.1	59.1	41.6	44.1	38.8	23.2	0.47
Liaohe		68 000	30 100	31 600	21 600	21 200	9300		457	152		34.3	56.7	38.2	30.0	19.6	0.12
Yalu		72 500	23 200	58 700	28 700	16 500	9700		404	231		44.6	42.3	40.2	40.0	24.5	0.15
Tumen		72 100	20 100	69 600	25 600	20 000	8700		456	351		46.6	55.0	43.6	34.6	24.4	0.12
Global Average	255 000	67 600	30 800	44 000	17 000	12 500	7430	4090	755.2	163.7	122.4	76.4	50.1	45.4	45.0	22.2	0.19
Seaward flux (10^3 tons year^{-1})[b]																	
Changjiang	123 000	33 900	11 200	19 900	8450	7570	2920	2390	369	67.8	56.8	31.6	20.4	20.7	16.0	9.992	0.14
Heilongjiang		3260	1000	2490	1290	765	498	334	34.4	8.06	4.02	2.73	1.32	1.92	1.77	0.969	0.01
Zhujiang	3100	1400	165	767	232	128	43.6	95.3	13.0	2.98	1.48	1.16	0.733	0.588	0.629	0.312	0.01
Minjiang		479	31.7	246	79.4	26.8	24.0	25.1	5.25	1.42	0.521	0.274	0.221	0.186	0.374	0.105	
Huanghe	243 000	57 800	46 400	28 400	14 700	14 200	8470	2930	554	96.0	69.8	55.1	21.8	33.6	16.6	10.96	0.15
Huaihe		525	195	391		97.8		45.0	7.73	0.76	0.839	0.806	0.219	0.419	0.219	0.226	
Qiantangjiang		150	14.6	66.1	38.4	23.4	10.2	9.26	1.88	0.477	0.196	0.118	0.083	0.088	0.078	0.046	
Liaohe		333	147	155	106	104	45.6		2.24	0.742		0.168	0.278	0.187	0.147	0.096	
Yalu		290	92.8	235	115	66.0	38.8		1.61	0.924		0.178	0.169	0.161	0.160	0.098	
Tumen		216	60.3	209	76.8	60.0	26.1		1.37	1.05		0.140	0.165	0.131	0.104	0.073	
Global Total[c]	4460 000	1180 000	539 000	770 000	298 000	219 000	130 000	71 600	13 200	2860	2140	1340	877	794	788	388	3.33

[a]Chen J and Wang F (1996) Chemical composition of river particulates in eastern China. *GeoJournal* 40: 31–37.

[b]The flux of each river is calculated from the concentration data in the first half of the table and the TSS flux data from **Table 1**.

[c]The global total flux is calculated from the global average concentration data and an estimate of 17 500 × 10^6 tons year^{-1} of global TSS flux.

loess (5–20%) in the middle reaches of the river. The concentrations of most trace elements (e.g., Cu, Pb, Zn, Cd, Cr) in TSS show a general trend in the increase from north to south, coincident with the high density of nonferrous mineral deposits in southern China. An anthropogenic impact is also possible but can be masked by the generally high sediment yields.

Due to the very high TSS load and Ca content, the TSS from the Huanghe alone accounts for 6.5% of the global TSS flux of particulate Ca. In total, the 10 Eastern Asian rivers listed in **Table 3** account for 5.2–11.1% of the global TSS flux of the elements studied.

Human Impacts

As is true around the world, the health of the river systems in Eastern Asia has been increasingly impacted by human activities since the Anthropocene. The situation is most challenging in China due to the regional imbalance of water resources, the large population base, and a rapid industrializing economy and society. Although southern China enjoys a bountiful freshwater supply, northern and northwestern China has one of the lowest per capita water resources in the world and faces severe water shortages. Large scales of damming and canal building have taken place along many large, medium and small rivers to facilitate irrigation, flooding control, water diversion to other regions, and electricity generating. Groundwater has been taken at an unsustainable rate in many areas in northern China. The need for such a development and its socioeconomic benefits are obvious, but in most cases their environmental impacts are not fully studied, understood, and mitigated. In addition to being increasingly regulated by dams and reservoirs, most rivers have been increasingly contaminated and polluted by domestic and industrial wastewaters, agricultural activities, solid wastes, and atmospherically transported chemicals. On top of all these is climate change which is likely to alter the hydrology and quality of the rivers further. As a result, not only has the quality of most small, medium, or urban rivers been heavily degraded, many large river systems in the continent have also been significantly modified. As discussed earlier, the dramatic decrease in water discharge of the Huanghe, the early acidification signs in the Changjiang, and the alkalinization in the Songhuajiang can be all related to, if not primarily caused by, human activities.

The Huanghe probably serves as the most alarming example of human impacts on large river systems in China and in the world. The Huanghe basin is regarded as the cradle of the Chinese culture, and has been inhabited by people since at least 1 Mya. At present more than 100 million people live in the basin, with a mean population density of 130 people per square kilometer. Irrigated agriculture has been the major economic development in the basin for more than 2000 years. The amount of irrigation water taken from the Huanghe has more than doubled in the past 50 years, from 12.5 $km^3 year^{-1}$ in the 1950s to more than 30 $km^3 year^{-1}$ in the 1990s. In addition, the hydrology of the Huanghe has been greatly regulated by thousands of dams and reservoirs. The total storage capacity of the reservoirs in the Huanghe basin was nearly 60 km^3 in 1993, almost twice that of its annual water discharge at the river mouth. Since 1972, water from the Huanghe has frequently been diverted to the City of Tianjin which is located outside of the Huanghe basin. Together with the decreased atmospheric precipitation, the increasing amount of water withdrawal and loss has resulted in a dramatic decrease in water, TDS and TSS discharge (**Figure 2**), one of the most severe in the world's history and only surpassed by the Colorado (USA/Mexico) and the Amu Darya (Turkmenistan/Uzbekistan). At present, the Huanghe only delivers 12.3 $km^3 year^{-1}$ of water and 155×10^6 tons $year^{-1}$ of TSS at the downstream station Lijin, a reduction of more than 75% and 87%, respectively, from its corresponding levels during 1950–1969.

The physical, ecological, and socioeconomic impacts of the sharp decline in water, TDS and TSS discharge of the Huanghe are expected to be profound and remain to be fully understood and appreciated. For instance, the reduced water discharge will further severe the water shortage problem in northern China, as well as make the lower reaches of the river more prone to contamination. The frequent dry-ups will increase the precipitation of calcite cement on the riverbed which will harden the riverbed and continuously raise the riverbed level thus threatening more the floodplain population, and increase water evaporation which will salinize soils and groundwater in the surrounding areas. The reduced sediment discharge to the delta region will slow down the growth of the delta and make the shoreline prone to seawater erosion and invasion. Ecologically, many aquatic biota in the lower Huanghe and coastal Bohai rely on the dissolved salts from the Huanghe as their nutrients. A declined or stopped flux of those salts and nutrients could result in dramatic changes in the ecosystem structures in the area.

Unless drastic actions are taken, the decrease in the water, TDS and TSS discharge of the Huanghe is likely to continue and probably become worse due to climate variations, population increase, and economical development. If this were the case, the Huanghe could become a new endorheic river cutoff from the ocean, similar to the Colorado River, and what have been experienced in the lower Colorado and the Gulf of California could be repeated in the Huanghe and the Bohai Bay.

Concluding Remarks

Two mega hydrological projects are currently ongoing in China which are likely to have major impacts on the river systems in China and on regional and global biogeochemical cycles. The first one is the Three Gorges Project to dam the main-channel Changjiang in its middle reaches to control flooding and generate electricity. The construction started in 1994 and the reservoir began filling in 2003. By its completion in 2009, the Three Gorges Reservoir will have a designed water storage capacity of $40\,km^3$. The second one is the South-to-North Water Diversion (SNWD) project which was initiated in 2003 to solve the serious water shortage problem in northern China. It is the largest water diversion project in the world. Under the SNWD project, three massive cross-basin canal networks (Eastern, Central, and Western lines) will be constructed to divert water from the Changjiang to the Huanghe and to northern China. The proposed capacity of the SNWD is $44.8\,km^3\,year^{-1}$, which is about 5% of the annual water discharge of the Changjiang, and 140% of the annual water discharge of the Huanghe. The project is expected to be completed by 2050 and will connect the drainage basins of the Changjiang, the Huanghe, the Huaihe, the Haihe and many other smaller river systems and essentially create a single mega pseudo-basin. Long-term and high density monitoring, and rapid analysis and interpretation of the monitoring data are needed now to understand and manage any dramatic beneficial and adverse effects of projects of this scale.

Further Reading

Chen J and Wang F (1996) Chemical Composition of River Particulates in Eastern China. *GeoJournal* 40: 31–37.

Chen J, Wang F, Xia X, and Zhang L (2002) Major Element Chemistry of the Changjiang (Yangtze River). *Chemical Geology* 187: 231–255.

Chen J, Wang F, Meybeck M, He D, Xia X, and Zhang L (2005) Spatial and Temporal Analysis of Water Chemistry Records (1958–2000) in the Huanghe (Yellow River) basin. *Global Biogeochemical Cycles* 19GB3016, doi:10.1029/2004GB002325.

Dadson SJ, Hovius N, and Chen HG (2003) Links between erosion, runoff variability and seismicity in the Taiwan orogen. *Nature* 426: 648–651.

Meybeck M (2003) Global Analysis of River Systems: From Earth system controls to Anthropocene syndromes. *Philosophical Transactions Of the Royal Society of London B* 358: 1935–1955.

Zhao S (1986) *Physical Geography of China*. New York: Wiley.

Asia – Monsoon Asia

M Meybeck, Université Pierre et Marie Curie, Paris, France

General Features of South and Southeast Asian Drainages

The relief, climate, and lithological features of this region are summarized in **Table 1** for 17 aggregated coastal catchments that link river networks to the coasts of the Indian Ocean and of the South China Sea (**Figure 1**). The northern hydrographic limits of this drainage are the anti-Taurus (3500 m) in Turkey, the Zagros Range in Iran (4070 m), the Belutschistan mountains and Hindu Kush range in North Pakistan (7690 m), the Karakorum (8600 m) and Himalaya (8840 m) ranges, the Tibetan Plateau (6000 m), and for Southeast Asia the Yunnan mountains in Southwest China. As a whole, these rivers are therefore characterized by headwaters of very high elevation with extended snow cover and permanent ice in areas exceeding 5000 m in the Hindu Kush and Karakorum (Indus basin), in the Himalayans (Ganges and Brahmaputra basins), and in some parts of the Tibet Plateau. These headwaters snow and ice covers provide most of the water resources from Anatolia to the Indus Valley and function as water towers in providing water to lower elevations.

The lithology of this region, a major controlling factor of river chemistry, is characterized by mixed sedimentary and crystalline rocks folded in the alpine ranges. An important volcanic plateau in Central India (the Deccan Traps) must be mentioned; the rest of Deccan and Sri Lanka have crystalline rocks originating from the former Gondwana continent. Sedimentary detrital rocks are found in all lowlands as in the Shatt el Arab, Indus, Ganges, Irrawaddy, and Mekong plains.

Climates of South and Southeast Asia are greatly dependent on elevation, which controls both temperature and precipitation, and by the presence of the Asian monsoon, which affects the central and eastern parts of this region from the Indus Valley to the South China Sea basin. The monsoon is characterized by winter–spring dryness and summer rain (June and August) that accounts for 50–80% of the annual rain. In India, the monsoon starts abruptly in June in the Western Ghats highlands, south of Mumbai, and progresses eastward during summer. The period of maximum precipitation depends on station location, ranging from July (e.g., Mumbai) to November (Chennai). The annual rainfall varies between 660 (Delhi) and 1700 mm year^{-1} (Hanoi) with a local maximum at 3000 mm year^{-1} in the Western Ghats. World extreme rainfall is recorded in the Khasi Hills (10 800 mm year^{-1}), north of Bangladesh in the upper Meghna basin, a tributary of the Ganges/Brahmaputra common estuary (also termed Padma). Another precipitation maximum (4800 mm year^{-1}) is observed in the southeastern Salween basin (Myanmar). In the southern part of the South China Sea basin (Malaysia, Indonesia, Philippines), near the equator, the rainfall is nearly equally distributed throughout the year and reaches 3200 mm in North Borneo. In the Middle East from the Indus Valley to the Persian Gulf Basin, annual precipitation reaches 500–800 mm year^{-1} in the mountains but barely exceeds 200 mm year^{-1} in the Shatt El Arab Valley (160 mm year^{-1} in Baghdad) and in the Indus plain (180 mm year^{-1}).

In this region, river temperature, a major control of aquatic species distribution, is very variable, including along the greatest basins that cross several climatic zones (**Table 1**). In high mountains, from the Anatolian Plateau to Tibet, rivers are commonly frozen during winter, while in equatorial catchments, from Sri Lanka to Philippines, thermal variations are limited, and monthly temperatures may exceed 22 °C. The greatest catchments, such as Shatt el Arab, Indus, Ganges, Brahmaputra, Irrawaddy, Salween, and Mekong, are characterized by the greatest spatial and temporal thermal variability.

Drainage Network and River Discharge Regimes

Because of the predominance of mountainous regions and of active tectonics, the drainage network of South Asia has not generated very large river basins compared with those found on other continents. The largest river drainage area does not exceed 1.05 million km^2 (Ganges), while on islands (Sri Lanka, Indonesia, Philippines), peninsulas (Malaysia), and narrow coasts (Iran, Oman, W. Deccan, Annam), river basins do not generally exceed 20 000 km^2. In the Arabian Peninsula and on the Iranian coast, medium-sized river networks are found, but they are mostly dry or flow only occasionally (desert wadis).

The major river systems from East to West are: the Shatt el Arab, draining the Anatolian Plateau to the Persian Gulf; the Indus, which reaches the Arabian Sea through one of the world's largest deltas; the Ganges (Ganga), which drains the southern side of the Himalayas; the Brahmaputra (named Tsang Po in

Table 1 General characteristics of South and Southeast Asian coastal catchments

Sea basin name	Code (a)	Principal basins	Sea basin area (Mkm^2)	Runoff (mm $year^{-1}$)	Population density/ sea basin (people km^{-2})	Sediment yield (t km^{-2} $year^{-1}$)	Relief			Climate							Geology			
							% low	% mid	% high	% polar	% cold	% temperate	% dry < 3 mm (b)	% dry ≥3 mm (b)	% tropical <680 mm (c)	% tropical ≥ 680 mm (c)	% plutonic metam.	% volcanic	% carbon.	% other rock type
East South China Sea	28	Kapuas, Rajang, Aguson	0.35	1824	115	512	21.7	77.3	1.0	0.0	0.0	10.3	0.0	0.0	0.0	89.7	8.3	15.5	4.5	24.8
West South China Sea	19	Mekong, Chao Phraya, Song Koi	1.49	1038	94	280	40.7	52.2	7.1	6.0	0.8	21.6	0.0	0.0	36.5	35.2	11.2	7.8	37.2	21.5
Sunda Strait	30	Barito, Mahakam	0.55	1220	122	278	49.0	51.0	0.0	0.0	0.0	6.3	0.0	0.0	8.3	85.4	4.8	27.2	5.4	26.9
Sulu-Celebes Sea	31	Mindanao	0.36	1194	107	605	14.8	85.2	0.0	0.0	0.0	1.3	0.0	0.0	9.1	89.6	4.2	34.6	16.9	9.6
Banda Sea	32	No important rivers	0.19	800	105	326	4.8	95.2	0.0	0.0	0.0	3.1	0.0	0.0	34.3	62.6	20.0	43.0	14.0	8.0
South Timor Coast	33	No important rivers	0.01	560	214	66	0.0	100.0	0.0	0.0	0.0	0.0	0.0	0.0	100.0	0.0	0.0	25.0	49.9	0.0
Java Trench	34	Cinamuk, Citanduy	0.17	1306	128	557	2.3	97.7	0.0	0.0	0.0	4.5	0.0	0.0	17.7	77.9	0.0	48.9	30.2	9.3
Andaman Sea	35	Irrawaddy, Salween, Kelantan, Musi	0.96	1106	79	652	18.7	64.1	17.1	16.1	1.5	37.6	2.5	2.1	10.9	29.3	12.7	9.7	35.0	14.6
Bengal Gulf	36	Ganges, Damodar, Pennar	1.82	820	261	612	48.7	23.6	27.6	22.2	2.2	60.9	3.3	3.4	5.0	3.0	24.7	3.1	17.7	40.7
East Deccan Coast	37	Godavari, Krishna, Mahanadi, Cauweri, Brahmani	1.12	273	315	355	45.8	54.2	0.0	0.0	0.0	11.2	17.0	12.0	55.5	4.3	53.9	23.0	8.4	10.6
Laccadive Basin	38	Ponnani, Payaswani, Kalinadi	0.12	695	377	732	32.5	67.5	0.0	0.0	0.0	0.0	0.0	6.0	45.5	48.5	83.8	0.0	0.0	13.5
West Deccan Coast	39	Narmada, Tapti, Mahi, Rabarmati	0.34	498	297	441	47.3	52.7	0.0	0.0	0.0	16.6	6.2	33.6	26.4	17.2	21.4	51.3	6.1	19.4
Indus Delta Coast	40	Indus	1.39	111	191	195	47.2	23.0	29.8	12.4	7.4	10.5	58.4	9.7	1.6	0.0	22.7	4.5	11.9	44.6
Oman Gulf	41	No important rivers	0.26	2	45	10	9.7	90.3	0.0	0.0	0.0	0.0	87.9	12.1	0.0	0.0	6.6	5.5	19.9	6.7
Persian Gulf	42	Shatt el Arab, Dawasir	2.47	48	28	122	45.1	52.3	2.7	0.0	6.4	10.0	79.4	4.2	0.0	0.0	13.2	3.4	50.6	29.6
South Arabian Coast	43	Muqshin	0.80	2	18	1	34.5	64.0	1.5	0.0	0.0	0.0	97.6	2.4	0.0	0.0	9.2	2.5	59.1	27.1
East Red Sea	44	No important rivers	0.44	0	36	1	0.7	97.1	2.2	0.0	0.0	0.0	100.0	0.0	0.0	0.0	52.9	23.3	4.0	8.8

(a) See location in **Figure 1.** (b) Annual runoff. (c) Annual precipitation.

Meybeck M, Dürr HH, and Vörösmarty CJ (2006) Global coastal segmentation and its river catchment contributors: a new look at land-ocean linkage. *Global Biogeochem. Cycles*, **20**, GB IS 90, doi 10.1029/2005 GB 002540.

Figure 1 Major rivers South Asia and South-East Asia and regional coastal catchments as defined by Maybeck *et al.* (2006).

Tibet, Jamuna in Bangladesh), which drains the northern side of the Himalayas and forces its way eastward across this range near the Namtcha Barwa Peak (7758 m) between Tibet and Assam, and finally flows south and shares the Bengal Delta with the Ganges, forming the Padma River or estuary; the Meghna, a relatively small basin in Bangladesh with an enormous discharge (3500 m^3 s^{-1} for only 80 000 km^2); the Irrawaddy (Ayerarwady in Myanmar); and the Salween (Thanlwin in Myanmar), which originates from the Tibet Plateau at 6000 m, where it is named Nag Chu then Lukiang, both of which have a very narrow and elongated basin caused by tectonic forcing; the Mekong (Dza Chu in China), also originating from the Tibetan Plateau with an extremely narrow upper course down to Vientiane (Laos) then a wider lower and middle course; the Song Koi or red river (Yunnan, China, and Vietnam). Other middle-sized basins include the Godavari, Krishna, Mahanadi, Narmada, Tapti in Deccan, and the Chao Phraya in Thailand. In deserts (Arabian Desert, Rajasthan Desert between Pakistan and India), ephemeral or occasional rivers (wadis) flow only during rare rain events. Internal regions in Iran and Afghanistan also have very limited river networks, excepted for the Helmand River (Afghanistan).

The Mekong river system has a unique hydrological feature, the Tonle Sap, or Great Lake, in Cambodia. This important lake is fed by the Mekong during its high water stage (June to November). At this period the lake progressively deepens and extends to a maximum size of 15 000 km^2. Then the connection between the Tonle Sap with the Mekong is reversed and the lake recedes to 3000 km^2 (annual depth variation: 9 m). A very diverse and abundant aquatic life is in equilibrium with this pulsing system, which also provides essential food and fibre resources to the riparian populations well adapted to this changing environment (e.g., floating villages). Other examples of such seasonal wetlands can also be observed on other continents as for the Niger, Nile, Senegal, Okavango in Africa and upper Parana in South America, although the Tonle Sap is the greatest pulsing lake.

Natural hydrologic regimes, as described by monthly long-term specific runoff ($l\ s^{-1}\ km^{-2}$), reflect climatic control factors. They are presented in **Table 2** on the basis of river discharges prior to damming according to the earliest records and from west to east. The natural Shatt el Arab regime, here reconstituted from Tigris plus Euphrates discharges, is controlled by winter rains and snowmelt from the

Table 2 Flow regimes of South Asian rivers before damming presented from west to east

River	Station	Country	Jan ($l\ s^{-1}\ km^{-2}$)	Feb ($l\ s^{-1}\ km^{-2}$)	Mar ($l\ s^{-1}\ km^{-2}$)	Apr ($l\ s^{-1}\ km^{-2}$)	May ($l\ s^{-1}\ km^{-2}$)	Jun ($l\ s^{-1}\ km^{-2}$)	Jul ($l\ s^{-1}\ km^{-2}$)	Aug ($l\ s^{-1}\ km^{-2}$)	Sep ($l\ s^{-1}\ km^{-2}$)	Oct ($l\ s^{-1}\ km^{-2}$)	Nov ($l\ s^{-1}\ km^{-2}$)	Dec ($l\ s^{-1}\ km^{-2}$)	Year ($l\ s^{-1}\ km^{-2}$)	A ($10^3\ km^2$)	Period
Euphrates	DS Baghdad	Iraq	3.5	4.8	6.95	10.35	11.1	6.1	2.65	1.5	1.12	1.18	1.6	1.6	4.42	398	1932/66
Narmada	Garudeshwar	India	1.75	1.3	0.90	0.66	0.41	2.2	24.0	52.5	62	16.8	4.9	2.3	14.1	89.3	1949/62
Damodar	Rhondia	India	1.8	1.8	1.4	1.1	1.75	10.7	39	59.8	49.8	22.0	4.8	2.5	16.5	19.9	1934/61
Mahamadi	Kaimundi	India	1.07	0.80	0.76	0.78	0.57	1.7	31.3	58.3	43.1	12.1	4.5	1.4	12.9	132.1	1947/61
Godavari	Dowlaishwaram	India	0.81	0.65	0.47	0.38	0.24	3.1	26.6	39.8	35.5	14.2	3.7	1.35	10.6	299	1901/60
Krishna	Vijayawada	India	0.44	0.27	0.17	0.15	0.77	2.38	21.9	26.1	17.3	10.9	4.0	1.0	6.9	251	1961/60
Pennar	Nellore	India	0.47	0.24	0.28	0.17	0.82	0.28	0.88	1.74	3.2	4.4	5.2	3.5	1.8	53.3	1934/47
Kelani	Nagalagam	Sri Lanka	51.8	48.0	51.3	63.3	123.7	129.4	88.7	77.2	76.7	127.5	109.3	70.5	84.9	2.085	1924/60
Chao Phraya	Nakohn S.	Thailand	2.1	1.4	1.0	0.86	1.61	4.3	5.94	11.8	19.6	26.5	15.3	4.75	7.9	111.4	1905/66
Mekong	Kratie	Cambodia	4.45	3.25	2.5	2.4	4.5	14.1	38.5	58.7	64.8	40.0	19.8	9.7	22.0	646	1933/53
Se Ban	Ban Komphun	Cambodia	9.9	6.8	5.2	5.6	14.0	25.3	58.5	83.4	74.7	55.2	25.7	17.2	31.7	48.2	1961/66
Kelantan	Guillemard	Malaysia	77.5	44.3	33.0	27.5	36.8	33.4	28.1	29.2	39.5	55.9	69.6	87.4	47.0	11.9	1949/64
Agusan	Poblacion	Philippines	139	173	105.9	60.7	53.7	57.7	54.0	60.6	62.1	66.2	65.0	125	81.9	7.39	1955/63

UNESCO (1969) Discharges of selected rivers of the world. Studies and reports in Hydrology, no. 5, p. 70. Paris: UNESCO.
UNESCO (1996) Global river discharge database (Riv Dis), Technical Documents in Hydrology, p. 41. Paris: UNESCO.
A: River basin in area.

Anatolian Plateau (maximum discharge in April and May, low flows from July to December). The upper Indus and its tributaries (Jhellum, Ravi, Sutlej, and Chenab, which altogether represent 40% of the drainage basin, are mainly fed by the Hindu Kush and Karakorum snow and ice melt. After their confluence, the Indus River crosses a very dry region and does not receive any other important tributary; it is a typical 'allogenic' river in which the mountain water tower allows the river to make its way through desert, as is also true for the Nile River. The upper Indus basin now has several major dams in Kashmir and Pakistan used for hydropower, irrigation, and water supply of downstream users.

The Narmada River, south of Mumbai, is characteristic of the West Deccan monsoon regime. From June to August the average specific discharge increases from 2.2 to 53.8 $l\ s^{-1}\ km^{-2}$ and 82% of the annual flow is discharged from July to September. East Deccan regimes, from the Damodar River in the North Bengal Gulf to the Pennar in the south (**Table 2**), are much dryer but also very seasonal. It must be noted that the average specific discharge in monsoon-fed rivers can be quite variable: 1.8 $l\ s^{-1}\ km^{-2}$ for the Pennar compared with 31.7 $l\ s^{-1}\ km^{-2}$ for the Se Ban (a lower Mekong tributary) and much more in some parts of the Meghna and in the lower Salween basins.

Ganges, Brahmaputra, Salween, and Mekong headwaters are fed by late spring snowmelt and by ice melt for their higher tributaries, after which their middle and lower basins receive the monsoon rains (July–September). The result is a mixed regime with a well-marked summer maximum ('monsoon period') and an extended low flow period from December to May ('pre-monsoon period'), as for the Mekong (**Table 2**). River basins near the equator, in western Sri Lanka (e.g., Kelani), Malaysia (e.g., Kelantan), the Philippines (e.g., Agusan), and Indonesia are characterized by minor seasonal discharge variations and by very high annual specific runoff (84.9, 47, and 81.9 $l\ s^{-1}\ km^{-2}$, respectively, **Table 2**), which is among the world's highest. Regional variability can be important: there is a 50-fold difference between the dry Pennar basin of Southeast Deccan and the Kelani basin, and a 250-fold difference between their minimum monthly runoff (0.17 and 48 $l\ s^{-1}\ km^{-2}$, respectively) over a distance of only 600 km.

Suspended Loads and Water Chemistry

South Asian rivers are much more turbid and erosive than world average because their headwaters are located in areas of very high relief, and because monsoon and snowmelt regimes are very seasonal, generating peak runoff with high velocities. Also, in some regions basin rock types are very erodible (e.g., volcanic ashes in Java).

The average concentrations of suspended particulate matter (SPM), calculated as the ratio of annual sediment loads over annual water volumes at the river mouth, are here presented for a selection of large and medium rivers, prior to damming (**Table 3**). All documented rivers have SPM concentrations higher than the world median of about 200 $mg\ l^{-1}$. SPM concentrations are much higher in headwaters, as in the upper Mekong, the upper Brahmaputra, and upper Indus, or in the Himalayan tributaries of the Ganges River, such as the Kosi River.

Small coastal rivers in tectonically active regions (e.g., Indonesia and the Philippines) also are very turbid and erosive (e.g., Cimanuk, Citanduy, and Citarum in Java, **Table 3**). As a result, the major part of the global sediment fluxes to oceans is generated by the South and East Asian rivers.

Dissolved solids in South and Southeast Asian rivers are generally close to or slightly higher than the world median, with a dominance of calcium and bicarbonate ions, as in most world rivers.

Because of the weathering of the volcanic Deccan Traps and the high evapotranspiration rate, Deccan rivers have a high ionic content (2000–5400 $\mu eq\ l^{-1}$ for the cation sum (TZ^+)), as in the Damodar, Godavari, Krishna, Narmada, and Tapti rivers (**Table 3**). Rivers draining the Himalayan and other alpine ranges (Indus, Ganges, Mekong, Brahmaputra, Irrawaddy) are less mineralized, with TZ^+ from 1180 (Mekong) to 2350 $\mu eq\ l^{-1}$ (Irrawaddy). The Salween is an exception, possibly linked to the presence of carbonate rocks.

In very wet regions, where the effect of evaporation on ionic content is quite limited, water chemistry is directly controlled by the presence or absence of easily weathered minerals. In the Mahakam River, which drains only crystalline rocks in Western Borneo, TZ^+ can be as low as 392 $\mu eq\ l^{-1}$, while it ranges from 800 to 1700 $\mu eq\ l^{-1}$ for rivers of Java. In central Thailand, the Chi and Mun rivers, which drain evaporitic rocks, are naturally very high in Na^+ and Cl^-.

Human Effects on South Asian Rivers

South Asian rivers, with the exception of the Irrawaddy, Salween, Brahmaputra, and Middle and Lower Mekong, are now extensively dammed. This greatly fragments the river courses, modifies the flow regimes, and decreases the downstream fluxes of particulate matter. In addition, reservoirs are generally associated with extensive irrigated areas where the water is lost by evaporation. As a result, the discharge

Table 3 Water chemistry and suspended sediments before damming for South Asian rivers

River	Station Country	L (km)	A (10^3 km^2)	Q (km^3 $year^{-1}$)	SPM ($mg\ l^{-1}$)	SiO_2 ($mg\ l^{-1}$)	Ca^{2+} ($mg\ l^{-1}$)	Mg^{2+} ($mg\ l^{-1}$)	Na^+ ($mg\ l^{-1}$)	K^+ ($mg\ l^{-1}$)	Cl^- ($mg\ l^{-1}$)	SO_4^{2-} ($mg\ l^{-1}$)	HCO_3^- ($mg\ l^{-1}$)	TZ^+ ($\mu eq\ l^{-1}$)
Brahmaputra	Bangladesh (a)	3000	580	510	1058	7.8	14	3.8	2.1	3.9	1.1	10	58	1200
Cauweri	India	800	88	20.9		19	28	24	60	5.5	50	32	177	6120
Chao Phraya	Thailand	1200	111.4	27.8	395	15.8	22	6.3			4.0	8.4	76	1615
Cimanuk	Java (Indonesia)		3.7	4.45	5600	19.2	16	6.2	9.0	2.0	4	19	81	1726
Citanduy	Java (Indonesia)		3.6	5.3	1800	13.3	6.7	3.2	4.3	1.1	3.3	4.7	36	812
Citarum	Java (Indonesia)		5.9	4.9		30	11	3.9	8	2	5	13	52	1267
Damodar	India		20	10	2800		22	8.2	15	3	9.7	45	62	2500
Ganges	India/Bangladesh	2525	1050	493	1055	11.7	23.2	6.5	9.6	2.6	5.0	8	119	2177
Godavari	India	1500	313	105	1619	21.1	30.2	2.4	8.1	2.2	14.1	10	105	2113
Indus	Pakistan	3180	916	90			26	6	9	2	7	26	90	2221
Irrawaddy	Myanmar	2300	410	486	535	10	10	6	30	2	18	5	120	2350
Krishna	India	1290	259	30	2130	5	27.5	13.5	42.5	3.0	37	63	125	4409
Mahakam	Borneo (Indonesia)		65.3	87		11.8	3.2	1.0	2.9	0.95	1.0	2.8	18.0	392
Mahanadi	India	858	141.6	66		9.0	10.4	9.5	10.2	1.5	30.9	15	60.9	1783
Mekong	Cambodia/Vietnam (b)	4850	795	467	321	8.9	14.2	3.2	3.6	2.0	5.3	3.8	57.9	1180
Musi	Sumatra (Indonesia)		57	80		24.5	3.2	1.1	4.6	1.1	4	5.2	15.5	470
Narmada	India	1300	99	40.7	3071		25	19.2	35		17	5	175	4350
Salween	Myanmar (c)	2820	325	211			46	16	10	1	20	1	212	4070
Shatt el Arab	Turkey/Syria/Iraq	2760	541	45.7		6.9	52	22	31	3	32	73	180	5830
Tapti	India	724	65	18			49.6	30.4	8.3	3	44	39.4	242	5415

Meybeck M and Ragu A (1996) River discharges to the oceans. An assessment of suspended solids, major ions and nutrients. Environnement Information and assessment Rpt., 250 p. Nairobi: UNEP (loadable from Gems Water:http://www.gemsstat.org/descstats.aspx)

A, basin area; *Q*, mean annual discharge; SPM, discharge weighted suspended particulate matter; TZ^+, sum of cations; *L*, river length.

(a) With headwaters in Tibet and India. (b) With headwaters in China and Laos. (c) With headwaters in China.

of many rivers in this region has decreased, as in the Shatt el Arab, the Indus, and the East Deccan rivers discharging to the Bay of Bengal. Impoundments are still created in India and in the upper Mekong Basin.

The Shatt el Arab headwaters (Tigris and Euphrates) also are extensively impounded in Turkey, the regional water tower of the Middle East (e.g., Attatürk Dam, one of the world's largest), and again in Syria and Iraq. The lower reaches of the Shatt el Arab and the Karun River in Iran form an extensive wetland that has been greatly affected by conflicts, oil drilling, and drainage. The restoration of this world-class wetland remains uncertain.

The upper Indus and its tributaries are impounded, and the Kotri dam in the lower reach also takes water for irrigation. As a result, the Indus water discharge to the Arabian Sea has been decreased by 80%, and its sediment load has been reduced even more. The upper Mekong, in China also, is now dammed in five places, as are some of its Thai tributaries. In the lower Mekong, a major tributary in Laos is being impounded. The water quality evolution influenced by irrigation is not much documented, but some increase of salt content is likely to occur in Pakistan and Iraq.

In South and Southeast Asia, the degradation of river quality is particularly caused by faecal contamination and by organic pollution, which depends on population density and on the degree of urban and industrial sewage collection and treatment. The pressure on water quality is therefore very high during the pre-monsoon period in densely populated regions where urban development is growing faster than sewage collection and treatment. Such a situation is found in most of the Indian subcontinent (Himalayan headwaters excepted), in Sumatra and Java, and in populated lowlands of Myanmar, Thailand, and Viet Nam, and is likely to occur in the lower Shatt El Arab.

Some of the largest and fastest growing cities are located in South Asia. Many of them are located inland on relatively small river courses with limited dilution or self-purification capacity, particularly during the pre-monsoon period, as is the case for Hyderabad (Pakistan) on the Indus; Lahore on the Ravi River; Hyderabad (India) on the Krishna River headwaters; Bangalore on Ponnayar River headwaters; New Delhi, Allahabad, and Varanasi on the Yamuna River, a Ganges tributary; Nagpur on the Godavari headwaters; Poona on the Bima headwaters, a branch of the Krishna River; Calcutta on the Hoogly, a branch of the Ganges Delta; Ahmedabad on the Sabarmati River; Dacca in the Bengal Delta; Baghdad on the Tigris; Ho Chi Minh City on the Saigon River; and Hanoi on the Song Koi (Red River).

Industrial activities are developing rapidly in the Indian subcontinent and in Southeast Asia. They may result in hotspots of pollution involving heavy metals already evidenced on the Yamuna River from Delhi to Varanasi, and on the Gomati River and on the Loni River at Lucknow, both Ganges tributaries, on Saigon River and delta branches of Song Koi River (Viet Nam).

Other possible threats are associated with conflicts (e.g., defoliant use in Viet Nam, leakage from destroyed industrial facilities in Iraq). However, the levels of toxic contaminants remain largely undocumented (see GEMS Water, a global program on river quality). The effect of rapid deforestation in Southeast Asia (e.g., Malaysia, Indonesia, Thailand) on water quality and aquatic species biodiversity should also be assessed.

Perspectives

River basins and water resources of South and Southeast Asia region are changing very rapidly in response to increasing water demand. From Turkey to Pakistan, shared basins, already sensitive to climate variations, are exposed to regional water management conflicts between water-rich upstream countries and water-poor downstream users. In this region, most of the river water is already being used for irrigation and water supply for large cities. The upper river courses are heavily fragmented by large dams and seasonal drought can now be observed in the lower courses, such as the lower Indus.

Although human influences on river quality are not yet adequately documented with regard to the water use and demand, the fast growth of urbanization, intensive agriculture, and industrialization could cause important problems in the future, particularly in densely populated floodplains (Ganges, Meghna, Irrawaddy, Chao Phraya, Mekong, Song Koi), and where water resources have already been much used (Shatt El Arab, Indus).

Glossary

Allogenic river – Any river the main stem of which is sustained almost entirely by waters derived from the uppermost regions of the drainage basin, because the upper part of the drainage basin is well watered, whereas the lower part of the basin is arid.

Water tower – In a hydrologic context, a water tower is a large source of water that feeds the lower reaches of a river, often because the upper reaches are mountainous and receive abundant precipitation.

Suspended particulate matter – Suspended particulate matter (SPM), which is also referred to as total suspended solids (TSS), consists of all material carried by water that can be separated from the water by use of a filter. SPM consists of both inorganic and organic particles.

TZ^+ – Sum of cations dissolved in water, expressed as microequivalents per liter ($\mu eq\ l^{-1}$). TZ^+ is an index of total salinity or total dissolved solids.

See also: Africa; Asia – Eastern Asia; Flood Plains; South America.

Further Reading

Ansari AA and Singh B (2000) Importance of geomorphology and sediment processes for the metal dispersion in sediments and soils of the Ganga Plain. *Chemical Geology* 162: 245–266.

Carbonnel JP and Guiscafré (1965) *Grand Lac du Cambodge, Sedimentologie et Hydrologie, 1962–1963,* 401pp. Paris: Rapport de Mission , Ministère des Affaires Etrangères.

Jennerjahn T, *et al.* (2004) Biogeochemistry of a tropical river affected by human activities in its catchment: Brantas River estuary and coastal waters of Madura Strait, Java, Indonesia. *Est. Coastal Shelf Science* 60: 503–514.

Korzun VI (ed.) (1978) *World Water Balance and Water Resources of the World. Studies and Reports. Hydrology* vol. 25, 663 pp. Paris: UNESCO (+atlas).

Le Thi Phuong Q, Garnier J, Billen G, Thery S, and ChanVan M (2007) Hydrological regime and suspended load of the Red River (Viet Nam), observation and modelling. *Journal of Hydrology* 334: 199–214.

Meybeck M, Dürr HH, and Vörösmarty CJ (2006) Global coastal segmentation and its river catchment contributors: A new look at land-ocean linkage. *Global Biogeochemical Cycles* 20, GB IS 90, doi 10.1029/2005 GB 002540.

Milliman JD and Syvitski JPM (1992) Geomorphic/tectonic control of sediment discharge to the oceans, the importance of small mountains. *Journal of Geology* 95: 751–762.

Salomons W, Kremer HH, and Turner RK (2006) The catchment to coast continuum. In: Crossland CJ, *et al.* (eds.) *Coastal Fluxes in the Anthropocene*, pp. 145–200. Springer.

UNESCO (1969) *Discharges of selected rivers of the world. Studies and reports in Hydrology*, no. 5, 70pp. Paris: UNESCO.

UNESCO (1996) *Global river discharge database (Riv Dis), Technical Documents in Hydrology*, 41pp. Paris: UNESCO.

Relevant Website

http://www.gemsstat.org/descstats.aspx – Gems Water program UNEP, global river water quality data.

Asia – Northern Asia and Central Asia Endorheic Rivers

M Meybeck, Université Pierre et Marie Curie, Paris, France

Rivers of Northern and Central Asia have been studied and surveyed by Soviet Geographers, hydrologists, and hydrobiologists during previous decades, which resulted in many publications poorly known outside the former USSR. They are synthesized in the unique World Atlas of Physical Geography and other specific surveys which provide a unique data set at a period when human impacts were still limited. Both the natural state of rivers and their *evolution* have been recently summed up by Russian scientists on whose work most of the information concerning this chapter is based. In addition to these sources of information concerning the hydrological networks, river regimes, sediment transport, water chemistry, and human impacts, the general description of these regions is also derived from a combination of global databases and their GIS applications to river basins and major coastal catchments of the world. The Volga River, the largest European river discharging to the Caspian, is not considered here. Many endorheic basins in Afghanistan (Helmand), Iran, China (Tarim), and Mongolia (Kerulen) could not be addressed here for lack of information. Major Arctic and Pacific rivers are addressed first, then the Central Asia rivers.

Arctic and Pacific Rivers

Present River Network Organization and Its Past Evolution

Siberian rivers are commonly grouped according to their final end-point in regional seas (**Figure 1**)

For Arctic Seas, from west to east:
Kara Sea: Ob, Pur, Taz, Yenisei, Pyassina, Taymyra
Laptev Sea: Khatanga, Anabar, Olenek, Lena, Omoloy, Yana
East Siberia Sea: Indigirka, Alazeya, Kolyma
Chukchi Sea: Amguema

For Pacific Seas:
West Bering Sea and Anadyr Gulf: Anadyr, Velikaya, Kamchatka
Okhotsk Sea: Penzhina, Amur

The general characteristics of these regional seas' catchments are listed in **Table 1** and the individual features of each river in **Tables 2** and **3**.

Four of these rivers (Ob, Yenisei, Lena, and Amur) are among the world's giants in terms of drainage area (1.8–3 M km^2), river length (2800–4400 km), and water discharge (10 400–19500 m^3 s^{-1}) (see comparative rankings in Chapter 21: The Kolyma, Olenek Indigirka, Anadyr, and Khatanga range between 200 000 and 660 000 km^2).

During the Quaternary glaciations, Siberian river catchments have been partly covered by several Arctic ice caps, as in North America, located on the lower Ob, the Taymyr Peninsula, the Indigirka-Kolyma region. Central Asia ice caps were also located in the mountainous upper courses of the Ob, Yenisei, and Lena. As a result the Pyassina, Taymyra, and Khatanga catchments were completely covered by ice as were large portions of the Indigirka and Kolyma catchments. Owing to ice caps damming, extended lakes covering hundreds of thousand km^2 were formed in the lower Ob (Pur-Mensi Lake), resulting in the present wetlands covering more than 100 000 km^2, one of the world's biggest riverine wetlands. At lower sea level periods, the catchment of many rivers, such as the Olenek and Lena was much greater, extending on the present continental shelf.

Temperature and River Discharge Regimes

Throughout Siberia the river regimes are essentially dependant on snowmelt, and locally on alpine glacier melt (**Table 2**). The temperature of the Siberian catchments, the world's coldest, is therefore the first control factor of river flows. Siberian rivers are exposed to polar and cold climate conditions, as defined by Ko¨ The temperate climate is completely absent (**Table 1**). There is a double temperature gradient: (i) warmer temperature from southern headwaters (+18 °C in July in the upper Ob), to northern lower courses (+6 °C in the lower Ob), (ii) colder temperature in Eastern Siberia with January temperature extending from −24 °C at the Ob River mouth to −45 °C in Central Kolyma basin. As a consequence, Siberian rivers are frozen for at least half of the year. The snowmelt starts early May in the Southern headwaters of the Ob then progressively reaches the most Northern and/or Eastern parts of the catchments: mid-June for the Amguema, Pyassina, and Khatanga. These rivers start to freeze again as early as mid-September and frost reaches the 60° N latitude in mid-October. In most northern and eastern parts of river catchments the freezing period can reach 9 months.

The resulting annual hydrographs of Siberian rivers are characterized by a short period (a few weeks) of very high flows during which 50–70% of the annual

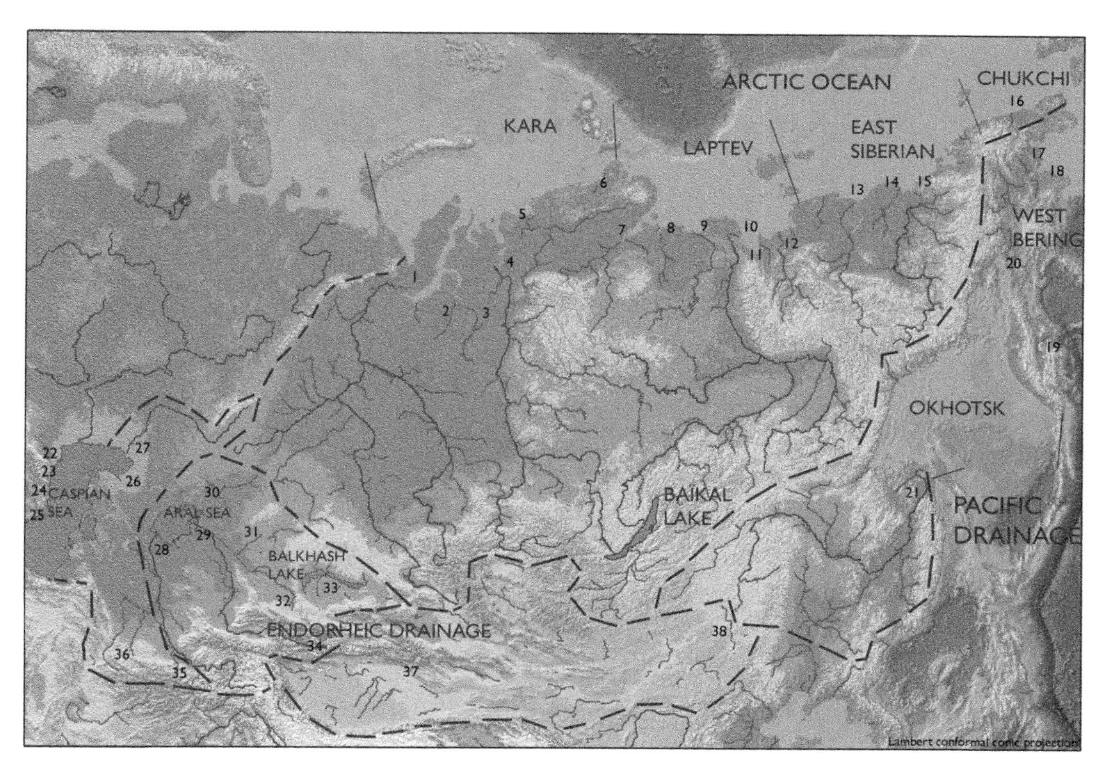

Figure 1 Northern and Central Asia rivers (river mouth location) and coastal catchments. Legend: 1 Ob, 2 Pur, 3 Taz, 4 Yenisei, 5 Pyasina, 6 Taymyra, 7 Khatanga, 8 Anabar, 9 Olenek, 10 Lena, 11 Omoloy, 12 Yana, 13 Indigirka, 14 Alazeya, 15 Kolyma, 16 Amguema, 17 Anadyr, 18 Velikaya, 19 Kamchatka, 20 Penzhina, 21 Amur, 22 Terek, 23 Sulak, 24 Samyr, 25 Kura, 26 Emba, 27 Ural, 28 Amu Darya, 29 Syr Darya, 30 Turgay, 31 Sary-Su, 32 Chu, 33 Ily, 34 Narym, 35 Murgab, 36 Tedzhen; 37 Tarim, 38 Kerulen.

flow is discharged. Peak flows occur from early May to mid-June depending on stations. The monthly maximum follows the monthly minimum discharge in most rivers (**Table 2**). In the Upper Ob and Upper Yenisei catchments summer rains (30–40% of annual precipitations) may generate a secondary discharge maximum. In the Kamchatka peninsula the high flow period is much more extended, from April to September (see Kamchatka, **Table 2**). For East Siberian Rivers (Yana, Indigirka, Kolyma, Amur) summer rains 54–70% of annual precipitations) are an important control of river flow. In North East Siberia the extreme negative temperatures result in complete freezing of rivers and cessation of base-flow as for Amguena and Yana Rivers (**Table 2**), which are explained by very limited groundwater inputs owing to permafrost extent.

In the large Siberian rivers, runoff is between 130 and 250 mm year^{-1}, i.e., half of the world's average (340 mm year^{-1} for exorheic regions connected to oceans). This corresponds to annual specific discharges between 4 and 10 l s^{-1} km^{-2}, for the Kara, Laptev, and East Siberia Sea catchments. The amount of precipitations, which is the major control of the runoff since evaporation is limited in this region, increases from South West to North East. As a result parts of the Ob sub-catchments from the Ubagal to the Childerty, are exposed to dry climate conditions and even to desertic conditions, defined here as annual runoff inferior to 3 mm year^{-1}. The West–East precipitation gradient results in maximum runoff (q >300 mm year^{-1}) for rivers draining to the Chukchi Sea (Amguema), Anadyr Gulf (Anadyr), and in Kamchatka where rivers are essentially rain-fed, with a high water period extending from May to September (600 mm year^{-1} for the Kamchatka River, more than 1000 mm year^{-1} in the tip of the peninsula).

River mouth average runoff may mask the catchment heterogeneity: in the Lena, it ranges from less than 50 mm year^{-1}, South of the Viluy, to more than 600 mm year^{-1} for the Middle Olekma sub-basin.

Relief, Sediment Transport and Suspended Solids

River catchment relief, a combination of mean altitude and relief rugosity defined at the medium resolution (30′ × 30′) (**Table 1**), is the major control factor of mechanical erosion and sediment transports: (i) in high relief portions (mean altitude > 2000 m) of catchments, erosion is maximum and sediment

Table 1 General characteristics of Northern and Central Asia regional sea catchments (Meybeck, Du¨o¨o¨

Regional Sea	*River*	*Area Mkm²*	*Runoff Mm year⁻¹*	*TSS mg l⁻¹*	*Pop. dens. p km⁻²*	*Relief (%)*			*Climate (%)*					*Lithology (%)*		
						Low	*Mid*	*High*	*Polar*	*Cold*	*Temp*	*Dry*	*Desert*	*Cryst.*	*Carb.*	*Other*
Arctic catchments																
Kara	Ob, Pur, Taz, Yenisei, Piasina, Taymyra	6.65	237	26	7	67.9	29.6	2.5	14.5	77.1	0	5.4	3	27.3	7.6	65
Laptev	Khatanga, Anabar, Olenek, Lena, Omoloy, Yana	3.61	176	40	1.5	47.4	52.6	0	26.5	73.5	0	0	0	23.7	4.8	71.5
East Siberia	Indigirka, Alezeya, Kolyma	1.32	140	170	1.3	31.6	68.4	0	59.6	40.4	0	0	0	16.8	3.8	79.2
Chukchi	Amguema	0.1	258	130	1.0	20	80	0	100	0	0	0	0	55	3	42
Pacific catchments																
W. Bering	Anadyr, Velikaya, Kamchatka	0.58	329	150	1.5	15	85	0	70	30	0	0	0	52.2	1.7	46
Okhostk	Amur, Penzhina	2.47	234	175	48	32.2	67.8	0	12.5	87.5	0	0	0	50.6	16.6	32.7
Endorheic catchments																
Caspian (Asia)[a]	Ural, Emba, Samyr, Sulak, Terek	1.43	66	1230	28	62.8	30.4	6.8	1.3	28.8	0.1	21.6	38.2	13.2	39.2	47.5
Aral	Amu Darya, Syr Darya, Turgay, Sary-Su, Chu	1.94	64	1000	22	66.1	16.9	17.0	9.6	26.5	3.3	23.9	36.7	18.7	17.4	63.8

[a]Caspian drainage without Volga catchment, and with Caucasus tributaries.

TSS: average total suspended solids for the catchment; cryst.: crystalline rocks; carb.: carbonate.

Source – Meybeck M, Du¨o¨o¨Global Biogeochemical Cycles 20, GB IS 90, doi 10.1029/2005 GB 002540.

Table 2 Natural regimes of North and Central Asia rivers: Monthly specific discharges at mouth (l s^{-1} km^{-2}) (from Unesco, 1969)

River	*Area Mkm²*	*Regional sea*	*Months*												*Year*
			I	*II*	*III*	*IV*	*V*	*VI*	*VII*	*VIII*	*IX*	*X*	*XI*	*XII*	
Amur	1.73	Okhotsk	1.12	0.67	0.50	1.68	8.32	9.77	9.65	11.62	12.54	10.35	3.97	1.49	5.95
Penzhina	0.0706	Okhotsk	0.42	0.35	0.30	0.32	9.83	66.29	16.15	14.05	8.13	3.94	1.10	0.67	10.13
Kamchatka	0.0456	Pacific	8.60	8.20	8.25	9.41	17.68	34.21	37.06	22.37	16.93	14.52	10.04	8.73	16.34
Anadyr	0.0473	Bering	0.32	0.24	0.23	0.22	2.30	68.92	21.99	14.88	7.10	2.79	1.00	0.53	10.04
Amguema	0.0267	Chuksi	0.01	0.00	0.00	0.00	0.58	54.68	33.63	22.81	10.79	2.49	0.73	0.13	10.49
Kolyma	0.361	East Siberia	0.31	0.21	0.17	0.15	5.84	28.53	14.63	11.39	9.11	2.63	0.83	0.55	6.20
Indigirka	0.305	East Siberia	0.13	0.07	0.04	0.03	0.95	18.30	18.62	13.41	6.98	1.57	0.43	0.27	5.08
Yana	0.216	Laptev	0.01	0.01	0.01	0.01	2.06	17.59	14.07	10.65	5.32	0.79	0.18	0.05	4.23
Lena	2.49	Laptev	0.76	0.55	0.40	0.33	1.31	21.52	11.40	7.87	7.11	4.20	0.95	0.79	4.75
Olenek	0.181	Laptev	0.04	0.02	0.01	0.01	0.89	30.99	9.61	5.19	4.99	1.38	0.34	0.10	4.46
Yenisei	2.44	Kara	1.98	1.85	1.72	1.63	12.38	31.15	11.43	7.75	7.34	6.02	2.51	2.00	7.30
Ob	2.95	Kara	1.48	1.22	1.05	1.09	4.92	10.88	9.86	7.46	4.61	3.49	2.06	1.68	4.14
Ural	0.19	Caspian	0.31	0.27	0.32	4.97	8.11	2.36	1.11	0.74	0.58	0.53	0.48	0.32	1.67
Amu-Darya	0.45	Aral	1.37	1.33	1.21	1.94	4.00	5.76	7.24	6.22	3.78	2.33	1.85	1.56	3.22
Syr-Darya	0.219	Aral	2.26	2.55	2.89	3.68	4.75	5.43	4.33	2.45	1.74	2.04	2.51	2.43	3.09
Kara-Turgay	0.0148		0.01	0.01	0.66	5.41	1.08	0.27	0.12	0.07	0.05	0.05	0.04	0.03	0.64

Source – Unesco (1969) Discharge of selected rivers of the world. *Studies and Reports in Hydrology*, No. 5, Paris, Unesco, p. 5.

Table 3 River basin characteristics and river chemistry for Northern Asia rivers (Meybeck and Ragu, 1996; Gordeev, 1998)

River	*L* *km*	*A* 10^3 km^2	*Qact* km^3 $year^{-1}$	*q* *mm* $year^{-1}$	*TSS* $mg\ l^{-1}$	*TDS* $mg\ l^{-1}$	SiO_2 $mg\ l^{-1}$	$N\text{-}NO_3^-$ $mg\ l^{-1}$	$P\text{-}PO_4^{3-}$ $mg\ l^{-1}$	*DOC* $mg\ l^{-1}$	*DIC* $mg\ l^{-1}$	Ca^{2+} $mg\ l^{-1}$	Mg^{2+} $mg\ l^{-1}$	Na^+ $mg\ l^{-1}$	K^+ $mg\ l^{-1}$	Cl^- $mg\ l^{-1}$	SO_4^{2-} $mg\ l^{-1}$	HCO_3^- $mg\ l^{-1}$	TZ^+ $\mu eq\ l^{-1}$
Ob	3650	2990	404	135	38	133	2.85	0.06	0.065	9.1	15.3	18.6	5.1	5.3	1.0	6.5	8.5	78	1604
Yenissei	3490	2580	630	244	10	110	3.0	0.02	0.008	7.4	11.3	16.5	3.7	5.2	1.0	9.5	10	57.3	1380
Khatanga	1636	364	85.3	234	20	110	3.2	0.03	0.006		9.4	12.5	3.6	9.7	1.0	12.5	5.6	47.9	1368
Anabar	939	78.8	13.2	168	24	52.8	2.6	0.03	0.003		6.1	9.8	2.3	0.2	1.0	1.73	4.0	31.2	713
Olenek	2270	219	34.3	156	31	113	2.7	0.030	0.003		14.3	20.1	4.2	3.2	1.0	4.8	4.8	72.6	1513
Lena	4 400	2490	532	213	34	112	4.2	0.030	0.004	6.6	10.4	17.1	5.1	4.5	0.7	12.0	13.6	53.1	1487
Omoloy		39	7.0	179	18														
Yana	872	238	30.7	129	103	49.7	2.2	0.01	0.001		4.1	6.1	1.5	2.8	1.0	2.3	9.0	20.7	572
Indigirka	1726	360	53.6	149	210	62.1	2.8	0.024	0.006		5.6	11.5	2.3	0.7	1.0	1.8	13.6	28.4	819
Alazeya		68	8.8	129	80	22.1													
Kolyma	2130	647	128	198	120	73.5	4.0	0.04	0.009		6.8	10.8	2.4	2.8	1.8	2.8	14.1	34.8	904
Amguema		29.6	9.2	311	6.0	18.4	5.9	0.025	0.012		1.5	3.3	0.6	1.2	1.0	1.1	3.3	7.6	292
Anadyr	1150	191	64.1	335	59	33.1						4.2	1.6	2.2	1.0	2.6	4.8	16.7	463
Kamchatka	704	55.9	33.1	592	90	102	12.6	0.1	0.075		9.5	9.4	4.7	7.2	1.0	4.7	14.1	48.6	1195
Penzhina	710	73.5	22.6	307	41	30.9	5.41	0.03	0.021		3.1	4.3	1.3	0.3	1.0	1.3	7.0	15.7	360
Amur	2820	1855	344	185	71	55	2.15	0.02	0.021		5.7	8.9	2.3	2.9	1.0	2.3	6.2	29.1	785

Qact: present average annual discharge; TZ^+: sum of cations.

Source – Gordeev VV, and Tsirkunov VV (1998) River fluxes of dissolved and suspended substances. In: Kimstach V, Meybeck M, Baroudy E. (eds.) *A Water Quality Assessment of the Former Soviet Union. E&FN Spon* 311–350.

Meybeck M and Ragu A (1996) *River Discharges to the Oceans. An Assessment of Suspended Solids, Major Ions and Nutrients*. Environment Information and Assessment Report, 250 pp. Nairobi: UNEP.

storage is very limited, (ii) in mid-relief portions (200–2000 m), erosion is important but sediment storage may occur, (iii) in low relief portions (< 200 m) erosion is limited, and sediment storage in flood plains is maximum.

The general relief of Siberia also presents a West–East gradient with a large prevalence of low-relief features (68%) in Kara Sea catchments to only 20% for the Chukchi Sea rivers, while the mid-relief features increase from 30 to 80%. Mid-relief is also dominant in the Pacific catchments, including the Amur (**Table 1**). High relief is only found in the Upper Ob and Yenisei catchments, in the Altaï Range and Tannu Mountains.

In Siberian rivers, the average total suspended solids (TSS), either obtained from direct surveys and discharge-weighted at gauging stations (**Tables 3** and **4**) or resulting from global models (**Table 1**), is low to medium. Kara and Laptev tributaries have TSS levels between 10 and 40 $mg\,l^{-1}$ while rivers characterized by higher relief (East Siberian Sea, Chukchi Sea and Pacific Ocean tributaries) have TSS levels between 50 and 250 $mg\,l^{-1}$. Higher TSS are likely to be found in the glacier-fed Ob headwaters.

Catchment Lithology and River Chemistry

In catchments with positive hydrological balance (precipitation > evaporation), the river chemistry mainly depends on their lithology (rock types): (i) crystalline rocks, either plutonic or volcanic, are moderately sensitive to chemical weathering, (ii) carbonate rocks (limestone, marl, dolomite) are easily weathered (5–10 times more than crystalline rocks), (iii) noncarbonated sedimentary rocks (most sandstones, shales) may be less weathered than crystalline rocks, (iv) maximum weathering corresponds to evaporitic rocks (gypsum, rock salt), however rare in most sedimentary basins. In dry regions where evaporation exceeds precipitation, ions are gradually concentrated and may reach saturation levels, as for calcite, thus changing the original ionic order. In most Siberian rivers, the lithology control is dominant, but the south-west part of the Ob catchment is under the same climate control as most of Central Asia endorheic basins (see later text).

The Ob, Yenisei and Lena catchments have similar lithologies and their chemistries are very close (**Table 3**) with a cation sum (TZ^+) between 1400 and 1600 $\mu eq\,l^{-1}$ and the following ionic order in eq/L: $Ca^{2+} \gg Mg^{2+} > Na^+ > K^+$ and $HCO_3^- \gg SO_4^{2-} > Cl^-$. Such chemistry is the most commonly found in world rivers. Other Siberian rivers are much less mineralized (TZ^+ from 300 to 800 $\mu eq\,l^{-1}$), and this can be due to local lithologies or to a larger occurrence of permafrost that greatly limits the chemical weathering. In the Kamchatka River, TZ^+ is much higher (1200 $\mu eq\,l^{-1}$), as well Mg^{2+}/Ca^{2+}, Na^+/Ca^{2+}, and SO_4^{2-}/HCO_3^- ratios in eq l^{-1}, typical of active volcanism regions.

Nutrients and Organic Carbon

Nutrients levels in Siberian rivers, either resulting from direct surveys or from global scale models, are among the lowest found in world rivers. Nitrate levels at river mouth (**Table 4**) are much lower than 0.1 mg N l^{-1} and phosphate levels are generally lower than 0.01 mg P l^{-1} –Ob excepted – both levels estimated to be representative of pristine rivers. About 50% of nitrogen is found as dissolved organic nitrogen (DON), ahead of nitrate plus ammonia (30%) and of particulate nitrogen (20%). Such proportions are characteristics of lowland pristine rivers and very different from those found in temperate rivers impacted by fertilizers (nitrate-dominated rivers) or by urban sewage and animal husbandry (ammonia-dominated rivers) (refer to 'See also' section).

Organic carbon is found mostly (80–90%) in dissolved form (DOC), except for a few rivers with TSS exceeding 100 $mg\,l^{-1}$. The DOC levels are somewhat higher than the world average due to the occurrence of wetlands and peat lands resulting from the deglaciation.

Dissolved silica levels are between 2.2 and 5 mg $SiO_2\,l^{-1}$, i.e., much lower than the world's average (9.4 $mg\,l^{-1}$). Chemical weathering is limited by permafrost occurrence; uptake and retention of silica in both terrestrial and aquatic vegetation is also possible. The Kamchatka River (12.6 mg $SiO_2\,l^{-1}$) is a noted exception, explained by its volcanic catchment (easily weathered minerals, hydrothermal inputs of silica): in the exported silica per km^2 is an order of magnitude higher than for other rivers.

Human Impacts

The population pressure is limited to a few big cities which are mostly located on river main courses where the dilution power is maximum, as Omsk, Novosibirsk on the Ob catchment, Krasnoiarsk and Irkoutsk on the Yenisei, Khabarovsk on the Amur and to mining or oil districts and their related industries as the Kusbass (on the Tom, an Ob tributary) and Norilsk, near the Pyasina River. The Ob and Yenisei rivers have a population density exceeding 10 p km^{-2}, the Amur catchment reaches 50 p km^{-2}: these levels do not reach the world average for exorheic rivers of around 60 p km^{-2} in 2000. In the rest of Siberia, from the Khatanga to the Penzhina Rivers, the population density corresponds to the world's minimum of 1 to

Table 4 River basin characteristics and water quality for the endorheic Asian rivers (Gordeev, 1998; Alekin and Brazhnikova, 1964)

River	*Location*	*L* km^2	*A* $10^3\ km^2$	*Q* $km^3\ year^{-1}$	*q* $mm\ year^{-1}$	*TSS* $mg\ l^{-1}$	*TDS* $mg\ l^{-1}$	Ca^{2+} $mg\ l^{-1}$	Mg^{2+} $mg\ l^{-1}$	Na^{+} $mg\ l^{-1}$	K^{+} $mg\ l^{-1}$	Cl^{-} $mg\ l^{-1}$	SO_4^{2-} $mg\ l^{-1}$	HCO_3^{-} $mg\ l^{-1}$	TZ^{+} $\mu eq\ l^{-1}$
Emba	Caspian basin		26	0.21	8		539	77.5	5.3	83.5	8.6	109.6	132.5	121.5	8150
Kura	Caspian basin		188	26.8	94	2000	418	51.4	16.5	39.9	4.3	26.7	68.7	210	5760
Samyr	Caspian basin		2.2	1.35	620	4000	180	26.8	7.1	11.6	1.6	4.7	34.1	94.4	2465
Sulak	Caspian basin		13.1	5.7	435	410	345	58.5	12.6	21.0	3.1	26.7	83.2	140	4940
Terek	Caspian basin		37.4	11.0	259	1940	361	56.7	13.7	200.2	3.1	17.6	87.5	162.5	4917
Ural	Caspian basin		236	3.43	42	390	404	37.0	15.6	45.5	5.9	60.9	62	177	5260
Amu Darya	Aral basin		227	49.4	218	4200	462	65.8	11.5	55.8	6.3	69.2	116.7	136.4	6815
Syr Darya	Aral basin		219	21.5	98	760	545	93.5	20.0	31.2	3.7	32.6	161.4	202.8	7765
Turgay	Aral basin		50.9	0.08	2		6180	103.5	68.2	2182	39.2	2833	695	261.5	106700
Chu	Kazakhstan	1190	27.1	1.06	39		408	49.1	12.5	48.8	4.7	22.4	95.5	175.2	5720
Ily	Kazakhstan		129	18.1	130		304	53.7	10.25	12.1	1.75	12.8	38.2	174.9	4097
Murgab	Turkmenistan		34.7	1.7	49		432	64.0	13.8	37.5	4.1	36.4	83.4	192.6	6067
Narym[a]	Kirgistan		58.4	12.5	213		294	50.9	9.4	16.2	2.35	17.1	55.3	143.1	4080
Tedzen	Turkmenistan	1124	70.6	1.0	14		868	79.6	41.1	131.3	6.7	144.1	222.7	242.5	13240

[a]Upper Syr Darya.

Source – Alekin OA and Brazhnikova LV (1964) *Runoff of Dissolved Substances from the USSR Territory* (in Russian), 228 pp. Moscow: Nauka.

Gordeev VV and Tsirkunov VV (1998) River fluxes of dissolved and suspended substances. In: Kimstach V, Meybeck M, and Baroudy E (eds.) *A Water Quality Assessment of the Former Soviet Union*, pp. 311–350. London: E&FN Spon.

$1.5\,\text{p}\,\text{km}^{-2}$. The human impact on most Siberian rivers is therefore still limited, even downstream of large cities. In the Chinese tributaries of the Amur River, higher urban and industrial pressures may resut in water quality deterioration (refer to 'See also' section).

The greatest threats come from mining and industrial settlements (kombinats) particularly when they are located on small catchments (Tsikurnov, 1998). This is the case of Norilsk, one of the world's greatest mines, or for the open-pit Myr diamond mine which discharges $0.5\,\text{Mt}\,\text{year}^{-1}$ of salts into the Viluy, a major tributary of the Lena. Oil and gas extraction can also have important impacts even on large rivers: in the Ob River at mouth, chloride has increased from $2\,\text{mg}\,\text{l}^{-1}$, its natural background level, to a maximum of $18\,\text{mg}\,\text{l}^{-1}$ in the 1960s. This 9-fold increase is not a major concern but it is certainly an indicator of major industrial inputs: it corresponds to an excess load of around $10\,\text{Mt NaCl year}^{-1}$. The associated impacts of such activities on heavy metals, hydrocarbon products, and other organic micropollutants is therefore likely although not much documented in Siberian rivers.

Large dams and some of the world's largest reservoirs have been established on the Ob – 13 dams totalling a volume of $75.2\,\text{km}^3$ – and on the Yenisei – 8 dams totalling $474\,\text{km}^3$ – their lower courses are therefore much regulated. In these regions, damming does not generate major water chemistry issues but it greatly modifies the aquatic habitat and the longitudinal connectivity of major rivers, critical for the migration of fish species. In Siberia, as in other parts of the Arctic Sea drainage (e.g., James Bay in Quebec, Sweden) damming can be regarded as the number one human impact.

Endorheic Rivers of Central Asia

As for Siberian rivers, the endorheic catchments (Caspian Sea, Aral Sea, Lake Balkash) and the Kazakhstan and Turkmenistan rivers have been very much studied by Soviet scientists. They can also be described through global river flux models and databases similar to the ones used for exorheic catchments (**Table 1**). All figures for the Caspian Sea drainage exclude the Volga catchment located in Europe.

Past and Present River Network Organization

From the Caspian Sea catchment to Mongolia, Central Asia is a succession of endorheic catchments. Many of them have potential interconnections that may develop during wetter climate periods, i.e., when the positive water balance generates a surficial overflow to the next downstream system. Their past connections are a major control of the present spatial distribution of many aquatic species in Central Asia. Moreover endorheic rivers fed by mountainous water towers are the only water resources in many of these regions and therefore have been used since Antiquity for irrigation (Amu Darya, Syr Darya, Tarim). This water use has been exponentially increased in the last 50 years, resulting in severe decrease of river flows. Therefore the catchment areas, river flows and river length (**Tables 1** and **4**) are only indicative.

Examples of interbasin connections include the following:

- Kerulen River (Mongolia) with the Amur River catchment, mapped in XIX atlases.
- Ili and Balkash Lake catchments.
- Chu, Sary Su, and Turgay catchments with Syr Darya catchment.
- Aral Sea with the Caspian Sea through the Uzboy Channel, still active in the Middle Age.
- Caspian Sea with the Black Sea through the Manych depression, active at the last glacial cycle.
- The Tarim River (in Sinkiang, China) was also draining tributaries originating from the Northern slope of the Tibet Plateau in the early Middle Age, corresponding to a catchment of $1\,\text{Mkm}^2$ eventually evaporated in the Lop Nor salt lake.

Hydrology and Climate Conditions

Climatic conditions over Central Asia are very different from those encountered in Northern Asia: the dry and desertic climates dominate (60%) in both Aral and Caspian catchments, Volga being set apart (**Table 1**). Temperate climate is not found in these regions located in the heart of continents. The desertic areas with less than $3\,\text{mm}\,\text{year}^{-1}$ of average runoff (considered here as the limit of arheism) cover more than a third of these catchments often covered by aeolian deposits. Most of the rivers are allogenic, i.e., they are fed only by a small portion of their drainage basin, the water towers being located in the highest and wetter mountain ranges where most of the water is stored as snow and ice: Pamir for Amu Darya, Tien Shan for Syr Darya, Pamir and Karakorum for Tarim, Afghanistan Plateau for Murgab and Tedzhen. 70% of the Amu Darya flow is during snow and ice melt (see **Table 2**). The Narym River, i.e., the Upper Syr Darya, has a runoff of $213\,\text{mm}\,\text{year}^{-1}$ and is well representative of these types of water towers (**Table 4**). The North Caucasus tributaries to the Caspian (Terek, Sulak, Samyr) are exposed to humid conditions with average runoff between 250 and $620\,\text{mm}\,\text{year}^{-1}$ while the Kura catchment ($178\,000\,\text{km}^2$) located between the Lesser and Greater Caucasus is much dryer ($q = 94\,\text{mm}\,\text{year}^{-1}$). In contrast the NW Caspian rivers, Ural ($q = 42\,\text{mm}\,\text{year}^{-1}$) and Emba

($q = 8$ mm year^{-1}), are lacking water towers and are much dryer; their maximum discharge is observed during the snow melt in April and May.

The Chu, Sary-Su and Turgay catchments, also located in Northern Kazakhstan (see Kara-Turgay, **Table 2**), do not benefit from extended water towers and are therefore very dry ($q = 2$ mm year^{-1}). The Ily River which originate from the Issyk Kul Lake, one of the greatest tectonic lakes in the world, has a higher runoff.

Relief, Sediment Supply and Suspended Solids

The two large Central Asia catchments consist of about two-thirds low-relief areas, 30% (Caspian) and 17% (Aral) of mid-relief areas and 7% (Caspian) and (17%) (Aral) of high-relief. Where high mountains are dominant, as in the Caucasus rivers, TSS levels are very high, from 2000 to 4000 mg l^{-1}, excepted for the Sulak (410 mg l^{-1}), and similar levels were found in natural conditions for the Upper Amu Darya (4200 mg l^{-1}) and Syr Darya (760 mg l^{-1}) (**Table 4**). Such levels are one or two orders of magnitude higher than those found in the Northern Asia rivers.

Lithology and Water Chemistry

The river chemistry of mountain rivers from Caucasus rivers (Terek, Samyr, Sulak, Kura) and from Tien Shan (Narym) is more mineralized than the world's average (cation sum TZ$^+$ from 2 500 to 5 800 µeq l^{-1}) with the following order (in eq l^{-1}) (**Table 4**):

$$Ca^{2+} > Na^+ \geq Mg^{2+} > K^+ \text{and } HCO_3^- \geq SO_4^{2-} > Cl^-$$

In such young mountain ranges folded sedimentary rocks are abundant, including some gypsum deposits and hydrothermal inputs that may explain the higher sulphate levels.

In all other catchments, the climate control evaporation then crystallization of $CaCO_3$ and then $CaSO_4$ is likely as for the Amu Darya and Syr Darya naturally exposed to evaporation. They are 50% more mineralized than mountain rivers (Narym): TZ$^+$ ranges from 5 700 (Chu) to 13 200 µeq l^{-1} (Tedzhen), which is among the highest level of TZ$^+$ found in world rivers. The ionic order is also different due to the gradual precipitation of carbonate minerals:

$$Na^+ = Ca^{2+} > Mg^{2+} > K^+ \text{and } SO_4^{2-} > HCO_3^- \geq Cl^-$$
$$\text{or } Cl^- > SO_4^{2-} > HCO_3^-$$

The analysis of Turgay River waters, north of the Aral Sea, (**Table 3**) shows one of the highest river mineralization levels ever reported for a nonpolluted catchment with TZ$^+$ reaching 106 700 µeq l^{-1}, i.e., a diluted NaCl solution (TDS = 6.18 g l^{-1}). This mineralization is more than two orders of magnitude higher than those of the Arctic/Pacific drainage. In addition to the evaporation control or chemistry (runoff $q = 2$ mm year^{-1}) the occurrence of evaporitic deposits in this catchment cannot be ruled out.

Human Impacts and Salinization

Human impacts on endorheic rivers include all impacts found on exorheic rivers (urban sewage inputs, eutrophication, damming, industrial and mining wastewaters, and use of agrochemicals). However the most specific impact is salinization caused by irrigation returns. Such evolution has been widely described for the Aral catchment where irrigated cotton and rice fields have markedly been developed since the 1950s in Kazakhstan, Uzbekistan, and Turkmenistan. Irrigation and diversions, such as the huge Kara Kum canal diverting waters from the Amu Darya to Turkmenistan, have also resulted in the overuse of the river flow, which has now dropped to zero for the Lower Amu Darya and to less than 5% of its original value for the Lower Syr Darya.

In 1912, the Syr Darya total dissolved solids (TDS) varied from 200 to 400 mg l^{-1}. In the late 1980s they reached an average of 1300 mg l^{-1} with peaks exceeding 2000 mg l^{-1}, well above the maximum acceptable concentration set up for potable waters. In the Amu Darya, average sulphate increased from 100 mg l^{-1} in the 1950s to peaks at 650 mg l^{-1} in 1985. The general ionic increases in the Terek, Kura, Murgab, Amu Darya, Syr Darya, Chu, and Ily between the 1940s and the 1990s range from 50% to 300% for $Na^+ + K^+$, SO_4^{2-} and Cl^-, and from 23% to 150% for TDS.

Urban and industrial sewage releases from fast developing cities located on minor tributaries, for example Boukhara, Samarkand, Tashkent, Alma-Ata, or on the lower courses of the Amu Darya (Chardzhou) and Syr Darya (Kzyl Orda), where river flow has been much reduced, generate important water quality impacts amplified by the lack of dilution capacity of receiving waters. The fast development of pesticides use in the Amu-Darya basin also generates a marked contamination of surface and groundwaters. Past industrial or army test sites on Kazakhstan and on Aral Sea island can be regarded as potential hot spots of contamination.

See also: Asia – Eastern Asia; Climate and Rivers; Streams and Rivers of North America: Overview, Eastern and Central Basins.

Further Reading

Alekin OA and Brazhnikova LV (1964) *Runoff of Dissolved Substances from the USSR Territory* (in Russian). 228 pp. Moscow: Nauka.

Crossland CJ, Kremer HH, Lindeboom HJ, Marshall-Crossland JI, and Le Tissier MDA (eds.) (2005) *Coastal Fluxes in the Anthropocene*, pp. 231. Berlin: Springer.

Dynesius M and Nilson C (1994) Fragmentation and flow regulation of river systems in the Northern third of the world. *Science* 266: 753–762.

Fedorov YA, Kulmatov RA, and Rubinova FE (1998) The Amu Darya. In: Kimstach V, Meybeck M, and Baroudy E (eds.) *A Water Quality Assessment of the Former Soviet Union*, pp. 413–433. London: E&FN Spon.

Gerasimov I (ed.) (1964) *Physical Geography Atlas of the World* (in Russian, with a special section on the Soviet Union). Moscow: Nauka.

Gordeev VV, Andreeva E, Lisitzin AP, Kremer HH, Salomon W, and Marshall-Crossland JI (2005) *Russian Arctic Basins: LOICZ Global Change Assessment.* LOICZ Reports and Studies no. 29. Netherlands: Texel.

Gordeev VV and Tsirkunov VV (1998) River fluxes of dissolved and suspended substances. In: Kimstach V, Meybeck M, and Baroudy E (eds.) *A Water Quality Assessment of the Former Soviet Union*, pp. 311–350. London: E&FN Spon.

Kimstach V, Meybeck M, and Baroudy E (eds.) (1998) *A Water Quality Assessment of the Former Soviet Union*, pp. 650. London: E&FN Spon.

Létolle R and Mainguet M (1993) *Aral.* p. 385. Berlin: Springer.

Rudoy A (1998) Mountain iced-dammed lakes of Southern Siberia. In: Benito G, Baker VR, and Gregory JK (eds.) *Paleo Hydrology and Environmental Change*, pp. 215–234. Chichester, UK: Wiley.

Salomons W, Kremer H, and Turner RK (2006) The catchment to coast continuum. In: Crossland CJ, *et al.* (eds.) *Coastal Fluxes in Anthropocene*, pp. 145–200. Berlin: Springer.

Skiklomanov IA, Georgievsky VY, and Polkanov MP (1998) Natural water resources. In: Kimstach V, Meybeck M, and Baroudy E (eds.) *A Water Quality Assessment of the Former Soviet Union*, pp. 2–23. London: E&FN Spon.

Tsirkunov VV, Polkanov MP, and Drabkova VG (1998) Natural composition of surface water and groundwaters. In: Kimstach V, Meybeck M, and Baroudy E (eds.) *A Water Quality Assessment of the Former Soviet Union*, pp. 25–68. London: E&FN Spon.

Tsirkunov VV (1998) Salinization. In: Kimstach V, Meybeck M, and Baroudy E (eds.) *A Water Quality Assessment of the Former Soviet Union*, pp. 113–136. London: E&FN Spon.

Relevant Websites

http://www.gemsstat.org/descstats.aspx – Gems Water program UNEP, global river water quality data.

http://webworld.unesco.org/water/ihp/db/shiklomanov/index.shtml – Global river discharge data set.

Australia (and Papua, New Guinea)

M Meybeck, Université Pierre et Marie Curie, Paris, France

Hydrological Network, Past and Present

At the last glacial cycle when the sea level was more than 100 m lower, the Australian continent (7.7 Mkm^2) and New Guinea (0.88 Mkm^2) were linked through the Torres Strait and separated from Southeast Asia by the Wallace line, and are therefore considered here together. This ensemble is actually quite heterogeneous in many ways (e.g., hydrology and climate, orography, population distribution). It is therefore decomposed into nine hydrological regions that correspond to different drainage systems to the Pacific and Indian Oceans and to the internal regions of Australia where rivers are discharging to internal basins (endorheism).

The Australian and New Guinean rivers discharge either to the Pacific Ocean or to the Indian Ocean, conventionally separated North by the Torres Strait, between New Guinea and the continent, and South by the Bass Strait between Tasmania and the continent. In Australia, due to the position of the Great Dividing Range, the cordillera located at the very eastern edge of the continent, the Pacific drainage area is actually very limited (0.5 Mkm^2) compared with the Indian drainage (5.0 Mkm^2). A large central position of the continent (2.2 Mkm^2) is presently without any overflow to the Indian Ocean. These endorheic catchments are numerous; the biggest one is the Lake Eyre catchment (1.2 Mkm^2), separated from the Indian Ocean by the Lake Torrens system, which is also endorheic most of the time.

New Guinean catchments are discharging to four regional seas: (i) the Bismarck Sea (e.g., Mamberano River), (ii) the Salomon Sea (Marham River), both on the Pacific Side, (iii) the Gulf of Papua (Fly, Kikori, Purari Rivers), and (iv) the Arafura Sea located in the most northern part of the Indian Ocean (Digul River).

In Australia, the southeast part of the Arafura Sea is the Carpentaria Gulf, a very shallow sea that receives the Mitchell, Gilbert, Flinders, Leinhardt, and Roper rivers. The Pacific drainage is described here as the North Coral Sea catchment, combined with the Gulf of Papua rivers, which extends south to Fraser Island (25° S). The Queensland catchments (Burdekin and East Fitzroy rivers) correspond to a narrow strip of land that crosses different climate zones. The Southern Coral Sea catchments (e.g., Snowy River) from Fraser Island to Tasmania are very limited in size compared to the other rivers of the continent. The southeast part of Australia is drained by the Murray-Darling system, one of the greatest basin areas (1.06 Mkm^2) and one of the longest river course (3490 km) in the world. On the southwest side, the Great Australian Bay is presently not fed by any river from the Nullabor Plain: the Great Victoria Desert is fragmented into numerous salt lakes without any present active river system (e.g., Barlee, Lefroy, and Moore Lakes). The western part of the continent is also very dry and is characterized by only one significant basin, the Swan-Avon on which Perth–Freemantle is located. The northwest part of Australia, from the Cape North-West to the Amiralty Gulf, is bordered by the Gibson Desert, also with multiple salt lakes (Disappointment, Percival, Mackay). The only active river systems are the Ashburton and the Fortescue, south of the Great Sand Desert and the Western Fitzroy (not to be confused with the other Fitzroy River, located in Queensland) that drains the relatively more humid Kimberley District. Then in the northwest side of the continent, from the Joseph Bonaparte Gulf to the Arhem Land, few active rivers are found (Ord, Victoria, and Daly).

The present hydrology is different from the one that occurred during past quaternary periods during more humid and/or lower sea level conditions:

- The Gulf of Carpentaria area was actually a large catchment (1.2 Mkm^2), probably endorheic and mainly fed by Australian rivers: the New Guinean contribution was very limited. As a consequence, the hydrological balance was negative and evidence of a large salt lake, the Carpentaria Lake, has been found in shelf sediments.
- The Lake Frome system was probably connected to the Lake Eyre system, and Lake Eyre itself was much bigger. Complex cascading connections between Lakes Frome, Eyre and Torrens, and the Spencer Gulf (SE Indian Ocean) are likely.

As in other continents, the present day endorheism is actually masking many river systems, some of them among the world's greatest. The numerous salt lakes found in Australia are the remnants of this past hydrological activity.

Hydroclimates and Hydrological Regimes

Out of the seven major hydroclimates types based on Köppen classification, Australia and New Guinea catchments cover five classes from north to south: wet tropical climate with annual runoff ($q > 680$ mm year^{-1}) exceeding twice the world's average, humid tropical climate ($q < 680$ mm year^{-1}), dry, temperate, and desertic defined here by an annual runoff $q < 3$ mm year^{-1}, a conventional limit that separates active river system, wich can flow permanently, seasonally, or occasionally, (rheic) from inactive system (arheic), i.e., with less that one flood event every 10 years (**Table 2**).

- New Guinea rivers are essentially (>80%) exposed to wet tropical regimes because of the very high rainfall (e.g., 5900 mm year^{-1} at Kikori on the Gulf of Papua). The average runoff ($q = 1380$ mm year^{-1}) provided by global scale models (**Table 1**), is four times higher than world's average. It may even be underestimated when considering individual rivers: $q = 2844$ mm year^{-1} for the Sepik, 2164 mm year^{-1} for the Tauri, 2530 mm year^{-1} for the Purari, 2190 mm year^{-1} for the Fly, and 3046 mm year^{-1} for the Kikori. Such runoff level for medium-sized catchments (20 000–50 000 km^2) is near the world's record. New Guinea river discharge regime (Sepik, Tauri, Purari, **Table 2**) is equatorial and characterized by a relatively stable annual mean hydrograph without marked low flows (minimum specific discharge q_{min} from 40 to 60 l s^{-1} km^{-2}) and high flows in February to May (maximum monthly discharge q_{max} at 100 l s^{-1} km^{-2}).
- In Australia, the North Coral Sea catchment is very heterogeneous with a dominance of the temperate climate. The wet tropical climate is much reduced to a narrow strip of coastal catchments around Cairns (15–18° S) (e.g., the Johnstone River, 1600 km^2, $q = 1750$ mm year^{-1}). North and south of this area, the runoff is much reduced, with only 86 mm year^{-1} for the Burdekin, the greatest catchment (130 000 km^2) that mixes humid coastal conditions and much dryer upland climate. The Burdekin and Eastern Fitzroy rivers (**Table 2**) are in opposition with the Papuan rivers with very moderate high flows in January–February (q_{max} = 12 and 5.3 l s^{-1} km^{-2}, respectively) and well-marked low flows close to complete desiccation, from July to November (q_{min} 0.09 and 0.04 l s^{-1} km^{-2}). The South Coral Sea tributaries such as the Snowy are under temperate climate conditions, including some snow cover on top of the Australian Alps, and have an average runoff of 150 mm year^{-1}.
- The Murray-Darling River runoff is low ($q =$ 39 mm year^{-1} according to models, 7 mm year^{-1} after water withdrawal for irrigation, according to actual gauging). More than half of this catchment is actually arheic, the other half is under dry temperate climate and fed by headwaters in Australian Cordillera. The Murray-Darling is also regulated by multiple reservoirs and used for irrigation. The high water period from September to November is very limited ($q_{max} = 0.5$ l s^{-1} km^{-2}) and the low flows are severe ($q_{min} = 0.08$ L s^{-1} km^{-2}).
- From the Gulf of Spencer to the Amiralty Gulf, most of the area draining to the Indian Ocean tributaries is presently arheic (80–95%, **Table 1**) with an annual runoff from 6 to 20 mm year^{-1} (7 mm year^{-1} for the Swan-Avon). Organized and active river systems are met again in the Arafura Sea–Carpentaria Gulf catchments, thanks to the humid tropical climate (37%, **Table 1**). Rivers draining the Kimberley district and Arhem Land (Western Fitzroy, Ord, Victoria, and Daly, **Table 2**) have a much contrasted regime: the high flow periods between January and March ($q_{max} =$ 11–19 L s^{-1} km^{-2}), then the river flow drops sharply under 0.5 l s^{-1} km^{-2} to near desiccations ($q_{min} <$ 0.1 L s^{-1} km^{-2}, except for the Daly ($q_{min} <$ 0.36 L s^{-1} km^{-2}).

The lake Eyre endorheic basin, is one of the greatest in the world, compared with those of Central Asia as the Aral Sea. It corresponds to the central-eastern part of the continent (**Figure 1**). It receives occasional flows – once every 6 years on average for the Cooper Creek – from its eastern tributaries, the Diamantina (250 000 km^2 at its mouth into Lake Eyre) and Cooper Creek (306 000 km^2 at mouth), which are fed by their headwaters in the Great Dividing Range. These rivers can totally stop flowing for more than three consecutive years. During the wettest years (annual runoff of 8 mm year^{-1} for Diamantina and of 2.1 mm year^{-1} for the Cooper Creek), their occasional flows can reach a maximum discharge of 4000 m^3 s^{-1} during a few days in their middle courses. Western Lake Eyre tributaries of similar catchment area originate from Mounts MacDonnell and Musgrave and flow even less frequently ($q = 1$ mm year^{-1}). Lake Eyre is completely dry most of the time: from 1930 to 1985 it was full during 1938, 1955, 1963, 1968, 1973, 1974, 1975, 1976, and 1984. At full stage, it covers 9300 km^2 and reaches an average depth of 2 m; its bottom is 9 m below sea level.

Sediment Transport

In New Guinea, the Pacific and Indian Ocean catchments are separated by a continuous mountain range exceeding 4000 m in altitude (maximum 5029 m), which was partially ice-covered at the Late Glacial

Table 1 General features of regional seas catchments, as delineated; in **Figure 1**, for Australia and New Guinea

Sea catchment	*Major river*	*A* *(Mkm²)*	*q* *(mm y⁻¹)*	*Pop density* *(p km⁻²)*	*Relief (%)*			*Climate (%)*						*Lithology (%)*		
					Low	*Mid*	*High*	*Cold*	*Temper*	*Desert*	*Dry*	*Wet* < *680**	*Humid* > *680**	*Cryst*	*Carb*	*Other*
Bismark–Salomon–N. Arafura (Pac.)	Mamberano, Sepik, Digul	0.691	1378	53.5	29.5	65.1	5.3	–	13.2	–	–	7.2	79.6	21.9	12.6	65.5
Gulf of Papua–N. Coral (Pac.)	Fly, Burdekin, Fitzroy E	0.659	633	3.9	66.8	31.6	2.4	1.0	47.2	27.3	–	7.0	17.5	29.7	35	35.3
South Coral (Pac.)	Snowy	0.290	144	24.3	30.7	69.3	0	–	100	–	–	–	–	30.8	48.9	20.3
Southeast Australia (Indian)	Murray	1.2	38.7	5.4	85.8	14.2	0	–	46.0	54.0	–	–	–	16.1	11.7	72.2
Nullabor Coast (Indian)	None	1.05	6.4	0.5	99	1	0	–	19.5	80.2	0.3	–	–	48.6	27.4	24.0
South West Australia (Indian)	Swan-Avon	0.47	19.8	3.4	99.4	0.6	0	–	18.5	81.3	0.2	–	–	65.8	17.4	16.8
North West Australia (India)	Ashburton, Fitzroy W	0.92	12.2	0.2	97.3	2.7	0	–	–	94.0	0.7	5.0	0.3	32.4	48.2	19.4
Timor–S. Arafura	Daly, Roper, Ord, Victoria, Flinders, Mitchell	1.1	110	0.3	99.2	0.8	0	–	3.7	53.3	6.4	37	0.3	49.5	18.0	31.5
Lake Eyre Basin	Cooper, Diamantina	1.17	0	0.2	99.4	0.6	0	–	–	100	–	–	–	9.5	34.7	55.8

A: drainage area; *q*: annual runoff; cryst: crystalline rocks; carb: carbonate rocks; p km^{-2}, people per square km; *annual precipitation on catchments.

Meybeck M, Dürr HH, and Vörösmarty CJ (2006) Global coastal segmentation and its river catchment contributors: a new look at land-ocean linkage. *Global Biogeochemical Cycles* 20: GB IS 90, doi 10.1029/2005 GB 002540.

Table 2 Hydrological regimes of Papuan and Australian rivers

River	*Ocean catchment*	*Area*[a] *(km²)*	*Annual runoff (mm y^{-1})*	*Mean monthly discharge (m³/s)*											
				Jan	*Feb*	*Mar*	*Apr*	*May*	*June*	*Jul*	*Aug*	*Sept*	*Oct*	*Nov*	*Dec*
Sepik	Bismark Sea (Pac.)	40 920	2894	4008	4210	4599	4562	4271	3728	2929	3129	2834	3708	3442	3677
Tauri	G of Papua (Pac.)	2410	2164	146	170	184	184	207	212	100	79.5	87.3	229	170	217
Purari	G of Papua (Pac.)	28 700	2530	1609	2659	2358	2880	2303	2552	2369	2386	2040	2395	1811	2288
Burkedin	N Coral Sea (Pac.)	129 660	86	1303	1599	701	109	263	38.2	17.0	11.9	22.7	8.2	28.9	222.5
Fitzroy (East)	N Coral Sea (Pac.)	135 900	43.9	464	713	298	36.5	368	108	38.3	23.5	16.4	5.6	29.4	199
Murray	SE Australia (Ind.)	991 000	8.2	159	114	81.5	99	143	187	238	271	426	508	483	372
Fitzroy (West)	NW Australia (Ind.)	16 800	144	180	328	312	83	10.2	4.2	4.2	0.4	0.8	0.0	4.6	11.8
Ord	Timor Sea (Ind.)	19 600	90	153	268	144	56	10.3	1.9	2.2	1.1	1.8	1.7	21.4	29
Victoria	Timor Sea (Ind.)	44 900	72	231	458	493	32	3.5	1.3	1.3	0.3	0.1	0.0	5.3	29
Daly	Timor Sea (Ind.)	47 000	133	288	823	900	165	40	29.6	24.4	20.6	17.1	17.1	22.8	66.2

UNESCO (1996) *Global River Discharge Database (Riv Dis)*, vol VI, Oceania. Technical Documents in Hydrology, p. 41. Paris: UNESCO.

[a]Drainage area at gauging station.

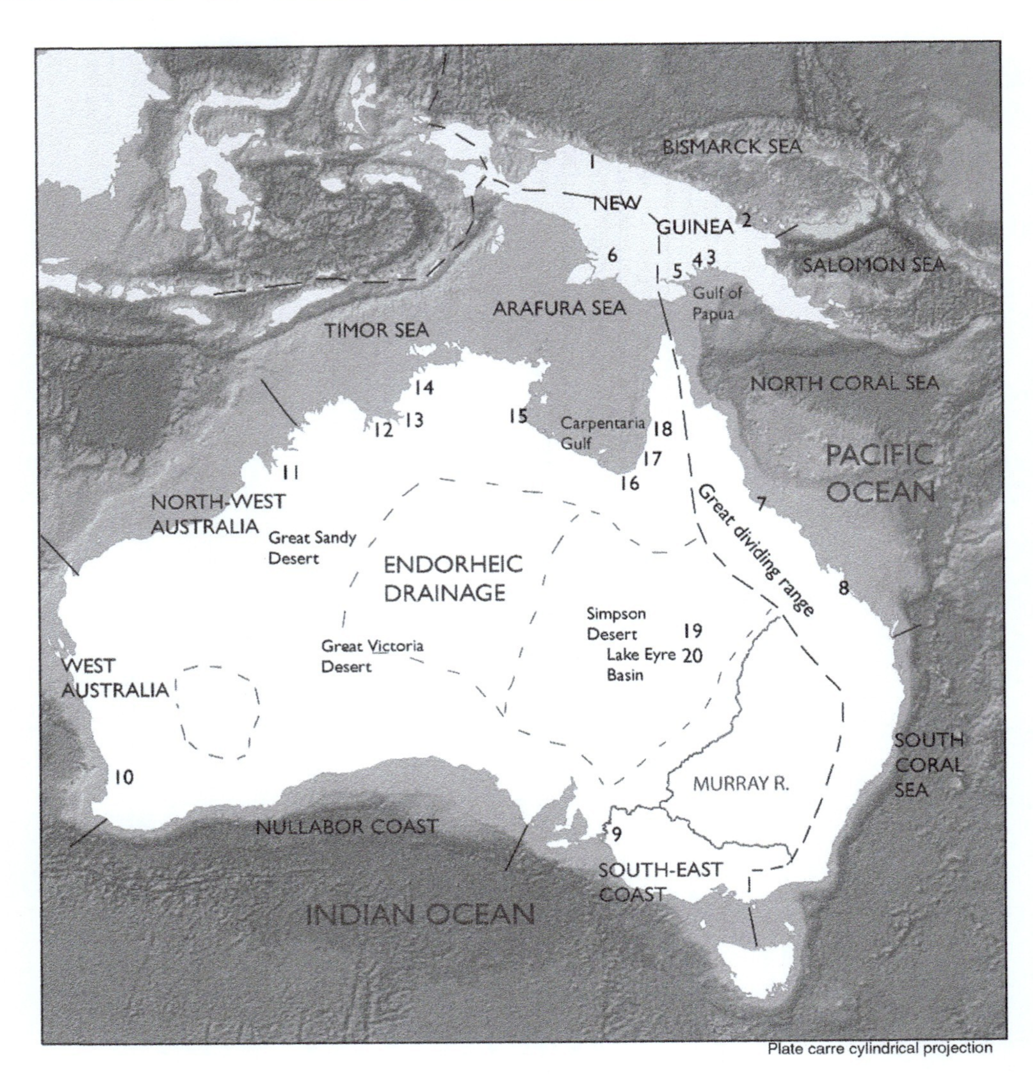

Figure 1 Rivers mouth and coastal catchment limits of New Guinea (1–6) and Australia (7–20). (1) Mamberano; (2) Sepik; (3) Purari; (4) Kikori; (5) Fly; (6) Digul (New Guinea); (7) Burdekin; (8) E.Fitzroy; (9) Murray; (10) Swan-Avon; (11) W.Fitzroy; (12) Ord; (13) Victoria; (14) Daly; (15) Roper; (16) Flinders; (17) Gilbert; (18) Mitchell; (19) Diamantina; (20) Cooper Creek (Australia).

Maximum, despite its position near the equator. The high-relief features (mean altitude > 2000 m) reach 5% of the whole island and the mid-relief features (altitude between 200 and 2000 m) are about 65% and low-relief patterns (30% altitude < 200 m) are mostly found in coastal plains of the Arafura Sea (Digul River) and of the Gulf of Papua (Fly River) (**Table 1**). Because of the steep slopes and highest river runoff, the sediment yields of New Guinean rivers are also close to the world maximum: 1800 t km^{-2} $year^{-1}$ for the Fly and 2600 t km^{-2} $year^{-1}$ for the Purari, with high levels of total suspended solids (TSS) exceeding 1000 mg l^{-1} most of the year.

By contrast, the Australian continent has no high relief; the mid-relief pattern is only found in the great Dividing Range (maximum altitude 2230 m between Melbourne and Sydney) and characterizes the Australian tributaries to the Coral Sea (about two-thirds of the catchments area). All other regions are dominated by lowlands: 86% for the Murray catchment, 99% for Lake Eyre catchment and for the rest of the continent. As a result, the sediment yields of Australian rivers are limited reaching 23.4 t km^{-2} $year^{-1}$ for the Burdekin. However, turbid pulses are likely to occur at the beginning of high flows and during occasional floods in endorheic catchments.

Ion Chemistry

New Guinean rivers have medium ionic concentration with cations sum ($TZ^+ = Ca^{2+} + Mg^{2+} + Na^+ + K^+$)

Table 3 River basin characteristics and water chemistry

River	L (km)	A (Mkm^2)	Q_{act} ($km^3\ y^{-1}$)	q ($mm\ y^{-1}$)	TSS ($mg\ l^{-1}$)	TDS ($mg\ l^{-1}$)	SiO_2 ($mg\ l^{-1}$)	$N\text{-}NO_3^-$ ($mg\ l^{-1}$)	$N\text{-}NH_4^+$ ($mg\ l^{-1}$)	$P\text{-}PO_4^{3-}$ ($mg\ l^{-1}$)	DIC ($mg\ l^{-1}$)	Ca^{2+} ($mg\ l^{-1}$)	Mg^{2+} ($mg\ l^{-1}$)	Na^+ ($mg\ l^{-1}$)	K^+ ($mg\ l^{-1}$)	Cl^- ($mg\ l^{-1}$)	SO_4^{2-} ($mg\ l^{-1}$)	HCO_3^- ($mg\ l^{-1}$)	TZ^+ ($\mu eq\ l^{-1}$)
(a) Swan-Avon	390	124	0.88	7															
(a) Mitchell	520	72	11.5	160		187	18				22.4	12.5	8.5	18.5	1.6	13.5		114	2169
(a) Burdekin	680	129	8.7	67	347	280	18.5				30.5	23	12.8	32.5	3.2	34	1.1	155	3697
(a) Fitzroy East	960	143	5.7	40		187	15.0				18.9	16	10	15	3	27	4.5	96	2351
(a) Murray	3490	1060	7.9	7		382	5.0	0.11	0.036	0.024	19.5	17	12	66	5.3	127	27	69	4840
(b) Diamentina[a]		115	1.42	12.3	440		32	0.18	0.10	0.21		6	4	14	5.7	1	12	58	
(b) Cooper[a]	1523	237	2.06	8.7			17	0.13	0.07	0.19		11	5	15	7.9	8	9	70	
(c) Fly	620	64.4	141	2190	815	116	9				15.4	21.3	1.76	2.33	0.43		2.67	78.3	1320
(c) Kikori		13.2	40	3046		177	8				24.6	37.3	4.0	1.3	0.2	0.3	1.1	125	2252
(c) Purari	630	30.6	84.1	2751	950	126	13.8	0.04	0.04	0.002	15.9	20.6	2.6	3.2	1.0	1.2	2.4	81	1407
(c) Sepik	700	78.7	120	1525	68	114	12.5				14.5	15.5	4.0	3.5	0.4		4.5	73.5	1265
(c) Mamberano		77.6	130	1675															

(a) For Australian exorheic drainage, (b) for Lake Eyre drainage, (c) for New Guinea drainage.

Meybeck M and Ragu A (1996) River discharges to the oceans. An assessment of suspended solids, major ions and nutrients. Environnement information and assessment Rpt, pp. 250, UNEP, Nairobi, (loadable from Gems Water http:// www.gemsstat.org/descstats.aspx).

Kotwicki V (1986) *Floods of Lake Eyre*, p. 99. Adelaide: Eng. Water Supply Department.

[a] 1978/1983 when flowing.

between 1250 and 2250 μeq l^{-1} (**Table 3**), with the following order (in eq l^{-1}):

$$Ca^{2+} \gg Mg^{2+} \geq Na^{+} + > K^{+} \text{ and } HCO^{3-} \gg SO_4^{2-} > Cl^{-}$$

Such analyses are representative of the chemical weathering of sedimentary rocks (78% in New Guinea catchments, including 12.6% of carbonate rocks). As the runoff is very high, the weathering rate is actually near to the world's maximum, particularly for dissolved silica, which ranges from 8 to 14 mg SiO_2 per liter, i.e., slightly higher than average.

Australian rivers draining to the Gulf of Carpentaria (Mitchell River) and to the North Coral Sea (Burdekin and E. Fitzroy) had in the mid-1970s (**Table 3**) medium-high TZ^{+}, between 2200 and 3700 μeq l^{-1} with a different ionic order:

$$Na^{+} \geq Ca^{2+} \geq Mg^{2+} \gg K^{+} \text{ and } HCO^{3-} > Cl^{-} > SO_4^{2-}$$

Such types of waters are not commonly found in world rivers and their relatively high Na^{+} and Cl^{-} levels may be a limit for irrigation use.

Similar ionic assemblages are also found in the Central Australian rivers, Diamantina and Cooper Creek, sampled during floods between 1978 and 1983 in their middle courses, i.e., with limited evaporation impact (**Table 3**): TZ^{+} are 1380 and 1810 μeq l^{-1}, respectively. Such quality is very different from the one observed in the 1950s in Central Asia endorheic rivers, which are 4–50 times more mineralized because of the evaporation/crystallization process and the occurrence of evaporitic deposits.

The analysis of the Murray river corresponds to a more recent average (1979–1987) with a higher TZ^{+} (4850 μeq l^{-1}) and a marked Na^{+} and Cl^{-} dominance:

$$Na^{+} > Mg^{2+} > Ca^{2+} \gg K^{+} \text{ and } Cl^{-} > HCO^{3-} > SO_4^{2-}$$

The salinization is enhanced by water use in this catchment as presented further below.

Human Impacts: Mining and Salinization

Human impacts are multiple, yet relatively limited or localized with regard to many other regions of the globe.

In New Guinea, the population density (5 people km^{-2}) is 10 times less than the global average; the urban impacts are mostly found on the coast. The on-going and recent deforestation is a major concern: it may result in local increase – by an order of magnitude – of the sediment supply, already very high. Mining districts, some of the world's greatest, also generate high sediment supply, as in the Fly River (Ok Tedi mine) or on the Arafura Sea catchment (Freeport mine), and to metal contamination.

In Australia, the population density (average 2 people km^{-2}) is near the world minimum. It ranges from 0.2 to 5 people km^{-2} in most regions, apart from the South Coral Sea coast between Melbourne and Sydney (25 people km^{-2}). Moreover, these two cities and others, like Adelaide and Perth, are located on the coast and do not contribute to river impacts. Canberra, the federal capital, located in the upper Murrumbidgee catchment, a tributary of Murray, is a noted exception. Mining activities in Australia are also important and their local impacts on rivers should be carefully surveyed for heavy metals (Cu, Pb, Zn, etc.) and for sediment inputs.

Yet the major water quality issue of the continent is the salinization of the Murray-Darling, which has been addressed in numerous official reports. The sedimentary cover of this river system is rich in evaporitic deposits. The related lenses of saline groundwaters and the salts stored in the capillary fringe zone, above the water table and in alluvial sediments, can be remobilized by the rising water table and the new distribution of hydrological pathways resulting from land clearing and irrigation. The river presently exports only half of the remobilized salt. The salt export now exceeds 6 t NaCl per square kilometre per year for some of the Upper Murrumbidgee tributaries, while the northern tributaries of the Darling branch, located in dryer catchments, export less than 1 t km^{-2} $year^{-1}$. The upstream/downstream increase of salinity of the Murray is 10-fold, exceeding 1000 mg l^{-1} in the lower course during dry years.

Nutrients increase and enhanced sediment supply have been reported in some tributaries of the Coral Sea, thus threatening the very fragile coral reef system with nutrient imbalance and sediment blanketing.

See also: Asia – Northern Asia and Central Asia Endorheic Rivers.

Further Reading

Brunskill GJ, Zagorski I, Pfitzner J, and Ellison J (2004) Sediment and trace element depositional history from the Ajkwa River estuarine mangroves of Irian Jaya (West Papua), Indonesia. *Continental Shelf Resarch* 24: 2535–2551.

Korzun VI (ed.) (1978) *World Water Balance and Water Resources of the World.* Studies and Reports. Hydrology 25, Paris: UNESCO (+atlas).

Kotwicki V (1986) *Floods of Lake Eyre*, p. 99. Adelaide: Eng. Water Supply Department.

Kotwicki V and Isdale P (1991) Hydrology of Lake Eyre, Australia. *Paleogeography, Paleoclimatology, Paleoecology* 84: 87–98.

MDBC (1997) Salt trends: historic trend in salt concentration and salt load in streamflow of the Murray-Darling. Murray-Darling Basin Commission, Canberra, Australia. Dryland Technical Report 1.

MDBMC (1999) The salinity audit of the Murray-Darling Basin: A 100-year perspective. Murray-Darling Basin. Ministerial Council, Canberra, Australia. p. 39.

Meybeck M, Dürr HH, and Vörösmarty CJ (2006) Global coastal segmentation and its river catchment contributors: A new look at land-ocean linkage. *Global Biogeochemical Cycles* 20: GB IS 90, doi 10.1029/2005 GB 002540.

Milliman JD (2001) River inputs. In Steele JH, Turekian KK, and Thorpe SA (eds.) *Encyclopedia of Ocean Sciences,* vol. 4, pp. 2419–2427. Academic Press.

Salomons W, Kremer H, and Turner RK (2006) The catchment to coast continuum. In Crossland CJ, *et al.* (eds.) *Coastal Fluxes in Anthropocene*, pp. 145–200. Springer.

Schulze RE (2005) River basin responses to global change and anthropogenic impacts. In Kabat P, *et al.* (eds.) *Vegetation, Water, Humans and the Climate*, pp. 338–374. Berlin: Springer.

Relevant Websites

http://www.gemsstat.org/descstats.aspx – Gems Water program UNEP, global river water quality data.

http://www.mdbc.gov.au/ – Murray Darling Basin Commission.

European Rivers*

K Tockner, Institute of Biology, Freie Universität Berlin, Germany
U Uehlinger, C T Robinson, R Siber, D Tonolla, and F D Peter, Swiss Federal Institute of Aquatic Science and Technology (Eawag), Duebendorf, Switzerland

Introduction

Rivers recognize no political boundaries. This is particularly true for Europe, which has more than 150 transboundary rivers. For example, the Danube is the 29th longest river globally, yet it flows through 18 countries and 10 ecoregions. Further, 8 of the 10 largest catchments in Europe are in the eastern plains of Russia and information on their present status is highly limited. Europe also has a long history in river training with most rivers being severely fragmented, channelized, and polluted. Recently, the European Union launched an ambitious program called the Water Framework Directive (WFD) that requires a catchment management plan for all major European rivers for achieving 'good ecological status' by 2015. In this chapter we provide a comprehensive overview of all major European catchments (**Figure 1**), starting with the biogeographic setting with an emphasis on physiography, hydrology, ecology/biodiversity, and human impacts.

Biogeographic Setting

Europe forms the northwestern physiographic constituent of the larger landmass known as Eurasia. Europe covers an area of ~11.2 million km^2 that includes the European part of Russia, parts of Kazakhstan (Ural River Basin), the Caucasus, Armenia, Cyprus, and Turkey (**Figure 1**). Armenia and Cyprus are considered as transcontinental countries; and Turkey is included because of political and cultural reasons. The average altitude of Europe is 300 m asl compared with 600 m asl for North America and 1000 m asl for Asia. Only 7% of Europe is above 1000 m asl. Europe has a highly extensive and deeply penetrating network of water bodies. Its 117 000 km convoluted coastline facilitated easy access to the interior, and it is this feature that contributed to the rapid development of its southern shores along the Mediterranean Sea.

Cultural and Socioeconomic Setting

There are distinct cultural, demographic, socioeconomic, and political gradients across Europe. Today's human population is 780 million with an average population density of 69 people per km^2. At the catchment scale, the population density ranges from <2 people per km^2 (Pechora Basin) to 313 people per km^2 (Rhine Basin). The annual Gross Domestic Product (GDP; US$/person) ranges over two-orders-of-magnitude, from 600$ (Dniester Basin in Moldova) to 65 000$ (Aare Basin in Switzerland). Human life expectancy ranges from 61 (Ural Basin) to 80 years (river basins in Iceland, Italy, Spain, Sweden, and Switzerland). More than 100 languages are spoken across Europe; with the greatest number (27 languages) spoken in the Caucasus region.

Hydrogeomorphic and Human Legacies

Last Glacial Maximum and Holocene Distribution of River Networks

Many European rivers have substantially changed in length, catchment area and flow direction over the past 20 000 years. For instance, at the onset of the last glacial maximum about 20 000 year BP, a paleo-river known as the 'Channel River' (located between the present France and Britain) extended across the raised continental margin. Most major rivers in northwestern Europe (e.g., Rhine, Meuse, Solent, and Thames) contributed to its waters. In addition, damming by the Fennoscandian ice sheet caused the development of southward-flowing melt-water valleys and ice-margin spillways running westward. These spillways collected proglacial waters from rivers east of the Elbe basin that drained into the Channel River. The Channel River was the largest river system that drained the European continent, thereby affecting the hydrology of much of Europe as well as that of coastal ecosystems.

The long-term evolution of European rivers during the Holocene can be placed into four regional categories:

1. Rivers recently developed on areas formerly covered by ice sheets and affected by isostatic uplift;
2. Rivers of the former periglacial zone partly influenced by ice sheets;
3. Rivers of the former periglacial zone with lower reaches influenced by eustatic sea-level changes; and
4. Rivers of southern Europe within the region of former cold steppe and forest-steppe.

* This text is a shortened version of the chapter 'Introduction to European Rivers' in Tockner K, Uehlinger U and Robinson CT (eds) *Rivers of Europe*. San Diego: Elsevier/Academic Press (2008).

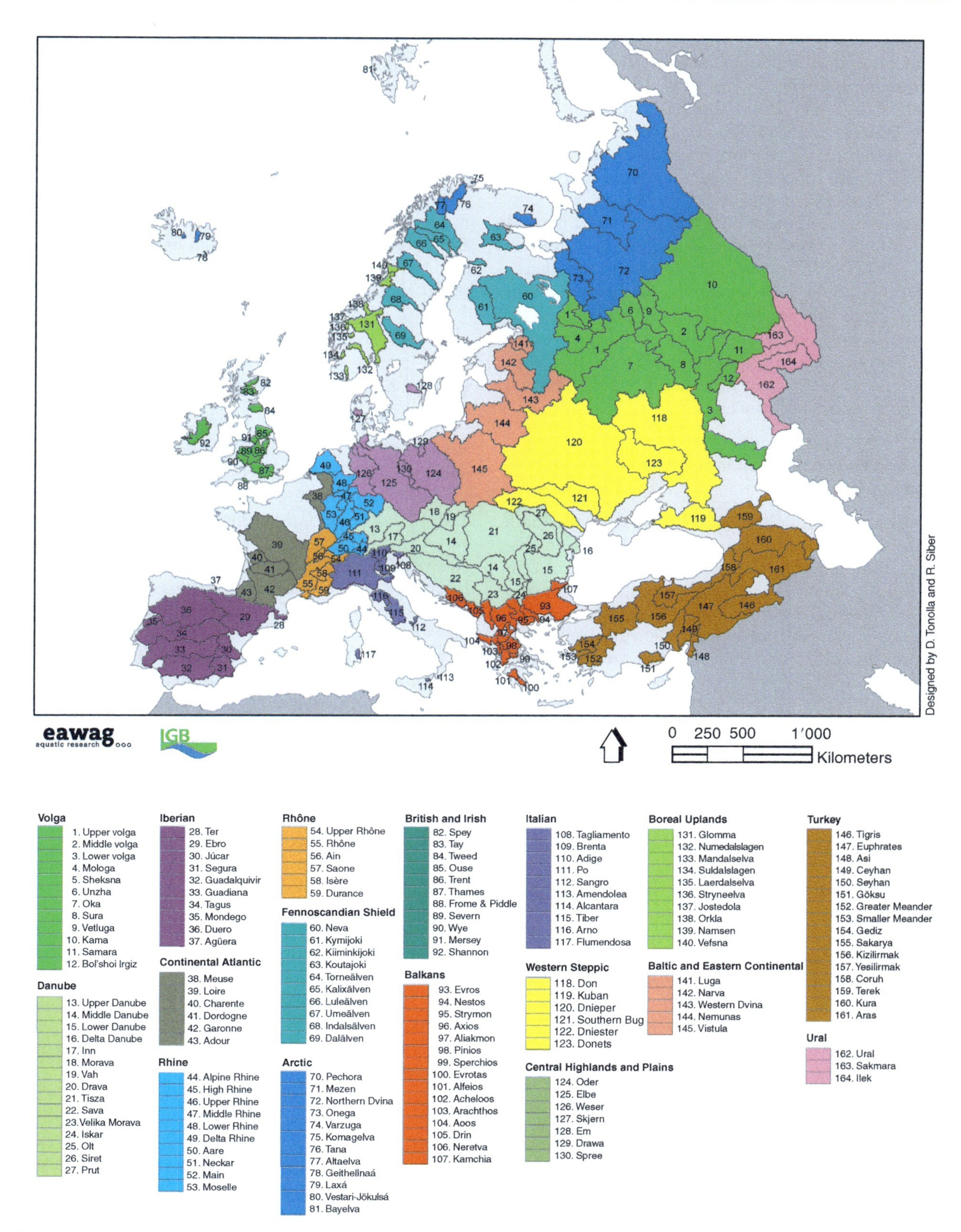

Figure 1 Spatial distribution of European catchments from 12 different geographic regions and subcatchments from the Volga, Danube, Rhine, Rhone, and Ural Rivers. Source – Tockner K, Uehlinger U, and Robinson CT (eds), *Rivers of Europe.* San Diego: Elsevier/Academic Press (2008).

Central European rivers show 3–6 separate alluvial fills that correlate well with stages of glacial advance and treeline lowering in the Alps. In the European lowlands, 2–3 fills are at times recorded, although major lateral channel shifts are more common as a consequence of fluctuations in discharge and sediment flux. In southern Europe, the tendency towards braiding has anthropogenic (deforestation and subsequent increases in sediment transport) and natural origins (higher flood frequency during the Little Ice Age, 1450–1850). In piedmont zones, the tendency towards braiding was repeated during cooler and moister stages. In the western Siberian river valleys, e.g., the Pechora and Mezen valleys, braided channels probably changed to a more meandering style after the retreat of permafrost and then towards braiding during permafrost advances.

Early and Recent Human Impacts of European Rivers

Deforestation and cultivation of soils were the main human activities that caused major changes in discharge and sediment transport. In southern and central Europe, distinct stages of sediment deposition have been recorded from the late Bronze Age and even more extensively in Roman times. General aggradation of European valley floors occurred in medieval times and is reflected in the rising of channel forms. Prior to the 11th century, river works were still primitive, consisting mostly of embankments built for flood control and land reclamation. There were two European centers of technological advance in river regulation during the medieval period. The Netherlands developed dredging technologies, designed floodgates, and built groynes and retaining walls. In Italy, land reclamation (*'La bonifica'*) was common with most large rivers being partly channelized by 1900. The greatest single engineering effort in the nineteenth century was the regulation of the lower Tisza River, the largest tributary of the Danube River, where 12.5×10^6 ha of flood plain marsh were drained and the river course shortened by 340 km.

Today, European catchments are highly fragmented by >6000 large dams. Reservoirs behind these dams can store about 13% of the mean annual runoff of Europe. The highest number of dams occur in Spain (1196) and Turkey (625). Of the 20 largest European rivers, only the Pechora River in Russia, draining to the White Sea, is considered free-flowing. Of the 165 catchments included in this chapter, only 30 have free-flowing large rivers. Most of these free-flowing rivers are in northern Europe (Arctic and Fenno-Scandinavian Shield) or drain relatively small catchments (e.g., Amendolea in Italy, Frome and Piddle in Great Britain, and Sperchios in the Balkans.

Among the major factors influencing water quality and quantity at the catchment scale is the change in land use intensity. Around 60% of the combined catchment area of the 165 examined rivers has been transformed into cropland and urban area. The proportion of developed area exceeds 90% for Central European and Western Steppic Rivers (**Table 1**). Over 70% of the European population lives in urban areas and the total number of cities with a population >100 000 is >360.

Water Availability, Runoff, and Water Stress

Water availability, defined as the annual long-term average renewable water resource derived from natural discharge including consumptive water use, shows a large spatial variation among river basins. Annual water availability ranges from >1000 mm $year^{-1}$ (western Norway, Britain's west coast, southern Iceland) to <100 mm $year^{-1}$ (parts of Spain, Sicily, large parts of the Ukraine, Southern Russia, large parts of Turkey). In most of Europe this reflects patterns of precipitation, whereas available water is transferred by rivers into more dry regions in other parts. Hungary, for example, attains most of its water from outside the country via the Danube and Tisza.

The total average runoff of European rivers is ~3100 km^3 $year^{-1}$ for 11 million km^2 (8% of the world average). The 20 largest rivers (total area: 5.9 million km^2) contributes more than 1/3 to the total continental runoff (**Table 1**). The average annual specific runoff ranges from 68 mm $year^{-1}$ (Asi River in southeast Turkey) to 1150 mm $year^{-1}$ (River Tay in Scotland). High seasonality in runoff is typical for rivers in southern Europe and Turkey such as the Guadalquivir (Iberian peninsula) and Upper Euphrates, and for Boreal and Arctic rivers such as the Glomma (Norway) and Pechora (Russia). Low runoff variability is characteristic for central European rivers (e.g., Elbe) and Steppic rivers (e.g., Dnieper) (**Figure 2**).

A recent assessment of Europe's environment by the European Environmental Agency indicated that high levels of water stress, both quantity and quality, exist in many areas throughout Europe and identified several significant ongoing pressures on water resources at the European scale. Total water withdrawal has generally increased in the last decades. By 1995, a total of ~476 km^3 water was being withdrawn

Table 1 The 20 largest catchments in Europe (including Turkey and the Caucasus)

	Area (km^2)	*Discharge (km^3 $year^{-1}$)*	*Relief a (m)*	*Population[b] (people per km^2)*	*Cropland and urban[c](%)*	*GDP[d] (\$ $year^{-1}$)*	*Protected[e] (%)*	*Fish (native)*	*Fish (nonnative)*
Volga	1 431 296	261.8	1536	45	58.5	2340	5.7	66	18
Danube	801 093	202.4	3651	102	65.3	7007	2.4	99	32
Dnieper	512 293	42.6	411	64	94.1	1388	3.2	29	5
Don	427 495	25.5	804	46	90.6	1508	3.2	64	7
N Dvina	354 298	107.5	422	5	10.2	2873	5.2	34	7
Pechora	334 367	150.9	1604	2	0.2	2928	12.2	34	3
Neva	281 877	79.1	390	17	25.9	6181	5.1	43	1
Ural	252 848	10.6	1094	15	61.6	2205	0.9	55	1
Kura	193 802	17.1	4816	74	58.6	1267	5.5	33	8
Vistula	192 980	32.9	2316	127	90.8	3789	2.6	54	18
Rhine	185 263	73.0	3786	313	76.4	31 822	0.4	46	25
Elbe	148 242	22.4	1456	164	83.6	14 068	4.3	38	8
Euphrates[f]	121 554	31.6	3557	57	43.0	1535	0.0	45	1
Oder	120 274	17.2	1468	132	91.3	5583	1.5	42	11
Loire	115 980	26.4	1704	67	88.1	22 196	1.5	32	26
Nemunas	98 757	17.0	354	52	93.0	2680	5.2	46	4
Rhône	98 556	53.8	4452	105	63.8	24 462	8.9	50	21
Duero	97 406	17.3	2359	37	75.6	15 058	1.2	18	13
Ebro	85 823	13.6	3104	34	63.4	19 587	1.5	29	19
W Dvina	83 746	13.6	307	32	87.6	2598	8.0	39	2

Relief: Calculated difference between highest and lowest point (resolution: 1000×1000 m) in catchment; Human population density: People per km^2; GDP: Annual Gross Domestic Product per person and year; Protected: National parks, Ramsar sites, nature reserves, and other nationally protected areas.
[a]http://edc.usgs.gov/products/elevation/gtopo30/gtopo30.html.
[b]http://gis.ekoi.lt/gis/.
[c]http://edcsns%2017.cr.usgs.gov/glcc/tablamberteuraseur.html.
[d]ESRI.
[e]http://sea.unep-wcmc.org/wdbpa/.
[f]only Turkey.

annually; 45% of this water is used for industry, 41% for agriculture, and 14% for domestic needs. There is a large difference between countries in how much and for what purpose water is withdrawn. Industrial uses dominate water withdrawals in most of Europe, whereas irrigation use is highest in southern and southeastern countries with low precipitation. Total withdrawal per catchment ranges from nearly zero (in the less populated areas of subpolar Scandinavia and Russia) to >400 mm $year^{-1}$ (in densely populated urban areas). In total, annual water withdrawal in Europe (excluding Turkey) is projected to rise from 415 km^3 today to ~660 km^3 by 2070. Although the annual total withdrawal in Western Europe will decrease from 236 km^3 to 190 km^3, it will increase considerably in eastern Europe from 180 to 470 km^3. In southeastern Europe, growth in water demand is complemented by reductions in water availability owing to climate change, which eventually will increase water stress. Overall, severe water stress is predicted to increase from 19% today to 34–36% by 2070. Since 1970, the total annual discharge of Balkan rivers already decreased by up to 70%, mainly due to water abstraction for irrigation.

Riverine Flood Plains

Owing to the development of agriculture in alluvial pains, the conversion of rivers for navigation, and the protection of settlements, flood plains have been 'trained' for centuries. Today, about 50% of the total European human population lives on former flood plains. As a consequence, ~50% of the original wetlands and up to 95% of riverine flood plains have been lost. In 45 European countries, 88% of the alluvial forests have disappeared from their potential range. The Seine River (France) shows the highest impact of all European rivers with 99% of its former riparian flood plains lost. Of the former 26 000 km^2 flood plain area along the Danube and its major tributaries, ~20 000 km^2 have been isolated by levees and have thus become 'functionally' lost; meaning that the basic attributes that sustain flood plains such as regular flooding and morphological dynamics are missing. Switzerland has lost about 95% of its original flood plains over the last two centuries. The remaining flood plains included in the inventory of 'flood plains of national importance' are far from being pristine, being heavily influenced by water abstraction, gravel

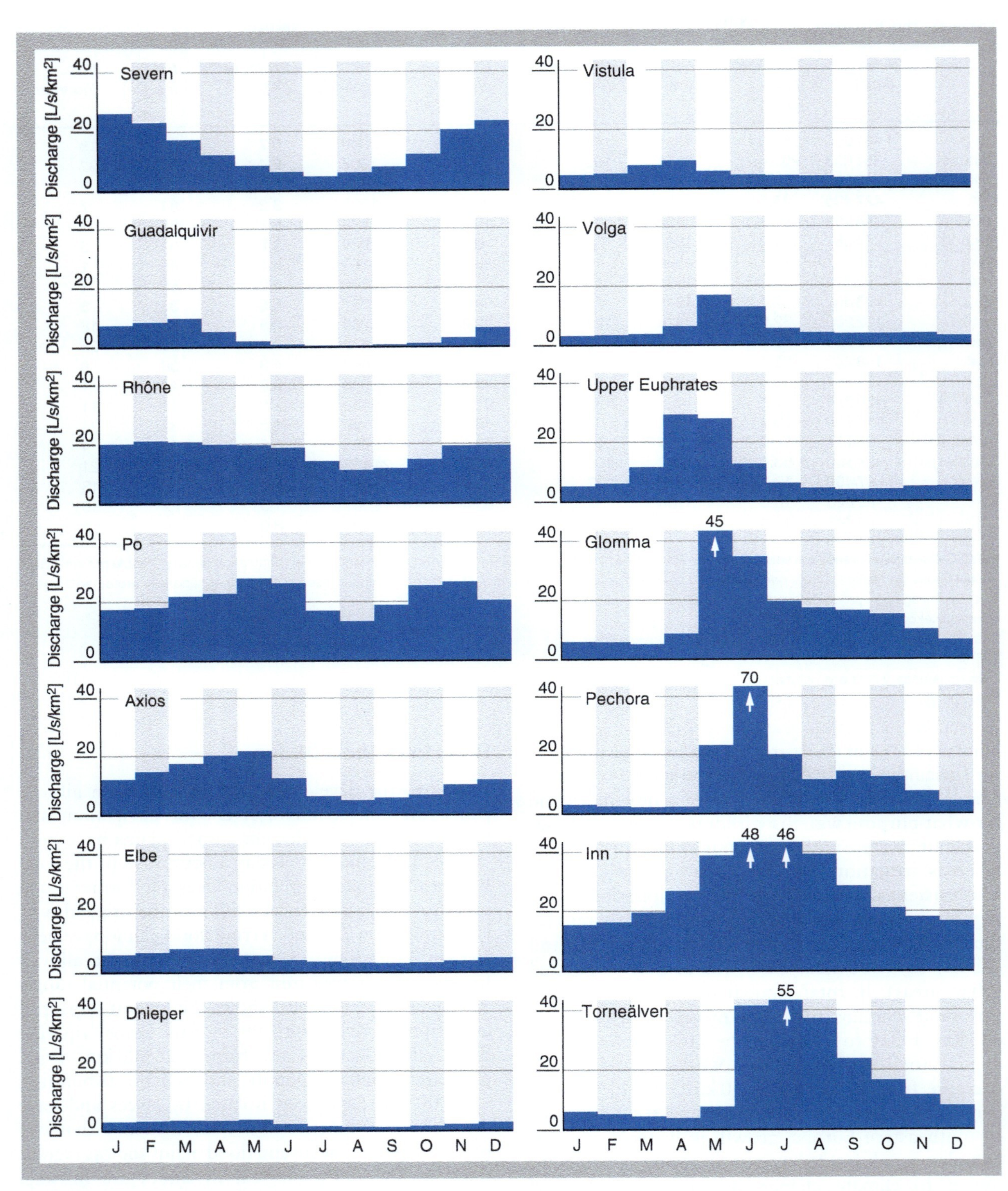

Figure 2 Seasonal distribution in catchment runoff (l s^{-1} km^{-2}) for selected rivers distributed across Europe. Runoff includes the difference between precipitation, evapotranspiration, and catchment topography. Data source: Global Water Runoff Data Center, GRDC, www.GRDC.bafg.de.

mining, and fragmentation. Today, the largest remaining flood plain fragment in Switzerland covers an area of only 3 km^2. Because most European flood plains are already 'cultivated,' even impacted systems that retain some natural functions, such as those along the Oder River (Poland/Germany), the Danube River, and eastern European river corridors (**Figure 3**), are extremely important to protect. This is especially true for the river corridors in Eastern Europe because of ever increasing pressures from development (gravel exploitation, damming, dredging for navigation, road constructions).

River Deltas

Deltas are integral features of many catchments, being important depositional landforms where the river mouth flows into an ocean, sea, or lake. The geometry, landform, and environment of deltas result from the accumulation of sediments added by the river and the reworking of these sediments by marine forces. Because many European rivers discharge into isolated and inland seas (Baltic, Black, and Mediterranean Seas), characterized by low tides and moderate wave powers, they can form extensive deltas (**Table 2**). The 35 major European deltas cover a total area of ~90 000 km^2. Despite their ecological and socioeconomic importance, European deltas are among the least investigated aquatic ecosystems.

Deltas are highly productive environments and, as a consequence, they have been extensively transformed into cropland and urban areas. Today, the human density in European deltas is often much higher than in the respective upstream catchment (**Tables 1** and **2**), although the opposite pattern can be found such as for the Danube catchment and the tributaries to the White Sea. Deltas formed by the Pechora, the N. Dvina (despite having a large seaport, Archangelsk, with a population of 350 000), and the Volga are among the few remaining relatively pristine deltas. Deltas are biologically diverse ecosystems, thus major efforts are underway to conserve and restore them. Several large deltas are already protected by the Ramsar Convention (e.g., Nestos, Axios, Kuban, Dnieper, Volga, Danube, Rhone). Around 90% of the Danube delta today is officially protected (Ramsar site and Unesco Biosphere heritage). Other large deltas such as those of the Ural and Terek Rivers in Russia, and the Seyhan and Kizilirmak Rivers in Turkey are not protected.

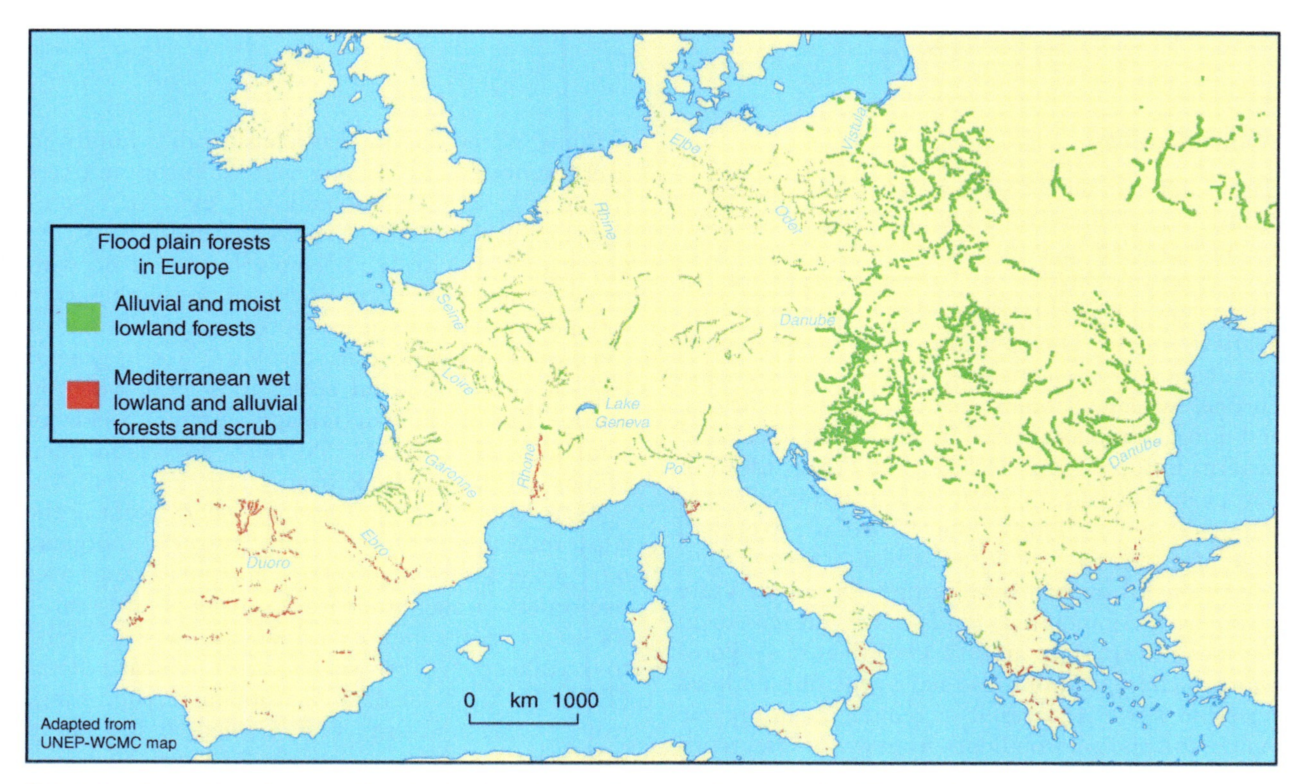

Figure 3 Distribution of the remaining riparian forests along European rivers (note the difference between eastern and western Europe. Hughes FMR (ed) (2003) *The Flooded Forest: Guidance for policy makers and river managers in Europe on the restoration of flood plain forests*. FLOBAR2, 96 pp. European Union and Department of Geography, University of Cambridge: UK, with permission.

Table 2 The 20 largest river deltas in Europe (including Turkey and the Caucasus)

	Area (km²)	*Average temperature[a] (°C)*	*Population[b] (people per km²)*	*Cropland[c] (%)*	*Protected[d] (%)*
Rhine	25 347	9.2	493	89.7	0.9
Volga	11 446	10.3	53	70.0	24.7
Ural	8586	9.1	24	13.6	0.0
Pechora	5490	−4.0	<1	<0.1	26.3
Kuban	5422	11.7	63	73.9	20.3
Danube	4560	10.7	34	56.0	89.1
Kura	4175	15.5	78	57.4	20.6
Terek	4026	11.6	46	86.3	3.3
Po	2878	12.8	119	86.9	10.0
Dnieper	2833	8.7	80	76.2	7.4
N Dvina	2229	0.6	118	3.4	5.9
Guadalquivir	2213	17.6	152	69.2	31.9
Seyhan	1903	17.1	116	86.0	0.0
Vistula	1858	7.7	187	93.5	0.0
Rhône	1783	13.5	64	63.7	59.7
Neman	1088	6.7	24	57.1	18.6
Don	604	10.1	541	71.9	80.8
Kizilirmak	474	11.1	126	84.6	0.0
Ebro	331	15.9	116	49.3	22.3
Nestos	319	12.5	53	83.3	14.6

Average annual temperature (1961–1990). Human population density: People per km². Protected: National parks, Ramsar sites, nature reserves, and other nationally protected areas.

[a]http://www.ipcc-data.org/obs/get30yr_means.html.

[b]http://gis.ekoi.lt/gis/.

[c]http://edcsns 17.cr.usgs.gov/glcc/tablamert_euras_eur.html.

[d]http://sea.unep-wcmc.org/wdbpa/.

Water Quality

European rivers show a wide variety of pollution problems. In Scandinavian rivers, acidification remains a major problem due to acid rain deposition that is not neutralized in the non-carbonated soils of the Fennoscandian Shield, while other contaminants are relatively minor. Eutrophication and nitrate deposition pose the greatest challenge in western and central Europe, whereas organic matter loads, pesticides, and nitrogen inputs are major issues in southern and eastern Europe. From 1992–1996, over 65% of European rivers had average annual nitrate concentrations exceeding 1 mg l^{-1} and 15% of the rivers had concentrations >7.5 mg l^{-1}. The highest nitrate concentrations are in northwest Europe where agriculture is intense. Ammonium levels have decreased in European rivers since around 1990 (**Figure 4**). Phosphorous concentrations also have generally declined since the 1990s as a result of reductions in organic matter and phosphorous loads from wastewater treatment plants and industry and of severe reduction or ban of phosphate detergents as in Switzerland and Germany.

While water quality has considerably improved over recent decades in many western European rivers (**Figure 4**), serious problems still exist in eastern and southern countries. For instance, 75% of the water in the Vistula, Poland's largest river with many seminatural flood plains, is unsuitable even for industrial use. The range of specific fluxes of river borne material (tons km^2 $year^{-1}$) is in general high at the continental scale; it is even wider in Europe due to human impacts. Annual yields of total suspended solids (TSS) range over more than two orders-of-magnitude from <1 ton km^2 $year^{-1}$ to >300 tons km^2 $year^{-1}$. Very high values occur in the Alps, reflecting natural erosion. Dissolved inorganic nitrogen yields from European catchments range over two orders-of-magnitude from <10 kg N km^2 $year^{-1}$ for rivers in the remote north (e.g., Finish rivers) to >2200 kg N km^2 $year^{-1}$ for the Rhine River (**Table 3**). Yields of dissolved organic carbon range from ~200 kg C km^2 $year^{-1}$ (Steppic Rivers) to >3000 kg C km^2 $year^{-1}$ in the Po River. Dissolved organic nitrogen, which primarily originates from anthropogenic sources in Western and Southern Europe, can reach 300 kg N km^2 $year^{-1}$. For all

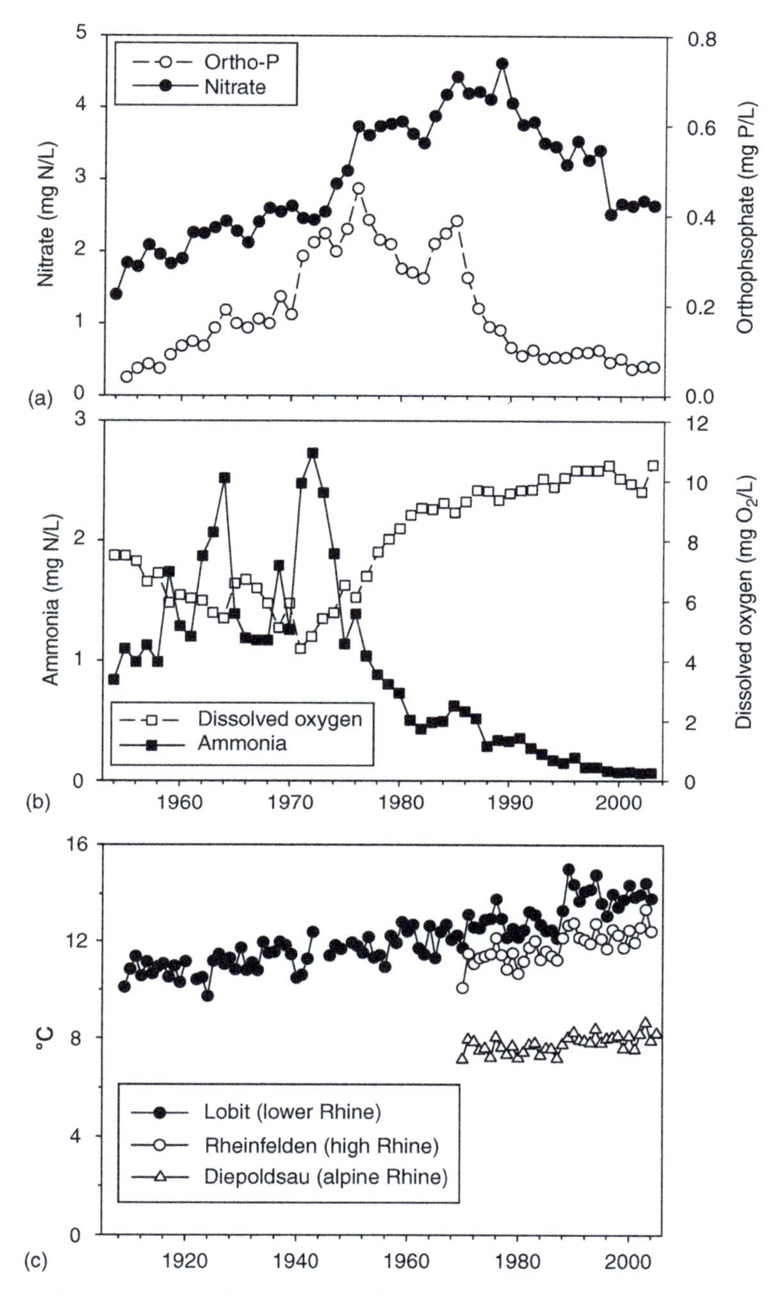

Figure 4 Temporal trends in (a) nutrients and (b) oxygen at Lobith, and (c) temperature (three stations) in the Rhine River. Note different scales of *x*-axes.

of Europe, estimated annual export rates were 1 Pg for inorganic suspended solids (ISS), 7 Tg for particulate organic carbon (POC), 1.1 Tg for particulate nitrogen (PN) and 0.3 Tg for particulate phosphorus (PP). Toxic substances as metals, polycyclic aromatic hydrocarbons (PAH) and polychlorinated biphenyls (PCBs) have reached some of the highest values ever recorded for Europe. Because their survey is costly, the recent trends are generally reconstructed from cores taken in deltas and flood plains. In Western Europe Rivers, bordering the Atlantic coast, very high levels of cadmium, mercury, lead, zinc, and of PAHs and PCBs have been recorded, with peaks from 1930 to 1970. Record contaminations are observed in rivers basins with high industrial and/or mining activities, megacities inputs (e.g., Paris, Berlin), often in combination with limited dilution by river sediments, as for the Seine, the Scheldt, Lot (France), Meuse,

Table 3 Predicted/measured catchment yields for particulate and dissolved organic matter and nutrients for selected European Rivers

	DIP (kg km^{-2} year^{-1})	*DIN (kg km^{-2} year^{-1})*	*DOC (kg km^{-2} year^{-1})*	*TSS (ton km^{-2} year^{-1})*	*POC (ton km^{-2} year^{-1})*	*PN (ton km^{-2} year^{-1})*	*PP (ton km^{-2} year^{-1})*
Volga	2.39	n.d.	n.d.	18	0.2	0.0	0.0
Danube	30.21	n.d.	1152	86	1.0	0.1	0.0
Dnieper	2.96	n.d.	570	5	0.2	0.0	0.0
Don	13.60	19.1	245	5	0.1	0.0	0.0
N Dvina	5.65	n.d.	1494	10	0.4	0.1	0.0
Pechora	5.94	64.7	1954	21	0.5	0.1	0.0
Vistula	36.98	371.8	n.d.	14	3.1	0.4	0.1
Rhine	119.32	2200.4	1388	21	2.4	0.4	0.1
Elbe	63.94	795.4	753	6	1.6	0.2	0.1
Oder	32.61	389.8	n.d.	1	0.4	0.1	0.0
Loire	30.95	n.d.	1065	4	0.7	0.1	0.0
Kuban	25.88	330.9	1044	120	1.0	0.2	0.0
Neman	9.42	74.1	n.d.	7	0.6	0.1	0.0
Ebro	2.34	n.d.	n.d.	217	1.6	0.2	0.1
Glama	16.06	191.8	n.d.	321	1.5	0.2	0.1
Kymjoki	3.95	n.d.	n.d.	3	0.3	0.1	0.0
Po	77.18	n.d.	3046	147	3.2	0.4	0.1
Seyhan	4.21	n.d.	n.d.	151	n.d.	n.d.	n.d.

TSS: Total Suspended Solidsl; n.d.: no data.

Sources

Beusen, AHW, Dekkers, ALM, Bouwman, AF *et al.* (2005) Estimation of global river transport of sediments and associated particulate C, N, and P. *Global Biogeochemical Cycles* 19, GB4S05.

Dumont E, Harrison JA, Kroeze C *et al.* (2005) Global distribution of dissolved inorganic mitrogen export to the coastal zone: Results from a spatially explicit global model. *Global Biogeochemical Cycles* 19, GB4S02.

Harrison JA, Caraco N, and Seitzinger SP (2005) Global patterns and sources of dissolved organic matter export to the coastal zobes: Results from a spatially explicit, global model. *Global Biogeochemical Cycles* 19, GB4S04.

Rhine, Idrija (Slovenia), Elbe, and Upper Vistula. In most cases the contamination levels have markedly declined since the 1970s, but they remain at high levels compared with natural levels. The heritage of this type of pollution, associated with particulate material, will last for decades and more.

Table 4 Global and European freshwater fauna species richness

Group	*World*	*Europe*	*Proportion of global (%)*
Bivalvia	1000	50	5
Gastropoda	4000	163	4
Ostracoda	2000	400	20
Copepoda	2085	902	43
Amphipoda	1700	350	21
Ephemeroptera	>3000	350	<10
Odonata	5500	150	3
Plecoptera	2000	423	21
Trichoptera	>10 000	1724	<17
Hemiptera	3300	129	4
Coleoptera	>6000	1077	<18
Diptera	>20 000	4050	<20
Lepidoptera	>1000	5	<1
Hymenoptera	>130	74	<56
Megaloptera	300	6	2
Pisces	>13 000	400	<3
Amphibia	5504	74	1
Ayes	1800	253	14
Total	*>82 500*	*10 580*	*<13*

Source – Lévéque C, Balian EV, and Martens K (2005) An assessment of animal species diversity in continental waters. *Hydrobiologia* 542: 39–67.

Freshwater Biodiversity

European freshwaters are relatively species-poor compared with other continents (**Table 4**). For example, continental waters provide habitat for <4% of the global freshwater fish fauna. The relative contribution of European freshwater fauna to global fauna is higher for groups with widespread species such as copepods and ostracods. It must be noted that the freshwater fauna (and flora) of Europe is much better described than the fauna of most other areas of the world. Around 25% of all European birds and 11% of all European mammals are dependent on freshwater for breeding or feeding, but only one species in each group is truly endemic to Europe (aquatic warbler, *Acrocephalus palustris*; southwestern water vole, *Arvicola sapidus*). Nine birds, 5 mammals and 25 fishes associated with European freshwaters are included in the International Union for the Conservation of Nature (IUCN) Red List of Globally Threatened Species, and two species are endangered by extinction (dalmatian pelican and ringed seal). One success story is the recent spread of the European beaver. At the beginning of the last century only a few hundred individuals survived in Norway, Germany, France, and the former Soviet Union. The population has now increased to at least half a million, and is attributable to large areas of suitable habitats and restricted hunting.

The European freshwater fish fauna includes 368 native species from 33 families (in the most recent inventory more than 500 species are listed including 59 species of the family Coregonidae. The most species (taxa) rich families are Cyprinidae (156 species), Gobiidae (40), Cobitidae (32), and Salmonidae (22; 64 species if all species and forms are considered). In comparison: North America contains ~1050 species, Africa >3000 species, and South America >5000 species. In Europe, a distinct west–east and north–south increase in species richness is found. The Danube River catchment has the highest diversity with ~100 fishes (~25% of the continental fauna). The mainstem of the Danube was unglaciated, and served as a 'refuge' during periods when the continental ice sheets advanced. As the ice sheets retreated, freshwater species expanded from this refuge to the rest of Europe (**Figure 5(a)**). Using area-corrected data (a power function of area and richness), the greatest diversity of fishes are in southeastern European catchments (Western Balkan, Turkey). River basins in Northern Europe, from Iceland to Northern Russia, have been covered by ice until 12 000 to 6000 years BP and therefore have low fish diversity.

At the continental scale, 13 fish species, including two fishes endemic to the River Drin (flowing into the Adriatic Sea in Albania/Croatia) and several salmon species are extinct. However, at the catchment scale, up to 40% of native fishes have disappeared, especially long-migrating species such as sturgeons, Allis shad (*Alosa alosa*) and lampreys. In contrast, 76 nonnative fishes belonging to 21 families have been introduced into European freshwaters, with ~50 of these having self-reproducing populations. Most nonnative fishes originated from North America (34 species) and from Asia (26 species), and between 30 and 50 fishes have been translocated within Europe. The proportion of nonnative fish exceeds 40% in some catchments, mostly in the Iberian Peninsula and the Atlantic region of France (**Figure 5(b)**). The highest proportion of irreplaceable fish (i.e., species with a limited geographic distribution), is found in the Iberian Peninsula, the southern Balkan, and Anatolia. These particular regions will face an even higher increase in water stress, pollution, and erosion in the near future.

The European Water Framework Directive

European catchments are under pressure of ever-increasing water stress and land-use change, especially those with high conservation value such as the Mediterranean area. The Water Framework Directive (WFD) creates a legislative framework to manage, use, protect, and restore surface water and groundwater resources in the European Union. The WFD approaches water management at the scale of the river catchment (river basin), which often includes several countries. The WFD requires the establishment of a 'river basin management plan' (RBMP) for each river catchment in the European Union. The RBMP is a detailed account of how environmental objectives (i.e., good ecological status of natural water bodies and good ecological potential of heavily modified and artificial water bodies) are to be achieved by 2015. For those countries that can demonstrate that this is not feasible without disproportionate economic and social costs, the WFD allows the possibility of delay to 2030. This sets a time scale for restoration of water bodies during which a considerable change in climate is expected. Although it is stated '*this Directive should provide mechanisms to address obstacles to progress in improving water status when these fall outside the scope of Community water legislation, with a view to developing appropriate Community strategies for overcoming them*' (WFD, Article 47), climate change and its possible impact on water bodies has been ignored in the scope of the WFD and the term 'climate' does not even appear in its text.

Knowledge Gaps

The catchment must be considered as the key spatial unit to understand and manage ecosystem processes and biodiversity patterns. However, biological information is mostly available at the country rather than the catchment level. In addition, available data are unevenly distributed across Europe and constrains potential comparability. Riverine flood plains and deltas are among the least studied ecosystems but yet the most threatened. As such, we need to identify and quantify the ecosystem services that these

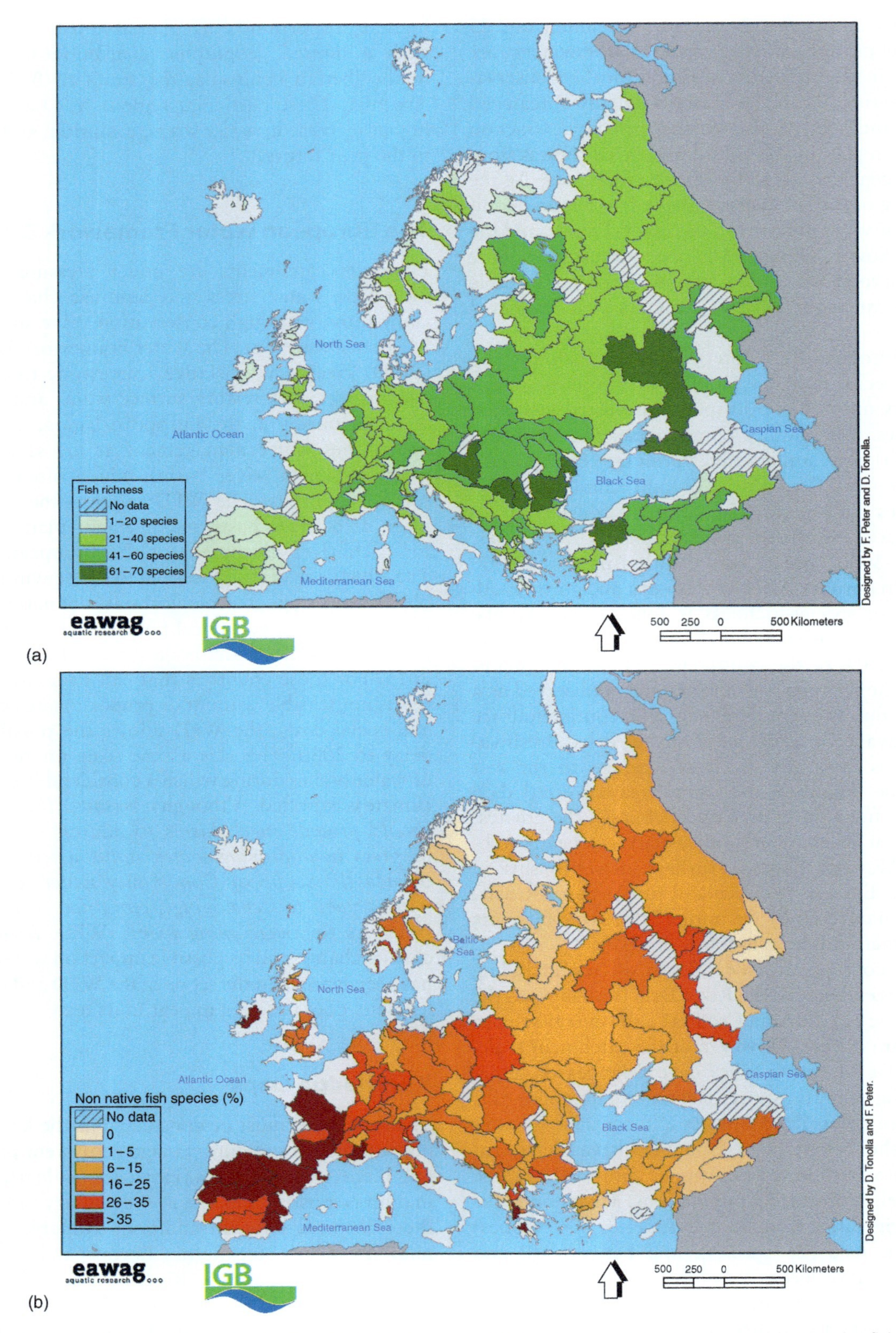

Figure 5 Species richness of fish in European catchments and sub-catchments: (a) native fish richness, (b) nonnative fish richness. Fabian P (2006) *Biodiversity of European Freshwater Fish – Threats and Conservation Priorities at the Catchment Scale*. Diploma thesis, 71 pp. Switzerland: University of Basel.

ecosystems provide in their natural state. Historic information and long-term data for freshwater organisms as well as key environmental drivers (e.g., temperature, habitat change) are rare, especially at the continental scale. While conservation planning is primarily driven by several native, endemic, and endangered species (so-called 'hot spot' areas), there is an urgent need to incorporate other ecosystem aspects such as the evolutionary potential of the system and its capacity to perform key ecological processes in conservation and restoration planning. Finally, there is an urgent need to establish a European network of 'reference' river systems against which human alterations can be assessed; and to better understand how rivers function in their (semi-)natural state. This provides pivotal baseline information for guiding future restoration and management programs.

Further Reading

Bacci ML (1998) *Europa und seine Menschen. Eine Bevölkerungsgeschichte.* München: Verlag C.H. Beck.

Beusen AHW, Dekkers ALM, Bouwman AF, Ludwig W, and Harrison J (2005) Estimation of global river transport of sediments and associated particulate C, N, and P. *Global Biogeochemical Cycles* 19: GB4S05.

Butlin RA and Dodgshon RA (eds.) (1998) *An Historical Geography of Europe.* Oxford: Clarendon Press.

CBD (Convention on Biological Diversity) (2003) *Status and Trends of Biodiversity of Inland Water Ecosystems.* CBD Technical Series No. 11. Montreal, Canada: Convention on Biological Diversity.

Dumont E, Harrison JA, Kroeze C, Bakker EJ, and Seitzinger SP (2005) Global distribution and sources of dissolved inorganic nitrogen export to the coastal zone: Results from a spatially explicit, global model. *Global Biogeochemical Cycles* 19: GB4S02.

Eisenreich SJ (ed.) (2005) *Climate Change and the European Water Dimension: A Report to the European Water Directors.* Ispra, Italy. http://natural-hazards.jrc.it.

Henrichs T and Alcamo J (2001) Europe's water stress today and in the future. (http://www.usf.unikassel.de/usf/archiv/dokumente/kwws/5/ew_5_waterstress_low.pdf).

Hughes FMR (ed.) (2003) *The Flooded Forest: Guidance for policy makers and river managers in Europe on the restoration of flood plain forests,* 96pp. UK: FLOBAR2, European Union and Department of Geography, University of Cambridge.

Hughes JD (ed.) (2001) *An Environmental History of the World.* London: Routledge.

Illies J (1978) *Limnofauna Europaea – Eine Zusammenstellung aller die europäischen Binnengewässer bewohnenden mehrzelligen Tierarten mit Angaben über ihre Verbreitung und Ökologie.* Stuttgart: Schweitzerbart'sche Vergasgesellschaft.

Kimstach V, Meybeck M, and Boroody E (eds.) (1998) *A Water Quality Assessment of the Former Soviet Union.* London: E & F Spon.

Klimo E and Hager H (eds.) (2001) *The Flood plain Forests in Europe.* Leiden, The Netherlands: Brill.

Kottelat M (1997) European freshwater fishes: An heuristic checklist of the freshwater fishes of Europe (exclusive of former USSR), with an introduction for non-systematists and comments on nomenclature and conservation. *Biologia* 52: 1–271.

Kottelat M and Freyhof J (2007) *Handbook of European freshwater fishes.* Berlin: Kottelat, Cornol and Freyhof.

Leppäkoski E, Gollasch S, and Olenin S (eds.) (1998) *Invasive Aquatic Species of Europe: Distribution, Impacts and Management.* Dordrecht: Kluwer.

Ménot G, Edouard Bard E, Rostek F, *et al.* (2006) Early reactivation of European Rivers during the last deglaciation. *Science* 313: 1623–1625.

Nilsson C, Reidy CA, Dynesius M, and Revenga C (2005) Fragmentation and flow regulation of the world's large river systems. *Science* 308: 405–408.

Parry M (ed.) (2000) *The Europe ACACIA project: Assessment of Potential Effects and Adaptations for Climate Change in Europe.* Norwich (UK): Jackson Environment Institute.

Peter F (2006) *Biodiversity of European Freshwater Fish – Threats and Conservation Priorities at the Catchment Scale,* 71pp. Switzerland: Diploma thesis. University of Basel.

Petts GE, Möller H, and Roux AL (eds.) (1989) *Historic Changes of Large Alluvial Rivers:* Western Europe. Chichester, UK: Wiley.

RIZA (2000) Ecological gradients in the Danube Delta lakes. RIZA report 2000.015, The Netherlands. ISBN 90.369.5309x.

Shiklomanov I (1997) *Assessment of Water Resources and Water Availability of the World.* Stockholm, Sweden: Stockholm Environment Institute.

Smith KG and Darwall RT (2006) *The Status and Distribution of Freshwater Fish Endemic to the Mediterranean Basin.* Gland, Switzerland: Report of The World Conservation Union (IUCN).

Stanners D and Bourdeau P (1995) *Europe's Environment – The Dobris Assessment.* London: Earthscan Publications.

Starkel L (1991) Long-distance correlation of the fluvial events in the temperate zone. In: Starkel L, Gregory KJ, and Thornes JB (eds.) *Temperate Palaeohydrology,* pp. 473–495. Chichester, UK: Wiley.

Starkel L, Gregory KJ, and Thornes JB (eds.) (1991) *Temperate Palaeohydrology.* Chichester, UK: Wiley.

Tockner K and Stanford JA (2002) Riverine flood plains: Present state and future trends. *Environmental Conservation* 28: 308–330.

Tockner K, Robinson CT, and Uehlinger U (eds.) (2008) *Rivers of Europe.* Burlington, MA: Academic Press/Elsevier.

UNEP (2004) Freshwater in Europe – Facts, Figures and Maps. *United Nations Environment Programme/Division of Early Warning and Assessment – Europe,* 92pp. Chatelaine, Switzerland: UNEP.

Van Dijk GM, Van Liere L, Bannik BA, and Cappon JJ (1994) Present state of the water quality of European rivers and implications for management. *The Science of the Total Environment* 145: 187–195.

Whitton BA (1984) *Ecology of European Rivers.* Oxford, UK: Blackwell.

Relevant Websites

http://ec.europa.eu/environment/index_en.htm – European Commission.
http://www.jrc.ec.europa.eu/ – EC Joined Research Center.
http://www.ecnc.nl – European Centre for Nature Conservation.
www.ecrr.org – European Center for River Restoration.
www.grid.unep.ch – United Nations Environment Programme/ Division of Early Warning and Assessment.
www.GWSC.bafgf.de – Global Runoff Data Center (GRDC).
http://www.gwsp.org – ESSP Global Water System Project (GWSP).
http://ec.europa.eu/environment/water/water-framework/index_en.html – Water Framework Directive (WFD).

South America

M E McClain and D Gann, Florida International University, Miami, FL, USA

Introduction and Overview

South America possesses distinctive and diverse river basins that are integral to the continent's natural, cultural, and socioeconomic identities. The geomorphologic and hydrological features of South American river systems are linked to the continent's unique tectonic and geographic settings. The Andes Cordillera delineates both the continent's western margin and its hydrologic divide, extending over 7500 km from 10° N of latitude to 55° S. As a result, enormous river basins, including the Amazon, Orinoco, and Paraná, drain the interior of the continent, and nearly all of the continent's runoff flows to the Atlantic Ocean – some through the Caribbean Sea. Moreover, the equator bisects South America near its widest point, and tropical moisture borne by the trade winds moves inland across the open Atlantic margin of the continent. This moisture falls on the Earth's largest expanse of remaining tropical rain forest, and due to the rain shadowing effect of the Andes and the southern deflection of Atlantic air currents, little Atlantic moisture is transferred to the Pacific side of the Cordillera. High rainfall rates across large interior basins make the Amazon, Orinoco, and Paraná Rivers three of the six highest discharge rivers in the world. Relatively small coastal rivers drain the Guyanas and much of the Atlantic Coast of Brazil, but the most remarkable coastal rivers drain the Pacific margin of the Andes. These Pacific Coastal rivers are generally less than 200 km in length and have very steep gradients, virtually falling into the Pacific Ocean from glacier-fed sources at 3000–6000 m above sea level (masl). To the extreme south of the continent, temperate rivers drain the Argentine pampas and Patagonia Steppe.

The diverse geographic and climatic settings of South American River Basins produce correspondingly diverse assemblages of riverine habitats and species. The large interior basins of the Amazon, Orinoco, and Paraná include floodplain and upland river reaches interconnected with large wetland complexes, such as the Amazonian floodplains and the Pantanal. Coastal rivers of South America occur in tropical, subtropical, and temperate zones, each with unique habitat characteristics, and the montane river habitats of the Andes range from arid to humid. The Amazon and Orinoco Basins support the world's richest freshwater biotas, with many species still to be described and a high prevalence of endemics. The Paraná and São Francisco River Basins also host rich assemblages of species.

Rivers are integral to the welfare and development potential of South America's population, providing services ranging from water supply (domestic, agricultural, and industrial) to important inland fisheries and hydroelectricity. Human pressures on the continent's river systems have lead to widespread contamination, disruption of natural flow regimes, land-cover conversion, and overexploitation of riverine resources. Human pressures threaten the extraordinary ecosystems of the continent and undermine the potential of South American rivers to continue to provide the services that benefit human communities. Improved river resource management and sustainable development are inextricably linked in South America.

Caribbean and North Atlantic Basins

South American rivers draining to the Caribbean and North Atlantic Basins include the Magdalena (Colombia), Orinoco (Colombia/Venezuela), Essequibo (Guyana), Corantijn (Guyana/Surinam), Maroni (French Guiana), and Oiapoque (French Guiana/Brazil). Together these river basins drain 1 504 241 km^2 or 8.54% (**Table 1, Figure 1**) of the continental area and account for ~20% of the continental runoff. The river basins fall entirely within the tropical zone and host rich assemblages of aquatic species (>100 species). Management issues in the basins include deforestation, water contamination, and flow regulation.

Geomorphology and Hydrology

The Caribbean and North Atlantic Basins, including smaller coastal catchments, drain an area of 2 122 707 km^2 or 12.06% of the continental area, including parts of the northern most cordilleras of the Andes, the Venezuelan Llanos (lowlands), and low mountains of the Guiana Shield. The Orinoco River is the largest river basin in the region, and third largest in all of South America, at 934 339 km^2 (5.31%). The average discharge of the Orinoco River is 30 000 m^3 s^{-1}, making it the second highest discharge river in South America and fourth in the world. The Orinoco originates in the Sierra Parima, on the Guiana Shield, at an elevation of ~1000 masl, although the basin also includes tributaries originating in the

Table 1 South American basins by region

Region	Area (km²)	% Continent	Basin	Name	Area (km²)	% Continent	Average Discharge ($m^3 s^{-1}$)
Caribbean and North Atlantic Basins (C)	2 122 707	12.06	C 1	Orinoco	934 339	5.31	30 000
			C 2	Magdalena	259 619	1.47	7500
			C 3	Essequibo	154 186	0.88	3 000
			C 4	Maroni	66 115	0.38	
			C 5	Corantijn	64 000	0.36	1085
			C 6	Oiapoque	25 982	0.15	
Amazon Basin (A)	5 888 270	33.45	A	Amazon	5 888 270	33.45	172 000
South Atlantic Basins (SA)	7 160 360	40.67	SA 1	Paraná	2 588 040	14.70	16 000
			SA 2	Tocantins	769 444	4.37	11 000
			SA 3	São Francisco	634 839	3.61	3000
			SA 4	Parnaiba	331 565	1.88	900
			SA 5	Colorado	294 076	1.67	130
			SA 6	Uruguay	265 786	1.51	8000
			SA 7	Chubut	95 416	0.54	50
Pacific Basins (P)	1 065 019	6.05	P 1	Loa	35 346	0.20	
			P 2	Guayas	32 538	0.18	
			P 3	Serrano	26 886	0.15	
Endorheic Basins (E)	1 368 406	7.77					

The source of all calculations except for Percent Runoff to Oceans is HydroSheds (WWF 2006). Discharge data were taken from the Global River Discharge Database (2007), selecting the station closest to the coastline (**Figure 1**).

Eastern Cordillera of the Andes. The river flows nearly 2500 km to its mouth at the Delta Amacuro (8°51′14″ N, 60°42′47″ W). Along its main channel the river ranges from 1 to 3 km in width with a narrow flood plain rarely exceeding 10 km in width.

The Magdalena River drains an area of 259 619 km² (1.47%) (**Table 1**, **Figure 1**) and is the ninth largest river basin in South America. It originates at elevations in excess of 5000 masl where the Central and Eastern Cordilleras of the northern Andes diverge. From its source the river flows northward 1500 km to its mouth (11°06′16″ N, 74°51′07″ W) at the City of Barranquilla, Colombia. The main channel of the Magdalena river transports seasonally high sediment loads and is meandering, anastomosing, and occasionally braided in its lower reaches. The Magdalena has an average discharge at Calamar of 7500 $m^3 s^{-1}$. Peak flows approach 10 000 $m^3 s^{-1}$ in October, while low flows drop to near 4000 $m^3 s^{-1}$ in February.

The major coastal rivers of the Guyanas, which are the Essequibo (154 186 km²), Corantijn (64 000 km²), Maroni (66 115 km²), and Oiapoque (25 982 km²) drain a combined area of 310 283 km² (**Table 1**, **Figure 1**). These rivers originate from a series of forested, low mountain ranges of the Guiana Shield. Their sources are generally less than 1000 masl, following low gradient, meandering paths, interrupted by occasional escarpments, which result in spectacular water falls. Kaieteur Falls (**Figure 3**) is a case in point, spilling over a 225 m escarpment on the Potaro River, a tributary of the Essequibo River. Annual discharge in the Essequibo River averages ~3000 $m^3 s^{-1}$ at Plantain Island, with high flows in June and July and low flows between October and February. Flows in the Corantijn River at Mataway average 1085 $m^3 s^{-1}$, with peak flows in excess of 2000 $m^3 s^{-1}$ during June and July.

Biology and Ecology

The Orinoco River, Magdalena River, and Guyana coastal rivers all exhibit high fish species richness by global standards. More than 300 fish species occur in the Orinoco River, more than 200 fish species occur in the Guyana Coastal Rivers, and ~150 species occur in the Magdalena River. The Guyana coastal rivers rank especially high in endemic species, with more than 75% of species occurring nowhere else. Research to identify and describe fish species in these rivers is still limited, so expectations are that continued research will reveal species richness even higher than currently documented.

The ecological status of the Orinoco and Guyana coastal rivers, with site-specific exceptions, is largely intact because human modifications and pressures remain low in large portions of the basins. The Magdalena River Basin is experiencing higher human pressures and its ecological status is correspondingly more impacted.

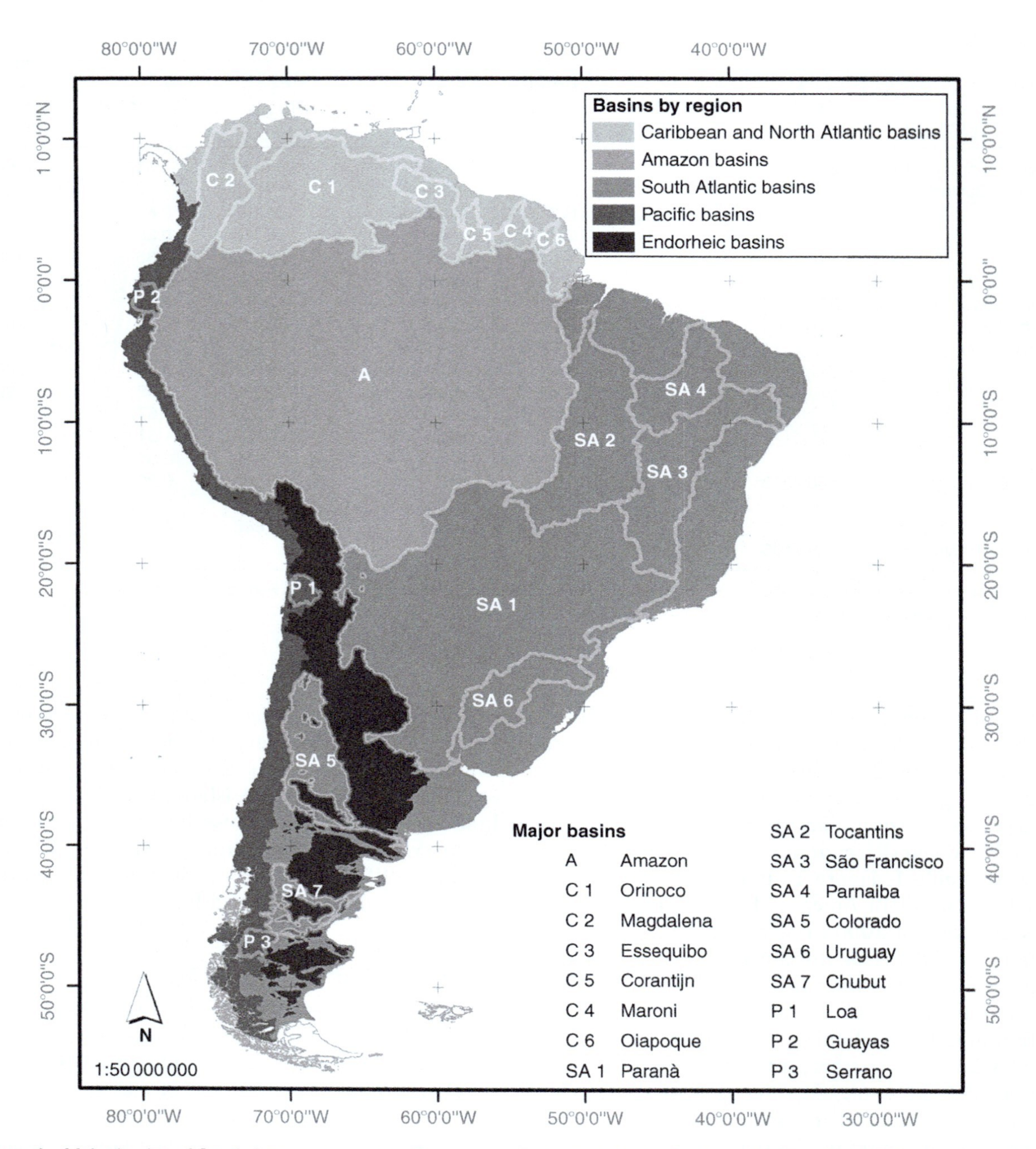

Figure 1 Major basins of South America by region. The source of all data is HydroSheds, delineated from SRTM 3 arc second data (WWF 2006) (Summary **Table 1**).

Management Issues

The most pressing management issues in the Caribbean and North Atlantic basins of South America lie in the Magdalena Basin and in the northern and coastal portions of the Orinoco and Guyanas river basins. The Magdalena is the most densely populated and heavily impacted of South America's major river basins. The basin includes 13 cities with populations greater than 100 000 people and a population density of 83 people per square kilometer (**Table 3**). Nearly 90% of the original forest cover of the basin has been cut and 60% of the basin is either urban or agricultural (**Table 2, Figure 2**). Consequently, soil erosion, domestic sewage, and agrochemicals are all serious contamination problems in the Magdalena. Human pressures remain low in the interior portions of the Orinoco and Coastal Guyanas rivers. Contamination and other management challenges are greatest in the

Table 2 Land cover by region and major basins within the regions

Basin	Name	Forest (km^2)	%	Grassland, Savanna, Shrubland (km^2)	%	Wetland (km^2)	%	Cropland (km^2)	%	Dryland (km^2)	%	Water (km^2)	%	Ice (km^2)	%	Urban (km^2)	%
Caribbean and North Atlantic Basin		**1 092 883**	**51.66**	**495 997**	**23.44**	**79 735**	**3.77**	**425 831**	**20.13**	**2705**	**0.13**	**16 951**	**0.80**	**0**	**0.00**	**1586**	**0.07**
C1	Orinoco	449 808	48.14	324 986	34.78	27 240	2.92	121 682	13.02	456	0.05	9 674	1.04	0	0.00	498	0.05
C2	Magdalena	48 197	18.57	51 763	19.94	221	0.09	156 734	60.38	1 298	0.50	1 314	0.51	0	0.00	70	0.03
C3	Essequibo	129 630	84.10	13 524	8.77	7 325	4.75	3 552	2.30	10	0.01	101	0.07	0	0.00	0	0.00
C4	Maroni	65 422	98.96	174	0.26	272	0.41	117	0.18	3	0.00	120	0.18	0	0.00	0	0.00
C5	Corantijn	62 188	97.18	883	1.38	596	0.93	190	0.30	5	0.01	133	0.21	0	0.00	0	0.00
C6	Oiapoque	25 227	97.44	105	0.41	491	1.90	58	0.22	0	0.00	8	0.03	0	0.00	0	0.00
Amazon Basin		**4 727 288**	**80.29**	**508 596**	**8.64**	**180 931**	**3.07**	**393 357**	**6.68**	**8916**	**0.15**	**67 075**	**1.14**	**1154**	**0.02**	**763**	**0.01**
South Atlantic Basins		**1 751 665**	**24.51**	**2 267 635**	**31.73**	**45 618**	**0.64**	**2 882 871**	**40.34**	**100 300**	**1.40**	**88 745**	**1.24**	**3659**	**0.05**	**5969**	**0.08**
SA 1	Paraná	808 135	31.23	674 094	26.05	2292	0.09	1 045 033	40.38	18 809	0.73	37 176	1.44	4	0.00	2 410	0.09
SA 2	Tocantins	163 346	21.23	281 468	36.58	5036	0.65	312 370	40.60	368	0.05	6815	0.89	0	0.00	20	0.00
SA 3	Sao Francisco	103 896	16.37	163 536	25.76	0	0.00	343 455	54.10	18 842	2.97	4678	0.74	0	0.00	445	0.07
SA 4	Parnaiba	118 379	35.71	58 484	17.64	377	0.11	150 075	45.27	2 568	0.77	1605	0.48	0	0.00	59	0.02
SA 5	Colorado	4786	1.63	253 946	86.34	0	0.00	681	0.23	32 615	11.09	1286	0.44	551	0.19	261	0.09
SA 6	Uruguay	51839	19.50	121 508	45.72	0	0.00	89 688	33.75	30	0.01	2708	1.02	0	0.00	5	0.00
SA 7	Chubut	4239	4.47	87 175	91.91	0	0.00	1453	1.53	873	0.92	1102	1.16	0	0.00	11	0.01
Pacific Basins		**235 806**	**22.57**	**439 475**	**42.06**	**2672**	**0.26**	**151 948**	**14.54**	**177 871**	**17.02**	**19 482**	**1.86**	**15 990**	**1.53**	**1549**	**0.15**
P1	Loa	7	0.02	8124	22.99	0	0.00	7	0.02	27 091	76.65	19	0.05	12	0.03	83	0.23
P2	Guayas	2228	6.86	7110	21.90	110	0.34	22 421	69.06	219	0.67	316	0.97	0	0.00	63	0.19
P3	Serrano	7187	26.73	14 205	52.84	0	0.00	575	2.14	162	0.60	3663	13.63	1092	4.06	0	0.00
Endorheic Basins		**61 403**	**4.49**	**822 370**	**60.13**	**143**	**0.01**	**201 947**	**14.77**	**253 576**	**18.54**	**26 346**	**1.93**	**1399**	**0.10**	**419**	**0.03**

The source of the summarized land cover is the Global Land Cover 2000 provided by the Global Vegetation Monitoring Unit of the Joint Research Center (**Figure 2**).

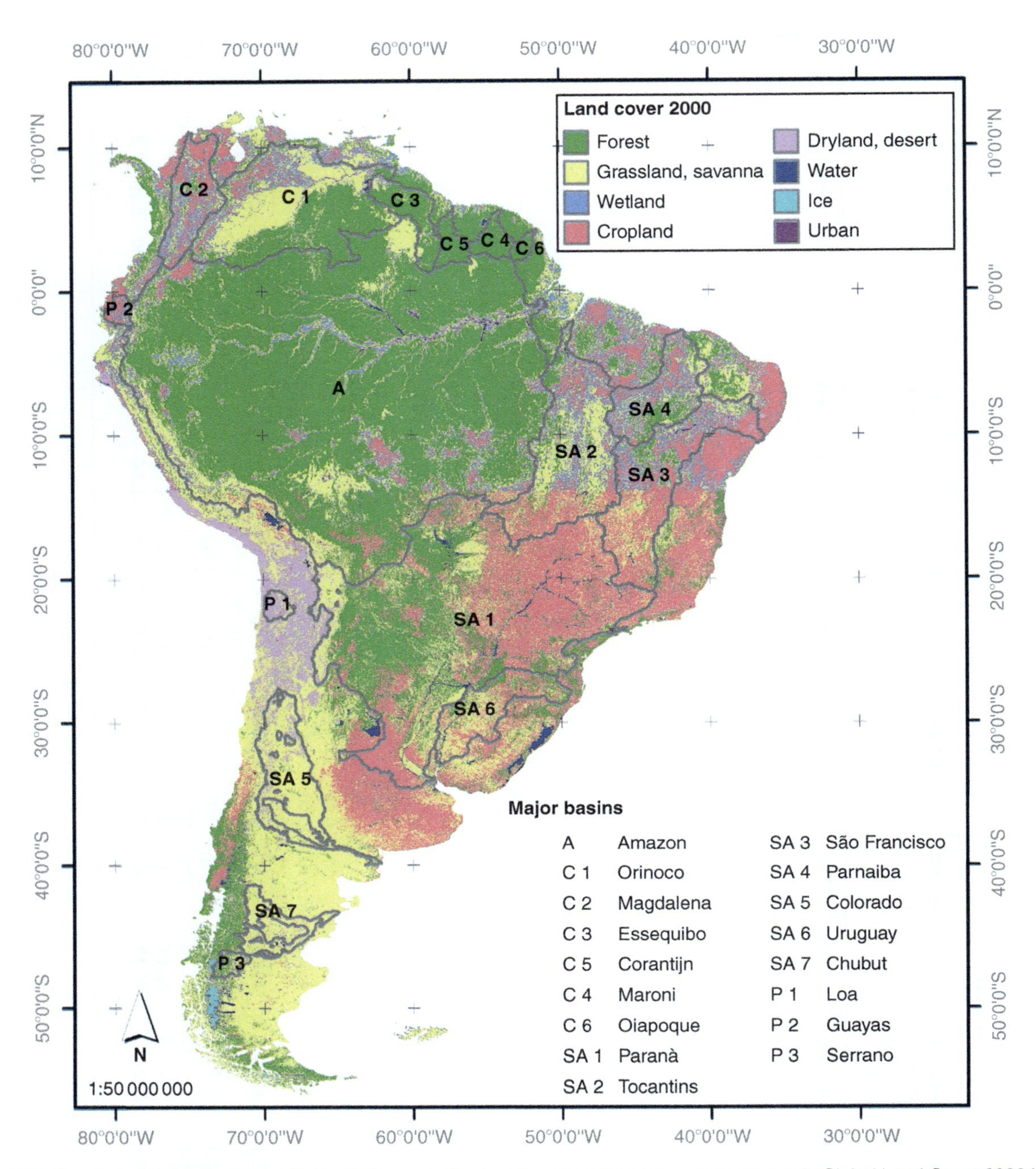

Figure 2 Land cover of major basins in South America. Source of summarized land cover classes is Global Land Cover 2000 by the Global Vegetation Monitoring Unit of the Joint Research Center (Summary **Table 2**).

Andean sections of the Orinoco River and near the populated coastline in the Guyanas. The Orinoco, Corantijn, and Oiapoque rivers are all transboundary rivers, so resolution of management issues requires cooperation between countries. Contamination linked to mining is an issue of concern in the Guyanas. River fragmentation because of dam construction is also of concern in the Magdalena River Basin and Caroní River, a major tributary to the Orinoco River in Venezuela.

Amazon Basin

The Amazon is Earth's largest river basin, covering an area of 5 888 270 km^2 (**Table 1**, **Figure 1**), and draining parts of Brazil, Peru, Bolivia, Colombia, Ecuador, and Venezuela. The basin includes a number of tributaries that are major rivers in their own right, including the Madeira, Negro, Xingu, Tapajos, Marañon, Purus, Ucayali, Japurá, Juruá, and Iça Rivers. The Amazon River Basin falls entirely within the tropical zone and hosts the largest diversity of fish species in the world, estimated to be more than 3000. The basin drains 33.45% of the continental area of South America and accounts for ~50% of its runoff. Many regions of the Amazon remain relatively pristine, but deforestation, over-fishing, untreated sewage discharges, and dams present management challenges in parts of the basin.

Figure 3 Kaieteur Falls, Guyana.

Geomorphology and Hydrology

The Amazon River Basin consists of four major geologic features. Although heavily weathered, the Precambrian Guyana and Brazilian shields still stand as highlands in the northern and southern sections of the basin, respectively, rising to elevations between 1000 and 3000 masl. The geologically young and still tectonically active Andean cordillera delineates the western margin of the basin and supplies the bulk of the sediment filling the shallow sections of the basin's two principal depositional features. The sub-Andean trough (foreland basin) accumulates sediments along the eastern margin of the Andes and subsequently serves as a secondary source for sediments carried further downstream and into the central Amazon trough. The elevation of both the sub-Andean and central Amazon troughs is below 500 masl and the topography is low-lying. The geologic configuration and climate patterns of the basin exert strong first-order controls on the distribution of soils and vegetation types. Well-drained and nutrient-poor oxisols dominate on the shields and upland portions of the central Amazon. Where annual rainfall exceeds 1500 mm on the shields and central trough, the dominant vegetation type is dense tropical forest adapted to the low nutrient status of the soils. Where annual rainfall is less than 1500 mm – mainly on the southeastern and northern fringes of the basin – forested and grassland savannas, or *cerrado*, occur. Oxisols give way to ultisols as the predominant soil type to the west and into the sub-Andean trough.

The average discharge of the Amazon River at Óbidos is 180 000 $m^3 s^{-1}$. This is far greater than that of any other world river. In fact, the Amazon discharge is equivalent to the combined discharges of the next seven largest rivers in the world, the

Table 3 Characteristics and human impact on major South American river basins

River basin	*Population density (per km²)*	*Number large cities (population exceeding 100 000)*	*Degree of fragmentation*	*Number of large dams*
Amazon	4	16	Medium	8
Parana	2	19	High	29
Orinoco	17	9	Medium	10
Tocantins	5	0	Medium	4
São Francisco	18	1	High	26
Colorado	6	2	–	10
Parnaiba	10	1	Medium	13
Uruguay	17	0	High	9
Magdalena	83	13	Medium	5
Chubut	1	0	–	2

The source of all information except for average discharge is the Water Resources eAtlas (WRI 2003). Discharge data were taken from the Global River Discharge Database (2007).

Congo, Yangtze, Orinoco, Brahmaputra, Yenisei, Rio de la Plata, and Mississippi. If the Negro River (45 300 $m^3 s^{-1}$) and Madeira River (32 000 $m^3 s^{-1}$) were considered individually, they would rank as the second and fourth highest discharge rivers in the world. The magnitude of discharge from the Amazon is the result of its large area and high average rainfall, which is ~2000 mm $year^{-1}$. The discharge of the Amazon River varies over the course of a year, resulting in distinct high water and low water periods. Water levels fall between June and November, reaching minimal flows at óbidos of approximately 100 000 $m^3 s^{-1}$. Water levels rise between December and the end of May, reaching normal peak flows between 200 000 and 250 000 $m^3 s^{-1}$. Along its main stem, the Amazon River follows a sinuous path, occasionally splitting into multiple channels that flow around riverine islands. During the high water period, the Amazon River floods a floodplain area of well over 100 000 km^2, and at its mouth (0°29′32″ N; 49°49′30″), the river forms a massive estuary ~140 km wide.

Biology and Ecology

The Amazon Basin hosts the most diverse assemblage of aquatic species and ecosystems of any river basin in the world. Ecosystems range from small, glacier-fed streams at high altitude to massive lowland rivers with deep channels and extensive flood plains. Amazon waters are classically separated into three groups with distinct limnological and ecological attributes. Blackwater rivers drain lowland areas of infertile sandy soils. They tend to be nutrient poor and rich in dissolved organic matter but are inhabited by rich assemblages of specially adapted aquatic species. The Negro River is the prime example, with more than 400 fish species, but many other black water rivers are found draining parts of the Andean foreland basin and Amazon trough. Clearwater rivers drain the Precambrian shields and exhibit intermediate nutrient levels and less diverse species assemblages. Finally, whitewater rivers originate in the Andes and carry large loads of sediments and associated nutrients. Whitewater rivers are often bounded by fertile flood plains and species richness is among the highest in the Amazon. In fact, the fertile flood plains built up along the corridors of whitewater rivers supported the development of early Amazonian civilizations like the Marajoara Culture centuries before European explorers entered the region.

Amazon ecosystems from the Andean foothills to the estuary are interconnected by the flow of water and nutrients and by the migrations of aquatic species. An illustrative example is the migration of large predatory catfish, which are hypothesized to spawn in the upper reaches of the basin. Larvae are then washed thousands of kilometers downstream to the estuary where juveniles develop. When they reach preadult and adult stages, these catfish begin migrating upriver, eventually reaching headwater spawning areas. The largest species migrating over these basin scales are of the genus *Brachyplatystoma*.

Management Issues

Much of the Amazon River Basin remains undeveloped (the average population density of the whole basin is ~1 person per square kilometer, near the world's minimum figure), but significant management challenges are emerging in areas subject to increasing population pressures and development activities. Road building and deforestation characterize the initial stages of development and are most extensive along the eastern, southern, and western margins of the basin. At present, ~80.3% of the basin remains forested, while 6.3% has been converted to agriculture and of the 8.6% of grassland a large part is used as pastures (**Table 2, Figure 2**). The basin contains 14 cities with populations in excess of 100 000, and it is in the vicinity of these cities that water contamination is greatest and where aquatic ecosystems are most impacted. Fish resources may also be degraded by poorly regulated commercial fisheries. An emerging management issue in the Amazon is dam building and river regulation. Currently, there are only eight large dams in the basin and no large dams on the Amazon mainstem or its major tributaries. There are, however, plans to build many additional dams, some of which are planned for the basin's major rivers. Fragmentation of the river could have severe impacts of the ecology of the system, as fluxes of nutrient rich sediments and migrations of many species will be impacted.

South Atlantic Basins

A number of important South American rivers drain to the South Atlantic Ocean, including the Parana River (Argentina, Brazil, Paraguay, Bolivia), Tocantins River (Brazil), Uruguay River (Uruguay, Brazil, Argentina), Parnaiba River (Brazil), São Francisco River (Brazil), Colorado River (Argentina), and Chubut River (Argentina). Many smaller coastal rivers also drain important parts of the continent. Rivers draining to the South Atlantic span tropical, subtropical, and temperate zones and host a corresponding diversity of species and ecosystems. Together these Southern Atlantic river basins drain 7 160 360 km^2 or 40.67% (**Table 1, Figure 1**) of the continental area. The major rivers named above account for 4 979 166 km^2 or 28.28% and account for ~25% of the continental runoff. Management challenges are great in these rivers because they drain some of the continent's most arid regions and some of its most populated and intensively developed regions. Consequently, demands and pressures on these rivers are high and include urban and agricultural pollution, large withdrawals for irrigation, and a high degree of regulation for navigation and flood control.

Geomorphology and Hydrology

The Parana River is the largest of the South American rivers draining to the South Atlantic. It drains an area of 2 588 040 km^2 (14.7%), making it the second most expansive river basin in South America after the Amazon. It originates in the Brazilian Highlands, south of the capital of Brasilia. Its major tributary, the Paraguay River, originates on the southern margin of the Brazilian Shield and flows into the Pantanal, an immense inland delta and one of Earth's largest freshwater wetland systems, covering more than 150 000 km^2. Western tributaries of the Paraguay River drain the Gran Chaco region of Bolivia, Paraguay, and Argentina. The Parana River, along with the Uruguay River, combine to form the Rio de la Plata near Buenos Aires, Argentina (34°11′40″ S; 58°13′50″ W). The average discharge of the Parana River at Corrientes is ~16 000 m^3 s^{-1}, making it the third largest discharge river in South America and one of the ten largest discharge rivers in the world. High flows in excess of 20 000 m^3 s^{-1} occur during January and February, and low flows below 15 000 m^3 s^{-1} occur in August. The Uruguay River lies to the east of the Parana Basin. It drains an area of 265 786 km^2 (1.51%) and has an average discharge at Puerto Salto of ~8000 m^3 s^{-1}.

The Tocantins River flows northward from the Brazilian capital of Brasilia to its mouth, 85 km west of the City of Belem (1°46′00″ S; 49°12′00″ W). The Tocantins River drains an area of 769 444 km^2 (4.37%), and its discharge averages 11 000 m^3 s^{-1}, with high flows exceeding 25 000 m^3 s^{-1} in February and March and low flows below 5000 m^3 s^{-1} from July through August.

The São Francisco and Parnaiba Rivers drain the arid and semiarid northeastern region of Brazil, the so-called Cerrado. The São Francisco River drains and area of 634 839 km^2 (3.61%) and has an average discharge at Traipu of ~3000 m^3 s^{-1}. High flows in excess of 5000 m^3 s^{-1} occur during February and March, and low flows below 2000 m^3 s^{-1} occur during July and August. The Parnaiba River drains an area of 331 565 km^2 (1.88%) and has an average discharge at Porto Formoso of ~800–900 m^3 s^{-1}. Peak flows of 1500 m^3 s^{-1} occur in February and March, and low flows below 500 m^3 s^{-1} occur in July and August.

The Colorado and Chubut Rivers originate on the eastern slopes of the southern Andes and drain the Pampas of Argentina. The Colorado River drains an area of 294 076 km^2 (1.67%). Its average discharge is ~130 m^3 s^{-1}, but peak flows during snowmelt in August and September regularly exceed 500 m^3 s^{-1}. The Chubut River drains and area of 95 416 km^2 (0.54%) and has an average discharge of ~50 m^3 s^{-1}. Like to Colorado River, the discharge in the Chubut is strongly influenced by spring snowmelt, and peak discharges in excess of 300 m^3 s^{-1} are normal.

Biology and Ecology

South American Rivers draining to the South Atlantic Ocean drain a wide range of landscapes and climatic zones and their biology and ecology are correspondingly varied. The most humid tropical river, the Tocantins, contains over 300 species of fish, while the temperate Colorado and Chubut Rivers in the far south contain fewer than 20 species. The highest numbers and percentages of endemic species occur in the tropical Tocantins and São Francisco Rivers.

The most remarkable aquatic ecological feature of the region is the Pantanal, which is flooded annually by the free-flowing Paraguay River. The Pantanal contains both a high diversity and high density of fauna. Nearly 300 species of fish, 650 species of birds, 160 species of reptiles, and nearly 100 species of mammals have been documented in the area, and it is likely that many new species remain to be discovered.

Management Issues

Rivers of this region are heavily used and subject to a number of challenging management issues. With relatively high population densities, major urban centers, and extensive irrigated agriculture there are large demands for water withdrawals and multiple sources of water contamination. Many rivers are highly regulated to store water, generate hydroelectricity, and facilitate river navigation. Nearly 100 large dams (>15 m high) occur in the seven river basins described in the region and many more are planned. The Parana, Uruguay, and Colorado River Basins have been almost completely deforested, and the Tocantins River Basin, on the margin of the Amazon, has lost 50% of its original forest cover. Loss of natural vegetation and increased erosion are also serious issues in the São Francisco and Parnaiba Basins.

Intensive use of basin and river resources poses significant management challenges for ecosystem conservation. The Pantanal wetland is threatened by a wide variety of issues. Upstream deforestation, agriculture, cities, and mining operations release contaminants into the Paraguay River, which flows into the wetland. Human development within and on the margins of the wetland destroys habitats, and increased population intensifies poaching and overfishing. Nearly 50 species in the Pantanal, including several large mammals, are listed as threatened or endangered with extinction. There are also plans for major river works to improve the navigability of the Paraguay River downstream of the wetlands.

Pacific Basins

The Pacific Coast of South America is drained by more than 100 river basins flowing side-by-side from the crest of the Andes to the coastline. These rivers cut steep canyons into the western slope of the Andes Cordillera and fertile valleys through the coastal plain. In the coastal deserts of northern Chile and Peru, river valleys form verdant corridors that have been lifelines for human communities since prehistoric times. The Andean Cordillera crests within 200 km of the coastline, limiting the extent of Pacific Coast river basins. Most basins are less than 20 000 km^2 in area. The largest of the Pacific Coast basins is the Chilean Loa at 35 346 km^2 followed by the Ecuadorian Guayas River at 32 538 km^2 and the Chilean Serrano at 26 886 km^2. All Pacific basins drain about 1 065 019 km^2 or 6.05% (**Table 1**, **Figure 1**) of the South American continental area and account for ~5% of the continental runoff. Pacific Coast rivers are heavily utilized by coastal populations and management issues linked to contamination and over withdrawals are common, especially in arid zones. Pacific Coast rivers are also sensitive to climate change, as many are fed by glaciers that are rapidly melting.

Geomorphology and Hydrology

Pacific Coast rivers of South America share many geomorphological and hydrological characteristics. They originate at high elevations in the Andes and are often glacier-fed. Their upper reaches are constrained in narrow valleys, while in their lower reaches they may be braided or meander through wide valleys with broad flood plains. Depositional fans develop at the base of the mountains where the rivers enter the coastal plain. In the southern temperate rivers of Chile annual rainfall ranges from 1000 to 3000 mm and peak flows coincide with high rainfall between May and October. North of 30° S latitude in Chile and along the entire Peruvian coast annual rainfall is generally less that 100 mm and rivers flow through desert landscapes. Here peak flows coincide with brief rains in January and February and during the remainder of the year river flows a very small and maintained by glacier melt and groundwater discharge in the higher elevations of the Andes. Along the humid Ecuadorian and Colombian Pacific Coasts annual rainfall is again in excess of 1000 mm and rivers enjoy relatively abundant flow year round.

Biology and Ecology

The biological and ecological characteristics of Pacific Coast river basins vary with both elevation and latitude. Pacific Coast river basins extend from sea level to elevations exceeding 5000 masl and aquatic and riparian biota vary accordingly. Species diversity is constrained by the relatively small size of these basins. Highest species diversity, approaching 100 fish species, occurs in the larger basins in the humid tropical zone. As mentioned previously, in the desert regions river valleys stand out as corridors of green and are of heightened ecological importance.

Management Issues

Because of their limited size and the high demands placed upon them, Pacific Coastal rivers of South America are especially vulnerable to mismanagement. Rivers flowing through the desert regions of Peru and northern Chile are under the most pressure to meet the demands of irrigated agriculture and domestic water needs. Where these desert rivers flow through large urban areas, contamination may be severe, as is the case with the Rimac River, where it flows through the City of Lima, Peru.

Endorheic Basins

A sizable portion of South America is drained by endorheic basins, which when combined cover 1 368 406 km^2 or 7.77% of the continent. This exceeds the area of the Pacific Basins (**Table 1**, **Figure 1**). A chain of endorheic basins stretches from Boliva and Peru, which include the Altiplano basin of Lake Titicaca as well as several salt lakes including Lake Poopó, Salar de Uyuni, and Salar de Coipasa, down to the Southern tip of Argentina. The major land cover of these basins is 60% grassland, savanna, and shrubland, followed by almost 15% cropland (**Table 2**, **Figure 2**).

See also: Flood Plains.

Further Reading

Biswas AK, Cordeiro NV, Braga BPF, and Torjada C (eds.) (1999) *Management of Latin American River Basins.* New York: United Nations University Press.

Global River Discharge Database (2007) Available at. http://www.sage.wisc.edu/riverdata/.

Goulding M, Barthem R, and Ferreira E (2003) *The Smithsonian atlas of the Amazon.* Washington, DC: Smithsonian Institution Press.

Joint Research Center of the European Commission. Brussels, Belgium

McClain ME (ed.) (2002) *The Ecohydrology of South American Rivers and Wetlands.* Wallingford: International Association of Hydrological Sciences Press.

Mittermeier RA, Harris MB, Mittermeier CG, *et al.* (2007) *Pantanal: South America's Wetland Jewel.* London: New Holland Publishers.

Pringle CM, Scatena FN, Paaby-Hansen P, and Núñez-Ferrera M (2000) River conservation in Latin America and the Caribbean. In: Boon PJ, Davies BR, and Petts GE (eds.) *Global Perspectives on River Conservation: Science, Policy, and Practice*, pp. 41–77. New York: John Wiley.

Sioli H (ed.) (1984) *The Amazon: Limnology and Landscape Ecology of a Mighty Tropical River and Its Basin.* Dordrecht: Dr. W. Junk Publishers.

Weibezahn F, Alvarez H, and Lewis WM Jr (eds.) (1990) *The Orinoco River as an Ecosystem.* Caracas: Impresos Rubel.

World Resources Institute (2003) Water Resources Atlas. Available at. http://earthtrends.wri.org.

WWF (2006) Hydrosheds. Available at. http://hydrosheds.cr.usgs.gov/.

Streams and Rivers of North America: Overview, Eastern and Central Basins

A C Benke, University of Alabama, Tuscaloosa, AL, USA

Introduction and Overview

The streams and rivers of North America are described in two articles. This article provides an overview of the continent's streams and rivers and then focuses on eastern and central river basins. East-central basins are divided into three mega-regions: Atlantic basins of the U.S. and Canada, Mississippi River basins, and U.S. Gulf of Mexico basins exclusive of the Mississippi River. Chapter 260 focuses on four remaining mega-regions of the continent: southwestern U.S., Mexico, Pacific Coast, and Arctic/subarctic. Both chapters borrow heavily from *Rivers of North America* (2005), edited by AC Beuke and Colbert Cushing in describing differences and similarities within regions, and major differences among regions in terms of landscape, climate, geomorphology, hydrology, biology, and ecology.

North America contains several million kilometers of flowing waters, beginning with small headwater streams and converging into ever larger streams and rivers. The continent includes a tremendous diversity of streams and rivers as a result of extremes in physiography, geomorphology, climate, and terrestrial vegetation. For example, the gradient (or slope) of small streams varies from very steep in mountains (>10 m drop in elevation per km of stream length) to very flat in coastal plains (<20 cm km^{-1}). Physiography can vary greatly both across and within river basins; e.g., streams of the Mobile River, Alabama, may flow over sandstones and shales of the Appalachian Plateau physiographic province, carbonate rocks of the Valley and Ridge, metamorphic rocks of the Piedmont, and finally over unconsolidated sands and gravels of the Coastal Plain before reaching the Gulf of Mexico. Climate varies from tropical rain forest (mean air temperature >25 °C and mean annual precipitation >180 cm) in southern Mexico to arctic (<−20 °C and <40 cm) in northern Canada and Alaska to desert (<30 cm) in northern Mexico and portions of the western United States. Variation in precipitation and temperature strongly affects the magnitude and seasonal pattern of stream flow, and also influences vegetation and aquatic animals.

The best known paradigm of river ecosystem function, the River Continuum Concept or RCC (1980, published in the *Canadian Journal of Fisheries and Aquatic Sciences*), was developed in North America by Robin Vannote and others, but has been evaluated for many rivers in North America and around the world. The RCC envisions the entire river system from headwater to mouth as an integrated series of physical gradients along with adjustments in associated plants and animals. In particular, it stresses the relative influence of terrestrial vegetation vs. instream photosynthesis as a food source for aquatic animals, the importance of upstream–downstream connectivity, and the shift in functional feeding groups of aquatic invertebrates. Tests of the RCC have been difficult in North America because of major human interventions, such as dams, in both medium-sized and large rivers.

North America's rivers and associated wetlands contain a relatively high biodiversity of aquatic animals and plants, a subject of major interest for both theoretical and conservation aspects. Joseph Nelson and others (2004, *Common and Scientific names of Fishes of the United States, Canada and Mexico*) have summarized all known fish species on the continent, including 1277 freshwater species, the great majority being found in streams and rivers. There are 521 from Mexico, 912 from the United States, and 212 from Canada, according to Paul Hudson and others (2005, in *Rivers of North America*). The highest diversity is in the central/southeastern U.S. and Mexico, the lowest in western and arctic states and provinces. Richard Merritt, Kenneth Cummins, and Martin Berg (2008, *An Introduction to the Aquatic Insects of North America*) have provided a comprehensive key to all known North American aquatic insect genera, and James Thorp and Alan Covich (2001, *Ecology and Classification of North American Freshwater Invertebrates*) have done the same for other freshwater invertebrates. North America probably contains >10 000 species of aquatic insects and >300 species each of crayfishes, freshwater snails, and freshwater mussels, as Colbert Cushing and the author mentioned in *Rivers of North America* (2005). The great majority of invertebrates are also likely found in streams and rivers or associated wetlands.

Unfortunately, many North American streams and rivers have been seriously degraded by pollution and exploitation. Domestic sewage remains a problem in many areas, even though significantly reduced by water quality laws. Industrial polluters have released a wide variety of persistent toxic chemicals that have accumulated in sediments and magnified through food chains. Nonpoint source pollution from agriculture, deforestation, and urbanization is the most widespread category and produces high levels of siltation, nutrients, and pesticides.

Exploitation involves deliberate physical changes such as obliteration of headwater streams, construction of dams and levees, channelization, and water extraction. Headwater streams of many areas have been destroyed or severely degraded by agriculture and urban development. In addition, water often is extracted for agricultural, domestic, or industrial uses, sometimes with disastrous ecological consequences from total dewatering. Medium to large rivers in many areas are strongly regulated by dams, channelization and levees. There were ~79 000 documented dams in the U.S. alone by early 2007 according to the National Inventory of Dams maintained by the U.S. Army Corps of Engineers. In an earlier paper (1990, published in *Journal of the North American Benthological Society*) I estimated there were only 42 reasonably natural free-flowing rivers of ≥200 km length remaining in the coterminous United States. Subsequent documentation of intensive fragmentation throughout North America has been provided by Mats Dynesius and Christer Nilsson (1994, published in *Science*) and by David Allan and the author (2005, in *Rivers of North America*).

The remainder of this chapter focuses on the Atlantic basins of the U.S. and Canada, the Mississippi River basins, and the U.S. Gulf of Mexico basins. Here are found the two largest rivers of the continent, the Mississippi, and St. Lawrence, as well as several large tributaries (**Table 1**, **Figure 1**). These three mega-regions include much of the low-relief areas of the continent (excluding Alaska and central Canada): the Coastal Plain of the southern states, and the Great Plains and Central lowlands of the continent's midsection. Also included are Appalachian and Ozark Mountains with substantially lower elevations (usually <2000 m asl) than mountains from the western part of the continent and Mexico (commonly >4000 m).

Atlantic Basins of the United States and Canada

Atlantic river basins are found along ~35° of latitude from Labrador to Florida. The major feature these basins have in common (besides draining into the Atlantic Ocean) is high precipitation distributed relatively uniformly throughout the year. As a result, they are also relatively similar in containing deciduous, coniferous, or mixed forests. Most of the rivers are small to medium in size, the major exception being the St. Lawrence River, which drains a large inland area that includes the Great Lakes. The largest of the rest include the Churchill, St. John, Hudson, Susquehanna, Santee, and Altamaha (**Figure 1**).

Landscape and Climate

From New Brunswick to Georgia, most rivers drain from the eastern mountains into foothills and through lowlands before emptying into the Atlantic Ocean (from St. John to Savannah). However, the

Table 1 Largest rivers in eastern and central North America exclusive of Arctic drainages, ranked by discharge, and including the largest tributaries of Mississippi and St. Lawrence basins (adapted from Benke and Cushing, 2005)

Rank	*River name*	*Discharge ($m^3\ s^{-1}$)*	*Basin area (km^2)*	*Relief (m)*	*General location or major basin*
1	Mississippi	18 400	3 270 000	>4300	Most of central U.S.
2	St. Lawrence[a]	16 800	1 600 000	1945	Eastern Canada and U.S.
4	Ohio[b]	8733	529 000	2300	Eastern Mississippi basin
8	Upper Mississippi[b]	3576	489 510	337	Northern Mississippi basin
14	Tennessee[c]	2000	105 870	1910	Eastern Mississippi basin
15	Missouri[b]	1956	1 371 017	4277	Northwestern Mississippi basin
16	Ottawa[d]	1948	146 334	911	Eastern St. Lawrence basin
17	Mobile	1914	111 369	1278	Southeastern U.S.
19	Churchill	1861	93 415	549	Eastern Canada
24	Saguenay[d]	1535	85 500	1130	Eastern St. Lawrence basin
Additional large basins					
	Rio Grande	~100	870 000	4272	Southwestern U.S.
	Arkansas[b]	1004	414 910	4340	Southwestern Mississippi basin

Rank indicates river's rank by mean annual discharge for the entire continent. Note that relief (highest peak in basin to mouth of river) is usually <2000 m, except for those rivers that arise in Rocky Mountains (>4000 m).

[a]Discharge includes flow just downstream of confluence with Saguenay.

[b]Tributary of Mississippi.

[c]Tributary of Ohio.

[d]Tributary of St. Lawrence.

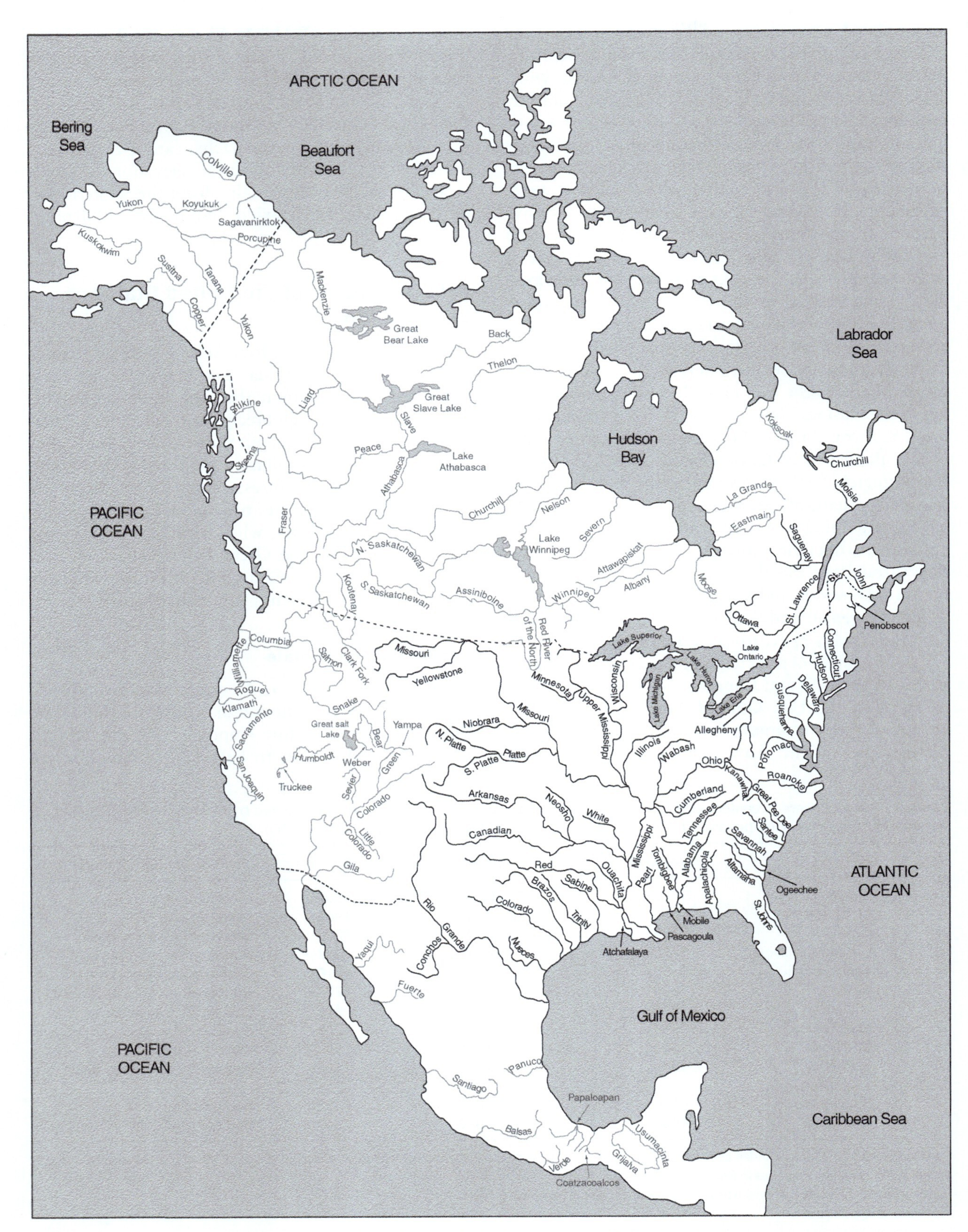

Figure 1 Major rivers of eastern and central North America, including the largest (**Table 1**), and selected other rivers.

Appalachian influence disappears in northern Georgia, with primarily Coastal Plain drainage from southern Georgia to Florida (e.g., Ogeechee, St. Johns). In Labrador and southeastern Quebec, drainage is from the Canadian Shield (e.g., Churchill, Moisie). In the interior St. Lawrence basin, Great Lakes tributaries from the south drain the Central Lowlands and tributaries from the north drain the Canadian Shield.

Precipitation within Atlantic basins is remarkably consistent considering the large range of latitude, typically with >100 cm distributed relatively evenly throughout the year (notice similarity of Ogeechee basin, Georgia, with Moisie basin, Quebec, **Figure 2**). On the other hand, mean annual air temperature varies greatly, from 20 °C (monthly summer means >25 °C) in the south to <0 °C (monthly winter means less than −10 °C) in Labrador. Thus, substantial amounts of frozen precipitation are stored throughout much of the northern winters.

Streams mostly drain temperate deciduous forest in the midsection of the Atlantic region from New Brunswick to the southern states. To the north, in southeastern Quebec and Labrador, vegetation shifts to coniferous Boreal Forest. Although deciduous forests extend well into the south, at least the lower reaches of rivers from New Jersey to Florida cross the relatively flat Coastal Plain typically dominated by conifers in uplands and hardwood swamps in lowlands. Agriculture is usually no more than 25% of basin land use in the south and substantially less in New England and Canada.

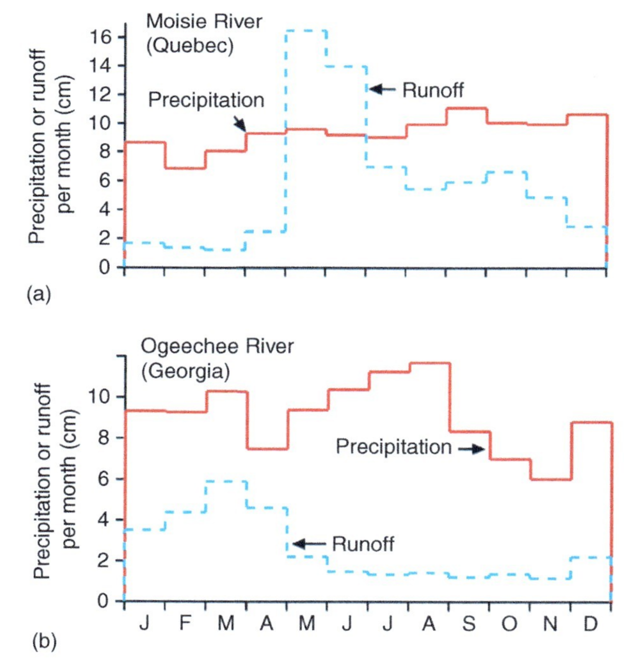

Figure 2 Mean monthly precipitation and runoff for rivers of the Atlantic coast of North America. (a) Moisie River of southern Quebec. (b) Ogeechee River of Georgia. Adapted from Cunjak and Newbury (2005, *Rivers of North America*) and Smock *et al.* (2005, *Rivers of North America*) with permission.

Geomorphology and Hydrology

Aside from the northern and southern extremes, the most common type of Atlantic river system begins at a relatively high gradient (≥2 m km^{-1}) in the Appalachian Mountains with riffle, run and pool morphology and a substrate often dominated by boulders, cobble and gravel (**Figure 3**). As the rivers drop into the Piedmont (from New Jersey to Georgia), gradients decrease to ~50 cm km^{-1} with sometimes large areas of bedrock, and also cobble, gravel, and sand. A marked change occurs when the river flows over a short, rocky, and steep transition zone (called the fall line) into the Coastal Plain, sometimes dropping at 2 m km^{-1}. Once in the Coastal

Figure 3 Stream below Anna Ruby Falls in southern Appalachian Mountains of Georgia. Stream flows into Appalachicola River system that empties into the Gulf of Mexico, but is like many other Appalachian streams that flow into Atlantic drainages. Note boulders and high gradient. Photo by Arthur Benke.

Plain, rivers typically have sand and silt beds and meander within a forested flood plain several kilometers in width and with gradients as low as 10 cm km^{-1} (**Figure 4**). Rivers traversing such a pathway include the Delaware, Potomac, James, Savannah, Great Pee Dee, and Santee (**Figure 1**).

The seasonal hydrology of Atlantic rivers is strongly influenced by climate. In spite of relatively uniform precipitation throughout the year, there are consistent patterns of high seasonal discharge influenced by two major factors. The first, most obvious from southeastern rivers, is that high evapotranspiration in summer results in considerably lower runoff than in winter (Ogeechee River, **Figure 2(b)**). This pattern is weaker in northern rivers, and is overridden when snow/ice storage greatly reduces winter flows until spring snowmelt (Moisie River, **Figure 2(a)**). For the many dammed rivers along the Atlantic coast, the characteristic monthly hydrographs are artificially smoothed, sometimes to a substantial degree (e.g., the remote Churchill River in Labrador). This can be deceptive, however, because actual discharge fluctuates much more than suggested by monthly averages, and hydropower releases often result in daily and weekly fluctuations of greater than ten-fold.

Figure 4 Aerial view of Ogeechee River, Georgia, meandering within a broad floodplain swamp. This is a low gradient river, with a sandy bed (note white sand bars) and darkly stained water. The Ogeechee is an unregulated river that begins in the Piedmont, but flows most mostly through the Coastal Plain. Photo by Arthur Benke.

Biology and Ecology

River systems that begin in the Appalachian Mountains are often reasonably consistent with the RCC model. Small high-gradient streams are heavily shaded by forest and receive large inputs of organic matter (leaves and wood) that function in habitat formation and as food sources (**Figure 3**). As streams become larger and more exposed to sunlight, production of algae, mosses, and aquatic vascular plants increases and provides habitat and food for a diverse and productive invertebrate assemblage. For rivers flowing into the Coastal Plain, ecological characteristics change greatly. With primarily a shifting sandy bottom, the major stable habitat for invertebrates is the wood of fallen trees (snags), typically anchored along the shore (unless removed by human snagging operations). Furthermore, the wide floodplain forest becomes a large aquatic habitat during flooding for several weeks or months of the year. Several small rivers from the mid-Atlantic to Florida originate in the Coastal Plain, have a strong floodplain influence throughout their length, often have low pH and alkalinity, and have a characteristic blackwater appearance, the result of high concentrations of dissolved humic acids (**Figure 4**).

Freshwater fish diversity is relatively high in rivers of the south Atlantic, commonly ~100 species. Diversity tends to dramatically decline north of the Delaware River, and drops to <50 in most rivers of New England and eastern Canada due to the strong influence of past glaciation. Invertebrate diversity can also be very high wherever water is unpolluted and there is a natural flow regime. Dams in most rivers have had a dramatic adverse effect on fish and invertebrate abundance, species composition and diversity. They have been especially devastating to migratory species such as Atlantic salmon in northeastern rivers. Introduction of nonnative species have also caused serious problems for native species.

Mississippi River Basins

The Mississippi River basin is the largest in North America by area (~3.3 million km^2) and total flow (18 400 $m^3 s^{-1}$) and the third largest by area in the world (**Table 1**). The Upper Mississippi flows from the north beginning in Minnesota and Wisconsin, but it is joined by two large tributaries, the Missouri and Ohio (**Figure 1**). The Missouri flows from as far west as western Montana (eastern slope of Rocky Mountains) and is the longest and largest in basin area of all Mississippi tributaries. The Ohio River flows from the east (as far as western New York and Pennsylvania) and has a higher discharge ($8733\,m^3\,s^{-1}$) than the Upper Mississippi and Missouri combined (**Table 1**). Four other large ($800–1000\,m^3\,s^{-1}$) tributaries flow from the west across the Great Plains (Arkansas, Red) or from the Ozark/Ouachita uplands (White, Ouachita).

Landscape and Climate

The most distant headwaters of the Missouri and Arkansas rivers are located in the Rocky Mountains, but most of their tributaries are within relatively arid and flat grasslands of the Great Plains and Central Lowlands. Precipitation and runoff from grassland tributaries are quite low, but increase in the wetter Central Lowlands and Ozark/Ouachita uplands. Precipitation in the Yellowstone basin (part of Missouri basin in Wyoming and Montana) typifies the pattern of the Great Plains, peaking in early summer ($5–10\,cm\,month^{-1}$) and declining substantially in winter (**Figure 5(b)**).

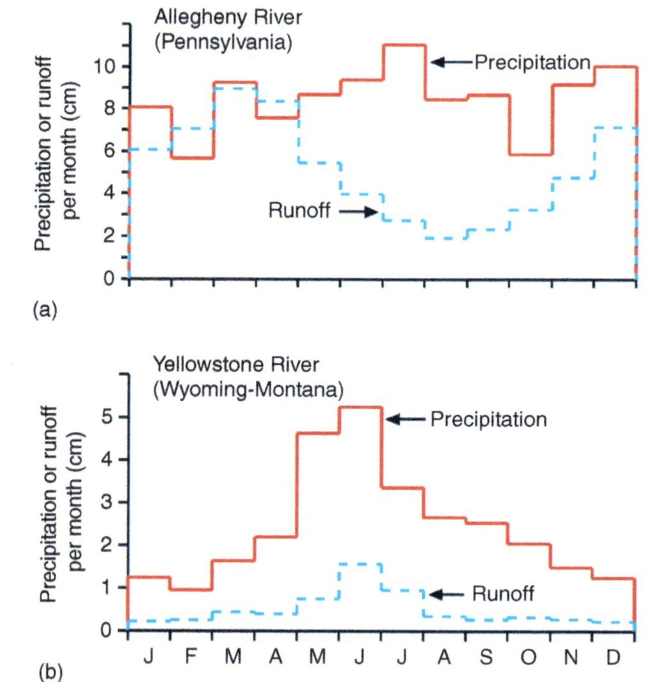

Figure 5 Mean monthly precipitation and runoff for rivers of the Mississippi River drainage of North America. (a) Allegheny River of western Pennsylvania. (b) Yellowstone River of northwestern Wyoming and Montana. Adapted from White *et al.* (2005, Rivers of North America) and Galat *et al.* (2005, *Rivers of North America*) with permission.

The Upper Mississippi River drains primarily from the Central Lowlands and includes grasslands in the western portion and deciduous forest in the east, although about 70% has been converted to agriculture. The basin retains some of the seasonal precipitation pattern of the Great Plains, with low precipitation in winter, but the relatively high precipitation ($\sim 10\,cm\,month^{-1}$) often extends for 7–8 months.

The Ohio River (including its large southernmost tributary, the Tennessee River) drains partly from the western slope of the Appalachian Mountains, but much of the drainage is from tributaries of the Central Lowlands and interior plateaus, most of which was originally deciduous forest. Although almost half is second-growth forest, most of the rest has been converted to agriculture. Precipitation patterns are much more similar to Atlantic basins than western basins of the Mississippi, with $8–12\,cm\,month^{-1}$ for most months (Allegheny River, Pennsylvania, **Figure 5(a)**).

Geomorphology and Hydrology

There is considerable diversity in river geomorphology and substrate within the Mississippi River basin. Streams arising on the western slopes of the Appalachians and on the eastern slopes of the Rockies have high gradients ($>5\,m\,km^{-1}$), often with large waterfalls, and coarse substrates of boulders and cobbles (**Figure 6**). Rivers that flow across the Great Plains and Central Lowlands, have much lower slopes ($<1\,m\,km^{-1}$) with substrate composition possibly with cobble/gravel and eventually shifting to gravel/sand/silt (**Figure 7**).

The hydrology of rivers in the Mississippi basin is strongly dictated by the seasonality and magnitude of precipitation and evapotranspiration which varies from west to east. The Yellowstone River illustrates the western pattern with the generally low runoff peaking in early summer, following the peak of precipitation and some degree of snowmelt influence (**Figure 5(b)**). In contrast are the much higher runoff and peak winter–spring flows seen for many tributaries east of the Mississippi and south of the Great Lakes (Allegheny River, **Figure 5(a)**), primarily the result of seasonal differences in evapotranspiration (and snow/ice storage in the north) rather than precipitation. The significance of east–west variation in precipitation and evapotranspiration is also illustrated by comparison of the Tennessee River (annual precipitation = 105 cm) with the Missouri River (annual precipitation = 50 cm). Although the Tennessee basin is only 8% of the drainage area of the Missouri, their mean discharge is almost identical ($\sim 2000\,m^3\,s^{-1}$, **Table 1**).

Figure 6 High gradient stream flowing from Glacier National Park, Montana, on the eastern slope of the Rocky Mountains and part of the Missouri River basin. Photo by Arthur Benke.

The hydrology and other natural ecological features of each major tributary of the Mississippi are strongly altered by regulation. There are 6 major dams on the Missouri, 5 on the Arkansas, 5 on the Ouachita, 4 on the White, 20 locks and dams on the Ohio, and 26 locks and dams on the Upper Mississippi. Many more dams occur on smaller tributaries. The lower Mississippi has been spared from large dams, but nonetheless has been extensively altered and degraded in an attempt to improve navigation and control floods. There are approximately 3000 km of levees over 9 m in height along the lower Mississippi according to Arthur Brown and others (2005, *Rivers of North America*).

Biology and Ecology

In general, smaller streams and rivers of the Great Plains have less shading by trees than seen in the eastern streams and represent a substantial variant of the RCC. With more open stream canopies, in-stream photosynthesis appears to be more important in the Great Plains than forested streams in the eastern Mississippi basin. These western streams are often subject to harsh physical conditions that include late summer drying, high temperatures, and high alkalinity, nutrients and suspended solids, providing a physical/chemical challenge for invertebrate and fish survival. Such stresses typically result in lowered animal diversity and productivity under the best of conditions. In addition to natural stresses, extensive

Figure 7 Niobrara River, northern Nebraska, which empties into the Missouri River upstream of Lewis and Clark Lake. This is one of the least altered rivers flowing across the Great Plains. Photo by Alexander Huryn.

water extractions for irrigation, municipal, and mining needs are widespread throughout the Great Plains, often resulting in greatly reduced flows that may cause even larger streams to dry. Channelization, dams, deforestation, snag removal and extensive damming have greatly changed the geomorphology of the main stem rivers, and degraded the biota. Although physical conditions within streams of the Upper Mississippi basin are less harsh than those of the Great Plains, human impacts are similarly severe; agricultural development has resulted in nutrient concentrations that are often extremely high; NO_3-N $>3\,mg\,l^{-1}$. Ecosystem characteristics of eastern streams (Ohio River basin and eastern portion of Upper Mississippi) are much more similar to those described for the Atlantic basins than those in the Great Plains, with terrestrial inputs being of major importance to streams with intact streamside vegetation. Furthermore, natural physical conditions of the eastern Mississippi drainage are less severe than in the Great Plains, although human alterations, such as dams and channelization have degraded many reaches.

Freshwater fish diversity throughout the basin reflects biogeography, extent of Pleistocene glaciation, and natural harshness. Rivers draining only from the Great Plains, with its harsh physical conditions, typically have <50 native fish species, whereas large rivers such as the Arkansas, with tributaries draining highly diverse Ozark/Ouachita uplands, can have >100. Rivers of the Upper Mississippi typically have fish diversity of ~100 species, even in areas that experienced glaciation. But the highest diversity is found in the Tennessee/ Cumberland rivers portion of the Ohio River basin, which did not experience glaciation. According to David White and others (2005, *Rivers of North America*), the Tennessee River alone has >225 fish species, with many endemics, and along with the Mobile River system to its south, has the highest fish diversity on the continent. This species richness also extends to some invertebrate groups, such as mussels and crayfishes. Unfortunately, the natural diversity of the entire Mississippi basin is seriously threatened by damming, channelization, navigation, levees, and pollution, and the Tennessee and Cumberland rivers are no exception. Arthur Brown and others (2005, *Rivers of North America*) report that the lower Mississippi River main stem is thought to have once had ~150 freshwater fish species, but only ~90 remain.

U.S. Gulf of Mexico Basins, Exclusive of Mississippi River

In addition to the main stem of the Mississippi River, there are many smaller U.S. rivers draining into the Gulf of Mexico (**Figure 1**). East of the Mississippi, rivers flow out of Florida, Georgia, Alabama and Mississippi. To the west are rivers draining portions of Louisiana, Texas, and northern Mexico, including the largest (in basin area) and westernmost river, the Rio Grande.

Landscape and Climate

A common feature of all U.S. Gulf coast rivers is that all or part of their drainage is within the Coastal Plain and they are characterized by low gradients. Although the largest basins in the eastern Gulf (Apalachicola and Mobile) have their headwaters in the lower end of the Appalachian Mountains or Piedmont, many drainages from Florida to eastern Texas flow only through the Coastal Plain (e.g., Suwannee, Pearl, Pascagoula, Sabine). The westernmost rivers of Texas (e.g., Trinity, Brazos, San Antonio/Guadalupe) typically have their upper portions in the Central Lowlands or Great Plains. The exception is the Rio Grande which begins its journey in the southern Rockies of Colorado and New Mexico, and drains portions of northern Mexico and western Texas. Along its path it has tributaries draining Basin and Range, Colorado Plateau, and Sierra Madre physiographic provinces.

These basins are primarily in the warm southern U.S., but there is a substantial decline in precipitation from east to west. Basins from Georgia to eastern Texas have high annual rainfall (>120 cm), and for many it is distributed relatively evenly throughout the year (Pearl River, Mississippi, **Figure 8(b)**), much like Atlantic river basins. Rivers in the westernmost Gulf, however, sometimes receive less than half such amounts (Nueces River, western Texas, **Figure 8(a)**).

Much of the Gulf drainage is characterized by Coastal Plain forests (unless converted to agriculture), but as rainfall declines from central to western Texas, basin vegetation changes from forests to grasslands (Great Plains). The Rio Grande, on the other hand, is one of the longest rivers in North America and begins in conifer forests of the Southern Rockies. Most of the basin, however, drains desert shrubland and grassland in New Mexico, northern Mexico, and western Texas, before reaching the Gulf.

Geomorphology and Hydrology

The headwaters of the largest U.S. Gulf rivers (by discharge), the Mobile and Apalachicola rivers of Alabama and Georgia, arise at moderately high gradients, with variable rocky substrates in the southern Appalachian Mountains, the Appalachian Plateau, the Valley and Ridge, or the Piedmont (**Figure 3**). At least the lower portions of these large rivers and most

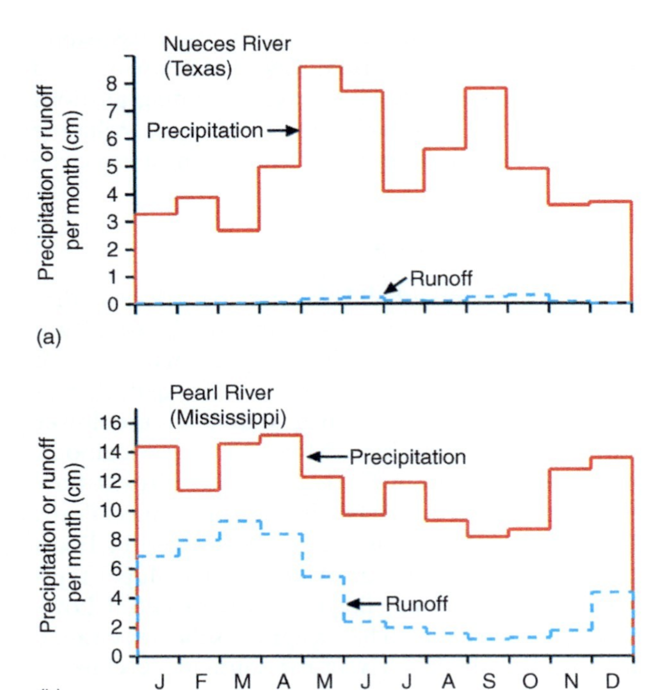

Figure 8 Mean monthly precipitation and runoff for rivers of the U.S. Gulf of Mexico coast. (a) Nueces River of western Texas. (b) Pearl River of Mississippi. Adapted from Dahm *et al.* (2005, *Rivers of North America*) and Ward *et al.* (2005, *Rivers of North America*) with permission.

of the other rivers pass primarily through the Coastal Plain with low gradients, as in the south Atlantic (**Figure 4**). Most of these Coastal Plain basins are quite unlike those envisioned for the RCC in having no mountainous headwaters and draining low-lying floodplain swamps from headwaters to mouth. Stream gradients are usually $<40\,\mathrm{cm\,km^{-1}}$, often much lower, and the bed is characterized by shifting sand or silt. Such rivers show considerable meandering among floodplain swamps and coastal wetlands, unless altered by channelization or dams.

In spite of the even distribution of rainfall throughout most of the U.S. Gulf, there is the same strong seasonal pattern of winter-spring high flows as shown for the south Atlantic rivers due to the influence of reduced evapotranspiration in winter (Pearl River, **Figure 8(b)**). These winter high flows are also reflected in extensive inundation of floodplain swamps, except where dams and channelization influence the hydrograph. In contrast, the Nueces River of western Texas has low precipitation and high evapotranspiration resulting in very low runoff (**Figure 8(a)**). Precipitation is even lower in the large Rio Grande basin ($21\,\mathrm{cm\ year^{-1}}$) and combined with naturally high evapotranspiration, mean annual discharge has been approximated as only $\sim 100\,\mathrm{m^3\ s^{-1}}$ prior to impoundments. By contrast, the Mobile River basin (precipitation $= 128\,\mathrm{cm\,year^{-1}}$) is only one-eighth the area of the Rio Grande but its virgin discharge is 19 times greater ($1914\,\mathrm{m^3\,s^{-1}}$). The combination of naturally low flows, damming, and water extractions for agriculture in the Rio Grande has resulted in extended periods (several months) of no-flow conditions that must be devastating to aquatic life.

Biology and Ecology

An essential feature of the natural ecology of these coastal rivers is a strong dependence on floodplain inundation during winter, much like rivers of the south Atlantic. The floods carry essential particulate organic matter to river invertebrates, and the inundated floodplain is the major breeding and feeding site for many fishes. Submerged wood is an important substrate for invertebrate diversity and production that represents an important food source for many fish species. This dependency of rivers on their flood plains and submerged wood is obvious in smaller tributaries and in the few natural rivers that remain, but it is much less apparent in the rivers that have been dammed, snagged and channelized. Rivers of central and western Texas are naturally stressed and exploited like Great Plains rivers that flow into the Mississippi River and suffer even more from the same types of exploitation in the eastern Gulf.

Rivers in the eastern portion of the U.S. Gulf typically have close to or >100 fish species, with the Mobile River system in Alabama having 236 species (178 strictly freshwater; the rest euryhaline or occasional marine intruders). All but about 10 species are native. Like the Tennessee/Cumberland rivers to the north, the high species richness of the Mobile River system is at least partially due to its physiographic diversity and lack of glaciation. Species richness of several other freshwater vertebrate and invertebrate groups is also extremely high in the Mobile River basin, and to a lesser extent for the smaller eastern Gulf basins. The more arid rivers in the western Gulf, however, typically have only 60–80 fish species.

Summary of Human Impacts

Eastern and central rivers of North America are among the most heavily fragmented rivers of the continent, and all of the largest rivers are strongly fragmented (**Table 2**). This is particularly true for the northeast Atlantic rivers and in the Ohio River basin. Although human population densities are highest ($>100\,\mathrm{ind.\ km^{-2}}$) in the smaller more urbanized basins of the eastern U.S. (e.g., Delaware, Hudson,

Table 2 Factors that reflect human influences on the largest rivers of eastern and central North America

River name	*Pop. density (ind. km^{-2})*	*% Agr*	*NO_3-N (mg l^{-1})*	*PO_4-P (mg l^{-1})*	*Fragmentation by dams*
Mississippi	22	50	1.40	0.13	2
St. Lawrence	54	20	<0.18	0.01	2
Ohio[a]	49	48	1.60	<0.17	2
Upper Mississippi[a]	54	70	3.20	0.19	2
Tennessee[b]	19	36	0.20	<0.01	2
Missouri[a]	8	37	0.02–3.20	0.01–0.23	2
Ottawa[c]	24	2	<0.17	<0.03	2
Mobile	44	18	0.27	0.02	2
Churchill	<1	0	NA	NA	2
Saguenay[c]	3	1	<0.10	0.01	2
Additional large basins					
Rio Grande	16	5	NA	NA	2
Arkansas[a]	15	45	0.25	0.02	2

Qualitative assessment of fragmentation by dams primarily based on Benke and Cushing (2005), but also Dynesius and Nilsson (1994): 0 – no or minor fragmentation, 1 – moderate fragmentation, 2 – strong fragmentation. % Agr. – % of basin used for agriculture. NA – not available.
[a]Tributary of Mississippi.
[b]Tributary of Ohio.
[c]Tributary of St. Lawrence.

Potomac), human densities in some of the largest basins (St. Lawrence, Ohio, Upper Mississippi) are roughly 50 ind. km^{-2} (**Table 2**). Concentrations of nutrients such as nitrate and phosphate are very high in those Mississippi River basins having agricultural land use of about 50% or higher (**Table 2**). Such high nutrients, particularly nitrates, are largely responsible for the growing 'dead zone' of the Gulf of Mexico. Although many streams and rivers of eastern and central North America have reaches of considerable beauty and high biodiversity, their waters are obviously in high demand for multiple uses, even along the Atlantic and eastern Gulf coasts where precipitation is plentiful. Such demands will continue to place a high stress on their ability to function as natural ecosystems and sustain their natural diversity and productivity. Relatively few of these east-central rivers receive protection as U.S. Wild and Scenic Rivers or Canadian Heritage Rivers.

Glossary

Discharge – The volume of water flowing in a stream, usually measured as cubic meters per second (m^3 s^{-1}) or cubic feet per second (ft^3 s^{-1}). Often presented as the average for a longer time; e.g., the average discharge for July is 510 m^3 s^{-1}.

Diversity – Generally used in the ecological literature to indicate the number of species within a biological community or ecosystem.

Ecosystem – A community of organisms and their nonliving environment interacting as a unit.

Evapotranspiration – The combined processes of evaporation of water and transpiration by plants, which together describe the loss of water to the atmosphere from an ecosystem.

Physiographic province – Subdivisions of the continent based on topographic features, rock type, and geological structure, and history: e.g., Coastal Plain, Great Plains, Basin and Range.

Production – The amount of algal or plant matter (primary production) or animal matter (secondary production) formed in an ecosystem over some period, such as a year (e.g., g m^{-2} year^{-1}).

Runoff – The amount of water draining from a basin presented as cm of water height for some interval of time (e.g., 120 cm year^{-1}). Calculated by dividing discharge (e.g., m^3 year^{-1}) by basin area (km^2).

See also: Africa; Algae of River Ecosystems; Asia – Eastern Asia; Asia – Monsoon Asia; Asia – Northern Asia and Central Asia Endorheic Rivers; Australia (and Papua, New Guinea); Benthic Invertebrate Fauna; Benthic Invertebrate Fauna, River and Floodplain Ecosystems; Benthic Invertebrate Fauna, Small Streams; Benthic Invertebrate Fauna, Tropical Stream Ecosystems; Biological Interactions in River Ecosystems; Climate and Rivers; Coarse Woody Debris in Lakes and Streams; Conservation of Aquatic Ecosystems; Currents in Rivers; Ecology and Role of Headwater Streams; European Rivers; Flood Plains; Floods; Geomorphology of Streams and Rivers; Hydrology: Rivers; Hydrology: Streams; Regulators of Biotic Processes in Stream and River Ecosystems; Restoration Ecology of Rivers; Riparian Zones; South America; Streams and Rivers as Ecosystems; Wetlands of Large Rivers: Flood plains.

Further Reading

Abell RA, Olson DM, Dinerstein E, *et al.* (2000) *Freshwater Ecoregions of North America: A Conservation Assessment.* Washington, DC: Island Press.

Benke AC (1990) A perspective on America's vanishing streams. *Journal of the North American Benthological Society* 9: 77–88.

Benke AC and Cushing CE (eds.) (2005) *Rivers of North America.* Burlington, MA: Academic Press/Elsevier.

Brown LR, Gray RH, Hughes RM, and Meader MR (eds.) (2005) Effects of urbanization on stream ecosystems. *American Fisheries Society Symposium 47.* Bethesda, MD: American Fisheries Society.

Cushing CE, Cummins KW, and Minshall GW (eds.) (1995) *Ecosystems of the World 22: River and Stream Ecosystems. Amsterdam: Elsevier.*

Dynesius M and Nilsson C (1994) Fragmentation and flow regulation of river systems in the northern third of the world. *Science* 266: 753–762.

Hughes RM, Wang L, and Seelback PW (eds.) (2006) Landscape influences on stream habitats and biological assemblages. *American Fisheries Society Symposium 48.* Bethesda, MD: America Fisheries Society.

Karr JR, Allan JD, and Benke AC (2000) River conservation in the United States and Canada. In: Boon PJ, Davies BR, and Petts GE (eds.) *Global Perspectives on River Conservation: Science, Policy and Practice*, pp. 3–39. Chichester, UK: Wiley.

Naiman RJ, DéCamps H, and McClain ME (2005) *Riparia: Ecology, Conservation, and Management of Streamside Communities.* Amsterdam: Academic Press/Elsevier.

Nelson JS, Crossman EJ, Espinosa-Pérez H, *et al.* (2004) *Common and Scientific Names of Fishes of the United States, Canada and Mexico.* Bethesda, MD: American Fisheries Society.

Palmer T (1996) *America by Rivers.* Washington, DC: Island Press.

Revenga C, Murray S, Abramovitz J, and Hammond A (1998) *Watersheds of the World: Ecological Value and Vulnerability.* Washington, DC: World Resources Institute.

Rinne JN, Hughes RM, and Calamusso B (eds.) (2005) Historical changes in large river fish assemblages of the Americas. *American Fisheries Society Symposium 45.* Bethesda, MD: American Fisheries Society.

Vannote RL, Minshall GW, Cummins KW, Sedell JR, and Cushing CE (1980) The river continuum concept. *Canadian Journal of Fisheries and Aquatic Sciences* 37: 130–137.

Wohl EE (2004) *Disconnected Rivers: Linking Rivers to Landscapes.* New Haven: Yale University Press.

Relevant Websites

http://www.daac.ornl.gov/RIVDIS/rivdis.html.
http://earthtrends.wri.org/maps_spatial/watersheds/ncamerica.php.
http://waterdata.usgs.gov/nwis/sw.
http://webworld.unesco.org/water/ihp/db/shiklomanov/index.shtml.
http://www.weatherbase.com/.
http://scitech.pyr.ec.gc.ca/climhydro/.
http://www.americanrivers.org/.
http://www.chrs.ca/.
http://www.nps.gov/ncrc/programs/rtca/nri/.
http://www.rivers.gov/.
http://www.sage.wisc.edu/riverdata/.
http://www.nationalgeographic.com/wildworld/terrestrial.html.
http://enpsychlopedia.org/psypsych/List_of_rivers_in_the_United_States%23State_river_lists.
http://ga.water.usgs.gov/edu/earthrivers.html.

Streams and Rivers of North America: Western, Northern and Mexican Basins

A C Benke, University of Alabama, Tuscaloosa, AL, USA

Introduction

This is the second of two articles on streams and rivers of North America. This article focuses on western, northern, and Mexican basins of the continent and is divided into four megaregions: arid southwestern basins of the United States, Mexican basins exclusive of the Rio Grande, Pacific basins of the United States and Canada, and Arctic and Subarctic basins. The first article provides an overview of the continent's streams and rivers and focuses on the three other megaregions in the east-central part of the continent. The brief descriptions of rivers in these articles borrow heavily from *Rivers of North America* (2005) edited by Colbert Cushing and me in describing differences and similarities within each region, and major differences among regions in terms of landscape, climate, geomorphology, hydrology, biology, and ecology.

Although the two larger rivers of the continent are found in the previous article (Mississippi and St. Lawrence), the next seven larger rivers flowing to the sea are described in this article, including the third largest, the Mackenzie (with several large tributaries), as well as the Columbia, Yukon, Fraser, Nelson, and Usumacinta/Grijalva (**Table 1; Figure 1**). These large rivers span the continent from the tropics of southern Mexico (Usumacinta/Grijalva) to the Arctic (Mackenzie).

The regions covered here include rivers draining the highest mountains of North America, such as Mts. McKinley in Alaska, Logan in the Yukon Territory, and Orizaba and Popocatépetl in Mexico. In contrast to the relatively low mountains (<2000 m asl) from eastern North America described in the previous article, many basins from Alaska to Mexico begin at elevations close to or >4000 m (**Table 1**), and many have had recent volcanic activity. Mountain gradients are thus characteristically very steep and several drain from glaciers. Furthermore, these regions include rivers that have carved the most spectacular canyons of North America, such as the Grand Canyon of Arizona and Copper Canyon of northwestern Mexico.

Arid Southwestern Basins of the United States (Colorado River, Great Basin)

The major rivers of the arid southwestern US are primarily found in the Colorado River basin, the Great Basin, and the Rio Grande basin (**Figure 1**). Only the first two are discussed here (see previous article for Rio Grande). The Colorado is the largest river in the southwest, has the seventh largest basin in North America, and drains from seven southwestern states. The Great Basin lies to the northwest (mostly in Nevada and Utah) and consists of several small rivers terminating inland, such as in the Great Salt Lake.

Landscape and Climate

The upper Colorado basin lies within the intermountain plateaus of the Middle and Southern Rocky Mountains of Wyoming, Colorado, and Utah. Many upper Colorado tributaries arise from mountains with peaks >3000 m asl before flowing into the Wyoming Basin or the Colorado Plateau physiographic provinces. The lower Colorado begins as the river enters Arizona, flows through the Grand Canyon, and enters the Basin and Range province on its path to the Gulf of California. The plateaus of the upper basin receive relatively little precipitation (<26 cm year^{-1}) compared to the mountains that provide most of the flow (Yampa River, **Figure 2(a)**) according to Dean Blinn and LeRoy Poff (2005, *Rivers of North America*). The lower basin receives even less; e.g., 17 cm year^{-1} for the Little Colorado basin (**Figure 2(b)**).

The Great Basin is a series of contiguous desert basins and mountain ranges between the Sierra Nevada and Cascades Ranges on the West, the Rocky Mountains on the east, the Snake River plain of Idaho on the north, and the Sonoran and Mohave deserts on the south. Precipitation is captured primarily by mountains, with some rivers (Weber and Bear) flowing mostly through conifer forests before reaching the Great Salt Lake. Others flow about half-way in mountains (Truckee) before flowing through desert shrub steppe, and still others (Humboldt and Sevier) arise in arid mountains before flowing mostly through desert before reaching their desert wetland lakes (now primarily dry because of water extractions).

In spite of the arid nature of the southwestern US, there is substantial use of the landscape for agriculture and grazing, often causing degradation of native vegetation and soils. Mean annual air temperatures are relatively cool at higher altitudes (<10 °C), but summer temperatures in the deserts of the lower Colorado basin exceed 40 °C.

Table 1 Largest rivers of western and northern North America, including Mexico (exclusive of Rio Grande), ranked by virgin discharge, and including largest tributaries of Mackenzie basin

Rank	*River Name*	*Discharge ($m^3 s^{-1}$)*	*Basin area (km^2)*	*Relief (m)*	*General location or major basin*
3	Mackenzie	9020	1 743 058	3620	Northwestern Canada
5	Columbia	7730	724 025	4392	Western U.S. and Canada
6	Yukon	6340	839 200	6200	Alaska and western Canada
7	Fraser	3972	234 000	3954	Southwestern Canada
9	Slave[a]	3437	606 000	3500	Mackenzie basin
10	Usumacinta/Grijalva	2678	112 550	3800	Southern Mexico
11	Nelson	2480	1 072 300	3370	Mostly southcentral Canada
12	Liard[a]	2446	277 000	2573	Mackenzie basin
13	Koksoak	2420	133 400	~600	Quebec
18	Kuskokwim	1900	124 319	>3550	Western Alaska
20	Copper	1785	63 196	>2500	Alaska and Yukon Territory
21	Skeena	1760	54 400	2755	British Columbia
22	La Grande	1720	96 866	~600	Quebec
23	Stikine	1587	51 592	>2900	British Columbia and Alaska
25	Susitna	1427	51 800	>4000	Southern Alaska
Additional large basin					
	Colorado	550	642 000	4100	Southwestern US

Rank indicates the river's rank by mean annual discharge for the entire continent. See **Figure 1**

Adapted from Benke AC and Cushing CE (eds.) (2005) *Rivers of North America*. Burlington, MA: Academic Press/Elsevier

[a]Tributary of Mackenzie.

Geomorphology and Hydrology

Many streams of the southwestern US begin in mountains at high gradients (>10 m drop in elevation per km of stream length) flowing over boulders and cobble as they quickly descend to the valleys and eventually into arid plateaus. In the Colorado basin, the headwaters generally lie in crystalline, granite bedrock, but the larger tributaries flow through erodible sedimentary deposits and sometimes carve enormous canyons.

Stream runoff can range from moderate to very low. Rivers with substantial mountain drainage, such as the Yampa River of the Colorado basin have fairly high runoff expressed in a strong snowmelt effect (**Figure 2(a)**). In contrast, some rivers such as the Little Colorado show extremely low runoff (**Figure 2(b)**). Mean monthly runoff from desert streams and rivers can be very misleading because they are subject to natural short-term extremes with both flash floods and complete surface drying (**Figure 3**).

The Colorado River has the highest virgin discharge of any river in the southwest (>550 $m^3 s^{-1}$), but is heavily impacted by >40 flow regulation dams and substantial extractions (within basin) and diversions (out of basin) according to Dean Blinn and LeRoy Poff (2005, *Rivers of North America*). Mean annual discharges of the Green River (172 $m^3 s^{-1}$) and the upper Colorado River (74 $m^3 s^{-1}$) are greater than actual annual discharge in the lower Colorado (~40 $m^3 s^{-1}$) in recent decades due to extractions and major diversions from the lower basin, particularly the California Aquaduct. Extractions and diversions from many smaller rivers (e.g., Gila River) severely alter their hydrology/ecology as well. Rivers within the Great Basin that once flowed to endorheic wetlands or lakes now are greatly reduced by extractions. The Sevier and Humboldt rivers typically yield negligible or infrequent flows to the now intermittently dried beds of Lake Sevier and the Humboldt Sink; reduced flows of the Truckee and Walker rivers have caused Pyramid Lake to fall 26 m and Walker Lake to fall 40 m over the past century according to Dennis Shiozawa and Russell Rader (2005, *Rivers of North America*).

Biology and Ecology

Studies of desert streams, particularly in Sycamore Creek (Gila basin), have been noteworthy in emphasizing how such streams differ from the River Continuum Concept (RCC, see first article) and have been summarized by Stuart Fisher (1995 in *Ecosystems of the World 22: Rivers and Stream Ecosystems*). These differences include the importance of instream production, intermittent flash flooding and drying, rapid recovery of algal and invertebrates communities, and the role of the hyporheic (subsurface) zone. Nonetheless, like the RCC, streamside vegetation is considered an integral part of natural stream ecology, from steep forested streams to an ever narrowing vegetation strip, as streams flow into increasingly dry landscapes (**Figure 3**).

Figure 1 Major rivers of western, northern, and Mexican North America, including the largest (**Table 1**), and selected other rivers.

Animal diversity of southwestern streams is a mixed story. On one hand, aquatic insects have been well studied in small to medium sized streams, with high diversity from mountains. On the other hand, fish diversity is low, even in unaltered systems. The Colorado River has 42 native fish species, but there is a high degree of endemism so that individual tributaries, such as the Green, typically have <15 species according the Dean Blinn and LeRoy Poff (2005, *Rivers of North America*). Furthermore, over 85% of all species are threatened from dams, extractions, pollution, and introduction of >70 non-native fishes. Native fish diversity in the Great Basin is also low, as described by Dennis Shiozawa and Russell Rader (2005, *Rivers of North America*) with individual rivers containing only 7–17 species and many endangered by the same problems, including more than a dozen non-native fishes per river.

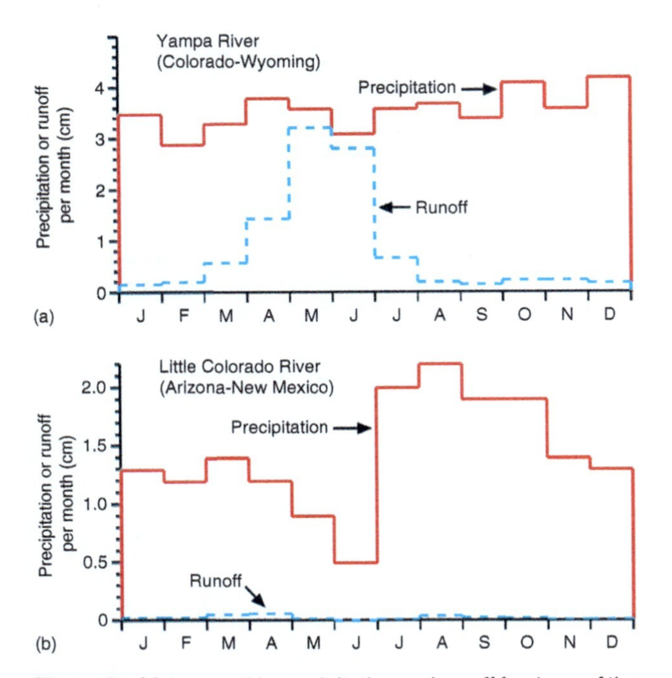

Figure 2 Mean monthly precipitation and runoff for rivers of the southwestern US. (a) Yampa River of southern Wyoming and northwestern Colorado. (b) Little Colorado River of Arizona. Adapted from Blinn and Poff (2005) In: Benke AC and Cushing CE (eds.) *Rivers of North America*. Burlington, MA: Academic Press/Elsevier, with permission.

Mexican Basins Exclusive of Rio Grande

Mexico's river basins are primarily located in the tropics and subtropics (ranging from 15° to 32° latitude) and thus all are found in areas of relatively warm climate. Nonetheless, there is great variation in these streams and rivers because of the strong influence of major mountain ranges and a significant increase in precipitation from north to south. Mexican rivers primarily drain into the Gulf of Mexico and the Pacific Ocean (**Figure 1**) and are the least studied of any region in North America as is made clear by Paul Hudson and others (2005, *Rivers of North America*).

Landscape and Climate

The Sierra Madre Occidental and Sierra Madre Oriental are major mountain ranges that divide

Figure 3 Sandy desert stream near Tucson, Arizona, with nearly dry bed. Note the narrow strip of streamside vegetation within an otherwise arid landscape. Currently, streams in this region rarely, if ever, reach their downstream tributaries of the Colorado River. Photo by Arthur Benke.

drainages into the Pacific and Gulf, as well as creating closed basins between them. The Basin and Range physiographic province extends southward from the United States as an arid plateau between these mountains (Mexican Altiplano), partly as the Chihuahuan Desert, and includes many endorheic drainages. The Basin and Range also extends to the west of the Sierra Madre Occidental as the Sonoran Desert. Thus, river systems in northern Mexico typically begin at high elevations but eventually flow across deserts to the east or west. The drainages in southern Mexico are influenced by other mountain ranges, particularly the east–west Trans-Mexican Volcanic belt (with Orizaba and Popocatépetl), the Sierra Madre del Sur, and the Sierra Madre de Chiapas. South-flowing rivers in central and southern Mexico typically begin at high elevations before crossing a narrow coastal plain and then emptying into the Pacific Ocean. East and north flowing rivers that primarily begin in mountains, flow across a coastal plain that becomes increasingly narrower and more tropical from north to south before emptying into the Gulf. In the karstic Yucatan Peninsula, there are only small rivers because of large subsurface drainage.

Temperatures are high throughout most of Mexico, except at high altitudes. Variation is less in the south, where mean monthly temperatures are >20 °C throughout the year. One of the larger differences among Mexican streams and rivers is the amount of precipitation received by their basins. In northern Mexico, precipitation is very low at lower elevations, and mountain precipitation typically provides much of the flow. For example, the Rio Conchos begins high in the pine–oak forests of the Sierra Madre Occidental and descends to the Chihuahuan Desert along a northeastern path before emptying into the Rio Grande. The Rio Fuerte drains the western slope flowing in the opposite direction. It descends from its forested headwaters, through the enormous Copper Canyon (Grand Canyon of Mexico) in its mid reaches, before crossing dry subtropical forest to the Gulf of California. In contrast, most of southern Mexico is tropical, with extremely high precipitation. Much was naturally covered with rain forest, particularly in the Usumacinta basin, although substantial areas have been converted to coffee plantations, cattle grazing, and agriculture. Regardless of whether precipitation is high or low, a seasonal pattern of rainfall occurs throughout Mexico, being highest in summer and early fall and lowest in winter (contrast the arid Yaqui with the moist Usumacinta–Grijalva, **Figure 4**).

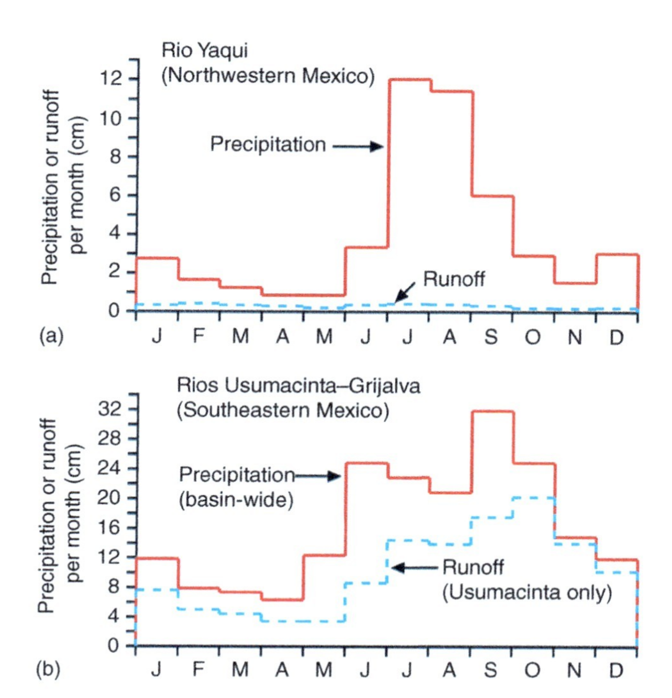

Figure 4 Mean monthly precipitation and runoff for rivers of Mexico. (a). Rio Yaqui of northwestern Mexico. (b). Rios Usumacinta–Grijalva of southeastern Mexico. Adapted from Hudson et al. (2005) In: Benke AC and Cushing CE (eds.) *Rivers of North America*. Burlington, MA: Academic Press/Elsevier, with permission.

Geomorphology and Hydrology

Geomorphology of Mexico's rivers is highly variable, especially within some of the larger systems. High stream gradients of several meters per kilometer are found in mountainous areas in the north (Fuerte), central (Panuco) and south (Usumacinta–Grijalva), often with spectacular waterfalls and gorges. In contrast, when rivers enter the Gulf Coastal Plain, they often meander within broad floodplains. Wetlands are particularly extensive along the coastal plain south of the Rio Panuco to the Rio Usumacinta (Tabascan lowlands), with seasonally flooded forests, marshes, and mangrove swamps.

The arid streams of northern Mexico resemble those of the southwestern US, being flashy but have low total runoff (Yaqui, **Figure 4(a)**). However, runoff from central rivers, such as the Panuco, can be moderate, and runoff from tropical rivers appears to be the highest in North America (Usumacinta, **Figure 4(b)**). Although damming is not as extensive as for many US rivers, strong regulation occurs widely, often reducing flows to negligible amounts in northern Mexican rivers so that stored waters can be used for irrigation, as described for the Rio Conchos by Paul Hudson and others (2005, *Rivers of North America*).

Biology and Ecology

Fish diversity of Mexico is relatively high with more than half (521) the species of the United States (912) of which 67% are endemic according to Paul Hudson and others (2005, *Rivers of North America*). Stream invertebrates and ecology are poorly studied in Mexican streams and rivers, although high diversity and unique species are likely. This lack of knowledge is unfortunate because many Mexican streams and rivers are highly exploited for hydropower and irrigation, and are also highly polluted. The Rio Usumacinta is Mexico's most natural and unregulated river, and its basin contains ancient Mayan ruins. Potential damming of this, the largest tropical river in North America, has been a highly contentious issue for decades.

Pacific Basins of the United States and Canada

The Pacific rivers of the United States and Canada cover about 30° of latitude from arid basins of southern California to subarctic basins of British Columbia and southern Alaska (**Figure 1**). Although they all drain into the Pacific Ocean, their basins differ greatly in both temperature and precipitation. The Columbia River is the largest and originates most deeply from within the continent. In general, there are more large rivers (>1200 $m^3 s^{-1}$) on the Pacific coast (e.g., Columbia, Fraser, Kuskokwim, Copper, Stikine, Skeena, and Susitna) than the Atlantic coast (St. Lawrence, Churchill) (**Table 1**).

Landscape and Climate

Rivers of the Pacific coast primarily drain from several extensive mountain ranges, some with tectonic activity, that dominate the landscape from California to southern Alaska (Sierra Nevada, Cascades, Coast, Alaska Range) including several with peaks >5000 m asl (**Figure 5**). In contrast, headwaters of the Columbia River drain primarily from the Rocky Mountains of southeastern British Columbia, western Montana, Idaho, and Wyoming (**Figure 6**), crossing the relatively flat Columbia Plateau (**Figure 7**) before cutting through the Columbia River Gorge of the Cascade Mountains to the Pacific. Other rivers with extensive lowland drainage include the Sacramento and San Joaquin rivers that flow through California's Central Valley after being fed by tributaries of the Sierra Nevadas and Oregon's Willamette River that is fed by tributaries from the Cascades (**Figure 8**). Also, while the Kuskokwim River of western Alaska drains a small portion of the Alaska Range, it flows primarily through relatively flat Tundra before reaching the Bering Sea.

Precipitation is generally lowest in southern California, the Central Valley, and the Columbia Plateau. It increases greatly along the coast from northern California into British Columbia, but is highly variable within mountain ranges, and declines substantially in northern British Columbia and southern Alaska (**Figure 9**). Mean air temperatures range from ~15 °C in southern California to <0 °C in southern Alaska. Mean summer temperatures are relatively mild (~15–20 °C along much of the coast from California to southern British Columbia), whereas mean midwinter temperature can be as low as −20 °C in northern British Columbia and southern

Figure 5 Susitna River, Alaska, being joined by Hurricane Creek in the Susitna Valley. Largely a pristine river, portions of the Susitna drain from Mt. McKinley, the tallest peak in North America and part of the Alaska Range. Photo by Jackson Webster.

Figure 6 Upper Snake River draining from Rocky Mountains, Wyoming. Downstream the Snake River enters the Columbia Plateau and is heavily impacted by a series of dams before joining the Columbia River. Photo by Arthur Benke.

Alaska. Precipitation, temperature, and elevation combine to affect basin vegetation which includes chaparral and grassland species at low elevations of south and central California, grasslands of the Columbia Plateau, and coniferous forests from the Rockies and along the coast from California to Alaska. At the highest altitudes, particularly in the north, alpine tundra, rocky slopes, ice fields, and glaciers are prominent.

Seasonality of precipitation changes dramatically from south to north. From southern California to Washington, precipitation is much higher during winter than summer as seen for the Rogue River of Oregon (**Figure 9(b)**). However, precipitation becomes progressively less seasonal to the north, and then shifts to late-summer peaks in river basins of southern Alaska as illustrated by the Kuskokwim (**Figure 9(a)**).

Geomorphology and Hydrology

In such a mountainous region, it is common for tributaries of major rivers to have gradients $>10\,\mathrm{m\,km^{-1}}$, falling to $<5\,\mathrm{m\,km^{-1}}$ in valleys. However, lower gradient streams and rivers are found in several areas such as the Columbia Plateau, the California Central Valley, the Willamette valley, and much of the Kuskokwim drainage. Fallen trees often play a major role in the geomorphology of both high gradient streams and low-gradient rivers, although much has been removed. Agriculture and grazing have been extensively developed in most areas of flat topography in the western US, often accompanied by channelization and water extractions.

Figure 7 John Day Dam on the Columbia River, between Washington and Oregon. It is located in the western portion of the arid Columbia Plateau, just east of the Columbia River Gorge. Photo by Arthur Benke.

Figure 8 McKenzie River draining the Cascade Mountains, western Oregon, before flowing into the Willamette River, a major tributary of the Columbia River. Photo by Arthur Benke.

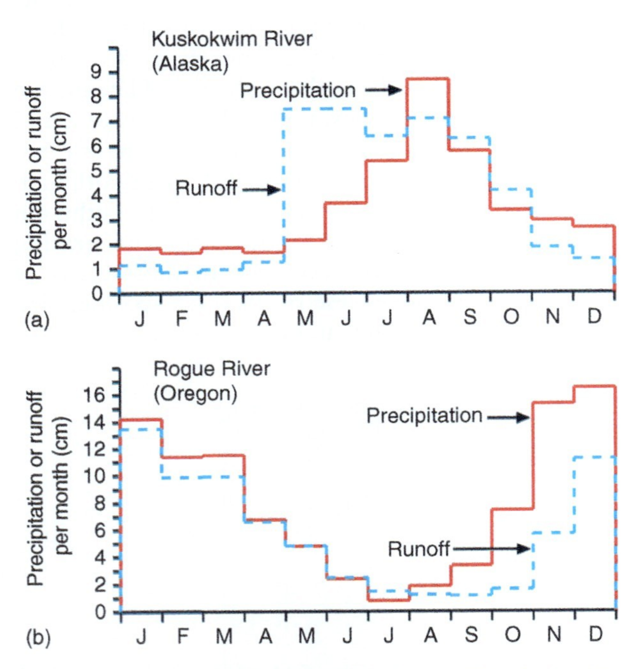

Figure 9 Mean monthly precipitation and runoff for rivers of the Pacific Coast of North America. (b). Kuskokwim River of western Alaska. (b). Rogue River of southwestern Oregon. Adapted from Richardson and Milner (2005) In: Benke AC and Cushing CE (eds.) *Rivers of North America*. Burlington, MA: Academic Press/Elsevier, and Carter and Resh (2005) In: Benke AC and Cushing CE (eds.) *Rivers of North America*. Burlington, MA: Academic Press/Elsevier, with permission.

Seasonal runoff from Pacific rivers is highly variable. From California to Washington, runoff is dominated by winter flows, a strong reflection of the seasonal precipitation pattern (**Figure 9(b)**). In contrast, rivers in British Columbia and Alaska have increasingly strong summer runoff that reflects not only the seasonal shifts from winter to summer precipitation, but winter snow/ice storage followed by spring/summer snowmelt (**Figure 9(a)**). Low evapotranspiration appears responsible for a high fraction of precipitation being captured in runoff for many rivers from northern California to Alaska, but excluding rivers of the drier Columbia basin.

Some basins, particularly the Columbia River, as described by Jack Stanford and others (2005, *Rivers of North America*) and the Sacramento–San Joaquin system in California, as described by James Carter and Vincent Resh (2005, *Rivers of North America*), are heavily dammed for hydropower and human uses such as agriculture. In contrast, Trefor Reynoldson and others (2005, *Rivers of North America*) describe the large Fraser River in southern British Columbia as relatively unregulated, and John Richardson and Alexander Milner (2005, *Rivers of North America*) found very little regulation within all the remaining Pacific rivers to the north.

Biology and Ecology

Pacific rivers include some of the wildest and some of the most ecologically damaged on the continent. The most pristine rivers in British Columbia and Alaska have enormous migrations of Pacific salmon (several species) and steelhead (**Figure 5**). The Fraser is one of the great salmon rivers of the world and is by far the longest river in the Canadian Heritage River system. In contrast, heavy damming in many Pacific rivers of the coterminous states has greatly reduced or eliminated many such migrations. Over 400 dams occur throughout the Columbia basin alone (15 huge dams on main stem, **Figure 7**) with all major tributaries but the Salmon River strongly regulated for hydropower and irrigation. Ironically, salmon cannot reach many Wild and Scenic rivers that drain the Rockies, Sierra Nevadas, and Cascades.

Although large salmon migrations are (or were) a remarkable feature of Pacific rivers, overall diversity of native fishes is relatively low in comparison to eastern rivers. Jack Stanford and others (2005, *Rivers of North America*) mention only 65 native freshwater fish species occur in the Columbia and Trefor Reynoldson and others (2005, *Rivers of North America*) indicate only 40 in the Fraser. Rivers south of the Columbia range from only 15 to 30 species based on information presented by James Carter and Vincent Resh (2005, *Rivers of North America*) and those north of the Fraser from about 25 to 35 according to John Richardson and Alexander Milner (2005, *Rivers of North America*). Impacts from regulation are often compounded by introduction of nonnative fishes. These same sources indicated about 53 nonnatives are in the Columbia, ~40 in the Sacramento, and ~20 for many other rivers south of the Columbia. In contrast, although the Fraser includes some nonnatives, they are virtually nonexistent to the north.

Invertebrates have been well studied in some Pacific rivers, especially mountain streams, and particularly in some tributaries of the Columbia, and coast rivers to the south. Many rivers, however, are poorly studied, especially north of the Fraser. Aquatic insect diversity in Pacific drainages is often quite high.

Important ecosystem studies have been done in certain Pacific rivers. The Columbia basin includes two of the original rivers in the RCC analysis – the Salmon in Idaho, and the McKenzie in Oregon – as described by Wayne Minshall and others (1983, published in Ecological Monographs). Both rivers generally showed the importance of terrestrial inputs in upper reaches and in-stream production in middle reaches. More recently, migratory salmon have been shown to represent a major source of nutrients to Pacific headwaters, but only in rivers of British Columbia and Alaska lacking dams.

Arctic and Subarctic basins

The northern portion of Canada and Alaska includes many rivers that flow primarily through subarctic and arctic regions (roughly north of 50° latitude). This section includes rivers that flow from the center of the continent, excluding Atlantic and Pacific drainages (**Figure 1**). This large area includes the third (Mackenzie) and sixth (Yukon) largest rivers on the continent, with the Mackenzie being more than half the area and about half the flow of the Mississippi River (**Table 1**). Tributaries from the larger of these basins (Mackenzie, Yukon, and Nelson) arise from deep within the continent, covering more than 20° of latitude. These large rivers, and many smaller ones, flow into the Arctic Ocean, Hudson Bay, the Bering Sea, or various connecting bays, straits, and channels of the Arctic (**Figure 1**).

Landscape and Climate

Several arctic northwestern drainages begin at high elevations, but also flow over vast low-relief landscapes (**Figure 1**). Tributaries of the Yukon River drain the northern slopes of the Alaska Range as well as the southern slopes of the Brooks Range and the western slopes of the Mackenzie Mountains. But the main stem of the Yukon and some of its larger tributaries flow through interior Alaska, with a relatively low relief and discontinuous permafrost. Many tributaries of the Mackenzie River drain either north or east from the Mackenzie Mountains and east from the Canadian Rockies. However, more than two-third of the Mackenzie flows through the low relief Great Plains and Canadian Shield to the east. Immediately south of the Mackenzie, tributaries from the Saskatchewan River (Nelson basin) also drain from the eastern slope of the Rockies, but the vast majority of the Nelson drainage is through the Great Plains, Central Lowlands, and Canadian Shield. One major tributary (Red River of the North) arises within the United States before flowing north to Lake Winnipeg. Other smaller arctic rivers drain from the Brooks Range and through the Arctic Plain (**Figure 10**), and many arctic/subarctic rivers arise within the Canadian Shield in northeastern and northcentral Canada, particularly those flowing into Hudson Bay.

Climate in the arctic/subarctic regions involves extremely cold temperatures. Mean monthly temperatures in mid-winter are often below −20 °C, and mean monthly temperatures in mid-summer may not go above 10 °C. As a result, permafrost is found at the most northern latitudes and has an important influence in basin vegetation. Annual precipitation is often low, typically ~30 cm, and tends to peak in summer (**Figure 11**). Winter precipitation is always frozen.

Vegetation across this vast northern area depends mostly on altitude, latitude, and precipitation. In northernmost regions with permafrost, tundra vegetation predominates, with hardy grasses, sedges, and dwarf willows and birch trees along water bodies. Tundra stretches from western Alaska to northern portions of Quebec and Labrador. Boreal forest, mostly white and black spruce, covers vast areas as

Figure 10 Ribdon River, a tributary of the Sagavanirktok River, drains from the Brooks Range on the Alaska North Slope. The ice in the foreground is called aufeis and is formed when spring and/or groundwater flows continue through the winter. Photo by Stephanie Parker.

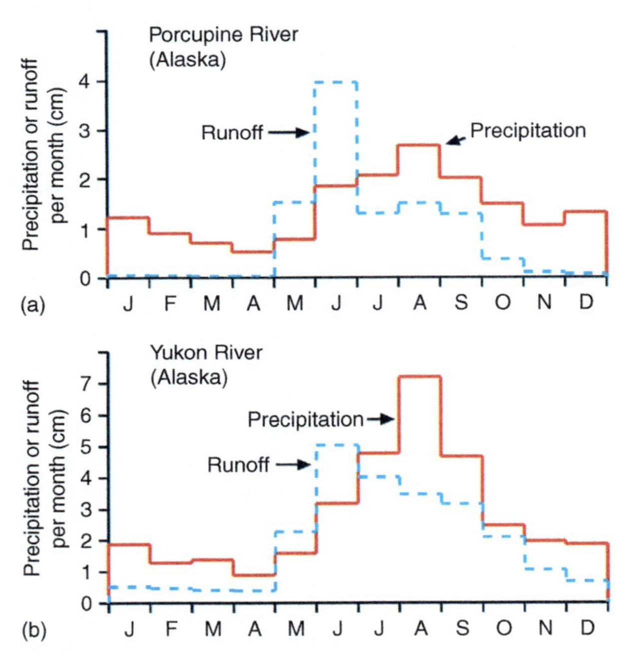

Figure 11 Mean monthly precipitation and runoff for Arctic rivers of North America. (a). Porcupine River of Alaska. (b). Yukon River of Alaska. Adapted from Bailey (2005) In: Benke AC and Cushing CE (eds.) *Rivers of North America*. Burlington, MA: Academic Press/Elsevier, with permission.

well. It is typically found south of the tundra, but reaching north into Central Alaska between the Brooks Range and the Alaska Range and stretching across Canada to Labrador and Newfoundland. The southern regions, especially within the Nelson basin, drain major portions of Great Plains and Central Lowlands with various prairie grasses.

Geomorphology and Hydrology

Headwaters of the larger rivers arise from some of the tallest mountains on the continent – Alaska Range, Canadian Rockies, Mackenzie Mountains – and thus are high gradient streams. However, because much of the drainage is across large plains, the majority of streams and most of the large rivers have relatively low gradients.

It is somewhat surprising that two of North America's larger rivers (Mackenzie and Yukon, **Table 1**) are in the Arctic because precipitation is relatively low. However, these are extremely large basins, and evapotranspiration is very low; thus a high fraction of precipitation eventually runs off to the sea (**Figure 11**). Two things contribute to the highest discharge occurring during summer in arctic streams and rivers. First, the peak in runoff is from melting of frozen precipitation that falls from early autumn through spring. Second, the summer runoff is sustained by the highest precipitation found in mid-to-late summer before freezing occurs (Yukon, **Figure 11(b)**). In many of the most northern tributaries, even large rivers such as the Porcupine (mean annual discharge $>400\,m^3\,s^{-1}$), water is almost entirely frozen from December through April (**Figure 11(a)**).

Biology and Ecology

Ecological characteristics in arctic drainages are highly variable because they include streams draining high altitude mountains, relatively arid low-relief plains, and the Canadian Shield. Furthermore, a great temperature range occurs from southern tributaries of the Nelson to small rivers draining directly into the Arctic. Most of these rivers are little studied, but knowledge of some streams and rivers is quite detailed. These include the small Kuparuk River, draining the north slope of the Brooks Range, the Moose River, flowing into the southern end of Hudson Bay, and tributaries of the Mackenzie (Peace) and Nelson (Saskatchewan). Studies of arctic systems have shown the powerful influence of freezing, with most biological activity in summer.

Fish diversity is strongly related to latitude in this region. Highest species richness occurs in the Nelson with >94 species, primarily due to its southernmost tributary, the Red River of the North according to David Rosenberg and others (2005, *Rivers of North America*). In contrast, Joseph Culp and others (2005, *Rivers of North America*) report only 52 species from the Mackenzie, and Robert Bailey (2005, *Rivers of North America*) could document only about 30 species from the Yukon. The fewest fish species per river (<15) were found by Alexander Milner and others (2005, *Rivers of North America*) in smaller rivers near the Arctic Circle. Diversity of invertebrates is also relatively low in arctic streams and rivers and thus food webs should be relatively simple compared to temperate and tropical streams.

Summary of Human Impacts

The rivers described in this article are far less fragmented than eastern and central rivers of the first article. Most unregulated rivers flow into the northern Pacific or Arctic oceans (**Table 2**). The major exceptions, however, are among the largest and most fragmented rivers of the continent (Columbia, Colorado, Nelson, La Grande). Furthermore, several large rivers are involved in enormous interbasin water transfers (Colorado to California Aquaduct, Churchill to Nelson, Koksoak, and Eastmain to La Grande). These transfers either greatly reduce (Colorado, Churchill, Eastmain, and Koksoak) or increase (Nelson, La Grande) natural flows (**Table 1**) with undoubtedly huge ecological consequences. Human population densities are generally lower in river basins of this article (most $<15\,km^{-2}$) than those in the first article. Concentrations of nutrients such as nitrate and phosphate are generally lower in the west than in Mississippi River basins because agricultural use is much lower (**Table 2**). However, where high-intensity agriculture is sustained in arid regions of the United States and Mexico (e.g., California Central Valley, San Joaquin River), heavy use of water for irrigation and pollution by nutrients/pesticides seriously

Table 2 Factors that reflect human influences on the largest rivers of western, northern, and Mexican river basins of North America

Rank	*River name*	*Pop. Density (ind. km^{-2})*	*% Agr*	*NO_3–N (mg l^{-1})*	*PO_4–P (mg l^{-1})*	*Fragmentation by dams*
3	Mackenzie	<1	<4	0.37	0.01	1
5	Columbia	13.8	15	0.26	<0.06	2
6	Yukon	<0.1	0	NA	0.02	0
7	Fraser	10.7	0.6	<0.21	<0.11	0
9	Slave[a]	<1	10.6	<0.40	<0.20	1
10	Usumacinta/Grijalva[b]	28	31	NA	NA	0/2
11	Nelson	5	51	<0.70	<0.03	2
12	Liard[a]	<1	1.6	<0.43	0.01	0
13	Koksoak	NA	NA	NA	NA	2
18	Kuskokwim	0.1	0	<0.08	<0.05	0
20	Copper	<0.05	0	NA	NA	0
21	Skeena	<3	<1	0.08	<0.06	0
22	La Grande	NA	NA	NA	NA	2
23	Stikine	<0.03	0	0.1	<0.2	0
25	Susitna	0.01	0	0.15	<0.12	0
Additional large basin						
	Colorado	7	67	0.12	0.04	2

Qualitative assessment of fragmentation by dams primarily based on Benke and Cushing (2005), but also Dynesius and Nilsson (1994): 0 = no or minor fragmentation, 1 = moderate fragmentation, 2 = strong fragmentation. % Agr. = % of basin used for agriculture. NA = not available

[a]Tributary of Mackenzie.

[b]Fragmentation: Usumacinta (0), Grijalva (2).

degrades river ecosystems. Although many more rivers are protected in western North America than in the east, specifically those with U.S. Wild and Scenic River status, future attempts to exploit pristine systems in the west and north can be anticipated.

Glossary

Discharge – The volume of water flowing in a stream, usually measured as cubic meters per second ($m^3 s^{-1}$) or cubic feet per second ($ft^3 s^{-1}$). Often presented as the average for a longer period of time; e.g., the average discharge for July is $510 m^3 s^{-1}$.

Diversity – Generally used in the ecological literature to indicate the number of species within a biological community or ecosystem.

Ecosystem – A community of organisms and their nonliving environment interacting as a unit.

Evapotranspiration – The combined processes of evaporation of water and transpiration by plants, which together describe the loss of water to the atmosphere from an ecosystem.

Physiographic province – Subdivisions of the continent based on topographic features, rock type, and geological structure and history; e.g., Coastal Plain, Great Plains, Basin and Range.

Production – The amount of algal or plant matter (primary production), or animal matter (secondary production) formed in an ecosystem over some period of time, such as a year (e.g., $g m^{-2} year^{-1}$).

Runoff – The amount of water draining from a basin presented as cm of water height for some interval of time (e.g., $120 cm year^{-1}$). Calculated by dividing discharge (e.g., $m^3 year^{-1}$) by basin area (km^2).

See also: Africa; Algae of River Ecosystems; Asia – Eastern Asia; Asia – Monsoon Asia; Asia – Northern Asia and Central Asia Endorheic Rivers; Australia (and Papua, New Guinea); Benthic Invertebrate Fauna; Benthic Invertebrate Fauna, River and Floodplain Ecosystems; Benthic Invertebrate Fauna, Small Streams; Benthic Invertebrate Fauna, Tropical Stream Ecosystems; Biological Interactions in River Ecosystems; Climate and Rivers; Coarse Woody Debris in Lakes and Streams; Conservation of Aquatic Ecosystems; Currents in Rivers; Ecology and Role of Headwater Streams; European Rivers; Flood Plains; Floods; Geomorphology of Streams and Rivers; Hydrology: Rivers; Hydrology: Streams; Regulators of Biotic Processes in Stream and River Ecosystems; Restoration Ecology of Rivers; Riparian Zones; South America; Streams and Rivers as Ecosystems; Streams and Rivers of North America: Overview, Eastern and Central Basins; Wetlands of Large Rivers: Flood plains.

Further Reading

Abell RA, Olson DM, Dinerstein E, *et al.* (2000) *Freshwater Ecoregions of North America: A Conservation Assessment.* Washington, DC: Island Press.

Benke AC and Cushing CE (eds.) (2005) *Rivers of North America.* Burlington, MA: Academic Press/Elsevier.

Brown LR, Gray RH, Hughes RM, and Meader MR (eds.) (2005) *Effects of Urbanization on Stream Ecosystems.* Bethesda, MD: American Fisheries Society. Symposium 47.

Cushing CE, Cummins KW, and Minshall GW (eds.) (1995) *Ecosystems of the World 22: River and Stream Ecosystems.* Amsterdam: Elsevier.

Dynesius M and Nilsson C (1994) Fragmentation and flow regulation of river systems in the northern third of the world. *Science* 266: 753–762.

Hughes RM, Wang L, and Seelback PW (eds.) (2006) *Landscape Influences on Stream Habitats and Biological Assemblages.* Bethesda, MD: America Fisheries Society. Symposium 48.

Karr JR, Allan JD, and Benke AC (2000) River conservation in the United States and Canada. In: Boon PJ, Davies BR, and Petts GE (eds.) *Global Perspectives on River Conservation: Science, Policy and Practice*, pp. 3–39. Chichester: John Wiley.

Naiman RJ, DéCamp H, and McClain ME (2005) *Riparia: Ecology, Conservation, and Management of Streamside Communities.* Amsterdam: Academic Press/Elsevier.

Nelson JS, Crossman EJ, Espinosa-Pérez H, *et al.* (2004) *Common and Scientific Names of Fishes of the United States, Canada and Mexico.* Bethesda, MD: American Fisheries Society. Special Publications 29.

Palmer T (1996) *America By Rivers.* Washington, DC: Island Press.

Palmer T (2006) *Rivers of America.* Harry N. Abrams.

Revenga C, Murray S, Abramovitz J, and Hammond A (1998) *Watersheds of the World: Ecological Value and Vulnerability.* Washington, DC: World Resources Institute.

Rinne JN, Hughes RM, and Calamusso B (eds.) (2005) *Historical Changes in Large River Fish Assemblages of the Americas.* Bethesda, MD: American Fisheries Society. Symposium 45.

Vannote RL, Minshall GW, Cummins KW, Sedell JR, and Cushing CE (1980) The river continuum concept. *Canadian Journal of Fisheries and Aquatic Sciences* 37: 130–137.

Wohl EE (2004) *Disconnected Rivers: Linking Rivers to Landscapes.* New Haven: Yale University Press.

Relevant Websites

http://www.daac.ornl.gov/RIVDIS/rivdis.html.
http://www.earthtrends.wri.org/maps_spatial/watersheds/ncamerica.php.
http://waterdata.usgs.gov/nwis/sw.
http://webworld.unesco.org/water/ihp/db/shiklomanov/index.shtml.
http://www.weatherbase.com/.
http://scitech.pyr.ec.gc.ca/climhydro/.
http://www.americanrivers.org/.
http://www.chrs.ca/.
http://www.nps.gov/ncrc/programs/rtca/nri/.
http://www.rivers.gov/.
http://www.sage.wisc.edu/riverdata/.
http://www.nationalgeographic.com/wildworld/terrestrial.html.
http://enpsychlopedia.org/psypsych/List_of_rivers_in_the_United_States#State_river_lists.
http://ga.water.usgs.gov/edu/earthrivers.html.

Subject Index

Notes

Cross-reference terms in italics are general cross-references, or refer to subentry terms within the main entry (the main entry is not repeated to save space). Readers are also advised to refer to the end of each article for additional cross-references - not all of these cross-references have been included in the index cross-references.

The index is arranged in set-out style with a maximum of three levels of heading. Major discussion of a subject is indicated by bold page numbers. Page numbers suffixed by T and F refer to Tables and Figures respectively. vs. indicates a comparison.

This index is in **letter-by-letter** order, whereby hyphens and spaces within index headings are ignored in the alphabetization. For example, acid rain is alphabetized after acidity, not after acid(s). Prefixes and terms in parentheses are excluded from the initial alphabetization.

Where index subentries and sub-subentries pertaining to a subject have the same page number, they have been listed to indicate the comprehensiveness of the text.

Abbreviations

DCAA - dissolved combined amino acids

DFAA - dissolved free amino acids

DIN - dissolved inorganic nitrogen

DOC - dissolved organic carbon

DOM - dissolved organic matter

DON - dissolved organic nitrogen

TDS - total dissolved solids

TSS - total suspended solids

A

B

C

D

E

F

I

J

K

L

M

N

O

P

Q

R

S

T

U

V

W

X

Y

Z